Green Energy and Technology

Climate change, environmental impact and the limited natural resources urge scientific research and novel technical solutions. The monograph series Green Energy and Technology serves as a publishing platform for scientific and technological approaches to "green"—i.e. environmentally friendly and sustainable—technologies. While a focus lies on energy and power supply, it also covers "green" solutions in industrial engineering and engineering design. Green Energy and Technology addresses researchers, advanced students, technical consultants as well as decision makers in industries and politics. Hence, the level of presentation spans from instructional to highly technical.

Indexed in Scopus.

Indexed in Ei Compendex.

Andreas Ch. Braun · Gabriela Espinosa Gutiérrez ·
Danny Tröger · Thomas Hirth
Editors

Eco-Industrial Development as an Industrial Strategy

Contributions from a German-Chilean Research Partnership

Springer

Editors
Andreas Ch. Braun
University of Kassel
Kassel, Germany

Danny Tröger
University of Kassel
Kassel, Germany

Gabriela Espinosa Gutiérrez
Institute of Regional Science
Karlsruhe Institute of Technology
Karlsruhe, Germany

Thomas Hirth
Karlsruhe Institute of Technology
Karlsruhe, Germany

ISSN 1865-3529 ISSN 1865-3537 (electronic)
Green Energy and Technology
ISBN 978-3-031-73575-2 ISBN 978-3-031-73576-9 (eBook)
https://doi.org/10.1007/978-3-031-73576-9

This Springer imprint is published by the registered company Springer Nature Switzerland AG
The registered company address is: Gewerbestrasse 11, 6330 Cham, Switzerland

Preface

This book is the result of a successful collaboration between four universities through the virtual Institute of Eco-Industrial Development (IECO): the Karlsruhe Institute of Technology, Universidad de Chile, Universidad de Concepcion, and Universidad Austral de Chile. The aim of this interdisciplinary project in research, higher education, and knowledge transfer was to identify sustainable solutions for industrial development in Chile, taking into account the challenges and opportunities in both the north and south of the country.

Of the many crossroads that Chile faces for its future, sustainable industrial development is one of the most important. Another major challenge is the ongoing process of decentralization which is both recent and incipient, especially in relation to the different potentials of the regions, the sustainability standards that will define these activities, and the benefits that each community and territory will obtain from them. This is why the theoretical development and case reviews described in this book are so relevant and timely, presented in the form of analyses by experts but also with a special concern for informing policymakers.

The eco-industrial development approach of IECO that is outlined in this book offers a conceptual framework to address the key challenges in this matter: that is, developing production chains sustained by circularity and based on science, integrating community perspectives, and constantly correcting towards the goal of process optimization. Industrial ecology is a broad field of study that encompasses the entire industrial system and its environmental impacts. According to Thomas Graedel, a prominent scholar in the field of industrial ecology, it is "the study of the physical, chemical, and biological interactions and interrelationships both within and between industrial and ecological systems" (Graedel, 1994). This definition emphasizes the need to integrate environmental considerations into the design and operation of industrial systems in order to achieve sustainable outcomes. Eco-industrial development is an approach that seeks to create sustainable industrial systems by integrating principles of industrial ecology, circular economy, and social equity. This approach aims to achieve economic growth while minimizing negative environmental and social impacts.

As a country rich in natural resources, Chile has significant opportunities for renewable production of energy and biomaterials. The country is known for its high potential for solar, wind, geothermal, and hydropower energy, as well as massive agricultural and forestry biomass production. These resources offer opportunities to reduce greenhouse gas emissions, replace fossil-derived raw materials, and improve energy security. Furthermore, biomaterials and renewable energy production can provide economic benefits through job creation, local manufacturing, and increased energy exports. In addition, Chile has a highly developed agricultural and forestry industry, producing by-products on which a circular bioeconomy with very promising economic, environmental, and social projections can be established. These opportunities also face important challenges, such as their impact on territories and local communities, which must be properly addressed in order to ensure a sustainable energy industry.

On the one hand, this book presents the findings of the projects run by the network of researchers from Chile and Germany who are behind the IECO, including chapters on renewable energy production, circular economy, sustainable agriculture, and social and environmental impact assessment, among others. On the other hand, this publication highlights the importance of international academic collaboration in order to achieve sustainable transformations in industry, while also providing insights into the particular challenges and opportunities of eco-industrial development in Chile.

We are proud to be active members of this binational consortium of experts and explorers of the potential of circular economy for the benefit of their communities and the country. In particular, we celebrate that the researchers represent a wide range of disciplines, from engineering to the social sciences, thus contributing to the multi- and interdisciplinary approach that is essential in this type of collaboration network. We as authors hope that this book will serve as a valuable resource for academics, policymakers, and practitioners interested in eco-industrial development and sustainable industrial practices. Our ultimate goal is to contribute to the ongoing efforts to achieve a more sustainable and equitable industrial development in Chile and beyond. We believe that the insights and recommendations presented in this book will inspire further research and action towards a sustainable future for all.

By including regional universities far beyond the traditional centers of knowledge production, we are recognizing the value of an academic community engaged in decentralization efforts, as well as the importance of research and technology transfer with territorial relevance in that process. In this sense, we believe that the analyses proposed in this publication, especially on topics such as urban development, biodiversity, and circular production, represent a substantive contribution to the design of regional development strategies at a time when they are highly required by stakeholders in the public sector and in businesses.

Inspired by nature and its processes but guided by the responsibility to offer effective and sustainable production alternatives with the aim of generating zero waste, the authors and editors of this publication offer an invaluable instrument for the generation of public policies that contribute to the productive diversification of Chile, in an endeavor of binational academic cooperation that should be applauded and replicated.

Kassel, Germany Andreas Ch. Braun
Karlsruhe, Germany Gabriela Espinosa Gutiérrez
Kassel, Germany Danny Tröger
Karlsruhe, Germany Thomas Hirth
Santiago, Chile Francisco Martínez Concha
Concepción, Chile Andrea Rodríguez Tastets
Valdivia, Chile Mylthon Jiménez Castillo

Reference

Graedel, T. (1994). Industrial ecology: definition and implementation. Industrial ecology and global change, 5, 23–41. In. Socolow, R., Andrews, C., Berkhout, F., Thomas, V. (1994). *Industrial ecology and global change*, Cambridge University Press, Cambridge, UK.

Contents

Project and Book Overview (Type: Introduction)

International Scientific Cooperation for Sustainable Development—The German-Chilean Institute of Eco-Industrial Development (IECO)

Andreas Ch. Braun, Irma Mantilla, Klaus Rümmele, Pascale Kohler, and Thomas Hirth

Abstract This chapter outlines the establishment and development of the virtual Institute of Eco-Industrial Development (IECO), a collaborative initiative between German and Chilean research institutions, including KIT, UCH, UdeC, and UACh. The IECO aims to promote sustainable resource management and energy usage through a multidisciplinary approach, focusing on eco-industrial development as a framework for research, teaching, and innovation. Following a successful funding proposal to Germany's Federal Ministry of Education and Research (BMBF), the institute received support for six years, later extended due to the Covid-19 pandemic, to enhance its research infrastructure. This funding catalyzed the formation of new partnerships and strengthened existing ties within the German-Chilean collaboration landscape. The IECO's vision encompasses the sustainable use of resources like water and biomass, urban mining, and the circular economy, aiming to regenerate urbanized areas and transform industrial residues into valuable materials. The document also acknowledges the contributions of various stakeholders, including funding agencies and partner universities, highlighting the importance of collaborative efforts in addressing global sustainability challenges. Through its initiatives, the IECO seeks to position itself as a significant player in fostering sustainable development practices and enhancing cooperation between academia and industry in both countries.

A. Ch. Braun (✉) · T. Hirth
Kassel Institute for Sustainability, University of Kassel, Kassel, Germany
e-mail: andreas.braun@uni-kassel.de

T. Hirth
e-mail: thomas.hirth@kit.edu

I. Mantilla · K. Rümmele · P. Kohler · T. Hirth
International Affairs Business Unit (INTL), Karlsruhe Institute of Technology, Karlsruhe, Germany
e-mail: irma.mantilla@kit.edu

K. Rümmele
e-mail: klaus.ruemmele@kit.edu

P. Kohler
e-mail: pascale.kohler@kit.edu

A. Ch. Braun et al. (eds.), *Eco-Industrial Development as an Industrial Strategy*, Green Energy and Technology, https://doi.org/10.1007/978-3-031-73576-9_1

1 International Scientific Cooperation for Sustainable Development

The discourse around sustainable development has always been influenced by input from both the Global North and the Global South. The idea of transitioning from the Millennium Development Goals (MDGs) to the Sustainable Development Goals (SDGs), for example, originated from Paula Caballero Gómez, a Colombian delegate to the Rio + 20 conference (Fukada-Parr & Muchhala, 2020). It was apparently recognized early on that sustainable development is only possible in the context of international cooperation—even if concrete actions in the spirit of Agenda 21 are to be local (Kefalas, 1998). International cooperation has shaped global environmental governance in several aspects. Firstly, joint conferences such as UNCED or UNFCCC took place (Spector et al., 1994). Secondly, specific targets were defined during these conferences, that envision international collaboration (such as MDG 8 and SDG 17, see Fig. 1) (Sachs, 2012). Thirdly, *attempts* have been made to achieve sustainable development through joint implementation mechanisms (such as the clean development mechanism, joint implementation, emissions trading, or joint development) (Sijm et al., 2000). Moreover, numerous UN-organizations, NGOs etc. aim to strengthen international collaboration.

Fig. 1 The seventeen Sustainable Development Goals (SDGs) of Agenda 2030. The SDGs break down into 169 targets. SDG 17 emphasizes the importance of international collaboration in particular. *Source* United Nations Sustainable Development (n.d.)

2 The Purpose of International Scientific Cooperation in the Context of Sustainable Development: Four Rationales

But why exactly is international cooperation useful in the context of sustainable development? This question will be answered in the following sections with regard to four important aspects: first, in terms of our global challenges; second, in terms of institutions; and third, in terms of scientific theory. Finally, the power relations in science and international environmental governance need to be addressed. These perspectives show why cooperative, joint research endeavors such as that of the IECO—Institute of Eco-Industrial Development make sense.

2.1 Rationale One: Global Challenges

In terms of global change, there are several challenges that humanity has to solve in the Anthropocene. Earth system science has long recognized that there is only one Earth system, which consists of complex, globally coupled energy and material cycles. This Earth system regulates itself through internal feedback mechanisms. If these are exceeded, there are potentially irreversible, negative effects with global and long-term consequences (Kump et al., 2004). Humans have the capacity to violate the carrying capacity limits of this Earth system and have already done so in many ways, thus risking the occurrence of the previously mentioned effects (Rockström et al., 2009; Steffen et al., 2015). While boundary demarcations may have political relevance, the seamless character of the Earth system renders them meaningless to the function of this system (Biermann et al., 2010). Within this setting, the challenges of the Anthropocene are inherently global in impact.

Some of the most important challenges that can only be addressed through international cooperation are mitigation and adaptation to global climate change, such as through emissions reductions or creating negative emissions to meet Paris Accord goals (Honegger et al., 2021; Loftus et al., 2015). Limited and critical resources raise issues of global distribution and equity (Dushyantha et al., 2020; McCoy & Kinyua, 2012; Sadik-Zada, 2021; Vaccari, 2009). Climate change and sustainability are also related to migration processes that require international strategies (Marotzke et al., 2020; Rees, 2006). In addition, our individual health is also linked to global processes through the problem of emerging infectious diseases (Morens & Fauci, 2013, 2020). These and other connections highlight the necessity of international collaboration to face today's global challenges.

2.2 Rationale Two: Institutions

The second rationale for international collaboration on sustainability has its basis in institutional linkages. As previously mentioned, many environmental problems that result from a lack of sustainable management often have a transboundary character. The negative effects of environmental pressures often emerge in a different place and time than where they were initially triggered (Hoel, 1997; Müller-Mahn, 2013). Thus, they create a causal link between the trigger and affected party and, in institutional terms, require management that transcends national boundaries (Milman et al., 2020). Moreover, global sustainability governance strategies also require global acceptance. For the latter, it is imperative that these strategies have been negotiated internationally in a fair and participatory manner (Akerboom et al., 2020; Buhmann, 2017). Lastly, a transition to greater sustainability will only be possible by uniting global knowledge bases, because different countries have different best practices for solving the given problems (Nagendra et al., 2018; Zeigermann, 2021). This also aligns with the European Commission's Global Approach to Research and Innovation which is to be implemented by "mobilizing science, technology and innovation to accelerate sustainable and inclusive development." To "cooperate across borders on a scale never seen before to develop innovative solutions to deliver just green and digital transitions in line with the sustainable development goals" is considered to be crucial for future collaboration and the way it is politically and financially supported (European Commission, 2021).

2.3 Rational Three: Theory of Science

Finally, we will take a look at the meaningfulness of international cooperation from a philosophy of science perspective. As Gerrard (1996), Orosz and Crăciun (2019), and Zeigermann (2021) explain, many aspects suggest that international cooperation makes inter- and transdisciplinary research fields more powerful. First and foremost, different countries face different challenges with respect to the same transdisciplinary context (such as the energy transition). This also gives rise to different bodies of knowledge with regard to possible factual solutions, from which all participating scientists can learn. In addition, international cooperation also makes it possible to uncover culture-dependent differences in the understanding of central concepts (such as the concept of sustainability) and thus to find a common definition of central terms. There is also great potential for joint teaching. International teaching networks simply increase the human capital available for teaching, which can make education richer from the students' point of view and create the potential for cost reduction from an organizational point of view. These advantages definitely outweigh the risks that exist at the same time (e.g. danger of brain drain, loss of autonomy, or loss of conceptual diversity). Professional and academic discourse on the internationalization of higher education widely accepts the benefits of international cooperation in higher education and occasionally acknowledges the costs of such cooperative activities.

2.4 Rationale Four: Power Relations

The IECO was and is not a project critical towards international regimes (but see Braun (2024), Chap. 31 of this book). Nevertheless, in a discussion on the importance of international cooperation, a look at power structures should not be neglected. As stated at the beginning of this chapter, the discourse on sustainable development has always received influences from the Global North and the Global South—but these have never been treated with equal importance with respect to their influence on sustainability discourse and practices. For example, Albert Denk's (2023) remarkable dissertation shows that the 2030 Agenda was much more explicitly dominated by voices from the Global North while those from the Global South were under-represented. It would be naïve to claim that modern science is an entirely differently structured domain, for it too is of course subject to neo-colonial patterns (e.g., Dahdouh-Guebas et al., 2003). By no means is every form of international scientific collaboration symmetrical in terms of power relations among participants. A true symmetry would at least mean that both partners are equally involved in knowledge production (from ideation to publication), that the knowledge produced is embedded in the contexts of both partners (countries) and benefits both equally, but also that there is symmetry with respect to the distribution of resources; in addition, many cultural aspects (e.g., with respect to research languages) would be treated with importance. Many cooperation formats are far from such an ideal.

Nevertheless, in our view, international cooperation can contribute to a symmetrization of international power relations in the context of sustainable development. This is true for at least three reasons. First, international projects are arenas in which awareness of power asymmetries can take place, so that the scientists involved increasingly take them into account in their work. Second, these projects involve metacommunication, with project leaders in exchange with funding agencies and able to advise them on how to adapt to more equitable international cooperation (Hampel, 2014). Third, international cooperation—even in suboptimal formats—makes the respective perspectives on sustainable development visible and thus incrementally contributes to a fairer international science system.

3 Project History of the Institute of Eco-Industrial Development

Now that it has been established why international cooperation in the context of sustainability makes so much sense, the IECO—Institute of Eco-Industrial Development will be presented as an example of German-Chilean cooperation in the context of industrial sustainability in research, higher education, and innovation. First, the historical and institutional factors that led to the IECO's emergence will be explained, as well as why a partnership between Chile and Germany is so attractive from a scientific point of view. This will be followed by an explanation of the IECO's structural setup. Finally, we will discuss the IECO's possible contributions to the UN's 2030 Agenda.

3.1 Particularities of Chile as a Country

Chile lends itself to various areas of scientific research due to its unique geographical characteristics (see Fig. 2). Located on the southwestern side of the South American cone, Chile stretches over 4,300 km from north to south while having an average east–west extension of only 180 km. This unique shape means that the Chilean territory covers a variety of climatic zones, from cold deserts in the south (Southern Ice Fields), temperate zones in the country's middle (including Mediterranean zones and a zone of unique temperate rainforests), to heat desert in the north (Atacama Desert). There are regions of the country that have not experienced precipitation for decades, while others belong to the wettest spots on earth. Certain regions are characterized by a variety of landscapes and several distinct ecosystems, such as vast forests, agriculture, or mountains including volcanos (Andes). The central and southern zones are considered biodiversity hotspots where around 30,000 species can be found. The country is also located on a highly seismic and volcanic zone, where several tectonic plates meet and move against each other.

This diversity of landscapes and ecosystems accounts for Chile's potential in the development of renewable energies: The country has vast geothermal, solar, and wind energy resources, as well as small-scale hydropower and biomass energy resources. What's more, Chile's national policy framework favors the shift towards the use and exploration of renewable energies. Gradual energy policy reform over the last decade "has made the country a world leader in renewable energy development" (Madariaga & Allain, 2020). For all of these reasons, researchers at the Karlsruhe Institute of Technology (KIT) have shown heightened scientific interest in the country for the last three decades.

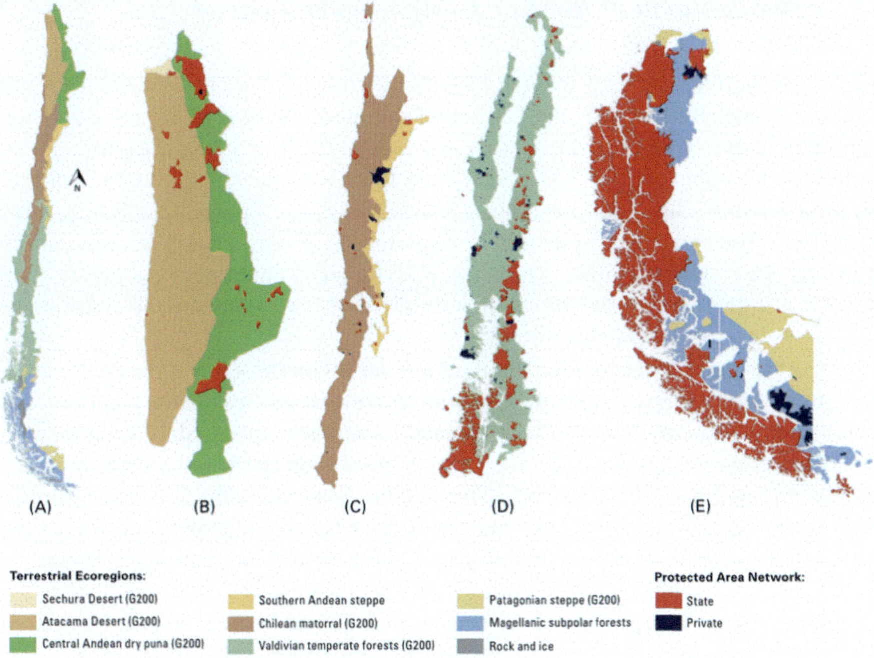

Fig. 2 Terrestrial ecoregions in Chile according to Schutz (2018): an example of the variety of geographical and ecological conditions in the country

3.2 Motivation for Cooperation

The IECO was born out of the desire for closer research cooperation between Germany and Chile. While Germany's KIT had held strong ties with Chilean universities for a long time these partnerships were isolated and conducted on a personal rather than institutional level, with individual researchers working together on papers, bilateral projects, joint field research, and data exchange. A double Master degree program between KIT and a Chilean university was established, as well as joint Ph.Ds and co-tutelle programs. In addition, there were lively student exchange programs between KIT and various universities in Santiago and other regions of Chile. Chilean alumni who had spent time at KIT as a student or researcher stayed in contact with KIT through their affiliation in the KIT alumni club, expanding the shared network between the two countries.

3.3 Establishment of the KIT Chile Cluster

In 2015, the regional strategy team at KIT International Affairs recognized the array of connections between KIT and Chile and initiated a workshop series to investigate the potential benefits of structuring and analyzing the existing activities. These workshops brought together KIT researchers to present their individual collaborative projects with Chile and to explore possible synergies. Approximately forty-five researchers from seventeen institutes or departments participated. The two leading questions were: Does it make sense for KIT as an institution to cooperate with Chile on a strategic level; is there sufficient interest? If so, what are the common or complementary research fields or topics?

The workshops were held on a regular basis and soon three key topics were identified: renewable energies, sustainable resource use, and environmental changes. The goals framing the collaboration were industrial ecology, sustainable development, and local economic growth. The researchers participating in the workshop series quickly initiated the KIT Chile Cluster, a network of individual research groups working on those topics. Prof. Dr. Andreas Braun became the spokesperson of the KIT Chile Cluster, with the organizational support of KIT International Affairs.

3.4 Institutional Process

The formation of the KIT Chile Cluster was supported by the institutional establishment of the vice-presidency in Innovation and International Affairs (today: Transfer and International Affairs) at KIT in 2016, emphasizing the importance of internationalization as a strategic advantage in academia. The newly appointed vice-president for Innovation and International Affairs closely followed the network meetings and with their support, the KIT Chile Cluster produced a first white paper outlining the relevance and uniqueness of Chile as a country of interest for cooperation in research, innovation, and higher education. The KIT Chile Cluster became one of the most visible networks both at KIT and in Chile with the main partner universities in Chile being Universidad de Chile (UCH), Universidad de Concepción (UdeC), and Universidad Austral de Chile (UACh).

3.5 Acquisition of Funding

At the end of 2016, Germany's Federal Ministry of Education and Research (BMBF) issued a call promoting research structures between German universities and research institutions with institutions in Argentina, Brazil, Chile, Colombia, and Mexico. The researchers working in KIT's Chile Cluster immediately decided to put together a solid proposal, an opportunity to further develop the existing ideas behind their initiative.

In the course of creating this proposal, it became clear that the preferred research infrastructure was that of a virtual institute, a way of turning the existing network into a highly institutionalized entity. The partners laying the foundations of this institute would be KIT, UCH, UdeC, and UACh (implementing partners). Eco-Industrial Development emerged as a framework topic, a relatively new multidisciplinary approach that encompasses a vast variety of topics from the use of resources (water and raw material) to climate to energy. This choice of focus would allow for German-Chilean collaboration in research, teaching, and innovation management and was broad enough to also include cooperation with the industrial sector in both countries.

In total, more than seventy proposals were submitted at the BMBF, of which nine were funded. As one of the winning proposals, the IECO was given funding for the next six years. (The project was prolonged due to the Covid-19 pandemic, so its total duration was seven years.) The funding would be directed almost entirely towards the development of the institute's research infrastructure, thus indirectly supporting the implementation of the institute's mission (research, teaching, innovation).

The funding approval bolstered the commitment of the four implementing partners and triggered numerous positive effects: Existing ties were strengthened and new ones were formed, various project proposals and joint papers were submitted, new strategic partners were added as associate members, and the network evolved into a significant player in the German-Chilean collaboration landscape, with the embassies, funding agencies, and entities of (economic) cooperation of both countries supporting the initiative.

4 Project Structure: The Five Academic Topics Within IECO

The vision of the IECO encompasses the development of concepts for the sustainable use and management of resources as well as sustainable energy usage and production in industry, including the interphase with urban areas. This is achieved by considering economic, ecological, and social concerns in equal measure and examining the entire value chain. In thematic terms, resources and energy are explored from five perspectives (topics) within the IECO. These five topics correspond to the existing partnerships, common interests, and expertise of the universities involved in the IECO and comprehensively cover the topic of eco-industrial development with regard to Chilean conditions under current global developments.

4.1 *Topic One: Sustainable Use of Water and Energy in an Urban Context*

The majority of humanity now lives in cities, which are responsible for around three-quarters of global resource use and the majority of emissions (e.g. Hayat, 2016). Therefore, sustainable urban development is essential for sustainable development, both globally and regionally. Management of energy usage is indispensable for maintaining the urban metabolism, with water being the greatest material flow in terms of quantity. In Chile, the water and energy sectors have until now been thought of and managed largely independently, despite exerting a considerable influence on one another.

The development of a sustainable urban environment is based on the integration of the water-energy nexus into the urban development process, which can only be possible if coordinated infrastructure is in place. For this reason, new, innovative technologies for the use of water and energy are easier to implement in new development areas than in existing buildings. Due to population and urban growth in Chile large numbers of new developments with a uniform building structure (mostly <5 house types) are being built, in contrast to Germany. These conditions could make it easier and cheaper to establish new technologies and infrastructures for integrated urban energy and water supply in Chile than would be possible in Germany, where infrastructure implementation requires costly conversions to existing buildings.

This first focal topic at the IECO, the sustainable use of water and energy in an urban context, addresses two strands of research:

(1) Reshaping the transport, use, and treatment of water and wastewater in new or existing building stocks, not only to improve the energy and water efficiency in settlements but also to improve the living conditions there, taking into consideration the climate, legal, techno-economic, as well as societal boundaries in the settlements under investigation. The ultimate goal is to achieve a local water-energy system that is sustainable.

(2) Reshaping the wastewater-energy-waste system, based on the idea of separating wastewater at its source and making use of the thermal and chemical energy content of greywater. This could contribute to the establishment of a sustainable local energy system.

4.2 *Topic Two: Sustainable Material and Energy Transformation of Biomass*

The production of forest biomass is one of Chile's most important economic sectors. It generates large profits, especially for some private-sector actors. However, biomass production, which is mainly carried out in large-scale plantations with monocultures, also raises questions about its sustainability. National actors who take a relatively small share of the sector's profits have to bear its social consequences, for example in the intensifying conflict between the forest industry and the state versus indigenous ethnic groups. These conflicts are exacerbated not least by the negative environmental consequences of this industry, whether in the form of gradual degradation (erosion) or rapid-onset natural hazards (mass movements) (e.g. Braun, 2021). With its rising demand for energy and high proportion of imported fossil fuels, Chile is increasingly faced with the question of ensuring a renewable and sustainable supply of energy. At the same time, large quantities of biogenic residues, such as forest residues, accumulate in the forestry and the wood-processing industries. Due to logistical obstacles, however, the provision of biogenic residues is often so costly that energy recovery does not appear to be economically viable.

IECO researchers began to address this second focal topic by studying a variety of issues. One issue was the transformation to a bioeconomy leading to a reorientation of production and logistics systems in agriculture and forestry, exemplified by the integration of various biorefinery concepts in the pulp and paper industry and the wood processing industry. Innovative technologies for semi-mobile biomass utilization were also developed, in which the utilization of biomass can be flexibly adapted to its spatial and temporal availability and thus have higher economic efficiency.

The research interests of the spokespeople and institutes involved in this topic currently go beyond the single utilization of biomass and include a more general view on sustainable materials within a circular economy in the future. Thus, the current research topics as a basis for joint activities are biomass utilization pathways and recycling of mixed plastic waste as well as the production, conversion, and transportation of green hydrogen.

4.3 Topic Three: Eco-Industrial Utilization of Subsurface Resources

The mining of economically relevant raw materials in Chile (e.g. Cu, Li, Ag, and Mo) requires approximately 61% of the country's energy with a total emission of 5.8 Mt of greenhouse gases. Chile's mineral and metallic resources are concentrated in paleo-geothermal systems to form mineable deposits dating back 250 million years, which are laid out in belts sub-parallel to the subduction zone in the north of the country. The easternmost and magmatically active belt contains the geothermal systems that are active today. The energy and water supply of the mining projects (which are located up to 4000 m above sea level) are usually sourced from the Pacific coast at a distance of 200 km. Based on the spatial proximity and size of the geothermal reservoirs, the aim of this topic area of the IECO is to achieve a decentralized, CO_2-neutral, baseload energy supply for mining processes and the desalination of water by working in collaboration with the mine operators (Goldberg et al., 2021).

To achieve this, the potential of desalination plants in Andean mining has been examined as well as the extraction of minerals from geothermal deep waters. Furthermore, concepts for an optimal decentralized energy supply have been developed and the implementation of an accompanied pilot plant if necessary. Specifically, the scientific progress in this topic area in recent years encompasses advances in the geothermal exploration of fractured geothermal reservoirs including the thermal source, and the development of an element-specific precipitation process to reduce the silicon dioxide concentration from geothermal waters.

4.4 Topic Four: Urban Eco-Industrial Development in South America

This topic involves investigations into potential circular economy approaches in the area of urban development. This includes developing process-oriented urban design approaches for space as a metabolism at different scales, or interpreting the analysis of urban metabolism and industrial processes and how their waste (material and energy) can be understood as a resource and used in urban design processes and vice versa. The interest is at the neighborhood scale, as an ambit to articulate regenerative design perspectives with dialogues and interventions.

For this purpose, the role of circular economy approaches related to resources such as space, water, materials, and energy (including storage and conversion systems) has been investigated at the neighborhood scale, and also for decentralized housing environment solutions. Additionally, approaches ensuring efficient use of local resources (especially water and energy) and reducing the amount of waste material in the object's life cycle have also been researched. The focus of the work in recent years was set on possible approaches and technologies to redesign neighborhoods while transforming the potential of industrial residues into new components and construction materials and recovering valuable materials from the urban settings (e.g. Tröger et al., 2023). The general objective was to construct knowledge on how circular economy and transition factors can spatially regenerate, redesign, and restructure urbanized areas, neighborhoods, and buildings (urban mining).

4.5 Topic Five: Land Use and Industrial Sustainability

This topic is focused on examining the sustainability of land use in Chile, concerning the forestry sector in particular. The forestry industry in Chile is one of the country's most important economic sectors and is responsible at the same time for triggering numerous negative environmental and social impacts. Many of these impacts are mediated through the spatial dimension of forest plantations. Plantations have spread rapidly in space, encompassing large spatial scales, increasingly approximating other land-use systems, and generating a mono-functionality of land use. Researchers working in this topic are focusing their efforts on studying the immediate environmental and social impacts of land use (impacts on regional ecosystems and rural livelihoods), on possible concepts for mitigating these impacts through governance and spatial planning, and on the possibilities for using eco-industrial strategies to generate added value for the population.

5 Conclusion: Expected Outcome of the IECO

As this chapter has shown, the IECO is a research institute that can address the issue of industrial sustainability in its full breadth. The concept of eco-industrial development—which in Europe and the US is mainly related to industrial symbioses in local eco-industrial parks (Ehrenfeld & Chertow, 2002)—is deliberately understood in a broader sense in the Latin American context. There are several reasons for this. First, extractive, rather than manufacturing, industries often predominate in Latin America (López-Morales et al., 2017). As pointed out by Kirchherr and van Santen (2019), there is a need to pay more attention to extractive industries in the context of eco-industrial development. For one, this means that eco-industrial development concepts need to be found that address this very situation, for example by focusing more on the regional, rather than the local, level (Deutz & Gibbs, 2008; Sterr & Ott, 2004). Braun and Hitzeroth (2024) have aimed to disentangle how political-economic factors influence land-use decisions in favor of extractive industries. Gibbs et al. (2005) provided the first attempts to explicitly relate eco-industrial development at the regional level. However, few publications have followed their foundations, which is why the role that eco-industrial development could play in regional development is still poorly understood. Second, it means that normative issues play a much larger role than in Western contexts, where industry triggers fewer spatial and resource use conflicts (Petras & Veltmeyer, 2014). Another reason is that eco-industrial strategies have only recently begun to be perceived as a sustainability strategy in Latin America (Tröger et al., 2023; Veiga & Magrini, 2009). Thus, the task of the IECO is more to show the potential of eco-industrial development in Latin America rather than to investigate developments that are already taking place.

The topics dealt with by the IECO therefore attempt to make generic contributions to the SDGs by asking how the environmental impacts of industry can be minimized by means of eco-industrial strategies at the regional level. The IECO's first topic area, Sustainable Use of Water and Energy in an Urban Context, contributes to SDG 6—Clean Water and Sanitation, SDG 7—Affordable and Clean Energy, and SDG 11—Sustainable Cities and Communities, while the second topic area (Sustainable Material and Energy Transformation of Biomass) develops strategies for SDG 7 and SDG 12—Responsible Consumption and Production. The third IECO topic, Eco-Industrial Utilization of Subsurface Resources, researches strategies for SDG 7, topic four (Urban Eco-Industrial Development in South America) is dedicated to SDGs 11 and 7, while topic five (Land Use and Industrial Sustainability) is interested in SDG 15—Life on Land as well as SDG 10—Reduced Inequalities. Overall, all of the IECO's topic areas contribute to SDG 9—Industry, Innovation and Infrastructure and all research that deals with mitigation also contributes to SDG 13—Climate Action.

With its focus on Latin America, a region where eco-industrial development strategies are just emerging, the IECO can also play a role in contributing to the discourse. The same applies to the interest in eco-industrial development strategies at the regional level. With the decision to work on eco-industrial development in a collaborative partnership between Germany and Chile in all three academic fields of action—research, teaching, and knowledge and technology transfer—the IECO is demonstrating its commitment to SDG 17—Partnerships for the Goals, Target 17.6 in particular: Knowledge sharing and cooperation for access to science, technology, and innovation.

Acknowledgements As the authors of this chapter, we would like to acknowledge the contributions that have shaped the virtual Institute of Eco-Industrial Development (IECO). We have received support from the German and Chilean side on countless occasions. Explicit mention should be made of the funding agencies at the BMBF, the DLR project management agency, the German and Chilean embassies, the DAAD, the German Chamber of Commerce CAMCHAL in Chile, the partner universities in Santiago de Chile, Concepción, and Valdivia, as well as universities in other locations such as Universidad Católica del Norte and the KIT Alumni Club. In addition, various people have participated in a coordinative role over the course of the project. Dr. Gabriela Espinosa, Jennifer Mesters, Danny Tröger, Dr. Fredy Rios, and Andrea Ramirez have—each in their own way—shaped this project, for which they are to be acknowledged.

References

Akerboom, S., Botzen, W., Buijze, A., Michels, A., & van Rijswick, M. (2020). Meeting goals of sustainability policy: CO_2 emission reduction, cost-effectiveness and societal acceptance. An analysis of the proposal to phase-out coal in the Netherlands. *Energy Policy, 138*, 111210.

Biermann, F., Betsill, M. M., Gupta, J., Kanie, N., Lebel, L., Liverman, D., Schroeder, H., Siebenhüner, B., & Zondervan, R. (2010). Earth system governance: A research framework. *International Environmental Agreements: Politics, Law and Economics, 10*(4), 277–298. https://doi.org/10.1007/s10784-010-9137-3

Braun, A. C. (2021). Encroached by pine and eucalyptus? A grounded theory on an environmental conflict between forest industry and smallholder livelihoods in Chile. *Journal of Rural Studies, 82*, 107–120.

Braun, A. C. (2024). Is eco-industrial development just "More of the Same"? A Hegelian dialectic view on an emerging concept. In A.C. Braun, G. Espinosa, D. Tröger, & T. Hirth (Eds.), *Eco-industrial development as an industrial strategy—Contributions from a German-Chilean Research Partnership*. Springer Nature.

Braun, A. C., & Hitzeroth, M. (2024). Relationship between political-economic factors and the spatial expansion of the plantation industry in the Coastal Region of Central Chile. In A.C. Braun, G. Espinosa, D. Tröger & T. Hirth (Eds.), *Eco-industrial development as an industrial strategy—Contributions from a German-Chilean Research Partnership*. Springer Nature.

Buhmann, K. (2017). *Power, procedure, participation and legitimacy in global sustainability norms: A theory of collaborative regulation*. Routledge.

Dahdouh-Guebas, F., Ahimbisibwe, J., Van Moll, R., & Koedam, N. (2003). Neo-colonial science by the most industrialised upon the least developed countries in peer-reviewed publishing. *Scientometrics, 56*, 329–343.

Denk, A. (2023). Nachhaltige Entwicklung und globale Ungleichheit. In *Eine wissenspolitologische Studie über die Entwicklungsagenda der Vereinten Nationen*. Nomos.

Deutz, P., & Gibbs, D. (2008). Industrial ecology and regional development: Eco-industrial development as cluster policy. *Regional Studies, 42*(10), 1313–1328.

Dushyantha, N., Batapola, N., Ilankoon, I., Rohitha, S., Premasiri, R., Abeysinghe, B., Ratnayake, N., & Dissanayake, K. (2020). The story of rare earth elements (REEs): Occurrences, global distribution, genesis, geology, mineralogy and global production. *Ore Geology Reviews, 122*, 103521. https://doi.org/10.1016/j.oregeorev.2020.103521

Ehrenfeld, J. R., & Chertow, M. R. (2002). 27. Industrial symbiosis: The legacy of Kalundborg. In R. U. Ayres & L. W. Ayres (Eds.), *A handbook of industrial ecology*, Edward Elgar.

European Commission. (2021). Communication from the Commission to the European Parliament, the Council, the European Economic and Social Committee and the Committee of the Regions on the Global Approach to Research and Innovation. Europe's strategy for international cooperation in a changing world.

Fukuda-Parr, S., & Muchhala, B. (2020). The Southern origins of sustainable development goals: Ideas, actors, aspirations. *World Development, 126*, 104706.

Gerrard, J. (1996). The importance of international cooperation in science and technology. *SRA Journal, 28*(1), 29.

Gibbs, D., Deutz, P., & Proctor, A. (2005). Industrial ecology and eco-industrial development: A potential paradigm for local and regional development? *Regional Studies, 39*(2), 171–183.

Goldberg, V., Winter, D., Nitschke, F., Rath, M., Held, S., Spitzmüller, L., et al. (2021). The potential of raw material extraction from thermal brines—Successful milestones of the BrineMine project. *Oil Gas, 1*, 26–33.

Hampel, A. (2014). *Fair cooperation: Partnerschaftliche Zusammenarbeit in der Auswärtigen Kulturpolitik*. Springer-Verlag.

Hayat, P. (2016). Smart cities: A global perspective. *India Quarterly, 72*(2), 177–191.

Hoel, M. (1997). Coordination of environmental policy for transboundary environmental problems? *Journal of Public Economics, 66*(2), 199–224.

Honegger, M., Poralla, M., Michaelowa, A., & Ahonen, H. M. (2021). Who is paying for carbon dioxide removal? Designing policy instruments for mobilizing negative emissions technologies. *Frontiers in Climate, 3*, 672996.

Kefalas, A. G. (1998). Think globally, act locally. *Thunderbird International Business Review, 40*(6), 547–562.

Kirchherr, J. W., & van Santen, R. (2019). Research on the circular economy: A critique of the field. *Resources, Conservation and Recycling, 151.*

Kump, L. R., Kasting, J. F., & Crane, R. G. (2004). *The earth system* (Vol. 432). Pearson Prentice Hall.

Loftus, P. J., Cohen, A. M., Long, J. C., & Jenkins, J. D. (2015). A critical review of global decarbonization scenarios: What do they tell us about feasibility? *Wiley Interdisciplinary Reviews: Climate Change, 6*(1), 93–112.

López-Morales, J. S., Ortega-Ridaura, I., & Ortiz-Betancourt, I. (2017). Strategies of corporate social responsibility in Latin America: A content analysis in the extractive industry. *AD-Minister, 31*, 115–135.

Madariaga, A., & Allain, M. (2020). Contingent coalitions in environmental policymaking: How civil society organizations influenced the Chilean renewable energy boom. *Policy Studies Journal, 48*(3), 672–699.

Marotzke, J., Semmann, D., & Milinski, M. (2020). The economic interaction between climate change mitigation, climate migration and poverty. *Nature Climate Change, 10*(6), 518–525.

McCoy, D., & Kinyua, K. (2012). Allocating scarce resources strategically-an evaluation and discussion of the global fund's pattern of disbursements. *Plos One, 7*(5), e34749.

Milman, A., Gerlak, A. K., Albrecht, T., Colosimo, M., Conca, K., Kittikhoun, A., Kovács, P., Moy, R., Schmeier, S., Wentling, K., Werick, W., Zavadsky, I., & Ziegler, J. (2020). Addressing

knowledge gaps for transboundary environmental governance. *Global Environmental Change, 64*, 102162. https://doi.org/10.1016/j.gloenvcha.2020.102162
Morens, D. M., & Fauci, A. S. (2020). Emerging pandemic diseases: How we got to COVID-19. *Cell*.
Morens, D. M., & Fauci, A. S. (2013). Emerging infectious diseases: Threats to human health and global stability. *Plos Pathogens, 9*(7), e1003467.
Müller-Mahn, D. (2013). *The spatial dimension of risk. The spatial dimension of risk. How geography shapes the emergence of riskscapes*. Routledge.
Nagendra, H., Bai, X., Brondizio, E. S., & Lwasa, S. (2018). The urban south and the predicament of global sustainability. *Nature Sustainability, 1*(7), 341–349.
Orosz, K., & Crăciun, D. (2019). *Dearth of evidence of value of international cooperation*. University World News.
Petras, J., & Veltmeyer, H. (2014). *Extractive imperialism in the Americas: Capitalism's new frontier*. Brill
Rees, W. E. (2006). Globalization, trade and migration: Undermining sustainability. *Ecological Economics, 59*(2), 220–225.
Rockström, J., Steffen, W., Noone, K., Persson, Å., Chapin, F. S., Lambin, E., Lenton, T. M., Scheffer, M., Folke, C., Schellnhuber, H. J., Nykvist, B., de Wit, C. A., Hughes, T., van der Leeuw, S., Rodhe, H., Sörlin, S., Snyder, P. K., Costanza, R., Svedin, U., et al. (2009). Planetary boundaries: Exploring the safe operating space for humanity. *Ecology and Society, 14*(2). http://www.jstor.org/stable/26268316
Sachs, J. D. (2012). From millennium development goals to sustainable development goals. *The Lancet, 379*(9832), 2206–2211.
Sadik-Zada, E. R. (2021). Political economy of green hydrogen rollout: A global perspective. *Sustainability, 13*(23), 13464.
Schutz, J. (2018). Creating an integrated protected area network in Chile: A GIS assessment of ecoregion representation and the role of private protected areas. *Environmental Conservation, 45*(3), 269–277.
Sijm, J. P., Ormel, F. T., Martens, J. W., Van Rooijen, S. N., Voogt, M. H., Van Wees, M. T., & de Zoeten-Dartenset, C. (2000). Kyoto mechanisms. The role of joint implementation, the clean development mechanism and emissions trading in reducing greenhouse gas emissions.
Spector, B. I., Sjöstedt, G., & Zartman, I. W. (1994). Negotiating international regimes: Lessons learned from the United Nations conference on environment and development (UNCED). Graham & Trotman/Martinus Nijhoff.
Steffen, W., Richardson, K., Rockström, J., Cornell, S. E., Fetzer, I., Bennett, E. M., Biggs, R., Carpenter, S. R., de Vries, W., de Wit, C. A., Folke, C., Gerten, D., Heinke, J., Mace, G. M., Persson, L. M., Ramanathan, V., Reyers, B., & Sörlin, S. (2015). Planetary boundaries: Guiding human development on a changing planet. *Science, 347*(6223). https://doi.org/10.1126/science.1259855
Sterr, T., & Ott, T. (2004). The industrial region as a promising unit for eco-industrial development—Reflections, practical experience and establishment of innovative instruments to support industrial ecology. *Journal of Cleaner Production, 12*(8–10), 947–965.
Tröger, D., Araneda, A. A. B., Busnelli, R., Yajnes, M., Williams, F., & Braun, A. C. (2023). Exploring eco-industrial development in the global south: Recognizing informal waste-picking as urban-industrial symbiosis? *Cleaner Waste Systems, 5*, 100096.
United Nations Sustainable Development. (n.d.). Communications materials. Retrieved April 10, 2023, from https://www.un.org/sustainabledevelopment/news/communications-material/
Vaccari, D. A. (2009). Phosphorus: A looming crisis. *Scientific American, 300*(6), 54–59.
Veiga, L. B. E., & Magrini, A. (2009). Eco-industrial park development in Rio de Janeiro, Brazil: A tool for sustainable development. *Journal of Cleaner Production, 17*(7), 653–661.
Zeigermann, U. (2021). Knowledge integration in sustainability governance through science-based actor networks. *Global Environmental Change, 69*, 102314.

Principles of Eco-Industrial Development

The Concept of Industrial Ecology

Witold-Roger Poganietz and Vanessa Bolivar Paypay

Abstract The core aim of Industrial Ecology is to overcome the linear production system by reorganizing the ways and means in which resources are transformed into products and services in order to reduce the environmental impact of human activities while improving living standards in the long term. To achieve this core aim, the field of Industrial Ecology addresses the challenges of economic agents as well as of entire societies at different scales and geographic boundaries, ranging from individual companies to global material flows. Although the focus is on material flows and their environmental impacts, economic and social science is gaining relevance in order to capture the embeddedness of modern economies in socio-economic developments. Although closing the loop is at the center of Industrial Ecology, the Circular Economy is a comparable concept that has gained prominence, leading to an intensive discussion of what type of relations both concepts have and how these could lead to improved answers to the challenges faced by modern societies.

1 Introduction

Since the beginning of the Industrial Revolution, starting in the 18th century, the industrial exploitation of fossil, metal, and mineral resources has been a central backbone of modern economies (Harari, 2015). This will not change in the foreseeable future even though many innovations aim at increasing resource efficiency, often by the introduction of composites. Higher resource efficiency allows for a better availability of products and services, resulting in further increases in demand for (still mostly primary) resources. While reuse and recycling are gaining in relevance, they still have not reached the necessary levels to achieve a noteworthy reduction

W.-R. Poganietz (✉) · V. B. Paypay
Institute for Technology Assessment and Systems Analysis (ITAS), Karlsruhe Institute of Technology (KIT), Karlsruhe, Germany
e-mail: witold-roger.poganietz@kit.edu

V. B. Paypay
e-mail: vanessa.bolivar@kit.edu

A. Ch. Braun et al. (eds.), *Eco-Industrial Development as an Industrial Strategy*, Green Energy and Technology, https://doi.org/10.1007/978-3-031-73576-9_2

in pressure on the exploitation of resources (OECD, 2019). Furthermore, composites challenge the recyclability of products, impeding potential closed-loop systems (Korniejenko et al., 2021).

In parallel, the global population will further increase to an estimated 9 billion by 2050, compared to 6.8 billion in 2015, raising the demands for energy, food, and water by between 50 and 60% (Kaza et al., 2018). The required fossil, metal, and mineral resources as well as the generation of waste will follow mostly in step with this, if the still-dominant system of value-creation within a linear production system—i.e. a system in which raw materials are used once and then disposed of—continues. The World Bank has estimated that global waste would rise by about 70% until 2050 compared to 2016, when about 2.01 billion tons of waste were generated (Kaza et al., 2018).

These figures describe the current and the expected future developments, calling into question the sustainability of linear production systems. These discussions had already begun as early as the 1960s. However, it took until 1989 before Industrial Ecology emerged as a field of study (Erkman, 1997).

The core aim of Industrial Ecology is to reorganize the ways and means in which resources are transformed into products and services in order to reduce the environmental impact of human activities while improving living standards in the long term (Erkman, 1997; Stiehl & Hirth, 2012). Generally, the transformation should occur in an environment of a market-based exchange of goods, taking into account the economic constraints of economic agents, such as a positive profit rate.

To achieve this core aim, Industrial Ecology addresses single companies' activities as well as specific material flows, sectors, or economies on a local, regional, national, or global scale (Graedel & Lifset, 2016). As a result the scope of Industrial Ecology is accordingly broad, leading to a plethora of methods to analyze systems as well as practical approaches to overcome linear production systems. The methods range from more engineering-influenced approaches like Material Flow Analysis (MFA) and Life Cycle Assessment (LCA) to more socio-economic ones such as Input–Output Approaches (IOA). Although in the last decades social science approaches have been gaining in relevance, MFA, LCA, and IOA are still the methodological backbones of Industrial Ecology, supplemented by Industrial Symbiosis as a practical application of Industrial Ecology (Graedel & Lifset, 2016).

Despite the broad scope of Industrial Ecology, the common ground and, therefore, the reason for the use of the term Industrial Ecology, is to transform regular, linear-based manufacturing processes or economies into closed-loop systems, at best mimicking ecological systems. In theory, ecological systems do not generate waste, achieving a complete cycling of material flows and using only solar energy (Graedel & Allenby, 2010). Consequently, Industrial Ecology makes use of different concepts from ecology, including material and energy cycles, complex adaptive systems, and ecological networks (Erkman, 1997).

While the idea of establishing circular production is a constituent part of Industrial Ecology, influenced by the same discussions on future viable production systems since the 1960s, in the last thirty years the concept of Circular Economy has gained relevant recognition in political and societal debates as well as in academia. Comparable to the ideas of Industrial Ecology, the Circular Economy looks for ways and means to transform the linear or semi-circular flows of material and energy into a closed-loop system in order to achieve sustainable benefits (Saavedra et al., 2018). This leads to the question of whether and to what extent collaborations between both approaches could lead to solutions to improve living standards in lasting and sustainable way.

The overarching aim of this chapter is to provide an overview of the status quo of Industrial Ecology today, while also addressing the roots of Industrial Ecology and the challenges it faces today and may face in the future. Due to the relevance of Circular Economy in current academic and societal discussions, the connection between Industrial Ecology and Circular Economy is also highlighted.

2 Concept and Origins

According to the International Society of Industrial Ecology (ISIE), the main international organization promoting Industrial Ecology, "*[i]ndustrial ecology is the study of systemic relationships between society, the economy, and the natural environment*" with the aim to reduce the environmental impact of human activities and to reconcile human development (https://doi.is4ie.org/about/what-is-industrial-ecology). In doing so, the focus of Industrial Ecology is on identifying and quantifying the use and transformation of materials and energy in society and their interchange with nature, by extraction, emitting, and disposing. Industrial Ecology seeks to achieve a harmonious integration of the industrial system with its surrounding environment. This has to be accompanied by an optimal utilization of the flow of materials and energy, considering a life cycle approach (Allenby & Cooper, 1994) (Fig. 1). However, Graedel and Allenby (1995) also emphasize the need to achieve economic, cultural, and technological growth while maintaining the carrying capacity of the ecosystem.

The driving force behind Industrial Ecology is the concern that the current way of extracting resources from the Earth, transforming these into products and services, and consuming and disposing of them is unsustainable and will lead in the long run to a deadlock, with undesirable outcomes for human beings worldwide. The supplies of fossil, metal, and mineral resources are limited with decreasing content per excavated material, raising the demand for additional financial, human, and material resources. Furthermore, the extraction of mineral and metal resources is often performed under inhumane conditions, while the income from such activities often constitutes the primary or only source of income for miners or of countries (Adeola et al., 2022). The mining, transformation, consumption, and disposal of resources are often connected with emissions that negatively affect the macro and micro climate as well as the regional environment and increasingly affect human health (Farjana et al., 2018).

These concerns have led to a search for sustainable ways to organize the use of fossil, mineral, and metal resources. The term "Industrial Ecology" derives from a

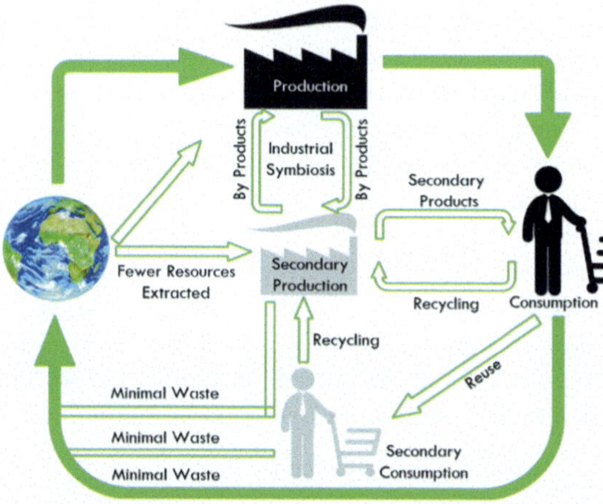

Fig. 1 Industrial ecology—Closing the loop. *Source* Abhishek and Biswas (2022)

recognition that economic systems (such as manufacturing processes) and ecosystems should be seen as similar. Erkman (1997, p. 1) highlighted the similarities between industrial and natural ecology in his definition of the field: "*The idea is first to understand how the industrial system works, how it is regulated, and its interaction with the biosphere; then, on the basis of what we know about ecosystems, to determine how it could be restructured to make it compatible with the way natural ecosystems function.*"

The basis of this connection between industrial and natural ecology can be dated back to the 19th century, with the establishment of metabolism as a description of the exchange of materials within the human body. In the meantime, the concept of metabolism spread into other scientific fields such as social sciences, agronomy, and ecology. In the 1960s, discussions began about the future of manufacturing systems. However, it wasn't until 1989 that Industrial Ecology emerged as a field of study (Graedel & Lifset, 2016).

The catalyst for Industrial Ecology as it is known and studied today is the paper "Strategies for Manufacturing" by Frosch and Gallopoulos (1989). At that time, the authors concluded that, taking into account the development of resource consumption and population, the traditional industrial mode of production was not sustainable in the long term. A transformation of the established production model to an industrial ecosystem would therefore be necessary and existential for securing and improving living standards. Such an industrial ecosystem should mimic the biological cycle system: plants synthesize minerals, herbivores feed on photosynthetic products, which in turn form the nutritional basis for carnivores, and the excrement of herbivores and carnivores as well as dead bodies in turn serve as nutrients for generations of plants (Erkman, 2007; Graedel & Allenby, 2010; Van Ewijk et al., 2023). Thus, nature mutates from a resource to a role model "*in dealing with substances, energy, information, space and time on the way to sustainable development*," assigning nature as an "*identity-forming characteristic of Industrial Ecology*" (Isenmann, 2007, p. 61; translations by the authors).

The original focus of Industrial Ecology was on industrial production and the interrelationship between production and its upstream and downstream processes. This interrelationship is often referred to by the term "industrial metabolism." In fact, this term can and should be expanded to include the "*metabolism of industrial societies as distinct from other forms of society*," which would include "*consumption processes and ultimately the way of life based on them*" (Fischer-Kowalski, 2007, p. 89; translations by the authors). Thus, the term "social metabolism" seems to be more appropriate.

The idea of mimicking ecological systems is disputed (Ayres, 2004; Bey, 2007). The design of economic systems follows other rules and is less straightforwardly organized in comparison to the functioning of ecological systems (Graedel & Allenby, 2010). Therefore, economic systems would fail if they would mimic ecological systems. In contrast to biospheres, economic systems make use of and require a plethora of energy carriers. There is no single energy carrier that could take over the unique role that photosynthesis plays in biospheres. The output of biospheres is organic waste and dead matter, whereas the output of economic systems is a heterogeneous mix of products, wastes, and residuals with a noteworthy share of non-organic wastes, challenging the possibilities for recycling. The principal concept of economic systems is based on the voluntary exchange of goods and services, with money as the central transaction medium. In the biosphere the exchange is (mostly) forced. The transition of economic systems is the result of innovations and new discoveries, in contrast to the random genome mutations that are the main driving force behind transition in the biosphere (Ayres, 2004).

Today's analyses within the field of Industrial Ecology consider different levels and scales. They address the challenges faced by companies to preserve the environment and to acknowledge the embeddedness of techno-economic activities in the life of local communities (Chertow, 2007; Frosch & Gallopoulos, 1989), as well as the challenges created by the need of modern economies for the flow of materials such as carbon (Uihlein et al., 2006), or challenges emerging within cities (Castan Broto et al., 2012; Graedel & Allenby, 2010).

Key to Industrial Ecology is to support decision-making processes with information about resources and the environmental impacts of industrial production processes. In recent decades the explicit recognition of the embeddedness of industrial activities in society has gained in relevance in research studies (Binder, 2007; Fuss et al., 2021), broadening the scope of potentially available information. In case of situations non-desired by societies, Industrial Ecology-based analysis also offers concepts for preferred or needed solutions. Industrial Ecology aims to produce concepts of sound systems and action knowledge, using a plethora of approaches and methods.

Traditionally, the main focus of Industrial Ecology is on technologies, i.e. how to develop and make use of a technology or a portfolio of technologies to reduce the environmental and resource footprint. This is achieved by combining knowledge from engineering and natural science, in particular environmental science and toxicology. However, knowledge from economics, sociology, and cultural sciences are gaining in relevance in helping to understand the current structure of material and energy flows and how to transform these to more sustainable futures. The inclusion of more disciplines allows for a broadening of perspectives on the challenges of human activities. Due to this, a more comprehensive understanding of sustainability paves the way forward, overcoming the understanding of sustainability that is more environment-focused.

3 Methods and Approaches

The driving intention of Industrial Ecology is to translate the general ideas of Industrial Ecology into practice. The core concept to realize this is Industrial Symbiosis. The concept translates the general ideas of Industrial Ecology into a practical approach. Industrial Symbiosis aims at physically connecting industries that are traditionally separated or independent—because they address different markets—through the exchange of energy, materials, water, and byproducts. Wastes and residuals of one company are used as input for another (Duchin & Levine, 2014; Hardy & Graedel, 2002). A pre-condition is seen in a specific proximity of the companies (Chertow, 2000). Prosman et al. (2017) see Industrial Symbiosis a key strategy to closing material flows claiming to mimic ecosystems. That means, material in the form of nutrients is cycled and energy cascades between actors of a natural system to the benefit of all (Hardy & Graedel, 2002).

Efforts to translate the concept of Industrial Symbiosis in practice led to the idea of establishing Eco-Industrial Parks (EIPs) like the one in Kalundborg, Denmark (Chertow, 2007; Ehrenfeld & Gertler, 1997). Characteristic of EIPs is the sharing or exchanging of resources between the businesses established on a common property (Lowe, 1997). Through joint management of resource and environmental issues, the companies endeavor to enhance their environmental, economic, and social performance (Halonen & Seppänen, 2019). Industrial Ecology places greater emphasis on the physical exchange of materials, while social science-oriented Industrial Ecology studies treat EIPs as a dynamic collaborative process developing over time (Boons et al., 2011, 2014).

Another approach to reduce environmental impacts is the "Design for the Environment (DfE)" approach, which aims to foster the design or redesign of products with better environmental performance (Graedel & Lifset, 2016).

The rapid broadening of the scope of Industrial Ecology at the start of the 1990s, for example by the inclusion of regional and national material flow analyses, led to the inclusion of multiple disciplines including natural sciences, engineering, and social sciences, which are often combined. This is mirrored in the variety of methods and approaches that are used in Industrial Ecology. Since the identification and quantification of material flows on different scales is of central interest, there is extensive literature devoted to Material Flow Analysis (MFA), addressing not only the scope of the MFA method but also providing required knowledge of material flows under investigation. The methodological discussions focus on setting systems boundaries, the required detail of modeling and analyzing, and generating required data to disclose the material and energy flows over the life cycle. MFA studies address companies and industrial plants, but also the global economy (Graedel & Lifset, 2016).

Although MFA is often used to assess the state of a system from a methodological perspective, it is not an assessment approach like Life Cycle Assessment (LCA). Comparable to MFA, LCA quantifies materials and energy "from the cradle to the grave," however the reference of the analysis is not a specific material but rather a product, service, or technology. In addition, it relates the material flows and energy to environmental flows, i.e. emissions, and the potential impacts on the environment in a broader sense, including impacts on human health as well. The aim of LCA is a comprehensive assessment of potential environmental impacts of products, services, or technologies during the entire life cycle. Ideally LCA is part of a decision-making process between alternative products, services, or technologies revealing possible consequences of human activities (Bauer & Poganietz, 2007; Espinosa & Casa, 2024). Due to the construction of LCA, which is to some extent comparable to an MFA, it can be smoothly connected with MFA, extending the scope of that analysis (Uihlein et al., 2006).

To disclose the material and energy flow-based interconnections between economic sectors, Input–Output Tables (IOT) are often used (Nakamura & Kondo, 2009). To quantify sector-specific environmental impacts, environmentally extended input–output analysis or connecting LCA with IOT are common approaches. To recognize waste flows and secondary resources in a more detailed manner the waste input–output table (WIOT) (Nakamura & Kondo, 2009) and EXIOBASE (Merciai & Schmidt, 2018) are other examples of the use of IOT-based research in Industrial Ecology.

Other approaches, such as socioeconomic metabolism for national economies, urban metabolism for cities, and the analysis of important socioeconomic factors such as consumer behavior, business models, and public policy are increasingly of relevance when it comes to achieving the goals of Industrial Ecology. For instance, urban metabolism has gained special attention as of late due to the rising proportion of the world's population living in cities. Based on the idea that urban systems can be studied analogously to the metabolism of organisms (Kennedy et al., 2011) and function similarly to ecosystems (Castan Broto et al., 2012), cities can be designed and planned in a more sustainable and self-sufficient way in order to build up resiliency to climate change and other global challenges. Studies of urban metabolism are mostly conducted with the use of MFA (Castan Broto et al., 2012) in order to quantify the urban flows and stocks e.g. for food, water, waste, or energy systems (Bristow & Kennedy, 2013; Serrao-Neumann et al., 2019). Regarding how consumer behavior can affect the life cycle of a product, especially during the use phase, studies that combine the use of Agent-Based Modelling and LCA have been recently conducted (Hicks, 2022; Lan & Yao, 2019).

In sum, Industrial Ecology makes use of and combines top-down and bottom-up approaches, with the ecology of industrial systems as the common base (Duchin & Levine, 2014).

4 Circular Economy in the Context of Industrial Ecology

Over the last three decades the concept of a Circular Economy has experienced increasing attention not only in academia, but also from companies, governments, and non-governmental organizations (Glöser-Chahoud & Schultmann, 2024). Although no widely accepted definition of Circular Economy exists (Kirchherr et al., 2017; Rizos et al., 2017), the broad idea of Circular Economy is to overcome the linear model of manufacturing, consumption, and disposal, comparable to Industrial Ecology. The aim is to establish a Circular Economy in which wastes and residuals are valorized through the so-called R strategies such as reuse and recycling, among others (Ellen MacArthur Foundation, 2013; Geng & Doberstein, 2008; Glöser-Chahoud & Schultmann, 2024). The guiding principle is to organize economics according to the principle "*everything is an input to everything else*" (Su et al., 2013). The reduction of wastes and utilization of renewable energy sources is also often seen as part of achieving a Circular Economy (Kirchherr et al., 2017). The main instruments discussed in Circular Economy with long-lasting impacts are design, maintenance, repair, reuse, remanufacturing, refurbishing, and recycling (Ellen MacArthur Foundation, 2013; Geng & Doberstein, 2008; Glöser-Chahoud & Schultmann, 2024; Gregson et al., 2015).

The Circular Economy concept aims at the reduction or even elimination of the conflicting aims of economic performance and environmental protection while stimulating economic growth. Transforming economies to a circular setting would allow for a decoupling of economic growth from natural resource depletion and environmental degradation (Glöser-Chahoud & Schultmann, 2024; Merli et al., 2018; Murray et al., 2017). Although the improvement of living standards should be the final goal of a Circular Economy, the available definitions of Circular Economy seldom incorporate the necessity of a systemic shift in the way an economy should be organized. Thus, any possible adverse impacts on the principles of market design (with possible market power shifts to the producer), on international trade, on social equity, and on future generations are not in the center of research. In addition, business models or consumers as possible drivers of a systemic shift are frequently not considered (Corvellec et al., 2022; Kirchherr et al., 2017), even though in practice business models play a crucial role in Circular Economy research (Baldassarre et al., 2019; Lüdeke-Freund et al., 2019). A prominent exception is the Ellen MacArthur Foundation, a US non-governmental organization promoting the implementation of Circular Economy. According to the foundation, a Circular Economy is "*an industrial system that is restorative or regenerative by intention and design. It replaces the 'end-of-life' concept with restoration, shifts towards the use of renewable energy, eliminates the use of toxic chemicals, which impair reuse, and aims for the elimination of waste through the superior design of materials, products, systems, and, within this, business models.*" (Ellen MacArthur Foundation, 2013, p. 7). However, societal impacts are still disregarded.

As previously mentioned, the idea of a Circular Economy is recognized in political discussions and activities, with Germany's "Circular Economy and Waste Act" (in German: Kreislaufwirtschafts-und Abfallgesetz (KrW/AbfG)), passed in 1994, as the first law worldwide to establish the idea of Circular Economy as a guiding principle for waste management. However, the act does not define the term "Circular Economy" but proposes a waste hierarchy to promote the establishment of a Circular Economy. This proposed hierarchy follows the common suggestions of modern waste management, starting with reduction, then reuse, recycling and other measures. China's "Circular Economy Promotion Law," on the other hand, defines Circular Economy as a "*generic term describing the activities of decrement, recycling, and resource recovery in production, circulation, and consumption*" (Article 2; National People's Congress of the People's Republic of China, 2008). Chile's "Extended Liability of the Producer" law (Ley 20920) follows the same intention (Biblioteca del Congreso, 2016). By making the producer liable for waste generated by their products, the producer shall be incentivized to reduce the generation of waste as well as to encourage its reuse, recycling, and other types of recovery. A broader understanding is stipulated by the European Commission: Circular Economy is seen as an economy "*where the value of products, materials, and resources is maintained in the economy for as long as possible, and the generation of waste minimized*" (European Commission, 2015, p. 2). Whereas Germany's "Circular Economy and Waste Act," China's "Circular Economy Promotion Law," and Chile's "Extended Liability of the Producer" follow the academic mainstream definitions, the EU at least implicitly addresses the systemic shift. However, societal impacts are not addressed at all.

Both concepts, Circular Economy and Industrial Ecology, follow the same idea of closing the loop of energy and material flows as a possible and feasible way to reduce the environmental impacts of economic activities, while improving economic development and striving for better, more sustainable standards of living of human beings (Ehrenfeld, 2004; Ellen MacArthur Foundation, 2013). Both concepts use the same literature as the starting point for discussions (Van Ewijk et al., 2023). The main methods are also similar, with LCA and MFA as prominent tools in the analyses (Al-Thani & Al-Ansari, 2021).

Despite the two most obvious similarities and some others such as addressing the company's level (Geisendorf & Pietrulla, 2017), the relationship between the two concepts is still under discussion (Puig, 2019). For example, whereas some scientists see Circular Economy as a broader concept compared to Industrial Ecology, with Industrial Ecology as one of many sources of Circular Economy, other scientists propose a completely reverse relationship (Puig, 2019). For example, according to Saavedra et al. (2018), the progress of Circular Economy as an approach to reorganize industrial systems occurred in the wake of Industrial Ecology, in particular due to the concept of Industrial Symbiosis and Eco-Industrial Parks. The authors identified conceptual, technical, and policy aspects where Industrial Ecology could contribute to the evolvement of Circular Economy (Saavedra et al., 2018; Van Ewijk et al., 2023).

With its research that is usually more engineer-oriented, Industrial Ecology is good at explaining the material and energy flow-related conditions necessary for the establishment of Industrial Symbiosis clusters as well as the long-run conditions for successful progress. Furthermore, the environmental impacts of such collaborations as well as the more systemic impacts are also often at the center of Industrial Ecology-based analysis. The growing relevance of social science-based Industrial Ecology research allows for a broadening of the scope of required conditions for establishing Industrial Symbiosis. In contrast to that, Circular Economy concentrates very much on the business logic of running Industrial Symbiosis clusters. Environmental, societal, or systemic implications such as path dependencies of chosen collaborations as well as possible lock-in situations are generally not considered (Baldassarre et al., 2019).

Although the overarching aims of both concepts are the same, i.e. to establish a closed-loop system, the general understanding of the functional aim of both concepts seems to differ. Circular Economy is principally a restorative system. Restorative systems are by definition resilient. Changes in the interrelation between systems components will not affect the guiding principle of the system under investigation. The idea of Circular Economy transforming the linear principle of production, i.e. of utilizing material and energy, to closed-loop systems aims to restore the natural base of economies (Ellen MacArthur Foundation, 2013). Industrial Ecology provides concepts which should be understood as more preventive. Preventive systems look more for measures to minimize the losses of activities and/or to optimize their handling. That is, Industrial Ecology would look for measures to minimize the negative environmental impacts of material use and energy conversion and/or to optimize their use, considering constraints given by economic agents such as profit orientation (Al-Thani & Al-Ansari, 2021).

Despite the general outline of both concepts, their research complements each other (Al-Thani & Al-Ansari, 2021). Combining the underlying, partly divergent understandings and intentions of both could lead to richer insights and support a more appropriate design and establishment of Industrial Symbiosis clusters (Bocken et al., 2017; Short et al., 2014). For example, Baldassarre et al. (2019) analyzed the conditions for designing Industrial Symbiosis clusters using the field of strategic design (Calabretta & Gemser, 2017) as an umbrella to include Industrial Ecology and Circular Economy.

5 Challenges

Currently, the world is experiencing several transformation processes, of which the process that aims to mitigate climate change is often seen as the most required one. In parallel or perhaps delayed in time, the framework in which the aforementioned transformation occurs is also transforming: the most significant changes are the increasing social conflicts in societies worldwide as well as the changes in the geo-political systems established after the collapse of the socialist countries at the beginning of the 1990s, of which the direction is still unclear. Currently, it seems that the transformation of the framework conditions will rather impede the desired transformation to mitigate climate change.

The changes to the global and national framework conditions underscore how the still-dominant techno-economic understanding of modern economies fails to find appropriate transformation pathways. Modern economies should be seen as socio-technical systems. Using the definition of Rohracher (2008, p. 147) for energy systems as a reference, modern economies are "*socio-technical configurations where technologies, institutional arrangements (*e.g. *regulation, norms), social practices and actor constellations (such as user-producer relations and interactions, intermediary organisations, public authorities) mutually depend on each other, and are embedded into broader contexts of cultural values, [and] socio-economic trends (globalisation, individualisation,* etc.*).*" The arrangements, social practices, and actor constellations determine the design of economic systems in which economic agents are acting, and thus the way the environment is used as a supplier of resources and as a sink for emissions.

Although at the beginning the focus of Industrial Ecology was strongly linked to the technical and environmental dimensions of modern economies (Graedel & Lifset, 2016), the embeddedness of human activities in an economic as well as social sphere was recognized (Fischer-Kowalski, 1998; Nakamura & Kondo, 2009). The introduction of Structural Agent Analysis by Binder (2007) is another example. Most recent research deals for example with the question of how behavioral change can play a vital role to reduce environmental impacts and therefore achieve environmental targets (Hennequin et al., 2023; Svenfelt et al., 2022). The role of gender when assessing human toxicity or the use phase of new technologies is also gaining relevance in Industrial Ecology research (Sobieraj & Krämer, 2020).

The broadening of the scope of research and required methods in Industrial Ecology is, however, a precondition for conducting a holistic sustainability assessment, i.e. for having a better understanding of which discussed solutions can support a sustainable transformation of the economic system. This means that the challenge, which is not new but increasingly challenging, is to "set the scene," i.e. to find societally relevant research question(s), to define required system boundaries appropriately, as well as to identify fitting components of the system under review. Furthermore, in case of a comprehensive sustainability assessment, there is a need to identify a sustainable reference system. Since an assessment not only follows strict laws of nature but is the outcome of "societal bargaining" of perceptions, it could be the case that there is not *one* sustainable solution. The challenge for Industrial Ecology would be to find ways and means to challenge the challenge. In doing so, one should be aware that the aim of Industrial Ecology is to address all relevant levels and scales of human activity, e.g. not only the levels of companies but also the sectoral, national, or global ones. Different geographical scales such as those of industrial zones or cities are also in the focus of Industrial Ecology.

Further leveraging the methods and approaches of Industrial Ecology to emerging concepts such as Circular Economy creates another challenge. Although there are studies that address many of these topics from an Industrial Ecology perspective, these still take place mainly at an academic level and in industrialized nations. The deployment of these tools at the level of industry and policy practitioners and their use in developing economies continues to present a challenge, due to a lack of financial resources for such studies, data availability or transparency, and changing policies, among other reasons (Graedel & Allenby, 2010; Li, 2018; Shenoy, 2016). If those challenges can be overcome, developing economies could profit from the application of Industrial Ecology in early phases of their development, for instance by reducing negative environmental impacts and resource consumption. The latter takes special relevance, considering that important stages of global value chains are taking place in developing countries, in particular for resource extraction, final disposal of goods, and increasing manufacturing overseas.

6 Conclusions

To overcome today's challenges with respect to living standards the core understanding of Industrial Ecology is to close the loop of material and energy flows, considering on the one hand the need for global economic growth to satisfy the demands of the still-growing population, and on the other hand the impacts of human activities on the environment and its adverse effects on living standards. The necessary transition shall follow the organizational principles of ecological systems (Isenmann, 2007), although this identity-forming characteristic is disputed, as shown above (Ayres, 2004; Bey, 2007).

Although the aim of Industrial Ecology is to improve living standards, originally societal aspects were not addressed explicitly, following the still-widespread perception that the changes created by new technologies or better organization of material and energy flows would automatically lead to sustainable development, a satisfied society, and acceptance of the transformation. However, the conflicts that are currently on the rise within societies and between nations on a global scale demonstrate the need for a better understanding of the interconnections between technology, environment, and societies (Blomsma & Brennan, 2017). The causes of social conflicts are manifold and induced not only by the challenges of climate change and other environmental and health risks, but also by socio-cultural developments in affected societies (Wieviorka, 2013). These social conflicts influence the space for potential actions taken and thus necessitate an understanding of that space, not only for politicians, but also for companies and societies. Additionally, in most cases future-oriented research is based on today's perception of the future. Though this cannot be changed, for obvious reasons, it is important to reflect on this by creating future-oriented scenarios with different perceptions and attitudes.

References

Abhishek, K., & Biswas, A. (2022). Understanding industrial ccology: An approach for sustainable industrial development. In R. Anbanandam & S. Rangnekar (Eds.), *Flexibility, innovation, and sustainable business. Flexible systems management* (pp. 299–306). Springer. https://doi.org/10.1007/978-981-19-1697-7_21

Adeola, A. O., Akingboye, A. S., Ore, O. T., Oluwajana, O. A., Adewole, A. H., Olawade, D. B., & Ogunyele, A. C. (2022). Crude oil exploration in Africa: Socio-economic implications, environmental impacts, and mitigation strategies. *Environment Systems and Decisions, 42*, 26–50. https://doi.org/10.1007/s10669-021-09827-x13

Allenby, B. R., & Cooper, W. E. (1994). Understanding industrial ecology from a biological systems perspective. *Environmental Quality Management, 3*(3), 343–354. https://doi.org/10.1002/tqem.3310030310

Al-Thani, N. A., & Al-Ansari, T. (2021). Comparing the convergence and divergence within industrial ecology, circular economy, and the energy-water-food nexus based on resource management objectives. *Sustainable Production and Consumption, 27*, 1743–1761. https://doi.org/10.1016/j.spc.2021.04.008

Ayres, R. U. (2004). On the life cycle metaphor: Where ecology and economics diverge. *Ecological Economics, 48*, 425–438. https://doi.org/10.1016/j.ecolecon.2003.10.018

Baldassarre, B., Schepers, M., Bocken, N., Cuppen, E., Korevaar, G., & Calabretta, G. (2019). Industrial symbiosis: Towards a design process for eco-industrial clusters by integrating circular economy and industrial ecology perspectives. *Journal of Cleaner Production, 216*, 446–460. https://doi.org/10.1016/j.jclepro.2019.01.091

Bauer, C., & Poganietz, W. R. (2007). Prospektive Lebenszyklusanalyse oder die Zukunft in der Ökobilanz. *Technikfolgenabschätzung – Theorie und Praxis, 16*(3), 17–23. http://www.itas.fzk.de/tatup/073/bapo07a.pdf

Bey, C. (2007). Grenzen der Kreislaufwirtschaft. In R. Isenmann & M. von Hauff (Eds.), *Industrial Ecology: Mit Ökologie zukunftsorientiert wirtschaften* (pp. 75–87). Elsevier.

Biblioteca del Congreso. (2016). Establece marco para la gestión de residous, la responsabilidad extendia del productor y fomento al reciclaje. https://www.bcn.cl/leychile/navegar?idNorma=1090894

Binder, C. R. (2007). From material flow analysis to material flow management part II: The role of structural agent analysis. *Journal of Cleaner Production, 15*, 1605–1617. https://doi.org/10.1016/j.jclepro.2006.08.017

Blomsma, F., & Brennan, G. (2017). The emergence of circular economy. A new framing around prolonging resource productivity. *Journal of Industrial Ecology, 21*(3), 603–614. https://doi.org/10.1111/jiec.12603

Bocken, N., Olivetti, E. A., Cullen, J. M., Potting, J., & Lifset, R. (2017). Taking the circularity to the next level: A special issue on the circular economy. *Journal of Industrial Ecology, 21*(3), 476–482. https://doi.org/10.1111/jiec.12606

Boons, F., Spekkink, W., & Jiao, W. (2014). A process perspective on industrial symbiosis: Theory, methodology, and application. *Journal of Industrial Ecology, 18*(3), 341–355. https://doi.org/10.1111/jiec.12116

Boons, F., Spekkink, W., & Mouzakitis, Y. (2011). The dynamics of industrial symbiosis: A proposal for a conceptual framework based upon a comprehensive literature review. *Journal of Cleaner Production, 19*, 905–911. https://doi.org/10.1016/j.jclepro.2011.01.003

Bristow, D. N., & Kennedy, C. A. (2013). Urban metabolism and the energy stored in cities. *Journal of Industrial Ecology, 17*(5), 656–667. https://doi.org/10.1111/jiec.12038

Calabretta, G., & Gemser, G. (2017). Building blocks for effective strategic design. *Journal of Design, Business & Society, 3*(2), 109–124. https://doi.org/10.1386/dbs.3.2.109_1

Castan Broto, V., Allen, A., & Rapoport, E. (2012). Interdisciplinary perspectives on urban metabolism. *Journal of Industrial Ecology, 16*, 851–861. https://doi.org/10.1111/j.1530-9290.2012.00556.x

Chertow, M. R. (2000). Industrial symbiosis: Literature and taxonomy. *Annual Review of Energy and the Environment, 25*(1), 31337. https://doi.org/10.1146/annurev.energy.25.1.313

Chertow, M. R. (2007). "Uncovering" industrial symbiosis. *Journal of Industrial Ecology, 11*(1), 11–30. https://doi.org/10.1162/jiec.2007.1110

Circular Economy and Waste Act (in German: Kreislaufwirtschafts- und Abfallgesetz (KrW/AbfG)) 1994. BGBl. I p. 2705. https://dejure.org/gesetze/KrW-AbfG

Corvellec, H., Stowell, A. F., & Johansson, N. (2022). Critiques of the circular economy. *Journal of Industrial Ecology, 26*, 421–432. https://doi.org/10.1111/jiec.13187

Duchin, F., & Levine, S. H. (2014). Industrial ecology. In B. Fath (Ed.), *Encyclopedia of ecology* (2 ed., pp. 352–358). Elsevier. https://doi.org/10.1016/B978-0-12-409548-9.09407-0

Ehrenfeld, J. (2004). Industrial ecology: A new field or only a metaphor? *Journal of Cleaner Production, 12*(8–10), 825–831. https://doi.org/10.1016/j.jclepro.2004.02.003

Ehrenfeld, J., & Gertler, N. (1997). Industrial ecology in practise. The evolution of interdependence at Kalundborg. *Journal of Industrial Ecology, 1*(1), 67–79.

Ellen MacArthur Foundation. (2013). *Towards the circular economy Vol. 1: An economic and business rationale for an accelerated transition.*

Erkman, S. (1997). Industrial ecology: An historical view. *Journal of Cleaner Production, 5*(1–2), 1–10.

Erkman, S. (2007). Historischer Überblick zur Industrial Ecology. In R. Isenmann & M. von Hauff (Eds.), *Industrial Ecology: Mit Ökologie zukunftsorientiert wirtschaften* (pp. 31–47). Elsevier.

Espinosa, G. G., & Casas, Y. L. (2024). Life cycle thinking: An essential framework for eco-industrial development. In A. C. Braun, G. G. Espinosa, D. Tröger & T. Hirth (Eds.), *Eco-industrial development as an industrial strategy—Contributions from a German-Chilean research partnership.* Springer Nature.

European Commission. (2015). *Closing the loop—An EU action plan for the Circular Economy.* COM 614. https://eur-lex.europa.eu/resource.html?uri=cellar:8a8ef5e8-99a0-11e5-b3b7-01aa75ed71a1.0012.02/DOC_1&format=PDF

Farjana, S. H., Huda, N., & Mahmud, M. A. P. (2018). Life-cycle environmental impact assessment of mineral industries. *IOP Conference Series: Materials Science and Engineering, 351*(1), 012016. https://doi.org/10.1088/1757-899X/351/1/012016

Fischer-Kowalski, M. (1998). Society's metabolism. *Journal of Industrial Ecology, 2*(1), 61–78. https://doi.org/10.1162/jiec.1998.2.1.61

Fischer-Kowalski, M. (2007). Multidisziplinäre Wurzeln der Industrial Ecology. In R. Isenmann & M. von Hauff (Eds.), *Industrial Ecology: Mit Ökologie zukunftsorientiert wirtschaften* (pp. 89–100). Elsevier.

Frosch, R. A., & Gallopoulos, N. E. (1989). Strategies for manufacturing. *Scientific American, 261*(3), 144–153. https://doi.org/10.1038/scientificamerican0989-144

Fuss, M., Barros, R. T. V., & Poganietz, W. R. (2021). The role of a socio-integrated recycling system in implementing a circular economy—The case of Belo Horizonte, Brazil. *Waste Management, 121*, 215–225. https://doi.org/10.1016/j.wasman.2020.12.006

Geisendorf, S., & Pietrulla, F. (2017). The circular economy and circular economic concepts—A literature analysis and redefinition. *Thunderbird—International Business Review, 60*, 771–782. https://doi.org/10.1002/tie.21924

Geng, Y., & Doberstein, B. (2008). Developing the circular economy in China: Challenges and opportunities for achieving 'leapfrog development.' *International Journal of Sustainable Development & World Ecology, 15*(3), 231–239. https://doi.org/10.3843/SusDev.15.3:6

Glöser-Chahoud, S., & Schultmann, F. (2024). Circular economy: An introduction. In A. C. Braun, G. G. Espinosa, D. Tröger, T. Hirth (Eds.), *Eco-industrial development as an industrial strategy—Contributions from a German-Chilean research partnership.* Springer Nature.

Graedel, T. E., & Allenby, B. R. (1995). *Industrial ecology* (2nd ed.). Prentice Hall.

Graedel, T. E., & Allenby, B. R. (2010). *Industrial ecology and sustainable engineering.* Prentice Hall.

Graedel, T. E., & Lifset, R. J. (2016). Industrial ecology's first decade. In R. Clift & A. Druckman (Eds.), *Taking stock of industrial ecology* (pp. 3–20). Springer International Publishing. https://doi.org/10.1007/978-3-319-20571-7_1

Gregson, N., Crang, M., Fuller, S., & Holmes, H. (2015). Interrogating the circular economy: The moral economy of resource recovery in the EU. *Economy and Society, 44*(2), 218–243. https://doi.org/10.1080/03085147.2015.1013353

Halonen, N., & Seppänen, M. (2019). Eco-industrial parks. In W. Leal Filho, A. Azul, L. Brandli, P. Özuyar & T. Wall (Eds.), *Responsible consumption and production. Encyclopedia of the UN sustainable development goals*. Springer. https://doi.org/10.1007/978-3-319-71062-4_5-1

Harari, Y. N. (2015). *Sapiens: A brief history of humankind*. Random House.

Hardy, C., & Graedel, T. E. (2002). Industrial ecosystems as food webs. *Journal of Industrial Ecology, 6*(1), 29–38. https://doi.org/10.1162/108819802320971623

Hennequin, T., Huijbregts, M. A. J., & van Zelm, R. (2023). The influence of consumer behavior on the environmental footprint of passenger car tires. *Journal of Industrial Ecology, 27*(1), 96–109. https://doi.org/10.1111/jiec.13334

Hicks, A. (2022). Seeing the people in LCA: Agent based models as one possibility. *Resources, Conservation & Recycling Advances, 15*, 200091. https://doi.org/10.1016/j.rcradv.2022.200091

Isenmann, R. (2007). Natur als Vorbild: Identitätsstiftendes Merkmal der Industrial Ecology. In R. Isenmann & M. von Hauff (Eds.), *Industrial Ecology: Mit Ökologie zukunftsorientiert wirtschaften* (pp. 61–74). Elsevier.

Kaza, S., Yao, L., Bhada-Tata, P., & Van Woerden, F. (2018). *What a waste 2.0: A global snapshot of solid waste management to 2050*. World Bank. https://doi.org/10.1596/978-1-4648-1329-0

Kennedy, C., Pincetl, S., & Bunje, P. (2011). The study of urban metabolism and its applications to urban planning and design. *Environmental Pollution, 159*(8–9), 1965–1973. https://doi.org/10.1016/j.envpol.2010.10.022

Kirchherr, J., Reike, D., & Hekkert, M. (2017). Conceptualizing the circular economy: An analysis of 114 definitions. *Resources, Conservation and Recycling, 127*, 221–232. https://doi.org/10.1016/j.resconrec.2017.09.005

Korniejenko, K., Kozub, B., Bak, A., Balamurugan, P., Uthayakumar, M., & Furtos, G. (2021). Tackling the circular economy challenges—Composites recycling: Used tyres, wind turbine blades, and solar panels. *Journal of Composites Science, 5*(9), 243. https://doi.org/10.3390/jcs5090243

Lan, K., & Yao, Y. (2019). Integrating life cycle assessment and agent-based modeling: A dynamic modeling framework for sustainable agricultural systems. *Journal of Cleaner Production, 238*, 117853. https://doi.org/10.1016/j.jclepro.2019.117853

Li, X. (2018). Challenges for applying industrial ecology and future development of industrial ecology. In X. Li (Ed.), *Industrial ecology and industry symbiosis for environmental sustainability: Definitions, frameworks and applications* (pp. 111–126). Springer International Publishing. https://doi.org/10.1007/978-3-319-67501-5_6

Lowe, E. A. (1997). Creating by-product resource exchanges: Strategies for eco-industrial parks. *Journal of Cleaner Production, 5*(1–2), 57–65.

Lüdeke-Freund, F., Gold, S., & Bocken, N. M. P. (2019). A review and typology of circular economy business model patterns. *Journal of Industrial Ecology, 23*(1), 36–61. https://doi.org/10.1111/jiec.12763

Merciai, S., & Schmidt, J. (2018). Methodology for the construction of global multi-regional hybrid supply and use tables for the EXIOBASE v3 database. *Journal of Industrial Ecology, 22*(3), 516–531. https://doi.org/10.1111/jiec.12713

Merli, R., Preziosi, M., & Acampora, A. (2018). How do scholars approach the circular economy? A systematic literature review. *Journal of Cleaner Production, 178*, 703–722. https://doi.org/10.1016/j.jclepro.2017.12.112

Murray, A., Skene, K., & Hynes, K. (2017). The circular economy: An interdisciplinary exploration of the concept and application in a global context. *Journal of Business Ethics, 140*, 369–380. https://doi.org/10.1007/s10551-015-2693-2

Nakamura, S., & Kondo, Y. (2009). *Waste input-output analysis. Concepts and application to industrial ecology*. Spiegel.

National People's Congress of the People's Republic of China. (2008). Circular economy promotion law. http://english.mep.gov.cn/Resources/laws/envir_elatedlaws/201712/t20171212_427823.shtml

OECD-Organization of Economic Cooperation and Development. (2019). *Global material resources outlook to 2060: Economic drivers and environmental consequences*. OECD Publishing. https://doi.org/10.1787/9789264307452-en

Prosman, E. J., Waehrens, B. V., & Liotta, G. (2017). Closing global material loops. Initial insights into firm-level challenges. *Journal of Industrial Ecology, 21*(3), 641–650. https://doi.org/10.1111/jiec.12535

Puig, R. C. (2019). The roles of industrial ecology and circular economy: A literature review and a classroom application. Universitat Poliytéchnica de Catalunya. https://upcommons.upc.edu/handle/2117/174079

Rizos, V., Tuokko, K., & Behrens, A. (2017). The circular economy. A review of definitions, processes and impacts. *CEPS Research Report* No. 2017/8. Retrieved August, 06, 2023, from https://www.ceps.eu/download/publication/?id=9969&pdf=RR2017-08_CircularEconomy_0.pdf

Rohracher, H. (2008). Energy systems in transition: Contributions from social sciences. *International Journal of Environmental Technology and Management, 9*(2–3), 144–161. https://doi.org/10.1504/IJETM.2008.019026

Saavedra, Y. M. B., Iritani, D. R., Pavan, A. L. R., & Ommetto, A. R. (2018). Theoretical contribution of industrial ecology to circular economy. *Journal of Cleaner Production, 170*, 1514–1522. https://doi.org/10.1016/j.jclepro.2017.09.260

Serrao-Neumann, S., Renouf, M. A., Morgan, E., Kenway, S. J., & Low Choy, D. (2019). Urban water metabolism information for planning water sensitive city-regions. *Land Use Policy, 88*, 104144. https://doi.org/10.1016/j.landusepol.2019.104144

Shenoy, M. (2016). Industrial ecology in developing countries. In R. Clift & A. Druckman (Eds.), *Taking stock of industrial ecology* (pp. 229–245). Springer International Publishing. https://doi.org/10.1007/978-3-319-20571-7_11

Short, S. W., Bocken, N., Barlow, C. Y., & Chertow, M. R. (2014). From refining sugar to growing tomatoes: Industrial ecology and business model evolution. *Journal of Industrial Ecology, 18*(5), 603–618. https://doi.org/10.1111/jiec.12171

Sobieraj, S., & Krämer, N. C. (2020). Similarities and differences between genders in the usage of computer with different levels of technological complexity. *Computers in Human Behavior, 104*, 106145. https://doi.org/10.1016/j.chb.2019.09.021

Stiehl, C., & Hirth, T. (2012). Vom additiven Umweltschutz zur nachhaltigen Produktion. *Chemie Ingenieur Technik, 84*, 963–968. https://doi.org/10.1002/cite.201200008

Su, B., Heshmati, A., Geng, Y., & Yu, X. (2013). A review of the circular economy in China: Moving from rhetoric to implementation. *Journal of Cleaner Production, 42*, 215–227. https://doi.org/10.1016/j.jclepro.2012.11.020

Svenfelt, Å., Bradley, K., André, E., Järbel, J., Kamb, A., Klintman, M., & Schoonover, H. (2022). *Sustainable consumption and beyond: Final report from phase 1 of the research program Mistra sustainable consumption: 2018–2021*. KTH Royal Institute of Technology. https://urn.kb.se/resolve?urn=urn:nbn:se:kth:diva-323102

Uihlein, A., Poganietz, W. R., & Schebek, L. (2006). Carbon flows and carbon use in the German anthroposphere: An inventory. *Resources, Conservation and Recycling, 46*, 410–429. https://doi.org/10.1016/j.resconrec.2005.09.001

Van Ewijk, S., Ashton, W. S., Berrill, P., Cao, Z., Chertow, M., Chopra, S. S., Fishman, T., Fitzpatrick, C., Heidrich, O., Leipold, S., Ritter, F., Sprecher, B., Yao, Y., & Myers, R. J. (2023). *Ten insights from industrial ecology for the circular economy*. International Society for Industrial Ecology (ISIE).

Wieviorka, M. (2013). Social conflict. *Current Sociology Review, 61*(5–6), 696–713. https://doi.org/10.1177/0011392113499487

Circular Economy: An Introduction

Simon Glöser-Chahoud and Frank Schultmann

Abstract The circular economy is an economic system that aims to minimize waste and make the most of resources by keeping products and materials in use for as long as possible. It contrasts with the traditional linear economy, which has a "take-make-dispose" approach. The transition towards a sustainable circular economy requires changes and innovation along the entire industrial value chain, starting with product design, material processing and manufacturing, consumption patterns, and the product use-phase, up to waste management and recycling. The practical implementation of circular value chains and closed-loop supply chains in most cases also requires an adaptation and modification of business models. In contrast to traditional linear business models, the use phase and the end-of-life treatment of products are brought to the fore in circular business models. As an introduction to circular economic principles, this chapter provides a general overview of the circular economy framework from a systemic perspective. After discussing basic strategies for sustainable development and underlining the severe need for the transition towards a circular economy, this chapter delivers a broad overview of the different areas of a circular economy including aspects of the materials-energy nexus. We introduce basic definitions and explain concepts along the entire circular value chain including business models and product-service systems (PSS). The subsequent discussions mainly focus on limits and challenges of the circular economy as a key element of sustainable industrial development.

S. Glöser-Chahoud (✉)
TU Bergakademie Freiberg, Chair of Corporate Sustainability and Environmental Management, Freiberg, Germany
e-mail: simon.gloeser-chahoud@bwl.tu-freiberg.de

F. Schultmann
Institute for Industrial Production (IIP), Karlsruhe Institute of Technology (KIT), Karlsruhe, Germany
e-mail: frank.schultmann@kit.edu

A. Ch. Braun et al. (eds.), *Eco-Industrial Development as an Industrial Strategy*, Green Energy and Technology, https://doi.org/10.1007/978-3-031-73576-9_3

1 A Circular Economy Framework

The continuously increasing demand for natural resources, the associated environmental impacts, and the acceleration of climate change foster the need for more sustainable economic development and a decoupling of welfare and resource consumption (Weizsäcker et al., 2010). In addition to measures of increased material efficiency and the greater use of renewable resources, the transition from a linear economy to a more circular economy is inevitable (Winning et al., 2017). While end-of-pipe emission reductions may decrease the environmental impact of industrial activities, circular-use resource structures are unavoidable if sustainable production and consumption patterns are to be achieved in the long term (Stiehl & Hirth, 2012). Inspired by the principles of natural ecosystems, the goal is to use various waste streams as material input for other processes, minimizing the losses of substances to the environment (Frosch & Gallopoulos, 1989). This idea of eco-industrial metabolism following the sustainability concept of consistency (consistency with natural material cycles) is extended by the circular economy framework combining aspects of waste prevention, product lifetime extension, reuse, remanufacturing, and subsequent recycling (Blomsma & Brennan, 2017). The systematic implementation of circular material use structures, in which obsolete products or waste streams form the basis for the production of new commodities, is a key issue for improving the sustainability and resource efficiency of industrial value chains. Circular economic systems are furthermore not limited to material cycles but also include natural cycles such as the carbon cycle by using bio-based renewable resources as a feedstock for renewable energy and for various chemical value chains. This forms the link between the circular economy and the bioeconomy, leading to the concept of a circular bioeconomy (see Fig. 1, based on the "butterfly diagram" by the Ellen MacArthur Foundation).

Recent research emphasizes the materials-energy nexus, drawing attention to a broad range of environmental impacts caused by the material intensity of an economic system (Masanet et al., 2021). Many materials and products have high ecological footprints that can be reduced through the systematic implementation of various renewable energy sources ranging from biomass utilization to renewable electricity through wind, solar, or hydropower. Hence, the use of renewable material and energy resources—in addition to the systematic circulation of technical materials—is a prerequisite for the circular economy. Last but not least, new technologies such as water electrolysis to produce green hydrogen from renewable electricity and subsequent chemical synthesis with e.g. CO_2 from direct air capture (Heinzmann et al., 2021), or the direct reduction of iron ore with green hydrogen in steel making (Bhaskar et al., 2020), will lead to an even stronger connection between the energy system and technical raw material supply.

Although some aspects of circular economy systems—such as the recycling of raw materials from waste—are not new but have been practiced since early human civilization (e.g. the remelting of obsolete metal products), circular economy research from a more systemic perspective has been an emerging topic in scientific literature in

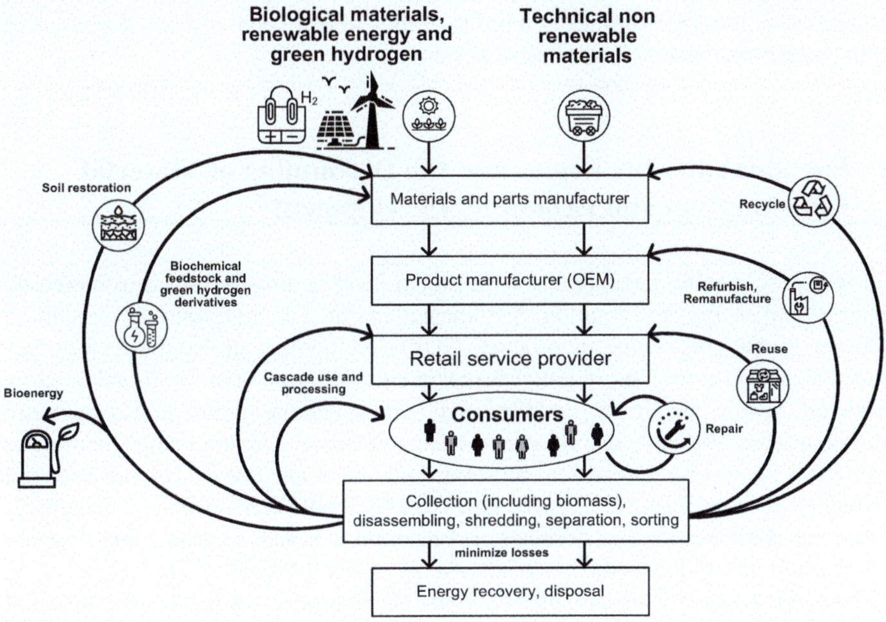

Fig. 1 Circular economy framework including both technical material cycles and the use of renewable bio-based materials. Figure adapted from the Ellen MacArthur Foundation (Winning et al., 2017)

recent years. A simple analysis of publications with the keyword "circular economy" listed in Scopus, the world's largest abstract and citation database of peer-reviewed literature, shows a strong increase in the number of manuscripts addressing circular economy aspects in recent years (see Fig. 2).

In the following sections, a general introduction to circular economy concepts and approaches will be given. Section 2 begins with a description of major challenges for sustainable development in the 21st century and the necessity of a more circular economy, followed by a summary in Sect. 3 of the major definitions and elements of a circular economy. Section 4 subsequently concludes this introduction with a

Fig. 2 Number of publications with "circular economy" as a keyword listed in Scopus, the world's largest collection of peer-reviewed academic literature

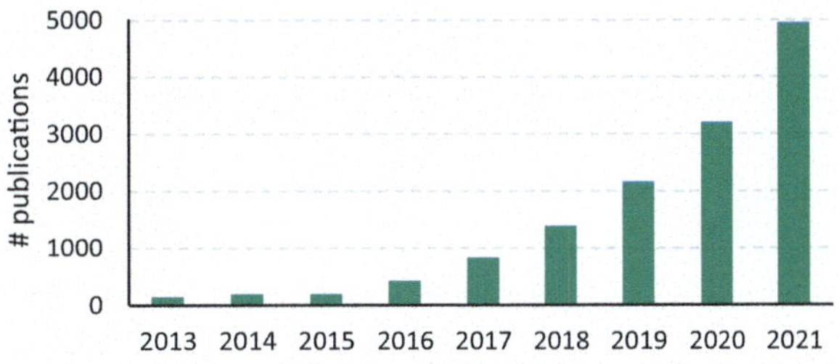

discussion of the opportunities and limits of the circular economy as a key element in the transition towards a sustainable economic system.

2 Sustainability Strategies and the Decoupling of Material Consumption and Economic Development

Currently, economic development on a global level is directly linked to increased material and energy consumption. As indicated in Fig. 3, there is a strong correlation between global gross domestic product (GDP) development, global material extraction, and carbon emissions, mainly caused by the consumption of fossil fuels such as crude oil, natural gas, lignite, and hard coal. Due to decreasing ore grades and more remote mineral deposits, the effort and energy required to extract respective materials tends to increase and so does the environmental impact caused by raw material extraction (Norgate et al., 2007). The analysis of embodied energy and related greenhouse gas (GHG) emissions of material consumption is subject to ongoing research and is often referred to as the materials-energy nexus (Masanet et al., 2021).

A key challenge for sustainable economic development will be the decoupling of economic development and material consumption on the one hand and a decoupling of material extraction and environmental impacts on the other (UNEP, 2011). Hence, decoupling material use and economic performance refers to the process whereby economic growth can be achieved without a corresponding increase in the consumption of natural resources. The goal of decoupling is to reduce the environmental impacts associated with economic activities while still allowing for development and the improvement of human well-being.

To successfully achieve decoupling, a combination of technological innovation, policy interventions, investments in sustainable infrastructure, changes in consumption patterns, and awareness and education initiatives are needed. This is represented by three key strategies for sustainable development, namely an increase in **efficiency**,

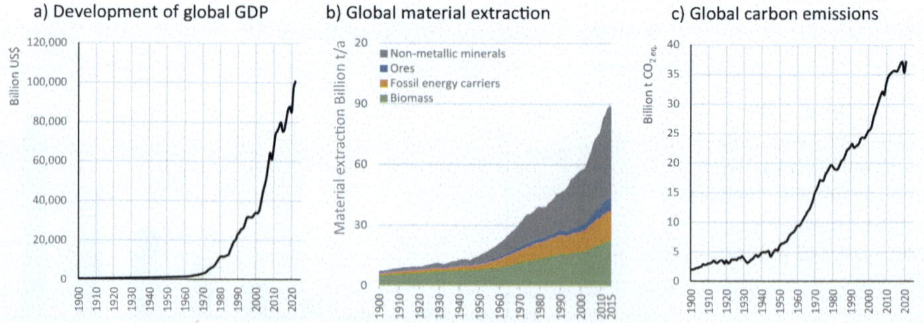

Fig. 3 Development of global GDP, global material extraction, and global carbon emissions (based on "Our World in Data" database and Krausmann et al. (2018)

consistency with natural material cycles, and **sufficiency**, questioning the scale and extent of consumption and production. These three terms capture the essence of sustainable development (Grunwald & Kopfmüller, 2012):

- Efficiency is about doing things right (optimizing processes). It refers to the achievement of a specific outcome using the least amount of resources or effort. From a sustainability perspective, efficiency is about reducing waste and optimizing the use of resources to minimize environmental impact.
- Consistency is about doing the right things (creating systems that are in harmony with nature). Consistency pertains to the creation of closed-loop systems in which resources are recycled, reused, or regenerated, mirroring natural ecosystems. It emphasizes the creation of systems in which waste from one process becomes the input for another, thereby minimizing the need for new resources.
- Sufficiency revolves around questioning the scale and extent of consumption and production. It entails understanding and determining what is enough to meet human needs and ensuring well-being without overshooting Earth's ecological limits.

While the basic framework of a circular economy as illustrated in Fig. 1 mainly builds upon the goal to be consistent with natural cycles where waste does not exist (consistency strategy), the different elements and fields of action within the circular economy address all three sustainability strategies including efficiency and sufficiency. Hence, the establishment of a circular economy is a prerequisite to overcome the aforementioned key environmental challenges in the 21st century and to pave the way towards a sustainable economic system.

3 Circular Economy: A Systems Approach

The basic principle of a circular economy is to minimize resource losses by increasing the material use intensity within a cyclic cascade use system. This includes aspects of waste prevention as well as the recovery of components and materials from waste streams (Hirth, 1997). Thereby, the circular economy concept builds upon the classical waste hierarchy (see Fig. 4) as e.g. implemented in the EU Waste Framework Directive (Directive 2008/98/EC, 2008). The highest level of the waste hierarchy is to reduce or prevent waste (such as through higher durability of products, avoidance of packaging, or reduction of fast-moving consumer goods), as waste that does not occur causes no environmental impacts and no material or resource losses.

The second stage of the waste hierarchy involves the reuse of products or components, as this directly keeps materials within the use cycle and further reduces the demand for new, resource-intensive products. The preparation for reuse can include remanufacturing or reconditioning of components, which is also referred to as refurbishment. The concept of product reuse may also include aspects of repurposing, which means that during the second life, the product or product component does not necessarily have to be used in the same field of application. A common example of

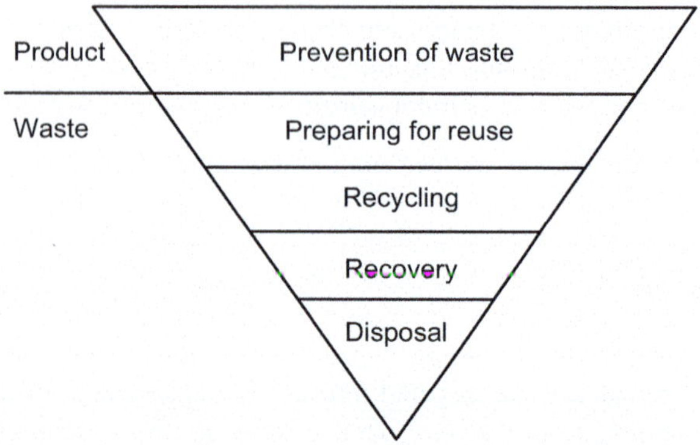

Fig. 4 Waste hierarchy as defined in Directive 2008/98/EC on waste (EU Waste Framework Directive) and implemented into national regulation by EU member states

repurposing is the reuse of old batteries or battery modules from electric vehicles as stationary storage systems in the electricity grid (Rosenberg et al., 2023).

If the reuse of components is not or no longer possible, the recovery of valuable materials through recycling forms the next level of the waste hierarchy. Recovery in the context of waste treatment does not refer to material recovery but to energy recovery, for example the combustion of mixed plastic waste for combined heat and electricity generation. The least-favorable waste treatment is disposal, as this does not include any aspects of circularity and corresponds to a linear use system of resources with high material losses and emissions to the environment.

The framework described by the waste hierarchy has been further extended and specified in academic literature. While the waste hierarchy generally addresses the concept "**r**educe, **r**euse, and **r**ecycle" and is therefore referred to as the 3R framework, further frameworks have been described ranging from the 5R framework (Korsunova et al., 2021) to the 9R framework illustrated in Fig. 5 (Kirchherr et al., 2017).

As depicted in Fig. 5, the concept of reusing obsolete products or key components in particular is distinguished in more detail in the 9R framework, which also includes aspects of direct reuse, refurbishment, remanufacturing, and repurposing. To specify these expressions, more detailed definitions are provided in the following sections. Furthermore, key elements of a circular economy and interdisciplinary research fields in the context of the transition towards a more circular economy will be listed and briefly described (Fig. 5).

3.1 Product Design as a Key Enabler of Circularity

During the design phase of a product, key properties such as material choice and composition, repairability, or recyclability are defined. Therefore, product design is relevant at all stages of circularity, ranging from concepts of "reduce" over "reuse" to the question of how efficiently recycling can be performed (see Fig. 5). The

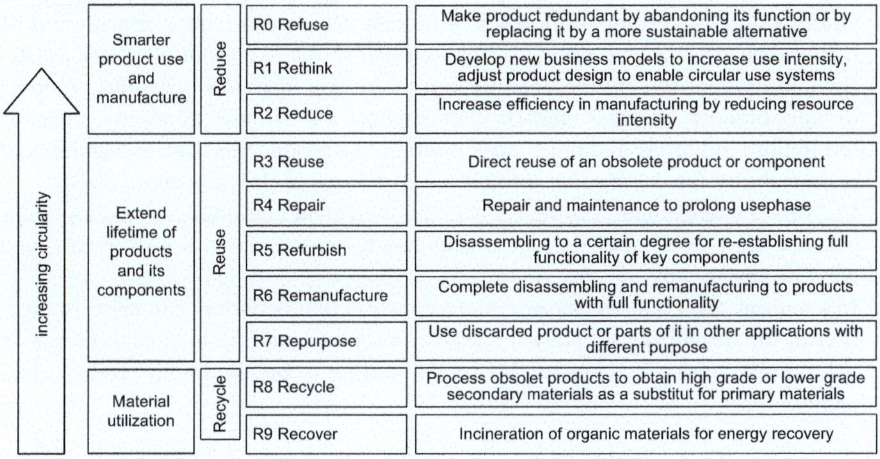

Fig. 5 9R framework for the circular economy as described by Kirchherr et al. (2017)

ways in which the design phase of a product is crucial for the circular economy are consequently manifold but may include the following aspects, among others (Hollander et al., 2017):

- Design for longevity: Products designed for durability and longevity reduce the need for frequent replacements. This results in fewer resources being consumed over time and reduces waste.
- Design for repair and maintenance: Products that are easy to repair or maintain can have their lifespans significantly extended. This means that rather than a broken product being discarded, it can be fixed and continue to be of service.
- Design for modularity: Modular designs allow individual components of a product to be replaced or upgraded without the entire product needing to be discarded. This can both extend product life and reduce waste (Machado & Morioka, 2021).
- Design for recycling: Designing products to be easily recyclable at the end of their life cycle ensures that materials can be reclaimed and reused. This decreases the demand for virgin materials and reduces the environmental impacts of extraction and processing.
- Design for reuse and remanufacturing: Some products can be designed to be easily disassembled and their parts used in the manufacturing of new products. This keeps the embedded energy and resources in the components circulating in the economy.
- Material selection: During the design phase, selecting materials that have a lower environmental footprint or are more easily recycled can significantly benefit the circular economy.
- Reduction of harmful substances: By avoiding or reducing the use of toxic or harmful substances in the design phase, products can be more safely recycled or decomposed at the end of their lifecycle.

- Optimization of resource efficiency: Efficient design can reduce the amount of material needed for a product, leading to lower resource consumption and waste.
- Business model integration: The design phase can also consider how products fit into broader business models that support the circular economy, such as leasing or product-as-a-service models where the producer retains ownership and responsibility for the product throughout its life cycle (Bocken et al., 2016).
- Consumer engagement: Products designed with clear communication about their circular attributes can encourage consumers to participate in return, repair, and recycling programs.
- Innovation: The challenges presented by the circular economy can drive innovation in the design phase. New materials, processes, and designs can emerge as solutions to these challenges, potentially leading to breakthrough products and technologies.

In essence, the design phase sets the foundation for how a product will interact with the environment and the economy throughout its life cycle. By making circular principles an integral part of the design process, businesses can drive sustainable practices and contribute positively to the environment and society.

3.2 Increasing Use Intensity and Material Effectiveness

As previously indicated, the durability of products plays a key role in the circular economy, as the longer use of products leads to a direct increase in materials' use intensity. Higher technical durability in the sense of higher quality, repairability, and the provision of spare parts may directly result in a longer use phase of a product. However, as obsolescence refers not only to the functional or technical properties of a product but also encompasses a psychological factor (meaning that the perceived age of a product might be different among different consumers), not only technical solutions but also the establishment of cascade use systems and respective business models are necessary (Rosenberg et al., 2022). Reuse in this context is strongly related to the concept of reselling used products, which are reintroduced into the same market and thus enter a further use phase. In many cases direct reuse is considered, meaning that no recovery processes lie in between the two use phases. This is particularly the case with products for which second-hand markets already exist and are well established, such as for used vehicles. In other cases, a refurbishment of the obsolete product may be necessary before it is reintroduced onto the market. With refurbishment, the product integrity is usually retained. The product is not fully disassembled and only various components such as the screen of a smartphone are upgraded or replaced. In the case that a product is fully disassembled and reassembled, this procedure would instead be called remanufacturing or reconditioning.

In some cases, recovery processes such as remanufacturing or reconditioning are necessary to allow for the reuse or repurposing of obsolete products or specific components. These recovery processes may differ concerning the following criteria (Glöser-Chahoud et al., 2021):

(a) the degree of disassembly of the product;
(b) the degree of replacement of components with new ones or used replacement parts that are retrieved from other used products (reconfiguration);
(c) the degree of product quality of the recovered product measured in comparison to the quality and functionality of a completely new product.

In the literature, multiple and sometimes conflicting ranking options of recovery concepts such as remanufacturing and reconditioning exist, especially concerning the criteria (c). The British Standards Institution (2009) for example defines remanufacturing as a standardized process in which the quality of the recovered product is equivalent to the original product, while during reconditioning only key components are replaced in order to achieve full functionality. Hence, remanufacturing and reconditioning can be distinguished on the basis of the quality of the secondary product as compared to a new product. Refurbishment often refers only to reestablishing the outer appearance of the product (Apra Europe, 2012). However, here there is also no clear distinction from reconditioning. If the remanufactured of refurbished product is not reused in its original application but in some different field where it gets a new function and replaces new products, this utilization pathway is called repurposing (e.g. using obsolete batteries from electric vehicles as stationary storage in the electricity grid).

3.3 *Recycling and Downcycling*

Although the concept of recycling to recover substances or materials from waste, which are then reintroduced into the production processes of new products, is the most common and best-known approach of material cycles, some basic definitions and distinctions are important. This is because there are often differences in the properties and quality of secondary materials as compared to primary, refined virgin materials. While in theory materials or substances are considered recycled if they replace primary materials, that is, if they provide a function that would otherwise be fulfilled by primary materials (Graedel et al., 2011), in practice there is a certain degree of "downcycling" in many cases. This means that the recycled material does not fully meet the properties, purity, or quality of primary alternatives. This does not mean that these secondary materials cannot be reused; however, there may be applications for which secondary materials do not meet required properties and therefore may not be applied. Nonetheless, the purity of secondary materials is strongly dependent on the recycling process and, similarly to primary processes, the level of purity is a question of technical processing and related costs or profitability. Helbig et al. (2022) define six different characteristics of downcycling, while several of these aspects may apply for one specific case of downcycling:

1. Reduction of material quality
2. Suitability only for less-demanding applications
3. Use in less-valuable products
4. Loss of alloying elements
5. Different material system with no functional use
6. Need for additional virgin material.

3.4 Circular Business Models and Product-Service Systems (PSS)

While circular value chains are often assessed from a systems perspective, such as by analyzing material flows and recycling efficiencies at a national or regional level, the transition from a linear to a circular economy will also strongly impact supply chains at a company level and their respective business models (Rosenberg et al., 2022). For most industries, the basic principle of business focuses on the production of goods and their subsequent sale while the classical supply chain encompasses material suppliers in the upstream value chain and customers in the downstream value chain, with different shareholders in between. In contrast, in a circular economy, the use phase and especially the end-of-life treatment of products becomes more important for equipment manufacturers as obsolete products are a source of materials and components. These circular supply chains are referred to as closed-loop supply chains (CLSCs), with a distinction made in comparison to open-loop supply chains (Schultmann et al., 2006). In a closed-loop supply chain, the obsolete components and recycled materials are reintroduced into their original value chain, while in an open-loop supply chain, recovered materials are reused in other processes or applications (Souza, 2013).

The transition towards a circular economy will only be successful if companies systematically adjust their business models and start including the use phase and end-of-life treatment of their products into their strategic decision making. This practical implementation of circular economy concepts at a business level requires new business models that are sometimes referred to as "circular business models." A circular business model refers to a strategic approach wherein a business creates and captures value through the regeneration, circulation, and extended use of products, components, and materials, which stands in contrast to the traditional linear "take-make-dispose" model (Geissdoerfer et al., 2020). It prioritizes reducing waste, optimizing resource efficiency, and enhancing the life cycle of products and services. In essence, a circular business model challenges the way value is created and delivered within the economy. These models often require businesses to rethink their relationships with consumers, suppliers, and resources, and to prioritize long-term sustainability over short-term profits. The adoption of such models can lead to benefits such as reduced costs, new revenue streams, enhanced customer loyalty, and a smaller environmental footprint. In many cases, circular business models are linked to so-called product-service systems (PSS). These are business models that combine products and services to provide value in a more sustainable manner. This approach can be seen as a subset of circular business models.

Several types of PSS can be distinguished based on the nature and extent of integration between the product and service elements (Tukker, 2015):

- Product-Oriented PSS: The product remains the primary source of value, but additional services are provided to enhance its use. For instance, a manufacturer might sell a washing machine and also offer maintenance services or training on optimal usage. Repair services are classical examples of product-oriented PSS. Advice and consultancy as an additional source of value creation are often included in the concept of product-oriented PSS. This is where the product is sold along with expert advice or consultancy to optimize its use. An example could be software sold with training or consulting sessions.
- Use-Oriented PSS: In this case, the manufacturer no longer sells the product but creates value from the use of the product. This includes leasing, sharing, and renting concepts. In the case of leasing, the product is owned by the provider and leased to the user for a certain period. An example would be office equipment that is leased instead of sold. Renting and sharing concepts are similar to leasing but usually refer to a shorter period of time. Car-sharing services or tool rental shops are examples. Sharing concepts might also be linked to pooling, in which multiple users share the product over its life cycle, either simultaneously or at different times. Carpooling or shared office spaces are typical examples.
- Result-Oriented PSS: The goal of result-oriented PSS is to replace the need for ownership of a specific product. Instead of buying the product, a customer buys the desired result, and the provider supplies the necessary products and services to achieve it. For instance, a company might not buy IT infrastructure but instead outsource its IT needs to a provider who manages the hardware, software, and support. The pay-per-service-unit model is a classical approach for creating value from result-oriented PSS. In this case, customers pay for the service output of a product rather than the product itself.

In practice, hybrid approaches combining different elements from various PSS concepts are used to create suitable business models. For instance, a company might lease a product and also charge based on its use or performance. The main aim of PSS models is to align the interests of providers and users, promoting more sustainable consumption and production patterns. For providers, this might mean designing durable, maintainable, or recyclable products, while for users, it can translate to reduced ownership responsibilities and access to the latest technology or best-performing solutions.

4 Discussion and Conclusions

The concept of a circular economy has gained significant traction over the last few years as a response to the traditional linear economic model, which follows a "take-make-dispose" pattern. The goal of this chapter was to provide a general introduction and framework of the circular economy, which emphasizes keeping products and materials in use for as long as possible, extracting the maximum value from them, and then recovering and regenerating products and materials at the end of their life cycle. However, there are limits to a fully circular economy and challenges and obstacles to tackle in the transition towards a more circular economic system. A major limit of the circular economy is its incompatibility with growing material consumption. Even though the total material requirement of an economy might decrease, there will be emerging technologies that require specific raw materials such as technology metals for the energy transition. For the provision of these materials but also for the compensation of efficiency losses during recycling and recovery processes, primary material extraction will remain a crucial component to overall supply in the future. In addition, there are economic and also ecological limits to recycling, as depicted in Fig. 6.

In some cases, the energy requirements and environmental footprint of recovering highly dispersed substances from waste streams may be higher than those of primary material extraction. This can lead to trade-offs between different environmental impact categories such as resource depletion and climate change when comparing recycling with primary material extraction processes.

Nonetheless, the transition towards a circular economy, driven by technological innovation, changing consumption and use patterns, new business models, as well as political regulation and societal education is a key requisite to reach sustainable production and consumption targets. Major challenges for this transition will be the significant changes in business models, initial investments in new technologies and

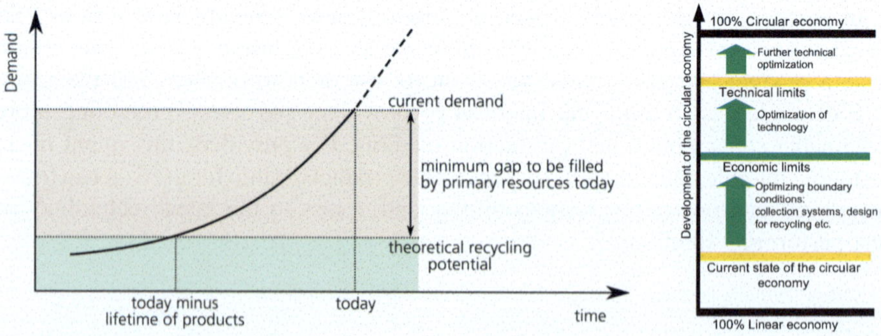

Fig. 6 Current limits to the circular economy include increasing demand for specific materials and the economic, technical, and in some cases also ecological limits of recycling processes in comparison to primary material extraction

infrastructures, and the increasing complexity of closed-loop supply chains. Furthermore, a new consumer mindset and shift in behavior will be necessary. Consumers are accustomed to disposable products and may need to adjust their buying habits and expectations for products that are designed to last longer or be refurbished. In addition, regulatory barriers for the circular economy need to be removed. For instance, existing laws and regulations may not be conducive to circular practices, such as product take-back, remanufacturing or repurposing.

On the other hand, the transition towards a circular economy also brings various opportunities, such as economic development and job creation through the implementation of new business models and markets for recycled and repurposed products. There are also clear gains in terms of resource efficiency and environmental benefits through the reduction of waste and the environmental impacts of primary material extraction. Furthermore, the need for circular solutions can drive innovation in product design, materials science, and waste management solutions. A circular economy can also contribute to risk mitigation as companies reduce their exposure to resource scarcity and price volatility by reusing and recycling materials.

In conclusion, while transitioning to a circular economy presents various challenges, primarily due to the significant paradigm shift required, the opportunities offered in terms of sustainability, economic development, and innovation are considerable. As the world grapples with environmental crises, global warming, and resource shortages, the implementation of a more circular economic system is a prerequisite for sustainable development.

References

Apra Europe. (2012). *Remanufacturing terminology remanufacturing: Term guideline.* https://cdn.ymaws.com/apra.org/resource/resmgr/European/Reman_Term_Guideline_2012-03.pdf

Bhaskar, A., Assadi, M., & Nikpey Somehsaraei, H. (2020). Decarbonization of the iron and steel industry with direct reduction of iron ore with green hydrogen. *Energies, 13*(3), 758. https://doi.org/10.3390/en13030758

Blomsma, F., & Brennan, G. (2017). The emergence of circular economy: A new framing around prolonging resource productivity. *Journal of Industrial Ecology, 21*(3), 603–614. https://doi.org/10.1111/jiec.12603

Bocken, N. M. P., de Pauw, I., Bakker, C., & van der Grinten, B. (2016). Product design and business model strategies for a circular economy. *Journal of Industrial and Production Engineering, 33*(5), 308–320. https://doi.org/10.1080/21681015.2016.1172124

British Standards Institution. (2009). *Design for manufacture, assembly, disassembly and end-of-life processing (MADE): Part 2: Terms and definitions.* British Standards Institution.

den Hollander, M. C., Bakker, C. A., & Hultink, E. J. (2017). Product design in a circular economy: Development of a typology of key concepts and terms. *Journal of Industrial Ecology, 21*(3), 517–525. https://doi.org/10.1111/jiec.12610

Directive 2008/98/EC of The European Parliament and of the Council of 19 November 2008 on waste and repealing certain Directives, OJ L31222.11.2008 Official Journal of the European Union 3 (2008).

Frosch, R. A., & Gallopoulos, N. E. (1989). Strategies for manufacturing. *Scientific American, 261*(3), 144–153.

Geissdoerfer, M., Pieroni, M. P., Pigosso, D. C., & Soufani, K. (2020). Circular business models: A review. *Journal of Cleaner Production, 277*, 123741. https://doi.org/10.1016/j.jclepro.2020.123741

Glöser-Chahoud, S., Huster, S., Rosenberg, S., Baazouzi, S., Kiemel, S., Singh, S., Schneider, C., Weeber, M., Miehe, R., & Schultmann, F. (2021). Industrial disassembling as a key enabler of circular economy solutions for obsolete electric vehicle battery systems. *Resources, Conservation and Recycling, 174*, 105735. https://doi.org/10.1016/j.resconrec.2021.105735

Graedel, T. E., Allwood, J., Birat, J.-P., Buchert, M., Hagelüken, C., Reck, B. K., Sibley, S. F., & Sonnemann, G. (2011). What do we know about metal recycling rates? *Journal of Industrial Ecology, 15*(3), 355–366. https://doi.org/10.1111/j.1530-9290.2011.00342.x

Grunwald, A., & Kopfmüller, J. (2012). *Nachhaltigkeit: Eine Einführung* (2., aktualis. Aufl.). *Campus Studium*. Campus-Verl. http://ub-mannheim.ciando.com/shop/book/short/index.cfm/fuseaction/short/bok_id/430402

Heinzmann, P., Glöser-Chahoud, S., Dahmen, N., Langenmayr, U., & Schultmann, F. (2021). *Techno-ökonomische Bewertung der Produktion regenerativer synthetischer Kraftstoffe*. Working Paper Series in Production and Energy.

Helbig, C., Huether, J., Joachimsthaler, C., Lehmann, C., Raatz, S., Thorenz, A., Faulstich, M., & Tuma, A. (2022). A terminology for downcycling. *Journal of Industrial Ecology*, (13289). https://doi.org/10.1111/jiec.13289

Hirth, T. (1997). Strategien für die Produktion im 21. Jahrhundert: 2. Internationales Symposium: Produzieren in der Kreislaufwirtschaft. *TATuP - Zeitschrift Für Technikfolgenabschätzung in Theorie Und Praxis, 6*(2), 80–81. https://doi.org/10.14512/tatup.6.2.80b

Kirchherr, J., Reike, D., & Hekkert, M. (2017). Conceptualizing the circular economy: An analysis of 114 definitions. *Resources, Conservation and Recycling, 127*, 221–232. https://doi.org/10.1016/j.resconrec.2017.09.005

Korsunova, A., Horn, S., & Vainio, A. (2021). Understanding circular economy in everyday life: Perceptions of young adults in the Finnish context. *Sustainable Production and Consumption, 26*, 759–769. https://doi.org/10.1016/j.spc.2020.12.038

Krausmann, F., Lauk, C., Haas, W., & Wiedenhofer, D. (2018). From resource extraction to outflows of wastes and emissions: The socioeconomic metabolism of the global economy, 1900–2015. *Global Environmental Change, 52*, 131–140. https://doi.org/10.1016/j.gloenvcha.2018.07.003

Machado, N., & Morioka, S. N. (2021). Contributions of modularity to the circular economy: A systematic review of literature. *Journal of Building Engineering, 44*, 103322. https://doi.org/10.1016/j.jobe.2021.103322

Masanet, E., Heeren, N., Kagawa, S., Cullen, J., Lifset, R., & Wood, R. (2021). Material efficiency for climate change mitigation. *Journal of Industrial Ecology, 25*(2), 254–259. https://doi.org/10.1111/jiec.13137

Norgate, T. E., Jahanshahi, S., & Rankin, W. J. (2007). Assessing the environmental impact of metal production processes. *Journal of Cleaner Production, 15*(8–9), 838–848. https://doi.org/10.1016/j.jclepro.2006.06.018

Rosenberg, S., Huster, S., Glöser-Chahoud, S., & Schultmann, F. (2022). Business models for second-life battery applications. In *Handbook on smart battery cell manufacturing: The power of digitalization* (pp. 391–408). World Scientific.

Rosenberg, S., Glöser-Chahoud, S., Huster, S., & Schultmann, F. (2023). A dynamic network design model with capacity expansions for EoL traction battery recycling—A case study of an OEM in Germany. *Waste Management (New York, N.Y.), 160*, 12–22. https://doi.org/10.1016/j.wasman.2023.01.029

Schultmann, F., Zumkeller, M., & Rentz, O. (2006). Modeling reverse logistic tasks within closed-loop supply chains: An example from the automotive industry. *European Journal of Operational Research, 171*(3), 1033–1050. https://doi.org/10.1016/j.ejor.2005.01.016

Souza, G. C. (2013). Closed-loop supply chains: A critical review, and future research. *Decision Sciences, 44*(1), 7–38. https://doi.org/10.1111/j.1540-5915.2012.00394.x

Stiehl, C., & Hirth, T. (2012). From end-of-pipe environmental protection to sustainable production. *Chemie Ingenieur Technik, 84*(7), 963–968. https://doi.org/10.1002/cite.201200008

Tukker, A. (2015). Product services for a resource-efficient and circular economy—A review. *Journal of Cleaner Production, 97*, 76–91. https://doi.org/10.1016/j.jclepro.2013.11.049

UNEP. (2011). *Decoupling natural resource use and environmental impacts from economic growth.*

Weizsäcker, E. U. von, Hargroves, K., & Smith, M. H. (2010). *Faktor Fünf: Die formel für nachhaltiges Wachstum.* Droemer.

Winning, M., Calzadilla, A., Bleischwitz, R., & Nechifor, V. (2017). Towards a circular economy: Insights based on the development of the global ENGAGE-materials model and evidence for the iron and steel industry. *International Economics and Economic Policy, 14*(3), 383–407. https://doi.org/10.1007/s10368-017-0385-3

Eco-Industrial Development: An Attempt to Define a Vague Concept

Danny Tröger and Andreas Ch. Braun

Abstract Due to industrialization, mankind is faced with an enormous quantity of products, most of which end up as "waste" in the environment. There is an urgent need to switch from this linear production process to more sustainable strategies, for example through circular economy approaches, industrial ecology and eco-industrial development. Most research on eco-industrial development relates to manufacturing industries; studies on countries with extractive industries are rare. Chile is an interesting country in this respect, as it provides the basis for a wide variety of extractive processes due to its diverse natural environments, and at the same time also offers many different renewable energy sources. This chapter describes Chile's special natural features, which represent both potentials and challenges for eco-industrial approaches. It also discusses the socio-economic context within which these strategies are or should be embedded. This is followed by a reference to the thematic focus and structure of the IECO project and this book. Finally, we will attempt to specify the general understanding of eco-industrial development for Chile and also formulate normative claims in order to contribute to an expansion of the definition.

1 The Demand for Sustainable Production Strategies and the Potential of Industrial Ecology

Industrialization has created enormous opportunities for the manufacture of huge quantities of complex substances and products. This is accompanied not only by an increasing complexity of inputs, but also of outputs of industrial production (Fig. 1). As a result, however, the consumption of non-renewable resources has increased and a growing amount of unusable and hardly self-degrading materials ends up in the environment as "waste" (Bilitewski & Härdtle, 2013).

D. Tröger (✉) · A. Ch. Braun
Kassel Institute for Sustainability, University of Kassel, Kassel, Germany
e-mail: danny.troeger@uni-kassel.de

A. Ch. Braun
e-mail: andreas.braun@uni-kassel.de

A. Ch. Braun et al. (eds.), *Eco-Industrial Development as an Industrial Strategy*, Green Energy and Technology, https://doi.org/10.1007/978-3-031-73576-9_4

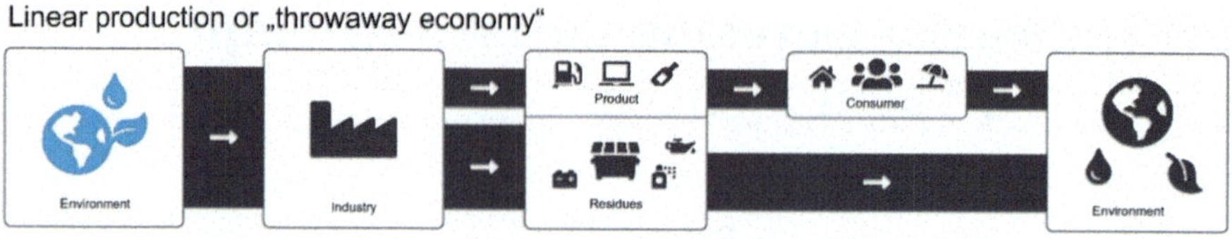

Fig. 1 Scheme of linear industrial production and end-of-life pathway. Graphic: Andreas Ch. Braun, Gabriela Espinosa, Danny Tröger, Gesa Siebert

Since the report of the Club of Rome in 1972 (Meadows et al., 1972), there have been increasing calls for a more sustainable industrial production model than the linear one depicted in Fig. 1. Three mechanisms are key to achieving more sustainable industrial production:

- Reuse—of waste or product by life cycle;
- Resource efficiency—optimization of the ratio of input per product (less waste); and
- Resource productivity—optimization of the achieved product per resource unit.

Larger frameworks for managing resource consumption and waste generation have been developed in recognition of the need for sustainable transformation. The concept of the circular economy (CE) contributes directly to some of the Sustainable Development Goals (SDGs) of the United Nations (UN), for instance SDG 6 (Clean Water and Sanitation), SDG 7 (Affordable and Clean Energy), SDG 8 (Decent Work and Economic Growth), SDG 12 (Responsible Consumption and Production), and SDG 15 (Life on Land) (Schroeder et al., 2019). However, the circular economy concept hardly addresses social aspects (Schöggl et al., 2020).

In contrast, industrial ecology (IE) approaches are designed to achieve the principles of a circular economy while being economically feasible at the same time and mitigating any social problems (Cohen-Rosenthal, 2004). The basic idea is that "there is potential to convert 'waste outputs' into 'resource inputs,' thereby reducing the economic, social, and environmental costs of waste disposal, while simultaneously preventing the need to utilize more raw materials for use in production" (McManus & Gibbs, 2008). As further outlined in chapter A5 of this book, industrial ecology additionally contributes to the SDGs 9 (Industry, Innovation, and Infrastructure), SDG 10 (Reduced Inequalities), SDG 11 (Sustainable Cities and Communities), and SDG 13 (Climate Action).

The term "industrial ecology" arose from the article "Strategies for Manufacturing" by Frosch and Gallopoulos (1989). The central idea is to consider industrial production systems in analogy to natural ecosystems by borrowing the terms, principles, and concepts of the latter (Hess, 2010; van Berkel, 2007; Yang & Lay, 2004). An important principle from nature is that of material cycles. Various system components transform substances in a process referred to as their metabolism. As different components work together in symbiosis, they complement each other in such a way that the initial substances are created again in the end (van Berkel, 2007).

In application, this means that *industrial symbiosis* is the study of the relationships between various actors in an industrial system that initiate certain characteristics of natural symbiosis such as reciprocity, constancy, balance, and interdependence to enhance efficiency (Hess, 2010). Each species or factory hereby has its own *metabolism*, which is the transformation and circulation of matter (materials and residues), (process) water, or energy (Fraccascia, 2018). *Industrial metabolism* refers to the measurement of material and energy flows resulting from industrial activities (Hess, 2010; Krausmann et al., 2008), while *urban metabolism* refers to the exchange processes that produce the urban environment, including municipal waste (Broto et al., 2012; Newell & Cousins, 2015). Figure 2 depicts some of the leverage points that industrial ecology offers to shape production in such a way that resource consumption is reduced.

When these concepts are applied in a targeted way on a common site or property, with companies from different industries or sectors cooperating in the form of mutual exchange of byproducts, water and energy, the result is called an eco-industrial park (EIP) (Cohen-Rosenthal, 2004; UNIDO, 2017). In the case of one of the oldest EIPs, the Kalundborg Symbiosis in Denmark (Fig. 3), individual economic benefits for the participating companies can be seen as well as enhanced environmental performance for the collective (Jacobsen, 2006).

Although this example of successful industrial symbiosis emerged on a small geographical scale in a bottom-up process, it nevertheless developed into a scientific approach with the potential success factors for upscaling. The geographical properties

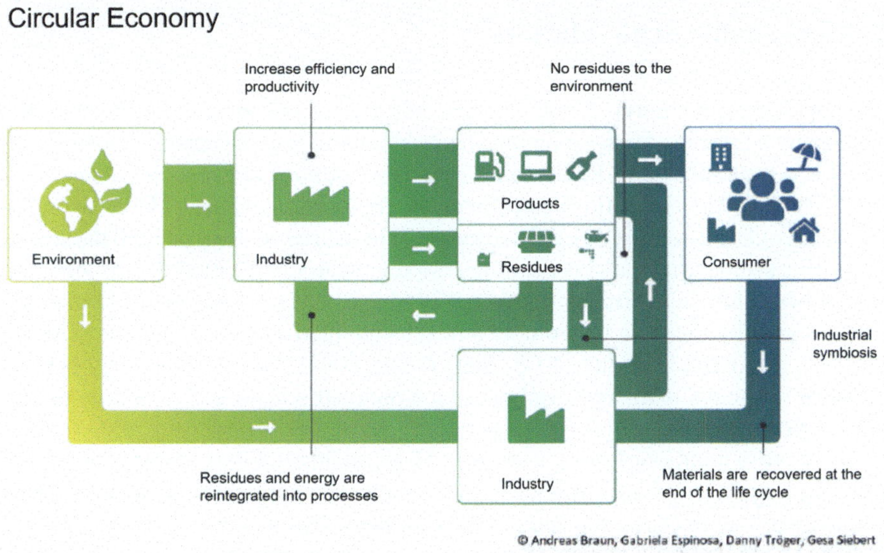

Fig. 2 Scheme of a circular economy with some leverage points of industrial symbiosis for reduced environmental impacts. Graphic: Andreas Ch. Braun, Gabriela Espinosa, Danny Tröger, Gesa Siebert

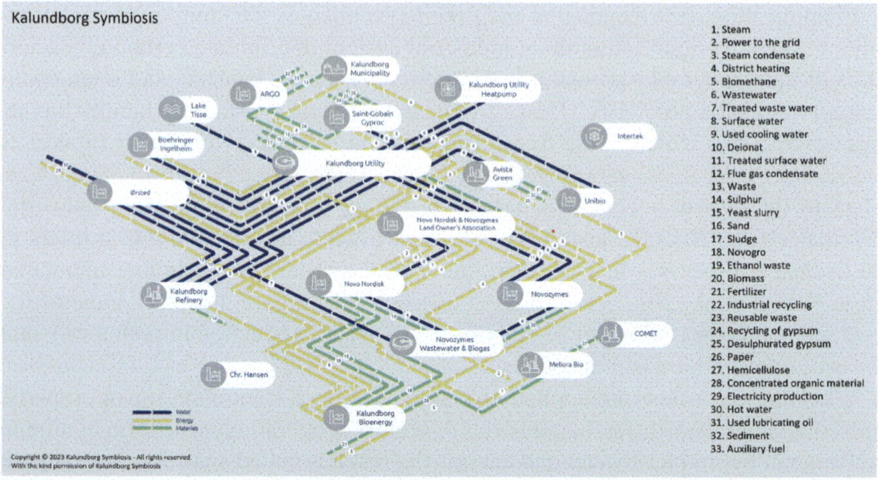

Fig. 3 The Kalundborg Symbiosis. Graphic: Kalundborg Symbiosis, 2023

of a given site are a decisive factor for the macro performance of an EIP (Raimbault et al., 2020; Sterr & Ott, 2004). This concerns the geographical scale as well as physical-geographical characteristics, as outlined in the following section.

2 Geographic Scale Matters

Geographical scale plays a role in eco-industrial development because as the complexity of the waste materials increases, the more specializations are required, which are generally more likely to be available within a larger area (Chen et al., 2012). Yet most specializations can already be found on the regional level, avoiding the need for global transport. A good example of industrial ecology beyond the spatial limits of eco-industrial parks is the industrial region around Heidelberg, Mannheim, and Ludwigshafen in Germany. Industrial symbiosis has been successfully initiated in this region by the state to promote an exchange platform for residual and waste materials (Sterr & Liesegang, 2003; Sterr & Ott, 2004). Industrial ecology beyond EIPs that considers industrial symbiosis on a regional scale is called *eco-industrial development* (Fig. 4).

There are a number of other examples of eco-industrial development strategies that have helped to reduce environmental impacts on a regional scale and are being used not only as an economic and ecologic strategy, but also integrated as a part of spatial planning and regional development (Deutz & Gibbs, 2008; Gibbs & Deutz, 2005, 2007; Gibbs et al., 2005). In the case of Chinese mega cities, for example, Liu et al. (2014) show that thinking of urban material flows as metabolisms and applying further approaches of industrial ecology strategies enabled urban planners

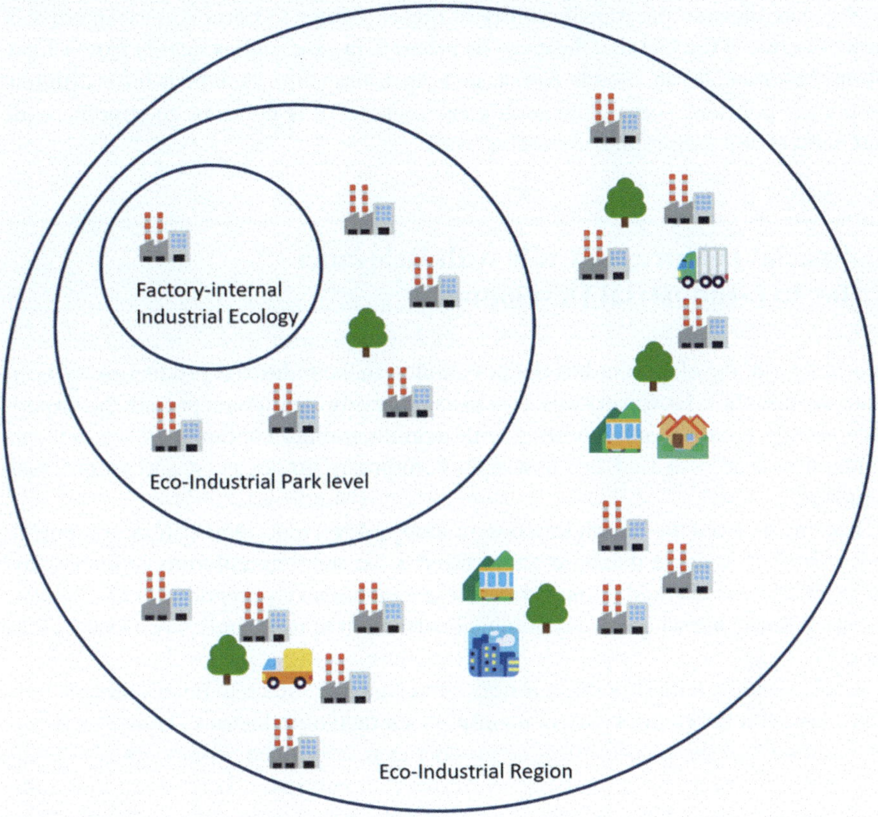

Fig. 4 Geographic scales in industrial ecology from factory-internal to regional. Graphic: Own graphic based on Cohen-Rosenthal (2017)

to develop more sustainable and resource-efficient urban pathways. In this application, an interesting and somewhat surprising insight is given by the example of Guiyang, a resource-dependent city with heavy industry in China, where resource saving and carbon dioxide emission reduction was achieved not by relocating industries outside the city but by the use of solid wastes and energy by local industries, that is to say putting an industrial-urban symbiosis in place (Dong et al., 2016). Simboli et al. (2019) illustrate the industrial ecology synergies of post-industrial cities in Italy by highlighting the various potential of material flows in industrial-residential-rural systems. In another Italian case study, Scarpellini et al. (2019) found that economic and social gains are achieved if CE activities are implemented at regional level. Gibbs and Deutz (2005), in contrast, conclude after analyzing case studies of eco-industrial development promoted as regional planning strategy that the integration of environmental, economic, and social aims remains difficult. This in itself suggests that eco-industrial strategies for regional development may theoretically but not automatically lead to win–win–win situations (Gibbs & Deutz, 2005).

We also question the transferability of these findings to Latin American cities in particular due to the fact that the analysis is based on case studies mainly from China, North America, Japan, South Korea, and Australia (Fig. 5), which have different economic, political, social, and ecological contexts. It is not only geographic scale that matters, but geographic context as well.

3 Special Features of Chile with Relevance for Eco-Industrial Development

Eco-industrial development has the potential to make industrial production systems more sustainable. However, this is a comparatively new concept and the experience mainly comes from countries with manufacturing economies. We conducted a literature search to find out how many empirical studies exist and from which countries.

For this we used the search function [country] AND [eco-industrial development] in the Web of Science database and created a list with the columns [country] and [number of case studies]. Connecting these to a country polygon in a GIS system results in a map showing the geographic distribution of case studies across the globe (Fig. 5).

Looking at the map (Fig. 5), it is noticeable that there seem to be few studies from countries with economies based mainly on extractive industries. However, it can be assumed that the possibilities for producing in a more environmentally friendly way differ between the processing and extractive industries. It is easy to imagine: residual materials and byproducts in the metal processing or chemical industries have

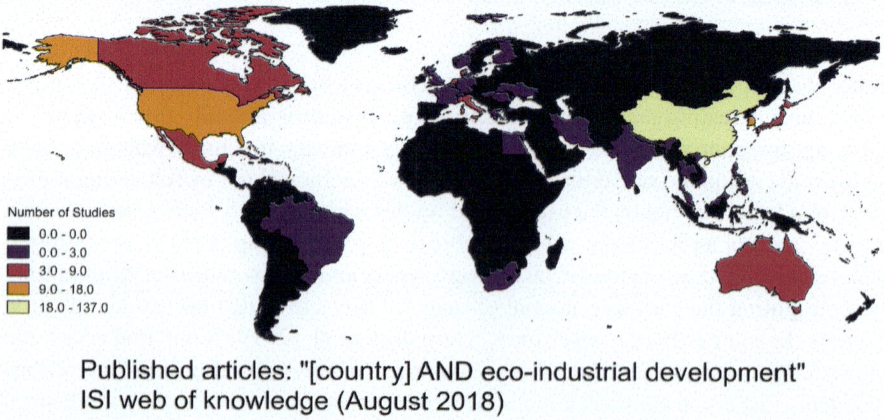

Fig. 5 World map of the number of empirical case studies on eco-industrial development cases in August 2018, before the IECO project started. Cartography: Andreas Ch. Braun, GIS data base: Opendatasoft.com

fundamentally different properties and potential uses than overburden or tailings from mining or wood chips from forestry. There were also hardly any empirical studies on eco-industrial development in Chile by the time of the IECO project's launch. In the following sections, we will outline some of Chile's environmental (Sect. 3.1) and socio-economic (Sect. 3.2) constraints which we consider relevant to eco-industrial development.

3.1 Natural Systems and Extractive Industries of Chile

Chile stretches along the western coast of South America for more than 4,300 km from 17°30′S to 56°0′S. The country spans several climatic zones and is bordered by the Pacific Ocean to the west and the Andes to the east. In the north, these mountains reach heights of nearly 7,000 m above sea level (the further south in the country, the lower the Andes generally are). This results in many different climatic zones, vegetation covers, and land uses (Zhao et al., 2016). The diversity of natural areas in Chile is the reason behind the large and varied potential for renewable energy supplies, green economies like bioeconomy, and other natural resources in the country.

In the north are some of the driest areas on earth such as the Atacama Desert, where hardly any vegetation can be found. The active mountain formation with its many volcanoes and high plateaus ensures the presence of mineral resources and geothermal energy. Due to the lack of water, mining faces particular challenges here (Carrasco et al., 2023). Traditional lithium mining, for example, is carried out by vaporizing mineral-rich groundwater. On the other hand, the high level of solar radiation provides enormous potential for solar energy.

In central Chile, more potential for renewable energy supplies can be found: orographically induced rainfall and the enormous differences in altitude offer opportunities for hydropower. In contrast to the north, agriculture is also possible at these latitudes. While Chile is best known for its wine-growing areas, other fruits like avocado also thrive here.

The Mediterranean climates of south-central Chile also offer good growing conditions for agricultural products as well as forestry (INFOR, 2021). This creates significant potential for a bioeconomy, which represents an attempt to replace fossil fuels and to produce industrial organic substances and products not on the basis of crude oil but by using wood—a renewable raw material—as a starting material (Hildebrandt et al., 2019).

The south of Chile is dominated by cold temperate climates, so that even inland ice fields can be found in Patagonia (the geographical name for the south of South America, divided between Chilean and Argentinian territory). Although the entire coast of Chile is rich in fish due to the cold and nutrient-rich Humboldt current, the Patagonian fjords are particularly suitable for fish farming in aquacultures due to the inflow of uncontaminated fresh water (Iizuka et al., 2016). The hinterland of these freshwater tributaries is characterized by extensive agriculture and sheep and cattle farming.

3.2 Historically Conditioned Peculiarities of Socio-economic Systems in Chile

Economic activities and regional development in Latin America have some path-dependent peculiarities. They are implemented in post-colonial organizational structures that are often very centralized (Andersson et al., 2009). This can lead to administrative mismatches (Garrick et al., 2022) and enormous inequalities in the effectiveness and efficiency of public administration (Nisar & Masood, 2021). Furthermore, this manifests in regional disparities and fragmentation in economic prosperity (Andersson et al., 2016), as well as disparities in access to and the quality of infrastructure (Blanco & Apaolaza, 2018; Gelman & Santilli, 2010).

Another phenomenon is widespread socio-spatial segregation with an increasing proportion of gated communities in urban areas (Haase et al., 2017; Michelini & Pintos, 2016) and in the—likewise socially exclusive—mono-functional use of land in rural areas (Braun, 2015, 2022; Braun & Koch, 2016). In the context of the region's colonial history, both can be seen as a consequence of the neoliberal economic policies established under the Pinochet dictatorship (Carcamo-Huechante, 2006; Centner, 2009; Haughney, 2007). At the same time, spatial planning is sometimes disconnected from actual land use, whether due to informal activities or telecoupling effects (Bolay et al., 2005; Gonzalez-Mathiesen & March, 2023).

Besides housing and land use, many other economic activities also take place in an informal way. The official and governmental handling of this issue ranges from support to disregard, criminalization, and suppression (Carenzo, 2020; Romero-Lankao et al., 2014). In the meantime, especially in the academic discourse, a strict duality between formal and informal practice is no longer drawn (Alfaro d'Alençon et al., 2018; Ezeah et al., 2013) and good experiences have been achieved by acknowledging and integrating existing social praxis by public authorities into their institutions (Carenzo, 2014; Gutberlet, 2015). In this regard, eco-industrial development in our understanding should provide an epistemic basis aiming for the inclusion of informal economic activities, for instance in the recycling economy (Purvis et al., 2023; Tröger et al., 2023; Valencia et al., 2023).

A problem area related to these phenomena arises from the fact that a prerequisite for the success of eco-industrial strategies is trust between stakeholders (Gibbs, 2003), which can be difficult in South American contexts. The necessary trust building is additionally challenged by differing risk perceptions regarding industrial development. Conflicting cultural worldviews, place attachment, and identity shape these concerns and are likely to be triggered by any project or strategy involving the link between urban and industrial metabolism as these often contain inherent aspects of spatial and territorial relevance (Masuda & Garvin, 2006). At the moment there are no known systematic ways to integrate spatial features when implementing IS synergies, leaving the appropriate strategies to achieve industrial symbiosis synergies highly case-specific (Fraccascia, 2018).

4 Eco-Industrial Development and the IECO Research Project

Poganietz and Bolívar-Paypay (2024) provided a general introduction and background knowledge on industrial ecology while chapter Glöser-Chahoud and Schultmann (2024) achieved the same on the topic of circular economy. The boundary conditions outlined in the previous section explain why the research content of the IECO project is particularly relevant for Chile, and which particular sustainability challenges are faced there in industrial development. Before we present our definition of eco-industrial development in this context, we would like to give an overview of how the structure of this book corresponds to those conditions.[1]

The second part of this book presents some insights from research into Topic One of the IECO project: sustainable use of water and energy in an urban context. This topic area "investigates the ramifications and implementation requirements of new urban infrastructures, to the purpose of achieving a circular economy regarding the resource utilization in the urban context. Specifically, this research line investigates alternatives to recover thermal energy from gray waters, increasing the efficiency of air conditioning mechanisms, and recovery of chemical nutrients and energy from dark waters, among others." (IECO, 2023a).

The third part of this book will address Topic Two: sustainable material and energy transformation of biomass. "This research line focuses on different aspects of biomass utilization and transformation. In particular, Topic Two investigates the transition to a bioeconomy through the reorganization of productive and logistic systems in the agroindustry. Additionally, the generation of biofuels from the waste of the forest industrial process is investigated as a means to increase economic and environmental efficiency." (IECO, 2023b).

Topic Three addresses the eco-industrial utilization of resources from the subsurface and is adressed in the fourth part of the book. "This research line investigates low and high enthalpy systems, to the purpose of identifying geothermal applications that improve the efficiency in different industrial processes. In the northern parts of Chile, geothermal energy can be used to enhance the sustainability of the mining and desalinization processes. In the southern part of the country the efficiency of energy-intensive industries, e.g. forest industrial or agricultural processes, can be increased and heating services can be provided to residential customers in urban and rural locations." (IECO, 2023c).

Topic Four of the IECO is focused on urban circularity and eco-industrial transition in South America. "This topic investigates possible circular economy approaches

[1] The topics are explained once more in Chapter A5 in the institutional context of their formation.

in the frame of urban development. This involves developing process-oriented urban design approaches of space as a metabolism at different scales, interpreting the analysis of urban metabolism and of industrial processes and how their waste (material and energy) could be understood as resourcefulness and used in urban design processes and vice-versa. The interest is at the neighborhood scale, as an ambit to articulate regenerative design perspectives with dialogues and interventions." (IECO, 2023d).

Research Topic Five, corresponding to part six of the book, has a certain special status. The topic of land use and industrial sustainability "examines the sustainability of land use in Chile—especially in relation to the forestry sector. The forestry industry in Chile is one of the most important economic sectors and at the same time triggers numerous negative environmental and social impacts. Many of these impacts are mediated through the spatial dimension of forest plantations. Plantations have spread rapidly in space, encompassing large spatial scales, increasingly approximating other land use systems and generating a monofunctionally of land use." (IECO, 2023e).

A fundamental prerequisite for minimizing the environmental impact of production systems, and thus realizing and implementing the principle of circularity and the creation of industrial symbioses, is to have the information about the energy, material, and water requirements of each production step fully available. This can be achieved by Life Cycle Assessments (LCA) or Material Flow Analysis (MFA). For this reason, LCA is presented by Espinosa and Casas Ledón (2024) in this book as a superordinate topic outside the topic structure of the IECO.

5 Towards an Understanding of Eco-Industrial Development

We argue here that the term "development" is used too uncritically and too conceptually in the context of eco-industrial development. Overall, the term is rarely used, and when it is, it is used almost synonymously with the term "industrial ecology." It is important to be careful and not to equate the existence of green industrial activities with eco-industrial development. To be more specific: *An* eco-industrial development—such as the building of a green industrial park *within* a region—and *the* eco-industrial development (e.g. of a region or even a country) are not the same thing.

At the same time, it is important to be careful about the context in which the term "development" is used. When it comes to the socio-economic development of "underdeveloped" countries, the term eco-industrial development is more closely associated with so-called leapfrogging strategies. Leapfrogging means not following the path of "catch-up development," i.e. imitating "developed" countries, but rather avoiding or skipping unfavorable path dependencies (hence the allusion to the leaping frog) and, for example, focusing on green industries etc. from the outset. Few examples are known where this has been completely successful. The city-state of Singapore could possibly serve as an example.

The term could also be used in connection with regional development, in which case regional planning criteria would have to be taken into account. There are few studies other than those by Gibbs et al. (2005) and Sterr and Ott (2004) that have systematically analyzed the relationship between eco-industrial development and regional development. It is then important to note that industrial ecology must be an essential and guiding part of a regional development strategy. The mere presence of an eco-industrial park in a region is just as insufficient for the diagnosis of an eco-industrial development strategy as the presence of a hospital suggests a health-oriented development strategy, or the presence of a university suggests an education-oriented development strategy.

Before any attempt at definition is made, it is also necessary to establish which development model forms the basis of "eco-industrial development" in the sustainability discourse. We are convinced that a classic growth-orientated economic model is intended here. Industrial ecology is a strategy that adheres to market and growth logic in the sense of green growth. According to numerous theorists—for example in the field of post-, de- or a-growth—this is the wrong approach. Instead, sufficiency strategies are proposed (Braun takes a critical look at this at the end of this book).

Based on the above, we believe that the following normative minimum criteria can be identified, all of which must be met in order to speak of eco-industrial development.

1. There are industries that actually endeavor to operate more sustainably, e.g. to improve their energy and resource efficiency, to reduce cycles, or to reduce emissions into the environment instead of merely greenwashing.
2. Green industries that focus on eco-industrial strategies make significant contributions to economic production and are not subordinate players.
3. Green industries that rely on eco-industrial strategies are (a) legitimate, (b) accepted, and (c) an integral part of the regional or national development strategy and not isolated projects of individual promoters.
4. The objective of such projects fulfils corporate social responsibility criteria, i.e. to make serious contributions to socio-economic welfare that are effective for the region and at the same time allow other (non-industrial) developments. (This criterion distinguishes them, for example, from so-called green special economic zones, in which the aim is simply to give industry as free a hand as possible in reaping corporate profits without "disturbing" them too much with social standards).

In our view, these criteria are a good first step in diagnosing eco-industrial development. At the same time, they are quite strong normative criteria. However, it is important to develop such strong criteria in order to prevent industries that have very negative environmental impacts but at the same time produce supposedly "green" products from being falsely labelled as "sustainability industries." One example would be the lithium industry, which could emphasize its applications in e-mobility in order to claim that it is an eco-industrial development. It would certainly violate criteria 1 and 2 in Chile. The forestry industry, which produces "green raw materials" but at the same time has massive environmental impacts, also violates at least criteria 2 and 3 in Chile. Such a demanding and critical-normative discourse aims to produce a demanding understanding of sustainability, fostering critical reflection on the developments described in this volume and setting the normative bar high, i.e. aiming at high standards that must be met in order to be able to speak of sustainable industries.

6 Conclusion

This chapter should be understood as an attempt to deal with the concepts of industrial ecology and eco-industrial development in terms of normative definitions. It should not (and cannot) be read conclusively as the "last word in wisdom," but is rather intended to sensitize people to the fact that these terms should be used more critically. Within the scientific community that deals with industrial sustainability, the terms industrial ecology and eco-industrial development are probably sufficiently clearly defined.

However, two things should be borne in mind. Firstly, as shown, the discourse is very much centered on the Global North. There needs to be more critical analysis of what eco-industrial development means in contexts in the Global South in order to understand whether and how this strategy can be implemented there. Secondly, the term "eco-industrial development" is related to other discourses such as socio-economic development, regional development, or sustainable development. We find that too little inter- and meta-discourse has taken place here so far, meaning that the literature on industrial ecology and eco-industrial development is still too isolated.

This chapter has provided some impetus with regard to both aspects. These impulses will help with the critical reading of this book and, additionally, should stimulate debates and subsequent systematic research.

References

Alfaro d'Alençon, P., Smith, H., Álvarez de Andrés, E., Cabrera, C., Fokdal, J., Lombard, M., Mazzolini, A., Michelutti, E., Moretto, L., & Spire, A. (2018). Interrogating informality: Conceptualisations, practices and policies in the light of the New Urban Agenda. *Habitat International, 75*, 59–66. https://doi.org/10.1016/j.habitatint.2018.04.007

Andersson, K., Gordillo, G., & van Laerhoven, F. (2009). *Local governments and rural development: Comparing lessons from Brazil, Chile, Mexico, and Peru (First paperback edition)*. University of Arizona Press.

Andersson, K., Lawrence, D., Zavaleta, J., & Guariguata, M. R. (2016). More trees, more poverty? The socioeconomic effects of tree plantations in Chile, 2001–2011. *Environmental Management, 57*(1), 123–136. https://doi.org/10.1007/s00267-015-0594-x

Bilitewski, B., & Härdtle, G. (2013). *Abfallwirtschaft: Handbuch für Praxis und Lehre* (4., aktualisierte und erweiterte Auflage). Springer Vieweg.

Blanco, J., & Apaolaza, R. (2018). Socio-territorial inequality and differential mobility. Three key issues in the Buenos Aires Metropolitan Region. *Journal of Transport Geography, 67*, 76–84. https://doi.org/10.1016/j.jtrangeo.2017.07.008

Bolay, J.-C., Pedrazzini, Y., Rabinovich, A., Catenazzi, A., & García Pleyán, C. (2005). Urban environment, spatial fragmentation and social segration in Latin America: Where does innovation lie? *Habitat International, 29*(4), 627–645. https://doi.org/10.1016/j.habitatint.2004.05.003

Braun, A. C. (in Press). A critical look: Is eco-industrial development just more of the same? In A. C. Braun, G. G. Espinosa, D. Tröger & T. Hirth (Eds.), *Eco-industrial development as an industrial strategy—Contributions from a German-Chilean research partnership.* Springer Nature.

Braun, A. C. (2015). Taxonomic diversity and taxonomic dominance: The example of forest plantations in South-Central Chile. *Open Journal of Ecology, 05*(05), 199–212. https://doi.org/10.4236/oje.2015.55017

Braun, A. C. (2022). Deforestation by afforestation: land use change in the coastal range of Chile. *Remote Sensing, 14*(7), 1686. https://doi.org/10.3390/rs14071686

Braun, A. C., & Koch, B. (2016). Estimating impacts of plantation forestry on plant biodiversity in southern Chile-a spatially explicit modelling approach. *Environmental Monitoring and Assessment, 188*(10), 564. https://doi.org/10.1007/s10661-016-5547-1

Broto, V. C., Allen, A., & Rapoport, E. (2012). Interdisciplinary perspectives on urban metabolism. *Journal of Industrial Ecology, 16*(6), 851–861. https://doi.org/10.1111/j.1530-9290.2012.00556.x

Carcamo-Huechante, L. E. (2006). Milton Friedman: knowledge, public culture, and market economy in the Chile of Pinochet. *Public Culture, 18*(2), 413–435. https://doi.org/10.1215/08992363-2006-010

Carenzo, S. (2014). Lo que (no) cuentan las máquinas: la experiencia sociotécnica como herramienta económica (y política) en una cooperativa de "cartoneros" del gran Buenos Aires. *Antípoda. Revista De Antropología Y Arqueología* (18), 109–135. https://doi.org/10.7440/antipoda18.2014.06

Carenzo, S. (2020). Contesting informality through innovation "from below": Epistemic and political challenges in a waste pickers cooperative from Buenos Aires (Argentina). *Tapuya: Latin American Science, Technology and Society,* 1–31. https://doi.org/10.1080/25729861.2020.1788775

Carrasco, S., Hernández, J., & Cariaga, V. (2023). The temporalities of natural resources extraction: Imagined futures and the spatialization of the lithium industry in Chile. *The Extractive Industries and Society, 15*, 101310. https://doi.org/10.1016/j.exis.2023.101310

Centner, R. (2009). Conflictive sustainability landscapes: The neoliberal quagmire of urban environmental planning in Buenos Aires. *Local Environment, 14*(2), 173–192. https://doi.org/10.1080/13549830802522079

Chen, X., Fujita, T., Ohnishi, S., Fujii, M., & Geng, Y. (2012). The impact of scale, recycling boundary, and type of waste on symbiosis and recycling. *Journal of Industrial Ecology, 16*(1), 129–141. https://doi.org/10.1111/j.1530-9290.2011.00422.x

Cohen-Rosenthal, E. (2004). Making sense out of industrial ecology: A framework for analysis and action. *Journal of Cleaner Production, 12*(8–10), 1111–1123. https://doi.org/10.1016/j.jclepro.2004.02.009

Cohen-Rosenthal, E. (2017). What is eco-industrial development? In E. Cohen-Rosenthal & J. Musnikow (Eds.), *Eco-industrial strategies: Unleashing synergy between economic development and the environment* (First ed., pp. 14–29). Routledge. http://www.greeneconomics.net/ei-developchap.pdf

Deutz, P., & Gibbs, D. (2008). Industrial ecology and regional development: Eco-industrial development as cluster policy. *Regional Studies, 42*(10), 1313–1328. https://doi.org/10.1080/00343400802195121

Dong, L., Fujita, T., Dai, M., Geng, Y., Ren, J., Fujii, M., Wang, Y., & Ohnishi, S. (2016). Towards preventative eco-industrial development: An industrial and urban symbiosis case in one typical industrial city in China. *Journal of Cleaner Production, 114*, 387–400. https://doi.org/10.1016/j.jclepro.2015.05.015

Espinosa, G., & Casas León, Y. (2024). Life cycle thinking: An essential framework for eco-industrial development. In A. C. Braun, G. Espinosa, D. Tröger & T. Hirth (Eds.), *Eco-industrial development as an industrial strategy—Contributions from a German-Chilean research partnership*. Springer Nature.

Ezeah, C., Fazakerley, J. A., & Roberts, C. L. (2013). Emerging trends in informal sector recycling in developing and transition countries. *Waste Management (New York, N.Y.), 33*(11), 2509–2519. https://doi.org/10.1016/j.wasman.2013.06.020

Fraccascia, L. (2018). Industrial symbiosis and urban areas: A systematic literature review and future research directions. *Procedia Environmental Science, Engineering and Management, 5*(2), 73–83.

Frosch, R. A., & Gallopoulos, N. E. (1989). Strategies for manufacturing. *Scientific American, 261*(3), 144–152. https://doi.org/10.1038/scientificamerican0989-144

Garrick, D., Alvarado-Revilla, F., de Loë, R., & Jorgensen, I. (2022). Markets and misfits in adaptive water governance: how agricultural markets shape water conflict and cooperation. *Ecology and Society, 27*(4). https://doi.org/10.5751/ES-13337-270402

Gelman, J., & Santilli, D. (2010). Crecimiento económico, divergencia regional y distribución de la riqueza: Córdoba y Buenos Aires después de la independencia. *Latin American Research Review, 45*(1), 121–147. http://www.jstor.org/stable/27919178

Gibbs, D. (2003). Trust and networking in inter-firm relations: The case of eco-industrial development. *Local Economy: The Journal of the Local Economy Policy Unit, 18*(3), 222–236. https://doi.org/10.1080/0269094032000114595

Gibbs, D., & Deutz, P. (2005). Implementing industrial ecology? Planning for eco-industrial parks in the USA. *Geoforum, 36*(4), 452–464. https://doi.org/10.1016/j.geoforum.2004.07.009

Gibbs, D., & Deutz, P. (2007). Reflections on implementing industrial ecology through eco-industrial park development. *Journal of Cleaner Production, 15*(17), 1683–1695. https://doi.org/10.1016/j.jclepro.2007.02.003

Gibbs, D., Deutz, P., & Proctor, A. (2005). Industrial ecology and eco-industrial development: A potential paradigm for local and regional development? *Regional Studies, 39*(2), 171–183. https://doi.org/10.1080/003434005200059959

Glöser-Chahoud, S., & Schultmann, F. (2024). Circular economy: An introduction. In A. C. Braun, G. Espinosa, D. Tröger, & T. Hirth (Eds.), *Eco-industrial development as an industrial Strategy—Contributions from a German-Chilean research partnership*. Springer Nature.

Gonzalez-Mathiesen, C., & March, A. (2023). Long-established rules and emergent challenges: Spatial planning and wildfires in Chile. *International Planning Studies, 28*(1), 37–53. https://doi.org/10.1080/13563475.2022.2136629

Gutberlet, J. (2015). More inclusive and cleaner cities with waste management co-production: Insights from participatory epistemologies and methods. *Habitat International, 46*, 234–243. https://doi.org/10.1016/j.habitatint.2014.10.004

Haase, D., Kabisch, S., Haase, A., Andersson, E., Banzhaf, E., Baró, F., Brenck, M., Fischer, L. K., Frantzeskaki, N., Kabisch, N., Krellenberg, K., Kremer, P., Kronenberg, J., Larondelle, N., Mathey, J., Pauleit, S., Ring, I., Rink, D., Schwarz, N., & Wolff, M. (2017). Greening cities—To be socially inclusive? About the alleged paradox of society and ecology in cities. *Habitat International, 64*, 41–48. https://doi.org/10.1016/j.habitatint.2017.04.005

Haughney, D. (2007). Neoliberal policies, logging companies, and mapuche struggle for autonomy in Chile. *Latin American and Caribbean Ethnic Studies, 2*(2), 141–160. https://doi.org/10.1080/17442220701489555

Hess, G. (2010). The ecosystem: Model or metaphor? *Journal of Industrial Ecology, 14*(2), 270–285. https://doi.org/10.1111/j.1530-9290.2010.00226.x

Hildebrandt, J., O'Keeffe, S., Bezama, A., & Thrän, D. (2019). Revealing the environmental advantages of industrial symbiosis in wood-based bioeconomy networks: An assessment from a life cycle perspective. *Journal of Industrial Ecology, 23*(4), 808–822. https://doi.org/10.1111/jiec.12818

IECO. (2023a). *Topic 1: Sustainable usage of water and energy in an urban context*. https://ieco-institute.com/topics/sustainable-use-of-water-and-energy-in-an-urban-context.html
IECO. (2023b). *Topic 2: Sustainable material and energy transformation of biomass*. https://ieco-institute.com/topics/sustainable-material-and-energetic-use-of-biomass.html
IECO. (2023c). *Topic 3: Eco-industrial utilization of subsoil resources*. https://ieco-institute.com/topics/eco-industrial-utilization-of-subsoil-resources.html
IECO. (2023d). *Topic 4: Urban circularity and eco-industrial transition in South-America*. https://ieco-institute.com/topics/urban-circularity-and-eco-industrial-transition-in-south-america.html
IECO. (2023e). *Topic 5: Land use and industrial sustainability*. https://ieco-institute.com/topics/land-use-and-industrial-sustainability.html
Iizuka, M., Roje, P., & Vera, V. (2016). The development of Salmon aquaculture in Chile into an internationally competitive industry: 1985–2007. In A. Hosono, M. Iizuka, & J. Katz (Eds.), *Chile's Salmon industry: Policy challenges in managing public goods* (pp. 75–108). Springer Japan.
INFOR. (2021). *Área de información y economía forestal: Estadisticas Forestales*. https://wef.infor.cl/sector_forestal/sectorforestal.php
Jacobsen, N. B. (2006). Industrial symbiosis in Kalundborg, Denmark: A quantitative assessment of economic and environmental aspects. *Journal of Industrial Ecology, 10*(1–2), 239–255.
Kalundborg Symbiosis. (2023, September 6). https://www.symbiosis.dk/en/
Krausmann, F., Fischer-Kowalski, M., Schandl, H., & Eisenmenger, N. (2008). The global sociometabolic transition. *Journal of Industrial Ecology, 12*(5–6), 637–656. https://doi.org/10.1111/j.1530-9290.2008.00065.x
Liu, H., Zhou, G., Wennersten, R., & Frostell, B. (2014). Analysis of sustainable urban development approaches in China. *Habitat International, 41*, 24–32. https://doi.org/10.1016/j.habitatint.2013.06.005
Masuda, J. R., & Garvin, T. (2006). Place, culture, and the social amplification of risk. *Risk Analysis, 26*(2), 437–454. https://doi.org/10.1111/j.1539-6924.2006.00749.x
McManus, P., & Gibbs, D. (2008). Industrial ecosystems? The use of tropes in the literature of industrial ecology and eco-industrial parks. *Progress in Human Geography, 32*(4), 525–540. https://doi.org/10.1177/0309132507088118
Meadows, D. H., Meadows, D., Randers, J., & Behrens, W. W. (1972). *The limits to growth: A report for the club of Rome's project on the predicament of mankind* (2nd ed.). Universe books.
Michelini, J. J., & Pintos, P. (2016). Metropolitan expansion and new socio-spatial segregation scenarios in contemporary Argentina. The case of Nordelta-Las Tunas (Buenos Aires). *Habitat International, 54*, 40–49. https://doi.org/10.1016/j.habitatint.2015.08.011
Newell, J. P., & Cousins, J. J. (2015). The boundaries of urban metabolism. *Progress in Human Geography, 39*(6), 702–728. https://doi.org/10.1177/0309132514558442
Nisar, A., & Masood, A. (2021). *Bureaucracy and the other: A systematic review of postcolonial scholarship in public administration*. Advance online publication. https://doi.org/10.2139/ssrn.3886409
Poganietz, W.-R., & Bolívar-Paypay, V. (2024). The concept of industrial ecology. In A. C. Braun, G. Espinosa, D. Tröger & T. Hirth (Eds.), *Eco-industrial development as an industrial strategy—Contributions from a German-Chilean research partnership*. Springer Nature.
Purvis, B., Celebi, D., & Pansera, M. (2023). A framework for a responsible circular economy. *Journal of Cleaner Production, 400*, 136679. https://doi.org/10.1016/j.jclepro.2023.136679
Raimbault, J., Broere, J., Somveille, M., Serna, J. M., Strombom, E., Moore, C., Zhu, B., & Sugar, L. (2020). A spatial agent based model for simulating and optimizing networked eco-industrial systems. *Resources, Conservation and Recycling, 155*, 104538. https://doi.org/10.1016/j.resconrec.2019.104538
Romero-Lankao, P., Hughes, S., Qin, H., Hardoy, J., Rosas-Huerta, A., Borquez, R., & Lampis, A. (2014). Scale, urban risk and adaptation capacity in neighborhoods of Latin American cities. *Habitat International, 42*, 224–235. https://doi.org/10.1016/j.habitatint.2013.12.008

Scarpellini, S., Portillo-Tarragona, P., Aranda-Usón, A., & Llena-Macarulla, F. (2019). Definition and measurement of the circular economy's regional impact. *Journal of Environmental Planning and Management, 62*(13), 2211–2237. https://doi.org/10.1080/09640568.2018.1537974

Schöggl, J.-P., Stumpf, L., & Baumgartner, R. J. (2020). The narrative of sustainability and circular economy—A longitudinal review of two decades of research. *Resources, Conservation and Recycling, 163*, 105073. https://doi.org/10.1016/j.resconrec.2020.105073

Schroeder, P., Anggraeni, K., & Weber, U. (2019). The relevance of circular economy practices to the sustainable development goals. *Journal of Industrial Ecology, 23*(1), 77–95. https://doi.org/10.1111/jiec.12732

Simboli, A., Taddeo, R., & Raggi, A. (2019). The multiple dimensions of urban contexts in an industrial ecology perspective: An integrative framework. *The International Journal of Life Cycle Assessment, 24*(7), 1285–1296. https://doi.org/10.1007/s11367-017-1411-y

Sterr, T., & Liesegang, D. G. (2003). Industrielle Stoffkreislaufwirtschaft im regionalen Kontext: Betriebswirtschaftlich-ökologische und geographische Betrachtungen in Theorie und Praxis. *Springer*. https://doi.org/10.1007/978-3-642-55666-1

Sterr, T., & Ott, T. (2004). The industrial region as a promising unit for eco-industrial development—Reflections, practical experience and establishment of innovative instruments to support industrial ecology. *Journal of Cleaner Production, 12*(8–10), 947–965. https://doi.org/10.1016/j.jclepro.2004.02.029

Tröger, D., Becerra, A. A. A., Busnelli, R., Yajnes, M., Williams, F., & Braun, A. C. (2023). Exploring eco-industrial development in the global south: Recognizing informal waste-picking as urban-industrial symbiosis? *Cleaner Waste Systems, 5*, 100096. https://doi.org/10.1016/j.clwas.2023.100096

UNIDO. (2017). *An international framework for eco-industrial parks*. https://www.unido.org/our-focus-safeguarding-environment-resource-efficient-and-low-carbon-industrial-production/eco-industrial-parks

Valencia, M., Solíz, M. F., & Yépez, M. (2023). Waste picking as social provisioning: The case for a fair transition to a circular economy. *Journal of Cleaner Production, 398*, 136646. https://doi.org/10.1016/j.jclepro.2023.136646

van Berkel, R. (2007). Industrial Ecology. In *Edward Elgar E-book archive. The international handbook on environmental technology management.* Edward Elgar Pub.

Yang, P.P.-J., & Lay, O. B. (2004). Applying ecosystem concepts to the planning of industrial areas: A case study of Singapore's Jurong Island. *Journal of Cleaner Production, 12*(8–10), 1011–1023. https://doi.org/10.1016/j.jclepro.2004.02.028

Zhao, Y., Feng, D., Yu, L., Wang, X., Chen, Y., Bai, Y., Hernández, H. J., Galleguillos, M., Estades, C., Biging, G. S., Radke, J. D., & Gong, P. (2016). Detailed dynamic land cover mapping of Chile: Accuracy improvement by integrating multi-temporal data. *Remote Sensing of Environment, 183*, 170–185. https://doi.org/10.1016/j.rse.2016.05.016

Life Cycle Thinking: An Essential Framework for Eco-Industrial Development

Gabriela Espinosa Gutiérrez and Yannay Casas Ledón

Abstract Life Cycle Thinking (LCT) seeks to identify potential improvements through all life cycle stages of a product or system in order to lower overall negative impacts (environmental, social, economic) and improve the global use of resources. It helps to create a better understanding of how decisions about a given product system could impact the economy, the environment, and society at all stages and related locations of a product system, becoming a central framework for numerous policies and instruments related to sustainable consumption and production, waste management, and improved product design. This article explains the interconnections of LCT with principles of the circular economy and eco-industrial development and elucidates several examples of successfully implemented LCT approaches for evaluating and improving production systems in Germany and Chile. Furthermore, it describes the incentives that currently exist at the regulatory and voluntary levels for companies to implement LCT in their organizations. Nevertheless, although the implementation of LCT helps provide more information on production systems and motivates innovation for more sustainable production, the behavior and decisions of consumers must also evolve towards a more responsible use of resources in order to achieve a circular economy and ensure global sustainability in the long term.

1 Introduction: Life Cycle Thinking

The life cycle of a product or service can be defined as the successive and interconnected stages in a product's life, from raw material extraction and conversion through to production (manufacturing), distribution, use, and disposal (Deutsches Institut für

G. Espinosa Gutiérrez (✉)
Infrastructure and Sustainability, TÜV SÜD Energietechnik GmbH Baden-Württemberg, Mannheim, Germany
e-mail: gabriela.espinosagutierrez@tuvsud.com

Y. Casas Ledón
Environmental Engineering Department, Faculty of Environmental Sciences-EULA Center, University of Concepción (UdeC), Concepción, Chile
e-mail: ycasas@udec.cl

A. Ch. Braun et al. (eds.), *Eco-Industrial Development as an Industrial Strategy*, Green Energy and Technology, https://doi.org/10.1007/978-3-031-73576-9_5

Normung [DIN], 2006). The succession of all interconnected processes related to the production, use, and disposal of a product is also called a product system. This may refer to a tangible product, but also to a service or even a human activity as long as the system fulfils a function and a specified performance is expected (Klöpffer & Grahl, 2014).

Life Cycle Thinking (LCT) seeks to identify potential improvements through all life cycle stages in order to lower overall negative impacts (environmental, social, economic) and improve the global use of resources. The key aspect of LCT is that it represents a holistic perspective for gathering information about potential environmental, social, and economic impacts through the complete life cycle (Joint Research Center of the European Union [JRC], 2010). LCT incorporates the gathered information into decision-making processes, with the aim of improving overall performance and preventing or minimizing impacts at any one life cycle stage or geographical region that may increase the impact at another life cycle stage, location, or impact category (United Nations Environment Programme [UNEP], 2012).

The roots of LCT emerged in the late 1960s and early 1970s out of concern for shortages of natural resources, particularly of fossil fuels, but also due to rising problems with the management of waste. The first studies adopting this approach were focused on energy analysis as well as resources and environmental modelling (JRC, 2010; Klöpffer & Grahl, 2014). Nowadays, needs and consumption patterns have changed, along with the environmental, policy, and social challenges. In this context LCT, also known as the "cradle-to-grave" approach, has become a central framework for numerous policies and instruments related to integrated product policies, sustainable consumption and production, waste management, and improved product design (JRC, 2010).

LCT helps create a better understanding of how decisions about a given product system could impact the economy, the environment, and society at all stages and related locations of the product system. It allows decision makers to focus decisions towards long-term perspectives and to identify opportunities for improving the overall system, not only a single stage (UNEP, 2012). Moreover, when applied to an early design stage of a given product or service, LCT helps to create sustainable products right from their point of creation. As LCT becomes more widely known outside the scientific community and political circles, its application and the communication of such practices as part of a company's philosophy help to improve the company's image and competitiveness, creating added value for shareholders and customers (JRC, 2010; UNEP, 2004).

LCT is also an inherent part of the circular economy. As described by Müller et al. (2020), a circular economy stands for a resource-efficient society with a closed-loop economy. Behind this are the principles of ecological economics and industrial ecology. Although the concept first became known in relation to waste-management approaches (e.g. reduce-reuse-recycle, zero-waste initiatives), a circular economy must actually consider all material and product life cycles, going far beyond waste management at the end of a product's life cycle (Müller et al., 2020). In an approach known as "cradle-to-cradle," products are designed in such a way that they do not waste resources at the end of their lifetime, but rather are reinserted into product systems as raw materials (Hauschild et al., 2017).

LCT and eco-industrial strategies are also related, as depicted in Fig. 1. Eco-industrial strategies, depicted at the center of Fig. 1, can be applied to any of the general life-cycle stages of a product and may contribute to a circular economy by increasing efficiency in resource use, minimizing waste, and, with the appropriate design, even enabling "cradle-to-cradle" approaches in support of a closed-loop economy.

Fig. 1 Linking eco-industrial development practices to life cycle thinking. Eco-industrial development strategies may be applied to any stage of the production life cycle to support a circular economy

2 Life Cycle Thinking: Tools

To support the implementation of LCT, several methodologies, tools, and software solutions have been developed, the best-known and most widespread methodology being Life Cycle Assessment (LCA). In the words of Hauschild et al. (2017), LCA is "a holistic, systemic, and rigorous environmental tool, which allows determining the potential environmental impacts quantitatively on human health, ecosystem quality, and resource depletion during all life cycle stages of a given product or goods."

LCA provides a normed framework for quantifying and assessing a given product's potential environmental impacts. The technical framework for the LCA methodology has been standardized by the International Standards Organization (ISO) in the norm ISO 14040:2006 (Life Cycle Assessment—Principles and Framework) and ISO 14044:2006 (Life Cycle Assessment—Requirements and Guidelines). According to these norms, LCA involves four phases: (i) Goal and Scope Definition, (ii) Inventory Analysis, (iii) Impact Assessment, and (iv) Interpretation.

The LCA phases are not strictly followed in a single sequence. It is an iterative process. Subsequent iterations can achieve increasing levels of detail analysis (from conceptual to detailed LCA) or lead to changes in the first phase prompted by the results of the last phase (DIN, 2006; Klöpffer & Grahl, 2014).

In the Impact Assessment phase of the LCA, the resulting mass flows from the Inventory Analysis phase are evaluated according to their potential impact[1] on human health, ecosystems, and resources. The indicators can be chosen either at the midpoint or endpoint level according to the required level of detail set in the Goal and Scope Definition phase of the LCA. The midpoint impacts translate potential impacts into environmental themes such as climate change, acidification, human toxicity, etc. (problem-oriented approach). Meanwhile, endpoint impacts, also called damage categories, translate environmental impacts into issues of concern such as human health, ecosystem diversity, and resource scarcity (damage-oriented approach) (Hauschild et al., 2017; Klöpffer & Grahl, 2014). The damage categories can be normalized and weighted to be expressed as a single score, which makes the product comparison and decision-making process easier. However, the process of aggregating and assigning severity scores for the different types of impacts is a difficult and uncertain task, which may reduce the comprehensiveness and consistency of the results (Hauschild et al., 2017). Figure 2 shows the relationship between inventory results and impact category levels of an LCA.

In the framework of LCT, several assessment variations exist based on the general LCA methodology. Each of them emphasizes a different aspect of sustainability (social, environmental, or economic). Table 1 summarizes some of the most important tools and LCT-based assessments. Figure 3 shows the arrangement of the LCT-based assessments according to the sustainability aspect they support. The Life Cycle

[1] Potential impacts do not necessarily translate into actual effects. For example, actual effects on an ecosystem may depend on preexisting concentrations of contaminants in the environment and the exceedance of pollutant thresholds may originate in multiple pollution sources converging at specific times and locations.

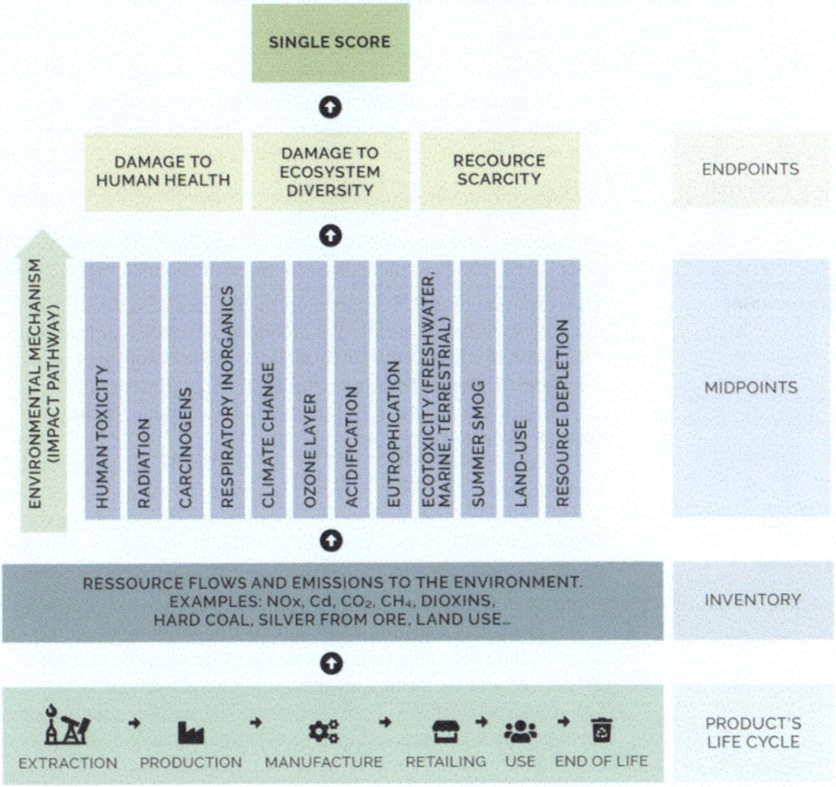

Fig. 2 Levels of impact category indicators: midpoint and endpoint levels. Adapted from Joint Research Centre—Institute for Environment and Sustainability (2010)

Sustainability Assessment (LCSA, in the middle of the figure) is the only one covering all aspects.

Currently, the ISO 14044 standard is widely accepted and used for LCA studies; nevertheless, its implementation may be complex. Table 2 provides a summary of the common general challenges encountered with the application of LCT in the industry. One of the significant challenges in the wider application of LCA studies is the development of regional Life Cycle Inventory (LCI) databases, particularly for Latin America and the Caribbean regions. Currently, there are operative databases for Mexico (81 datasets, Mexicaniuh) and clusters of datasets for Brazil and Chile (Sonnemann et al., 2016; Valdivia et al., 2017). For Chile in particular, the LCA database named Ecobase developed by *Fundacion Chile* is relevant (Valdivia et al., 2015). However, this database mainly focuses on export food products and building materials produced during 2015 (Sonnemann et al., 2016).

Table 1 Overview of selected tools and methodologies for the implementation of LCT

Tool	Description
Eco-efficiency assessment	Assesses the environmental performance of a product system in relation to its value (product system value/environmental impact) during the whole life cycle. The product system value is referred to functional, monetary, aesthetic aspects, among others (ISO, 2012)
Carbon footprint	This variation of an LCA accounts for the greenhouse gas emissions CO_2, CH_4, N_2O and chlorofluorocarbons and includes two approaches: (i) evaluation of the carbon footprint of organizations (ISO, 2018a) and (ii) the carbon footprint of products (ISO, 2018b)
Water footprint	This variation of an LCA considers the amount of water consumed directly or indirectly during the production of goods and services. Two methodologies are available: (i) the methodology developed by the Water Footprint Network (WFN) in 2002 (Hoekstra et al., 2011), which considers different methods depending on the source and destination of the water used (blue,[a] green,[b] and grey[c] water footprint) (Hoekstra et al., 2011), and (ii) the life cycle approach standardized by the ISO. It provides a more holistic perspective on water use and pollution than the first mentioned methodology, orienting its impacts on human health and the damage to ecosystem quality caused by water loss and quality degradation (ISO, 2014)
Ecological footprint	A metric that measures the biologically productive land and sea area required for a given population, product, region, or country to produce the renewable resources and ecological services it uses. The ecological footprint is expressed in global average bioproductive hectares[d] (gha) and considers six provisioning areas: cropland, grazing land, fishing grounds (marine and inland), forests, and infrastructure areas (Ewing, 2010). Borucke et al. (2013) provide a detailed description of this methodology
Life cycle costing (LCC)	This evaluation assesses the internal (e.g. private) and external (e.g. by pollution) costs related to a product or system during its life cycle. There are three LCC approaches: (i) conventional LCC,[e] (ii) environmental LCC, and (iii) social LCC. This evaluation is handy for monitoring costs in different scenarios, making it attractive to clients and the financial sector (Hauschild et al., 2017; Swarr et al., 2011)

(continued)

Table 1 (continued)

Tool	Description
Social life cycle assessment (S-LCA)	This methodology emphasizes the direct impacts that a product or a service could exert on stakeholders, correlating the companies' behavior with socioeconomic processes or social capital. UNEP/SETAC Life Cycle Initiative has developed a guideline following the standard of ISO 14040, this being one of the most comprehensive guidelines developed to date for future S-LCA (Benoit-Norris et al., 2013; UNEP/SETAC Life Cycle Initiative, 2013)
Life cycle sustainability assessment (LCSA)	This is the most holistic and transdisciplinary methodology, integrating the three sustainability pillars (environmental, economic, and social). It allows for an evaluation of all the positive and negative impacts of environmental, economic, and social criteria, making it a key tool for supporting decision making towards more sustainable products, sectors, and economies (Guinée et al., 2011; UNEP/SETAC Life Cycle Initiative, 2011)

[a] The blue water footprint refers to water catchment from surface water or groundwater that is either evaporated or incorporated into a product
[b] The green component corresponds to water coming from rainfall and melted snow, which is stored in the soil and evaporated back into the atmosphere
[c] The grey water footprint involves the water amount needed to assimilate pollutants coming from any process to meet local water-quality standards
[d] Bioproductive hectares are the area that is available for production; for example, cropland for provisioning of food and fiber products, or forest land for producing timber and forest products
[e] Referred to as traditional process economic analysis when applied to processes

3 Material Flow Analysis: Accounting Framework of Life Cycle Thinking

Putting Life Cycle Thinking into practice requires accounting the material and energy flows necessary for the production of a given good or the provision of a service. The purpose is to quantify the physical exchanges of such goods or services with the environment over their entire life cycle (JRC, 2010). Baccini and Bader (1996) refer to LCA as a scientific-technical procedure to quantify the energy and material flows resulting from the production, distribution, use, and disposal of a good or process, with the aim of then using the information gained for the most economical use of resources with the least possible environmental impact.

A Material Flow Analysis (MFA), defined as "a systematic assessment of the flows and stocks of materials within a system defined in space and time," helps to assess the mass and energy flows of a given system in quantitative terms while reducing the complexity of the system to a level helpful for decision making (Brunner & Rechberger, 2004). Typically, the general procedure for an MFA would include a mass and energy balance for a well-defined system in which several processes occur (Fig. 4). The flows entering and leaving the system (import and export flows) as well as the internal flows and accumulating stocks are accounted and calculated to obtain

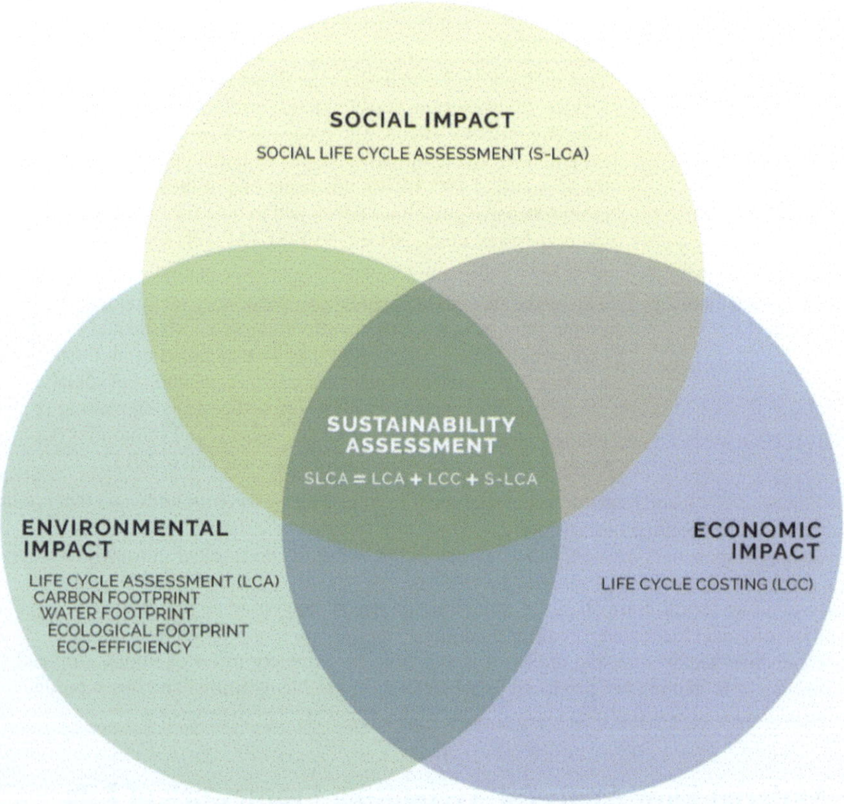

Fig. 3 Life Cycle Thinking tools organized according to the aspect of sustainability to which they relate (environmental, social, and economic)

an inventory of all flows relevant to the chosen system. The MFA (and LCI in its scientific part) is based on the laws of mass and energy conservation, the principles of thermodynamics, and stoichiometry (Klöpffer & Grahl, 2014).

A typical MFA focuses on the evaluation of the flows in a geographically well-defined region during a defined period of time, which could be a factory or even a unit process inside a factory, with the goal of a mass and energy balance in the system of a specific material or substance (e.g. water, copper, or nitrogen). The results are helpful for identifying bottlenecks and appropriate control and monitoring points (Brunner & Rechberger, 2004). Furthermore, by means of modelling scenarios and evaluation of their results, MFA helps in the search for improvements to the analyzed system (Espinosa, 2016).

The LCI, on the other hand, could be considered a type of MFA focusing on accounting for specific flows (substances, materials, energy) through a complete value chain. LCI focuses on the function of a product or the service it delivers, not

Table 2 Examples of challenges in the application of LCT in the industry, according to the stages of the LCA. Based on: (Guinée et al., 2011; Klöpffer & Grahl, 2014; Reap et al., 2008; Zamagni et al., 2008)

Challenges at stage "Goal and scope definition"	Challenges at stage "Life Cycle Inventory (LCI)"
• Defining the functional unit for the life cycle assessment is challenging for practitioners • The boundaries of the product system are difficult to define, as it requires deciding how broadly to expand the scope of the study to ensure a complete LCA is obtained without unnecessarily increasing the complexity of the analysis • Some simplifications to the system are often too restrictive, particularly for their application at the mesoscale (e.g. industrial sectors) and macroscale (e.g. complex technological systems). This can lead to a partial evaluation and a shifting of the environmental impacts due to the implementation of measures with impacts outside the system	• Clear agreement is lacking on the most appropriate impact allocation approach for different cases • The time variable is not considered in life cycle inventories; basically, the current LCI modelling is done statically, limiting future projections • More research is still needed on the reliability and uncertainty of the data, in particular for input–output analysis • There is a general lack of region-specific data availability and quality, specifically for non-developed countries • Freely accessible databases are limited
Challenges at stage "Life Cycle Impact Assessment (LCIA)"	*Challenges at stage "LCA Interpretation"*
• It is difficult for non-experts to choose the impact assessment method and specially to interpret the impacts, due to a variety of method approaches (midpoints/endpoints). The harmonization of the two approaches could solve this concern • Spatial differentiation and geographical allocation is required for several impact categories (e.g. acidification, eutrophication, human toxicity, among others). Current methods do not allow these to be depicted properly • New characterization methods and impact categories are needed, such as for risk assessment, noise pollution, etc. • Normalization of data sets being applicably only at a regional scale	• The assessment of the input data quality and the uncertainty analysis of the flows from the LCI implies a high degree of complexity • More comprehensive guidelines are needed for defining uncertainty in inventory data and impact assessment models

on the product itself. It accounts all required materials and energy inputs required to fulfil the function or service in question. The goal of LCI is not to provide a complete mass balance on the system, but to depict the resource consumption along the life cycle of the product. After accounting all the flows required to provide the service or function, these are correlated against the environmental, social, or economic impacts during the assessment and interpretation phase of the LCA (Klöpffer & Grahl, 2014). Often, an LCI will be conducted for several product alternatives in order to compare

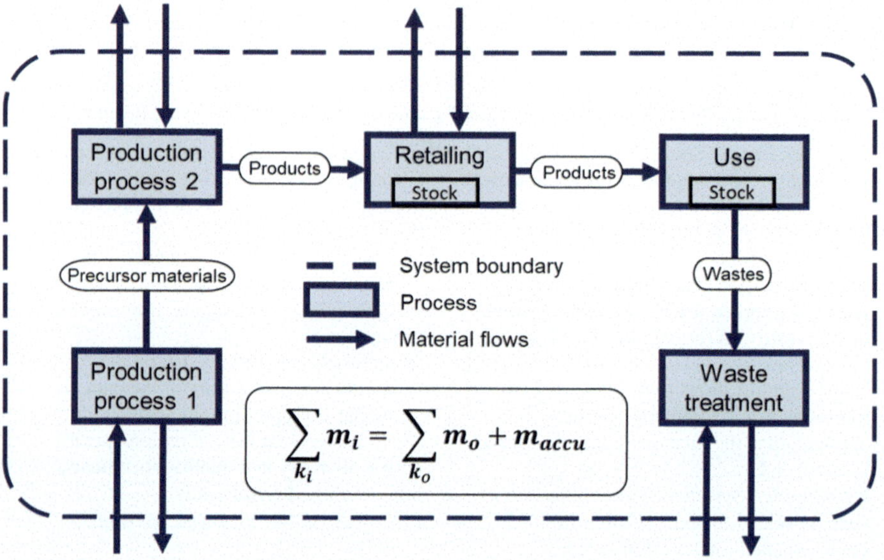

Fig. 4 Example flowchart for a Material Flow Analysis and a basis equation for mass balance. In the equation, *k* is the number of flows in a system, *m* refers to mass flows, *I* stands for inputs, *o* stands for outputs, and *accu* stands for the change in the stocks of the system, i.e., accumulation or depletion

their impacts and to support decisions with respect to purchasing, improvements, or design (Bouman et al., 2000).

Recent investigations support the combined use of MFA and LCI to effectively identify hotspots and assess the environmental sustainability of complex systems comprising numerous products and services, such as countries, cities, regions, economic sectors, and industrial or service polygons. The detailed knowledge of the production processes generated by the combination of these tools (MFA and LCI) was proven to increase resource efficiency in companies (Lopes Silva et al., 2015; Schmidt et al., 2017, 2018).

The information required for a complete MFA and LCI may not always originate from sources or measurements directly related to the evaluated product or system and literature values would be required. This introduces a certain uncertainty into the assessment, affecting the reliability of the results. However, well-established procedures for handling the data uncertainty of the input values and for calculating the propagation of their uncertainty into output flows by means of Gauss's law (for normally distributed variables) or Monte Carlo simulations (for not normally distributed variables, or if the deviations are too large) will be of help (Brunner & Rechberger, 2004). Additionally, the use of incomplete data sets can be maximized by means of data validations. This and model adaptations are useful tools to complete the flows inventory and can help provide a robust framework for decision making (Espinosa, 2016; Espinosa & Otterpohl, 2014).

4 Life Cycle Thinking: Application Examples

The application of LCT tools for eco-industrial development has gained popularity over the last decades. The number of scientific publications on the topic alone has experienced exponential growth since 1990 (Fig. 5), with geographical distribution of the publications strongly focused on developed countries. The United States, China, and European countries, followed by Canada, Australia, Brazil, and India, rank as the countries with the largest number of publications on the topic (Fig. 6).

A transfer of knowledge about LCT and LCT-related tools into an industrial and praxis-oriented environment has been gaining momentum over the past decades. The following examples describe some of the applications of LCT in the industry. The applications and benefits of LCT and related tools in the industry are numerous and as diverse as the industry sectors, with an ever-increasing complexity in the analyzed systems and product alternatives.

An example from the cotton industry shows how LCT helped in the successful implementation of a new production strategy, leading to savings in raw materials and energy and in the placement of a new product in the market. The German company *Gebrüder Otto GmbH & Co.* evaluated the impact of using processed cotton waste originating from various process stages of the textile value chain (e.g., from waste produced in the spinning mill and the cut edges of the knitting or weaving) for the fabrication of a new cotton textile product. The LCA of the current product and the alternative product focused on the water and energy consumption. The results

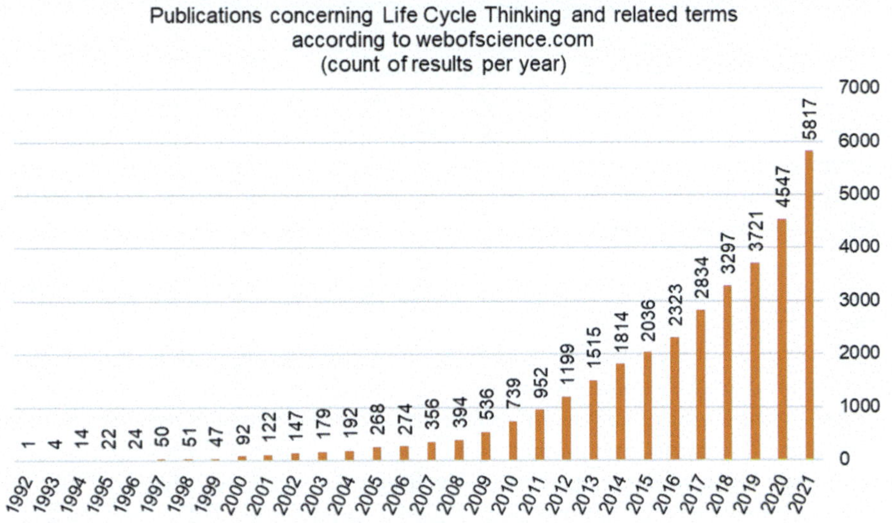

Fig. 5 Number of published articles related to LCT and related tools* according to web of science. *Terms searched: life cycle thinking, life cycle assessment, life cycle impacts assessment, life cycle impact assessment, water footprint, carbon footprint, life cycle costing, social life cycle assessment, and life cycle sustainability assessment

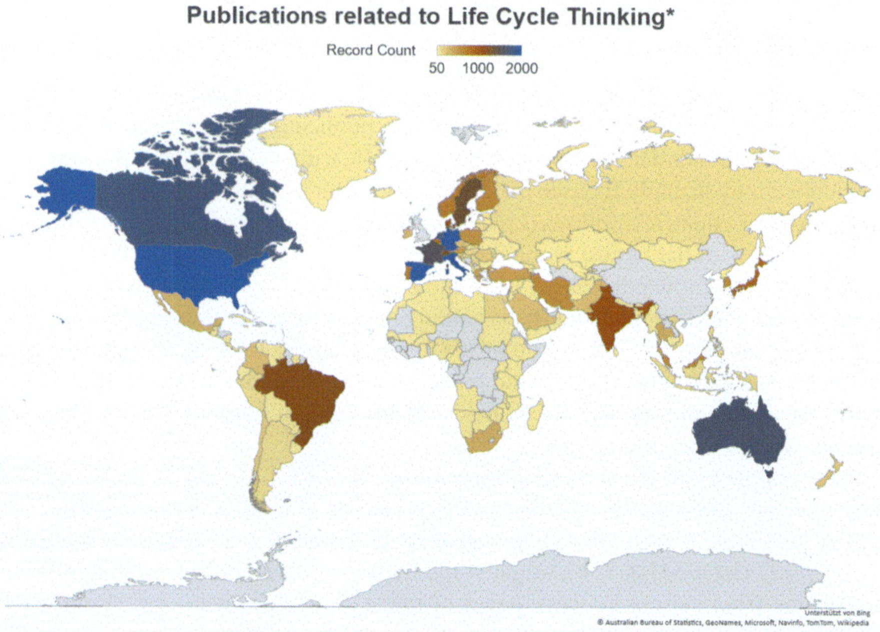

Fig. 6 Geographical distribution of publications (cumulative, up to 2022) related to LCT*, according to web of science. *Terms searched: life cycle thinking, life cycle assessment, life cycle impacts assessment, life cycle impact assessment, water footprint, carbon footprint, life cycle costing, social life cycle assessment, and life cycle sustainability assessment

showed that, besides savings in raw materials and chemical supplies, a mix rate of 50% recycled cotton fibers would create water savings on the cotton fields of up to 10,000 L/kg of raw cotton and energy savings of 20% at the production site. The introduction of the more sustainable product alternative with recycled cotton was made possible by the development of a new technology for processing cotton waste. From the production perspective this represented small economical savings due to the investment costs, but from the sales perspective the company gained significant leverage by improving their unique selling proposition with a more sustainable product (Schmidt et al., 2017).

The example of Uzin Utz AG (Germany) illustrates the application of LCT for a complex product. This company specializes in the manufacture of chemical product systems for the laying of all types of floor lining. With the help of LCA, they assessed the ecological footprint of a product system consisting of textile flooring adhesive, primer, and self-levelling cementitious levelling compound. All phases of the life cycle were analyzed for optimization potential: production of the raw materials, processing in the factory, product disposal, and all transport stages in between. The results of the LCA provided the company with the decision criteria to start a project for reducing the raw materials of synthetic origin used in their products and packaging. The outcome was an alternative packaging system with lower environmental impacts, consisting of a flexible plastic bag and an outer cardboard box instead of the traditional system of hard plastic containers. The new system uses 85% less plastic, generates 89% less CO_2 emissions, and requires 60% less energy compared to the traditional packaging system[2] without compromising performance or functionality. It also has a better transport performance given that more packages can be transported on one pallet. In this example the economic efficiency of the alternative packaging was the same as the traditional packaging, due to necessary adaptations in the process. Customer acceptance of the new packaging will be the deciding factor for the company to invest in a new facility with higher automation (Schmidt et al., 2018).

The Latin American context also provides examples of LCT application. In a reference project for the Guatemalan palm oil industry at the refining facilities of Las Palmas S.A., the life cycle, water footprint, and carbon footprint of the palm oil production were analyzed. The LCA covered the environmental impacts of the pre-nursery, nursery, palm plantation, crude palm oil extraction, and oil refining stages of production.[3] The results provided a basis for improving the environmental sustainability of the industry. They show that, for the analized life cycle stages, the emissions to air, water, and soil originate mostly from fertilizer and pesticide application, as well as from the use of fossil fuels. The highest overall impact originates from the agricultural stages, with more than 50% of the carbon footprint stemming from the manufacture of the fertilizers and pesticides. Furthermore, the carbon footprint increases by up to 270 kg of CO_2-equivalents per ton of crude palm oil when the production draws upon tropical forests and transforms them for palm cultivation. Several countermeasures to lessen the environmental impacts were identified. For example, biodigestion of the palm oil mill effluent for electricity generation reduces the carbon footprint by 20%. Additionally, the LCA showed that sustainable procurement and reduced consumption of fertilizer and pesticides together with reducing fuel and water consumption represent the most important actions for increasing the sustainability of the industry in Guatemala (Sacayón et al., 2018).

A Chilean example of applied LCT can be found in the wine industry. The carbon footprint measured for several Chilean wine companies in 2009 was 0.83–2.93 kg of CO_2-equivalents per bottle for wine destined for the North American market, and 0.87–2.97 kg of CO_2-equivalents per bottle for the European market. Later, detailed LCAs identified that the elaboration and use of the bottles were the main sources of carbon emissions, representing 32% of total emissions, followed by wine storage (28%), field operations (20%), and transport of the products (20%). This information was used to introduce measures for the reduction, compensation, or neutralization of the emissions. For example, the bottle weight was reduced and the usage of space in transport containers was optimized, changes in energy supply sources and waste

[2] The results originate from a cradle-to-gate analysis, from the extraction of raw materials to the finished container at the factory gate. This approach is often used when the analyzed product serves as a raw material for further products, making it excessively complicated, if not impossible, to follow this raw material until the end-product stage.

[3] Also a cradle-to-gate analysis.

recycling were introduced in addition to eco-design strategies (for new products, bottles, or processes). Additionally, the information was used as part of a National Sustainability Codex for the Chilean wine industry, defining sustainability parameters for the different life cycle stages within the framework of an accreditation system with annually increasing requirements. Accredited wineries are then entitled to a certificate and to use a distinctive seal on their labels or other advertising material. Furthermore, recommendations were drawn from the project policy and a calculator for the Chilean agricultural sector was developed (Frohmann & Olmos, 2013; Olmos, 2012).

Further examples from Latin America illustrate how LCT and LCT tools have helped companies from other branches to reduce their environmental impacts, such as by increasing energy efficiency, and reducing their carbon footprint and waste generation (including wastewater). The production sites and companies were also able to build brand value and provide more transparency to their customers by creating publicly available Environmental Product Declarations (EPD). In some cases, it was also demonstrated that LCT tools helped with the decisions to introduce new products while ensuring an environmentally sustainable supply chain for them (Wiche & Valdivia, 2014).

5 Incentives for the Implementation of Life Cycle Thinking in Chile and Germany

The growing importance of a life cycle perspective as a necessary element of eco-industrial solutions is underlined by the ever-increasing and advancing framework of regulations and norms around the topic. The incentives can be organized into three groups:

- regulatory frameworks
- voluntary standards
- scientific communities, practitioners' networks, and industrial lobby groups.

5.1 Regulatory and Strategic Framework in Chile and Germany

The regulatory and strategic framework in Germany directly and indirectly supporting the implementation of LCT is strongly correlated to the sustainability strategy and action plans of the European Union (EU). Much has been undertaken in this area in the past decade, starting with the 2015 EU action plan titled Closing the Loop, part of a circular economy package (Müller et al., 2020). This action plan evolved into the EU Circular Economy Action Plan published in 2020 which presents a holistic approach to the circular economy, incorporating the entire life cycle from product design, production processes, and sustainable consumption to waste management (European Commission, 2020a). This new action plan also recognizes the global dimension of such a holistic approach, given that European companies often integrate international supply chains (European Commission, 2020b). Derived from the EU models, the German laws and ordinances have also evolved, following the same philosophy as their precedents while complementing them with local adaptations and specifications.

The following German regulations can be highlighted for their implementation of LCT in the industry:

- The Circular Economy Law (*Kreislaufwirtschaftsgesetz-KrWG)*, approved in Germany in 2012 and last updated in 2021, was created as a response to the European Waste Framework Directive. Its aim is to promote the conservation of natural resources and the protection of people and the environment mainly through the promotion of a circular economy, the prevention of waste, and, above all, through the increased recycling of waste. In this law it is stated that the entire life cycle of the waste shall be taken as the basis for the consideration of its effects on humans and the environment. The law emphasizes the importance of eco-design as one possible measure to reduce the production of waste by systematically integrating environmental aspects into product design, with the aim of improving the environmental performance of the product throughout its entire life cycle. Furthermore, in order to bolster waste reduction, the already known and proven system of product responsibility[4] was expanded, requiring the extension of the useful lifetime of products and allowing waste disposal only as a last possibility. Additionally, the system of product responsibility includes a duty of transparency for the information that can best be addressed by applying LCT (Bundesministerium für Umwelt, Naturschutz und nukleare Sicherheit, 2021).
- Further laws and regulations in Germany complement and support the implementation of the Circular Economy Law and extended product responsibility in a detailed manner for certain products considered of major importance. This encompasses packaging products, motor vehicles, portable batteries, electric and electronic devices, and petroleum products (Bundesministerium für Umwelt, Naturschutz und nukleare Sicherheit, 2021). The corresponding laws are:
  - Law for placing on the market, withdrawal, and high-quality recycling of packaging materials (*Verpackungsgesetz*-VerpackG) from 2017, last updated in 2021
  - Ordinance on the transfer, withdrawal, and environmentally sound disposal of end-of-life vehicles (*Altfahrzeugverordnung*-AltfahrzeugV) from 1997, last updated in 2020

[4] The concept of product responsibility implies that manufacturers are responsible for the waste generated from their products. This concept was created to foster the implementation of improved product designs allowing for less waste generation and the recycling of products at the end of their useful lifetime.

- Law for placing on the market, withdrawal, and environmentally sound disposal of batteries and accumulators (*Batteriegesetz*-BattG) from 2009, last updated in 2020
- Law for placing on the market, withdrawal, and environmentally sound disposal of electrical and electronic equipment (*Elektro- und Elektronikgerätegesetz*-ElektroG) from 2015, last updated in 2021
- Waste oil ordinance (*Altölverordnung*-AltölV) from 1987, last updated in 2020
- Eco-design guideline (*Ökodesign Richtlinie*, also known as *Energieverbrauchsrelevante-Produkte-Richtlinie*, ErP-RL) from 2009 and updated in 2012, establishing a framework of eco-design requirements for energy consumption-relevant products.

The regulatory framework in Germany (as in Europe) requires the environmental aspects of a number of products to be communicated in obligatory environmental labels. An example is the energy consumption efficiency certificate for all domestic appliances (*Energieausweis*).

In Chile, the following regulations and strategic frameworks can be highlighted for their implementation of LCT in the industry:

- The National Program of Sustainable Consumption and Production (SCP), promoted by the Environmental Minister in 2016 (MMA, 2016b), explicitly incorporates LCT and its tools (LCA, carbon footprint, ecolabeling) in the SCP action plans.
- The Chilean Energy roadmap for 2015–2050, "Towards a Sustainable and Inclusive Energy," also promotes LCT by means of low-emission and cost-competitive energy development, considering energy infrastructure, transmission, and final users.
- The Extended Producer Responsibility law (REP, its acronym in Spanish; Law N° 20.920) promulgated from 2016 (MMA, 2016a), establishes a comprehensive framework for efficient waste management. Based on life cycle thinking, the law addresses the reduction of waste generation, the promotion of waste recovery, and valorization of six priority wastes: lubricant oils, electrical and electronic devices, batteries, containers, packaging, and tires.
- The National Strategy for Sustainable Construction (CDT, 2015) boosts a continuous and permanent improvement of building construction standards at the national level, advancing towards a more sustainable material construction market across the value chain. Moreover, the strategy is the basis for implementing different national certification systems, such as the Sustainable Building Certification (CES) and the Housing Energy Rating.
- The National Roadmap for a Circular Chile until 2040 (MMA, 2021) is aimed at developing a waste strategy that promotes organic waste recovery generated at the municipal level, focusing on four strategy lines such as circular innovation, circular culture, circular territories, and circular regulation.

5.2 *Voluntary Standards at the International Level*

Numerous documents and guidelines for the voluntary implementation of LCT exist at the international level. For example, the UN's Sustainable Development Goal of "ensuring sustainable consumption and production patterns" (SDG 12) includes a holistic approach based on LCT. Specifically, SDG 12 aims to achieve the environmentally sound management of chemicals and all wastes throughout their complete life cycle. In this way, it intends to reduce the release of this waste into the air, water, and soil in order to minimize adverse impacts on human health and the environment, to achieve the efficient use of resources, and to reduce waste generation. This SDG encourages companies to integrate sustainability information into reporting cycles and also aims at providing people with the relevant information and awareness needed for sustainable lifestyles (United Nations, 2015).

Another example of voluntary standards at the international level are the standards for environmental management issued by the International Organization of Standardization (ISO) under the series number ISO 14000, which are widely applied in the industrial context. Table 3 presents some of the most relevant ISO standards related to LCT.

Table 3 Standards from the International Organization of Standardization (ISO) with relation to LCT

Standard number	Content of standard
ISO 14006:2020	Guidelines for incorporating eco-design
ISO 14040:2006/AMD 1:2020, ISO 14044:2006/AMD 2:2020	Life cycle assessment—principles and framework jointly with Life cycle assessment—requirements and guidelines
ISO 14045:2012	Eco-efficiency assessment of product systems—principles, requirements, and guidelines
ISO 14046:2014	Water footprint—principles, requirements, and guidelines
ISO 14064-1:2018	Greenhouse gases, part 1: specification with guidance at the organizational level for quantification and reporting of greenhouse gas emissions and removal (ISO, 2018a)
ISO 14064-2:2019	Greenhouse gases, part 2: specification with guidance at the project level for quantification, monitoring, and reporting of greenhouse gas emission reductions or removal enhancements
ISO 14067:2018	Greenhouse gases, carbon footprint of products—requirements and guidelines for quantification
ISO 14075 (under development)	Principles and framework for social life cycle assessment

Additionally, several types of voluntary product labels exist that rely on information obtained from LCT approaches and which serve to inform the consumer as well as other companies about the sustainability of the products. Another important function is to allow the labelled products access to more competitive or restrictive markets in other countries, where environmental regulations are stricter than in the place of production. However, the application of voluntary product labels has sometimes shown to be misused on products with only a relative improved environmental performance without consideration of their complete life cycle (biased environmental-related advertising known as "greenwashing"). To counteract such misuse, the ISO 14020:2000 for environmental labels and declarations sets the basis framework for voluntary product-related environmental information at an international level and clearly states that all product labels and declarations must consider the complete life cycle of the product (Bundesministerium für Umwelt, Naturschutz, nukleare Sicherheit und Verbraucherschutz [BMUV], 2019). Of the voluntary environmental labels, the examples presented in Table 4 are worth mentioning due to their high integration of a life cycle perspective. They all require certification by a third party as required for "Type I labels" in ISO 14020. Figure 7 shows the corresponding logos of the mentioned ecolabels.

Additionally, Environmental Product Declarations (EPDs) are important communication documents of voluntary implementation. The ISO14025:2006 establishes

Table 4 Voluntary product labels with a high integration of a life cycle perspective

Label name (website)	Description
EU Ecolabel (https://www.ecolabel.eu)	Recognized in all countries belonging to the EU, this label identifies products and services with a lower environmental impact than the market average, considering their life cycle and without diminishing the product's functionality or safety
Blauer Engel (https://www.blauer-engel.de/en)	In use by the German federal government since 1978, the Blauer Engel (Blue Angel) is awarded to environmentally friendly products and services
Global Organic Textile Standard (GOTS) (https://global-standard.org/)	This label aims to identify textile products with a guaranteed application of sustainable standards along their lifespan: from the extraction of raw fibers to environmentally and socially responsible production and labeling of end products
The Blue Certification (Certificado Azul) (https://www.certificadoazul.cl/)	A recently released initiative in Chile (2021) that promotes the sustainable management of water resources for companies and product value chains
#ElijoReciclar (Mas de 80% RECICLABLE) (https://elijoreciclar.mma.gob.cl/)	A new ecolabel developed in Chile that provides information to consumers about the percentage of recyclable materials used to produce packaging

Fig. 7 Logos of ecolabels with a high integration of a life cycle perspective. (Images sourced from respective official websites, see Table 4)

the use of the ISO14040 series of standards (LCA standards) in the development of product declarations and focuses on the comprehensive presentation of all environmentally relevant data for a product along the complete life cycle without comparing the product's environmental performance to other alternatives on the market. It is mainly used for communication between companies and is not targeted towards end consumers due to its complexity. Special variations of EPDs are the Product Environmental Footprint, Product Water Footprint, and the Product Carbon Footprint.

5.3 Communities for LCT Practice in Chile and Germany

Non-governmental organizations such as scientific communities, practitioners' networks, and industrial lobby groups focusing on the application of LTC and LCA are also an important driver for the implementation of this approach to industry and for promoting interest on the consumer side. The UN Life Cycle Initiative is a good example of a community of LCT practitioners and interested stakeholders on a global scale. At the European level, the European Platform of Life Cycle Assessment promotes the integration of life cycle approaches in environmental policy, initiatives, and programs (Pombo et al., 2019). In Germany, this philosophy is supported by several communities from both the industrial and academic sectors. Examples include the German Society for Waste Management (Deutsche Gesellschaft für Abfallwirtschaft e.V., DGAW),[5] the German Association for Water, Wastewater,

[5] https://www.dgaw.de/.

and Waste (Deutsche Vereinigung für Wasserwirtschaft, Abwasser und Abfall e. V., DWA),[6] and the Forum for Sustainability through Life Cycle Innovation (FSLCI).[7]

In Latin America, the life cycle approach is gaining more interest, and its practical implementation is still in progress (Pombo et al., 2019; Valdivia et al., 2017). LCA collaboration networks have played an important role in the progress of the LCT approach. For example, the Ibero-American Network of Life Cycle Assessment[8] is aimed at integrating the different local networks existing in Latin America such as the existing local country LCA networks, promoting the potential of LCA for decision making in the production sector, business, and government. In 2007 the Chilean LCA network (RedACV)[9] was created, which is also a member of the Ibero-American Network. Its members are primarily academics and researchers, as well as some environmental consultants. Like the rest of the regional LCA networks, the RedACV aims to generate information and diagnoses in different areas related to life cycle thinking, bolstering sustainable production and consumption.

In addition to the communication of information, holding events also helps to generate interest and spread knowledge about LCT practices. For example, in 2009 the third edition of the International Life Cycle Congress (CILCA) was held in Chile. The event is one of the most important and recognized conferences of the Ibero-American Life Cycle Network (RICV) and the starting point for the establishment of Chilean conferences on LCA. The first edition took place in the city of Viña del Mar in 2016, after which new editions have been held every two years. Sustainable production and consumption, environmental product declarations, eco-innovation, as well as tools towards a circular economy are central topics of discussion. In Germany and at the European level, conferences and international meetings on topics related to LCT are numerous, especially taking into account all the events organized under the umbrella topics of circular economies and resource efficiency, where concepts related to LCT are very commonly on the program. The Life Cycle Management Conference (LCM) series, for example, is one of the world's leading forums for exchange on topics of environmental, economic, and social sustainability. Held every second year and organized each time by a leading research institution and industry in the domain, the 2021 edition of the conference was in Stuttgart, Germany. The annual Greener Manufacturing Show in Cologne and the Plastic Waste Free World Conference (also in Cologne for the 2021 edition) are further examples of international events in Germany under the LCT umbrella and in relation to circular economy approaches.

---

[6] https://en.dwa.de/en/.

[7] https://fslci.org/.

[8] http://rediberoamerica.byethost7.com/.

[9] https://redacvchile.wixsite.com/acvchile.

6 Final Remarks

The application of LCT approaches as an essential framework for a sustainable industry is experiencing accelerated growth. This has been recognized by the scientific community as a necessary step towards a sustainable society. The legislative bodies of numerous countries all over the world, including Germany and Chile, have recognized that the inclusion of a life cycle perspective into legislative frameworks is necessary to motivate producers to embrace such approaches. Furthermore, recent amendments to legislation and voluntary standards increase transparency across entire production systems, reinforcing and supporting the exchange of information and collaboration amongst industries. This also motivates supporting industries (such as waste and wastewater collection, treatment, and recycling industries) to innovate and upgrade their services and creates room in the market for new products for the conversion of waste materials into raw supplies.

Nevertheless, equally important to pressure from the "top" is also pressure from the "bottom," meaning the consumer side. Consumer behavior and decisions must also experience an evolution towards a more responsible use of resources in order to actually achieve a circular economy and ensure global sustainability in the long term. For this, education and raising awareness of consumers young and old about resource conservation is indispensable, along with more information about the environmental impacts of products and services. The information reaching consumers must be understandable and transparent to support them in their decision making. In this regard, ecolabeling with energy and environmental information (e.g. carbon footprint and water footprint) becomes a useful instrument for promoting more sustainable consumption patterns. Furthermore, it is equally important that consumers learn and use the appropriate disposal channels in order to reintegrate discarded packaging materials and used products at the end of their useful life back into production processes.

It is beyond doubt that the cost of products and services plays a decisive role in consumer behavior. It is for this reason that the development of new, cost-efficient technologies and innovative materials, designed for both resource conservation and improved environmental performance, has become a cornerstone in the road towards a sustainable industry. To ensure technological progress, important investments are necessary in research and technology transfer but also in the education of young professionals and in the advanced training of experienced staff and in raising awareness of the importance of LCT amongst consumers.

As legislative frameworks evolve on a national level and as experience with LCT increases and technological innovations are developed, new possibilities are emerging that enable sustainable products to be competitive on the market. The goal is that by implementing LCT and with the use of appropriate analysis tools and innovative technologies, new, sustainable products and services will become even more profitable than conventional alternatives in the market. This situation, coupled with a more educated consumer who chooses sustainable alternatives, is decisive in promoting eco-industrial development and advancement towards a circular economy around the globe.

Acknowledgements The support of the Chilean National Research and Development Agency (Agencia Nacional de Investigación y Desarrollo, ANID) by mean of projects ANID CENTROS BASALES ACE210012, ANID/FONDAP/15130015 and ANID/FONDAP/1523A0001 is gratefully acknowledged.

References

Baccini, P., & Bader, H.-P. (1996). *Regionaler Stoffhaushalt: Erfassung, Bewertung und Steuerung*. Spektrum Akad. Verl.

Benoit-Norris, C., Traverso, M., Valdivia, S., Vickery-Niederman, G., Franze, J., Mazijn, B., Azuero, L., Ciroth, A., & Aulisio, D. (2013). The methodological sheets for sub-categories in social life cycle assessment (S-LCA). *Pre publication-version. The methodological sheets for subcategories in social life cycle assessment (S-LCA)*. https://doi.org/10.1007/978-1-4419-8825-6

Borucke, M., Moore, D., Cranston, G., Gracey, K., Iha, K., Larson, J., Lazarus, E., Morales, J. C., Wackernagel, M., & Galli, A. (2013). Accounting for demand and supply of the biosphere's regenerative capacity: The National Footprint Accounts' underlying methodology and framework. *Ecological Indicators, 24*, 518–533. https://doi.org/10.1016/j.ecolind.2012.08.005

Bouman, M., Heijungs, R [Reinout], van der Voet, E., van den Bergh, J. C., & Huppes, G. (2000). Material flows and economic models: An analytical comparison of SFA, LCA and partial equilibrium models. *Ecological Economics, 32*(2), 195–216. https://doi.org/10.1016/s0921-8009(99)00091-9

Brunner, P. H., & Rechberger, H. (2004). *Practical handbook of material flow analysis* (Vol. 9). CRC Press.

Bundesministerium für Umwelt, Naturschutz und nukleare Sicherheit. (2021). *Gesetze, Vorschriften und Vereinbarungen*. https://www.bmu.de/service/gesetze-verordnungen

Bundesministerium für Umwelt, Naturschutz, nukleare Sicherheit und Verbraucherschutz (Ed.). (2019). *Umweltinformationen für Produkte und Dienstleistungen: Anforderungen – Instrumente – Beispiele*.

CDT. (2015). *Corporación de Desarrollo tecnólogico. Guía Desarrollo Sustentable de Proyectos Inmobiliarios*. Santiago de Chile. http://mainichi.jp/feature/interview/news/p20150907mog00m040015000c.html

Deutsches Institut für Normung (DIN). (2021). *Ökobilanz – Anforderungen und Anleitungen (ISO 14044)*.

Espinosa, G. (2016). *Material flow analysis of the urban water system in Tepic Mexico*. Dissertation. *Hamburger Berichte zur Siedlungswasserwirtschaft: Band 93* [XVI, 210 Seiten].

Espinosa, G., & Otterpohl, R. (2014). Assessing material flows in urban systems: An approach to maximize the use of incomplete data sets. *Water Science and Technology: A Journal of the International Association on Water Pollution Research, 70*(6), 1135–1142. https://doi.org/10.2166/wst.2014.352

European Commission. (2020a). *2020 Circular economy action plan: For a cleaner and more competitive Europe*. https://op.europa.eu/en/publication-detail/-/publication/45cc30f6-cd57-11ea-adf7-01aa75ed71a1

European Commission. (2020b). *2020 Circular economy action plan: International aspects*. https://op.europa.eu/s/pnKX

Ewing, B. (2010). Calculation methodology for the national Footprint accounts 2010 Edit I on Authors. *Global Footprint Network Report*.

Frohmann, A., & Olmos, X. (2013). *Huella de carbono, exportaciones y estrategias empresariales frente al cambio climático*. Cepal. www10.iadb.org/intal/intalcdi/PE/2013/13146.pdf%0A. https://repositorio.cepal.org/bitstream/handle/11362/4101/S2013998rev1.pdf?sequence=6&isAllowed=y

Guinée, J., Heijungs, R., Huppes, G., Zamagny, A., Masoni, P., Buonamici, R., Ekvall, T., & Rydberg, T. (2011). Life cycle assessment: Past, present, and future. *Environmental Science & Technology, 45*(1), 90–96. https://doi.org/10.17654/JPHMTFeb2015_029_042

Hauschild, M. Z., Rosenbaum, R. K., & Olsen, S. I. (Eds.). (2017). *Life cycle assessment: Theory and practice*. Springer.

Hoekstra, A. Y., Chapagain, A. K., Mekonnen, M. M., & Aldaya, M. M. (2011). *The water footprint assessment manual: Setting the global standard*. Routledge.

ISO. (2012). *Environmental management—Eco-efficiency assessment of product systems—Principles, requirements and guidelines* (ISO 14045:2012).

ISO. (2014). *Environmental management—Water footprint—Principles, requirements and guidelines* (ISO 14046:2014).

ISO. (2018a). *Greenhouse gases—Part 1: Specification with guidance at the organization level for quantification and reporting of greenhouse gas emissions and removals* (ISO 14064-1:2018).

ISO. (2018b). *Greenhouse gases—Carbon footprint of products—Requirements and guidelines for quantification* (ISO 14067:2018).

ISO. (2020). *DIN EN ISO 14040:* Ökobilanz: Grundsätze und Rahmenbedingungen.

Joint Research Center of the European Union. (2010). *Making sustainable consumption and production a reality: A guide for business and policy makers to life cycle thinking and assessment*. https://doi.org/10.2779/91521

Joint Research Centre-Institute for Environment and Sustainability. (2010). *International reference life cycle data system (ILCD) handbook: General guide for life cycle assessment—Detailed guidance*. https://doi.org/10.2788/38479

Klöpffer, W., & Grahl, B. (2014). *Life cycle assessment (LCA): A guide to best practice*. Wiley-VCHVerlag GmbH & Co.

Lopes Silva, D. A., Oliveira, J. A. de, Saavedra, Y. M., Ometto, A. R., Rieradevall i Pons, J., & Gabarrell Durany, X. (2015). Combined MFA and LCA approach to evaluate the metabolism of service polygons: A case study on a university campus. *Resources, Conservation and Recycling, 94*, 157–168. https://doi.org/10.1016/j.resconrec.2014.11.001

MMA. (2016a). Ley 20920. Establece Marco Para la Gestión de Residuos, la Responsabilidad Extendida del Productor y Fomento al Reciclaje. *Diario Oficial*, 17–20.

MMA. (2016b). *Ministerio del Medio Ambiente. Programa Nacional de Consumo y Producción Sustentables*. Gobierno de Chile. https://mma.gob.cl/wp-content/uploads/2016/07/ccps_1307 2016_alta.pdf

MMA. (2021). *Ministerio del Medio Ambiente*. Hoja de Ruta para un Chile Circular al 2040. Gobierno de Chile. https://economiacircular.mma.gob.cl/hoja-de-ruta/

Müller, F., Kohlmeyer, R., Krüger, F., Kosmol, J., Krause, S., Dorer, C., & Röhreich, M. (2020). *9 Principles for a circular economy*. https://www.umweltbundesamt.de/en/publikationen/9-principles-for-a-circular-economy

Olmos, X. (2012). *La huella de carbono en el comercio internacional: el caso de las viñas chilenas*. https://www.uchile.cl/documentos/estudio-de-caso-completo_84936_0_5938.pdf

Pombo, O., Rivela, B., & Neila, J. (2019). Life cycle thinking toward sustainable development policy-making: The case of energy retrofits. *Journal of Cleaner Production, 206*, 267–281. https://doi.org/10.1016/j.jclepro.2018.09.173

Reap, J., Roman, F., Duncan, S., & Bras, B. (2008). A survey of unresolved problems in life cycle assessment. Part 1: Goal and scope and inventory analysis. *International Journal of Life Cycle Assessment, 13*(4), 290–300. https://doi.org/10.1007/s11367-008-0008-x

Sacayón, E. E., Vionnet, S., & Rodriguez, L. (2018). *Life cycle and water footprint assessment of a palm oil company in Guatemala: Pilot Study.*

Schmidt, M., Haubach, C., Preiß, M., Spieth, H., & Bauer, J. (2018). *100 Betriebe für Ressourceneffizienz - Band 2: Praxisbeispiele und Erfolgsfaktoren.* Springer. https://doi.org/10.1007/978-3-662-56712-8

Schmidt, M., Spieth, H., Bauer, J., & Haubach, C. (2017). *100 Betriebe für Ressourceneffizienz – Band 1: Praxisbeispiele aus der produzierenden Wirtschaft.* Springer. https://doi.org/10.1007/978-3-662-53367-3

Sonnemann, G., Strothmann, P., Weyand, S., & Valdivia, S. (2016). Opportunities for national life cycle network creation and expansion around the world. *United Nations Environment Programme, 84.*

Swarr, T. E., Hunkeler, D., Klöpffer, W., Pesonen, H. L., Ciroth, A., Brent, A. C., & Pagan, R. (2011). Environmental life-cycle costing: A code of practice. *International Journal of Life Cycle Assessment, 16*(5), 389–391. https://doi.org/10.1007/s11367-011-0287-5

UNEP/SETAC Life Cycle Initiative (2011). Towards a life cycle sustainability assessment. Making informed choices on products. *International Journal of Life Cycle Assessment.*

UNEP/SETAC Life Cycle Initiative. (2013). *Guidelines for social life cycle assessment of products.*

United Nations. (2015). *Transforming our world: The 2030 Agenda for Sustainable Development: A/RES/70/1.* https://www.un.org/ga/search/view_doc.asp?symbol=A/RES/70/1&Lang=E

United Nations Environment Programme (Ed.). (2004). *Why take a life cycle approach?* https://sustainabledevelopment.un.org/content/documents/846Why_take_a_life_cycle_approach_EN.pdf

United Nations Environment Programme (Ed.). (2012). *Greening the economy through life cycle thinking.* https://www.lifecycleinitiative.org/wp-content/uploads/2013/03/2012_LCI_10_years_28.3.13.pdf

Valdivia, S., Quispe, I., & Mila i Canals, L. (2015). Regional Stakeholder Consultation on LCA Databases in Latin America.

Valdivia, S., Sonnemann, G., & Milà i Canals, L. (2017). LCA mainstreaming conditions in Latin America—Based on learnings from 2005 to 2014. *The International Journal of Life Cycle Assessment*, *22* (4), 485–491. https://doi.org/10.1007/s11367-016-1142-5

Wiche, P., & Valdivia, S. (2014). *Life cycle thinking in Latin America: 12 case studies of LCA and LCM approaches of companies in.* https://www.lifecycleinitiative.org/wp-content/uploads/2015/01/LCA-LCM-company-case-studies-in-Latin-America.pdf

Zamagni, A., Buttol, P., P.L., P., R., B., P., M., Guinee, J., Heijungs, R., Ekvall, T., Bersani, R., Bienkowska, A., & U. Pretato (2008). Critical review of the current research needs and limitations related to ISO-LCA practice—Deliverable D7 of work package 5 of the CALCAS project. I*Co-ordination Action for Innovation in Life-Cycle Analysis for Sustainability (CALCAS),* (037075), 1–106. http://www.calcasproject.net/

Sustainable Use of Water and Energy in an Urban Context—Introduction

[illegible]

[illegible]

Challenging the Current Wastewater Treatment Management System—A Water-Energy-Waste Perspective

Witold-Roger Poganietz and V. Alfredo Erlwein

Abstract Seeing the climate crisis as a water crisis, the question arises as to how current wastewater treatment management could be changed to mitigate the effects of climate change while contributing to slow down the increase in the average temperature. An innovative wastewater management system could not only reduce water demand, but would also provide an additional, locally available energy source as well as a better recovery of nutrients compared to a conventional wastewater treatment management system. However, the gain from an innovative wastewater management system depends on a set of different factors that are partly given, like the local climate, but also the outcome of socioeconomic developments in the settlements under investigation. The following four Chapters "Adapting the German Bioenergy Village Concept to Los Ríos Region in Southern Chile: Integrating Community, Water,Energy and Organic Waste Management–Circular Bioeconomy in Chile and Germany" outline the main features of the innovative wastewater management system.

1 Introduction

The climate crisis is a water crisis, as global warming has disastrous impacts on the water cycles (Sköld, 2020). Global warming increases the number and size of extreme weather events, while average precipitation is decreasing. For example, most parts of Chile are experiencing a long-lasting drought (Garreaud et al., 2020), and

W.-R. Poganietz (✉)
Institute for Technology Assessment and Systems Analysis (ITAS), Karlsruhe Institute of Technology (KIT), Karlsruhe, Germany
e-mail: witold-roger.poganietz@kit.edu

V. A. Erlwein
Department of Agriculture Engineer and Soil Science, IIAS, Faculty of Agrarian Sciences, Universidad Austral de Chile, Valdivia, Chile

Transdisciplinary Centre of Environmental Studies and Sustainable Human Development, CEAM, Universidad Austral de Chile, Valdivia, Chile

A. Ch. Braun et al. (eds.), *Eco-Industrial Development as an Industrial Strategy*, Green Energy and Technology, https://doi.org/10.1007/978-3-031-73576-9_6

in many parts of Germany the soil is drying out (Umweltbundesamt, 2021). At the same time, to slow down climate change, the transformation of the energy system to a more environmentally friendly one is most urgent, as around 75% of greenhouse gas emissions are energy-related (Global Change Data Lab, 2022). However, not only the energy system should be the target of a transformation process. It is widely accepted that the processes of extracting and using resources should be transferred to a recirculation system (European Commission, 2020). The treatment of nutrients is highly relevant, as the provision of these is generally energy intensive as well as environmentally questionable. The question arises, how current wastewater treatment management could be changed to become adapted to and mitigate the effects of climate change in terms of both water availability and GHG emissions.

Centralized wastewater treatment plants are commonly used to treat wastewater in most developed cities worldwide. The treated wastewater is then discharged into water bodies while the sludge is either incinerated, disposed of in landfills, or used as a fertilizer if the sludge fulfills specific quality conditions (Gutjahr & Müller-Schaper, 2018). The acceptance of these systems by users is high because of the relatively simple and robust collection of wastewater, while reducing the occurrence of diseases. It has to be said that in the Latin American context, where many areas still do not have water treatment plants, such centralized plants are an important contribution to reducing organic contamination and increasing the sanitation of water bodies. Although those systems contribute to the primary aims of wastewater treatment, i.e., the removal of wastewater with pathogens from settlements, the main disadvantage could be seen in the insufficient recovery of energy and nutrients. Furthermore, the increasing share of micropollutants in the wastewater will require significantly improving existing wastewater treatment plants, which cannot be guaranteed with the conventional technologies.

In the literature, several strategies are discussed and have been partially implemented on a pilot scale. Following the conventional setting of the wastewater treatment system, implementing advanced treatment technologies are often proposed. Such advanced technologies enable cleaning the wastewater to achieve higher quality effluent, thus reducing the demand for potable water, while nutrients can be recovered, thus cutting the requirements to produce mineral fertilizer. It is important to note that some nutrients are usually obtained from extraction processes so that they are currently non-renewable, which means that they may not be sufficiently available for a period of decades, with the consequent risk of food security (Vaccari et al., 2014). A special case is phosphorus, one of the most restrictive nutrients in the volcanic soils of southern Chile in particular, which is currently obtained via mining processes.

However, an alternative sees a rather complete overhaul of the current wastewater treatment management by separating the wastewater streams at the source, i.e., in buildings (Erlwein, 2024; Poganietz & Friedrich, 2024; Suryandari & Bolivar Paypay, 2024; Vergara-Araya et al., 2024).

The most straightforward sorting of wastewater separates it into black water and gray water. A simpler design would require a second pipeline system in the building and to the black water treatment site (Vergara-Araya et al., 2024). Gray water, coming from showers, kitchens, and washing machines, is less contaminated

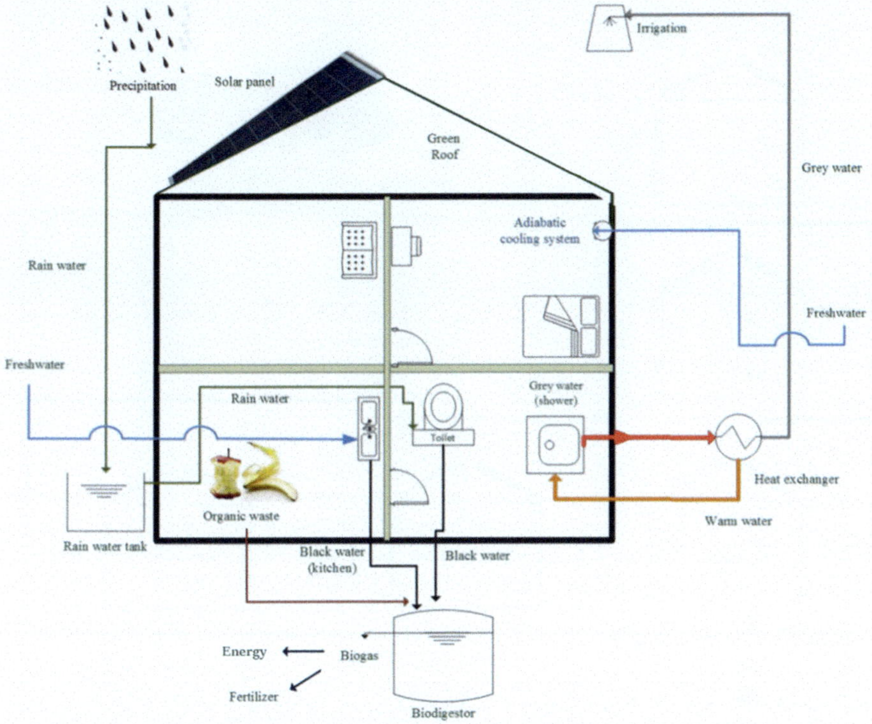

Fig. 1 Sketch of a system with separated wastewater streams and biodigestion of black water

and quite warm; black water is nutrient rich but cold compared to gray water. To increase the nutrient content of black water, the addition of organic municipal solid waste is often suggested (e.g., Vergara-Araya et al., 2024), enhancing the chances of proper biodigestion (see Fig. 1).

A more sophisticated sorting is to separate black water into brown water, i.e. only feces, and yellow water, i.e. only urine. The advantage is a better recovery of nutrients (Erlwein, 2024). Another important factor in designing optimized systems is the additive effect of collective real estate and housing projects. Such increases in magnitude allow for more sophisticated technologies and different uses due to the volumes of resources involved. A reference example is shown in Fig. 2.

The challenges in transforming a conventional wastewater system are manifold and address nearly every aspect of constructing, operating, and using such a system. In the center stands the technical viability of the system inside and outside the building. As shown in the following Chapters "Adapting the German Bioenergy Village Concept to Los Ríos Region in Southern Chile: Integrating Community, Water,Energy and Organic Waste Management–Circular Bioeconomy in Chile and Germany", implementing such a system would need, among other things, separated

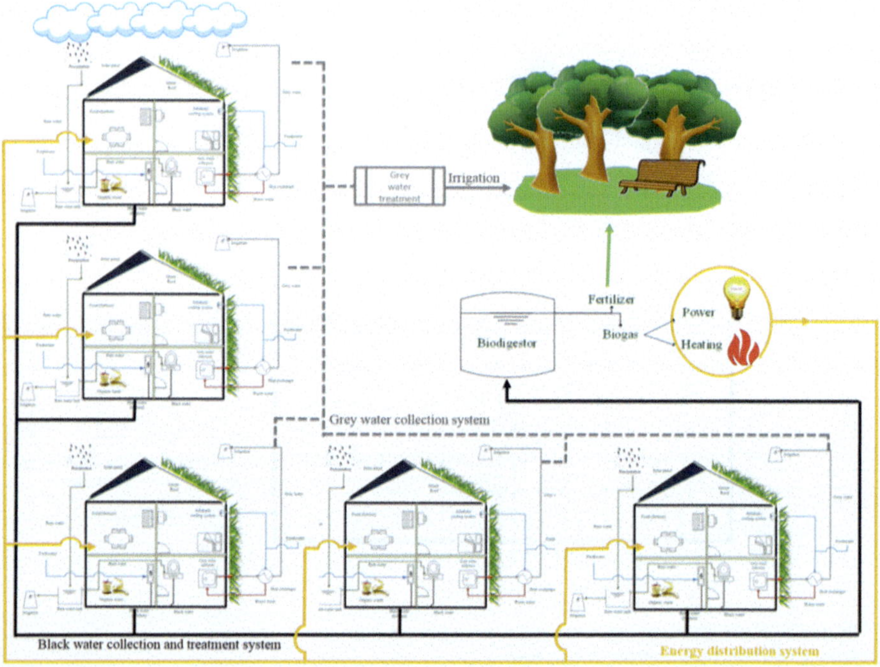

Fig. 2 Collective scale of housing projects enable a wider range of different technologies and resource uses, i.e. cogeneration of power and heat via biogas combustion, green facades and roofs, or communal land irrigation, to name but a few

pipelines for each wastewater stream, specific toilet systems, like vacuum or urine-diverted toilets, and stations for the separated black water if the water content of black water should be minimized. Also, shredder systems for grinding organic municipal solid wastes to be added to the black water in order to include that waste in the system increase the energetic value of the black water and enhance the biodigestion process. Furthermore, wastewater treatment plants have to be adapted to the changed composition of wastewater, i.e. since black water is treated elsewhere, only gray water and storm water will enter wastewater treatment plants with a comparable low load of solid wastes and pathogens, if these are still necessary (Poganietz & Friedrich, 2024). This situation could also be the opposite if the black water went to sewage (because there is no treatment facility, or no near agricultural field available for fertilizer application) and gray water were treated and reused on-site in projects that implement some technologies mentioned in e.g., Erlwein (2024). In these cases, the black water can be treated in water treatment plants, improving the performance of the plant due to the concentration effect. However, the plants should also be adapted to the new wastewater composition anyway (Erlwein, 2024). As can be appreciated, the optimal system depends on the design concepts used. In the following chapters, several concepts are presented following that criterion. Of full relevance is also the technical resilience of the system as a pre-condition of a long-run acceptance by operators and users of such a system.

Whereas the technical feasibility is a required precondition for any transformation, two further aspects are often in the center of discussions: the costs of transforming and the acceptance of the new system. The complexity of constructing and operating an integrated system should be higher, since additional technologies and infrastructures are implemented. Thus, at least higher investment costs could be expected, leading to the question regarding the willingness as well as the bearer's capacity to pay for the cost of the system (Poganietz & Friedrich, 2024).

An overhaul of the current system could and should affect the living conditions of users of the system. If recycled gray water cannot be used for agriculture, e.g., due to contamination concerns, it could fuel blue-green areas in the neighborhood with water. Blue-green areas in the middle of built environments could influence the microclimate of the environment positively, e.g., cooling in hot summers, as well as being used for recreational activities. An important side effect of this is a reduced demand for potable water, which is commonly used for irrigation.

However, the most different designs of alternative systems would require integrating important features into the living environment of users. For example, the anaerobic digester (Erlwein, 2024; Poganietz & Friedrich, 2024; Vergara-Araya et al., 2024) as well as the facility to convert black water into nutrients (Suryandari & Bolivar Paypay, 2024) should be in proximity. Other examples are required, such as new toilet systems (Erlwein, 2024; Suryandari & Bolivar Paypay, 2024) or a shredder in the kitchen sink (Poganietz & Friedrich, 2024; Suryandari & Bolivar Paypay, 2024). All these features would confront the user with the new wastewater treatment, overcoming the current approach of "hiding" wastewater transport and treatment.

Typically, water-wastewater systems, energy systems and waste systems are run independently, although crossovers are also known, like incinerated waste or biogas offered to the utilities or directly to energy demanders. In any case, the system under discussion should increase the amount of available energy from a local source, affecting the local energy system and its management (Vergara-Araya et al., 2024). The integration of the currently independently operated system will change the relationship between the sectors. The question arises, how to organize the changed relationships to achieve the desired results of a transformed system (Vergara-Araya et al., 2024).

The challenges sketched above are only a selection; some, like climate conditions, were not mentioned, but could have some impact on the required design. In addition, the relevance of the challenges sketched depends on a large number of different conditions, addressing, e.g., climate, societal, environmental, and economic circumstances that differ between regions and urban settlements, partly even within a city. That means the precise design of a transformed integrated water-energy-waste system is determined by the layout of the above-mentioned conditions, challenging the identification of a concept that "fits for all".

2 Outline of the Chapters "Transforming the Urban Water-Energy-Waste Nexus in Chile and Germany"

The following Chapters "Opportunities and Challenges of Integrating Water, Waste, and Energy Management Systems in the Conditions of Curauma, Chile–Closing the Loop of Nutrients by an Urban-Rural Partnership—A German Project" will substantiate the ideas sketched above and address the challenges to specific contexts.

Chapter "Opportunities and Challenges of Integrating Water, Waste, and Energy Management Systems in the Conditions of Curauma, Chile", by Vergara et al., aims at: (a) comparing a system integrating wastewater, energy, and wastes with a conventional system under the conditions of Curauma, a settlement within the Municipality of Valparaíso, Chile, and (b) providing initial insights into the institutional challenges of implementing an integrated water-energy-waste management system. With respect to energy and water, an integrated system would favor the use of biogas as a relevant energy source for electricity and heat, whereas the demand for potable water would decrease considerably. The challenges of implementing such a system could be noteworthy, as different markets with different regulations are affected, leading to the question of whether the current regulations would be sufficient.

Chapter "Adapting the German Bioenergy Village Concept to Los Ríos Region in Southern Chile: Integrating Community, Water, Energy and Organic Waste Management", "Adapting the German Bioenergy Village concept to Los Ríos Region, southern Chile: Integrating Community, Water, Energy and Organic Waste Management" from Erlwein focuses on testing the German approach of Bioenergy Villages applied to southern Chile in order to understand its feasibility and what kinds of modifications need to be done in order to optimize it in this new scenario. Results show that the Bioenergy Village concept is technically feasible, with abundant energy supply stocks. As southern Chile has temperate rainy weather, the same technologies and even biomass sources can be replicated from the original concept. However, economic and environmental restrictions related to the local price of energy sources and the need for energy crops development seriously limit the application of the concept in southern Chile. However, some modifications to the original concept have been proposed and tested, including higher efficiency of wood biomass-based heating, biogas from domestic organic waste used for cooking, and a wastewater separation system that includes constructed wetlands. According to the results of the modeling, the "Chilean" Bioenergy Village concept shows relevant improvements in energy, water, and nutrient recycling efficiency in relation to the current local standards, suggesting the promising approach of integrating a systemic view in resource management.

Chapter "Alternative Urban Water-Energy-Waste Nexus Options—Sustainable Assessment in an Existing Settlement", "Alternative urban Water-Energy-Waste Nexus options—Sustainable assessment in an existing settlement" written by Poganietz and Friedrich, analyzes three different system alternatives with the status quo under the conditions of an existing settlement in Germany using the Integrative Concept of Sustainability as a methodological backbone. The system alternatives differ in the technological design of treating separated wastewater flows. The least invasive system change sees a separation of wastewater at the source while the current pipeline system and central treatment plant are still in use; the most elaborate one takes the current wastewater system fully out of operation. The most elaborate option shows the highest potential to enhance the resource and energy efficiency; however, this option could face the least acceptance of all the options, including the status quo, and compared to the status quo, higher costs.

Chapter "Closing the Loop of Nutrients by an Urban-Rural Partnership—A German Project", "Closing the loop of nutrients by an urban-rural partnership—a German project" of Suryandari and Bolivar Paypay, sets the focus on the chances of closing the nutrient loop in the context of a German region. Using the Stuttgart Process to treat wastes allows for the following materials to be recovered from the blackwater: struvite containing nitrogen and phosphor, zeolite containing nitrogen and potassium, as well as biopolymers as raw materials for foils and tarpaulins. The focus of the contribution is to present and discuss the required indicators for conducting a sustainable assessment, using the Integrated Concept of Sustainable Development as the methodological backbone.

3 Concluding Remarks

Key to all options presented in the Chapters "Adapting the German Bioenergy Village Concept to Los Ríos Region in Southern Chile: Integrating Community, Water, Energy and Organic Waste Management–Circular Bioeconomy in Chile and Germany" is the management of waste in terms of water, energy, organic residues, and other resources in order to reuse, recycle, or recirculate them as a core implementation concept of Industrial Ecology (Poganietz & Bolivar Paypay, 2024).

The different options to improve the current wastewater treatment management system can contribute to mitigating the effects of climate change on the water cycle by reducing the demand for potable water. However, regulations as well as the acceptance of using recycled wastewater or gray water is crucial. Depending on the design of the system, enhanced nutrient recovery is possible, leading to a further reduction in the carbon footprint, as well as the biochemical contamination of water bodies. The same can be true with respect to energy, but the gain and thus the relevance for the local energy system is influenced by a wide range of factors, demanding further research.

Thus, the necessity of including users and operators in designing an enhanced system is obvious. The success of an innovative system depends deeply on the willingness of those stakeholders to make behavioral changes. The more disruptive the changes in the current wastewater management system are (and thus, the more traditional behavior is challenged), the more important the participation of the stakeholders is. Also very relevant is the regulation of the wastewater treatment, the energy provision, and the use of recovered nutrients, because such regulations set the design boundaries for any innovation system.

To understand in depth the impacts of these systems of redesigned material and energy flows, a rigorous inventory of the current as well as of future wastewater management systems, consistent with the contributions here presented, is a sine qua non condition for any recommendations. However, the different contributions make it clear that a comprehensive sustainability assessment demands an extension of the methodological base of the analysis to a broader spatial, cultural, and temporal scale.

References

Erlwein, A. (2024). Adapting the German Bioenergy Village concept to Los Ríos Region, southern Chile: Integrating Community, Water, Energy and Organic Waste management.

European Commission. (2020). A new Circular Economy Action Plan for a cleaner and more competitive Europe. COM/2020/98 final.

Garreaud, R. D., Boisier, J. P., Rondanelli, R., Montecinos, A., Sepúlveda, H. H., & Veloso-Aguila, D. (2020). The Central Chile Mega Drought (2010–2018): A climate dynamics perspective. *International Journal of Climatology, 40*, 421–439. https://doi.org/10.1002/joc.6219

Global Change Data Lab. (2022). Our world in data. Emissions by sector. Retrieved March, 25, 2022, from https://ourworldindata.org/emissions-by-sector#energy-electricity-heat-and-transport-73-2

Gutjahr, M., & Müller-Schaper, J. (2018). Treatment of sewage sludge in Europe: Status quo and perspectives of energy and phosphorus recovery—Illustrated by the example of selected countries. In S. Thiel, E. Thomé-Kozmiensky, F. Winter, & D. Juchelková (Eds.), *Waste management—Waste-to-energy* (Vol. 8, pp. 371–384). Thomé-Kozmiensky Verlag.

Poganietz, W. -R., & Bolivar Paypay, V. (2024). The concept of industrial ecology.

Poganietz, W. -R., & Friedrich, J. (2024). Alternative urban water-energy-waste nexus options—Sustainable assessment in an existing settlement.

Sköld, M. (2020). 5 Reasons why the climate crisis is a water crisis. Retrieved March, 25, 2022, from https://siwi.org/latest/5-reasons-why-the-climate-crisis-is-a-water-crisis/

Suryandari, D., & Bolivar Paypay, V. (2024). Closing the loop of nutrients by an urban-rural partnership—A German project.

Umweltbundesamt. (2021). Trockenheit in Deutschland. Retrieved March, 25, 2022, from https://www.umweltbundesamt.de/themen/trockenheit-in-deutschland-fragen-antworten

Vaccari, D. A., Mew, M., Scholz, R. W., & Wellmer, F. W. (2014). Exploration: What reserves and resources? In R. Scholz, A. Roy, F. Brand, D. Hellums, & A. Ulrich, A. (Eds.), *Sustainable Phosphorus Management*. Springer, Dordrecht. https://doi.org/10.1007/978-94-007-7250-2_2

Vergara-Araya, M., Poganietz, W. -R., & Bolivar Paypay, V. (2024). *Opportunities and challenges of integrating water, waste, and energy management systems in the conditions of Curauma, Chile.*

Sustainable Use of Water and Energy in an Urban Context—Case Studies from Germany and Chile

Opportunities and Challenges of Integrating Water, Waste, and Energy Management Systems in the Conditions of Curauma, Chile

Mónica Vergara-Araya, Witold-Roger Poganietz, and Vanessa Bolivar Paypay

Abstract Wastewater management systems are intended to protect people's health and natural sources of water. However, modern wastewater treatment systems do not always exhibit satisfactory energy and resource efficiency from a circular economy perspective. Separating wastewater at the source and recognizing organic municipal solid waste as a co-substrate to produce biogas and energy are options to improve efficiency. This study has two goals: first, to compare a system integrating wastewater, energy, and wastes with a conventional system in the conditions of Curauma, a community within the Municipality of Valparaíso, Chile. And second, to provide initial insights into the institutional challenges of implementing an integrated water-energy-waste management system. With respect to energy, water, and waste, an integrated system would favor the use of biogas as a relevant energy source for electricity and heat, and its nutrient-rich subproducts, e.g. biosolids for agriculture, and would decrease the demand for potable water considerably. The institutional implementation of such a system could be carried out via the establishment of respective markets or through the installation of one company that would be responsible for the entire value-added chain. However, the challenges could be noteworthy, as different markets with different regulations are affected.

M. Vergara-Araya (✉)
Department for Water, Environment, Construction and Safety, Magdeburg-Stendal University of Applied Sciences, Magdeburg, Germany
e-mail: monica.vergara-araya@h2.de

W.-R. Poganietz · V. B. Paypay
Institute for Technology Assessment and Systems Analysis (ITAS), Karlsruhe Institute of Technology (KIT), Karlsruhe, Germany
e-mail: poganietz@kit.edu

V. B. Paypay
e-mail: vanessa.bolivar@kit.edu

A. Ch. Braun et al. (eds.), *Eco-Industrial Development as an Industrial Strategy*, Green Energy and Technology, https://doi.org/10.1007/978-3-031-73576-9_7

1 Introduction

The next few decades will experience ongoing urbanization (UN DESA, 2018). Accompanied by increasing living standards, the demand for energy and water will follow suit. To satisfy the associated needs in a sustainable way, a more efficient use of both energy and water has become one of the most important premises for sustainable development (Dai et al., 2018).

Energy and water consumption are to some extent mutually interconnected, i.e. the former is required to provide the latter, but the latter could also support the supply of the former (Vergara-Araya et al., 2020). The relevance of the water-energy nexus in resolving the growing demands and the associated dilemmas has been highlighted by techno-economic (Mo et al., 2014) as well as policy or societal (DeNooyer et al., 2016) perspectives. Several studies extend the analysis of the nexus by considering, e.g., carbon emissions (Venkatesh et al., 2014) and food (Villarroel Walker et al., 2014).

Common to all these studies is the analysis of the opportunities for the technological progress of available technologies without questioning the appropriateness of the given water and energy infrastructures in satisfying future demands. A possible, widely discussed option is to separate wastewater into greywater and blackwater to make use of the chemical and thermal energy of wastewater (Otterpohl et al., 2002). Greywater is lightly polluted and warm, coming from showers, bathtubs, or washing machines. Blackwater consists of human excreta and urine plus flushing water (Otterpohl et al., 2002). Blackwater covers between half to two-thirds of the carbon compounds and between 80% and more than 90% of all valuable plant nutrients contained in domestic wastewater (Kleckers & Haberkamp, 2021).

The consumption of potable water and the energy demand of households satisfied by external sources could be reduced and the energy efficiency of biogas production could be increased if a mixture of blackwater with organic waste as a feedstock for an anaerobic digestion plant is used, and greywater is treated and recycled for reuse. Additionally, the greywater can be treated with less sophisticated technologies (i.e. close to natural treatment systems) or with less energy consumption, as fewer organic pollutants and nutrients are present. This approach is widespread in the literature, usually under the name new alternative sanitation systems (NASS) (Schütze et al., 2019), as an alternative way to treat wastewater, closing nutrients, and water cycles.

The reuse of wastewater and the more efficient treatment of wastewater is an important part of the water cycle, and a potentially affordable and sustainable source of water, energy, nutrients, organic matter, and other useful byproducts (UNESCO, 2017). For example, in many regions with water scarcity, wastewater reuse is one of the main aspects that could allow for sustainable water management.

The idea of separating wastewater at the source has been set in practice through facility managers and city developers. Single buildings (e.g. Arminplatz (Berlin), Nolde & Grunow, 2013) as well as smaller communities (e.g. Stockholm Sea Port (Sweden), Skambraks et al. (2017)) are equipped with unbundled wastewater treatment systems, although they differ in their technical layout. Some of them recover thermal heat from greywater and reuse recycled greywater for non-hygienic purposes, like toilet flushing (Nolde & Grunow, 2013) or for the irrigation of gardens or recreation areas (Skambraks et al., 2017).

Using the organic fraction of municipal solid waste (OFMSW) as an additional source of substrates for anaerobic digesters, the energy efficiency of the process could be further improved (Mata-Alvarez et al., 2000), contributing not only to biogas and energy production, but also reducing the transportation cost to disposal sites. Such decentralized solutions could be of special interest in isolated regions, or regions with poor or a lack of infrastructure, where wastewater and waste transportation is either difficult or even nonexistent.

The overarching aim of this contribution is twofold. First, it is to analyze the impacts of implementing an integrated water, waste, and energy management system on water and energy flows. The proposed integrated system consists of separating domestic wastewater at the source, and co-digesting blackwater with OFMSW to generate electricity and heat via biogas. Secondly, it is to discuss the challenges in implementing an integrated water-energy-waste system from an institutional perspective.

Both goals will make use of the situation in Curauma, Chile. Curauma is a fast-growing community under the administration of the Municipality of Valparaíso in Chile.

Material flow approaches are a common instrument to analyze material and energy flows (Allesch & Brunner, 2015; Kennedy et al., 2011), revealing the relevance of single elements for urban metabolism, but they could also take into account the nexus between multiple elements (Chen & Lu, 2015; Madrid-López & Giampietro, 2015).

Although these methods can help to reveal water and energy flows and their interconnections, they fail to describe different designs of systems where detailed technological characteristics are needed. In this study the material flow approach used is based on a detailed modeling of the technologies and processes using life cycle inventory (LCI) data (Uihlein et al., 2006).

The following Sect. 2 will present the idea of separating wastewater at the source and integrating OFMSW flows within an urban community to disclose the underlying assumptions of the following analysis. The selected system under review is outlined in Sect. 3, while Sect. 4 gives an overview of the model used and the scenarios developed for the analysis. The main findings are presented and discussed in Sect. 5. The institutional challenges of transforming the current system into an integrated management system can be found in Sect. 6. Section 7 offers some concluding remarks.

2 Wastewater Treatment Systems

State-of-the-art urban wastewater systems (WWS) in developed countries connect gravity-driven sewers with mechanical-biological treatment plants, with sludge as the most important solid byproduct of wastewater treatment plants (WWTP). To stabilize

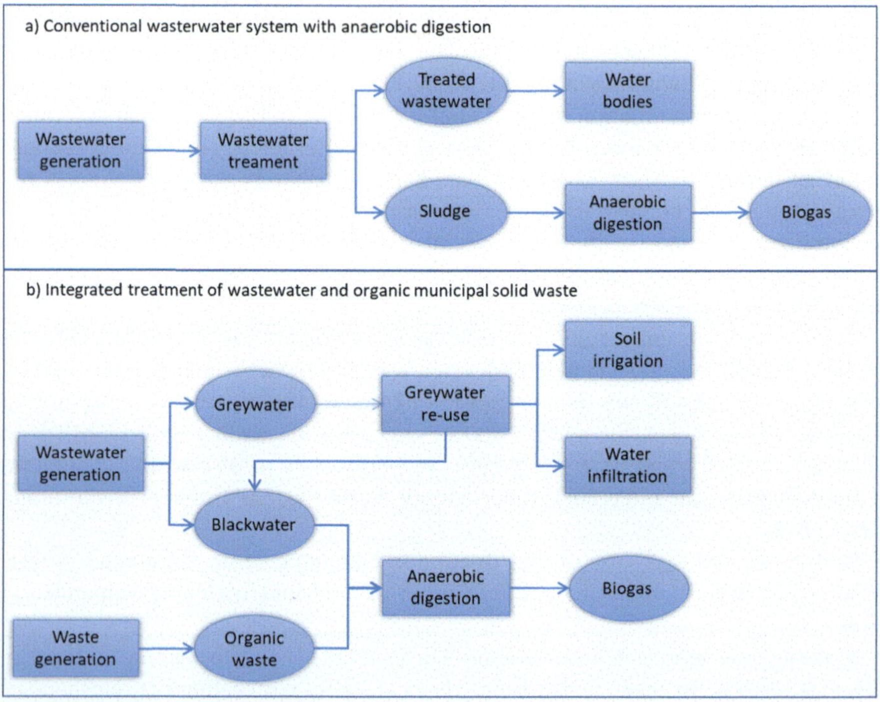

Fig. 1 Wastewater treatment management systems. *Source* Vergara-Araya et al. (2020

sludge as a pre-condition for further treatment, anaerobic stabilization in large scale WWTPs (i.e. >30,000 population equivalents or PE) is the most efficient way in terms of energy and resource recovery potential. Anaerobic stabilization results in biogas and a nutrient-rich bio-solid in auxiliary digestion plants (Fig. 1a).

However, if the WWTP performance is poor, or if wastewater treatment is limited to organic matter removal like in Curauma, the waters receiving wastewater are affected by organic material and nutrients, thus polluting water bodies. Moreover, the treated wastewater is often not reused, which is water crucial for arid or semiarid regions. It is estimated that in Chile, approximately 10% of the national water gap could be filled by the reuse of wastewater from marine sewage outfalls. Marine sewage is mechanically (or partially) treated wastewater that is discharged into the ocean (Fundación Chile, 2017).

If nutrients are not removed during wastewater treatment, or if the nutrient-rich sludge is not used, e.g. in agriculture, a loss of nitrogen and phosphorus available to the plants can be observed. An uncontrolled discharge of nitrogen, and especially of phosphorus, drives eutrophication, which is a leading cause of damage in many freshwater and coastal marine ecosystems worldwide (Chislock et al., 2013). Phosphorus is also a finite and scarce resource, whereas phosphorus-based fertilizers are a pillar of modern agriculture. Consequently, many countries use the sewage sludge directly as fertilizer (e.g. Chile), while many developed countries (e.g. Austria, Germany, etc.) push or have already set an obligation to recover phosphorus from wastewater treatment byproducts within a transition period.

Wastewater treatment is usually an energy-intensive process, in fact, it is estimated that WWTPs represent about 1–3% of the overall energy use of a country (Capodaglio & Olsson, 2020). Without biogas-based energy production and proper energy efficiency measures, the energy balance of WWTPs is negative due to high energy demand from the several cleaning steps, e.g. aeration, dewatering, pumping, etc. (Guerrini et al., 2017).

To overcome the sometimes poor resource and energy efficiency of the entire wastewater treatment system and to reduce the negative environmental impacts, separating wastewater in different flows complemented with OFMSW is suggested as an alternative (Lange et al., 1997). A simple system would separate wastewater between greywater and blackwater at the source of wastewater generation (Fig. 1b).

The main sources of greywater in a traditional domestic wastewater collection system are bathtubs, showers, bathroom sinks, laundry machines and kitchen sinks (Fane, 2013; Henze & Comeau, 2008). The share of greywater in wastewater varies worldwide, generally between 50 and 80% (Allen et al., 2010). The amount of greywater depends on the lifestyle, fixtures, type of community, and climatic conditions (Oteng-Peprah et al., 2018), but also on the efficiency of water use. Treated greywater could be reused for toilet flushing, water infiltration, or irrigating non-edible plants. The decentralized treatment of greywater is possible, reaching the required water quality levels, e.g. $BOD_5 < 5$ mg/L (Nolde & Grunow, 2013).

The reuse of treated wastewater has a huge potential to preserve fresh water resources. Different studies estimate fresh water savings of ca 29–35% (Ghisi & Ferreira, 2007), and up to 80% of the WC flush water (Dixon et al., 1999) with the reuse of greywater. However, the main use for treated wastewater worldwide is irrigation. For example, it is estimated that Europe reused 964 million m^3/a of treated wastewater in 2006, mostly in agriculture. However this is only 2.4% of the urban wastewater effluents (BIO by Deloitte, 2015). In Israel, in 2011 already 40% of the water used for irrigation came from reclaimed wastewater (UNESCO, 2017).

Blackwater, consisting of human excreta, urine and flushing water, contains between 50% and nearly 100% of all valuable plant nutrients or carbon compounds in domestic wastewater (Kleckers & Haberkamp, 2021; Otterpohl et al., 2002) (Fig. 2).

Blackwater, which is rather undiluted, could enhance the energy efficiency of anaerobic digestion, as no additional thickening steps are required. An efficient anaerobic digestion process requires substrates with a high concentration of total solids, usually higher than 4–5% and up to 8% for wet digestion, or even higher for dry digestion systems. Otherwise, the electrical (e.g. for mixing) and thermal demands would not match. Moreover, due to the long solids retention times (SRT) needed in anaerobic processes, the necessary reactor dimensions would be too large—and expensive. To feed anaerobic digesters with sewage sludge from WWTPs, usually an energy-intensive sludge thickening process is required, which usually demands more than 10% of the energy in the wastewater treatment process.

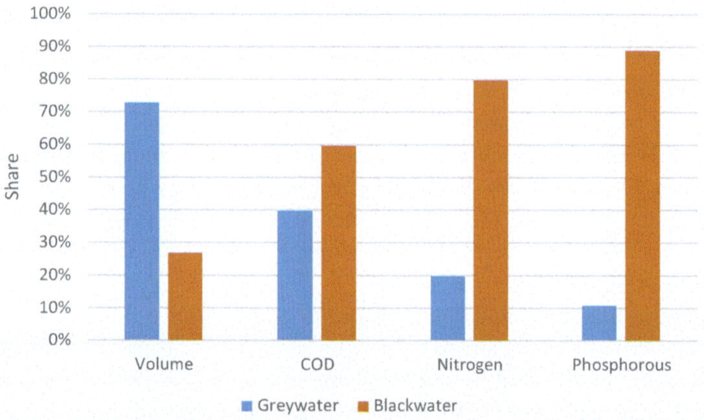

Fig. 2 Distribution of organic components and nutrients in different wastewater fractions. *Source* Kleckers and Haberkamp (2021), translated from German

The efficiency of the anaerobic digester could be enhanced by adding OFMSW to blackwater. Co-digestion with a mixture of two or more substrates raises energy efficiency, as it makes use of positive synergisms established in the digestion medium and the supply of missing nutrients by the co-substrates (Aguilar et al., 2017; Mata-Alvarez et al., 2000). However, the potential gains and losses of adding co-substrates depend crucially on a number of parameters, the most relevant factors being the composition of the inlet flow and the feed ratio.

Making the biogas produced available to combined heat and power plants (CHPs), both electricity and heat can be produced. This energy can be used for operating the biogas production system and the WWTP, or be offered to the community. However, to use the excess energy outside CHP and WWTP and thus, to improve the energy efficiency of the entire wastewater-energy-waste system, the CHP has to be connected with a local heat network and grid system. The former is not common everywhere. Moreover, treating greywater would decrease the overall energy consumption, allowing treatment facilities to move closer to an autarkic energy supply. It should be noted that the operational phase of a WWTP dominates with around 94% the life cycle energy use (Racoviceanu et al., 2007).

The system requires changes to the wastewater infrastructures. Within buildings, separated pipe systems for each type of wastewater transportation are needed. For the transportation and digestion of blackwater, a low dilution flush system for toilets would be beneficial (Stauffer, 2018).

Another requirement is the separation of solid waste flows at the source, e.g. within private households, at least between organic and non-organic MSW. This can be critical, since large amounts of plastics, metals, and other (large) objects can seriously disrupt the treatment process, not only from a biological perspective, where pollutants can potentially inhibit the process, but also from a mechanical point of view, blocking pumps and other equipment or impairing mixing and heating. Separating solid household waste at the source is known both in principle and in many countries' practices. However, it is not exempt from problems, such as miss-sorting and contamination of waste, demanding adequate policies (Muralikrishna & Manickam, 2017).

3 Methods

The analysis of the conventional system as well as of an integrated system will make use of a process-based material flow approach (MFA) that combines MFA with life cycle inventory (LCI) (Uihlein et al., 2006). For the process-based MFA analysis, openLCA v.1.7.4 and the database ecoinvent v.3.5 (www.ecoinvent.org) are used. However, environmental impacts will be not considered in this study; these were investigated in Vergara-Araya et al. (2020), although for a different time period. The study assumes only flows induced by private households in residential areas.

The process-based MFA consists of three parts, which form the system boundary:

- water/wastewater system, which depicts the provision of water, wastewater generation, wastewater collection and its treatment;
- generation of MSW, transport and treatment; and
- provision of and demand for electricity and heat.

Whereas in the conventional system the parts are treated separately (Fig. 3), the integrated system will combine these, according to the discussion in Sect. 2 (cf. Sect. 4.3).

At each step the relevant technologies are modeled (see Sect. 4).

To quantify the impacts of an integrated system, two scenarios are set up. The first scenario, called "conventional system" and could be seen as a reference scenario, describes the current system. The second scenario, called "integrated system," will assume an integrated water, waste, and energy management system, as featured

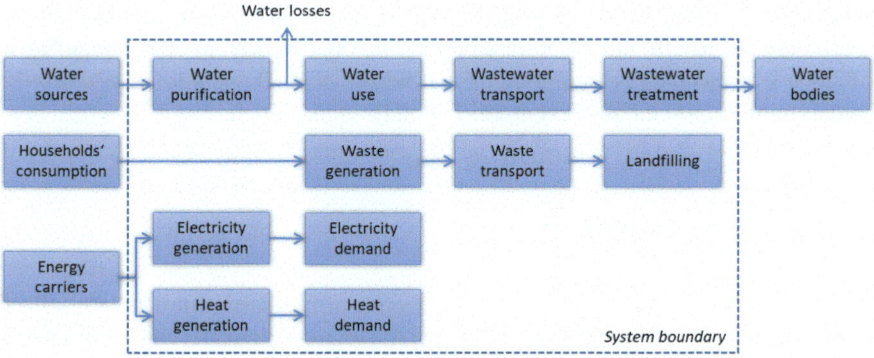

Fig. 3 Conventional water, waste and energy management systems. Adapted from Vergara-Araya et al. (2020)

in Sect. 4. In both scenarios the year 2015 is the reference; the final year under consideration is the year 2040.

The scenarios are constructed as "as-if-scenarios" (Börjeson et al., 2006): the analysis aims to answer the question, "What happens with respect to resource demand if the conventional system runs from 2015–2040 or if, as an alternative, the integrated system is implemented right from the start of the community's establishment and also operates between 2015 and 2040?".

4 Models

In the following, the main features and assumptions of the two models required to analyze the scenarios are described, which draws heavily on Vergara-Araya et al. (2020). Further details and the mathematical structure of the models can be found there.

4.1 General Conditions

The geographical systems boundary of the system under review is Curauma (Fig. 4). Curauma is a site for different real estate development projects in the region of Valparaíso, which borders the Metropolitan Region of Santiago de Chile, to foster the rapidly increasing demand for living areas in the region (ENACO, 2011; Mansilla & Fuenzalida, 2010).

The main common driver for the impacts of both systems under review is population development. Since the other relevant factors, such as the energy mix, differ between the two scenarios, they are described in the Sects. 4.2 and 4.3.

Using a variety of official data and available data from press media from 2007–2017 (3V Chile, 2012; Álvarez & Durán, 2008; "Curauma: Todas Las Ventajas De Una Ciudad Satélite," 2014; "Proyectan Millonaria Inversión En Curauma," 2011), the estimated population in 2015 was approx. 30,500; for the year 2040 a population of 87,800 is projected, assuming an annual inflow of 2,100 people adjusted to the population of 39,500 inhabitants in 2017 (Poduje et al., 2018).

4.2 Conventional System

Current resource management in Curauma follows traditional lines in centralized systems. The water management, municipal waste management, and local energy systems are separated (Fig. 3), and carried out by different companies and municipal departments.

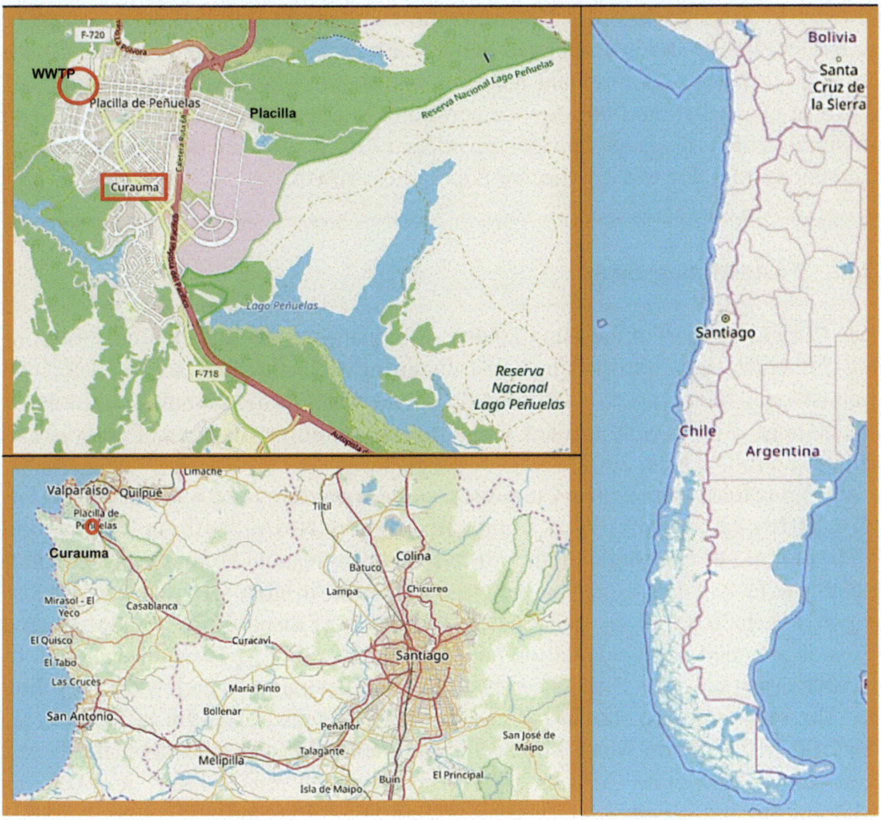

Fig. 4 Location of Curauma in Chile. *Source* Vergara-Araya et al. (2020)

4.2.1 Water and Wastewater Management

In Curauma, the private company Esval S.A. is responsible for the drinking water supply, sewerage, and sewage treatment. In the region of Valparaíso, water comes mainly from the Aconcagua River (about 1.074 million m^3/a) and subterranean water sources (about 0.276 million m^3/a (both 2012)). Water loss is set to the current national average value of 35% (ECONSSA CHILE S.A., 2014; MMA, 2012). Average water consumption equals 156 L/(PE·d) (own calculation based on SISS (2014a) and ANDESS (2013)). The water demand per person per day as well as the water loss is assumed not to change until 2040. Investments in the current and future pipelines are not taken into account in the analysis due to a lack of reliable data.

The wastewater from Curauma is treated in the wastewater treatment plant Placilla de Peñuelas (also called Placilla V). In 2015 the sewage water collection coverage was 93.1% (own calculation based on Esval (2017)). The collection rate is expected to improve annually by 0.5% per year, reaching complete coverage in 2029. Treatment in the WWTP comprises a completely mixed aerated pond with sedimentation (SISS,

2014b), with a design flow of 1.2 L/s and a daily maximum of 92.3 L/s (Esval, 2009). The treated wastewater is discharged to the nearby Las Cenizas creek (Esval, 2009). The average energy requirement accounts for about 0.097 kWh/m^3 for pumping (Matos et al., 2014), and the one for aeration around 2 W/m^3 (DWA, 2005). The computation of the analyzed impacts considers the necessity of expanding the existing treatment system due to the expected population growth.

4.2.2 Waste Management

Since Curauma is part of the Municipality of Valparaíso, the municipality is responsible for collecting, transporting, and disposing of residential solid waste, and commissioned the company COSEMAR (UCV radio, 2015). Currently, solid waste management is practically limited to landfilling, aiming to dispose of all waste in sanitary landfills (BCN, 2005).

Assuming that one person in Curauma generates about 0.97 kg waste per day (SGS SIGA, 2018), of which 53.3% is of organic origin (CNMA, 2010), in 2015 around 10,710 t OFMSW are available; the waste composition should remain constant until 2040 due to lack of data to capture the possible decline in share (Hoornweg & Bhada-Tata, 2012). In 2015 about 86% of the OFMSW was disposed of to landfills, and 14% was recycled (own calculation based on MMA (2018)). The future amount of available organic waste is influenced by the growing population and by the recycling rate. The recycling rate should increase by 5% per year to recognize the enforcement of Law 20.920 on Waste Management, Extended Producer Responsibility and Promotion of Recycling (MMA, 2016). The collection rate of MSW in 2015 was 100% with no indication of a change until 2040.

4.2.3 Energy Management

With the establishment of Sistema Eléctrico Nacional (SEN) as the result of a merger of SIC (Sistema Interconectado Central) with SING (Sistema Interconectado del Norte Grande) in November 2017, the electricity mix in the central region of Chile looks at follows: hydropower (share 27%; −14% compared to the mix between 2015 and 2017, which is recognized in the modeling for these years), renewable energies (21%; +13%), coal (share 21%; +6%), natural gas (19%; +7%) and diesel (13%; −8%) (CNE, 2015, 2018). The electricity mix of 2018 will be held constant until 2040. The possible provision of private households and small businesses are neglected, assuming that the current minor relevance for the electricity system will not change. As of April 2018 only 16 MW were installed nationwide, compared to the total installed capacity of 23,315 MW (MMA, 2018).

In the Curauma region the average annual household demand for electricity (excluding electricity used for heating) equals 1,867 kWh (2015) (own calculation based on Agostini et al. (2012)). In the 20 years before 2015 the annual growth rate was 2.9%, showing the influence of increasing income on energy demand (World

Energy Council, 2013). With a growing saturation the trend is expected to slow down in the future, and we estimate electricity demand per household in 2040 of 3,050 kWh.

In the respective region, the main energy sources for heating are wood (share 65.3%), liquefied petroleum gas (LPG; 24.0%), and kerosene (7.5%). Electricity (3.0%) and charcoal (0.2%) are rather unimportant (own calculation based on CChC (2010)). The heating mix should not change before 2040. The annual energy requirement for heating per household is currently about 0.9 MWh and should remain constant. No significant changes in housing construction materials or laws enforcing higher energy efficiency standards are expected in the next few years.

4.3 *Integrated System*

The integrated water, waste, and energy management system scenario under review combines the different resources and energy flows (Fig. 5). Wastewater is separated within the household into greywater and blackwater. Greywater is treated and then reused, e.g. for toilet flushing, while blackwater is transported to an anaerobic digestion plant. Together with OFMSW, blackwater is fermented into biogas, which is converted into electricity and heat in a CHP. The nutrient-rich digestate could be converted into fertilizer for agriculture, which has not been considered further.

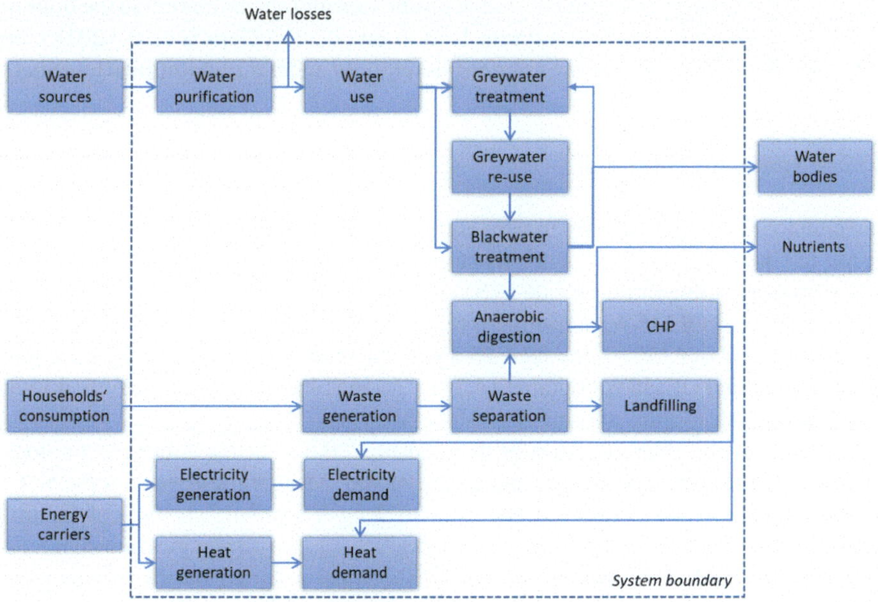

Fig. 5 Integrated water, waste and energy management system. Adapted from Vergara-Araya et al. (2020)

4.3.1 Water and Wastewater Management

The provision of drinking water, including water losses and infrastructure, will not be affected by the integrated management system. However, the installation of, e.g., vacuum toilets and the reuse of greywater could reduce the need for potable water. Depending on the design of the decentralized water treatment systems with water reuse, the requirement to provide potable water could decrease by up to 50% (Friedrich et al., 2020; Hernández Leal et al., 2007). Since no special technologies using small flushing volumes are foreseen, in this study a reduction of 30% in drinking water consumption is assumed, resulting in a freshwater demand of 109 L/(PE·d).

Of the wastewater generated, about 70% is assigned as greywater, and the rest as blackwater (Fane, 2013; Henze & Comeau, 2008). The share of greywater and blackwater in wastewater is assumed not to change in the period under investigation. The initial concentration of blackwater is relatively low, i.e. 1% total solids (TS), since no technologies using small flushing volumes are implemented. Therefore, a primary settling in the existing WWTP is planned to reach a 3% TS concentration. The supernatant water collected in this settling is treated with the greywater, representing 22% of the total treated volume.

The treatment of greywater and recirculated wastewater from the primary settling will take place in the existing aerated pond with sedimentation. By treating only the greywater fraction and the recirculated wastewater, the organic load is significantly reduced, from 60 to 34 g BOD/(PE·d) for pure greywater (DWA, 2017), extending the lifetime of the existing pond due to sludge production and accumulation in the bottom of the pond. Due to the lower nutrient load in greywater (Ziemba et al., 2018), no additional nitrogen or phosphorus removal is required under the conditions described.

4.3.2 Anaerobic Digestion

The composition of OFMSW can vary significantly (Deublein & Steinhauser, 2011; Eisted & Christensen, 2011). For this study, total solids (TS) content of 35% with 50% of fermentable volatile solids (VS) was assumed.

The collected OFMSW, assuming a recovery rate of 80%, is mixed with blackwater, sharing about 75% of the feed TS and is then fed into the reactor. The mixture of blackwater and OFMSW contained 4.2% TS—the blackwater significantly dilutes the organic waste—and 78% volatile solids (VS).

Considering the characteristics of the feedstock, the wet (low solids) technology for anaerobic digestion was assumed in the model. Mesophilic digesters (~35 °C) with an organic loading rate (OLR) of up to 2 kg VS/(m^3 d) are typical for this purpose. A biogas yield of 0.97 Nm^3 biogas/kg $VS_{removed}$ with a composition of 55% CH_4 was assumed (Tchobanoglous, 2003). Technical progress on the biogas yield was not considered, so as not to overestimate biogas production. The methane loss accounted for 1 g CH_4/kg feed (TS) (Pipatti et al., 2006).

The anaerobic digestion (AD) plant demands electricity and heat for pumping and mixing, as well as to heat up the feedstock and the anaerobic digesters to 35 °C, maintaining the desired mesophilic temperature. It is estimated that the AD process requires 25% of the heat and 4.4% of the electricity produced by the CHP.

4.3.3 Combined Heat and Power Plant (CHP)

For the adjunct combined heat and power plant (CHP), an electrical efficiency of 30% and a thermal efficiency of 45% (FAO, 2009) were assumed. Taking into account the caloric value of methane (15 kWh/kg), its density (0.716 kg/m^3), and the methane content of biogas (55%), 1.49 kWh/m^3 biogas electricity and 2.23 kWh/m^3 biogas heat were generated by the CHP.

4.3.4 Energy Mix

The remaining electricity and heat can penetrate the respective markets, changing the energy mix.

5 Results and Discussion

In both scenarios, population growth and the increasing per capita demand for electricity are driving the demand for electricity. Irrespective of the scenario, Curauma's demand for electricity will increase to about 89,300 MWh in 2040, compared to about 19,200 MWh in 2015 (Fig. 6; left scale), i.e. the demand will grow at a factor of 4.65.

While in the conventional system no biogas is available to produce electricity, in the integrated system biogas could contribute up to one-fifth to the electricity production. However, this share is only achieved in 2015 with a rather low per capita demand. Although with the population growth the availability of blackwater and OFMSW also increases, the share of biogas in the electricity mix would decline until 2040 from 19.1% (2015) to 12.6%, although the decline slows down slightly (Fig. 6; right scale). The availability of blackwater and organic MSW grows in the period from 3,664 MWh/year (net of own use in the AD and CHP) in 2015 at a factor 3.08. Thus, the growth rate reaches just two-thirds of the growth rate of the electricity demand.

The relevance of biogas-based electricity depends crucially on the accessibility of blackwater and OFMSW and the per-capita demand for electricity. A decline in available blackwater and OFMSW by 10% each would reduce the share of biogas in electricity production by about 8.2% (blackwater) and by 2.2% (OFMSW). OFMSW contributes to biogas production with 19.2% in 2015 and with 18.1% in 2040. A 10%

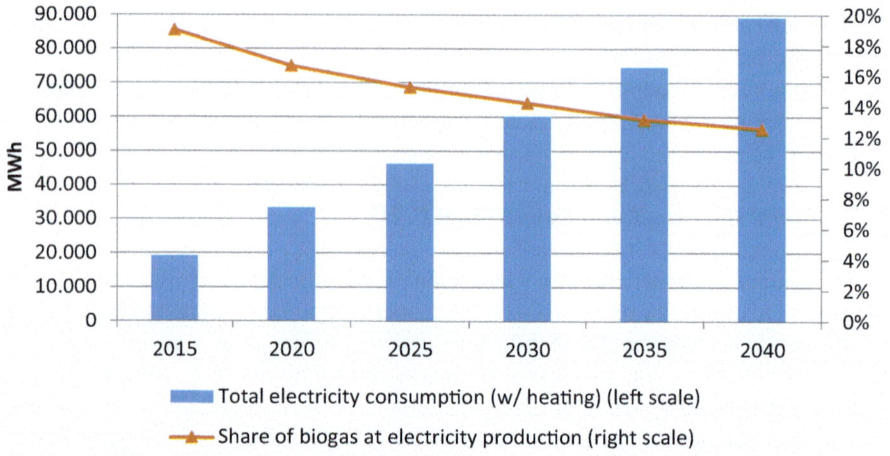

Fig. 6 Estimated electricity consumption in households per year and the share of biogas at electricity production in the scenario Integrated System

higher per capita demand for electricity would cause a decline in the share of biogas in electricity generation by 9.0%.

A slightly different picture can be observed in the case of heat. In both scenarios the heat demand is only determined by the growth of the population, resulting in an increase from about 8,900 MWh in 2015 to about 25,900 MWh in 2040 (Fig. 6; left scale). But the share of biogas-based heat is around 50% in the integrated system (Fig. 7; right scale), whereas in the conventional system no biogas is produced. In contrast to electricity production, the share of biogas-produced heat increases slightly until 2030 from 48.0% (2015) to 50.8% (2030), with no change in the share until 2040. This pattern is the result of the growing sewage coverage until 2029, then reaching 100%.

Like in the case of electricity, the relevance of biogas-based heat depends crucially on the accessibility of blackwater and OFMSW and the per capita demand for heat. A decline in available blackwater and OFMSW by 10% each would reduce the share of biogas in heat production by 7.9% (blackwater) and 2.1% (OFMSW). As mentioned before, OFMSW contributes to biogas production with 19.2% in 2015 and with 18.1% in 2040. A 10% higher per capita demand for heat would cause a decline in the share of biogas in heat production by 8.8%.

The growth of the population will have no effect on the biogas production share, and thus on neither electricity nor heat generation. The larger population will generate a higher demand for electricity and heat, but will also generate more blackwater and OFMSW.

In contrast to the demand for electricity and heat, which is identical for both scenarios, the water demand in both scenarios differs significantly. The integrated system is technically defined by the re-use of the treated greywater, reducing the provision of potable water by 30% compared to the conventional system (Fig. 8).

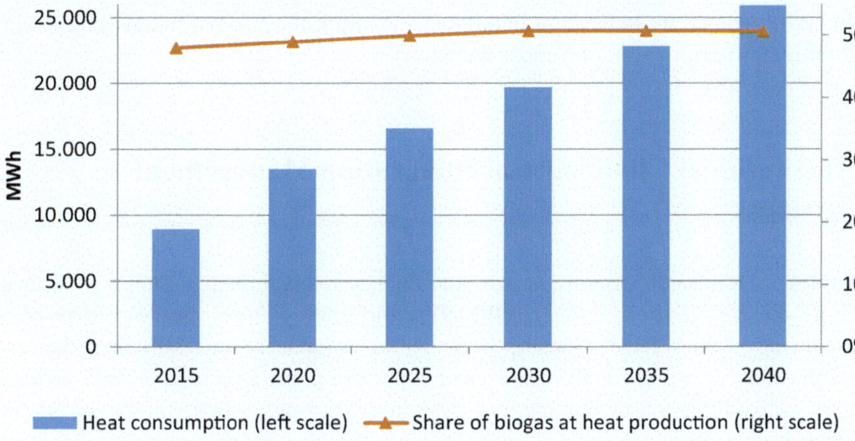

Fig. 7 Estimated heat consumption in households per year and the share of biogas at heat production in the scenario integrated system

Irrespective of this, according to the calculation the resource demand in 2020 in the integrated system scenario should have overturned the potable water requirement in the conventional system scenario in 2015. The reason for this is the growing population.

A 10% larger population would increase the demand for water in both scenarios by 10%. Reducing the reuse rate of greywater by 10% would cause a higher water provision by 4.3%. The reviewed variables regarding drinking water consumption

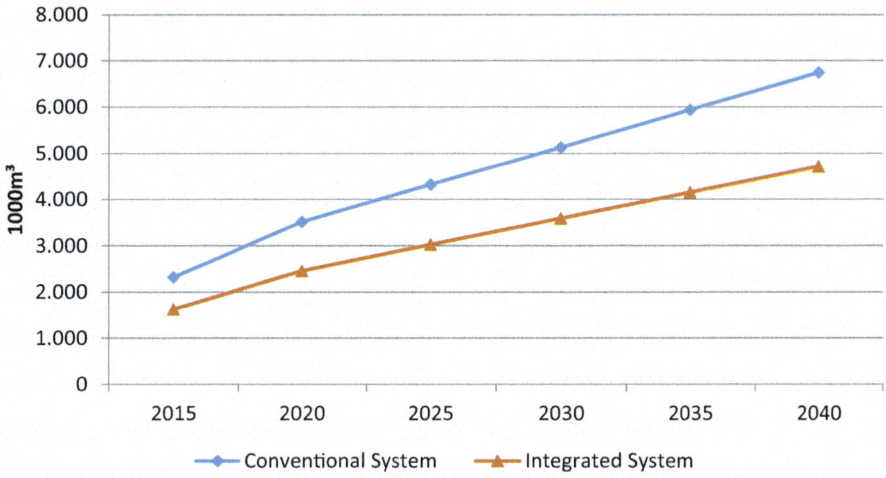

Fig. 8 Water provision in the conventional system scenario and the integrated system scenario

could be influenced by individual behavior, e.g. cultural customs, but also by water availability and political-economical decisions.

6 Institutional Challenges of Integrating Management Systems

Integrating wastewater, organic waste, and energy management systems will have not only an impact on the water, OFMSW, and energy flows, but would need a restructuring of the interrelationship between the companies and municipal departments that currently provide the services or supervise them. In principle, two options exist to reach an integrated management system: (a) through establishing markets for different products along the value-added chain, or (b) through integrating the different steps of the value-added chain into one company. Both options could be combined, however, the combination will be only touched on, but not further discussed, for reasons of simplifying the discussion.

Integration via markets would require the identification of the respective markets with their (potential) market participants, as well as the legal conditions under which they have to act. Assuming for the following discussion a compulsory separation of the waste flows, in principle the integrated system would offer market opportunities for OFMSW, sludge, biogas, heat and electricity, which are established products, and for recycled greywater, which would be a new product.

The supplier of sewage sludge for converting energy would be the operator of the WWTP, in Curauma currently Esval S.A. The demander for the sludge could be any AD, which has to compete with another AD or other user of the sludge, e.g. agriculture. The AD would also demand OFMSW. Following current Chilean law (BCN, 2016, 2020) the supplier of OFMSW would be the generator of waste, i.e. in our case the Municipality of Valparaíso. Nevertheless, they could commission a company, e.g. the current collector COSEMAR, to trade OFMSW on its account or on the account of the municipality. The demander of the OFMSW could be any AD, but also composter. The supplier of biogas is the operator of the AD, the supplier of electricity, and heat the operator of the CHP, and, finally, the demander of the heat and electricity households and companies. The operator of the AD or the CHP could be any company or non-profit organization, but also the owner of the WWTP, to secure a channel to distribute the sludge, or any power generation utility like the ones associated to the Sistema Eléctrico Nacional (SEN), the current provider of electricity in the region of Curauma. The WWTP could also offer treated greywater for reuse.

A market will form when conditions in the conditions of trading allow a desired return of investment, which the market participants must determine for themselves. The achievable return of investment depends on the earnings and costs of providing the product, which are crucially influenced by technological constraints, and on the legal conditions, which, however, could influence the economic variables of a company.

Market solutions have to take into account the specifics of the products to be offered. The transport of biogas, heat, electricity and greywater are grid-bounded. Biogas could be also transported with trucks as long the transported amount is low, even though it is rather exceptional. If the infrastructures belong to the provider of the transported product, the flexibility to change the trading partner is limited, if it is possible at all. Heat from CHP would be network-bounded with heat losses during transport, limiting the transport distances and thus, restricting the potential demander to the near neighborhood. In contrast to that, the transport of electricity is comparably easy to manage, as the grid system is typically available. Therefore, either the trading conditions are regulated by the government, e.g. by setting the sales price, or the transport infrastructure is unbundled from the provider of the product, in principle allowing everybody to use the infrastructure for trading. But even unbundling would require a sufficient number of suppliers to achieve a competitive market. Looking at the products, this seems to be unlikely, as their production is connected with high investments. Dewatered blackwater and organic MSW can be transported with trucks and thus, could be offered to different ADs as a potential demander. However, this would require appropriate logistics, challenging the possible advantages of long-haul transportation. The impacts of the existing transferability of water rights with the change in the water intake location (Molinos-Senante, 2018) on the functioning of such market solutions needs additional research.

Due to the limited opportunities to transport and trade the products described, the number of potential market participants would be rather low. Thus, presumably the market exchange will be characterized by regional monopolies in combination with a regional oligopsony, i.e. a small number of demanders, leading to the question how the pricing process between market participants should be organized to achieve a maximum economic welfare to all participants, including the final demander. The low flexibility of grid-bounded markets increases the risk of delivery shortfalls, e.g. the delivery of substrates, leading in practice to long-term contracts over more than ten years. An alternative would be vertical integration (see below). Here, additional research would also be necessary.

Regulations have to control the conditions under which market participants can enter the market and produce and trade the respective goods, which directly or indirectly affects the earnings and costs of a company. On the one hand, they should support the public interests, e.g. sustainable provision of water. On the other hand, the rules should allow the trading partners to achieve a positive return on their investments and the end consumer an affordable price for decent services. A specific challenge is dealing with the integration of private and public interests, which could be conflicting. For example, the planned new Chilean Water Code reinforces the character of water as a public good, requiring that the water concessions are linked to availability and sustainability of the water source (Senado de la Republica de Chile, 2021). The requirement could impact the provision of water, and the treatment of wastewater and trading of the byproducts. Currently, there is a complex and sometimes overlapping institutional structure, which often lacks expertise on water issues, to settle disputes and to define responsibilities (Vergara & Rivera, 2018).

According to Chilean Law 382, concessions for water supply and a sewerage system are granted in urban or developable areas. In cities like Valparaíso, it is compulsory to supply drinking water and treat wastewater (BCN, 2014). The provision of water and the treatment of wastewater by a water company is to be separated from water rights, i.e. the right to extract water from surface water bodies (BCN, 2014; Vergara & Rivera, 2018). The Chilean Water Code demands that all treated wastewater has to be returned to its source, whether directly or indirectly, if e.g. "canalistas" use the treated water for irrigation (BCN, 2014; Vergara & Rivera, 2018). The ownership of treated greywater is currently regulated. A published law aims at limiting the reuse of greywater to the premises of the polluter of the wastewater (BCN, 2018). The required regulation to implement the law has been recently published (BCN, 2024).

The current law demands the WWTP operator dispose of the sludge, for which it currently has to pay to the waste management company. These costs are then included in the utility bill to be balanced by the consumer. Comparable to that, ADs can sell the biogas using sludge as a feed. The earnings or savings have to be considered in the utility bill (BCN, 2018).

The discussion presented assumes that wastewater and its byproducts can be treated as economic goods with a positive price, which has to exceed the costs of supplying. The current situation not only in Chile implies that this could be a huge assumption. Maybe steady public financial support could be crucial for establishing the respective markets, as the experience with the German biogas market shows. The German biogas market depends crucially on the financial support of the government. Among other reasons, the feed is too expensive to compete with other energy sources (Harenbrock, 2020), leading to the question of whether non-economic reasons could justify lasting support.

The alternative to the market solution would be to integrate the management system into one company.

The merging of the different services within one company could be understood as a vertical integration. The entire value chain, beginning with the provision of water, ending with recycling and supply of greywater and provision of heat and electricity, or parts of it, is organized within one company. The main gain is the ability of the company to secure the supply chain and to synchronize the supply and demand along the value chain. Potentially, the transaction costs could be lower (Perry, 1989).

Due to the different legal frameworks for each product along the chain, a complete merger will be unlikely. That means each service will be provided by rather independent departments or subsidiaries, with their own legal and business demands to be fulfilled, which could conflict with each other. Typical for vertical integrated companies is weaker motivation for good performance at the start of the supply chain, since sales are guaranteed or efforts are not rewarded (Perry, 1989).

Even though the entire value chain in one region is organized within one company, such a company will compete on different markets with competitors if regional monopolies are not implemented, which is not discussed here. This will influence the possible market power of such a company and thus the price setting on the different markets, within the boundaries set by the regulations on each market. It should be noted that competing markets at different stages of a value-added chain impede the ability to cross-subsidize parts of a company.

Recycled greywater would compete with potable water for some water applications, like the irrigation of parks, gardens, and industrial uses. The water scarcity observed in the region could be a driver for economic or political incentives for wastewater reuse, increasing its market attractiveness or enforcing its realization. If the planned regulations are set into law, a market for greywater will not be possible. However, a water company could offer the knowledge to construct and operate greywater recycling treatment plants or provide on-site service. The provision of heat would compete for new customers without access to a heat grid with alternative heat-related energy sources. The flexibility of existing customers that are connected to the grid is rather low, since changing the energy source would result in high transaction costs. The flexibility on the electricity market could be highest if the market offers alternatives.

The points raised focus very much on the economic relations between potential market participants or a vertical integrated company, ignoring possible social and environmental impacts. It is obvious that these shall be considered in the organization of markets, or if a vertical company is seen as an option.

In Chile, the arrangement of water rights is a long-term topic of discussion (Vergara & Rivera, 2018). The legal reform of water rights in 1981 tried to establish a water market with the goal of improving the efficiency of the water supply, and thus the welfare provided by the water system. However, the law ignored many peculiarities of water management, e.g. aquifers as an additional water source, leading to an inefficient water regime enforcing an unsustainable use of water bodies in Chile. The flaws of the 1981 law were not resolved by the 2005 reform (Donoso, 2018; Morán, 2014; Peña, 2018). The question is whether a market solution would enforce the flaws of the current water regime, with an overstretched water outtake or reduce these defects. Or whether an organizational solution would allow for the needed welfare impacts.

7 Concluding Remarks

The analysis presented gives some indication of the possible impacts of implementing an integrated water, waste, and energy management system under the conditions of Curauma on energy and water flows, but also of the institutional challenges to implement such a system. The analysis of the impacts is based on a comprehensive approach, i.e. on a process-based MFA analysis, but considers mainly the impacts of flows. Construction expenses are not considered, although they would be necessary to implement the integrated system, such as in-house pipelines to separate wastewater at its source, as well as a network system for heat or gas transport within the community. Other potential benefits, such as using heated greywater as an energy source, were also not modeled. The discussion of the institutional challenges provides some insights into implementing an integrated system. Additional research is therefore needed to get a better picture of the impacts and challenges, but also to find solutions to implement a water-energy-waste system, which would improve the current system.

The opportunities to transform the current system depend very much on the economic soundness and the acceptance of the users. The integrated system, which has been analyzed in this contribution, would require, among other things, a change in the behavioral pattern, such as separating solid waste at home. As discussed in Sect. 6, the implementation of an integrated system also demands the reorganization of the provision of water and energy, as well as the treatment of wastewater and waste, which are currently operated by four different organizations, affecting the economic soundness of the entire system. However, not only are the entire costs to establish an integrated system relevant, but also the allocation of the costs (and benefits) to the different stakeholders.

Using Curauma as a case study, the findings are specific to the conditions of that community. The valuation of the different advantages of an integrated system will vary according to climate and economic and cultural conditions, among others. For example, in semi-arid or arid regions, the focus will be more on water-related impacts, whereas in humid regions this could be of minor relevance. Moreover, in warmer (e.g. tropical) regions, the potential benefit of providing heat must be carefully studied.

Furthermore, in cases where the system is installed in communities with existing wastewater treatment plants, the environmental and economic costs of investment have to be balanced with the environmental and societal benefits—which could differ considerably by region. For example, the removal of the black water stream could negatively influence some wastewater treatment systems with targeted nitrogen removal (e.g. with upstream denitrification) due to the reduced organic matter content in the greywater in comparison with raw wastewater. Moreover, in order to enable the reuse of treated wastewater, it is possible that the disinfection requirements for treated wastewater may be increased, thereby increasing the resources involved and the operational costs, an aspect that has not been taken into account in this study.

If a separate flow system is integrated into a solid waste collection and treatment system, additional benefits could be realized because, e.g., solid waste without the organic fraction is easier to treat (Bilitewski et al., 2018). However, in the context of Chilean solid waste management, this could have a negative impact, as an enlarged share of inorganic waste in sanitary landfills will slow down its degradability, also influencing changes in the solid waste structure and other characteristics (e.g. compatibility, flammability, etc.). Additional research is necessary to better understand the relevance of site-specific impacts for the overall findings.

References

3V Chile. (2012). *CURAUMA: Un lugar que conjuga calidad, entorno natural y las comodidades de una ciudad de futuro.*

Agostini, C. A., Cecilia Plottier, M., & Saavedra, E. H. (2012). Residential demand of electricity in Chile [La demanda residencial de energía eléctrica en Chile]. *Economia Chilena, 15*(3), 64–83. https://www.scopus.com/inward/record.uri?eid=2-s2.0-84871695181&partnerID=40&md5=9466589a244d2644af0ac895cda29cae

Aguilar, M. C., Wang, Y. D., Roskilly, T., Pathare, P. B., & Lamidi, R. O. (2017). Biogas from anaerobic co-digestion of food waste and primary sludge for cogeneration of power and heat. *Energy Procedia, 142*, 70–76. https://doi.org/10.1016/j.egypro.2017.12.012

Allen, L., Christian-Smith, J., & Palaniappan, M. (2010). *Overview of greywater reuse: The potential of greywater systems to aid sustainable water management*. Pacific Institute.

Allesch, A., & Brunner, P. H. (2015). Material flow analysis as a decision support tool for waste management: A literature review. *Journal of Industrial Ecology, 19*(5), 753–764. https://doi.org/10.1111/jiec.12354

Álvarez, O., & Durán, M. (2008, August 20). Nuevo centro cívico une Placilla y Curauma. *El Mercurio De Valparaíso.* http://www.mercuriovalpo.cl/prontus4_noticias/site/artic/20080830/pags/20080830003401.html

ANDESS. (2013). *Informe de Gestión de la Sequía 2014 Industria Sanitaria en Chile, Regiones de Atacama al Biobío.* Asociación Nacional de Empresas de Servicios Sanitarios. http://www.andess.cl/descargas/noticias/201401-INFORME-GREMIAL-SEQUIA-ANDESS-ENERO-2014.pdf

BCN. (2005). *Reglamento Sobre Condiciones Sanitarias Y De Seguridad Básicas En Los Rellenos Sanitarios.*

BCN. (2014). *DFL 382. Ley general de servicios sanitarios: Versión: Última Versión - 10-OCT-2014.* Última modificación: 09-ENE-2014 - Ley 20720. http://bcn.cl/29kpk

BCN. (2016). *Ley 20920: Establece marco para la gestión de residuos, la responsabilidad extendida del productor y fomento al reciclaje.* Ministerio del Medio Ambiente. http://bcn.cl/297lc

BCN. (2018). *Ley 21075. Regula la recolección, reutilización y disposición de aguas grises.* http://bcn.cl/2eoxk

BCN. (2020). *Decreto 2385: Fija texto refundido y sistematico del decreto ley num. 3.063, de 1979, sobre renta municiplaes.* http://bcn.cl/2ep5t

BCN. (2024). *Decree 40. Approves regulation on basic sanitary conditions for the reuse of grey water.* https://bcn.cl/3jve0

Bilitewski, B., Wagner, J., & Reichenbach, J. (2018). *Best Practice Municipal Waste Management: Information pool on approaches towards a sustainable design of municipal waste management and supporting technologies and equipment.* Dessau-Roßlau. Umweltbundesamt.

BIO by Deloitte. (2015). *Optimising water reuse in the EU: Final report prepared for the European Commission (DG ENV).* Part I. Luxembourg. https://ec.europa.eu/environment/water/blueprint/pdf/BIO_IA%20on%20water%20reuse_Final%20Part%20I.pdf

Börjeson, L., Höjer, M., Dreborg, K.-H., Ekvall, T., & Finnveden, G. (2006). Scenario types and techniques: Towards a user's guide. *Futures, 38*(7), 723–739. https://doi.org/10.1016/j.futures.2005.12.002

Capodaglio, A., & Olsson, G. (2020). Energy issues in sustainable urban wastewater management: Use, demand reduction and recovery in the urban water cycle. *Sustainability, 12*(1), 266. https://doi.org/10.3390/su12010266

CChC. (2010). *Estudio De Usos Finales Y Curva De Oferta De Conservación De La Energía En El Sector Residencial De Chile.* Cámara Chilena de la Construcción. http://antiguo.minenergia.cl/minwww/export/sites/default/05_Public_Estudios/descargas/estudios/Usos_Finales_COC_Sector_Residencial_2010.pdf

Chen, B., & Lu, Y. (2015). Urban nexus: A new paradigm for urban studies. *Ecological Modelling, 318*, 5–7. https://doi.org/10.1016/j.ecolmodel.2015.10.010

Chislock, M. F., Doster, E., Zitomer, R. A., & Wilson, A. E. (2013). Eutrophication: Causes, consequences, and controls in aquatic ecosystems. *Nature Education Knowledge, 4*(4), 10.

CNE. (2015). *Anuario Estadístico De Energía 2005–2015.* Comisión Nacional de Energía, Ministerio de Energía. https://www.cne.cl/wp-content/uploads/2016/07/AnuarioCNE2015_vFinal-Castellano.pdf

CNE. (2018). *Capacidad Instalada. Energía Abierta BETA.* Comisión Nacional de Energía, Ministerio de Energía. http://energiaabierta.cl/visualizaciones/capacidad-instalada/

CNMA. (2010). *Primer Reporte Del Manejo De Residuos Sólidos En Chile.* Comisión Nacional del Medio Ambiente. http://www.sinia.cl/1292/articles-49564_informe_final.pdf

Curauma: Todas las ventajas de una ciudad satélite (2014, July 1). *El Mercurio De Valparaíso.* http://tematicas.cl/ee/wp-content/uploads/2014/06/CURAUMA-en-baja.pdf

Dai, J., Wu, S., Han, G., Weinberg, J., Xie, X., Wu, X., Song, X., Jia, B., Xue, W., & Yang, Q. (2018). Water-energy nexus: A review of methods and tools for macro-assessment. *Applied Energy, 210*, 393–408. https://doi.org/10.1016/j.apenergy.2017.08.243

DeNooyer, T. A., Peschel, J. M., Zhang, Z., & Stillwell, A. S. (2016). Integrating water resources and power generation: The energy–water nexus in Illinois. *Applied Energy, 162*, 363–371. https://doi.org/10.1016/j.apenergy.2015.10.071

Deublein, D., & Steinhauser, A. (2011). *Biogas from waste and renewable resources: An introduction* (Second, revised and expanded edition). Wiley-VCH. http://lib.myilibrary.com/detail.asp?id=337047 https://doi.org/10.1002/9783527632794

Dixon, A., Butler, D., & Fewkes, A. (1999). Water saving potential of domestic water reuse systems using greywater and rainwater in combination. *Water Science and Technology, 39*(5), 25–32. https://doi.org/10.2166/wst.1999.0218

Donoso, G. (2018). Overall assessment of Chile's water policy and its challenges. In G. Donoso (Ed.), *Global issues in water policy. Water policy in Chile* (Vol. 21, pp. 209–219). Springer International Publishing. https://doi.org/10.1007/978-3-319-76702-4_14

DWA. (2005). *Grundsätze für Bemessung, Bau und Betrieb von Abwasserteichanlagen. DWA-Regelwerk: A 201.* Deutsche Vereinigung für Wasserwirtschaft Abwasser und Abfall e. V.

DWA. (2017). *30. Leistungsvergleich kommunaler Kläranlagen.* Deutsche Vereinigung für Wasserwirtschaft Abwasser und Abfall e. V.

ECONSSA CHILE S.A. (2014). *Agua no facturada en sistemas de agua.* http://www.econssachile.cl/pdf/estudios/ANF_web.pdf

Eisted, R., & Christensen, T. H. (2011). Characterization of household waste in Greenland. *Waste Management (New York, N.Y.), 31*(7), 1461–1466. https://doi.org/10.1016/j.wasman.2011.02.018

ENACO. (2011). *Curauma: Inversión, crecimiento y calidad.* https://sge.lclark.edu/wp/wp-content/uploads/2012/01/Curauma-Suplemento-edicion-especial-2011.pdf

Esval S.A. (2009). *Solictitud de Ampliación de Concesiones Sanitarias Según Ley N 20307 para un terreno urbano Localidad Placilla de Peñuelas, Valparaíso.* http://www.siss.cl/appsiss/repositorio_siss/conc_tram/V/PLacilla%20de%20Pe%C3%B1uelas,%20Com.%20Valpara%C3%ADso/1%20SOLICITUD%20DE%20CONCESI%C3%93N.pdf

Esval S.A. (2017). *Memoria Anual 2017.* http://portal.esval.cl/wp-content/uploads/2018/04/Memoria-201720181.pdf

Fane, S. (2013). *Wastewater reuse—Your Home—Australia's Government.* http://www.yourhome.gov.au/sites/prod.yourhome.gov.au/files/pdf/YOURHOME-5-Water-3-WastewaterReuse-(4Dec13).pdf

FAO. (2009). *ETo-Calculator.* http://www.fao.org/nr/water/eto.html

Friedrich, J., Poganietz, W.-R., & Lehn, H. (2020). Life-cycle assessment of system alternatives for the water-energy-waste nexus in the urban building stock. *Resources, Conservation and Recycling, 158*, 104808. https://doi.org/10.1016/j.resconrec.2020.104808

Fundación Chile. (2017). *Desafíos del Agua para la Región Latinoamericana.*

Ghisi, E., & Ferreira, D. F. (2007). Potential for potable water savings by using rainwater and greywater in a multi-storey residential building in southern Brazil. *Building and Environment, 42*(7), 2512–2522. https://doi.org/10.1016/j.buildenv.2006.07.019

Guerrini, A., Romano, G., & Indipendenza, A. (2017). Energy efficiency drivers in wastewater treatment plants: A double bootstrap DEA analysis. *Sustainability, 9*(7), 1126. https://doi.org/10.3390/su9071126

Harenbrock, K. (2020, October 19). *Germany's biogas plants face bankruptcy.* DW. https://www.dw.com/en/germanys-biogas-plants-face-bankruptcy/av-55321525

Henze, M., & Comeau, Y. (2008). *Wastewater characterization.* http://ocw.unesco-ihe.org/pluginfile.php/462/mod_resource/content/1/Urban_Drainage_and_Sewerage/5_Wet_Weather_and_Dry_Weather_Flow_Characterisation/DWF_characterization/Notes/Wastewater%20characterization.pdf

Hernández Leal, L., Zeeman, G., Temmink, H., & Buisman, C. (2007). Characterisation and biological treatment of greywater. *Water Science and Technology: A Journal of the International Association on Water Pollution Research, 56*(5), 193–200. https://doi.org/10.2166/wst.2007.572

Hoornweg, D., & Bhada-Tata, P. (2012). *What a waste. A global review of solid waste management.*

Kennedy, C., Pincetl, S., & Bunje, P. (2011). The study of urban metabolism and its applications to urban planning and design. *Environmental Pollution (Barking, Essex: 1987), 159*(8–9), 1965–1973. https://doi.org/10.1016/j.envpol.2010.10.022

Kleckers, J., & Haberkamp, J. (2021, May 4). *Einfluss von Stoffstromtrennung auf den Betrieb einer Modellkläranlage.* 27. SIMBA-Anwendertreffen, Magdeburg.

Lange, J., Otterpohl, R., & Steger-Hartmann, T. (Eds.). (1997). *Ökologie aktuell. Abwasser: Handbuch zu einer zukunftsfähigen Wasserwirtschaft* (1. Aufl.). Mall-Beton-Verl.

Madrid-López, C., & Giampietro, M. (2015). The water metabolism of socio-ecological systems: Reflections and a conceptual framework. *Journal of Industrial Ecology, 19*(5), 853–865. https://doi.org/10.1111/jiec.12340

Mansilla, P., & Fuenzalida, M. (2010). Procesos de desarrollo urbano-regional y exclusión territorial: Nuevas formas de urbanización en el área metropolitana de Valparaíso. Estudio de caso ciudad de Curauma. *Revista INVI, 25*(69). https://doi.org/10.4067/S0718-83582010000200003

Mata-Alvarez, J., Macé, S., & Llabrés, P. (2000). Anaerobic digestion of organic solid wastes. An overview of research achievements and perspectives. *Bioresource Technology, 74*(1), 3–16. https://doi.org/10.1016/S0960-8524(00)00023-7

Matos, C., Pereira, S., Amorim, E. V., Bentes, I., & Briga-Sá, A. (2014). Wastewater and greywater reuse on irrigation in centralized and decentralized systems—An integrated approach on water quality, energy consumption and CO2 emissions. *The Science of the Total Environment, 493*, 463–471. https://doi.org/10.1016/j.scitotenv.2014.05.129

MMA. (2012). *Informe del estado del medio ambiente 2011: Residuos.* Ministerio del Medio Ambiente. http://www.mma.gob.cl/1304/w3-article-52016.html

MMA. (2016). *Ley 20920. Establece Marco para la Gestión de Residuos, la Responsabilidad Extendida del Productor y fomento al Reciclaje.* Ministerio del Medio Ambiente. https://www.leychile.cl/N?i=1090894&f=2016-06-01&p=

MMA. (2018). *Cuarto Reporte del Estado del Medio Ambiente 2018.* Ministerio del Medio Ambiente.

Mo, W., Wang, R., & Zimmerman, J. B. (2014). Energy-water nexus analysis of enhanced water supply scenarios: A regional comparison of Tampa Bay, Florida, and San Diego, California. *Environmental Science & Technology, 48*(10), 5883–5891. https://doi.org/10.1021/es405648x

Molinos-Senante, M. (2018). Urban Water Management. In G. Donoso (Ed.), *Global issues in water policy. Water policy in Chile* (Vol. 21, pp. 131–150). Springer International Publishing. https://doi.org/10.1007/978-3-319-76702-4_9

Morán, J. I. (2014). *Water legislation in Chile: The need for reform.* https://content.next.westlaw.com/9-547-7926?__lrTS=20210203103555733&transitionType=Default&contextData=(sc.Default)&firstPage=true

Muralikrishna, I. V., & Manickam, V. (2017). Solid waste management. In *Environmental management* (pp. 431–462). Elsevier. https://doi.org/10.1016/B978-0-12-811989-1.00016-6

Nolde, E., & Grunow, P. (2013). *Dezentrale Abwasserwärmerückgewinnung in Kombination mit einer Grauwasserrecyclinganlage.*

Oteng-Peprah, M., Acheampong, M. A., & deVries, N. K. (2018). Greywater characteristics, treatment systems, reuse strategies and user perception-a review. *Water, Air, and Soil Pollution, 229*(8), 255. https://doi.org/10.1007/s11270-018-3909-8

Otterpohl, R., Braun, U., & Oldenburg, M. (2002). *Innovative technologies for decentralised wastewater management in urban and peri-urban areas.* IWA Small2002.

Peña, H. (2018). Integrated water resources management in Chile: Advances and challenges. In G. Donoso (Ed.), *Global issues in water policy. Water policy in Chile* (Vol. 21, pp. 197–207). Springer International Publishing. https://doi.org/10.1007/978-3-319-76702-4_13

Perry, M. K. (1989). Chapter 4 Vertical integration: Determinants and effects. In *Handbook of industrial organization. Handbook of Industrial organization Volume 1* (Vol. 1, pp. 183–255). Elsevier. https://doi.org/10.1016/S1573-448X(89)01007-1

Pipatti, R., Silva Alves, J. W., Gao, Q., Mareckova, K., Oonk, H., Scheehle, E., Sharma, C. A., Smith, A., Svardal, P., & Yamada, M. (2006). Waste. Biological treatment of solid waste. In *2006 IPCC Guidelines for national greenhouse gas inventories: Prepared by the National Greenhouse Gas Inventories Programme* (Vol. 5, 4.1–4.8).

Poduje, I., Ruiz, M., Iribarne, C., & Vergara, J. (2018). *Monitor diagnóstico: El despoblamiento de Valparaíso – Análisis de Resultados: CENSO 2017.*

Proyectan millonaria inversión en Curauma (2011, May 18). *El Mercurio De Valparaíso.* http://www.mercuriovalpo.cl/prontus4_noticias/site/artic/20110518/pags/20110518000434.html

Racoviceanu, A. I., Karney, B. W., Kennedy, C. A., & Colombo, A. F. (2007). Life-cycle energy use and greenhouse gas emissions inventory for water treatment systems. *Journal of Infrastructure Systems, 13*(4), 261–270. https://doi.org/10.1061/(ASCE)1076-0342(2007)13:4(261)

Schütze, M., Wriege-Bechtold, A., Zinati, T., Söbke, H., Wißmann, I., Schulz, M., Veser, S., Londong, J., Barjenbruch, M., & Alex, J. (2019). Simulation and visualization of material flows in sanitation systems for streamlined sustainability assessment. *Water Science and Technology : A Journal of the International Association on Water Pollution Research, 79*(10), 1966–1976. https://doi.org/10.2166/wst.2019.199

Senado de la Republica de Chile. (2021, August 4). *Nuevo Código de Aguas que refuerza carácter de bien de uso público fue despachado por el Senado* [Press release]. https://www.senado.cl/noticias/agua/nuevo-codigo-de-aguas-que-refuerza-su-caracter-de-bien-de-uso-publico

SGS SIGA. (2018). *Informe 1 "Diagnóstico De La Situación Por Comuna Y Por Región En Materia De Rsd Y Asimilables. Capítulo 4–Diagnóstico De La Situación Por Comuna Y Por Región En Materia De Residuos Sólidos Domicliarios Y Asimilables. Subsecretaría De Desarrollo Regional Y Administrativo (Subdere): Programa Nacional De Residuos Sólidos.*

SISS. (2014a). *Facturación mensual.* Superintendencia de Servicios Sanitarios. http://www.siss.gob.cl/577/w3-propertyvalue-3447.html

SISS. (2014b). *Plantas de tratamiento de aguas servidas - V Región.* Superintendencia de Servicios Sanitarios. http://www.siss.gob.cl/577/w3-article-5124.html

Skambraks, A.-K., Kjerstadius, H., Meier, M., Davidsson, Å., Wuttke, M., & Giese, T. (2017). Source separation sewage systems as a trend in urban wastewater management: Drivers for the implementation of pilot areas in Northern Europe. *Sustainable Cities and Society, 28*, 287–296. https://doi.org/10.1016/j.scs.2016.09.013

Stauffer, B. (2018). *Vacuum toilet.* https://sswm.info/water-nutrient-cycle/water-use/hardwares/toilet-systems/vacuum-toilet

Tchobanoglous, G. (2003). *Wastewater engineering: Treatment and reuse* (4. ed.). *McGraw-Hill higher education.* McGraw-Hill. http://www.loc.gov/catdir/description/mh021/2001053724.html

Uihlein, A., Poganietz, W.-R., & Schebek, L. (2006). Carbon flows and carbon use in the German anthroposphere: An inventory. *Resources, Conservation and Recycling, 46*(4), 410–429. https://doi.org/10.1016/j.resconrec.2005.09.001

UN DESA. (2018). *2018 Revision of World Urbanization Prospects*. UN Department of Economic and Social Affairs.

UNESCO. (2017). *Wastewater: The untapped resource: The United Nations world water development report* (Vol. 2017). OECD.

Venkatesh, G., Chan, A., & Brattebo, H. (2014). Understanding the water-energy-carbon nexus in urban water utilities: Comparison of four city case studies and the relevant influencing factors. *Energy, 75*, 153–166. https://doi.org/10.1016/j.energy.2014.06.111

Vergara, A., & Rivera, D. (2018). Legal and institutional framework of water resources. In G. Donoso (Ed.), *global issues in water policy. Water policy in Chile* (Vol. 21, pp. 67–85). Springer International Publishing. https://doi.org/10.1007/978-3-319-76702-4_5

Vergara-Araya, M., Lehn, H., & Poganietz, W.-R. (2020). Integrated water, waste and energy management systems—A case study from Curauma, Chile. *Resources, Conservation and Recycling, 156*, 104725. https://doi.org/10.1016/j.resconrec.2020.104725

Villarroel Walker, R., Beck, M. B., Hall, J. W., Dawson, R. J., & Heidrich, O. (2014). The energy-water-food nexus: Strategic analysis of technologies for transforming the urban metabolism. *Journal of Environmental Management, 141*, 104–115. https://doi.org/10.1016/j.jenvman.2014.01.054

World Energy Council. (2013). *Energy efficiency indicators: Average electricity consumption per electrified household*. https://www.wec-indicators.enerdata.eu/household-electricity-use.html

Ziemba, C., Larivé, O., Reynaert, E., & Morgenroth, E. (2018). Chemical composition, nutrient-balancing and biological treatment of hand washing greywater. *Water Research, 144*, 752–762. https://doi.org/10.1016/j.watres.2018.07.005

Adapting the German Bioenergy Village Concept to Los Ríos Region in Southern Chile: Integrating Community, Water, Energy and Organic Waste Management

Alfredo Erlwein

Abstract German bioenergy villages (BEVs) achieve high energy efficiency, a low carbon footprint, optimal organic waste management, and community integration. This research found that the German approach applied to Los Ríos Region is technically feasible, but economic and environmental constraints make its application unfeasible. However, modifications to the original concept are proposed as a new BEV concept adapted for southern Chile. Firstly, in urban neighborhoods of cities saturated by air pollution or in rural villages, district heating networks fueled by wood biomass could be effectively applied thanks to the efficiency gained with centralized boilers and their low emissions of fine dust. Similar efficiency could be reached in houses using pellet stoves. Secondly, management of domestic organic waste could supply enough biogas for cooking, allowing energy savings and the replacement of traditional low-performance stoves, reducing the demand for firewood and air pollution. Finally, a specific proposed system for organic waste and water management would allow a better functioning of biodigesters for black waters and organic waste, and constructed wetlands for grey waters. This combination could optimally recover over 80% of the total wastewater and 90% of the total nutrients and provide a new source of energy, in addition to enhancing the biodiversity of the place.

A. Erlwein (✉)
Department of Agriculture Engineer and Soil Science, Faculty of Agrarian Sciences, Universidad Austral de Chile, IIAS-UACh - Campus Isla Teja, Valdivia, Chile
e-mail: aerlwein@uach.cl

Transdisciplinary Centre of Environmental Studies and Sustainable Human Development (CEAM), Universidad Austral de Chile, Valdivia, Chile

Interdisciplinary Centre for Sustainable Development (IZNE), Universität Göttingen, Göttingen, Germany

A. Ch. Braun et al. (eds.), *Eco-Industrial Development as an Industrial Strategy*, Green Energy and Technology, https://doi.org/10.1007/978-3-031-73576-9_8

1 IZNE and the First Bioenergy Village in Germany

Based on the principles of sustainability science (Schmuck et al., 2013), the Interdisciplinary Center for Sustainable Development (IZNE, Interdisziplinäres Zentrum für Nachhaltige Entwicklung) at the University of Göttingen, Germany, developed the first project of an electricity-and-heat autonomous bioenergy village in Germany. The village Jühnde has 800 inhabitants and is located in Lower Saxony (Ruppert et al., 2008). The energy supply is based on a biogas reactor that uses crop silage from the surrounding farmland and manure from livestock. Energy crops (mainly maize silage) provide approximately 90% of the biogas, and manure only 10%. The biogas goes to a central heat-and-power cogeneration unit (CHP) where it is burned in a combustion engine powering an electric generator. The heat released during that process is transferred to a hot water pipeline, which brings thermal energy to 75% of the houses in the village. In winter, additional heat is provided by a central heating facility that burns wood chips. Figure 1 shows the conceptual structure of a bioenergy village.

There are many advantages to these combined heat and power biogas systems (Ruppert et al., 2008; USDA, 2014; WBGU, 2008; Wüste & Schmuck, 2012) even if applied to non-industrialized countries. Some benefits include the following: high energy efficiency through cogeneration of electricity and heat (85%), decentralized structure, autonomy, integration of fermentable biological waste, complete nutrient recycling without the need of additional fertilizers, tested simple-safe-clean commercial technology, high social impact and community strengthening due to its participative characteristics. The biogas produced from energy crops in that area provides two to three times more electricity than the population of Jühnde (800 inhabitants) and

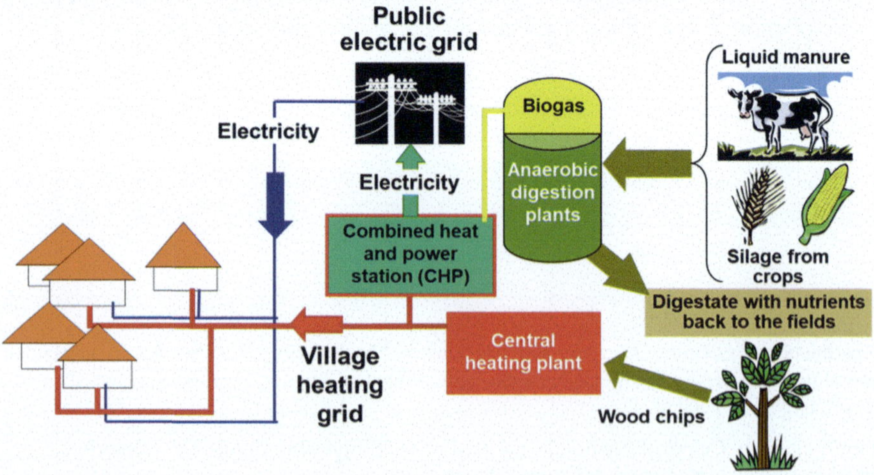

Fig. 1 Heat and electricity production and distribution in a bioenergy village. Modified from Schmuck et al. (2013)

supplies local businesses' needs. A disadvantage of such an approach is primarily the use of land for energy crops, which in the case of Jühnde reaches approximately 300–320 ha, which is over 30% of the agricultural land of the village. Additionally, bioenergy villages need the agreement of many people involved in the system with regard to the production of biomass, the distribution and feed-in system for heat into households, etc. Sometimes such a complex agreement is difficult to achieve.

Given such an interesting model of energy management, it would be highly interesting to know if such approach is transferable to Chile, and especially to Los Ríos Region, which has similarly temperate weather conditions as central Germany. For epistemological reasons that cannot be developed here due to the space limitations of this article, this complex multivariable-transdisciplinary question has no direct answer. However, some technical criteria can be used to assess the feasibility of implementing the German bioenergy village concept for Los Ríos Region.

2 Bioenergy in Rural Areas of Los Ríos Region

Chile has a wide variety of climates due to its north–south configuration and its extraordinary length of almost 5,000 km. Four main climate zones can be found from north to south: desert, Mediterranean (central north, dry), temperate rainy (central south, wet) and Patagonian (south, cold and dry). Los Ríos Region is located in central south Chile, where forests and biomass productivity are high and irrigation is generally not necessary. The main income source for the small rural households in the region comes from animal and firewood production. Most countryside households and rural villages of Los Ríos Region greatly depend on wood biomass for heating, hot water, and cooking. As shown in Fig. 2, there is a direct latitudinal relationship between the availability of forests and the use of firewood, and an inverse relationship regarding the use of liquefied gas for domestic thermic use.

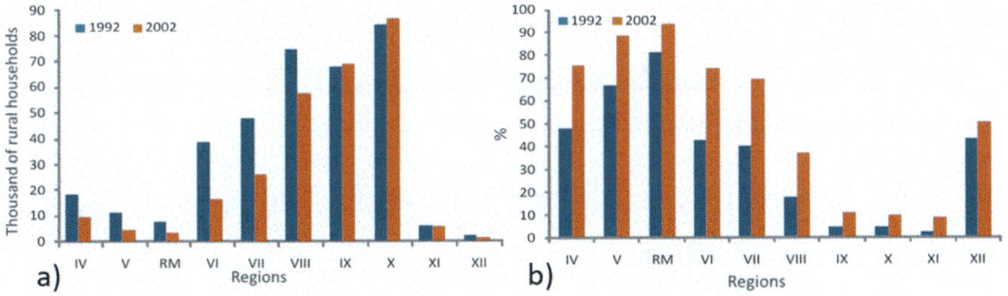

Fig. 2 Main household thermal energy sources in Chile with north–south distribution of regions from left to right, from the censuses of 1992 and 2002 (Reyes & Neira, 2012). Los Ríos Region (named Region XIV in 2007) was contained in Region X until 2007. **a** The number of rural households that consume wood fuels for cooking and water heating. **b** The percentage of households that consume liquefied gas for cooking and heating water

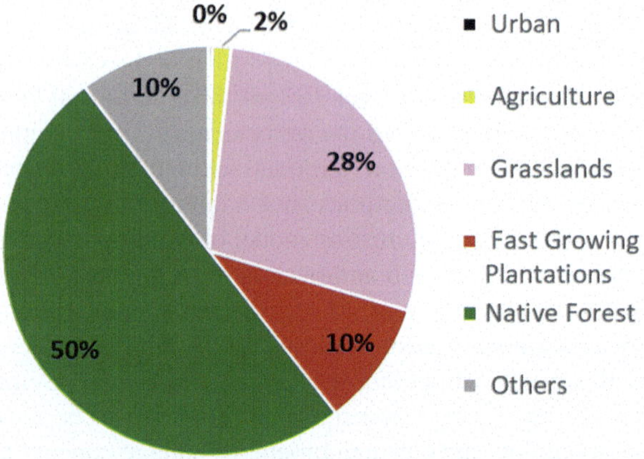

Fig. 3 Distribution of land uses in Los Ríos Region, in percentage

Although the substitution of native forests by fast growing tree plantations of exotic species for the forestry industry was for decades the most important degrading factor of native forests, today firewood extraction is critical (Echeverria et al., 2006). In a non-formalized firewood market fed by a mainly unmanaged forest resource through very low-capacity operators in the sector, firewood extraction today is the principal factor for regional forest degradation and deforestation. From the 15.2 million m^3 of wood annually extracted in Chile from native forests, the portion extracted for firewood is 9.3 million m^3 (AIFBN, 2011). However, southern Chile possesses vast areas of native forests of more than 14 million hectares. More than 6 million hectares are younger, successional forests that can be restored and managed for multiple integral uses and services, such as wood, energy, biodiversity, and water production. As shown in Fig. 3, half of the region is covered by native forests and almost one third by grasslands. With less than half a million inhabitants, the region has abundant bioenergy resources for the population, representing a promising energy future for their development.

3 Testing the German Bioenergy Village (BEV) Concept in Los Ríos Region

3.1 Technical Viability: Modelling the German BEV Concept Adapted to Los Ríos

In order to test the feasibility of the German bioenergy village concept, Los Ríos Region's rural villages were studied in terms of their surrounding land use and biomass availability. As can be seen in Fig. 1, the BEV approach needs biomass (energy crops and slurry) for biogas production and wood residues for extra thermal energy. Therefore, the regional potential for biomass in terms of energy crops (maize silage) and wood were estimated. Regarding energy crops, site-specific biomass potential was modelled using the software BioSTAR (Bauböck, 2013). For the model operation, a GIS-based database that included soils and climate districts of the region was developed, applied, calibrated, and validated for Los Ríos Region (Erlwein-Vicuña, 2016). The potential of sustainable forest biomass was modelled with the *Explorador de Biomasa Forestal* (forestry biomass explorer) developed by Universidad Austral (UACh, 2013). Based on energy management logic, the biomass supply from local land areas depends on the biomass productivity and available areas, and must correspond to the local energy demand of the population for electricity and heat.

To establish this, an area with a 5 km radius from the center of the villages was specified based on Bauböck (2013). This is the maximum distance to avoid economic losses due to the transportation costs for green biomass. As it is more dense and dry, wood allows for distances of up to 50 km for cost-effective transportation (CCA, 2008), so that distance is a less restrictive parameter. This way, a circle with a 5 km radius translates to a total area of 7,850 ha (non-agricultural land included). The population of the villages in the region is variable. For a comparative analysis with the German BEV, a population size of 800 inhabitants per village is assumed (apart from the regional capital of Valdivia, the other cities in the region have an average of 10,000 inhabitants). In Jühnde, the ratio of surface area for energy crops per capita necessary to cover the population's electricity needs is roughly 0.2 ha per person (160 ha for 800 people). Regarding heat, Jühnde has surplus heat in the summertime, but in the winter, heat from wood burning is necessary. 72% of households are connected to the hot water pipeline. Therefore, considering that the maize silage yield in Los Rios region is similar or higher than in Jühnde, and assuming that the rural people in Los Rios region consume the same amount of electricity as a typical Jühnde resident, a village of 800 people requires 160 ha of surface area for energy crops under local conditions. Regarding the modelling, this surface can be satisfied by only 2% of the whole area defined by the 5 km radius. In any case, this would mean an important re-allocation of land use, as currently only 2% of the regional land is used in agriculture. However, agricultural land use has experienced dramatic exponential growth due to climate change and better prices for agricultural products during the last half decade. In Fig. 4, the potential for energy crops and firewood are calculated for every village and presented by assuming the current land use. In Fig. 5 regional villages are shown on a map with both wood and bioenergy potential highlighted. With the combined information provided by the two models, it is possible to state that the majority of the regional villages can satisfy their demand for energy, as most of them have far more than 5% viable bioenergy potential in the surrounding area of 5 km.

Therefore, from a technical point of view, without considering environmental restrictions for bioenergy development in the area, it seems clear that the German concept of a bioenergy village is certainly applicable. However, some other factors exist that may restrict the application of such a concept, which will be discussed next.

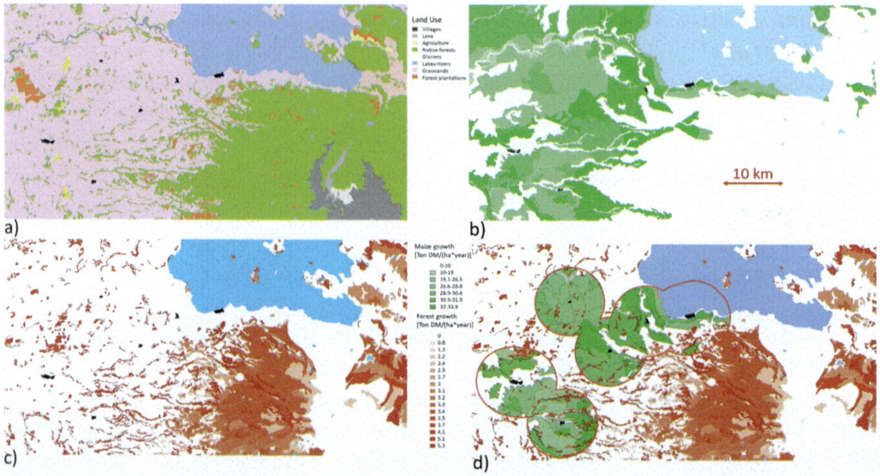

Fig. 4 Biomass potential for several rural villages (in black) in the area surrounding Lake Ranco in Los Ríos region. **a** current land use, **b** maize potential, c sustainable native forest potential, **d** agriculture and forest potential. Agricultural biomass is restricted to a 5 km radius due to transportation cost-effectiveness

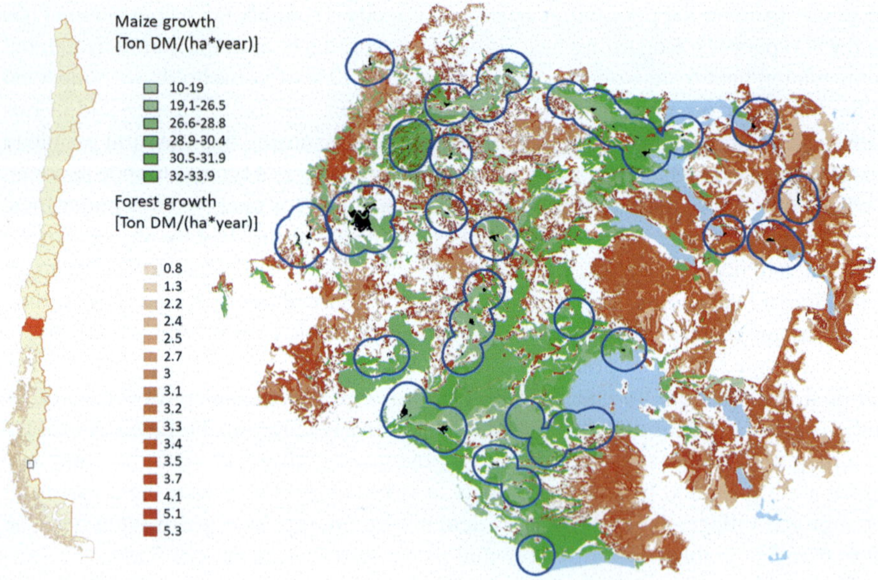

Fig. 5 Los Ríos region biomass potential for maize modeled with BioStar (Erlwein-Vicuña, 2016) and for wood from native forest modeled with *Explorador de Biomasa Forestal* (UACh, 2013). Circles (blue) define areas around cities and villages (black) with a 5 km radius from the center. Chilean map from Wikipedia

3.2 *Sustainable Viability: Critical Economic, Social, and Environmental Parameters for the Transfer of the German BEV Concept to Los Ríos Region*

As mentioned, the results of the modeling show great potential for energy and technical prospects. However, other relevant parameters should be considered that restrict the practical application of the German BEV concept in Los Ríos Region.

Economic restrictions

Since Los Ríos Region is already a bioenergy region (as thermal energy is mainly obtained from firewood), the main reason for adapting the German concept is not to increase the use of bioenergy, but to install technologies and infrastructures to secure/guarantee its more efficient and environmentally-friendly use. In the context of local economic scarcity, that would be to allow households to have access to modern energy in parallel with organic waste management. However, in Latin America an improvement of technologies is restricted by its costs.

Since wood is so abundant in the region, it is also very inexpensive, especially in the countryside where it is produced. Price details are shown in Table 1.

According to the data in Table 1, the conditions for achieving a similar scenario in Los Ríos as in Jühnde are restricted, as it is not economically feasible. In 2008, Jühnde had a 19% higher heat price than in Los Ríos, (even by using 3 times higher thermal efficiency in the calculation), and charged nearly three times as much for electricity, even without the EEG subsidy. Furthermore, it is probable that the potential efficiency of providing heat via a hot water pipeline in Chile would be lower than in Germany, due to cultural and technical factors. Besides, Chile has projected a future subsidy for renewable energies, but this is not possible or available within a short time period.

Additionally, rural villages in Chile have different conditions than German villages:

- The consumption of electricity by the population is very low due to low incomes and the availability of firewood.
- Due to the economic standard of living and the local culture, people use less energy than in urban areas, and definitely less than in Germany.

Cultural differences

Secondly, it is important to consider that the German concepts do not consider local cooking devices, such as the traditional kitchen wood stove, which is culturally the central hearth of the rural household in Chile. There is no cultural adaptation to electric cooking devices, in part because the price of electricity and their related devices are too high. In fact, the abundance of forests is very high in the region and most of the household energy consumption is spent on thermic energy, as shown in Fig. 6. Therefore, wood biomass is the culturally-prioritized base energy source in the region, which is different from the German concept of a bioenergy village.

Table 1 Comparison of energy prices in Germany and Los Ríos Region in €[1]

Heat—Los Ríos	
Price of firewood logs in Los Ríos Region[2] in €/m^3	At source farm: 25 At villages: 33 At cities: 41
Wood density (Kg/m^3)	600
Wood energy content (KWh/kg)	4
Heat price (villages, €/KWh)	0.0137
Traditional wood stoves' thermal efficiency (AIFBN, 2008)	25%
Usable heat price (villages, €/KWh)	0.0548
Heat—Jühnde	
Heat price in Jühnde (€/KWh) (Ruppert et al., 2008)[3]	0.049
Heat net thermal efficiency	75%
Usable heat price (€/KWh)	0.0653
Electricity—Los Ríos	
Farmers' buying price (€/KWh) [4]	0.163
Farmers' selling price (€/KWh)	0.0746
Electricity—Jühnde	
Selling price (€/kWh)	0.205 (without EEG)[5]

[1] 1 € corresponds to 730 Chilean pesos ($CP)
[2] Prices of certified wood. More than 90% of the market is informal with no certification, which implies at least a 19% lower price
[3] The heat in the summer is nearly free (surplus heat), but in wintertime, wood chips have to be burned in separate boilers
[4] Selling prices for electricity used are between $CP 76 and 180 depending on the load factor given by the national energy authority according to the temporal abundance of energy sources
[5] EEG = Erneuerbares Energiengesetz = Renewable Energy Act. Linked to subsidies

It is also important to consider that in general, the educational level in rural areas is lower than in urban areas of the region, and even more so in comparison to Germany. That makes any technological transition more complex, as there is neither much experience in new technologies, nor technical capacity to implement them. Finally, there is limited interest in ecology and little environmental consciousness in rural areas.

Food-to-energy opposition

Finally, there is worldwide but also local public opposition toward the use of agricultural land for energy production (FAO, 2008; Giampietro & Mayumi, 2009; Tomes et al., 2011). Although the concept of bioenergy in Chile is still in a very early development stage, the eruption of fast-growing plantations of exotic trees like Eucalyptus over the previous decades has set a bad precedent, especially as these plantations were originally established through environmental arguments such as providing erosion control and renewable energy production (through carbon sequestration). Over time,

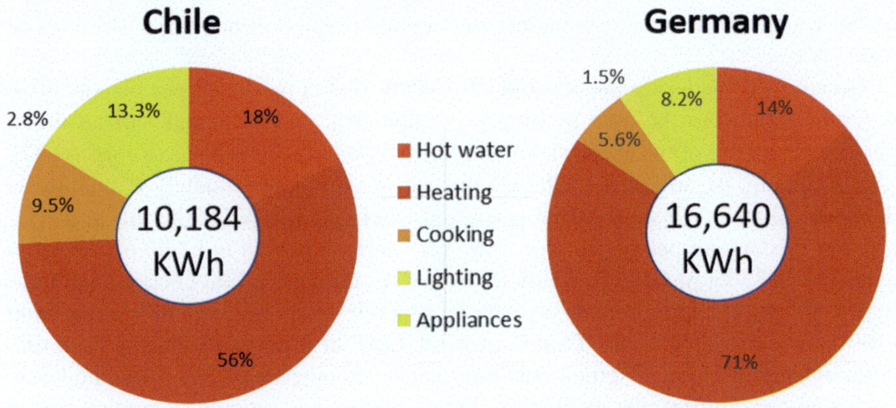

Fig. 6 National average per household of final energy consumption for Chile (CDT, 2010) and for Germany (Statistisches Bundesamt, 2018) in KWh per family per year. Yellowish colors: electricity; Reddish colors: thermal energy. In general terms, an average German household comprises 2 people, whereas the Chilean household 4 people. This way, the consumption of energy per household member in Germany is about 3 times larger than in Chile

the Chilean forestry background of fast growing trees has shown several weaknesses (Erlwein et al., 2007; Little et al., 2009). Additionally, global concerns over the costs and availability of food versus gasoline have influenced Chilean society more than concerns over ecological consciousness. Citizen opposition to bioenergy concepts may in fact be a valid reflection of the country's social inequality, which only allows a powerful minority of firms to access the energy business.

3.3 *Searching for an Alternative Local BEV Concept*

Considering these sociopolitical complications, the direct application of the German concept of a bioenergy village to Los Ríos Region seems to not be feasible at present. However, three main modifications to the original German BEV approach may have better prospects in the region. The necessary modifications are developed in the following subsections.

3.3.1 Variation 1: Replacing Energy Crops with Domestic Organic Waste

As was analyzed in the previous section, there are meaningful arguments to avoid energy crops, both from an ecological and social point of view. Therefore, one way to avoid such disadvantages but maintain the advantages of the bioenergy approach is to replace the biomass provided by the energy crops with any local source of organic waste, which in urban areas is mainly the organic fraction of municipal solid waste (OMSW).

However, considering the amount of OMSW that is produced per person, such a system would not be able to supply the domestic energy demand of an urban person, regardless of the urban size: village, town, or city. Indeed, according to CDT (2010, see Fig. 6), an average Chilean household with approximately 4 inhabitants demands 10.184 kWh yearly. That is approximately 2.550 kWh of thermic and electric energy demand per person per year. Yet yearly, a person in Chile produces an average of 233 kg of OMSW (AMRR, 2015). Assuming the best energy efficiency, that would yield approximately a total of 117 kWh yearly in terms of cogeneration (40% electricity, 45% heat). **That is around 4.6% of a person's energy demand.** Although certain human settlements have access to other sources of organic waste (i.e. from agriculture), the availability, type, quality, and amount of organic waste varies significantly among different places, so that no generalization is possible. This way, the concept of an OMSW-based BEV should be rejected if household energy self-sufficiency is the goal. If implemented, such a setting should ideally entail a public facility where the bioenergy is used publicly, like energizing public buildings or public transportation. Citizens at home should separate the OMSW, and therefore they should share the benefits of such management in order to achieve commitment and support for the system. Another way to engage people could be the development of an incentive-based pricing system.

3.3.2 Variation 2: Replacing the Biogas Cogeneration of Heat and Electricity with Wood Resources for Heat and Biogas for Cooking

With variation 2, technological complexity, cost, and land use can be significantly reduced in comparison with the original German BEV, and the energy efficiency of traditional Chilean systems can be significantly improved.

A. Heat from wood-generated thermal energy

Thermal energy can be produced from wood resources instead of from biogas cogeneration. Wood is a local, abundant, cheap, and near carbon neutral source. Wood thermal energy can be obtained with different levels of efficiency, from district heating, pellet stoves, to less efficient devices. Figure 7 shows the thermal efficiencies of different heating devices currently used in southern Chile.

(a) *Urban neighborhood with district heating network*

This variation simply means to install a central heating system and district network similar to the original German BEV, but produce heat from high-efficiency communitarian wood boilers instead of from biogas cogeneration. Heating networks can be more expensive and complicated than the traditional, individual system in some respects. However, a lot of efficiency can be gained in the process, which compensates for the higher implementation costs, especially in cities where wood is more

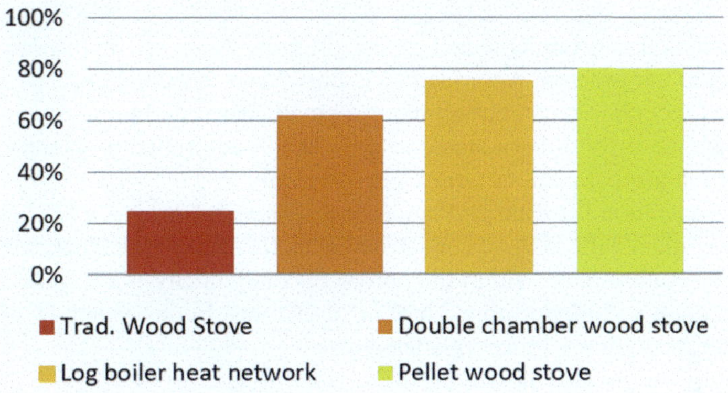

Fig. 7 Thermal efficiencies of different heating and cooking devices. The efficiency of double-chamber wood stoves might vary significantly according to their use in each household

expensive. Moreover, wood burning is critical in cities with air pollution problems. A central heating system can radically reduce noxious gas emissions, and because of its automation, it implies comfortable operation for homeowners. In fact, in Chile some new projects using this concept are slowly beginning to appear. Such initiatives are shown in the following box.

Cases of district heat networks in Chile

A. One case that has already been built is the residential condominium project Frankfurt in the neighborhood north zone of Temuco city. It is the first locality that has set up a district heat network for the distribution of geothermal heat in the country. The project is unique in its type, but faces many problems with Chilean regulations: http://www.frankfurt.cl/wp-content/uploads/2013/08/geotermica.pdf

B. The Pilot Project of district heating in Coyhaique (Proyecto Piloto De Calefaccion Distrital En Coyhaique), in the neighboring region to the south, developed for the Ministry of Energy (Renewable Energies Centre) and the Ministry of the Environment. Given the climatic conditions of southern Chile and the abundance of forests, the government of Chile has decided to study the feasibility of developing a heating plant district based on wood to deliver heat and hot water to new social service housing and middle-class houses: http://sig.goreaysen.cl/pedze/161_Calefaccioon_Distrital_Coyhaique.pdf

   http://www.ebpchile.cl/servicios/recursos-naturales-y-cambio-climatico/aktuell/sistema-de-calefaccion-distrital-a-base-de-biomasa.html

C. Biomass Center in Coyhaique, where development of the next generation of products and services for heating and biomass energy is being researched and implemented, realized by the Agrarian Innovation Fund (Fondo Innovación Agraria, FIA): http://www.ebpchile.cl/servicios/recursos-naturales-y-cambio-climatico/aktuell/centro-de-biomasa-en-coyhaique-desarrollo-de-la-siguiente-generacion-de-productos-y-servicios-de-biomasa-para-calefaccion-y-energia.html

D. Co-generation Bio-Energy Plant *Bioenergía de Los Ríos*. A 9 MW electric power plant with 35 t/h of condensed steam for a district heat network from a gasification plant within the border of Valdivia city is being designed. The plant would consume 2,500 m^3/d of forest biomass and 71 m^3/h of water: http://www.bioenergiadelosrios.cl/

According to the CCA (2008), it is important to note that price signals are not only inaccurate and misleading, but also prevent or limit designing sound public policy for clean energy. Today's prices exclude concomitants in regard to the use of firewood such as deteriorating air quality in large cities or a loss of environmental services, plus the uncertainty about wood's caloric content, moisture content, physical characteristics, dispersion qualities, origins, among other concerns. The absence of measures to clarify real prices and standardize the quality of firewood makes resource management difficult and reduces the chance of incorporating innovative and environmentally sustainable technological options like district heating, even though local conditions suggest that they are promising for the region.

(b) *Pellet Stoves*

Another alternative to a district heating system is the technologies that use pellets, which reach high efficiency even in single houses (see Fig. 7). This resource can achieve a similar efficiency to a district heating system. The combustion efficiency is only a little lower than an industrial wood boiler, but it does not have the loss of energy of a heat network. Although the pellet furnace and the pellets supplied are relatively expensive in caloric terms, it can be compensated by saving the investment of a heat network. Additionally, the individual devices do not suffer the constraints of a collective management system.

(c) *Double-chamber wood stoves*

Double-chamber wood stoves are standard and presently the most traded heating device. They are not so expensive and are commonly widespread in southern Chile. It is important to consider that the efficiency of double-chamber wood stoves can be significantly low, especially if the wood is not dry or the device is not properly adjusted. This can be a frequent situation when the population has not been properly introduced to its use. Because of the latter, this technology is less recommended for urban contexts where air quality is an important issue.

Fig. 8 Existing heating devices in Los Ríos Region: **a** traditional wood stove (cooking, heating, and hot water), **b** double-chamber wood stove (with or without heat exchanger for hot water), **c** heat exchanger that can be installed on either type of stove to heat water from the residual heat of the exhaust gas

Biogas/biomethane network

A complement to obtaining heat from wood-generated thermal energy could be the implementation of a gas net in the village specifically to power cooking devices.[1] That would allow for the replacement of the traditional kitchen wood stove in rural villages with biogas/biomethane stoves. The gas can be provided by the fermentation of organic residues complemented with biomass. Considering the efficiencies given in Fig. 7, some of the reasons to adapt such a combination are presented next:

- This model would also work for single countryside houses.
- Traditional wood stoves have low efficiency. Most of the present-day rural kitchens use traditional wood stoves.
- Firewood and other fuels are increasing in price as their availability decreases and population increases. The costs for most heating fuels has increased in the last few decades (Reyes & Neira, 2012).
- Gas stoves are the most common devices in urban Chile. LPG and city gas (natural gas networks) are industry standards. Many cities and villages in Chile use a gas network infrastructure for cooking, heating water, and heaters, so there are firms and technical services available. General knowledge of this technology exists.
- Using biogas cookers would allow traditional wood stoves to be replaced by double-chamber wood stoves as sources of heat. Double-chamber wood stoves have more than double the efficiency when compared to traditional wood stoves. Additionally, the stoves can be complemented with heat exchangers that enable the heating of water. In the region, there are a variety of small firms that offer the installation of heat exchangers on the chimney tube where exhaust gasses are released, recovering part of the residual heat from combustion. Stoves and the heat exchanger are shown in Fig. 8.

[1] In this concept, hot water and other thermic consumptions are obtained by wood-generated thermal energy.

- Doubling the efficiency means that half of the firewood is required. Additionally, cooking with biogas means even lower firewood demand. This can be accomplished in a rural house by replacing traditional wood stoves with a double-chamber wood stove or even a more efficient device for heating and hot water. The savings can reach up to 0.8 ha of harvested forest per family annually. Figure 11 shows the effect of the proposed replacement on the reduction in firewood demand, and therefore forest surface area necessary to satisfy the heat demand.
- As mentioned before, an alternative to the previous technology would be pellet boilers. The efficiency improvement would be even higher, but the device and pellets are more expensive, which is a restriction in rural areas where purchasing power is usually low and wood logs are abundant in the countryside.
- Electricity should preferably be obtained through other renewables. According to the German Academy of Science *Leopoldina* (Anton & Steinicke, 2012), "*ideally the local electricity supply should be covered by other renewable sources, as bioenergy can produce fuels that are otherwise scarce, so that the conversion of biomass should be concentrated on biofuels for such uses as heavy vehicles, airplanes, and large ships that most likely cannot be powered by electricity.*"
- Biogas is compatible with current agriculture waste management in Los Ríos Region. *Leopoldina* also suggests that energy crops should be used for biogas production only as far as this is needed for the stabilization and optimization of the overall process of utilizing agricultural wastes and for stabilizing fluctuating energy demand. This is precisely the function that energy crops play in the proposed concept.
- Biogas is also compatible with grasslands. According to the large extension of grasslands in Los Rios Region, there is clear bioenergy potential for grass, although it is more land demanding than energy crops, as grass has a lower biomass yield per hectare. This is a limiting factor, since harvesting grass needs more distance to be traveled and surface to be covered to gather the same amount of biomass from energy crops. As fresh, green biomass is both voluminous and humid (high water content), its transport is rather expensive per unit of dry mass, limiting the distances between the field and the biodigestor. In the case of the bioenergy village in Jühnde, the critical distance is a radius of about 5–7.5 km around the biodigestor (Bauböck, 2013). However, the use of perennial grasslands for bioenergy production seems to have some relevant advantages, such as the following:
  - The GHG reduction is much higher than with energy crops because (WBGU, 2008):

    Less fossil energy is needed (low inputs, no tillage)
    No indirect land use change emissions are produced[2]
    High soil carbon sequestration (no tillage)

  - Low-cost biomass production

[2] When iLUC are included, grassland GHG emissions can be 1/3 compared to energy crops.

- Higher biodiversity of plants and animals
- Compatibility with land management
- Lower landscape impact.

Regarding land management, there are important land areas available in the region that are "sub-utilized," which means traditionally "left to rest ('resago')," a type of set aside between harvests. These land areas are usually covered with a species of grass and bushes of low feed value for animal production that could be used for bioenergy production as a part of an overall land use management plan.

- Small amounts of biomass from crops or grass are needed. In this proposed variation, hot water is heated with heat exchangers so that the heat energy comes mainly from wood sources. Although there are plentiful biomass resources available to satisfy energy demands (as shown in Fig. 5), cooking represents only 9.5% of total household energy demand (in Fig. 6), and therefore, not much biomass is needed for biogas supply. Moreover, organic residues from the village can be used, further reducing the need for energy crop land leading to less environmental impacts. When domestic organic waste is used,[3] a very small amount of extra biomass is needed to satisfy the cooking demand. Estimates are shown in Table 2 based on data from Erlwein and Sotomayor (2016).

As can be observed, for an 800-inhabitant rural village only 5.1 ha of maize are needed for self-sufficient cooking with gas. This is a very significant reduction in comparison with the 150 ha mentioned for Jühnde, although the necessary hectares in Jühnde rely on electricity production, not heat. In any case, this replacement might have a positive impact on the standard of living as well, as the simplicity and time of cooking with gas does not require the long waiting period of warming up a wood stove, nor the constant feeding process with wood to avoid letting the fire go out. Unlike with firewood stoves (even the sophisticated ones), a biogas cooker is detached from the warming-cooling cycle, being able to be turned on and off quickly, cleanly, and efficiently.

To implement such a system in human settlements, a village gas network would be required, which implies additional costs. There are current studies for methane gas networks (city gas) in the region that are compatible with biomethane, but no projects have been implemented yet. In any case, the emission of harmful particles (PM2.5) and associated noxious organic and inorganic substances from burning wood would decrease tremendously. The air quality would improve immensely if biogas would replace wood. Burning gas is the cleanest option to produce bioenergy. Split wood logs have the worst environmental performance, followed by wood chips and to a lower extent, by pellets (see Seidel et al., 2013).

It is also possible to apply this concept without the need of a village gas network, but this would require the transformation of biogas into biomethane, which could be delivered in gas cylinders as is done today with LPG. But this may be a more complex and expensive solution because biomethane is not as easy to liquefy as LPG and considerable energy is necessary to transform biogas into pure methane.

[3] Surrounding rural organic waste is probably also available but is not considered in the calculations.

Table 2 Estimation of biomass needed for cooking with biogas by a self-sufficient family of four persons, and savings of energy as a substitute, based on Erlwein and Sotomayor (2016)

Biogas demand for cooking and savings of other fuels		
Total biogas (m^3/(family * year))	241	
Savings of liquefied gas (kg/year)	104	
Savings of firewood (m^3/year)	8.1	
Biogas production from available biomass		
A. Family organic waste		
Domestic organic waste (kg FM/(person * day))	0.5	
Human waste (kg FM/(person * day))	0.3	
Total biogas from waste (m^3/(family * year))	150.4	
B. Fresh biomass (grass)		
Average grassland biomass (ton DM/(ha * year))	7.0	
Fresh biomass (30% DM) (kg/m^2)	2.3	
Biogas/(kg fresh biomass) (m^3/kg)	0.18	
Biomass area necessary for self-sufficient biogas cooking	Biomass to be used	
Required area for an average family (m^2)	Grass	
Without household waste	574	
With household waste	**216**	
Required area for an 800-inhabitant village (hectares)	Grass	Maize silage
Without household waste	45.9	13.5
With household waste	17.2	**5.1**
Required area for a 10,000-inhabitant city (hectares)	Grass	Maize silage
Without household waste	574	169
With household waste	216	**63**

Also, the concept of a village gas network is compatible with the production of biomethane from lignocellulosic gasification-methanation. Considering the increasing problem of low air quality in Valdivia, such a concept might be attractive not only for cooking but for heating as well, radically lowering harmful emissions.

This concept could also be applied to countryside houses and farms if the location had a biodigester. Although this would conceptually imply positive impacts previously explained, family-scale biodigesters bear certain risks. First, every family should know how to manage a biodigester, which demands certain technical and theoretical skills. Secondly, the efficiency of domestic biodigesters is commonly lower than big-automatized ones due to the higher technology and control of the digestion parameters. Finally, materials and the quality of construction of domestic digesters are very important elements to prevent the emission of the greenhouse gas methane. The assumption that biogas reduces a household's carbon footprint by 90% is in relation to fossil fuels, so that a loss of only 4.5% of the methane produced in a family-scale biodigester could entail a total loss of the attempted reduction of the greenhouse gas footprint, since methane has a global warming potential 20–30 times stronger than carbon dioxide (CO_2).

Finally, according to previous calculations (Erlwein-Vicuña, 2016), the proposed stove replacement from wood stoves to double-chamber wood stoves implies an annual savings of 0.78 ha per family for wood exploitation, which means a 62% reduction in wood demand and forest land needed. That amount can even be increased by new technology reaching higher efficiency such as wood boilers or pellet stoves (69% reduction). Such a reduction can definitely diminish the pressure on native forests and their corresponding degradation. The BEV approach proposed in this analysis for heating is based on wood biomass, which is a locally-available, close to net-zero emissions fuel.

3.3.3 Variation 3: Biodigestor and Constructed Wetland for Separated Wastewater

The previous BEV variations can be significantly enhanced, with a high degree of synergy, if the houses alone or in a villages have a roof rainwater catchment system and a wastewater separation system, so that the black water can be treated in a biodigester and the grey water in a constructed wetland. Figure 9 shows the proposed concept of the integrated management of thermal energy, wastewater, and organic waste. Calculations were made with MS Excel and the dynamic systems model Stella© 7.0.1 (see diagram in Fig. 10).

<u>Grey-black wastewater separation system</u>

The proposed wastewater separation system considers two water circuits and the separation of the kitchen sink wastewater, as follows:

1. Grey waters that include water from showers, washbasin sink, washing machine and second chamber (detergent rinse) for the kitchen sink are directed to a constructed wetland.
2. Concentrated black waters that include the fecal content of urine-diverting toilets and the first chamber of the kitchen sink (soaking dishes with leftover food for washing) and/or the whole dishwasher wastewater are directed to a biodigester. In addition, domestic solid organic waste and some eventual extra garden biomass, after passing through a grinder, may also be transferred to the biogas reactor.

Metaphorically, this separation of circuits can be imagined as the separation between the liquid and solid body wastes of a person. The constructed wetland, as the kidneys, processes very low organic matter, deals with higher dilution, and uses physical processes for depuration. Biodigesters on the other hand, as the large intestine, deal with very high organic loads and metabolic processes of depuration mediated mainly by associated microorganisms. Through the proposed wastewater management system concept, significant synergies can be achieved, as it is detailed next:

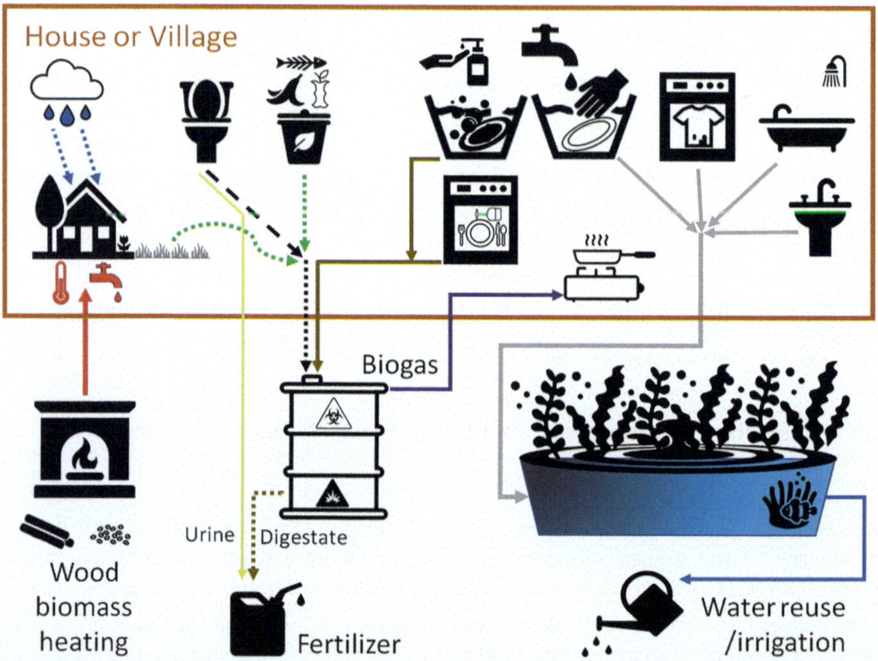

Fig. 9 Proposed concept of BEV for southern Chile: integration of solid and liquid organic waste management considering water efficiency and energy production. The wastewater separation system treats black and grey waters in a biodigester and a constructed wetland respectively. Continuous line arrows denote pipelines

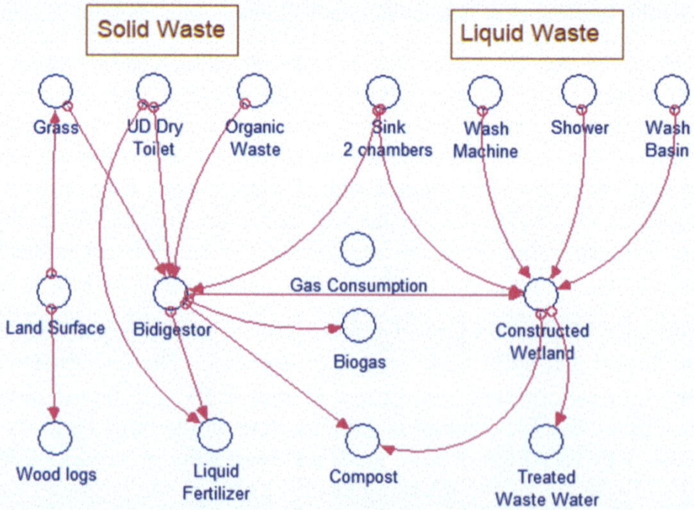

Fig. 10 Modeling diagram for the proposed concept of BEV in software Stella©

Biodigesters and Constructed Wetlands

- Biodigesters are suited for black water treatment, as they require relatively high organic loads, avoid methane emissions, and produce energy as biogas. They mineralize nutrients and significantly reduce pathogens. However, organic matter in black waters is generally too diluted, restricting their technical and economic viability.
- Constructed wetlands, on the contrary, are much cheaper to operate per m^3 of treated wastewater, and they work well with low organic loads, favoring biodiversity development. These wetlands are ideal for tertiary treatment, which enables more water recycling. In this way, constructed wetlands show synergic performance with simultaneous improvements in water quality, biodiversity richness, and landscape aesthetics. They use low energy and no chemicals in the process, being a nature/photosynthesis-based solution with low carbon emissions.
- In terms of constructed wetland-biodigestor interactions, the biomass of constructed wetlands, with low lignin content, is a good substrate for biodigestion with relevant energetic potential (Avellan et al., 2017; Hassan et al., 2021; Kochi et al., 2020). Extracting some biomass as part of the wetland maintenance to feed a biodigestor is also an effective way to extract nutrients from the wetland system, increasing its removal efficiency, producing biogas in the process and increasing the nutrient recovery into the liquid fertilizer (digestate). In the opposite direction, treating the liquid fraction of digestate in the constructed wetland is not intended as a way of normal functioning in the concept presented here, as it searches for no pathogens and minimum nutrients in grey water. However, if the liquid fraction of digestate exceeds the accumulation capacity, the constructed wetland can be a palliative treatment option. Evidence shows that biodigestion before the constructed wetland step can improve the treatment, avoid clogging, and reduce the wetland surface required (Álvarez et al., 2008; Barros et al., 2008).

Urine-diverted toilets (UD toilets)

Diverting urine can significantly achieve positive impacts on human excreta and wastewater management. An average person flushes the toilet approximately 6 times a day, but only one carries feces. Using UD toilets presents several synergic advantages:

- It implies an approximate reduction of 5/6 of the water needed for the toilet, and 1/3 of the overall water demand of one person. Therefore, it proportionally reduces the wastewater generation.
- It concentrates the black water in terms of organic matter.
- Diverting urine is an important issue. Roughly 70% of the nutrients that a person's body excretes come from urine. Unlike organic matter, nutrients are very difficult to remove from water once diluted, demanding larger treatment systems that frequently use chemical reagents. On the other hand, urine produces no biogas and can increase the ammonia volatilization in the digester.
- Toilet flushing water is too cold for the biodigestor. If few or no toilet flushes are made, and the kitchen sink/ food chamber is used in the biodigestor (the thermal

energy of the kitchen waste water would add heat) the biodigestor can be kept warm, which is a requirement for its proper functioning.

Here, an important distinction must be made for the concept presented. There are two kinds of UD toilets: urine-diverted dry (composting) (UDD toilet) or urine-diverted flush (UDF toilet). A UDD toilet does not need water, which implies less water demand, a higher concentration of black waters, less digestate production (relevant for store & application), and no cold water to the biodigestor (higher temperature). Also, the system is more flexible, as without a biodigester, substrate can simply be composted. On the contrary, a UDF toilet allows no feces manipulation and less odors or vectors, a much simpler biodigestor load system (through a pipeline), and a system more similar to conventional toilets (less behavioral change needed). In such toilets, just one flush a day (instead of 6) per person is feasible. In eco-toilets, one flush uses 6 L of water.

Separation of chambers of the kitchen sink

- As can be observed in the daily life of many homes, the first chamber is commonly used for physically removing food residue, washing the dishes—hydration of the organic matter, and for the application of detergent. The second chamber can then be used for rinsing the detergent. This way, the performance of the whole system is enhanced through the separation of the wastewater of the two chambers, assuming that:

  a. More than 80% of the organic matter goes in the first chamber
  b. More than 80% of the water volume goes in the second chamber.

Overall synergies of the separation-concentration system

- Through this system, black waters can attain significantly lower volume and inversely proportional higher solids concentration. Such volume reduction implies a proportional reduction in the size of the biodigestor, digestate storing tanks, and digestate transport to field.
- Additionally, this solid concentration in black waters significantly enhances biodigestor performance. Optimal wet biodigestion requires a 10% of dry mass concentration, which is affordable after this concentration process.
- After digestion treatment, digestate becomes a concentrated liquid fertilizer that can be stored and then used directly on soil together with collected urine, allowing high nutrient recycling and avoiding the elimination of this wastewater into sewerage, subsoil, or watersheds.
- In parallel, the opposite happens to grey water: significantly lower organic matter and higher volume can be afforded. Additionally, practically all the pathogen content is absent due to the separation system, enabling more recycling uses. Such reduction in organic load allows a smaller, simpler, cleaner, and safer water treatment system, enabling a good performance from the constructed wetland and an increase in biodiversity.

- Although in this approach there are no heat exchangers, it is thought that thermal energy from wastewaters (from kitchen, shower, etc.) is used to enhance the biological process, either in the biodigestor or in the constructed wetland.

It is important to note that a conventional house with a good wastewater treatment system can recycle the treated water for flushing the toilet, saving a good 34% of the total water demand. However, there are not many other uses that can be implemented with this recycled water, especially if there is the chance of pathogens in it. In any case, such systems demand an additional water circuit for the toilets, which also request specific practices due to its remaining pathogenic content. Additionally, in such systems black waters are diluted more than 10 times in comparison to the concept presented here, making a well-functioning biodigestor very improbable.

3.3.4 Overhaul Performance Assessment

Regarding the energy performance of the system, figures from Sect. 3.3.2 were used, using a pellet stove as the new wood combustion source compared to a house with a traditional wood stove for heating, hot water, and cooking.

Regarding the system's water performance, for rainwater calculations, the monthly rainfall of the region was considered, differentiating the 3 driest months (with an average of 85.7 mm/month) from the other 9 months (with an average of 237.6 mm/month). For rainwater harvesting, an average house roof surface of 25 m^2/person was used. For water storage, 2.5 m^3/person was considered, which for a conventional person would be enough to supply 1/3 of the monthly water demand, but for the concept presented here, would supply half of the monthly water demand during the 3 driest months of the year, and the entire water demand throughout the rest of the year.

For modeling organic waste flows, the parameters used were based on Castillo-Avalos and Rovira (2013), Franco (2007), Spångberg et al. (2014) and Edwin et al. (2014). Regarding the sink chamber's specifications, for the first sink chamber (food removal) 80% of the whole organic matter and 20% of the overall volume of the sink was used. For the second sink chamber (rinse, detergent removal), the remaining amounts were used: 20% and 80%, respectively. For washing machines, a water load of 49 L utilized once a week per person was used. For the dishwasher a water load of 13.5 L utilized every two days was used.

For the calculation, a combination of the cheapest devices was considered, which is a UDD toilet plus sink chambers separation. However, possible alternatives were calculated. A similar performance, but more comfortable and expensive scheme would be to use a UDF toilet plus a dishwasher. The optimal performance, in terms of water use and organic matter concentration, would be a UDD toilet plus a dishwasher. In this last scheme a tank of 2,500 L would be enough to store the digestate that a person produces in one year. Parameters used are detailed in Table 3.

In Table 3 it can be noticed that using the UDD toilet plus sink/food chamber for the biodigestor allows it to reach 5.6% DM in the reactor, instead of 0.44% of

Table 3 Daily flows of organic waste and wastewater of an average Chilean. Comparison between a conventional management and the proposed BEV concept

	Volume (L)	Total solids (kg DM)	Concentration [weight-%]
Conventional			
Solid waste			
Domestic organic waste (A)	–	0.28	–
Liquid waste			
Conventional flush toilet	74	0.110	0.15%
Kitchen sink, chamber food (B)	12	0.048	0.40%
Kitchen sink, chamber rinse (C)	48	0.012	0.03%
Washing machine (D)	7	0.014	0.20%
Shower (E)	75	0.015	0.02%
Wash basin sink (F)	19	0.0095	0.05%
Alternatives			
Dishwasher (G)	6.8	0.060	0.89%
Total conventional wastewater	235[a]	0.21	0.09%
BEV proposed concept			
Solid waste			
A		0.28	
Grass cuts[b]	–	0.24	–
UDD toilet (H)	–	0.10	–
Liquid waste (B + C + D + E + F)	161	0.099	0.06%
Alternatives			
UDF toilet (I)	6	0.100	1.67%
TOTAL BEV wastewater (A + B + C + D + E + F + H)	161[c]	0.71	0.44%
Alternatives			
A + D + E + F + G + H	108	0.71	0.66%
A + D + E + F + G + I	114	0.71	0.63%
A + B + C + D + E + F + I	167	0.71	0.43%
BEV flow to biodigestor (B + H)[d]	**12**	0.66	**5.6%**
Alternatives			
G + H	**6.8**	0.66	**9.8%**
G + I	12.8	0.66	5.2%
B + I	18	0.66	3.7%
BEV flow to constructed wetland (C + D + E + F)	**149**	0.051	**0.034%**

(continued)

Table 3 (continued)

	Volume (L)	Total solids (kg DM)	Concentration [weight-%]
Alternatives			
D + E + F	**121**	0.051	**0.042%**
BEV final (only grey, treated) wastewater	149	0.010	0.007%

[a] Also corresponds to the water demand
[b] 0.75 kg of fresh grass is added to fulfill the daily biogas demand for cooking, from the surface calculated in Table 2 (216 m^2 for the whole year) and assuming a grassland yield of 4 ton of DM/ha
[c] Also corresponds to the water demand
[d] This flow ends as liquid fertilizer (digestate)

a conventional wastewater. Also, the effect of UDD instead of flushing toilet avoids a 6.2 times higher dilution in black water, which allows a volume reduction of the biodigestor, digestate storing tanks, and digestate amount transported to field by a factor of 6.

For modeling the treatment performance, a constructed wetland with 80% removal of organic matter is assumed (Oteng-peprah & Nanne, 2018). Regarding nutrient recycling, as the wetland biomass is not considered for biodigestion in the calculations due to its low quantity, a removal of only 50% in nutrient content after the wetland was considered. Nevertheless, it has to be noticed that as part of the treatment process, a constructed wetland can retain up to 90% of the nutrients, especially in the first years of operation. Therefore, harvesting and feeding the biodigestor with that wetland biomass can significantly increase the amount of nutrients recycled, allowing it to exceed 90% nutrient recycling in this system. In any case, it was also assumed that irrigation with treated grey water, as a form of nutrient recycling, is realized only during the 3 driest months of the year, then recycling ¼ of the nutrients of the treated grey water. The results of the modeling are shown in Fig. 11. Recycling is expressed as a reduction in nutrient loss from all the organic waste in comparison with a conventional household, that to the greatest extent eliminates solid organic waste from landfills and liquid organic waste from the sewer, so that the nutrients are currently not recycled.

As the separation process reduces the amount of total solids that go into the wetland up to 76%, in comparison to a conventional domestic wastewater, the required size for treatment can be significantly reduced, especially if practically no pathogens are present in the grey water. However, as a biological process, the treatment does follow a non-linear dynamic, so that the wetland size cannot be reduced in the same proportion (76%). A range between 1/3 and 1/2 of the conventional size of a constructed wetland can be considered for design purposes. For the evaluation of the performance of a constructed wetland with the proposed wastewater separation system, a constructed wetland was designed and implemented for a house of 4 people according to principles of ecological design (Todd & Josephson, 1996). Figure 12 shows some details of the constructed wetland.

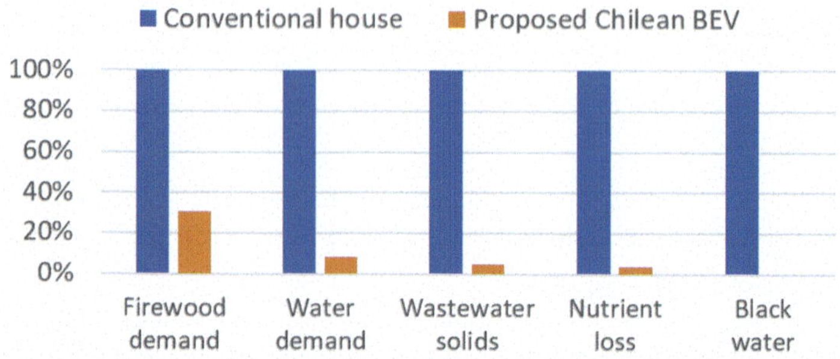

Fig. 11 Performance of some parameters of the proposed BEV approach for southern Chile, expressed as percentages related to a current average household in the region

Fig. 12 Experimental constructed wetland (for a house of 4 people) for evaluating the quality and treatability of the grey wastewater separation system. *Left*: The 3-pond wetland, a view of the two separated wastewater pipe circuits, and a visual comparison between the inflow and outflow of grey wastewater from the wetland. *Right:* Biodiversity attracted by and developed in the wetland. Bottom right: frog tadpoles preying on mosquito larvae as part of the self-organized development of the food web complexity of this ecosystem

Another relevant positive aspect of the proposed BEV approach is its flexibility in terms of maintaining good performance even if some components are not implemented:

- If there is no district heating, similar thermal efficiency can be reached through pellet boilers on individual houses, or even with second chamber woodstoves with lower efficiency.

- In strict terms, there is no mandatory need for composting in this system: as all the organic solid waste goes to the biodigestor, the whole digestate can be disposed of directly in the field as nutrient-rich sludge, a semi-liquid fertilizer like slurry, which includes the solid fraction.
- If there is no biodigestor, in the case of UDD toilet, the whole solid organic waste (including feces) can be simply composted, reaching almost the same nutrient recovery and allowing the relevant water savings, plus avoiding pathogens in the wastewater.
- If there is no use of UD toilets, a separated wastewater system is a relevant improvement anyway. When grey water is separated from sewage, the amount of black water is considerably reduced with a proportional increase in its dry matter concentration. That can be very relevant for rural septic tanks, or for water treatment plants in the case of more urban towns, or even allowing the implementation of municipal biodigesters as their operating conditions improve. When diverting urine systems are included, urine can be recovered, and the number of toilet rinses and the volume of black water decrease by a factor of 6 (Spångberg et al., 2014).

The reduction in emissions of greenhouse gasses is not calculated in this work. However, important reductions can be made in terms of savings on municipal water treatment systems, methane emissions in (locally open) municipal landfills from domestic organic waste, and in the case of urban houses using gas stoves, the use of biogas instead. As a reference, the daily 710 g of dry organic matter from wastes feeding the biodigestor (Table 3) should yield a range between 200 and 350 L of biogas for a person daily, depending on the optimization of the parameters of the process. According to Table 2, a person demands 165 L of biogas per day (241 m^3 yearly for a family of four). As biogas is a very versatile fuel, its surplus can be used for many purposes, like heating, pumping or lighting, among others.

Some negative aspects of the proposed BEV approach are:

- It depends on external wood resources for satisfying thermal energy demand
- It does not produce electricity
- It implies more complex/unpleasant waste management, as manipulation of human excreta (in the case of the UDD toilet)
- It implies additional work for household chores: dry toilet, biodigestor and constructed wetland operation.

Finally, as the Chilean BEV approach includes different components for the system proposed, the appropriateness and feasibility of those components may change depending on the context. A referential qualitative assessment of the technical feasibility of the different components is presented in Table 4. The main criteria for its elaboration were access to agricultural land for fertilizer disposal, sanitary conditions for organic waste management, air quality relevance, complexity of collective management, technical and economic restrictions, among others.

The concepts presented here share many aspects with the work developed in Germany by Friedrich et al. (2020), also presented in this book. Among the most relevant similarities are rainwater catchment and the separation between black and

Table 4 Qualitative assessment of technical feasibility of the components of the Chilean BEV approach regarding different contexts

	Rural house	Rural village	Towns
District heating	x	+-	√
Pellet boiler	±	√	√√
Double chamber wood stove	√	±	x
Biodigestor	√	√	±
Biogas/biomethane network	x	√	√
Composting dry (UDD) toilet	√√	√	x
Urine-diverted (UD) toilet	√	√	±
Human excretes biodigestion	√√	√	x
Liquid fertilizer collection/use	√	√√	±
Composting organic waste	√	√	√
Constructed wetland	√√	√√	√

grey waters, the introduction of the biodigestion of black waters, and domestic organic waste. However, in the German model biogas is used for heating, not for cooking. Also, its technological level is higher, including a vacuum toilet system and heat exchangers. In any case, both approaches show very significant and coherent improvements in the environmental performance concerning energy, water, and organic waste.

Finally, it is important to note that the concepts presented here are an attempt to find a systemic solution for an integrated energy, water, and organic waste management system for a specific cultural background. However, as a systemic topic, they deal with complex and transdisciplinary problematics, so that many aspects have not yet been studied in this work. This way, for implementing the concepts presented, more research and pilot projects transferring these ideas into practice are needed.

4 Conclusions

German bioenergy villages (BEV) are models of energy management that achieve a low carbon footprint, high energy efficiency, organic waste management, and social integration. This research found that the German approach of bioenergy villages is technically feasible for Los Ríos Region. However, it would be not feasible in terms of sustainability, as Chilean prices for buying heat and selling electricity are much lower than in Germany, and therefore, heating networks and biogas electricity cogeneration are too expensive for the regional village reality. Additionally, energy crops are facing increasing critics worldwide due to their environmental and social impacts. Nevertheless, modifications to the original BEV concept seem to be very promising for the region and are presented as a new concept of BEV adapted for southern Chile. Firstly, in medium or larger cities, where atmospheric pollution is problematic, district heating networks using firewood or derivates can be successfully applied in discrete urban areas, reaching high energy efficiency and low particulate emissions through centralized boilers. Secondly, replacing energy crops with the domestic organic solid waste for biogas production can only supply less than 5% of the total energy demand of a household. However, it can supply enough gas for cooking, therefore allowing energy savings and technological replacement of low efficiency stoves. This reduces firewood demand, improves air quality, and improves organic waste management. The Chilean BEV does not supply electricity, but 100% of the thermal energy (including for cooking), which is over 80% of the total domestic energy demand. That number can be reached by using only very little agricultural land area for energy plant production for biodigestion. Finally, this model can also meaningfully improve water efficiency through a wastewater separation system, ideally with the incorporation of urine-diverting dry toilets that alone can save 1/3 of the water demand. These toilets concentrate 85% of total organic waste in only 14% of the total wastewater volume compared to normal toilets. This enables better performance of biodigestion for the black water. The dilution of grey water enables good performance of constructed wetlands for the grey water. This way, a high degree of synergy is reached. At its optimum, this concept makes it possible to efficaciously recover 86% of the wastewater that has no contact with human excreta (no black water), and to recover up to 90% of the household waste nutrients, besides enhancing the biodiversity and landscape quality of the place. This model could work in villages or urban neighborhoods, but also in individual rural houses, although with lower efficiency due to the size and technology of the familiar biodigestor. In any case, such a model requires a high level of ecological advocacy of the users and stakeholders due to the demanding practices required for its operation.

Acknowledgements This research was financially supported by a scholarship from Becas Chile/Conicyt and DAAD. I thank Professor Hans Ruppert, Dept. of Sedimentology/Environmental Geology, for his valuable help with relevant comments on the manuscript, and the support of the whole team of the Interdisciplinary Center for Sustainable Development (IZNE), all from the Universität Göttingen in Germany. I also would like to acknowledge the support of Professor Teodoro Kausel, Professor Victor Sandoval, the Agriculture Engineering and Soil Science Department of the Faculty of Agrarian Sciences, and the Direction of Services, all from Universidad Austral de Chile, besides Johannes Horstmann, and the Municipality of Valdivia.

References

AIFBN. (2008). *Manual of Environmental Education; Use of wood preserving the forest and polluting less (Usar leña conservando el bosque y contaminando menos. Manual de Educación Ambiental). Agrupación de Ingenieros Forestales por el Bosque Nativo.*

AIFBN. (2011). *Hacia un Nuevo Modelo Forestal. Propuestas para el desarrollo sustentable del Bosque Nativo y el Sector Forestal en Chile.* Agrupación de Ingenieros Forestales por el Bosque Nativo.

Álvarez, J. A., Ruíz, I., & Soto, M. (2008). Anaerobic digesters as a pretreatment for constructed wetlands. *Ecological Engineering, 33*(1), 54–67.

Association of Muncipalities of Los Ríos Region. (2018).*Sustainable Management of Residential Solid Waste Project (Proyecto Manejo Sustentable de Residuos Sólidos Domiciliarios),* Valdivia, Chile

Anton, C., & Steinicke, H. (2012). *Bioenergy - Chances and Limits. Deutsche Akademie der Naturforscher Leopoldina.* http://www.leopoldina.org/nc/en/publications/detailview/?publication[publication]=433&publication[action]=show&cHash=52f32691fc9e320f41f1faacd7608d96

Avellan, C. T., Ardakanian, R., & Gremillion, P. (2017). The role of constructed wetlands for biomass production within the water-soil-waste nexus. *Water Science and Technology, 75*(10), 2237–2245. https://doi.org/10.2166/wst.2017.106

Barros, P., Ruiz, I., & Soto, M. (2008). Performance of an anaerobic digester-constructed wetland system for a small community. *Ecological Engineering, 32*(2), 142–149.

Bauböck, R. (2013). Modelling site-specific biomass potentials. In H. Ruppert, M. Kappas, & J. Ibendorf (Eds.), *Sustainable bioenergy production—An integrated approach.* Springer Science+Business Media Dordrecht. https://doi.org/10.1007/978-94-007-6642-6

Castillo-Avalos, Y., & Rovira, A. (2013). Eficiencia Hídrica en la Vivienda (Hidric Efficiency on Household). *Tecnología y Ciencias Del Agua (Technology and Water Sciences), IV*(4), 159–171.

CCA. (2008). *Informe Final Estudio de "Análisis del Potencial Estratégico de la Leña en la Matriz Energética Chilena". Corporación Chile Ambiente.* http://www.ced.cl/ced/wp-content/uploads/2012/03/potencial-de-la-lena-en-matriz-energetica.pdf

CDT. (2010). *Estudio de Usos Finales y Curva de Oferta de Conservación de la Energía en el Sector Residencial de Chile. Informe final y resumen ejecutivo. Realizado para el Programa País de Eficiencia Energética del Ministerio de Energía del Gobierno de Chile. Corpora.*

Echeverria, C., Coomes, D., Salas, J., Lara, A., & Newton, A. (2006). Rapid deforestation and fragmentation of Chilean temperate forests. *Biological Conservation, 130*, 481–494. https://doi.org/10.1016/j.biocon.2006.01.017

Edwin, G. A., Gopalsamy, P., & Muthu, N. (2014). Characterization of domestic gray water from point source to determine the potential for urban residential reuse: A short review. *Applied Water Science, 4*(1), 39–49. https://doi.org/10.1007/s13201-013-0128-8

Erlwein-Vicuña, A. (2016). *Interactions between land use and bioenergy potentials on designing a bioenergy area in Los Ríos Region.* Universität Göttingen.

Erlwein, A., Lara, A., & Pradenas, A. (2007). Industria de Celulosa en Chile; Un Modelo de Desarrollo no Sustentable. In J. A. . A. P. y E. L. MONJEAU (Ed.), *Temas de ecología y filosofía.* Ediciones de la Universidad Atlántida Argentina- Fundacao O Boticario - Universidade Federativa do Paraná.

Erlwein, A., & Sotomayor, E. (2016). Estimación de la producción de biomasa, generación de biogás y su potencial energético. In *Biogás de Residuos Agropecuarios en la Región de Los Ríos; Aspectos Generales, Experiencias Y Potencial De Producción. INDAP – GORE Los Ríos, Osorno.*

FAO. (2008). State of Food and Agriculture (SOFA) 2008—Biofuels: Prospects, risks and opportunities. In *FAO Agriculture Series.* Food and Agriculture Organization. http://www.fao.org/publications/sofa/2008/en/

Franco, M. V. (2007). *Tratamiento y Reutilización de Aguas Grises en Chile.* Universidad de Chile.

Friedrich, J., Poganietz, W. R., Lehn, H. (2020). Life-cycle assessment of system alternatives for the water-energy-waste Nexus in the urban building stock. *Resources, Conservation & Recycling, 158*(104808). https://doi.org/10.1016/j.resconrec.2020.104808

Giampietro, M., & Mayumi, K. (2009). The biofuel delusion: The fallacy of large-scale agro-biofuel production. *Earthscan.* https://doi.org/10.4324/9781849770200

Hassan, I., Chowdhury, S. R., Prihartato, P. K., & Razzak, S. A. (2021). Wastewater treatment using constructed wetland: Current trends and future potential. *Processes, 9*(11), 1–27. https://doi.org/10.3390/pr9111917

Kochi, L. Y., Freitas, P. L., Maranho, L. T., Juneau, P., & Gomes, M. P. (2020). Aquatic macrophytes in constructed wetlands: A fight against water pollution. *Sustainability (Switzerland), 12*(21), 1–21. https://doi.org/10.3390/su12219202

Little, C., Lara, A., Mcphee, J., & Urrutia, R. (2009). Revealing the impact of forest exotic plantations on water yield in large scale watersheds in South-Central Chile. *Journal of Hydrology, 374*, 162–170. https://doi.org/10.1016/j.jhydrol.2009.06.011

Oteng-peprah, M., & Nanne, K. (2018). Greywater characteristics, treatment systems, reuse strategies and user perception—A review. *Water, Air, and Soil Pollution, 229*(255), 1–16.

Reyes, R., & Neira, E. (2012). Leña, energía renovable para la conservación de los bosques nativos de Chile. In *Agrupación de Ingenieros Forestales por el Bosque Nativo, MIRA Ediciones.*

Ruppert, H., Eigner-Thiel, S., Girschner, W., Karpenstein-Machan, M., Roland, F., Ruwisch, V., Sauer, B., & Schmuck, P. (2008). *Paths for bioenergy village - Guidelines for a separate power and heat supply based on biomass in rural areas (Wege zum Bioenergiedorf – Leitfaden für eine eigenständige Strom- und Wärmeversorgung auf Basis von Biomasse im ländlichen Raum).* Fachagentur Nachwachsende Rohstoffe e.V. (FNR). http://www.wege-zum-bioenergiedorf.de/fileadmin/bioenergiedorf/dateien/Leitfaden_Wege_zum_Bioenergiedorf.pdf

Schmuck, P., Eigner-Thiel, S., Karpenstein-Machan, M., Sauer, B., Ruppert, H., Girschner, W., & Roland, F. (2013). Bioenergy Villages in Germany: Applying the Göttingen approach of sustainability science to promote sustainable bioenergy projects. In H. Ruppert, M. Kappas, & J. Ibendorf (Eds.), *Sustainable bioenergy production—An integrated approach.* Springer Science+Business Media Dordrecht. https://doi.org/10.1007/978-94-007-6642-6

Seidel, T., Orasche, J., Ruppert, H., Schnelle-Kreis, J., & Hartmann, H. (2013). Emissions of organic and inorganic pollutants during the combustion of wood, straw and biogas. In H. Ruppert, M. Kappas, & J. Ibendorf (Eds.), *Sustainable bioenergy production—An integrated approach.* Springer Science+Business Media Dordrecht. https://doi.org/10.1007/978-94-007-6642-6

Spångberg, J., Tidåker, P., & Jönsson, H. (2014). Environmental impact of recycling nutrients in human excreta to agriculture compared with enhanced wastewater treatment. *Science of the Total Environment, 493*, 209–219. https://doi.org/10.1016/j.scitotenv.2014.05.123

Statistisches Bundesamt. (2018). *Energieverbrauch privater Haushalte für Wohnen 2017 erneut gestiegen.* Retrieved April 5, 2022, from https://www.destatis.de/DE/Presse/Pressemitteilungen/2018/10/PD18_378_85.html

Todd, J., & Josephson, B. (1996). The design of living technologies for waste treatment. *Ecological Engineering, 6*, 109–136.

Tomes, D., Lakshmanan, P., & Songstad, D. (2011). *Biofuels: Global impact on renewable energy, production agriculture, and technological advancements.* Springer.

UACh. (2013). *Evaluación del Mercado de Biomasa y su Potencial. Informe Final de Proyecto. Para Ministerio de Energía y CONAF.*

USDA-EPA-USD of Energy. (2014). *Biogas opportunities roadmap* (Issue August). https://www3.epa.gov/climatechange/Downloads/Biogas-Roadmap.pdf

WBGU. (2008). Future bioenergy and sustainable land use. German advisory council on global change. In *Engineer*. Earthscan. https://doi.org/10.4324/9781849774505

Wüste, A., & Schmuck, P. (2012). Bioenergy villages and regions in Germany: An interview study with initiators of communal bioenergy projects on the success factors for restructuring the energy supply of the community. *Sustainability, 4*(2), 244–256. https://doi.org/10.3390/su4020244

Alternative Urban Water-Energy-Waste Nexus Options—Sustainable Assessment in an Existing Settlement

Witold-Roger Poganietz and Jasmin Friedrich

Abstract Sophisticated water-wastewater systems contribute significantly to public health, but their operation is tied to significant energy consumption and their construction to material-intensive, long-lasting infrastructures. The systems are also characterized by an inability to effectively and efficiently recover nutrients. Separating wastewater into warm and lightly polluted greywater and nutrient-rich blackwater at the source, supplemented with organic municipal solid wastes, is often seen as an option to enhance resource and energy efficiency. This study compares three different system alternatives with the status quo, using the Integrative Concept of Sustainability as a methodological backbone. The system alternatives differ in the technological design of treating the separated wastewater flows. The least invasive system change focusses on the separation of wastewater at the source while the current pipeline system and central treatment plant are still in use. The most elaborate option takes the current wastewater system out of operation completely and shows the highest potential to enhance resource and energy efficiency. However, this option could face the lowest acceptance of all of the options, including status quo, and, compared to the status quo, has higher costs.

1 Introduction

Sophisticated water-wastewater systems as they exist in developed cities are an important factor in promoting public health (Daigger, 2007), but their operation is tied to a noteworthy amount of energy consumption and their construction to material-intensive, long-lasting infrastructures (U.S. Department of Energy, 2013). The systems are also characterized by a lack of effective and efficient nutrient

W.-R. Poganietz (✉)
Institute for Technology Assessment and Systems Analysis (ITAS), Karlsruhe Institute of Technology (KIT), Karlsruhe, Germany
e-mail: witold-roger.poganietz@kit.edu

J. Friedrich
Heidelberg, Germany

A. Ch. Braun et al. (eds.), *Eco-Industrial Development as an Industrial Strategy*,
Green Energy and Technology, https://doi.org/10.1007/978-3-031-73576-9_9

recovery (Arcadis, 2016). For example, in Europe only 53% (2015) of the nutrient-rich sewage sludge is distributed to agriculture or to compost facilities (Gutjahr & Müller-Schaper, 2018).

Technical options to reduce the demand for materials and energies without deteriorating the quality of the services offered are being widely discussed. Separating wastewater into greywater and blackwater at the source is a commonly proposed option (Larsen et al., 2009; Skambraks et al., 2017; UN WWAP, 2017; Vergara-Araya et al., 2020). Greywater is generally warm and lightly polluted; the thermal energy it contains can be recovered, while recycled greywater can be reused mainly for non-hygienic uses, like toilet flushing. Blackwater is nutrient-rich, and co-digesting it with organic municipal solid waste (MSW) could enhance the efficiency of biogas plants as well as their nutrient recovery (Vergara-Araya et al., 2020). Different variations of these ideas can be seen in the first pilot plants, ranging from home solutions for single buildings (e.g., Arminplatz, Berlin) (Nolde, 2013) to larger neighborhoods (e.g., Jenfelder Au, Hamburg) (Stadt Hamburg, 2017).

Transforming urban wastewater systems with separate wastewater streams could have far-reaching impacts on the environment, economies of providing services, but also on the daily lives of users, demanding a comprehensive assessment of possible systems alternatives.

The small number of studies assessing entire systems with separate wastewater flows and not only single components focus mostly on the environmental impacts (Hiessl et al., 2010; Remy, 2010; Vergara-Araya et al., 2020). Hiessl et al. (2010) and Remy (2010) also refer to the costs.

The overarching aim of this contribution is to comprehensively analyze and assess different system alternatives that would transform the current system of treating wastewater and organic municipal solid waste while providing energy to an integrated water-energy-waste system. Whereas the current system follows a more linear understanding of providing and treating water, all system alternatives aim to close several loops, increasingly introducing circular elements to the status quo system.

Using a multi-functional approach (Friedrich et al., 2020), the study compares three different system alternatives with the status quo. All system alternatives assume the separation of wastewater into blackwater and greywater as well as the co-digesting of blackwater with organic MSW. The system alternatives differ in the technological design of the treatment of the separate wastewater flows. The least invasive system change focuses on the separation of wastewater at the source, while the current pipeline system and central treatment plant are still in use. The most elaborate one takes the current wastewater system fully out of operation.

This chapter is organized as follows: the following section, Sect. 2, describes the systems under review, followed by Sect. 3 where the applied methods to carry out a sustainability assessment are presented. The results are shown and discussed in Sect. 4. Concluding remarks are offered in Sect. 5.

2 Status Quo and System Alternatives

2.1 System Boundaries

The main function of the current water-wastewater system is to provide drinking water and to suitably transport wastewater and its associated pathogens and harmful substances out of the settlement, and to treat wastewater accordingly to ensure healthy and dry living conditions in the settlement. Although the current system provides nutrient-rich sludge and energy, these functions are of minor importance in the current system. Contrary to that, the system alternatives analyzed see the provision of energy to the settlement and of nutrient-rich sludge as a source for fertilizer as an additional objective for the water-wastewater system. To allow for a comparison of the alternatives with the current system, the "new" primary functions are also included in the analysis of the current system. That means the primary function of all of the systems analyzed is comprised of wastewater and waste management, and the energy supply, i.e., electricity and heat. It is a prerequisite that the system alternatives need to fulfill the primary functions without any limitations to security, reliability, and service comfort. The systems under review are meant to protect public health in urban areas as well as the water bodies affected by sewage effluents, which places an important constraint on the transformation pathways.

Local and regional characteristics determine the framework of an urban water-energy-waste-waste system. This is not only true for the resources and technologies required to install and maintain the respective systems, but also for the composition of wastewater and organic MSW (Hiessl et al., 2010). To adequately consider these specifics, the analysis is restricted geographically to a neighborhood in the city of Heidelberg, Germany, with around 5,081 inhabitants. Residential buildings of different sizes and a school with about 1692 students characterizes the neighborhood (Friedrich et al., 2020).

2.2 Status Quo[1]

The current system is characterized by a centralized water supply and treatment system, which includes the treatment of rainwater. The drinking water provided is intended for all household uses regardless of any specific quality requirements (Stadt Heidelberg, 2017). Wastewater is discharged into a mixed sewer and flows to a sewage treatment plant. Nutrients are removed during treatment but not recovered. Sewage gas is generated and covers the energy demand of the treatment plant. The treated wastewater is discharged into the local river (AZV, 2019). Heat is provided by a central district heating plant with coal as the main energy carrier with a share of

[1] This and the following section draw heavily on Friedrich et al. (2020) (cf. Friedrich, 2020).

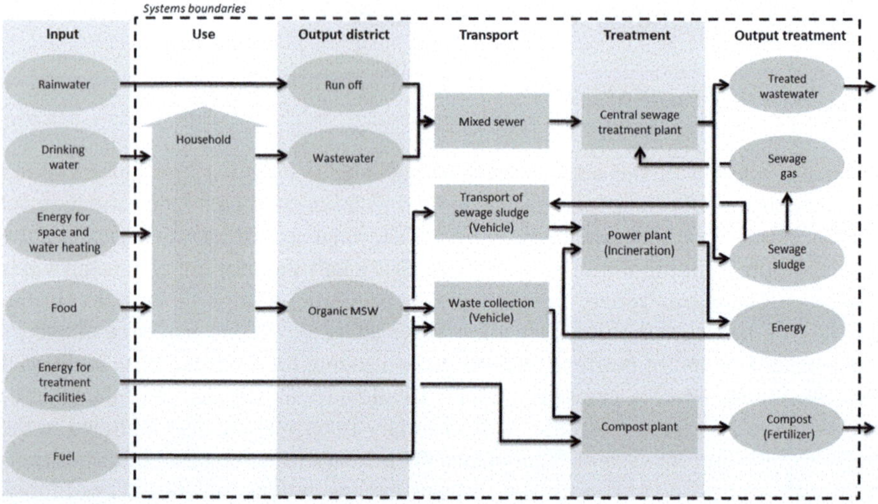

Fig. 1 Flow chart of the status quo. *Source* Friedrich et al. (2020)

about 75%. Organic MSW is collected by a municipal waste collection service and processed in a centralized composting plant (Stadt Heidelberg, 2017) (Fig. 1).

2.3 System Alternatives

The different system alternatives can be understood as possible ranges of transformation: system alternative 1 (SYAL1) describes the least invasive intervention in the current system, SYAL3, the complete overhaul of the current system, while SYAL2 could be understood as a middle course between SYAL1 and SYAL3. Common to all system alternatives is the separation of wastewater into greywater and blackwater. Blackwater, coming from toilets, is nutrient-rich but highly contaminated. Co-digesting it with organic MSW could significantly increase the biogas yield and the provision of organic fertilizer (Han et al., 2016; Mata-Alvarez et al., 2000). Greywater, coming from e.g., showers and dishwashers, is warm and contains relatively low levels of contamination. Recycled greywater can be reused for non-hygienic purposes, like the irrigation of non-edible plants (Hiessl et al., 2010). The heat from the greywater can be recovered and used as an additional energy carrier in buildings. The system alternatives also aim to reduce the amount of rainwater reaching the sewer system (the so-called "run-off") through measures for more evaporation and infiltration (Matzinger, 2017).

In SYAL1, greywater is separated from blackwater but not treated for reuse, addressing a situation with no imminent water scarcity but still the need for a more efficient treatment of wastewater and use of energy, which is observable in Heidelberg. The heat is recovered from the greywater via a heat exchanger combined with a

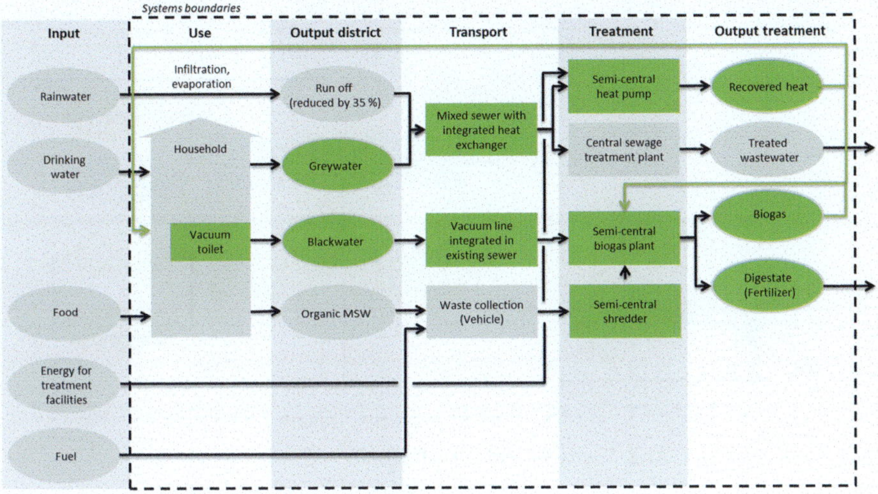

Fig. 2 Flow chart of SYAL1. *Source* Friedrich et al. (2020)

heat pump in the main sewer of the neighborhood and used for heating in the neighborhood. The greywater is treated together with rainwater in the centralized treatment plant and disposed of in the local river. It is assumed that the rainwater reaching the treatment plant is reduced by 35% via measures for more evaporation and infiltration (Morandi et al., 2018). Figure 2 shows SYAL1. System changes compared to the status quo are marked in green.

To minimize the water content of blackwater, which influences the biogas yield, vacuum toilets are installed in households (Staben, 2008). The vacuum sewer is implemented in the existing sewer to make use of already existing assets, transporting the blackwater to a biogas plant in the neighborhood. The organic MSW collected is delivered to a semi-centralized shredder and after pretreatment is mixed with the blackwater, to be used as the feed in a biogas plant. Together with the recovered heat from greywater, the biogas produced replaces the previous non-renewable energy supply for heating. The nutrient-rich residues of the digestion process can be used after conditioning as organic fertilizer (Fig. 2).

In SYAL2, the greywater is recycled in the buildings directly and replaces some of the drinking water for non-hygienic purposes. The heat contained in the greywater is immediately recovered in decentralized treatment plants. The unused treated greywater is discharged into the local river together with the runoff, which accounts for 30% of the rainwater. As all greywater is treated in the households, the central treatment plant is no longer needed. Figure 3 depicts SYAL2. Changes compared with the status quo system and SYAL1 are marked in blue.

In SYAL2, organic MSW is added to the vacuum system right in the households via shredding food waste in the kitchen sink. Therefore, no municipal waste transport is needed. As in SYAL1, in combination with the recovered heat from the greywater,

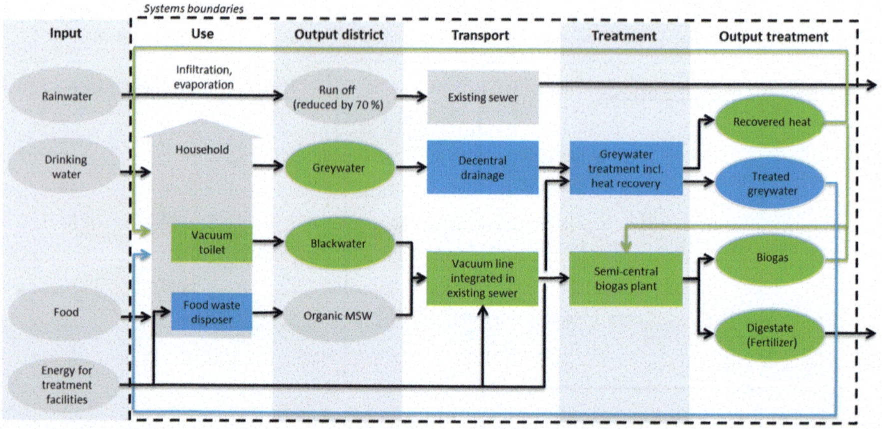

Fig. 3 Flow chart of SYAL2. *Source* Friedrich et al. (2020)

the biogas generated replaces the previous non-renewable energy supply for heating in the neighborhood (Fig. 3).

SYAL3 has all the elements SYAL2 has but could be considered the most extreme system alternative, as it represents the idea of a "sewerless society" (Martin et al., 2015). By implementing a vacuum line outside the existing sewer and infiltrating, retaining, and evaporating all rainwater locally, the sewers are taken completely out of the operation. Figure 4 gives an overview of SYAL3. The changes compared with the other systems are marked in orange.

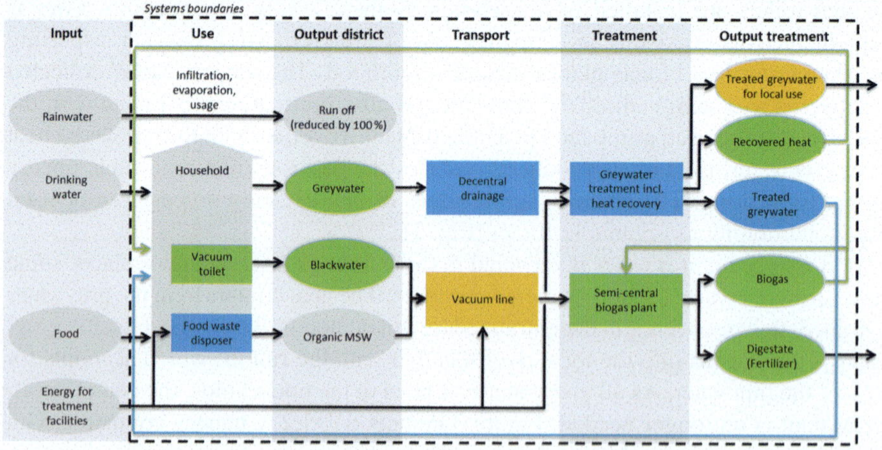

Fig. 4 Flow chart of SYAL3. *Source* Friedrich et al. (2020)

3 Method

3.1 Overview

The interest in and the necessity of a sustainability assessment of technologies or systems is extensive (Grunwald, 2016; Nakamura et al., 2019), which was pushed recently by the adoption of the Sustainable Development Goals (SDGs) through the resolution adopted by the general assembly on September 25, 2015 (Transforming our world, 2015).

In general, sustainability assessments are oriented toward the already classic pillar concept, with the three dimensions of ecology, economy, and society (Purvis et al., 2019). The main disadvantage of the pillar concept is that the paradigms of sustainable development as formulated by the Brundtland Commission are inadequately covered (Grunwald & Kopfmüller, 2012; Kopfmüller et al., 2001). The implementation of sustainable development should pay attention to (a) the global perspective, both in the analysis of problems and in the development of solution strategies, (b) the linkage between environmental and development perspectives, and (c) the consideration of intra- and intergenerational justice (Brundtland, 1991).

The Integrative Concept of Sustainable Development (ICoS) that is applied in this study (Grunwald, 2016; Kopfmüller et al., 2001), allows for an integrated and comprehensive assessment of systems addressing the above-mentioned demands. Using, among others, the claims of the Brundtland Commission, three sustainability goals were defined: (a) the securing human existence, (b) the maintaining of society's productive potentials, and (c) the preserving society's options for development and action (Kopfmüller et al., 2001; Suryandari & Bolivar Paypay, 2024).

Central for sustainable development is the safeguarding of human existence. This is captured through the protection of human health, the guarantee of a basic supply, and the possibility of securing one's own livelihood through a freely undertaken activity. Additionally, the distribution of the opportunities to use environmental resources should be fair, and the differences in income and wealth in the global perspective should not be extreme (Grunwald, 2016; Grunwald & Kopfmüller, 2012).

Intergenerational justice demands comparable opportunities for present and future generations to satisfy their needs, which requires the preservation of society's productive capacity over time. This means a sustainable use of the environment as a resource and as a sink, and avoidance of major technical risks that would massively endanger society's productive potential. The preservation of productive societal potential also requires the safeguarding and improvement of economic performance through the corresponding development of physical, human, and knowledge capital (Grunwald, 2016).

An essential condition for achieving intra- and intergenerational equality, taking into account the global orientation as well as an enlightened anthropocentrism, is the permanent preservation of social development and possibilities for action. For this, equal opportunities for and the participation of the members of the society, the preservation of cultural diversity, and the cultural functions of nature as well as social resources are essential (Grunwald & Kopfmüller, 2012).

In order to conduct a sustainability assessment for each above-mentioned goal, five rules were defined (Kopfmüller et al., 2001; Suryandari & Bolivar Paypay, 2024), which are further operationalized by context-specific indicators. However, after defining the system under review, the very first step should be to identify the appropriate rules, since not all of the rules always need to be considered (see below).

In the following, the selected rules and indicators are described in detail.

3.2 *Securing Human Existence*

Considering the system analyzed, three rules were identified to address the goal "securing human existence": (a) protection of human health, (b) ensuring satisfaction of basic needs, and (c) just distribution of opportunities to use natural resources (Table 1). The other two rules proposed by ICoS, "autonomous subsistence based on own income" and "reduction of extreme income or wealth inequalities" are topics to focus on that are not affected by a transformation of the water-energy-waste system.

Activities involving direct or indirect contact with water in rivers that are not designated as swimming areas pose a health risk to people. They can trigger illnesses that are associated with fever, diarrhea, and vomiting (Schaum, 2018). This could happen if sewer overflows discharge untreated wastewater into the river, polluting the water body (Gantner & Barjenbruch, 2012). The hygienic load is measured in CFU/100 ml for E.coli.

Urban blue-green spaces are defined as all unpaved areas in urban space that contain water and greenery, e.g., park-like, near-surface rainwater retention measures (Ramboll, 2018). They can help to mitigate heat effects during the summer, as they heat up less than the built environment, cool down significantly more at night, and provide evaporative cooling and shade (SRU, 2018). It is predicted that the number of days with heat stress in Baden-Württemberg will double by 2050, and the number

Table 1 Securing human existence

Rule	Indicator	Unit
Protection of human health	Hygienic load of local streams	E.coli: CFU/100 ml
	Provision of urban blue-green spaces with treated wastewater	m^3/a of available treated wastewater
Ensuring satisfaction of basic needs	Recovered nutrients	t/a per nutrient
	Diversification of water provision	Number of types of water sources
Just distribution of opportunities to use natural resources	Global warming potential	t CO_{2eq}/a

of tropical nights will even quadruple (LUBW, 2017). Urban blue-green spaces also provide an incentive for all residents for outdoor activities in the midst of built-up surroundings (BBSR, 2016). Available treated wastewater could reduce the demand for drinking water and could thus increase the incentive to install additional blue-green spaces or to supply an existing one with additional water. The availability of treated wastewater is estimated as the sum of treated greywater and rainwater less re-used greywater through the households in m^3/a.

The nutrients nitrogen (N), phosphorus (P), and potassium (K) are essential plant nutrients and an essential component of NPK fertilizers (Steinmetz & Morandi, 2017). The deficiencies of these nutrients have significant negative effects on plants and affect food security. The production of nitrogen is energy-intensive (Gellings & Parmenter, 2009), the supply of phosphorous is globally limited (Heckenmüller et al., 2014), and the extraction of potassium from geogenic deposits is associated with negative environmental impacts (BGR, 2016). These limits create an urgency to recover the available nutrients, e.g., in wastewater (Vaneeckhaute, 2019). The recovered nutrients are counted by t/a, which differs between the nutrients.

Dry summer periods, which are expected to increase in the next few decades, leads to a decrease in groundwater recharge (LUBW, 2019), while the demand for groundwater supplies should grow as a result of augmented demand for drinking water and agricultural irrigation, potentially resulting in regional shortages (Rechenberg, 2019). For the city of Heidelberg, the first local shortages are becoming apparent (Stadtwerke Heidelberg, 2019). In order to reduce the demand on groundwater reserves, the diversification of local water resources is therefore gaining in importance. If recovered wastewater could be used for non-hygienic purposes, the overall demand for drinking water would decrease, relieving water sources (Arcadis, 2016). The unit used to estimate this indicator is the number of types of water sources, only differing between drinking water sources and reused greywater and rainwater.

Global warming caused by greenhouse gasses has negative physical effects on all globally accessible environmental goods (atmosphere, water, biodiversity, etc.) (BMUB, 2016), damaging the living conditions of current and future generations locally and globally (UN DESA, 2016, 2017). Climate change is thus above all also a justice issue (Northcott, 2018; Posner & Weisbach, 2010). The unit used to estimate global warming potential is t CO_{2eq}/a (Goedkoop et al., 2013).

3.3 Maintaining Society's Productive Potentials

To address the goal "maintaining society's productive potentials", the systems discussed will be assessed with the help of four rules: (a) sustainable use of renewable resources, (b) sustainable use of non-renewable resources, (c) sustainable use of the environment as a sink, and (d) sustainable development of man-made, human, and knowledge capital (Table 2). The rule "avoiding technical risks with potentially catastrophic impacts" is not included in the assessment, because a breakdown of the investigated water-energy-waste system would not endanger society's productive potentials.

Although Germany is not yet characterized by water shortages throughout the country, as discussed above, in the region water shortages have occurred (Stadtwerke Heidelberg, 2019). Careful use of drinking water resources and conscious drinking water consumption in households is therefore important for the conservation of water resources (Umweltbundesamt, 2018). The unit to measure the requirements is m^3/a.

Wastewater is a source of renewable energy that demands effective and efficient treatment (UN Vergara-Araya et al., 2020; WWAP, 2017). The unit to measure the provided energy is MWh/a.

The construction, operation, and maintenance of the technical infrastructures that currently shape the water-energy system are energy- and resource-intensive (U.S. Department of Energy, 2013). Since energy and material resources are limited worldwide and their extraction often takes place under unsustainable conditions, it is imperative that they are used sparingly (Umweltbundesamt, 2016). The units used to measure the demands are t oil_{eq} and t FE_{eq} (Goedkoop et al., 2013).

Organic micropollutants consist of a large number of substances from, for example, pharmaceuticals, cleaning agents and cosmetics, which are not or not completely degraded or retained by conventional wastewater systems (LAWA, 2016). The possible effects of micropollutants are metabolic changes and the delayed development of organisms in, e.g., flowing waters (Behrendt, 2018). The large number of substances makes it difficult to select a single substance to assess the problem. With the neighborhood as an urban residential area, pharmaceutical residues are particularly relevant (LAWA, 2016). For the assessment, the most frequently sold painkiller (analgesic) in Germany—diclofenac—was selected (WIdO, 2013). The German Environmental Agency (Umweltbundesamt) has assessed the environmental relevance of diclofenac to municipal wastewater systems as high and its harmful effect on aquatic ecosystems has been identified in corresponding concentrations (Schwaiger et al., 2004; Umweltbundesamt, 2018). The indicator shall disclose the load of diclofenac in the local water bodies.

Table 2 Maintaining society's productive potentials

Rule	Indicator	Unit
Sustainable use of renewable resources	Drinking water requirements	m^3/a
	Provision of renewable energy	MWh/a
Sustainable use of non-renewable resources	Demand for fossil fuels	t oil_{eq}
	Demand for metal resources	t FE_{eq}
Sustainable use of the environment as a sink	Entry of organic micropollutants in local streams	Not available
	Global warming potential	t CO_{2eq}/a
Sustainable development of man-made, human, and knowledge capital	Contribution to the knowledge of inhabitants with respect to water and energy challenges	Not available
	Operating and investment costs	million €

Global warming negatively impacts the carrying capacity of the environment (Myhre et al., 2013). The potential is measured with t CO_{2eq}/a (Goedkoop et al., 2013).

An environmentally sound and resource-conserving household behavior requires knowledge of the functioning of the water-wastewater system as well as the consequences on the environment (Hummel et al., 2004). The indicator is intended to disclose the availability of the respective knowledge among households.

The rule "sustainable development of man-made, human, and knowledge capital" shall include economic conditions comprehensively, since, like material and energy constraints, economic conditions could reduce the welfare of a region, if e.g., non-economic advantages are overruled by the lack of sufficient financial resources. The common suggestion is the estimation of the value added. When no competitive markets are achievable or a market is highly regulated, like water provision, costs could be a good proxy for estimating the value added. The cost analysis shall consider the full costs of implementing, maintaining, and operating the system, which provides the services under consideration, irrespective of the bearer of the costs. The costs are measured in million € (Poganietz et al., 2021).

3.4 *Preserving Society's Options for Development and Action*

Two rules will address the goal "preserving society's options for development and action": (a) participation in societal decision-making processes and (b) conservation of social resources (tolerance, solidarity, etc.) (see Table 3). The rules "equal access for all people to information, education, occupation", "conservation of cultural heritage and cultural diversity", and "conservation of the cultural function of nature" will not be included in the assessment. These rules deal with topics, which are not affected by the transformation of the system.

The participation of users in respect to infrastructure is seen as an opportunity to identify their wishes and concerns as well as to identify the most appropriate infrastructure design. This can enhance the acceptance of a transformation of the system under discussion (see below). Additionally, participation leads to improved user knowledge with respect to water and energy challenges.

User acceptance tackles the question under which conditions a user would accept a specific technology as part of their daily life. A new technology is accepted when

Table 3 Preserving society's options for development and action

Rule	Indicator	Unit
Participation in societal decision-making processes	Initiation of opportunities in respect to participation	Qualified statement
Conservation of social resources (tolerance, solidarity, etc.)	Acceptance by user	Qualified statement
	Promotion of societal responsibility	Qualified statement

it fulfills the expected demands and the provision of the service that is connected with, at least, a marginal improvement compared to the established one. Acceptance is not only driven by possible costs, but also whether and to what extent known and accepted daily procedures and the urban living environment are interfered with (Hegger et al., 2008; van Vliet & Spaargaren, 2010; Wüstenhagen et al., 2007).

Social responsibility points to the behavior of users when making use of an infrastructure. Individual behavior influences the efficiency of an infrastructure and thus, the living standard of others. The relevance of social responsibility increases with the intensity of social or technological interlinkages between neighbors and beyond. For example, the current system tolerates user misconduct as long as it is negligible (Hegger et al., 2008), allowing "a low level of responsibility" (Hummel et al., 2004, p. 70).

3.5 Comprehensive Assessment

The comprehensive assessment sees three steps: (1) the estimation of the identified indicators, (2) the transfer of the different types of results into a common assessment scheme, and (3) the assessment of the results, considering the different goals, rules, and indicators.

For the estimation of each indicator, the study uses different methods: environmental and resource related indicators are calculated with the help of a regional Life Cycle Assessment (r-LCA), the costs by estimating the net value added from investing and operating the systems under review and the residual indicators by carrying out interviews with experts and stakeholders, as well as literature research (Table 4).

The indicator values differ in respect to the units and to the type of result, differing in the latter between quantitative and qualitative findings. Therefore, the different indicator values and units are transferred into a common assessment scheme or rating scale. The base of the rating of each option is the rank of an option compared to the others. The rating scale ranges from 1 to 4, defining 1 as the lowest position in the ranking, i.e., the worst outcome of a system in respect to the indicator analyzed compared to the other systems. The value 2 indicates that exactly one system option shows an inferior outcome, whereas the value 3 implies that the outcome of two other options are worse. The value 4 is the best one, i.e., no other options show a better or equal outcome.

The essence of the selected rating scale is to rank the options without qualifying the difference between the options and whether the best option achieves a sustainable outcome. The former means for the assessment, it is irrelevant whether the difference between the achieved values of an indicator are small, e.g., less than 1%, or large, e.g., more than 100%. The latter is the consequence of not defining sustainable targets.

ICoS tackles different goals and rules, addressing the different needs and aims. The weighting of the goals and rules, which reveals the preference structure of a society or a decision-maker, will have an impact on the recommended characteristics of a system or even on the system finally put into effect (Dusseldorp, 2017; Robinson,

Table 4 Overview about the used methods

Indicator	Method	Description in …
Hygienic load of local streams	Literature research	Friedrich (2020)
Provision of urban blue-green spaces with treated wastewater	r-LCA	Friedrich et al. (2020)
Recovered nutrients	r-LCA	Friedrich et al. (2020)
Diversification of water provision	Literature research	Friedrich (2020)
Global warming potential	r-LCA	Friedrich et al. (2020)
Drinking water requirements	r-LCA	Friedrich et al. (2020)
Provision of renewable energy	r-LCA	Friedrich et al. (2020)
Demand for fossil fuels	r-LCA	Friedrich et al. (2020)
Demand for metal resources	r-LCA	Friedrich et al. (2020)
Entry of organic micropollutants in local streams	Literature research	Friedrich (2020)
Contribution to the knowledge of inhabitants with respect to water and energy challenges	Interviews and literature research	Friedrich (2020)
Operating and investment costs	Net present value method	Poganietz et al. (2021)
Initiation of opportunities in respect to participation	Interviews and literature research	Friedrich (2020)
Acceptance by user	Interviews and literature research	Friedrich (2020)
Promotion of societal responsibility	Interviews and literature research	Friedrich (2020)

2004). Thus, the assessment should take into account the sensitivity of the findings with respect to the assumed weighting.

4 Sustainability Assessment

In the following Sects. 4.1–4.3, the results as well as the rating of the results are presented along with the sustainability goals. Section 4.4 presents the comprehensive assessment.

4.1 Securing Human Existence

At the time of our analysis, the local river has not been declared safe for swimming, since the number of detected E.coli, which account for more than 900 CFU/100 ml, exceed the allowed benchmark (Untersteller, 2017). In all system alternatives, the treated feces containing black water should not be discharged into the local river to avoid any fecal contamination. Thus, the hygienic burden of the hygienic load of the system alternatives is set to approximately zero. Since the treatment of wastewater is in all system alternatives the same, outperforming the status quo, they are rated alike, assigning them a 2. The status quo scores 1 (Table 5).

In the status quo, the urban blue-green spaces are typically supplied by drinking water. Rainwater is polluted via its collection in a mixed sewer, impeding other uses. An appropriate joint treatment of greywater and rainwater would allow use for non-hygienic purposes in blue-green spaces. The available amount of such water is highest in SYAL 1 with 318,000 m^3/a, since all treated greywater and rainwater could be directed to blue-green spaces. Due to the partial reuse of greywater by households in SYAL 2 and 3, the available amount is lower, achieving 186,000 m^3/a and 102,000 m^3/a, respectively. The lowest availability of SYAL3 is due to the complete infiltration, retention, and evaporation of rainwater. The rating follows accordingly (Table 5).

Separating blackwater from greywater allows for a more efficient recovery of nutrients. Since the treatment of blackwater is in all system alternatives the same, outperforming the status quo, they are rated with a 2; the status quo achieves 1 (Table 5).

The reuse of greywater for non-hygienic purposes would reduce the demand for drinking water. Characteristic for SYAL2 and 3 is the re-transport of treated greywater to households, allowing for a direct reuse. In the status quo and in the SYAL1, treated wastewater and greywater, respectively, are dispersed into the local

Table 5 Securing human existence

Rule	Indicator	Unit	SQ	SYAL1	SYAL2	SYAL3
Protection of human health	Hygienic load of local streams[1]	E.coli: CFU/100 ml	>900	~0	~0	~0
		Rating	1	2	2	2
	Provision of urban blue-green spaces with treated wastewater	m^3/a	0	318,603	186,332	101,558
		Rating	1	4	3	2
Ensuring satisfaction of basic needs	Recovered nutrients[2]	t/a	N: 1.7 P: 0.4 K: 1.1	N: 23.0 P: 3.2 K: 7.0	N: 23.0 P: 3.2 K: 7.0	N: 23.0 P: 3.2 K: 7.0
		Rating	1	2	2	2
	Diversification of water provision	Number of water sources	1	1	2	2
		Rating	1	1	3	3
Just distribution of opportunities to use natural resources	Global warming potential[2]	t CO_{2eq}/a	7,923	434	309	280
		Rating	1	2	3	4

Note Calculation of the figures compare further [1]Friedrich (2020) and [2]Friedrich et al. (2020)

river. Consequently, SYAL2 and 3 are rated with 3, whereas SQ and SYAL1 with 1 (Table 5).

The transformation from the status quo to the system alternatives leads to a reduction in greenhouse gas emissions by up to 96.5% (SYAL3). The main reason is the switch from the coal-fired central heating plant in the status quo to biogas generation in the system alternatives. The lower impacts of SYAL2 and SYAL3 in comparison to SYAL1 can be associated with the reduced wastewater treatment and heat recovery, as well as to the biogas plant (only in case of SYAL3) (Friedrich et al., 2020). The estimated greenhouse gas emissions determine the rating of each option (Table 5).

4.2 *Maintaining Society's Productive Potentials*

The installation of vacuum toilets reduces drinking water requirements by about 27.2% compared to the status quo. The reuse of recycled greywater in SYAL2 and 3 would lead to an additional decrease in drinking water demand by 18.5% (Friedrich et al., 2020). The rating of the options is determined by this outcome (Table 6).

The co-digesting of blackwater and organic MSW in a biogas plant as well as the recovering of the thermal heat of greywater enhance the energy efficiency compared to the status quo by a factor 6.6. The slight difference between SYAL 1 and the other two system alternatives results from higher heat losses in SYAL1. In SYAL1, the greywater is carried to a semi-central heat-exchanger; in SYAL2 and 3 the heat exchanger is in the building (Friedrich et al., 2020). The rating of the options is determined by this outcome (Table 6).

The transformation of the status quo to the systems alternatives leads to a noteworthy decline in the demand for fossil fuels and metal resources between 68.0% (SYAL1, metal resources) and 95.7% (SYAL3, fossil fuels). The most prominent factor for the drop in fossil fuels is the switch from the coal-fired central heating system to biogas generation. In the case of metal demand, the improvement stems mainly from the switch from co-incineration of sludge to the co-digestion of blackwater and organic MSW, driving at least 87.7% (SYAL3) of the change. The differences between the system alternatives can be associated with the reduced wastewater treatment and heat recovery, as well as to the biogas plant (only in the case of SYAL3) (Friedrich et al., 2020). The calculated demands set the rating of the options (Table 6).

The current system cannot remove the diclofenac analyzed and consequently, the load is discharged into the river. In the system alternatives, the blackwater carrying the diclofenac should not be allowed to be discharged into the local river. Fermentation of blackwater reduces the content of diclofenac, but not completely (Graaff, 2010). Soil has to function as a sink (Facey et al., 2018). Comparing the different options, the system alternatives perform better than the status quo (Table 6).

As presented above (Sect. 4.1), with respect to global warming potential, SYAL3 shows the best performance and the status quo the worst, rating the options accordingly (Table 6).

Table 6 Maintaining society's productive potentials

Rule	Indicator	Unit	SQ	SYAL1	SYAL2	SYAL3
Sustainable use of renewable resources	Drinking water requirements[1]	m^3/a	195,768	142,523	106,378	106,378
		Rating	1	2	3	3
	Provision of renewable energy[1]	MWh/a	7,271	47,648	48,082	48,082
		Rating	1	2	3	3
Sustainable use of non-renewable resources	Demand for fossil fuels[1]	t oil_{eq}	2,227	123	101	96
		Rating	1	2	3	4
	Demand for metal resources[1]	t FE_{eq}	146	47	39	31
		Rating	1	2	3	4
Sustainable use of the environment as a sink	Entry of organic micropollutants in local streams		Yes	No	No	No
		Rating	1	2	2	2
	Global warming potential[1]	t CO_{2eq}/a	7,923	434	309	280
		Rating	1	2	3	4
Sustainable development of man-made, human, and knowledge capital	Contribution to the knowledge of inhabitants with respect to water and energy challenges		No	Yes	Yes	Yes
		Rating	1	2	2	2
	Operating and investment costs[2]	million €	37.4	57.7	62.7	54.7
		Rating	4	2	1	3

Note Calculation of the figures compare further [1]Friedrich et al. (2020) and [2]Poganietz et al. (2021)

The current system tolerates a rather uninformed user, who must recognize only a minimum of regulations (Hummel et al., 2004). The implementation of the system alternatives is partly motivated to reduce the environmental load of the current system and to install an improved source for energy and nutrients. Thus, they incorporate knowledge-transferring design elements. In addition, due to the semi- and decentralized system components of the system alternatives, the structures that shape the water-energy-waste system are per se much more present in the residents' environment than in the status quo system (Michalek et al., 2016; van Vliet & Spaargaren, 2010). Therefore, the system alternatives are better rated than the status quo. Since the decentralized greywater treatment plants in SYAL2 and 3 bring an additional system component into the immediate living environment, both options are more highly rated compared with SYAL1 (Table 6).

The entire implementation and operating costs of the system alternatives are higher by far compared to the current systems, ranging between +46.2% (SYAL3) and + 67.7% (SYAL2). The main reason is the comparably high investment costs, which are higher by at least a factor of 2.3 (SYAL1) than the one of the status quo (Poganietz et al., 2021). The ratings follow the ranking accordingly (Table 6).

4.3 Preserving Society's Options for Development and Action

The current system is the outcome of the long history of development of the water-wastewater system with no participation by the user. Constituent to the system alternatives is the participation of the user, since the final design of the future system has to be identified between the stakeholders. Thus, the system alternatives are rated higher, i.e., with 2, than the status quo (Table 7).

The proven functionality of the current system leads to high acceptance, even though the newly emerging challenges, like micropollutants, are known to at least some users (Centrum für Soziale Investitionen & Innovationen, 2019). Characteristic of the system alternatives is the implementation of technical components that challenge traditional conventions as well as the immediate urban living environment (van Vliet et al., 2010). At least as long as the new components are not established and their functionality not proven in practice, users will be reluctant to accept these technologies and the accompanying changes in behavior. Users demand clear evidence of the improvement in the quality of the service offered, affecting their own lives in particular (Centrum für Soziale Investitionen & Innovationen, 2019). Thus, the acceptance of the system alternatives could be rated below the one of the status quo. The challenges of SYAL2 and 3 could be evaluated even higher compared to SYAL1, since the changes are more demanding, e.g., reuse of treated greywater. Therefore, the status quo is rated highest, followed by SYAL1. SYAL2 and 3 get the lowest rating (Table 7).

The status quo is characterized by a low need for societal responsibility, since it allows incorrect usage to some extent (Hummel et al., 2004). In contrast, the system alternatives demand a more active user than the current system, such as misconduct,

Table 7 Preserving society's options for development and action

Rule	Indicator	Unit	SQ	SYAL1	SYAL2	SYAL3
Participation in societal decision-making processes	Initiation of opportunities in respect to participation		No	Yes	Yes	Yes
		Rating	1	2	2	2
Conservation of social resources (tolerance, solidarity, etc.)	Acceptance by user		Accepted	Reduced acceptance	Lowest acceptance	Lowest acceptance
		Rating	4	3	1	1
	Promotion of societal responsibility		No	Enhanced	Highest	Highest
		Rating	1	2	3	3

e.g., disposing of solid waste via the vacuum toilet will result in more severe disruptions of the entire system (Hegger et al., 2008). It is necessary that the residents be made aware of their new role as "prosumers". Residents stay "consumers" of the system in a traditional understanding, and at the same time they provide necessary resources. The demands for residents are higher in SYAL2 and 3 compared to SYAL1, as the former places shredder for organic kitchen waste below the sink. The options are rated accordingly (Table 7).

4.4 *Comprehensive Assessment*

As a reference for the comprehensive assessment of the systems under review, the goals, the rules within a goal, and the indicators within each rule, if required, are treated alike. Figure 5 summarizes the ranking of the options under the conditions of the reference.

In respect to the goal "securing human existence", SYAL3 shows the highest ranking regarding two rules. SYAL2 is second, and SYAL1 third. Only in the case of "protection of human health" does SYAL1 outperform the other options, followed by SYAL2 and 3. The main reason for this outcome is the availability of treated greywater and rainwater for urban blue-green spaces. For all rules, the status quo has the lowest ranking. This entire pattern is robust; any other weighting scheme would not change it, only the "distance" between the options would vary.

In contrast to the first goal, regarding the second goal, "maintaining society's productive potentials", SYAL3 outperforms all other options in respect to all rules considered, with the exception of "sustainable use of renewable resources". Here,

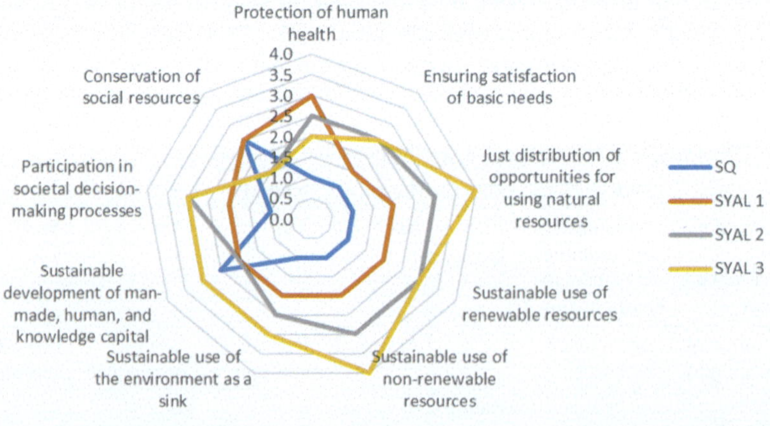

Fig. 5 Rating the rules. *Note* The first three rules, starting clockwise from the top, belong to the goal "securing human existence"; the next four rules define the goal "maintaining society's productive potentials"; finally, the last two address the goal "preserving society's options for development and action"

SYAL3 and 2 are ranked alike. Under the assumption of equal weighting of all four rules, SYAL1 and 2 are only better than the status quo regarding three of four rules. In respect to the rule "sustainable development of man-made, human, and knowledge capital", the status quo achieves the second rank. However, the outcome of this rule depends crucially on the weighting of the indicators. Reinforcing the relevance of costs would favor the current system; a weighting of 0.7 or higher would outpace SYAL3. Costs (or economic considerations in general) are often seen as a crucial aspect of infrastructure-related decisions. Besides this rule, any other change in the weighting would not change the ranking between the options, indicating a robust pattern with respect to the first three rules.

An unclear pattern can be observed with respect to the third goal "preserving society's options for development and action". The rule "participation in societal decision-making process" favors SYAL2 and 3, whereas the rule "conservation of social resources" supports the status quo and SYAL1. The outcome of the second rule depends crucially on the chosen weightings. Even a small change in the weights to the advantage of the indicator "acceptance by user" would back the status quo unequivocally.

To sum up, the weighting of two indicators, which are often in the center of public discussions, could challenge the findings of the reference, with an even ranking of the goals, the rules within each goal, and the indicators within each rule.

5 Discussion and Conclusions

The overarching aim of this contribution is to analyze whether different system alternatives for urban infrastructures are favorable from a sustainability point of view, i.e., whether the transformation of the current water-energy-waste system to an integrated one will improve the current situation.

A big picture view of the results indicates that SYAL3 has the highest potential for the desired improvement. Realizing SYAL3 would mean the most intense change of the current system, leading to a more or less "sewerless" society. With respect to the first goal, SYAL3 dominates two of three rules, but trails in respect to the "protection of human health" behind SYAL1 and 2. Regarding the second goal, SYAL3 shows the best performance of all of the rules. Only the last goal gives no clear pattern, although SYAL3 could achieve the best outcome of one rule. The main weaknesses of SYAL3 could be seen in its potentially high costs that affect the "sustainable development of man-made, human, and knowledge capital" and in the "conservation of social resources", or to be more specific, in the presumably low acceptance of SYAL3 due to the required changes.

However, the findings do not claim that SYAL3 or any other option achieves a sustainable outcome. The chosen approach to compare the indicators and rules allows only the ranking between the options, indicating an improvement in the direction of sustainability (or not). A precondition to declaring one or more options as sustainable would demand the respective knowledge regarding each selected indicator requiring additional research.

The decision whether and which system alternatives should be realized will ultimately depend on how society evaluates the goals, rules and indicators. The reference system indicates a specific preference structure of the "society" that is not entirely robust. Changes in the weighting could challenge the findings. For example, if acceptance and costs are valued sufficiently higher than in the reference, this would intensify possible conflicting aims, impeding the transformation of the water-energy-waste system.

The presented study assumes a society with homogeneous preferences. However, the decision process faces different challenges, like diverse preferences between decision-maker and user, between the beneficiaries and losers of a transformation, and the possible discrepancy between the willingness to pay for the implementation of a system alternative and the ability to pay for it.

The advantage of the rating scheme used is the simplicity and transparency. But such a system does not recognize improvements and deterioration compared to the current situation, valuing each change alike. A possible alternative would have been to apply, e.g., Risk Assessment Matrixes or Leopold Matrixes (Leopold et al., 1971; Pastakia & Jensen, 1998). Both approaches value changes in relation to a reference situation. Such an approach would require at least a semi-quantification of specific indicators and rules based on defined criteria and not in comparison to alternative options. This would need additional research in order to apply it.

The study focuses on the conditions of a neighborhood in Heidelberg, a medium-large city in southwest Germany, which determine the selected rules and indicators. The conditions in the neighborhood also set the extent of the required changes to achieve a specific system design, influencing the potential gain of a transformation and thus, the final design of a transformed system. Additional research under different conditions is needed to validate the presented findings.

To sum up, the sustainability assessment shows the advantages and applicability of an integrated wastewater-waste system even in an existing building stock. But the exercise also shows the system alternatives discussed have to overcome some hurdles, like costs and user acceptance, before they can be implemented successfully. Thus, a main challenge is to learn more about the nature of the hurdles and about potential solutions to mitigate their relevance in a decision process.

Acknowledgements The study presented bases partly on the Ph.D. thesis of Jasmin Friedrich, which was prepared within the framework of the cooperative college Energy Systems and Resource Efficiency (ENRES) (https://www.hs-pforzheim.de/forschung/kooperative_promotionskollegs/enres_energiesysteme_und_ressourceneffizienz). Jasmin Friedrich was supported by a scholarship of the Landesgraduiertenstiftung of the State of Baden-Württemberg. The funds are gratefully acknowledged.

References

Arcadis. (2016). *Sustainable cities water index*. https://www.arcadis.com/media/4/6/2/%7B462EFA0A-4278-49DF-9943-C067182CA682%7DArcadis%20Sustainable%20Cities%20Water%20Index_003.pdf

AZV. (2019). *Aufgabe des Abwasserzweckverbands Heidelberg*. http://www.azv-heidelberg.de

BBSR. (2016). *Anpassung an den Klimawandel in Stadt und Region: Forschungserkenntnisse und Werkzeuge zur Unterstützung von Kommunen und Regionen*. Bonn. Bundesinstitut für Bau-, Stadt- und Raumforschung im Bundesamt für Bauwesen und Raumordnung (BBSR).

Behrendt, S. (2018). *Mikroschadstoffe in Gewässern*.

BGR. (2016). *Salze in Deutschland*. Bundesanstalt für Geowissenschaften und Rohstoffe (BGR).

BMUB. (2016). *Leitfaden Nachhaltiges Bauen: Zukunftsfähiges Planen, Bauen und Betreiben von Gebäuden*. Bundesministerium für Umwelt, Naturschutz, Nukleare Sicherheit und Bauen (BMUB).

Brundtland, G. H. (1991). *Our common future* (13. impr). *Oxford paperbacks*. University Press.

Centrum für Soziale Investitionen und Innovationen. (2019). *Bericht zur Untersuchung der Einstellung zu NASS in Heidelberg*. University of Heidelberg.

Daigger, G. T. (2007). Wastewater management in the 21st century. *Journal of Environmental Engineering, 133*(7), 671–680. https://doi.org/10.1061/(ASCE)0733-9372(2007)133:7(671)

UN Desa. (2016). *Climate change resilience: An opportunity for reducing inequalities*. United Nations Department of Economic and Social Affairs.

Dusseldorp, M. (2017). *Zielkonflikte der Nachhaltigkeit*. Dissertation [XI, 256 Seiten]. Springer Fachmedien Wiesbaden GmbH.

Facey, S. J., Nebel, B. A., Kontny, L., Allgaier, M., & Hauer, B. (2018). Rapid and complete degradation of diclofenac by native soil microorganisms. *Environmental Technology & Innovation, 10*, 55–61. https://doi.org/10.1016/j.eti.2017.12.009

Friedrich, J. (2020). *Nachhaltigkeitsbewertung von Systemalternativen zur Transformation des Wasser-Energie-Nexus im städtischen Gebäudebestand* [Dissertation]. Karlsruhe Institute of Technology.

Friedrich, J., Poganietz, W.-R., & Lehn, H. (2020). Life-cycle assessment of system alternatives for the water-energy-waste nexus in the urban building stock. *Resources, Conservation and Recycling, 158*, 104808. https://doi.org/10.1016/j.resconrec.2020.104808

Gantner, K., & Barjenbruch, M. (2012). *Reduzierung des Frachteintrags aus Mischwasserentlastungen*. https://www.siwawi.tu-berlin.de/fileadmin/fg118/120326Abschlussbericht_Frachteintrag_downl.pdf

Gellings, C. W., & Parmenter, K. (2009). Energy efficiency in fertilizer production and use. In C. W. Gellings (Ed.), *Efficient use and conservation of energy—Volume II* (pp. 123–136). Eolss Publishers.

Goedkoop, M., Heijungs, R., Huijbregts, M., Schryver, A. de, Struijs, J., & van Zelm, R. (2013). *ReCiPe 2008. A life cycle impact assessment method which comprises harmonised category indicators at the midpoint and the endpoint level*. https://35f23ee4-a-62cb3a1a-s-sites.googlegroups.com/site/lciarecipe/file-cabinet/ReCiPe_main_report_MAY_2013.pdf

Graaff, M. S. de. (2010). *Resource recovery from black water* [Ph.D thesis]. Wageningen University.

Grunwald, A. (2016). *Nachhaltigkeit verstehen: Arbeiten an der Bedeutung nachhaltiger Entwicklung*. oekom.

Grunwald, A., & Kopfmüller, J. (2012). *Nachhaltigkeit: Eine Einführung* (2., aktualisierte Auflage). Campus Verlag.

Gutjahr, M., & Müller-Schaper, J. (2018). Treatment of sewage sludge in Europe: Status Quo and perspectives of energy and phosphorus recovery—Illustrated by the example of selected countries. In S. Thiel, E. Thomé-Kozmiensky, F. Winter, & D. Juchelková (Eds.), *Waste management—Waste-to-energy* (Vol. 8, pp. 371–384). Thomé-Kozmiensky Verlag.

Han, W., Zhao, Y., & Chen, H. (2016). Study on biogas production of joint anaerobic digestion with excess sludge and kitchen waste. *Procedia Environmental Sciences, 35*, 756–762. https://doi.org/10.1016/j.proenv.2016.07.089

Heckenmüller, M., Narita, D., & Klepper, G. (2014). *Global availability of phosphorus and its implications for global food supply: An economic overview* (Kiel Working Paper No. 1897). Kiel Institute for the World Economy (IfW).

Hegger, D., van Vliet, B., & Spaargaren, G. (2008). *Decentralized sanitation and reuse in Dutch society: Social opportunities and risks.* Wageningen University.

Hiessl, H., Hillenbrand, T., Gandenberger, C., Klug, S., Peter, A., & Tettenborn, F. (2010). *DEzentrales Urbanes InfrastrukturSystem DEUS21: Abschlussbericht.* Fraunhofer ISI.

Hummel, D., Hertler, C., Niemann, S., Lux, A., & Janowicz, C. (2004). *Versorgungssysteme als Gegenstand sozial-ökologischer Forschung: Ernährung und Wasser*. Frankfurt / Main

Kopfmüller, J., Jörissen, J., Brandl, V., Paetau, M., Banse, G., Coenen, R., & Grunwald, A. (2001). *Nachhaltige Entwicklung integrativ betrachtet – Konstitutive Elemente, Regeln, Indikatoren.* Edition Sigma.

Larsen, T. A., Alder, A. C., Eggen, R. I. L., Maurer, M., & Lienert, J. (2009). Source separation: Will we see a paradigm shift in wastewater handling? *Environmental Science & Technology, 43*(16), 6121–6125. https://doi.org/10.1021/es803001r

LAWA. (2016). *Mikroschadstoffe in Gewässern.* Bund/Länder-Arbeitsgemeinschaft Wasser (LAWA).

Leopold, L. B., Clarke, F. E., Hanshaw, B. B., & Balsley, J. R. (1971). *A procedure for evaluating environmental impact* (Geological Survey Circular No. 645). Geological Survey.

LUBW. (2017). *Planungsempfehlungen für die (stadt-)klimawandelgerechte Entwicklung von Konversionsflächen* (KLIMOPASS-Berichte: Klimawandel und modellhafte Anpassung in Baden-Württemberg). Landesanstalt für Umwelt Baden-Württemberg (LUBW).

LUBW. (2019). *Zustandsbericht über die Grundwasservorräte.* https://guq.lubw.baden-wuerttemberg.de/GuQWeb.dll/p79197.html

Martin, B., Curddas, P., & Hutchings, P. (2015). *Imagining a sewerless society* (Nexus Network Think Piece Series No. 5). http://www.thenexusnetwork.org/wpcontent/uploads/2014/08/MartinCruddasandHutchings_SewerlessSocietyNexusThinkpiece2015.pdf

Mata-Alvarez, J., Macé, S., & Llabrés, P. (2000). Anaerobic digestion of organic solid wastes. An overview of research achievements and perspectives. *Bioresource Technology, 74*(1), 3–16. https://doi.org/10.1016/S0960-8524(00)00023-7

Matzinger, A. (2017). *Zielorientierte Planung von Maßnahmen der Regenwasserbewirtschaftung: Ergebnisse aus dem Projekt KURAS.* http://www.kuras-projekt.de/fileadmin/Dokumenten_Verwaltung/pdf/20170428_Leitfaden_Regenwasser_full_final_med_res.pdf

Michalek, G., Meran, G., Schwarze, R., & Yildiz, Ö. (2016). Nudging as a new 'soft' tool in environmental policy. An analysis based on insights from cognitive and social psychology. *Zeitschrift für Umweltrecht & Umweltpolitik, 39*, 169–207.

Morandi, C. G., Wasielewski, S., Mouarkech, K., Minke, R., & Steinmetz, H. (2018). Impact of new sanitation technologies upon conventional wastewater infrastructures. *Urban Water Journal, 15*(6), 526–533. https://doi.org/10.1080/1573062X.2017.1301502

Myhre, G., Shindell, D., Bréon, F.-M., Collins, W., Fuglestvedt, J., Huang, J., & et al. (2013). Anthropogenic and natural radiative forcing. In T. F. Stocker, D. Qin, G.-K. Plattner, M. Tignor, S. K. Allen, J. Boschung et al. (Eds.), *Climate change 2013: The physical science basis. Contribution of working group I to the fifth assessment report of the intergovernmental panel on climate change* (pp. 659–740). Cambridge University Press.

Nakamura, M., Pendlebury, D., Schnell, J., & Szomszor, M. (2019). *Navigating the structure of research on sustainable development goals.* Institute for Scientific Information.

Nolde, E. (2013). *Hohe Energie- und Wassereffizienz durch Grauwasserrecycling mit vorgeschalteter Wärmerückgewinnung.* https://www.fbr.de/fileadmin/Daten/Artikel_aus_wsp/Artikel-Arnimplatz_GW_fbr_1_13.pdf

Northcott, M. S. (2018). Neoliberal (Mis)Management of earth-time and the ethics of climate justice. In C. Murphy, P. Gardoni, & R. McKim (Eds.), *Climate change and its impacts* (pp. 233–250). Springer International Publishing. https://doi.org/10.1007/978-3-319-77544-9_14

Pastakia, C. M., & Jensen, A. (1998). The rapid impact assessment matrix (RIAM) For EIA. *Environmental Impact Assessment Review, 18*(5), 461–482. https://doi.org/10.1016/S0195-9255(98)00018-3

Poganietz, W.-R., Friedrich, J., & Lehn, H. (2021). Eco-efficiency of system alternatives of the urban water-energy-waste nexus. *NachhaltigkeitsManagementForum | Sustainability management forum*. https://doi.org/10.1007/s00550-021-00517-5

Posner, E. A., & Weisbach, D. (2010). *Climate change justice*. Princeton University Press. https://press.princeton.edu/books/hardcover/9780691137759/climate-change-justice

Purvis, B., Mao, Y., & Robinson, D. (2019). Three pillars of sustainability: In search of conceptual origins. *Sustainability Science, 14*(3), 681–695. https://doi.org/10.1007/s11625-018-0627-5

Ramboll. (2018). *Blau-grüne Infrastruktur*. https://de.ramboll.com/services/stadtplanung-und-gestaltung/blau-grueneinfrastruktur

Rechenberg, J. (2019). *Deutschland droht Kampf ums Wasser*. https://www.tagesschau.de/inland/wasser-131.html

Remy, C. (2010). *Life cycle assessment of conventional and source-separation systems for urban wastewater management [Dissertation]*. TU Berlin.

Robinson, J. (2004). Squaring the circle? Some thoughts on the idea of sustainable development. *Ecological Economics, 48*(4), 369–384. https://doi.org/10.1016/j.ecolecon.2003.10.017

Schaum, C. (2018). *Abwasserbehandlung der Zukunft: Gesundheits-, Gewässer- und Ressourcenschutz* [Dissertation]. TU Darmstadt.

Schwaiger, J., Ferling, H., Mallow, U., Wintermayr, H., & Negele, R. (2004). Toxic effects of the non-steroidal anti-inflammatory drug diclofenac: Part I. Histopathological alterations and bioaccumulation in rainbow trout. *Aquatic Toxicology, 68*, 141–150.

Skambraks, A.-K., Kjerstadius, H., Meier, M., Davidsson, Å., Wuttke, M., & Giese, T. (2017). Source separation sewage systems as a trend in urban wastewater management: Drivers for the implementation of pilot areas in Northern Europe. *Sustainable Cities and Society, 28*, 287–296. https://doi.org/10.1016/j.scs.2016.09.013

SRU. (2018). *Wohnungsneubau langfristig denken: Für mehr Umweltschutz und Lebensqualität in den Städten*. Geschäftstelle des SRU.

Staben, N. (2008). *Technische Möglichkeiten der alternativen Gestaltung städtischer Wasser- und Abwasserinfrastruktur: Eine Technikrecherche im Rahmen des Projekts Transformationsmanagement für eine nachhaltige Wasserwirtschaft*. Deutsches Institut für Urbanistik. netWORKS-Paper.

Stadt Hamburg. (2017). *Jenfelder Au am Wasser Zuhause*. http://www.jenfelderau-info.de/index.php/Broschueren-Plaene.html

Stadt Heidelberg. (2017). *Der Bioabfall: Aus Abfall wird neue Erde*. http://www.heidelberg.de/hd,Lde/HD/Rathaus/Bioabfall+und+Kompost.html

Stadtwerke Heidelberg. (2019). *Trinkwasser-Umstellung am Hang in Handschuhsheim und Neuenheim*. https://www.swhd.de/de/SWH/Presse/Meldungslisten-SWH/Meldungsliste-SWH-2/Trinkwasser-Umstellung-am-Hang-in-Handschuhsheim-und-Neuenheim.html

Steinmetz, G., & Morandi, C. (2017). Ressourceneffizienz in der Abwasserwirtschaft. *Ingenieurspiegel, 4*, 69–72.

Suryandari, D., & Bolivar Paypay, V. (2024). Closing the loop of nutrients by an urban-rural partnership—A German project. In A. C. Braun, G. Espinosa, D. Tröger, & T. Hirth (Eds.), *Eco-industrial development as an industrial strategy—Contributions from a German-Chilean research partnership*.

Transforming our world: The 2030 Agenda for Sustainable Development (2015 & rev. A/Res/70/1).

UN DESA. (2017). *Climate change and social inequality* (DESA Working Paper No. 152). United Nations Department of Economic and Social Affairs.

U.S. Department of Energy. (2013). *Water heating*. https://www.energy.gov/energysaver/heat-and-cool/water-heating

Umweltbundesamt. (2016). *Die Nutzung natürlicher Ressourcen: Bericht für Deutschland 2016*. Dessau-Roßlau.

Umweltbundesamt. (2018). *Antibiotika und Antibiotikaresistenzen in der Umwelt: Hintergrund, Herausforderungen und Handlungsoptionen*. Dessau-Roßlau.

Untersteller, F. (2017). *Baden im Neckar: Möglichkeiten, Hindernisse und Chancen*. Stuttgart. Landtag des Landes Baden-Württemberg. https://www.landtag-bw.de/files/live/sites/LTBW/files/dokumente/WP16/Drucksachen/2000/16_2950_D.pdf

UN WWAP. (2017). *The United Nations World Water Development Report 2017: Wastewater: The untapped resource*. United Nations World Water Assessment Programme (WWAP). http://unesdoc.unesco.org/images/0024/002471/247153e.pdf

van Vliet, B., & Spaargaren, G. (2010). Sense and sanitation. In B. van Vliet, G. Spaargaren, & P. Oosterveer (Eds.), *Social perspectives on the sanitation challenge* (1st ed., pp. 31–47). Springer Netherlands. https://doi.org/10.1007/978-90-481-3721-3_3

van Vliet, B., Spaargaren, G., & Oosterveer, P. (Eds.). (2010). *Social perspectives on the sanitation challenge* (1. Aufl.). Springer Netherlands. http://gbv.eblib.com/patron/FullRecord.aspx?p=993591

Vaneeckhaute, C. (2019). Towards quality by design and process analytical technology for enhanced nutrient recovery from wastewaters. *NPJ Clean Water, 2*(1). https://doi.org/10.1038/s41545-019-0038-x

Vergara-Araya, M., Lehn, H., & Poganietz, W.-R. (2020). Integrated water, waste and energy management systems—A case study from Curauma, Chile. *Resources, Conservation and Recycling, 156*, 104725. https://doi.org/10.1016/j.resconrec.2020.104725

WIdO. (2013). *GKV-Arzneimittelindex*. Springer.

Wüstenhagen, R., Wolsink, M., & Bürer, M. J. (2007). Social acceptance of renewable energy innovation: An introduction to the concept. *Modeling Transport (Energy) Demand and Policies, 35*(5), 2683–2691. https://doi.org/10.1016/j.enpol.2006.12.001

Closing the Loop of Nutrients by an Urban-Rural Partnership—A German Project

Dinar Suryandari and Vanessa Bolivar Paypay

Abstract The Rural-Urban Nutrient Partnership (RUN) project offers a solution to close the nutrient cycle in a German urban settlement by introducing an innovative technical and social approach. The technical concept of RUN considers the recovery of nutrients and biopolymers from a mixture of blackwater and kitchen waste through the development and combined application of technologies, such as struvite chemical precipitation, usage of natural zeolites for the sorption of nitrogen and potassium, biopolymer, and biochar production. The selection and operational principles of these technologies are described and discussed in this chapter. In addition, the non-technical concept as part of the societal approach is also comprehensively discussed, including the real laboratory and experience space. Moreover, a sustainability assessment of this project will be conducted to evaluate the concept holistically. The indicator-based sustainability assessment, called the Integrative Concept of Sustainable Development (ICoS), was chosen in this purpose. The proposed indicators that represent the sustainability in RUN are also included. Finally, this sub-chapter is expected to provide insights into the sustainable use of wastewater treatment and nutrient recovery technologies from a research perspective. It can also be seen as an innovative solution to address the sustainability challenges that countries in the world face, including Chile.

1 Introduction

1.1 Motivation and Aim of the Study

The urban area has become an essential part of a country in implementing sustainable development. It has unique capabilities in creating economic activities and shaping society while causing environmental degradation. In 2018, about 55% of the world's

D. Suryandari (✉) · V. B. Paypay
Institute for Technology Assessment and Systems Analysis (ITAS), Karlsruhe Institute of Technology (KIT), Karlsruhe, Germany
e-mail: dinar.suryandari2@kit.edu

A. Ch. Braun et al. (eds.), *Eco-Industrial Development as an Industrial Strategy*, Green Energy and Technology, https://doi.org/10.1007/978-3-031-73576-9_10

population lived in urban areas, and it is predicted that that number will be even higher in the future (United Nations, 2018). It is a challenge for the urban practitioner to meet the population's basic needs without compromising the environment, for example, the extensive use of groundwater in drought areas, or inadequate treatment of wastewater. The urban-rural partnership is seen as a promising approach to improve urban developments by creating a regional circular economy with respect to nutrient recovery and use.

The new concept of the German research project Rural-Urban Nutrient Partnership (RUN) (https://www.run-projekt.de/) aims to introduce a partnership concept between rural and urban areas to close nutrient cycles. Resource recovery from waste streams is considered for addressing the urban challenge and achieving sustainability within the project. It creates an opportunity to reuse the embedded resources while the treatment activities are carried out. The innovative technologies within the RUN concept are designed to recover nutrients from wastewater and organic kitchen waste generated in urban areas. The wastewater is separated directly at the source into greywater and blackwater, and the latter is mixed with organic kitchen waste and processed into fertilizer. The fertilizer is used in rural areas to produce food that urban residents can later consume (RUN Consortium, 2020).

Most of the research on nutrient recovery technologies for the design of nutrient-specific fertilizers, e.g. struvite, has focused mostly on conventional wastewater treatment plants without wastewater separation (Umweltbundesamt, 2019). While in some previous experiences with alternative on-source wastewater separation techniques, nutrient recovery with the use of advanced technologies was intended but not always implemented (Skambraks et al., 2017), in many cases biosolids from blackwater were only used to produce soil improvers such as compost (Schimpf & Mildenberger, 2021). Fewer research projects similar to RUN include the production of high-quality fertilizers such as struvite and/or nitrogen rich zeolites from on-source separated blackwater (Larsen et al., 2009; Schimpf & Mildenberger, 2021; Winker et al., 2012). Building upon some of these experiences, the RUN technical concept includes the combination of innovative nutrient and biopolymer recovery technologies, which realize synergies within the urban-rural system of RUN.

The sustainability assessment of the RUN concept is conducted to ensure that the concept can be a sustainable treatment and resource recovery technology option. The indicator-based sustainability assessment, called the Integrative Concept of Sustainable Development (ICoS), evaluates the RUN concept with a holistic approach. Unlike other sustainability assessment concepts, ICoS provides a multidimensional approach with three goals of sustainability. It does not look at the individual dimensions of sustainability, namely ecology, social, and economy, but at the comprehensive and holistic approach (Kopfmüller et al., 2001).

Based on the above motivation, this chapter aims to provide a concise overview of innovative technical possibilities to manage household wastewater in a different and more efficient way, e.g. through wastewater separation at the source and recovery of valuable nutrients from wastewater and waste streams. In addition, this paper allows the reader to gain insights into the unique technical approach for resource recovery from separately collected household blackwater and biowaste of a study case in Germany. The latter includes the development of a technical system that maximizes the recovery of nutrients and biopolymers through the development and use of different innovative technologies that will be tested afterwards at pilot scale in a German neighborhood. In addition, the transformation process of society in participating and accepting the new technology itself is comprehensively presented by conducting a sustainability assessment. The sustainability assessment is presented to summarize the preliminary findings from the technical and non-technical approaches and to explore the possibility of implementing the sustainability of the RUN concept. It could also help decision makers to decide whether the project can contribute to sustainable development and serve as a reflection medium for the project consortium to improve the project. However, as the project is still ongoing, the focus of this sub-chapter is on discussing the conceptual framework as well as the identified rules and indicators. It is expected that this chapter can help broaden the research perspective in the field of the sustainable use of wastewater treatment and nutrient recovery. Therefore, it could help countries around the world, including Chile, to address future sustainability challenges.

2 Latest Development of Current Wastewater Treatment Technologies

Several studies have already been conducted in sustainable wastewater treatment and resource recovery that can provide scientific references in developing the RUN. These other similar works are explained in detail in this section.

2.1 Wastewater Treatment with Source Separation

Source separation is an approach to dividing wastewater into different streams. The goal is to minimize the dilution of domestic wastewater and thus maximize nutrient recovery. It also has the advantage of efficiently treating wastewater streams according to their characteristics (Otterpohl et al., 2004). Source separation often separates domestic wastewater into different streams, such as wastewater from the toilet (blackwater) and other domestic wastewater, such as water from the shower, washing machine, kitchen, etc. (greywater). It is possible to add organic kitchen waste to the blackwater (Skambraks et al., 2017). In another source separation objective, the blackwater could be split again into urine and feces with flush water (brown water) (Larsen et al., 2009; Tilley, 2014).

Urine source separation is one of the examples of wastewater treatment with source separation. The aim is to use the urine directly for agriculture with minimal treatment. By separating urine and brown water, dilution can be avoided. Thus, nutrients in both wastewater streams can be efficiently preserved. Urine separation is accomplished by installing a special toilet called a urine diversion toilet (UDT) that stores urine and brown water in different chambers (Larsen et al., 2009). From here, the urine is treated to achieve some objectives such as hygienization, volume reduction, stabilization, and removal of micropollutants (Maurer et al., 2006). The brown water could be treated with various treatment technologies to recover the nutrients or added to the biogas digester for energy production (Tilley, 2014).

A project in which wastewater is separately collected into two streams—blackwater and greywater—similar to the RUN approach, has also been implemented in the Netherlands. The project in the Noorderhoek residential area is part of the Desah and Waterschoon project in the Netherlands and runs on a scale of 232 houses, where wastewater separation and treatment of wastewater into useful products is carried out, see Fig. 1. The blackwater is collected using a vacuum toilet with less water and mixed with organic kitchen waste from the kitchen shredder and treated together in a biogas reactor. The nutrients from the blackwater are recovered in the form of fertilizer. The greywater is treated at the local wastewater treatment plant and the heat is extracted for residential heating (Kisser et al., 2020). Separate collection with a vacuum transport system for household wastewater was also implemented and operated at small pilot scale in a multifamily building in Freiburg-Vauban (Coates, 2013) and in an ecological housing state in Lübeck-Flintenbreite (Bauer, 2017). Most of the previous experiences with wastewater separation into streams such as greywater and blackwater considered the use of the solid byproducts as fertilizers or soil improvers, for instance as compost. Unlike the RUN project, fewer include both the development of high-quality fertilizers such as struvite and nitrogen rich zeolites.

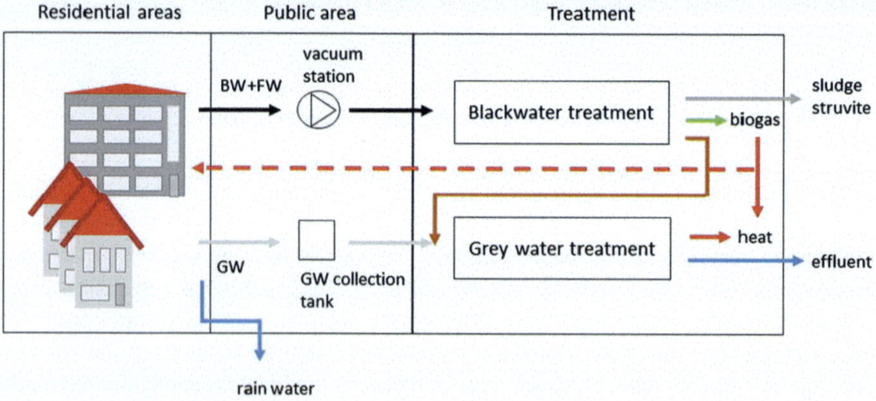

Fig. 1 Concept of wastewater separation at source in Noorderhoek, Netherlands. Figure was redrawn and adapted by Kisser et al. (2020) from (Wit et al., 2018)

2.2 *Nutrient Recovery Technologies from Organic Substrates*

Recovery of nutrients from organic substrates such as raw sewage sludge, digested sludge, supernatant liquor and animal manure have been widely investigated and implemented at pilot and full scale. In contrast, experiments with blackwater were conducted mostly at lab scale (Hiessl & Hillenbrand, 2010; Morandi et al., 2016). This section focuses mostly on the nutrient's recovery technologies from the liquid phase (after solid–liquid separation) or the digested sludge from conventional wastewater treatment plants. The RUN technical concept builds upon the operational principles of some of these technologies, with the difference that the RUN approach tries to optimize and maximize the recovery of not only one nutrient but three of them—nitrogen, phosphorus and potassium—with a unique sequential arrangement of nutrient recovery technologies that has not been applied for blackwater management so far.

Chemical Precipitation of Struvite with Addition of $MgCl_2$

Struvite is a crystalline mineral with the chemical formula $MgNH_4PO_4 * 6H_2O$. Phosphorus and nitrogen recovery in the form of struvite has been implemented at full scale in some wastewater treatment plants (WWTP) in Europe and the United States. For instance, in the Berliner Wasserbetriebe (Water and Wastewater Utility of Berlin) (Berliner Wasserbetriebe, 2021), struvite is successfully separated and produced from digested sludge on a large scale and is used later as fertilizer. In this process, struvite crystallization with the addition of $MgCl_2$ occurs directly from the sludge, and the optimal pH value is reached by stripping CO_2, using sludge aeration. Another example is the Canadian company Ostara Nutrient Recovery Technologies, Inc., which commercializes the technologies WASSTRIP (Waste Activated Sludge Stripping to Recover Internal Phosphate) and Pearl for struvite recovery in the wastewater treatment sector. As a practical example (OSTARA, 2016), in June 2016, the Dutch Waterboard Vallei & Veluwe (The Waterboard) officially opened Europe's first commercial nutrient recovery facility in Amersfoort, in partnership with Ostara Nutrient Recovery Technologies and Eliquo Water & Energy. The RUN concept also considers the chemical precipitation of struvite with the addition of $MgCl_2$. However, unlike the existing approaches, within the RUN project, blackwater mixed with biowaste is used as an input substrate. In addition, RUN includes a pre-acidification step prior struvite precipitation that improves phosphorus recovery.

Electrochemical Precipitation of Struvite

The electrochemical phosphorus recovery ePhos (Fraunhofer IGB) is a novel process to recover ammonium (NH_4^+) and phosphate (PO_4^{3-}) from filtrate wastewater, which was developed and patented by Fraunhofer IGB in Germany (Fraunhofer IGB, n.d.). In this electrochemical process with a magnesium electrode, NH_4^+ and PO_4^{3-} are precipitated as struvite. In comparison to other struvite recovery technologies, there is no need to manipulate chemicals such as NaOH for the precipitation process, which can also result in an easier operation. However, in comparison to RUN, it lacks a pre-bioacidification step to increase the dissolution of phosphates, which usually results in higher phosphorus recovery rates and is unique to the RUN approach.

Membrane Technologies

Some of the more common membrane technologies (MT) for phosphorus removal are tertiary membrane filters such as reverse osmosis (RO) systems and membrane bioreactors (MBRs) (Sengupta et al., 2015). However, MTs usually have a high phosphorus retention rate, which is undesirable for phosphorus recovery. Therefore, this is still an active field of research. For instance, in 2016, Thong et al. (2016) fabricated a nanofiltration membrane (NF) that was desirable for phosphorus recovery at 10 bar and using a feed solution of pH 2. Up to 90% of the feed phosphorus could be recovered using this newly developed NF membrane at a permeate recovery of 90%. Still another alternative is the use of membranes such as electro-membrane filtration (Bargeman et al., 2002) or forward osmosis (Ansari et al., 2017) as a phosphorus pre-concentrating technique, where the phosphorus rich brine (retentate) can be processed with the use of another phosphorus recovery technology to produce struvite afterward. This means that at the end of the process, a RUN-like approach is additionally required for recovering the phosphates in a marketable product. Apart from adding more complexity to the overall nutrient recovery process, the use of membrane technologies usually demands high energy consumption while membrane fouling is also considered a risk that should not be ignored.

Physical–Chemical Sorption and Ion Exchange

The phosphate removal can occur at the surface but also at the interior of a sorbent medium, therefore the term sorption instead of adsorption (as cited in other sources (Sengupta et al., 2015)) will be used here. There are five mechanisms of phosphate sorption (Loganathan et al., 2014): (a) ion exchange, (b) ligand exchange, (c) hydrogen bonding, (d) surface precipitation, and (e) diffusion into the interior structure of the sorbent. The sorbents for phosphate removal can be either inorganic (e.g. zeolites (Ma et al., 2011)), organic (e.g. ion exchange resins (Johir et al., 2011)), industrial by product (e.g. slag material, fly ash (Lu et al., 2009)) or biological waste (e.g. oyster shells (Lee et al., 2009)). The sorbent, once saturated with phosphate ions contained in the liquid phase of an organic substrate or wastewater, can also be regenerated (desorption) and reused repeatedly, and phosphorus can be recovered from the regenerant in a usable form such as struvite through chemical precipitation (Sengupta & Pandit, 2011). Alternatively, the phosphorus rich sorbent can also be directly used as a source of phosphorus if it has no contaminants (Loganathan et al., 2014).

However, certain sorbents, e.g. zeolites, are also used to recover nitrogen and potassium instead of phosphorus, which is the case of the RUN approach. A common drawback of the use of sorption processes either for phosphorus, nitrogen or potassium removal is the existence of other competing ions in the solution, which might affect the final removal efficiency. To overcome this gap, RUN project uses a selective ion exchanger known as clinoptilolite for nitrogen removal.

Nitrogen Recovery Through Ammonia Stripping

Ammonia stripping is a pH-dependent process where at a high pH value, ammonium nitrogen from solution converts to ammonia gas. Therefore, in the process, lime or sodium hydroxide are usually applied to keep the pH around 10.8–11.5 (Sengupta et al., 2015). This nitrogen removal technology has been implemented so far in municipal WWTPs, as well as in the livestock sector for slurry and manure treatment. Regarding nitrogen recovery, many studies have been conducted. For instance, McLeod et al. (2014) came up with a novel scheme where ammonia-rich wastewater was used as an absorbent to selectively extract CO_2 from biogas during its upgrading to biomethane in a packed bed mode. As a result of this study, ammonia in the form of NH_4HCO_3 was recovered. Jiang et al. (2014) recovered 90% ammonia from anaerobically digested dairy manure as $(NH_4)_2SO_4$ (ammonium sulfate). Both products are considered high quality nitrogen fertilizers. A limitation of this technology in comparison with the use of natural zeolites, as applied in RUN, is the addition of lime to control pH, which may create operational and maintenance concerns.

Nitrogen Recovery in the Form of Ammonium Sulfate Through the Use of Contactor Membranes

Studies mostly at the lab scale (Dittmann, 2016; Hasanoğlu et al., 2010) have been conducted to test the recovery of nitrogen in the form of ammonium from municipal wastewater and animal manure with the use of contactor membranes, resulting in nitrogen removal rates of over 99%. The contactors' membranes allow a gaseous phase (which contains ammonia from a nitrogen rich wastewater) and a liquid phase (which contains sulfuric acid) to come into direct contact with each other. So then the nitrogen rich ammonia reacts with sulfuric acid to form ammonium sulfate that can be commercialized for agriculture. In comparison to the use of zeolites for ammonium removal (e.g. RUN project), this process allows us to conduct two steps in one, removal and recovery of nitrogen in one step, reducing the technical complexity. Within RUN project, it is still to be defined whether the ammonium contained in the zeolites requires further processing prior to application in agriculture.

3 RUN Concept

Rural-Urban Nutrient Partnership (RUN) is an inter- and transdisciplinary research project that aims to build and practically test the future vision of a partnership between farmers and urban residents. The project aims to close nutrient and resource loops between regional urban and rural areas and to make a valuable contribution in increasing the resiliency and resource efficiency of agricultural systems by treating blackwater and organic kitchen waste. A representation of the RUN concept is presented in Fig. 2. Technology solutions are assessed in detail for ecological and economic sustainability to ensure that the project goal is achieved. The social concept promotes intensive communication between the actors, where the customers

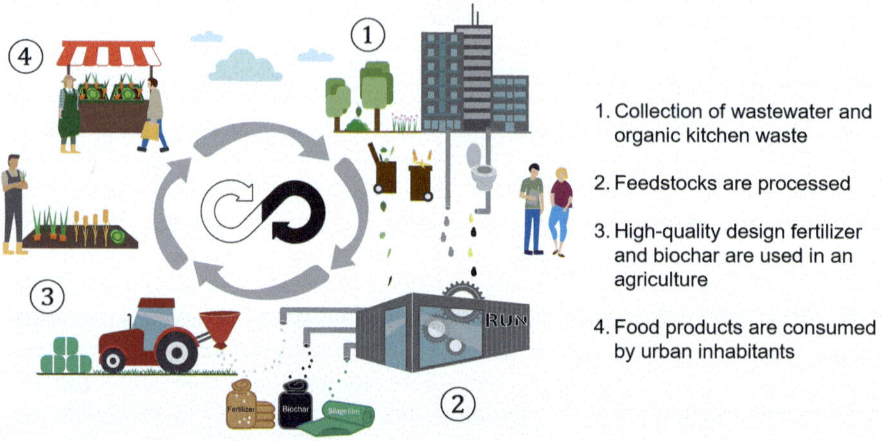

Fig. 2 Illustration of RUN concept (RUN Consortium, 2020)

and producers are actively involved in addressing the challenges (RUN Consortium, 2020).

In the structure of the RUN project, the concept is divided into two different approaches. The first one is the technical concept, where innovative technologies have been developed and integrated in one system to maximize the recovered resources such as nutrients for the production of different NPK fertilizers, as well as biopolymer for bioplastic production. The second one is the non-technical concept, demonstrating the technological concept understandably and practically in society.

The pilot plant with the capacity to treat the blackwater from around 100–200 inhabitants will be built in the area of Birkenfeld to put this project into practice. In addition, physical experience spaces will be installed close to the pilot plant as a real-world laboratory. It aims to illustrate the whole RUN concept to the audience and provide an experience with innovative technologies. It is expected that the experience spaces could induce a dialogue between stakeholders to discuss the RUN concept and its sustainability. Furthermore, the RUN concept will also include a scenario analysis, a life cycle assessment, and a sustainability assessment. The goal of implementing these methods is to evaluate the concept in a holistic approach. Later, the results will be used to analyze the transferability of the RUN concept to different regions and the possible future development, including the transformation process (RUN Consortium, 2019).

3.1 Technical Concept of RUN

Based on the principle of wastewater and waste separation at the source, household greywater and blackwater, as well as biowaste, are collected separately with the use of vacuum toilets and food grinders, respectively. The valuable resources contained in the blackwater and biowaste streams, such as biogenic carbon and nutrients, e.g. phosphorus (P), nitrogen (N) and potassium (K), are then later recovered through innovative technologies to produce fertilizer with a specific nutrients' ratio for agriculture, as well as biochar and biopolymers for the production of bioplastics. The technical component of the RUN Project (RUN Consortium, 2019) consists mainly of the development and combination of these innovative resource recovery technologies both at the conceptual and process engineering level.

Figure 3 shows the technical concept for integrated resource recovery from blackwater and biowaste, as it could be implemented in the pilot plant. The output flows (marked in red) represent the valuable sub-products. The testing and determination of the final combination of process technologies that maximize the nutrients (NPK) and biopolymer recovery is one of the research topics of the RUN project.

In a first step, blackwater and previously shredded biowaste are mixed and acidified. The use of acidic leaching prior to phosphorus recovery from digested sewage sludge was a technical approach initially followed in the so-called Stuttgart Process (Meyer et al., 2019), which was developed by the Institute of Sanitary Engineering, Water Quality and Waste Management of the University of Stuttgart. Unlike the Stuttgart Process, the acidification process of the substrate blackwater-biowaste within RUN does not require the use of chemicals such as sulfuric acid and is

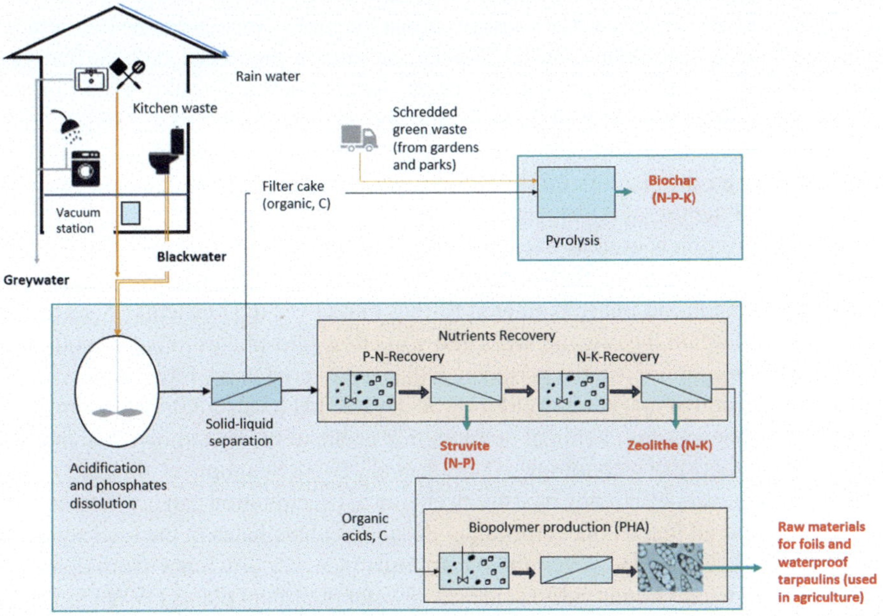

Fig. 3 Example of a possible concept for integrated nutrient and further material recovery from municipal wastewater and waste streams (RUN Consortium, 2021)

conducted under controlled anaerobic conditions up to a step previous to the production of biogas. On one side, acidification aims to optimize phosphorus dissolution from the solid to the liquid fraction, which enhances the phosphorus recovery efficiency. On the other side, it aims to convert proteins, fats and carbohydrates contained in the blackwater-biowaste mixture into short-chain fatty acids, also known as volatile fatty acids (VFA) (Uhrig et al., 2020). The latter are the raw materials for the production of polyhydroxialkanoates (PHA), which are biodegradable polymers that will be used later for manufacturing bioplastics.

After acidification, a solid–liquid separation step takes place. The remaining solid phase is then mixed with garden waste that could be later pyrolyzed. Pyrolysis is a thermochemical conversion process in which biomass is heated up to 400–900 °C under anoxic conditions (Franck & Schröder, 2015). In the last few years, pyrolysis has gained special attention as a carbon sequestration technology for climate change mitigation. The use of this technology also allows the reduction of biomass volumes and the elimination of pathogens and other toxic compounds contained in the biomass (Vali et al., 2021). In addition, pyrolysis products include a high energy pyrolysis gas and a chemical biological stable biochar. The latter can be potentially used as soil improver especially in low organic carbon soils (Jatav et al., 2021). Nevertheless, there are some limitations for its use. According to the Düngemittelverordnung (DüMV) (German fertilizer regulation), biochar from sewage sludge is still not allowed to be used for direct application to the soil. The ongoing debate focuses on the presence of heavy metals such as cadmium in certain types of biochar. Even though recent research studies have shown that heavy metals can be potentially separated from ash contained in the biochar, additional processes such as flue gas cleaning might lead to additional costs for product preparation, which according to experts ended in economic problems for pyrolysis plants in the past (Gleis, 2012; Vali et al., 2021). Therefore, the context-specific market demand of the recycling products among other factors might play an important role in the decision-making process toward the use of pyrolysis within RUN.

The ultrafiltration is used as a technology where the solid–liquid separation takes place. As mentioned above, the solid phase is transferred to the pyrolysis treatment facility while the liquid phase is treated further to recover the nutrients. A selective membrane will retain the organic acids and will allow phosphates to pass through it. The phosphates contained in the permeate will be recovered in the form of struvite by controlled addition of magnesium chloride at elevated pH levels >8.0. As exemplified in part 1.2.2., phosphorus chemical precipitation is one of the best known and market ready nutrient recovery technologies (Lorick et al., 2020; Sengupta et al., 2015). The phosphorus recovery efficiency of struvite chemical precipitation can lay between 85 and 97% compared to the concentration of dissolved phosphates in the feed solution (Sengupta et al., 2015). However, the phosphorus recovery efficiency from digested sludge or supernatant liquor in full scale wastewater treatment plants (WWTPs) with enhanced biological phosphorus removal (EBPR) yield recovery rates between 5 and 40% related to the total phosphorus load in raw sludge (Umweltbundesamt, 2019). With the technology known as the Stuttgart Process, which also includes a pre-step of phosphorus dissolution, up to a 50% phosphorus recovery rate can be achieved in WWTPs with either chemical or biological phosphorus removal. For the RUN project, the phosphorus recovery rate related to substrate input is still under investigation. Recovered phosphate in the form of struvite is expected to be used later as a high-quality soil fertilizer in the area of influence of the RUN project.

Following struvite precipitation, the remaining nitrogen- and potassium-rich stream undergoes adsorption in a reactor loaded with natural zeolites. Preliminary tests with biochar (by product from pyrolysis) as a sorption medium were also conducted within RUN. Results showed an insufficient adsorption rate of biochar for nitrogen and potassium removal (RUN Consortium, 2021), and therefore the use of ammonium and potassium selective natural zeolites was preferred (RUN Consortium, 2021). In recent years, wastewater treatment using zeolite has been examined by many researchers as a low-cost absorbent, which has the potential to remove heavy metals, ammonium, phosphorus, COD, dissolved organic matter and other cations (Kotoulas et al., 2019). For instance, other pilot studies with the use of natural zeolite clinoptilolite resulted in ammonium removal rates of around 70% for municipal wastewater (Beebe et al., 2013) and 85% for a mixture of blackwater with yellow water (Castro et al., 2021). However, the specific nutrient's removal efficiency of zeolites for the RUN project will depend on the influent characteristics, e.g. nutrien's load and pH as well as the design parameters e.g. type of zeolite, contact time, number of sorption cycles, etc. The further treatment steps of the loaded zeolites, prior to their use as fertilizer, also still need to be defined.

The now phosphorus-, nitrogen- and potassium low stream is sent to an aerobic second step where the VFA (organic acids that were obtained during the bioleaching) are used as a substrate of PHA-producing bacteria and stored in their cells in the form of PHA. The technical feasibility of this process within RUN has been conducted by the Technical University of Kaiserslautern. In this process, PHA-producing bacteria are first selected and enriched by a feast-famine process. Subsequently, PHA will be produced, harvested and finally chemically extracted for the purpose of being used later as a raw material for bioplastic production (Uhrig et al., 2020). Scientists of the same university have previously conducted lab tests with brown and blackwater, where these and other wastewater streams were acidified in order to obtain the substrate (VFA) for PHA production. Based on these results, they estimated that blackwater in Germany has the potential to produce 50,313 tons of PHA per year (Uhrig et al., 2020).

The production of bioplastics from the PHA produced within RUN is outside the system boundaries of the project. However, the recovered PHA, could potentially be commercialized for the production of foils and waterproof tarpaulins that can be used in agriculture.

3.2 Non-technical Concept of RUN

As briefly mentioned earlier, the RUN concept includes innovative technologies and a non-technical concept to make it more applicable to the current social situation.

The non-technical concept will help bridge the gap between technology and society. In this case, experimentation under actual conditions is required to understand the concept in a more practical way. In the RUN concept, this is demonstrated by an actual laboratory that includes the pilot plant, the information and experience space, and the integration of the user perspective. The pilot plant is to be set up in Birkenfeld that can serve as a demonstrator for the specific processes of nutrient recycling in which the system is to be tested. The pilot plant is thus the laboratory test on a larger scale. In order to provide residents with a real-life experience, an information and experience room will be set up near the pilot plant. Here, interested visitors can learn about the project and glimpse how the pilot plant works through "transparent production." Some installations such as vacuum toilets and a kitchen shredder will be set up to get to know the users' perspective on the new technologies. Local people, especially decision-makers and directly affected residents, will be involved explicitly through dialogue events. Their perspective is a significant input for planning and establishing nutrient partnerships in the region. The partnerships should develop into a long-term alliance. Therefore, the perspectives of relevant future users of the new technologies will be given special consideration.

4 Sustainability Assessment Approach

It is necessary to conduct a sustainability assessment to ensure that the RUN concept can provide practitioners and researchers with a sustainable treatment option. Therefore, the following section will explain the concept of sustainability assessment used in RUN Project in more detail.

4.1 Systems Under Investigation

The aim of the sustainability assessment is to compare the RUN system with a currently existing one. As a reference for the current system, the one operated in the City of Heidelberg is used (Friedrich et al., 2020). The wastewater waste system of Heidelberg is characterized by a central wastewater system without source separation of wastewater. Organic solid waste is collected separately from wastewater and is treated in a composting facility. Furthermore, the current production of fertilizer and cultivation of food is also included in the reference system. The assessment of the reference system will include changes in the economic and demographic conditions in the next decades as well as the implementation of currently know new regulations, e.g. for recovering phosphorus (AbfKlärV, 2017).

The RUN system sees source separation of wastewater, the integrated feed of kitchen wastes to blackwater, treatment of blackwater with organic wastes in RUN facilities, supplying fertilizer as well as bioplastic polymer PHA (see Sect. 3.1 and Fig. 3). In addition, the cultivation of food using RUN products is also included. The economic and demographic conditions assumed for the reference systems are also implemented in the RUN system.

4.2 The Assessment Approach (ICoS)

The Integrative Concept of Sustainable Development (ICoS) is one of the sustainability assessments commonly used to evaluate new technologies. It is a concept developed by the Helmholtz Association in Germany. Unlike other sustainability assessment concepts, the ICoS provides a multidimensional approach with three goals of sustainability. ICoS does not follow the structure of individual sustainability dimensions: ecology, social, and economy, but uses the comprehensive and holistic approach (Kopfmüller et al., 2001). ICoS has translated the key documents on sustainable development, such as the Brundtland Report and the Rio Declaration, Agenda 21, and formulated its constitutive elements from them. Therefore, it can give equal weight to all three pillars of conventional sustainability assessment and make a fair judgment (Rösch et al., 2017). The consecutive elements are first translated into three general sustainable development goals. Then, the goals are concretized by formulating several substantial rules. Later, these rules are concretized by corresponding indicators (Kopfmüller et al., 2001). The detail of the goals and rules in the ICoS is shown in Table 1.

In this study, the ICoS rules are translated into different indicators representing the whole RUN concept. It compares the current system, the city of Heidelberg in this case, which will serve as a benchmark with the alternative system of RUN to give more options for sustainable wastewater treatment. Furthermore, the sustainability

Table 1 Goals and rules within ICoS

Securing human existence	Maintaining society's productive potential	Preserving society's options for development and action
Substantial rules		
Protection of human health	Sustainable use of renewable resources	Equal access for all people to information, education, occupation
Ensuring satisfaction of basic needs	Sustainable use of non-renewable resources	Participation in societal decision-making processes
Autonomous subsistence based on own income	Sustainable use of the environment as a sink	Conservation of cultural heritage and cultural diversity
Just distribution of opportunities for using natural resources	Avoiding technical risks with potentially catastrophic impacts	Conservation of the cultural function of nature
Reduction of extreme income or wealth inequalities	Sustainable development of man-made, human and knowledge capital	Conservation of social resources (tolerance, solidarity, etc.)

of the transformation process of the society towards a new system alternative is also evaluated.

4.3 Identification of the Indicators

The application of ICoS in the RUN concept was carried out in two steps. First, the discussion on the substantial rules was conducted to select the most suitable rules for the RUN concept. That is because not all rules fit the context of the RUN concept. Once the substantial rules were selected, they were translated into measurable indicators. Indicators are a typical tool for any sustainability assessment. They are very useful for communicating ideas and effective for making the information more available to decision-makers (Rösch et al., 2017).

The development of sustainability indicators for the RUN concept was an iterative process, see Fig. 4. The first step of the process was an extensive literature review to translate the sustainability rules into indicators. References used to select indicators included the topic of wastewater treatment and agricultural systems. Since most of the literature still follows the standard dimension of sustainability assessment, it is important in this process to assign the pre-selected indicators to the corresponding substantial rules of ICoS. The second step was an expert review of the pre-selected indicators. The experts here are the partners in the RUN consortium, consisting of interdisciplinary expertise. The partners in the RUN consortium consist of experts in the wastewater and waste treatment system, the agricultural, the social, the regional, and the evaluation system. Therefore, it was determined that the review by the RUN consortium is sufficient to select the indicators that can represent the RUN concept. Discussions were held extensively with expert partners within the RUN concept. Decisions were made as to whether a particular indicator applied to the RUN concept. Some pre-selected indicators were eliminated because they did not fit the context of the RUN concept, and some were modified. At the end of the iterative process, the specific set of indicators for the RUN concept was available and could be used for the sustainability assessment.

4.4 Identified Indicators

As previously mentioned, the sustainability indicators in the RUN concept are assigned to each goal and substantial rules in ICoS. In this subchapter, the discussion about selected indicators will be explained in detail.

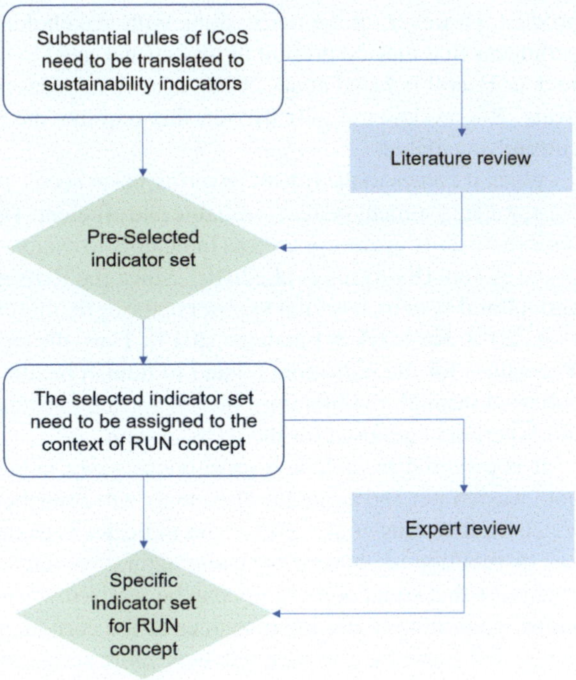

Fig. 4 Iterative process on the selection of sustainability indicator set

4.4.1 Indicators for the Goal Securing Human Existence

The first goal stated in ICoS and to be achieved in the RUN project is to secure human existence. This goal is considered the most important core of sustainability, since sustainability means that conditions must be maintained for present and future generations. The indicators of this goal are used to assess whether the basic requirements for securing human existence, such as health protection, basic infrastructure and food supply, can be achieved.

As urged in the Sustainable Development Goals (SDGs), protecting human health from waterborne diseases is crucial (United Nations, 2015), especially in the RUN concept. Some indicators could be presented to address the rule of human health protection. Some included waterborne disease outbreaks (Hellström et al., 2000) and the others included the degree of pathogen removal (Balkema et al., 2002; Muga & Mihelcic, 2008). Some studies explicitly used fecal coliform as a pathogenic parameter (Kazora & Mourad, 2018; Sartorius et al., 2019) because it is a common parameter used in wastewater studies (Blumenthal et al., 2000). Since the final product from the RUN treatment plant will be a design fertilizer, the consideration focuses on the fertilizer and the food product. Based on the German Fertilizer Ordinance, the commonly used microbiological parameter to assess hygiene is Salmonella (Düngemittelverordnung-DüMV, 2012). Therefore, it was decided that for the RUN concept, the indicator will be salmonella concentration in design fertilizer and food product. However, in the discussions with stakeholders, it became clear that other pathogens that may be present in wastewater need to be included in the assessment, such as E. coli (Chahal et al., 2016), as well as micropollutants such as pharmaceuticals (Rogowska et al., 2020); yet, this requires the provision of data that are not currently available.

When it comes to the rule of ensuring basic needs, it is essential to ensure that the wastewater treatment system is reliable and efficient. The number of interruptions and reserve capacity are considered as indicators to evaluate the reliability and efficiency of the system (Sartorius et al., 2019). Since the RUN concept will also consider the agricultural system, it is vital to observe the agricultural productivity (Fallah-Alipour et al., 2018; Królczyk & Latawiec, 2015). Thus, the crop yield indicator is used here. The values for the indicators related to human health and ensuring basic needs are collected through the literature review for the existing system, while the results of the experiment are used for the RUN system.

It is essential to include a greenhouse gases indicator in the assessment, since both wastewater and agricultural systems will contribute to global warming (DLG e. V., 2013; Sartorius et al., 2019). The indicator is evaluated according to the rule of just distribution of chances for using natural resources. The idea behind this rule is to express that equal opportunity can be realized when the future generation has the same opportunity to use the resources as the current generation. Under the premise that global warming is expected to have more severe impacts in the future, it hinders the opportunity of the future generation to use the resources freely (Kopfmüller et al., 2001). Hence, it is necessary to minimize the contribution to global warming at the present time. The value for this indicator will be estimated by using Life Cycle Assessment (LCA) (REF) both for existing and RUN system (Table 2).

Table 2 Selected indicators for the goal securing human existence

Sustainability rule	Sustainability Indicator	Unit
Protection of human health	1. Total salmonella in the design fertilizer	CFU/mL
	2. Total salmonella in the food product	CFU/g
Ensuring satisfaction of basic needs	3. Interruption of the system in the building	Incident/a
	4. Capacity reserve of the treatment facility	% from normal capacity
	5. Crop yield	kg/ha
Just distribution of chances for using natural resources	6. Greenhouse gases equivalent	t CO_2 eq/a

4.4.2 Indicators for the Goal Maintaining Society's Productive Potentials

The second goal is to preserve the productive potential of society to ensure that resources, not only environmental but also economic and knowledge resources, can be passed on to the next generation. In this way, the future generation can satisfy its own needs, which may be different from the needs of the present generation. The indicators could be used to maintain stocks and balances to minimize resource depletion. To achieve this goal, a total of four rules have been defined, which are listed in Table 3. The first two rules concern the sustainable use of renewable and non-renewable resources as a prerequisite for a society to generate added value. The third rule recognizes that the environment is used as a sink for human activities, but also that carrying capacity is limited. The fourth rule addresses the question of how human-created human and knowledge capital can be secured or even accelerated as a further prerequisite for sustaining society's productive potential (Kopfmüller et al., 2001).

To measure whether renewable resources are being used sustainably, three indicators are proposed: water footprint, energy demand from renewable resources (Balkema et al., 2002), and nitrogen recovery rate (Sartorius et al., 2019), included in Table 3. Since proper water use is at the heart of any technology, the inclusion of a water footprint is obvious. In addition, the provision of water, treatment of wastewater and organic waste requires energy, which should come from renewable sources, as this could reduce the environmental impact of the system under consideration. Moreover, the core idea of RUN is to establish a circularity for the nitrogen used. For non-renewable resources, the indicators focus on the recovery rate of phosphorus and potassium, in addition to the energy demand, as they are the core of the RUN vision. The construction and operation of the current system, as well as the RUN system, will initiate the depletion of fossil fuels and metal sources.

Human activities affect the environment in different ways, acidification and eutrophication being the most common, along with global warming. Specifically, for the RUN system, ammonia emissions are monitored, mainly from agriculture, as well as soil balance in terms of nutrients (Muga & Mihelcic, 2008; Zhang & Ma, 2020). The last rule refers to the value creation of the entire system, i.e. the value creation of both systems is compared. Value creation is a prerequisite for the acceleration of man-made human and knowledge (DLG, 2013; Kazora & Mourad, 2018). The construction and operation of facilities also induces resource- and environmentally-relevant activities in the upstream and downstream process steps. To capture these, a life cycle approach is used where applicable, i.e. resource requirements and emissions over the entire life cycle of the resources are estimated using LCA as a general approach; impacts are calculated using the ReCiPe 2016 Midpoint Impact Categories (Huijbregts et al., 2017).

Table 3 Selected indicators for the goal maintaining society's productive potential

Sustainability rule	Sustainability Indicator	Unit
Sustainable use of renewable resources	7. Water footprint	m^3/a
	8. Energy demand from renewable resources	kWh/a
	9. Conversion rate of nitrogen in conventional fertilizer to design fertilizer	% of conversion
Sustainable use of non-renewable resources	10. Energy demand from non-renewable resources	kWh/a
	11. Conversion rate of phosphor in conventional fertilizer to design fertilizer	% of conversion
	12. Conversion rate of potassium in conventional fertilizer to design fertilizer	% of conversion
	13. Fossil fuel depletion	t oil eq/a
	14. Metal depletion	t Cu eq/a
Sustainable use of the environment as a sink	15. Total ammonia emission	kg NH_3/a
	16. Eutrophication in water bodies	kg P eq/a
	17. Acidification	kg SO_2 eq/a
	18. Phosphorus balance in the soil	kg P/ha
	19. Nitrogen balance in the soil	kg N/ha
	20. Potassium balance in the soil	kg K/ha
Sustainable development of man-made, human and knowledge capital	21. Value creation of the entire system	euros (€)

4.4.3 Indicators for the Goal Preserving Society's Options for Development and Action

The goal aims to preserve the immaterial resources for the future generation to complement the material resources that have already been addressed in the previous goal. This goal plays an important role in sustainability because without these immaterial resources, such as education, culture, and social environment, the material resources cannot be used efficiently and effectively. Therefore, a total of three rules were identified as suitable for achieving this goal (Table 4): the first addresses how the RUN system affects equal access to information, education, and employment for all people, as well as access to the social infrastructure. The second deals with the participation of society in decision-making processes, and the last with the preservation of the cultural function of nature (Kopfmüller et al., 2001).

The equal access rule is translated into the household cost as a share of the household income indicator to assess whether the new system is affordable for low-income households (Kazora & Mourad, 2018; Yli-Viikari, 1999). This is to ensure that the

Table 4 Selected indicators for the goal preserving society's options for development and action

Sustainability rule	Sustainability Indicator	Unit
Equal access for all people to information, education, occupation	22. Household costs as a share of household income (of low-income households)	%
Participation in societal decision-making processes	23. Residents' acceptance of vacuum toilet and kitchen shredder	Ordinal scale
	24. Farmer acceptance of design fertilizers	Ordinal scale
	25. Acceptance of food products from agriculture using design fertilizers	Ordinal scale
Conservation of the cultural function of nature	26. Visual nuisance due to the installed equipment	Ordinal scale
	27. Noise pollution from the treatment plant	dB (A)
	28. Odor nuisance from the treatment plant	ouE/m^3

transformation of the system does not affect access to basic infrastructure for the low-income population. The establishment of the RUN technology will have some impact on the living environment of households as well as the conditions under which farmers will cultivate their crops. A successful implementation of the RUN system would require the acceptance of these stakeholders by involving the stakeholders by implementing the RUN systems in practice, as well as by using it. While the former can be demanded but is no constituent of the technology, the latter can be influenced by the design of the RUN system. The acceptance level is meant to show the commitment of society to the new system. The acceptance is divided into three different indicators: the acceptance of the new wastewater system, the design fertilizer, and the food product that uses the design fertilizer. Some studies have included acceptance as a sustainability indicator in the wastewater and agriculture system (Zhen & Routray, 2003). Conservation of the cultural function of nature tackles with demand that the transformation of an existing system will not deteriorate the opportunities to use the nature among others as a recreational sphere. Visual disturbances, unwanted noises as well as odors could reduce the functional value of nature, demanding to include these as indicators in the assessment (Balkema et al., 2002; Molinos-Senante et al., 2014). For the values of the indicators, literature research and direct contact with the research subject are used here. It is planned to conduct interviews and questionnaires and to observe the perspective of the users in the experience space.

5 Conclusion and Outlook

Closing the nutrient cycle in systems with wastewater separation at the source has been in most cases limited to the use of the nutrient-rich sub-products in the form of soil improvers or fertilizers, e.g. compost or urine. The RUN project fills this gap in the research field of alternative sanitation systems through the recovery of phosphorus, nitrogen, and potassium in the form of high-quality designed fertilizers. On the other hand, a myriad of advanced technologies for the recovery of nutrients have been investigated and applied at pilot and large scale in some conventional combined wastewater treatment plants. Some of them such as struvite chemical precipitation can be potentially transferred to applications with separately collected blackwater, as demonstrated within RUN. In addition, special importance has been given to the recovery of a specific nutrient, e.g. either phosphorus and/or nitrogen, while other valuable resources such potassium, the biogenic carbon and the volatile fatty acids contained in the sub products remained unrecovered. In this context, RUN was conceptualized under a holistic approach that maximizes not only the recovery of nutrients, but also of other valuable resources such as biopolymers, by following a sub-sequential order of different advanced technologies.

In addition, closing the nutrient cycle requires not only the development of innovative technologies at lab or pilot scale, but also the realization of synergies within urban and rural systems. Therefore, the technical concept of RUN follows an iterative approach that links the development of technologies embedded in an urban area based on the demands and feedback of the rural systems.

However, not only the technological solution introduced in the RUN project, but also the social aspect is strongly integrated into the project. In order to bridge the gap between the technological solution and the society, a real laboratory and an experience space are set up in the area. The prototype household installations, such as vacuum toilet and kitchen grinder, will be provided in the experience room for the future residents to try out. The acceptance study for the new facilities is being conducted in two steps. Before the installation of the new facilities, the experiences from other similar projects have been studied. In the second step, after the installation of the new facilities, a direct survey of the future residents of the residential area is planned. The project will be shared with the community to provide transparency, and their perspective and participation are expected to enhance the project. This will be done through dialogue between stakeholders and partnership with the local community.

The RUN project is also evaluated from a sustainability point of view using a sustainability assessment method called ICoS. The development of ICoS is based on the key documents of sustainable development such as the Brundtland Report, the Rio Declaration and Agenda 21. In ICoS, the content of this report is translated into three different sustainable goals, each goal also containing several sustainability rules. In the RUN sustainability assessment, these ICoS rules are translated into different indicators that can represent the sustainability aspect in the project. The indicators are developed in an iterative process to obtain a comprehensive list of indicators. First, a comprehensive literature review was conducted to develop a preliminary list of indicators. Then, the indicators were shared with the experts and adapted to be consistent with the concept of RUN. In the end, a total of 29 indicators were selected to be used for the sustainability assessment of the RUN project.

It is expected that the whole concept of the RUN project, which includes technical and non-technical innovations, could broaden the research perspective in wastewater and waste treatment, resource recovery, urban-rural partnership, sustainable agriculture, and the transformation process of society toward new technologies. The future vision of the study is that the transferability of the RUN concept to different regions and cases will help accelerate global sustainable development.

Acknowledgements The authors would like to thank the RUN Consortium (https://www.run-projekt.de/) for providing data as a base for writing this contribution. The work of the RUN Project is funded by German Federal Ministry of Education and Research (BMBF) through a funding program called Agrarsysteme der Zukunft (Agricultural System of the Future). The authors gratefully acknowledge the funds received. The authors also want to express their gratitude to Witold-Roger Poganietz for his revisions and valuable advice in the process of writing this contribution.

References

Ansari, A. J., Hai, F. I., Price, W. E., Drewes, J. E., & Nghiem, L. D. (2017). Forward osmosis as a platform for resource recovery from municipal wastewater—A critical assessment of the literature. *Journal of Membrane Science, 529*, 195–206. https://doi.org/10.1016/j.memsci.2017.01.054

Balkema, A. J., Preisig, H. A., Otterpohl, R [Ralf], & Lambert, F. J. (2002). Indicators for the sustainability assessment of wastewater treatment systems. *Urban Water, 4*(2), 153–161. https://doi.org/10.1016/S1462-0758(02)00014-6

Bargeman, G., Houwing, J., Recio, I., Koops, G.-H., & van der Horst, C. (2002). Electro-membrane filtration for the selective isolation of bioactive peptides from an alpha(s2)-casein hydrolysate. *Biotechnology and Bioengineering, 80*(6), 599–609. https://doi.org/10.1002/bit.10419

Bauer, N. (2017). *Modell-Siedlung Flintenbreite in Lübeck: Auch Abwasser lässt sich trennen.* https://www.deutschlandfunkkultur.de/modell-siedlung-flintenbreite-in-luebeck-auch-abwasser.1001.de.html?dram:article_id=389816

Beebe, D. A., Castle, J. W., & Rodgers, J. H. (2013). Treatment of ammonia in pilot-scale constructed wetland systems with clinoptilolite. *Journal of Environmental Chemical Engineering, 1*(4), 1159–1165. https://doi.org/10.1016/j.jece.2013.08.034

Berliner Wasserbetriebe. (2021). *Wir recyceln Phosphor zu Dünger.* https://www.bwb.de/de/4951.php

Blumenthal, U. J., Mara, D. D., Peasey, A., Ruiz-Palacios, G., & Stott, R. (2000). Guidelines for the microbiological quality of treated wastewater used in agriculture: Recommendations for revising WHO guidelines. *Bulletin of the World Health Organization, 78*(9), 1104–1116.

Castro, C. J., Shyu, H. Y., Xaba, L., Bair, R., & Yeh, D. H. (2021). Performance and onsite regeneration of natural zeolite for ammonium removal in a field-scale non-sewered sanitation system. *The Science of the Total Environment, 776*, 145938. https://doi.org/10.1016/j.scitotenv.2021.145938

Chahal, C., van den Akker, B., Young, F., Franco, C., Blackbeard, J., & Monis, P. (2016). Pathogen and particle associations in wastewater: Significance and implications for treatment and disinfection processes. *Advances in Applied Microbiology, 97*, 63–119. https://doi.org/10.1016/bs.aambs.2016.08.001

Coates, G. J. (2013). The sustainable urban district of Vauban in Freiburg, Germany. *International Journal of Design & Nature and Ecodynamics, 8*(4), 265–286. https://doi.org/10.2495/DNE-V8-N4-265-286

Dittmann, M. (2016). *Untersuchungen zur Ammoniumrückgewinnung aus flüssigen Reststoffen mittels eines Membrankontaktors im Labormaßstab und Auslegung einer Pilotanlage.* Fraunhofer IGB.

DLG e. V. (2013). *Nachhaltiger Ackerbau:* Effizienz steigern, Image pflegen, Ressourcen schonen (DLG-Merkblatt 369).

Düngemittelverordnung-DüMV. (2012). Verordnung über das Inverkehrbringen von Düngemitteln, Bodenhilfsstoffen, Kultursubstraten und Pflanzenhilfsmittel.

Fallah-Alipour, S., Boshrabadi, H. M., Mehrjerdi, M. R. Z., & Hayati, D. (2018). A framework for empirical assessment of agricultural sustainability: The case of Iran. *Sustainability, 10*(12), 4823. https://doi.org/10.3390/su10124823

Franck, J., & Schröder, L. (2015). Zukunftsfähigkeit kleiner Klärschlammverbrennungsanlagen. In K. J. Thomé-Kozmiensky & M. Beckmann (Eds.), *Energie aus Abfall* (pp. 457–474).

Fraunhofer IGB. *ePhos – Electrochemical phosphorus recovery.* Retrieved June 09, 2021, from https://www.igb.fraunhofer.de/en/research/resource-recovery-nutrients-metals-biogas/ephos-electrochemical-phosphorus-recovery.html

Friedrich, J., Poganietz, W.-R., & Lehn, H. (2020). Life-cycle assessment of system alternatives for the Water-Energy-Waste Nexus in the urban building stock. *Resources, Conservation and Recycling, 158*, 104808. https://doi.org/10.1016/j.resconrec.2020.104808

Gleis, M. (2012). Gasification and Pyrolysis—Reliable options for waste treatment? In K. J. Thomé-Kozmiensky & S. Thiel (Eds.), *Waste management* (403–310).

Hasanoğlu, A., Romero, J., Pérez, B., & Plaza, A. (2010). Ammonia removal from wastewater streams through membrane contactors: Experimental and theoretical analysis of operation parameters and configuration. *Chemical Engineering Journal, 160*(2), 530–537. https://doi.org/10.1016/j.cej.2010.03.064

Hellström, D., Jeppsson, U., & Kärrman, E. (2000). A framework for systems analysis of sustainable urban water management. *Environmental Impact Assessment Review, 20*(3), 311–321. https://doi.org/10.1016/S0195-9255(00)00043-3

Hiessl, H., & Hillenbrand, T. (2010). *Dezentrales Urbanes Infrastruktur System DEUS 21: Abschlussbericht*. Fraunhofer ISI.

Huijbregts, M. A. J., Steinmann, Z. J. N., Elshout, P. M. F., Stam, G., Verones, F., Vieira, M., Zijp, M., Hollander, A., & van Zelm, R. (2017). ReCiPe2016: A harmonised life cycle impact assessment method at midpoint and endpoint level. *The International Journal of Life Cycle Assessment, 22*(2), 138–147. https://doi.org/10.1007/s11367-016-1246-y

Jatav, H. S., Rajput, V. D., Minkina, T., Singh, S. K., Chejara, S., Gorovtsov, A., Barakhov, A., Bauer, T., Sushkova, S., Mandzhieva, S., Burachevskaya, M., & Kalinitchenko, V. P. (2021). Sustainable approach and safe use of biochar and its possible consequences. *Sustainability, 13*(18), 10362. https://doi.org/10.3390/su131810362

Jiang, A., Zhang, T., Zhao, Q.-B., Li, X., Chen, S., & Frear, C. S. (2014). Evaluation of an integrated ammonia stripping, recovery, and biogas scrubbing system for use with anaerobically digested dairy manure. *Biosystems Engineering, 119*, 117–126. https://doi.org/10.1016/j.biosystemseng.2013.10.008

Johir, M., George, J., Vigneswaran, S., Kandasamy, J., & Grasmick, A. (2011). Removal and recovery of nutrients by ion exchange from high rate membrane bio-reactor (MBR) effluent. *Desalination, 275*(1–3), 197–202. https://doi.org/10.1016/j.desal.2011.02.054

Kazora, A. S., & Mourad, K. A. (2018). Assessing the sustainability of decentralized wastewater treatment systems in Rwanda. *Sustainability, 10*(12), 4617. https://doi.org/10.3390/su10124617

Kisser, J., Wirth, M., de Gusseme, B., van Eekert, M., Zeeman, G., Schoenborn, A., Vinnerås, B., Finger, D. C., Kolbl Repinc, S., Bulc, T. G., Bani, A., Pavlova, D., Staicu, L. C., Atasoy, M., Cetecioglu, Z., Kokko, M., Haznedaroglu, B. Z., Hansen, J., Istenič, D., & Beesley, L. (2020). A review of nature-based solutions for resource recovery in cities. *Blue-Green Systems, 2*(1), 138–172. https://doi.org/10.2166/bgs.2020.930

Kopfmüller, J., Jörissen J, Brandl, V., Paetau, M., Banse, G., Coenen, R., & Grunwald, A. (2001). *Nachhaltige Entwicklung integrativ betrachtet – Konstitutive Elemente, Regeln, Indikatoren.*

Kotoulas, A., Agathou, D., Triantaphyllidou, I., Tatoulis, T., Akratos, C., Tekerlekopoulou, A., & Vayenas, D. (2019). Zeolite as a potential medium for ammonium recovery and second cheese whey treatment. *Water, 11*(1), 136. https://doi.org/10.3390/w11010136

Królczyk, J. B., & Latawiec, A. E. (2015). 10. Sustainability indicators for agriculture in the European Union. In A. Latawiec & D. Agol (Eds.), *Sustainability indicators in practice* (pp. 182–204). De Gruyter Open Poland. https://doi.org/10.1515/9783110450507-015

Larsen, T. A., Alder, A. C., Eggen, R. I. L., Maurer, M., & Lienert, J. (2009). Source separation: Will we see a paradigm shift in wastewater handling? *Environmental Science & Technology, 43*(16), 6121–6125. https://doi.org/10.1021/es803001r

Lee, C. W., Kwon, H. B., Jeon, H. P., & Koopman, B. (2009). A new recycling material for removing phosphorus from water. *Journal of Cleaner Production, 17*(7), 683–687. https://doi.org/10.1016/j.jclepro.2008.11.019

Loganathan, P., Vigneswaran, S., Kandasamy, J., & Bolan, N. S. (2014). Removal and recovery of phosphate from water using sorption. *Critical Reviews in Environmental Science and Technology, 44*(8), 847–907. https://doi.org/10.1080/10643389.2012.741311

Lorick, D., Macura, B., Ahlström, M., Grimvall, A., & Harder, R. (2020). Effectiveness of struvite precipitation and ammonia stripping for recovery of phosphorus and nitrogen from anaerobic digestate: A systematic review. *Environmental Evidence, 9*(1). https://doi.org/10.1186/s13750-020-00211-x

Lu, S. G., Bai, S. Q., Zhu, L., & Shan, H. D. (2009). Removal mechanism of phosphate from aqueous solution by fly ash. *Journal of Hazardous Materials, 161*(1), 95–101. https://doi.org/10.1016/j.jhazmat.2008.02.123

Ma, J., Lenhart, J. H., & Tracy, K. (2011). Orthophosphate adsorption equilibrium and breakthrough on filtration media for storm-water runoff treatment. *Journal of Irrigation and Drainage Engineering, 137*(4), 244–250. https://doi.org/10.1061/(ASCE)IR.1943-4774.0000291

Maurer, M., Pronk, W., & Larsen, T. A. (2006). Treatment processes for source-separated urine. *Water Research, 40*(17), 3151–3166. https://doi.org/10.1016/j.watres.2006.07.012

McLeod, A., Jefferson, B., & McAdam, E. J. (2014). Biogas upgrading by chemical absorption using ammonia rich absorbents derived from wastewater. *Water Research, 67*, 175–186. https://doi.org/10.1016/j.watres.2014.09.010

Meyer, C., Preyl, V., Steinmetz, H [Heidrun], Maier, W., Mohn, R.-E., & Schönberger, H. (2019). The Stuttgart process (Germany). In H. Ohtake & S. Tsuneda (Eds.), *Phosphorus recovery and recycling* (pp. 283–295). Springer. https://doi.org/10.1007/978-981-10-8031-9_19

Molinos-Senante, M., Gómez, T., Garrido-Baserba, M., Caballero, R., & Sala-Garrido, R. (2014). Assessing the sustainability of small wastewater treatment systems: A composite indicator approach. *Science of the Total Environment, 497–498*, 607–617. https://doi.org/10.1016/j.scitotenv.2014.08.026. Epub 2014 Aug 27. PMID: 25169875.

Morandi, C. G., Mouarkech, K., Wasielewski, S., Minke, R., & Steinmetz, H. (2016). *Phosphorrückgewinnung mittels MAP-Fällung: BMBF-Projekt TWIST++ (2013–2016): Transitionswege WasserInfraSTruktursysteme: Anpassung an neue Herausforderungen im städtischen und ländlichen Raum.* ISWA Universität Stuttgart.

Muga, H. E., & Mihelcic, J. R. (2008). Sustainability of wastewater treatment technologies. *Journal of Environmental Management, 88*(3), 437–447. https://doi.org/10.1016/j.jenvman.2007.03.008

OSTARA. (2016). *Dutch-waterboard-vallei-en-veluwe-and-ostara-to-produce-high-value-fertiliser-crystal-green-at-amersfoort-wwtp.* https://ostara.com/project/dutch-waterboard-vallei-en-veluwe-and-ostara-to-produce-high-value-fertiliser-crystal-green-at-amersfoort-wwtp/

Otterpohl, R., Braun, U., & Oldenburg, M. (2004). Innovative technologies for decentralised water-, wastewater and biowaste management in urban and Peri-urban areas. *Water Science and Technology, 48*(11–12), 23–32. https://doi.org/10.2166/wst.2004.0795

Rogowska, J., Cieszynska-Semenowicz, M., Ratajczyk, W., & Wolska, L. (2020). Micropollutants in treated wastewater. *Ambio, 49*(2), 487–503. https://doi.org/10.1007/s13280-019-01219-5

Rösch, C., Bräutigam, K.-R., Kopfmüller, J [Jürgen], Stelzer, V., & Lichtner, P. (2017). Indicator system for the sustainability assessment of the German energy system and its transition. *Energy, Sustainability and Society, 7*(1). https://doi.org/10.1186/s13705-016-0103-y

RUN Consortium. (2019). *Gesamtkonzept für das Projekt: Nährstoffgemeinschaften für eine zukunftsfähige Landwirtschaft - Rural-Urban Nutrient Partnership (RUN)* [Vorhabenbeschreibung].

RUN Consortium. (2020). *RUN—Rural Urban nutrient partnership: Nutrient communities for sustainable agriculture.* RUN Project. https://en.run-projekt.de/das-ziel

RUN Consortium. (2021, September 28). *Nährstoffgemeinschaften für eine zukunftsfähige Landwirtschaft: Rural Urban Nutrient Partnership.* Statusseminar "Agrarsysteme der Zukunft", Webex.

Sartorius, C., Hillenbrand, T., & Niederste-Hollenberg, J. (2019). Multikriterielle Bewertung von Wasserinfrastruktursystemen im Kontext der SDGs. In W. Leal Filho (Ed.), *Aktuelle Ansätze zur Umsetzung der UN-Nachhaltigkeitsziele* (pp. 271–289). Springer. https://doi.org/10.1007/978-3-662-58717-1_15

Schimpf, G.-C., & Mildenberger, G. (2021). *Sekundäranalytische Ermittlung der Vorerfahrungen an anderen Standorten: RUN - Bericht.* (Unpublished). CSI, Universität Heidelberg.

Sengupta, S., & Pandit, A. (2011). Selective removal of phosphorus from wastewater combined with its recovery as a solid-phase fertilizer. *Water Research, 45*(11), 3318–3330. https://doi.org/10.1016/j.watres.2011.03.044

Sengupta, S., Nawaz, T., & Beaudry, J. (2015). Nitrogen and Phosphorus recovery from wastewater. *Current Pollution Reports, 1*(3), 155–166. https://doi.org/10.1007/s40726-015-0013-1

Skambraks, A.-K., Kjerstadius, H., Meier, M., Davidsson, Å., Wuttke, M., & Giese, T. (2017). Source separation sewage systems as a trend in urban wastewater management: Drivers for the implementation of pilot areas in Northern Europe. *Sustainable Cities and Society, 28*, 287–296. https://doi.org/10.1016/j.scs.2016.09.013

Thong, Z., Cui, Y., Ong, Y. K., & Chung, T.-S. (2016). Molecular design of nanofiltration membranes for the recovery of phosphorus from sewage sludge. *ACS Sustainable Chemistry & Engineering, 4*(10), 5570–5577. https://doi.org/10.1021/acssuschemeng.6b01299

Tilley, E. (Ed.). (2014). *Compendium of sanitation systems and technologies* (2. rev. ed.). EAWAG.

Uhrig, T., Zimmer, J., Rankenhohn, F., & Steinmetz, H. (2020). Biopolymerproduktion aus Abwasser-strömen für eine kreislauforientierte Siedlungswasserwirtschaft. *Wasser Und Abfall, 22*(6), 13–18. https://doi.org/10.1007/s35152-020-0229-2

Umweltbundesamt. (2019). *Ökobilanzieller Vergleich der P-Rückgewinnung aus dem Abwasserstrom mit der Düngemittelproduktion aus Rohphosphaten unter Einbeziehung von Umweltfolgeschäden und deren Vermeidung (PHORWÄRTS): Abschlussbericht.*

United Nations. (2015). *SDG Goal 3 (Targets and Indicators): Ensure healthy lives and promote well-being for all at all ages.* https://sdgs.un.org/goals/goal3

United Nations. (2018). *68% of the world population projected to live in urban areas by 2050, says UN.* United Nations. https://www.un.org/development/desa/en/news/population/2018-revision-of-world-urbanization-prospects.html

Vali, N., Åmand, L.-E., Combres, A., Richards, T., & Pettersson, A. (2021). Pyrolysis of municipal sewage sludge to investigate char and phosphorous yield together with heavy-metal removal—Experimental and by thermodynamic calculations. *Energies, 14*(5), 1477. https://doi.org/10.3390/en14051477

Verordnung über die Verwertung von Klärschlamm, Klärschlammgemisch und Klärschlammkompost (Klärschlammverordnung - AbfKlärV) (2017).

Winker, M., Schröder, E., Saadoun, A., & Kilian, F. (2012). *Factsheets of the SANIRESCH project in Eschborn, Germany—Factsheets of the MAP (struvite) reactor, greywater and brownwater treatment plant.* Eschborn. Deutsche Gesellschaft für Internationale Zusammenarbeit GmbH.

Wit, J. de, Graaf, R. de, Elzinga, N., Debucquoy, W., Lam, G. P. T, Rodenhuis, J., Stutterheim, E., & Piekema, L. (2018). *Evaluatie nieuwe sanitatie Noorderhoek/Waterschoon 2. Stowa rapport: 2018–63.* STOWA.

Yli-Viikari, A. (1999). Indicators for sustainable agriculture—A theoretical framework for classifying and assessing indicators. *Agricultural and Food Science, 8*, 265–283.

Zhang, J., & Ma, L. (2020). Environmental sustainability assessment of a new sewage treatment plant in China based on infrastructure construction and operation phases emergy analysis. *Water, 12*(2), 484. https://doi.org/10.3390/w12020484

Zhen, L., & Routray, J. K. (2003). Operational indicators for measuring agricultural sustainability in developing countries. *Environmental Management, 32*(1), 34–46. https://doi.org/10.1007/s00267-003-2881-1

Sustainable Material and Energy Transformation of Biomass—Introduction

Circular Bioeconomy in Chile and Germany

Simon Glöser-Chahoud, Alex Berg, and Frank Schultmann

Abstract The circular bioeconomy is an economic approach that merges the sustainable transformation of renewable biological materials into high-value products with circular economy principles. These principles strive to minimize waste, maximize resource efficiency, and establish self-sustaining production and consumption cycles. Circular bioeconomy is an emerging global research area and Chile as a country with both high biomass potentials and renewable energy capacities may become an important player for the production of green chemicals. This potential is showcased through four detailed case studies in the subsequent chapters. These studies span topics from the evaluation of lignocellulosic fuels using thermochemical methods to the innovative use of chemical synthesis in producing alternate fuels and eco-friendly base chemicals. The case studies also touch upon the complexities of gathering biomass and the logistics involved. Within this scope, there is an in-depth analysis of semi-mobile units designed for local biomass value addition, comparing different technologies and biomass types. As an illustration of introducing bio-derived products to commercial markets, one case study focuses on the launch of Nanofibrillated Cellulose (NFC) in the Chilean market. This present chapter provides some basic definitions regarding bioeconomy and circular bioeconomy and serves as an introduction to the following in-depth case studies.

S. Glöser-Chahoud (✉)
TU Bergakademie Freiberg (TU BAF), Chair of Corporate Sustainability and Environmental Management, Freiberg, Germany
e-mail: simon.gloeser-chahoud@bwl.tu-freiberg.de

A. Berg
Unidad de Desarrollo Tecnológico (UDT), Universidad de Concepción. Avenida Cordillera 2634, Parque Industrial Coronel, 419996 Coronel, Chile
e-mail: a.berg@udt.cl

F. Schultmann
Karlsruhe Institute of Technology (KIT), Institute for Industrial Production (IIP), Karlsruhe, Germany
e-mail: frank.schultmann@kit.edu

A. Ch. Braun et al. (eds.), *Eco-Industrial Development as an Industrial Strategy*, Green Energy and Technology, https://doi.org/10.1007/978-3-031-73576-9_11

1 From Bioeconomy to Circular Bioeconomy

The global economy depends heavily on fossil hydrocarbons including crude oil, hard coal, and lignite as well as natural gas to meet its material and energy requirements. This dependence causes serious environmental problems including global warming and air and water pollution, leading to ecosystem degradation and impacts on human health (Dorian et al., 2006). A key requisite for the transition towards a more sustainable economy will be to reduce this dependence on fossil fuels. In addition to renewable energy sources from wind, solar, and hydropower as well as the use of geothermal energy sources, biomass is an important source of both renewable energy and renewable hydrocarbon materials for chemical industries that currently source their raw materials and energy primarily from crude oil and petrochemicals (Li & Anastas, 2012). As previously described by Glöser and Schultmann (2024) in this book, the systematic use of renewable, bio-based materials is a key component of a sustainable circular economy. Hence, circulatory economic systems are not limited to material cycles but also include natural cycles such as the carbon cycle by using bio-based renewable resources as a feedstock for various chemical value chains (Ellen MacArthur Foundation, 2019). This forms the link between circular economy and bioeconomy, leading into the concept of a circular bioeconomy (Stegmann et al., 2020).

In the following subsections, basic definitions of the concepts of a bioeconomy and a circular bioeconomy are provided. This is intended as an introduction to the case studies in the following chapter, which provide illustrative examples of how bio-based value chains in the context of a circular bioeconomy can be realized in practice using the example of Chile, a country with high biomass and renewable energy potential.

1.1 Bioeconomy

The term "bioeconomy" refers to an economic model that emphasizes the production of renewable biological resources and their conversion into value-added products, including food, feed, bio-based products, and bioenergy (Birner, 2018). The concept of the bioeconomy originated as a policy idea from the European Commission during the early 2000s (McCormick & Kautto, 2013). Alongside the idea of substituting fossil resources, it also introduced the notion of advancements in biotechnology and life sciences. This concept has displayed considerable success, prompting other nations such as Chile to develop strategies centered around the bioeconomy, including initiatives related to bioenergy. However, given the substantial and fundamental reliance of modern society on fossil fuels, even a gradual transition towards a bioeconomy implies profound shifts in both societal and economic landscapes.

From an economic perspective, particularly material availability in the form of agriculture, marine, and forestry output as well as reasonable resource allocation and

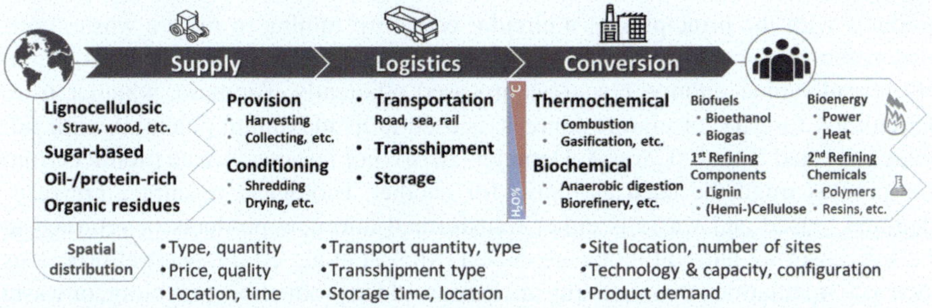

Fig. 1 General depiction of a bioeconomy value chain consisting of biomass provision, collection, transportation, and conversion

the consideration of alternative utilization pathways is of utmost importance when regarding bioeconomy potential. Since the main purpose of agriculture is to supply food for an increasing global population and farmland is a limited resource, sustainable concepts for the use of bio-based value chains should avoid direct competition with food production and focus more strongly on agricultural and forestry residues or on biomass from low-fertility areas that are not used for food production. The fact that sustainable bioeconomy value chains will have to be implemented in existing agricultural systems without compromising food production poses additional challenges to the planning of bio-based value chains (Diehlmann et al., 2019). For a range of distributed biomass types, different utilization pathways and processing technologies need to be considered while taking into account potential revenues from bio-based products on the one hand and various costs for biomass provision, transportation, and conversion on the other. A general bio-based value chain is depicted in Fig. 1.

Nevertheless, as the inherent sustainability of a bioeconomy is limited, efforts are underway to expand and enhance the concept. This involves integrating circular economy principles, promoting the idea of utilizing biomass in multiple stages, and situating it within the broader context of a green economy powered by renewable energy sources.

1.2 Circular Bioeconomy

The term "circular bioeconomy" merges the concepts of the bioeconomy and the circular economy. While the bioeconomy emphasizes the use of renewable biological resources for the production of food, materials, chemicals, and energy, the circular economy emphasizes the efficient use and reuse of resources, minimizing wastes and maximizing the value extracted from them (Stegmann et al., 2020). Thus, the circular bioeconomy can be defined as an economic model that integrates the sustainable production and conversion of renewable biological resources into value-added products with the principles of a circular economy, aiming to reduce waste, optimize resource use, and create closed-loop systems of production and consumption. In a circular bioeconomy, resources are used efficiently: Biomass, whether from agriculture, forestry, or marine sources, is used to its maximum potential, with full value extracted from it (Carus & Dammer, 2018). For instance, waste products from one process might be used as inputs for another. Biological resources are recycled and reused and waste is either transformed into new products or returned to the soil, ensuring nutrient loops are closed (Muscat et al., 2021). Natural processes such as fermentation or composting are harnessed for production, recycling, or waste management. The extraction and use of biological resources are managed in a sustainable manner to maintain the health of ecosystems, ensuring they remain resilient and productive over the long term. The economic model of a circular bioeconomy takes a holistic approach that considers not just economic viability but also environmental health and social well-being. As illustrated in Fig. 2, in addition to circulatory use systems of technical materials, the use of renewable resources as a feedstock for chemical industries, for example, is a key element of the circular bioeconomy.

In addition to its relevance to bio-based renewable carbon cycles, the material-energy nexus is of research interest as it merges the use of renewable hydrocarbons from biomass and further renewable energy sources. In this context, the production and processing of green hydrogen via electrolysis and the potential subsequent chemical synthesis with CO_2 from various sources will play a key role in the decarbonization of industrial value chains. An important and broadly available renewable carbon source will always be biomass, which makes the joint analysis of biomass utilization and the production of e-fuels and base chemicals through chemical synthesis with green hydrogen a promising alternative to fossil fuel-based hydrocarbons (König et al., 2019). This is demonstrated in the last case study of this section (Glöser-Chahoud et al., 2024).

An analysis of academic literature clearly shows the emerging importance of combining the concept of a bioeconomy with aspects of the circular economy. The outcome of an analysis in Scopus, the world's largest collection of academic literature, of documents that use the term "circular bioeconomy" as a keyword is presented in Fig. 3. In addition to the significant increase in publications in this field (Fig. 3a), Figs. 3b and c also emphasize the international relevance of the topic as represented by the number of trans-national academic collaborations. While Brazil met expectations by being the strongest player in Latin America in the study of circular bioeconomy, Chile is also listed in the analysis, mainly through collaborations with European countries (Fig. 3c).

In this context, a number of joint Chilean-German research activities have been conducted in recent years. This book chapter provides reports on these activities in the form of case studies regarding the production of sustainable materials and the provision of biomass-based energy (Berg & Guzmán, 2024; Glöser-Chahoud et al., 2024; Müller & Segura, 2024 ; Zimmer et al., 2024). The final case study (Glöser-Chahoud et al., 2024) incorporates the use of green hydrogen for chemical synthesis, which can be seen as an extension of bio-based renewable materials and fuel supply chains.

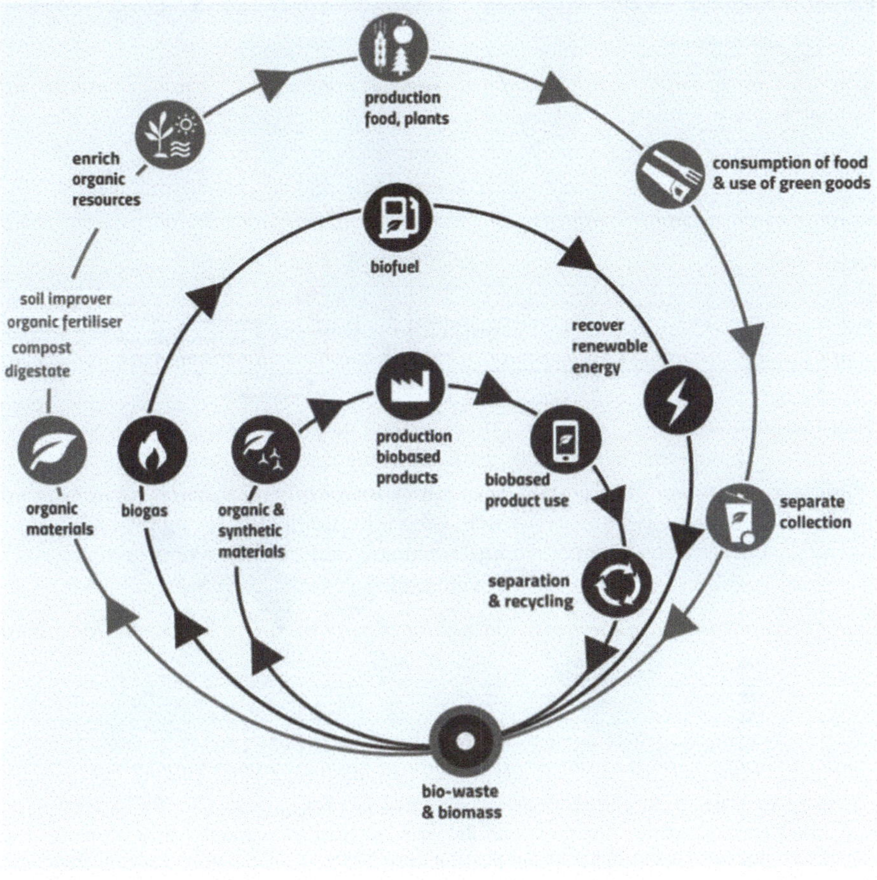

Fig. 2 Basic concept of a circular bioeconomy merging the use of bio-based material and energy with major aspects of the circular economy, mainly regarding the cascade use und recyclability of materials (Figure based on European Compost Network, ECN 2023)

2 Case Studies for Circular Bioeconomy in Chile

This section provides a brief overview of the various case studies that will be described in greater detail in the following chapter. Overall, these case studies clearly demonstrate the potential of Chile to become an important player in the international circular bioeconomy, particularly when combining the potential of biomass utilization with the availability of renewable energy from photovoltaics and wind turbines.

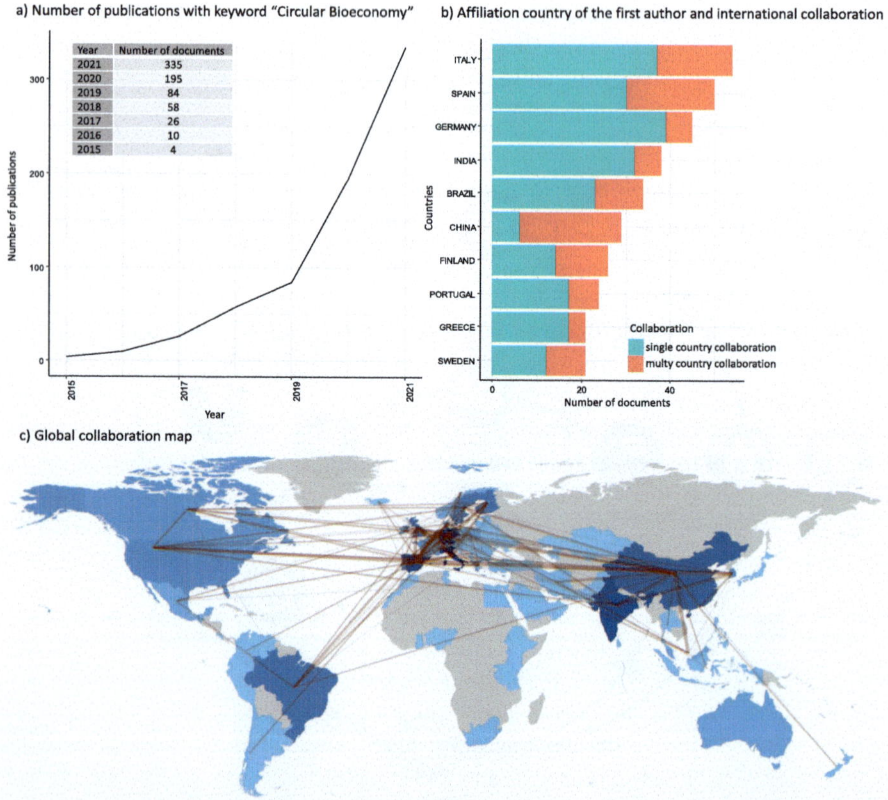

Fig. 3 Analysis of academic literature with the term "circular bioeconomy" as a keyword. The analysis was conducted using data from Scopus, the world's largest collection of academic literature, using the R-package Bibliometrix (Aria & Cuccurullo, 2017)

2.1 Case Study 1: Perspectives for the Development of Lignocellulosic Fuels by Thermochemical Processes

One of the core concepts of the bioeconomy revolves around substituting fossil fuels with renewable, bio-based resources. While the current focus on decarbonizing the electricity sector in Chile through wind energy and photovoltaics has downplayed the significance of bioenergy, the actual role of biomass in the primary energy landscape remains substantial, accounting for 23% of Chile's total generated energy in 2019. Around 53% of this biomass is channeled into electricity generation, primarily in combined heat and power plants within the forestry sector. Industries use 25% while the remaining 22% serves the public, commercial, and residential sectors. In southern Chile, the use of firewood sourced from native forests for heating and cooking still prevails. Biomass, particularly in pellet form, constitutes a small fraction, yet the domestic demand for pellets is experiencing robust growth, surpassing industrial demand due to government subsidies promoting the transition to pellet stoves in order to mitigate winter air pollution.

Given this context, the inquiry arises regarding the potential contribution of thermochemical processing of lignocellulosic biomass to a circular bio-based economy in Chile. Lignocellulose based biomass such as wood or straw is of specific interest is Chile as there are significant amounts of unused residues in forestry and agriculture. From an eco-industrial standpoint, the existing strategies aimed solely at improving the efficiency of individual companies are insufficient for advancing sustainability objectives. Analyzing the flows of materials and energy through industrial systems underscores the significance of local clusters and collaborative associations among companies to drive progress towards these objectives. In this context, case study 1 by Müller and Segura (2024) analyzes the perspectives for the development of lignocellulosic fuels by thermochemical processes in Chile.

2.2 *Case Study 2: The Path to Introducing Nanofibrillated Cellulose (NFC) to the Chilean Market*

Presently, Chile boasts a forested area exceeding 16 million hectares, with natural forests constituting 86% of that area and plantations making up the remaining 14%. The predominant species in these plantations encompass Pinus radiata (comprising 56% of the plantations), Eucalyptus globulus, and Eucalyptus nitens. These three species, in fact, cover an impressive 92% of the total planted area, owing to their adaptability to Chile's growth conditions and their ability to yield multiple cycles of growth at high rates. For instance, Pinus radiata demonstrates an average growth rate of approximately 25 m3/ha/year, while Eucalyptus species achieve around 40 m3/ha/year. Such production levels have facilitated a consistent increase in forestry exports over recent years. By 2018, forestry product exports had reached 6.84 million US $, marking a notable 27% upsurge since 2015. These figures position Chile as one of the globe's top twenty exporters of forest products. Noteworthy segments within the realm of forest exports encompass pulp and sawn timber, collectively accounting for more than 60% of all exports. Other significant exported items include woodchips, plywood, various paper types (such as paperboard, newspaper, corrugated paper, wrapping paper, printing and writing paper), tissue products (including toilet paper, paper towels, paper napkins, and facial tissue), as well as sanitary items like diapers and sanitary pads.

Within this case study, we delve into the trajectory pursued by our research team thus far, focusing on the introduction of a particularly fascinating and versatile product into the Chilean market, one that capitalizes on these byproducts: nanofibrillated cellulose. We explore the potential of this material and subsequently narrow down its promising market opportunities. Subsequently, we detail the effective methods we've employed to adapt its inherent traits, enhance its compatibility with diverse products, and elevate its performance in specific applications.

2.3 Case Study 3: Techno-Economic Analysis of Mobile Production Plants for Bioenergy Carriers

In Chile, substantial quantities of lignocellulosic byproducts are generated through forestry and agricultural activities, encompassing residues from forests and cereal straw. While some of these biomass resources are minimally utilized for bioenergy, a notable portion is discarded without exploitation, often through practices like stubble burning. The chief hurdle in effectively utilizing these biomass residues lies in their low density, resulting in modest truckloads and elevated transportation expenses. A strategic solution involves deploying mobile or containerized facilities at the point of harvest. These facilities can transform the biomass into densified bioenergy carriers, facilitating straightforward transportation. This study focuses on an economic evaluation of mobile pre-treatment plants for forest residues in Chile. Two dynamic configurations—fully mobile units and satellite plants—are contrasted against larger, conventional establishments.

The findings reveal that while mobile plants may not match the economies of scale demonstrated by the largest commercial facilities, their modular design proves advantageous at smaller capacities. In the Chilean context, these mobile units would be particularly suitable for remote locations and decentralized utilization strategies. In the ensuing subchapter, an assessment of the potential of mobile plants in producing densified bioenergy carriers is conducted through a techno-economic analysis. This evaluation encompasses elucidating the structure of a biomass supply chain with decentralized pre-treatment, presenting technologies for energetically compacting biomass, and outlining concepts for mobile or containerized production. Lastly, a succinct case study founded on economic considerations will exemplify the capacity of mobile plants to harness residues from Chilean forest plantations.

2.4 Case Study 4: Production of Alternative Fuels and Green Chemicals Based on Hydrogen and Various Carbon Sources

The economic viability of generating green hydrogen through electrolysis using renewable electrical power sources, primarily solar and wind energy, is profoundly influenced by local circumstances related to the potential for renewable energy generation. Chile boasts one of the world's most substantial capacities for harnessing both solar and wind energy, rendering it an appealing hub for green hydrogen production and subsequent processes like chemical synthesis that enhance value creation and enable more efficient transportation using conventional oil tankers. Chile stands poised to emerge as a global hub for green hydrogen supply or hydrogen-derived alternative fuels and green chemicals. Presently, German industry collaborators Siemens Energy AG and Porsche AG have established a pilot facility in southern Chile for producing green methanol using wind energy, with plans to scale up to industrial capacities in the near future.

Moreover, Chile possesses considerable biomass potential encompassing agricultural and wood residues, which could complement the process of green hydrogen production. Recent evaluations of alternative fuel production have demonstrated optimal economics when coupling biomass gasification with supplementary green hydrogen and subsequent chemical synthesis. This synergy enhances the carbon efficiency of the process and facilitates the stoichiometric conditioning of synthesis gas, yielding maximal outputs of synthetic fuels and chemicals. The presented case study delivers an overview of different value chains for the production of so-called e-fuels and green hydrocarbons through chemical synthesis using different carbon sources.

3 Summary and Conclusions

Circular bioeconomy can be defined as an economic model that integrates the sustainable production and conversion of renewable biological resources into value-added products (concept of the bioeconomy) with the principles of a circular economy, aiming to reduce waste, optimize resource use, and create closed-loop systems of production and consumption. While the idea of a circular bioeconomy is an emerging research topic at the global level, Chile with its biomass potentials on the one hand and its high renewable energy potential on the other possesses promising preconditions to become a global player in the field of green chemistry. This is demonstrated in the following chapters through the example of four case studies.

These case studies include analyses of lignocellulosic fuels by thermochemical processes, the application of chemical synthesis for the production of alternative fuels and green base-chemicals combining biomass gasification and green hydrogen production through electrolysis to optimize the stoichiometric ratio between hydrogen and carbon monoxide in the synthesis gas. Further aspects addressed in the case studies are the challenges of biomass collection and logistics. In this context, a techno-economic assessment of semi-mobile plants for decentralized biomass valorization is conducted using the example of different technologies and various biomass feedstocks. Finally, a case study for the introduction of nanofibrillated cellulose (NFC) to the Chilean market is discussed as an example of bringing bio-based products to market maturity.

References

Aria, M., & Cuccurullo, C. (2017). Bibliometrix: An R-tool for comprehensive science mapping analysis. *Journal of Informetrics, 11*(4), 959–975. https://doi.org/10.1016/j.joi.2017.08.007

Birner, R. (2018). Bioeconomy concepts. In I. Lewandowski (Ed.), *Bioeconomy: Shaping the transition to a sustainable, biobased economy* (pp. 17–38). Springer Nature. https://doi.org/10.1007/978-3-319-68152-8_3

Carus, M., & Dammer, L. (2018). The circular bioeconomy—concepts, opportunities, and limitations. *Industrial Biotechnology, 14*(2), 83–91. https://doi.org/10.1089/ind.2018.29121.mca

Diehlmann, F., Zimmer, T., Glöser-Chahoud, S., Wiens, M., & Schultmann, F. (2019). Techno-economic assessment of utilization pathways for rice straw: A simulation-optimization approach. *Journal of Cleaner Production, 230*, 1329–1343. https://doi.org/10.1016/j.jclepro.2019.04.369

Dorian, J. P., Franssen, H. T., & Simbeck, D. R. (2006). Global challenges in energy. *Energy Policy, 34*(15), 1984–1991. https://doi.org/10.1016/j.enpol.2005.03.010

Ellen MacArthur Foundation. (2019). *Completing the Picture: How the Circular Economy Tackles Climate Change.* https://www.solvay.com/sites/g/files/srpend221/files/2022-10/completing%20the%20picture%20-%20how%20the%20circular%20economy%20tackles%20climate%20change.pdf

Glöser-Chahoud, S., & Schultmann, F. (2024). Circular economy: An introduction. In A. C. Braun, G. Espinosa, D. Tröger, & T. Hirth (Eds.), *Eco-industrial development as an industrial strategy-contributions from a German-Chilean research partnership.* Springer Nature.

Glöser-Chahoud, S., Heinzmann, P., Langenmayr, & Schultmann, F. (2024). Production of alternative fuels and green chemicals based on hydrogen and various carbon sources. In A. C. Braun, G. Espinosa, D. Tröger, & T. Hirth (Eds.), *Eco-industrial development as an industrial strategy-contributions from a German-Chilean research partnership.* Springer Nature.

König, A., Ulonska, K., Mitsos, A., & Viell, J. (2019). Optimal applications and combinations of renewable fuel production from biomass and electricity. *Energy & Fuels, 33*(2), 1659–1672. https://doi.org/10.1021/acs.energyfuels.8b03790

Li, C.-J., & Anastas, P. T. (2012). Green chemistry: Present and future. *Chemical Society Reviews, 41*(4), 1413–1414. https://doi.org/10.1039/C1CS90064A

McCormick, K., & Kautto, N. (2013). The bioeconomy in Europe: An overview. *Sustainability, 5*(6), 2589–2608. https://doi.org/10.3390/su5062589

Muscat, A., de Olde, E. M., Ripoll-Bosch, R., van Zanten, H. H. E., Metze, T. A. P., Termeer, C. J. A. M., van Ittersum, M. K., & de Boer, I. J. M. (2021). Principles, drivers and opportunities of a circular bioeconomy. *Nature Food, 2*(8), 561–566. https://doi.org/10.1038/s43016-021-00340-7

Müller, N., & Segura, C. (2024). Chilean case study 1: Perspectives for the development of lignocellulosic fuels by thermochemical processes. In A. C. Braun, G. Espinosa, D. Tröger, & T. Hirth (Eds.), *Eco-industrial development as an industrial strategy-contributions from a German-Chilean research partnership.* Springer Nature.

Stegmann, P., Londo, M., & Junginger, M. (2020). The circular bioeconomy: Its elements and role in European bioeconomy clusters. *2590-289X, 6*, 100029. https://doi.org/10.1016/j.rcrx.2019.100029

Sustainable Material and Energy Transformation of Biomass—Case Studies from Chile and Germany

Perspectives for the Development of Lignocellulosic Fuels by Thermochemical Processes

Niels Müller and Cristina Segura

Abstract An overview of the status and prospects of lignocellulosic fuels in Chile is given and placed in the broader context of biorefineries and the bioeconomy. The possible role thermochemical processes could play as a pretreatment step, production of energy carriers and biorefinery platform is reviewed. From a conceptual point of view and a practical assessment of the status quo and government policies in Chile, the large-scale production of only lignocellulosic fuels is unlikely to be implemented. The future contributions of lignocellulosic biomass to a more circular and bio-based economy in Chile are expected to be in material applications rather than fuels. Opportunities for application of thermochemical processes will come mainly from the recycling of industrial wastes and sludges, as well from the gradual implementation of the Extended Producer Responsibility Act which will oblige some producers to explore recycling technologies for their products.

1 Introduction

One of the main ideas of the bioeconomy is the substitution of fossil resources with renewable resources. In view of the remarkable and fundamental dependence of modern society on fossil fuels, crude oil in particular, as a cheap and dense net energy source, even a gradual replacement implies profound societal and economic transformations. The path to this new post-industrial society remains largely unknown and marked by controversies, such as the narrow focus on technology and technological alternatives, underlying sustainability constraints, and the insufficient involvement of societal stakeholders (Priefer et al., 2017).

The bioeconomy was foremost a policy concept formulated by the European Commission in the early 2000s. In addition to the paradigm of fossil resource substitution, a perspective of innovations in biotechnology and life science was put forward. The bioeconomy was further labeled "knowledge-based," emphasizing a vision of

N. Müller (✉) · C. Segura
Unidad de Desarrollo Tecnológico, Universidad de Concepción. Avenida Cordillera 3624, Parque Industrial Coronel, 4191996 Coronel, Chile
e-mail: n.muller@udt.cl

A. Ch. Braun et al. (eds.), *Eco-Industrial Development as an Industrial Strategy*, Green Energy and Technology, https://doi.org/10.1007/978-3-031-73576-9_12

high-technology industries supporting economic growth (Birner, 2018). The bioeconomy concept spread globally helped by the oil price peak prior to the 2008 financial crisis, the need for climate change mitigation, and other pressing environmental problems. The concept has proven to be remarkably successful, prompting other countries, including Chile, to prepare bioeconomy or related strategies, such as bioenergy. However, as a bioeconomy is not inherently sustainable per se, the concept is being expanded and complemented with circular economy principles, the idea of cascading biomass use, and placed in a broader context of a green economy driven by renewable energy sources (Birner, 2018).

Nevertheless, a concrete bioeconomy requires a technological base for biomass processing at an industrial scale. Initially, it was envisioned that new biorefineries analogous to those of the petroleum industry would process complete plants providing fuels, chemicals, and materials. In contrast to oil-derived hydrocarbons, lignocellulosic materials are solid with a comparatively low energy density and high oxygen content. To obtain competitive liquid fuels, oxygen must be removed from the biomass molecules, mainly as CO_2, which lowers the mass yield of lignocellulosic-derived fuels. Considering also the distributed nature of the resource, the low conversion of solar energy into lignocellulosic biomass and, therefore, the large area required to achieve economy of scale, the low energy return on investment (EROI) is not surprising. The net energy obtained for society by producing transport fuels from lignocellulosic biomass represents one of the lowest EROI compared with other conventional and renewable energy sources. Therefore, contrary to the initial enthusiasm, lignocellulosic biofuels are of little help in offsetting the declining EROI of crude oil, which is threating to undermine current lifestyles, without even discussing the sustainability issues of these biofuels (Hall et al., 2014; King & van den Bergh, 2018). It is worth noting at this point that solid biomass fuels do have an attractive EROI.

In general, thermochemical conversion, the deconstruction of biomass by thermal energy at high temperatures, is the less technologically demanding route compared to biochemical conversion at near-ambient conditions or chemical conversion at intermediate temperatures. It is also less sensitive to changes in biomass composition and can be used as a pre-processing step, platform technology, or in combination with other routes (Loeffler et al., 2018).

Thermochemical biomass conversion technologies have been the focus of the bioenergy research group at the Technical Development Unit (UDT) at Universidad de Concepción, Chile, for the last 15 years, with an initial emphasis on fast pyrolysis and bio-oil as a fossil fuel substitute (Fig. 1). In 2013, the group elaborated recommendations to the Chilean Ministry of Energy for the development of a National Bioenergy Strategy 2013–2020; a biofuel strategy was not among them, as it was already clear that such a development would not be competitive without subsidies in the short to medium term. In the meantime, the high development potential of renewable energies together with the commercial boom in solar and wind power generation has boosted the share of these energies in the national energy matrix, without the need for state subsidies. The installed capacity of non-conventional renewable energies in Chile (i.e. without major hydropower) reached 8.398 MW in 2021, a contribution of

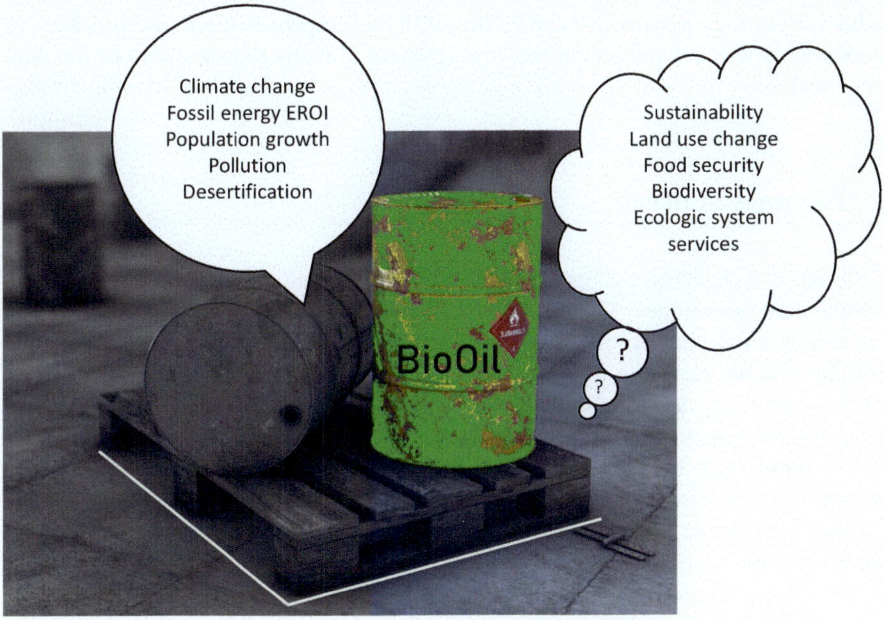

Fig. 1 Thinking within a box: substitution of fossil fuels by bio-oil or biofuel (Adapted from "oil barrels and stairs pallet" by Remco van der Meer, used under CC BY 2.0 DEED)

26% to the country's power generation (ACERA, 2021), with an additional capacity of 4.516 MW under construction (CNE, 2021).

The current emphasis on the decarbonization of the electricity matrix and the new strategy for the development of a green hydrogen economy have contributed to the low profile of bioenergy. However, at 23% (2019) the importance of biomass in the primary energy matrix has remained high (CNE, 2020). About 53% is used for electricity generation, mostly in combined heat and power (CHP) plants of the forestry sector, 25% by industries, and 22% by the public and the commercial and domestic sectors (Electricidad, 2019). In southern Chile, the use of firewood for heating and cooking, mainly from native forests, remains very common. Biomass for energy in the form of pellets accounts for only a small percentage, but domestic demand for pellets is growing strongly and has long exceeded industrial demand mainly due to state subsidies for switching to pellet stoves to reduce air pollution in winter.

In this context, the question arises as to what contribution thermochemical processing of lignocellulosic biomass could make to a circular bio-based economy in Chile and what role eco-industrial strategies could play. From an eco-industrial perspective, i.e. a systemic point of view, current efficiency strategies limited to individual companies are not advancing sustainability goals any further. By studying material and energy flows through industrial systems, the importance of local clusters and the associativity of companies to advance those goals becomes evident. An industrial ecology perspective can be useful in this respect as the future success of a bioeconomy will depend on the new arrangement of many industry sectors and their value chains.

2 Biomass Sources

Chile has a well-developed export-oriented forestry and agricultural industry with comparative advantages. The production of pulp, paper, cardboard, timber, panels, veneers, among other products, from an annual harvest of 45 million cubic meters of industrial wood, is run in many large-scale and efficient processing plants with little potential for capturing underutilized biomass residues. Of the industrial wood, 98% comes from pine and eucalyptus plantations totaling 2.3 million ha, established mainly on eroded soils, and only the remaining 2% stems from native forests. The annual area of forest plantations in 2019 was 107,000 ha and nearly all were reforestation (INFOR Instituto Forestal, 2021). An estimated distribution of areas with relevant biomass production is given in Fig. 2.

The initial focus for new lignocellulosic biorefineries was the production of biofuels with an eye on dedicated forest and perennial crop plantations, grown on marginal lands that do not compete with food and feed production. However, the geophysical and socioeconomic constraints of marginal soils, such as limited infrastructure, low yield, and logistical problems, are also conditions for low-profit margins (Priefer et al., 2017). The potential surface area for short-rotation wood crops in the central-south of Chile has been conservatively estimated to be only 380,000 ha, accounting for hydric and agricultural reserves (Gerding et al., 2013). In addition, a change in land use by introducing extensive monocultures in areas with poor soil threatened by erosion and irregular water supply not only results in a loss of biodiversity and multiple ecological services of these areas, it can also impact

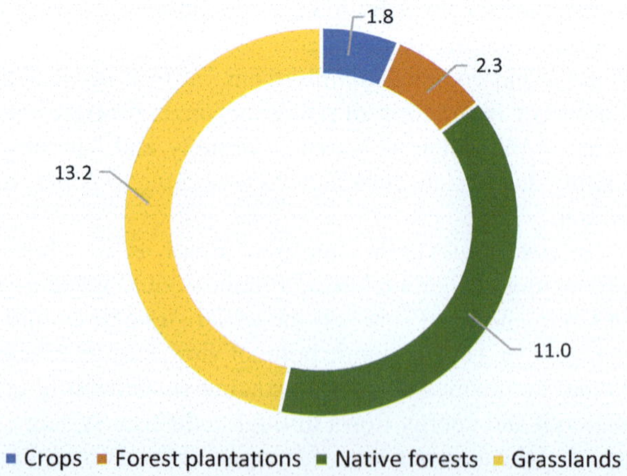

Fig. 2 Areas (millions of ha) with relevant biomass in Chile. Estimated values collected from multiple sources representing the years 2007–2020 (Total area 28.3, country area 75.7)

nearby cultivable land, for example by altering the water supply (Priefer et al., 2017). Therefore, the real potential of biomass from marginal land for energy use has often been overestimated, not only from a theoretical, economically top-down approach, but also by neglecting social aspects, i.e. the willingness and motivation of small farmers to establish plantations for bioenergy (Lewandowski, 2015).

Numerous studies have been conducted on the potential of wood crop plantations in the central-south of Chile (Acuña et al., 2018; Baettig et al., 2010). For example, plantations on drylands with severe growth restrictions of agronomic and forest species revealed technical potential but with the obstacles mentioned above (Santelices-Moya et al., 2019). Biomass productivity must be high enough to compensate for higher maintenance costs and low biomass market prices. Short-rotation wood crops often use genetically optimized plants and involve more intensive activities than traditional forest management, including fertilization, irrigation, and weed control, which can cause undesirable effects on the environment (Morales et al., 2017).

The main biomass sources for new biorefineries in Chile are mixed native forests not used for firewood, agricultural residues, and underutilized forest residues (Fig. 3). The power generation potential from forest plantation residues, has been estimated at 300–500 MW for a utilization of 50–75% of the residues, and was shown to be attractive only for plantation owners, i.e. if residues were available free of charge (Bertrán & Morales, 2008). Around 1.6 million tons of cereal straw are produced in Chile each year, with much of it burned on site. At present, there are only two co-located large power plants (Comasa, Lautaro) of 24 and 22MWe that can also burn cereal straw. This is the only case of large-scale bioenergy production where the entire supply comes from third parties and different industrial sectors (forestry and agriculture). With an economic transport radius of about 70 km, there is substantial technical potential to install up to five similar plants in the region. However, falling electricity prices and conflicts with local communities, mainly indigenous, have prevented their materialization (Almonacid, 2018).

The distributed nature and low energy density of lignocellulosic biomass and the seasonal nature of agricultural residues cause supply chain challenges if used on a

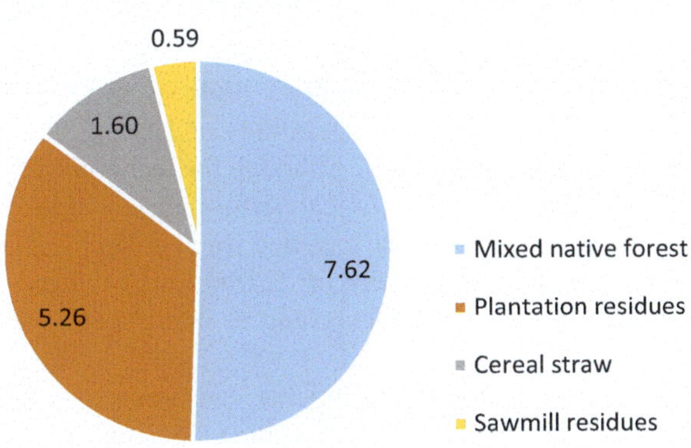

Fig. 3 Biomass potential (millions of tons/y) for lignocellulosic fuels (Zimmer et al., 2017)

large scale and require densification and homogenization strategies. As the major cost of these materials is transportation and handling, supply is often volume-driven while demand is energy-oriented. The current business is largely dominated by a few big players that own plantations and processing plants, with well-developed logistics that enable them to leverage the thermal output of their cogeneration plants.

On a smaller scale, other sources of biomass are municipal solid waste, wastewater treatment sludges, livestock manure, and recycling waste among others, which are of potential interest especially for thermochemical processing.

3 Thermochemical Processing

Thermochemical processes can be divided into dry processes (carbonization, pyrolysis, gasification) and those carried out in an aqueous medium (hydrothermal liquefaction, carbonization, gasification). Here we are interested in the dry and more conventional pyrolysis and carbonization processes applied to lignocellulosic biomass, providing a distribution of gaseous, liquid (bio-oil), and solid (biochar) products (Fig. 4). Depending on the reactor temperature, the heating rate, and the removal rate of vapor and liquid products, the distribution of the product changes. A mild pyrolysis, called torrefaction, delivers mainly a roasted hydrophobic solid that chemically resembles the feedstock, with a slightly higher heating value and that is easy to grind. Fast pyrolysis is primarily aimed at maximizing bio-oil production, while slow pyrolysis produces mainly biochar. All these relatively simple thermal treatments performed in a largely inert atmosphere have been proposed as an energy densification step of lignocellulosic biomass in decentralized plants, addressing biomass variations, seasonality, blending of different types of biomass, and reduction of water content to improve supply chain and transportation costs and achieve better economy of scale at central processing sites.

3.1 Solid Fuels and Lignocellulosic Energy Carriers

Pellets of torrefied lignocellulosic biomass and bio-oil by fast pyrolysis have been found to be economically feasible for heating residential spaces within the EU (Schipfer & Kranzl, 2019). Although the technologies have a high technology readiness level, scaling up, technical performance, and production cost are still subject to uncertainties. There is one torrefied wood pellet producer in Chile (Traiguen Energy) with a production capacity of 22,000 tons/y but no fast pyrolysis plant. Slightly torrefied wheat straw pellets could be interesting as an industrial fuel (Azócar et al., 2019).

A slurry of bio-oil and biochar can be used as an energy carrier. In the Bioliq process (Dahmen et al., 2012) a slurry of bio-oil and char is obtained locally by fast pyrolysis of cereal straw and forest residues and then processed centrally in (very)

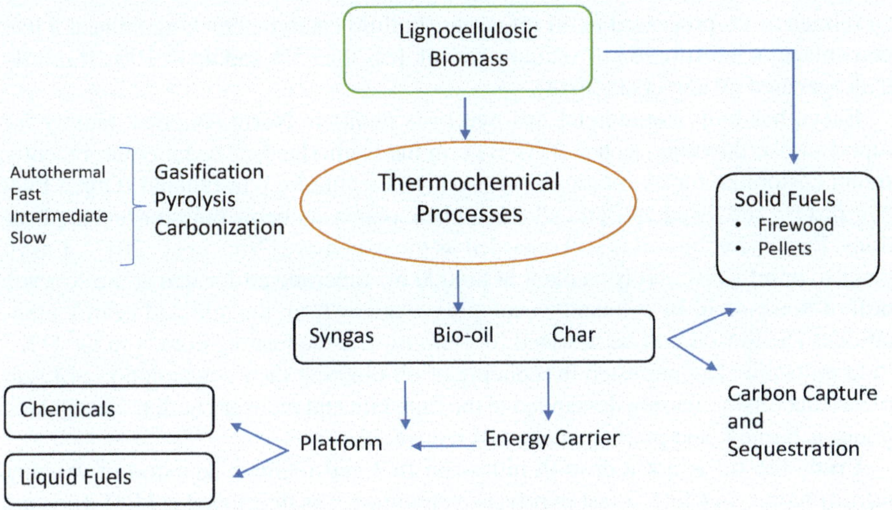

Fig. 4 Thermochemical processes, products, and potential applications of lignocellulosic biomass

large-scale facilities using gasification and Fischer–Tropsch synthesis to produce synfuels. The supply chain for this Biomass-to-Liquid (BtL) concept has been simulated for Chile, taking into account established utilization pathways for CHP plants and domestic consumers (Zimmer et al., 2017). Configuration of the BtL supply chain was strongly affected by current and projected feedstock demand of these pathways. Synfuel costs were calculated below €1/L, mainly because of the comparatively low cost of biomass in Chile. The simulated scenarios supported only one gasification plant and considered a very high percentage of the potential biomass supply, mainly cereal straw, mixed native forest, and some plantation forest residues.

Nevertheless, decentralized pretreatments also have drawbacks. Mass loss of biomass due to microbial contamination, size reduction, storage, and transport, as well as high energy requirements for drying and size reduction, small-scale operations, and low annual load factors are to be mentioned. Techno-economical and life cycle analysis have not shown a clear advantage of decentralized energy densification and homogenization (Lan et al., 2020).

3.2 *Bio-Oil Platform*

The development of fast pyrolysis into a commercial technology that converts solid biomass into a petroleum-like bio-oil has been an endeavor for more than 40 years now (Oasmaa et al., 2021). Bio-oil is a non-distillable biocrude with properties very different from crude oil. It can be used in boilers with special burners and auxiliary fuels for startup and shutdown or upgraded to transport fuels using hydrogen. It can be refined or co-processed in oil refineries, in fluid catalytic cracking units at a low percentage of substitution of vacuum gas oil, less than 5% and up to 10%, if mildly hydrogenated (Wang et al., 2018).

Ensyn has built commercial fast pyrolysis plants in North America, mainly for liquid smoke flavoring, as has BTG-BTL in the Netherlands. The latter has recently found customers for its rotating cone pyrolysis technology in Finland (Green Fuel Nordic Oy) and Sweden (Pyrocell AB), where plants are being built for fuel applications. Elsewhere, bio-oil is still regarded as too expensive (Ning et al., 2013). Ensyn plans to build a fast pyrolysis plant of 80,000 m^3 biocrude co-located at the Aracruz mill of Souzano in Brazil using eucalyptus residues from the mill and nearby plantations. The bio-oil will be destined for refining and the heating market in the U.S.[1] Pulp mills are also interested in reducing or eliminating their consumption of fossil fuels (fuel oil #6), mainly demanded in the lime kiln and recovery boiler, where black liquor is burned, and pulping chemicals recovered.

Crude bio-oil is not a drop-in industrial fuel and requires adaption of existing infrastructure. In Chile, a fast pyrolysis technology was developed at UDT 10 years ago. However, bio-oil as an industrial fuel is not an interesting business model for attracting venture capital to further develop the new technology. Some local forestry investors interested in the fast pyrolysis technology eventually opted for a low-risk conventional cogeneration plant.

Maximizing liquid yield as in fast pyrolysis is not necessarily the optimal strategy for fuel application. A high liquid yield can only be achieved with small and dry wood particles, while larger particles, bark, green forest residues, and agricultural and other residues are all detrimental to liquid yield. An alternative is intermediate pyrolysis with longer solid residence times followed by Thermal Catalytic Reforming (TCR) using the produced char. TCR bio-oil is more thermally stable, easier to upgrade and hydrodeoxygenate to conventional fuels (Jäger et al., 2016), and can be used as an industrial fuel without blending with conventional fuels (Broumand et al., 2021).

Bio-oil can also serve as a special platform for fuels, chemicals, and materials. Thermal decomposition of lignocellulose results in a complex mixture of many compounds with a wide range of molecular weights and multiple chemical functionalities, which contributes to the instability and corrosiveness of bio-oil. Simple fractionation of bio-oil in a phenolic (pyrolytic lignin) and an aqueous fraction has been studied, mainly for applications without further purification. For example, components in the aqueous fraction can be fermented to obtain bioethanol (Chi et al., 2019), oxidized into formic acid (Müller et al., 2016), or separated, e.g. to obtain acetic acid (Sarchami et al., 2021). The phenolic fraction, especially using bark as raw material, can have applications in the production of wood resins and as a wood preservative (Pinto et al., 2018). As resins for wood panels have only a limited lifetime, they are produced locally. The replacement of synthetic oil-derived phenol in resins with renewable resources such as tannin from pine bark or the phenolic fractions of bio-oil has been extensively studied with some success (Effendi et al., 2008). The lower reactivity and number of binding sites of the oligomeric compounds in the phenolic

[1] https://www.ensyn.com/brazil---suzano.html.

fractions of bio-oil used in resins require higher temperatures and longer pressing times to reach panel specifications, making them less attractive as components in wood resins. At the same time, the effects of variability of biomass and bio-oil production on properties of these fractions are also a concern for resin producers.

3.3 Biochar as Fuel, Soil Amendment, Carbon Capture, and Sequestration Path

At the lower end of the technology spectrum is the direct combustion or co-firing of raw or pretreated biomass, for example by low-temperature pyrolysis (330 °C). Although major ash-forming metals are essentially retained in the biochar and intensify slagging and fouling problems, thermal efficiency and environmental benefits increase during combustion of pyrolytic biochar compared to raw pine wood (Liu & Han, 2015). Biochar-to-fuel systems using woody wastes and crop residues can be economically viable negative-emission technologies, but hidden environmental impacts, e.g. from indirect land use change or additional fertilizer used on crop land, should not be neglected (Lee et al., 2020).

Soil sustainability is an important long-term criterion for bioenergy feedstocks. Biochar from the pyrolysis of biomass returned to soil can sequester the CO_2 taken up by plants from the atmosphere and improve soil health, fertility, and structure, as well as retain water and improve nutrient availability (Roberts et al., 2010). Biochar temporarily accelerates soil N transformations by stimulating soil microbial activity, thereby increasing N bioavailability (Nelissen et al., 2015). In addition to improving soil quality, biochar has been used to remove environmental pollutants, in water treatments, and in gaseous emission mitigation. Although the sorption capacity has been found to be lower than that of activated carbon, the spent product can be sold as a soil amendment (Hwang et al., 2018).

4 Discussion and Conclusions

In general, the technological challenges for new biorefineries are mainly related to the heterogeneity and natural variations in biomass quality, supply chains, pre-processing, the establishment of new technologies, and diverse environmental impacts. In addition, as lignocellulosic biomass production is limited by available land, its use in a resource-efficient context should resemble a cascade from high value-added material uses to energy recovery at the end. This is already practiced to some extent in the forestry sector, where parts of wood that cannot be profitably processed into other products are used for energy production. An improvement in this area would be the separation of a lignin stream for material applications in existing

Kraft pulp mills, as an example of an evolutionary strategy for more efficient resource use.

Unlike forest management in other parts of the world, in Chile the forestry sector almost exclusively exploits monoculture of pine and eucalyptus plantations, which are more like open-air production systems than natural forests. These plantations have an impact on biodiversity, water supply, erosion, and sediment transport in addition to social impacts and so far have not been required by law to undergo an environmental impact assessment (Salinas, 2020). Rural development is often mentioned as an objective of bioenergy policies, though further employment and value creation is found for biomaterials in the manufacturing industry. Large-scale deployment of monocultures for bioenergy is likely to negatively affect sustainability in one or more dimensions (Calvin et al., 2021) and will not improve the current situation of neighboring rural communities, which have a significant poverty gap in comparison to the country's average values (Cerda et al., 2020).

Lignocellulosic biomass is an important contribution to renewable heating. However, apart from air pollution problems, the carbon neutrality of bioenergy is not self-evident and depends on forest management and development of carbon stocks and greenhouse gas emissions of the bioenergy system. Thermochemical processing of lignocellulosic biomass with carbon capture and sequestration as biochar or CO_2 can play an important role in a carbon-negative economy (Brown, 2021). This is especially true for autothermal fast pyrolysis that could dispense with the use of biochar in the process and produce higher-value energy products than slow pyrolysis.

Common to circular economy, industrial ecology, and process integration is the aim of minimizing resource and energy consumption by integration, recycling, and recovery. Nevertheless, contexts and scales differ. The involved disciplines, i.e. chemical and process engineering, ecology, and economy, study a range of contexts from processes and industrial sites up to entire economies with little overlap. At the process scale, design tools and binding concepts are well developed, while at larger scales methods are largely descriptive. At the process level, many of the principles of integration, recycling, and recovery are commonplace; for practical implementation at larger scales, non-technical aspects such as sustainable business models, multi-stakeholder cooperation, policy, and regulatory framework become important (Walmsley et al., 2019).

Circular material flows involve close dependencies between different companies that only function on the basis of trustful cooperation. The potential synergies and environmental benefits of material and energy flows between companies are often offset by constraints that these dependencies impose on the management and independent operation of each business. The establishment of industrial ecosystems only seems possible with the management and financial support of large corporations. Something similar occurs with the introduction of bio-based alternatives to fossil-derived commodities that are currently not cost competitive. The financing of biorefinery projects is challenging. Biorefinery projects are very capital intensive and involve many risks related to implementation, scaling, feedstock variability, and logistics, among others. The typical financial assessment of banks and the financial sector does not consider the long-term benefits of biorefinery projects (Vadlani, 2020).

In Chile, forest biorefineries are already in place but are tailored to pulp, timber, and derived materials. Forest and wood-processing residues are widely used for cogeneration and wood drying. From the supply side, significant expansion of forest plantation appears limited by land availability and conflicts of land use. The forest business is dominated by big players and interest in biofuel biorefineries is limited, as capital requirements are high and current profits would suffer. The amount of biomass needed to reach the necessary economy of scale for biofuel production using forest and agricultural residues is quite large and competes with already established supply chains and energetic uses of these residues. Value creation and employment are higher in the forest products sector than in the bioenergy sector. Governmental bioenergy policies focus on regulating solid fuels for heating and the sustainable management of native forests to improve the use of firewood, or switching to electricity altogether in urban centers to improve air emissions.

Therefore, the future contributions of lignocellulosic biomass to a more circular and bio-based economy in Chile are likely to be in material applications rather than fuels, particularly in the construction sector where energy-intensive materials such as cement and steel can be replaced. On the other hand, opportunities for thermochemical processes may arise from the recycling of industrial wastes and sludge from municipal and industrial wastewater treatments, and in the short term from the gradual implementation of the Extended Producer Responsibility Act, which obliges companies in certain sectors to organize the recycling of their products and meet recycling quotas. This will give a boost to waste-recycling technologies for the treatment of residues such as packaging waste, plastics, and used tires.

References

ACERA. (2021). *Boletín Estadísticas* [Reporte Noviembre 2021]. ACERA. https://acera.cl/estadisticas-generacion-de-energia/

Acuña, E., Rubilar, R., Cancino, J., Albaugh, T. J., & Maier, C. A. (2018). Economic assessment of Eucalyptus globulus short rotation energy crops under contrasting silvicultural intensities on marginal agricultural land. *Land Use Policy, 76*, 329–337. https://doi.org/10.1016/j.landusepol.2018.05.028

Almonacid, F. (2018). Bioenergy in an Agroforestry Economy under Crisis: Complement and Conflict. La Araucanía, Chile, 1990–2016. *Sustainability, 10*(12), 4478. https://doi.org/10.3390/su10124478

Azócar, L., Hermosilla, N., Gay, A., Rocha, S., Díaz, J., & Jara, P. (2019). Brown pellet production using wheat straw from southern cities in Chile. *Fuel, 237*, 823–832. https://doi.org/10.1016/j.fuel.2018.09.039

Baettig, P. R., Yáñez, A. M., & Albornoz, A. M. (2010). Cultivos dendroenergéticos de híbridos de álamo para la obtención de biocombustibles en Chile: estado del arte. *Bosque (Valdivia), 31*(2). https://doi.org/10.4067/S0717-92002010000200002

Bertrán, J., & Morales, E. (2008). *Potencial de generación de energía por residuos del manejo forestal en Chile.*

Birner, R. (2018). Bioeconomy Concepts. In I. Lewandowski (Ed.), *Bioeconomy: Shaping the transition to a sustainable, biobased economy* (pp. 17–38). Cham: Springer International Publishing. https://doi.org/10.1007/978-3-319-68152-8_3

Broumand, M., Khan, M. S., Yun, S., Hong, Z., & Thomson, M. J. (2021). The effect of thermo-catalytic reforming of a pyrolysis bio-oil on its performance in a micro-gas turbine burner. *Applications in Energy and Combustion Science, 5*, 100017. https://doi.org/10.1016/j.jaecs.2020.100017

Brown, R. C. (2021). The role of pyrolysis and gasification in a carbon negative economy. *Processes, 9*(5), 882. https://doi.org/10.3390/pr9050882

Calvin, K., Cowie, A., Berndes, G., Arneth, A., Cherubini, F., Portugal-Pereira, J., Grassi, G., House, J., Johnson, F. X., Popp, A., Rounsevell, M., Slade, R., & Smith, P. (2021). Bioenergy for climate change mitigation: Scale and sustainability. *GCB Bioenergy, 13*(9), 1346–1371. https://doi.org/10.1111/gcbb.12863

Cerda, R., Gallardo-Cobos, R., & Sánchez-Zamora, P. (2020). An analysis of the impact of forest policy on rural areas of Chile. *Forests, 11*(10), 1105. https://doi.org/10.3390/f11101105

Chi, Z., Zhao, X., Daugaard, T., Dalluge, D., Rover, M., Johnston, P., Salazar, A. M., Santoscoy, M. C., Smith, R., Brown, R. C., Wen, Z., Zabotina, O. A., & Jarboe, L. R. ('). Comparison of product distribution, content and fermentability of biomass in a hybrid thermochemical/biological processing platform. *Biomass and Bioenergy, 120*, 107–116. https://doi.org/10.1016/j.biombioe.2018.11.006

CNE. (2020). *Anuario estadístico de energía.*

CNE. (2021). *Reporte Mensual ERNC* (Nr. 64). Comision Nacional de Energía. https://www.cne.cl/wp-content/uploads/2021/12/RMensual_ERNC_v202112.pdf

Dahmen, N., Henrich, E., Dinjus, E., & Weirich, F. (2012). The bioliq® bioslurry gasification process for the production of biosynfuels, organic chemicals, and energy. *Energy, Sustainability and Society, 2*(1). https://doi.org/10.1186/2192-0567-2-3

Effendi, A., Gerhauser, H., & Bridgwater, A. V. (2008). Production of renewable phenolic resins by thermochemical conversion of biomass: A review. *Renewable and Sustainable Energy Reviews, 12*(8), 2092–2116. https://doi.org/10.1016/j.rser.2007.04.008

Electricidad. (2019). Biomasa: potencial al alza. *Electricidad*(232), 4–8.

Gerding, V., An Der Fuhren, F., Holmqvist, C., Almonacid, B. N., & Ceballos, R. (2013). *Evaluación del mercado de biomasa y su potencial: Estimación de superficie disponible para el establecimiento de plantaciones dendroenergéticas.* Universidad Austral de Chile. https://doi.org/10.13140/RG.2.1.2247.0561

Hall, C. A., Lambert, J. G., & Balogh, S. B. (2014). EROI of different fuels and the implications for society. *Energy Policy, 64*, 141–152. https://doi.org/10.1016/j.enpol.2013.05.049

Hwang, O., Lee, S.-R., Cho, S., Ro, K. S., Spiehs, M., Woodbury, B., Silva, P. J., Han, D.-W., Choi, H., Kim, K.-Y., & Jung, M.-W. (2018). Efficacy of different biochars in removing odorous volatile organic compounds (VOCs) emitted from swine manure. *ACS Sustainable Chemistry & Engineering, 6*(11), 14239–14247. https://doi.org/10.1021/acssuschemeng.8b02881

INFOR Instituto Forestal. (2021). *El sector forestal chileno 2020.*

Jäger, N., Conti, R., Neumann, J., Apfelbacher, A., Daschner, R., Binder, S., & Hornung, A. (2016). Thermo-catalytic reforming of woody biomass. *Energy & Fuels, 30*(10), 7923–7929. https://doi.org/10.1021/acs.energyfuels.6b00911

King, L. C., & van den Bergh, J. C. J. M. (2018). Implications of net energy-return-on-investment for a low-carbon energy transition. *Nature Energy, 3*(4), 334–340. https://doi.org/10.1038/s41560-018-0116-1

Lan, K., Park, S., Kelley, S. S., English, B. C., Yu, T.-H.E., Larson, J., & Yao, Y. (2020). Impacts of uncertain feedstock quality on the economic feasibility of fast pyrolysis biorefineries with blended feedstocks and decentralized preprocessing sites in the Southeastern United States. *GCB Bioenergy, 12*(11), 1014–1029. https://doi.org/10.1111/gcbb.12752

Lee, M., Lin, Y.-L., Chiueh, P.-T., & Den, W. (2020). Environmental and energy assessment of biomass residues to biochar as fuel: A brief review with recommendations for future bioenergy systems. *Journal of Cleaner Production, 251*, 119714. https://doi.org/10.1016/j.jclepro.2019.119714

Lewandowski, I. (2015). Securing a sustainable biomass supply in a growing bioeconomy. *Global Food Security, 6*, 34–42. https://doi.org/10.1016/j.gfs.2015.10.001

Liu, Z., & Han, G. (2015). Production of solid fuel biochar from waste biomass by low temperature pyrolysis. *Fuel, 158*, 159–165. https://doi.org/10.1016/j.fuel.2015.05.032

Loeffler, M., Hinrichs, J., Moß, K., Henkel, M., Hausmann, R., Kruse, A., Dahmen, N., Sauer, J., & Wodarz, S. (2018). Processing of biobased resources. In I. Lewandowski (Hrsg.), *Bioeconomy* (S. 179–230). Springer International Publishing. https://doi.org/10.1007/978-3-319-68152-8_7

Morales, M., Quintero, J., & Aroca, G. (2017). Environmental assessment of the production and addition of bioethanol produced from Eucalyptus globulus to gasoline in Chile. *The International Journal of Life Cycle Assessment, 22*(4), 525–536. https://doi.org/10.1007/s11367-016-1119-4

Müller, N., Romero, R., Grandón, H., & Segura, C. (2016). Selective production of formic acid by wet oxidation of aqueous-phase bio-oil. *Energy & Fuels, 30*(12), 10417–10424. https://doi.org/10.1021/acs.energyfuels.6b01345

Nelissen, V., Rütting, T., Huygens, D., Ruysschaert, G., & Boeckx, P. (2015). Temporal evolution of biochar's impact on soil nitrogen processes-a 15 N tracing study. *GCB Bioenergy, 7*(4), 635–645. https://doi.org/10.1111/gcbb.12156

Ning, S.-K., Hung, M.-C., Chang, Y.-H., Wan, H.-P., Lee, H.-T., & Shih, R.-F. (2013). Benefit assessment of cost, energy, and environment for biomass pyrolysis oil. *Journal of Cleaner Production, 59*, 141–149. https://doi.org/10.1016/j.jclepro.2013.06.042

Oasmaa, A., Lehto, J., Solantausta, Y., & Kallio, S. (2021). Historical review on VTT fast pyrolysis bio-oil production and upgrading. *Energy & Fuels, 35*(7), 5683–5695. https://doi.org/10.1021/acs.energyfuels.1c00177

Pinto, O., Romero, R., Carrier, M., Appelt, J., & Segura, C. (2018). Fast pyrolysis of tannins from pine bark as a renewable source of catechols. *Journal of Analytical and Applied Pyrolysis, 136*, 69–76. https://doi.org/10.1016/j.jaap.2018.10.022

Priefer, C., Jörissen, J., & Frör, O. (2017). Pathways to shape the bioeconomy. *Resources, 6*(1), 10. https://doi.org/10.3390/resources6010010

Roberts, K. G., Gloy, B. A., Joseph, S., Scott, N. R., & Lehmann, J. (2010). Life cycle assessment of biochar systems: Estimating the energetic, economic, and climate change potential. *Environmental Science & Technology, 44*(2), 827–833. https://doi.org/10.1021/es902266r

Salinas, F. (2020). *La necesaria discusión sobre la evaluación ambiental de plantaciones forestales industriales*. Heinrich-Böll-Stiftung-Santiago de Chile. https://cl.boell.org/es/2020/10/06/la-necesaria-discusion-sobre-la-evaluacion-ambiental-de-plantaciones-forestales

Santelices-Moya, R., Espinoza-Meza, S., Cabrera-Ariza, A., & Medina-Urbina, A. (2019). Potential of some species for biomass energy production in the mediterranean dryland of Chile. *Ciência Florestal, 29*(2), 844–853. https://doi.org/10.5902/19805098

Sarchami, T., Batta, N., & Berruti, F. (2021). *Production and separation of acetic acid from pyrolysis oil of lignocellulosic biomass: A review*. Vorab-Onlinepublikation. https://doi.org/10.1002/bbb.2273

Schipfer S. & Kranzl, l. (2019). Techno-economic evaluation of biomass-to-end-use chains based on densified bioenergy carriers (dBECs). *Applied Energy*, 239, 715–724. https://doi.org/10.1016/j.apenergy.2019.01.219

Vadlani, P. V. (2020). Financing strategies for sustainable bioenergy and the commodity chemicals industry. In A. A. Vertès, N. Qureshi, H. P. Blaschek, & H. Yukawa (Hrsg.), *Green energy to sustainability* (S. 569–586). Wiley. https://doi.org/10.1002/9781119152057.ch23

Wang, C., Venderbosch, R., & Fang, Y. (2018). Co-processing of crude and hydrotreated pyrolysis liquids and VGO in a pilot scale FCC riser setup. *Fuel Processing Technology, 181*, 157–165. https://doi.org/10.1016/j.fuproc.2018.09.023

Walmsley, T. G., Ong, B. H. Y., Klemeš, J. J., Tan, R. R., & Varbanov, P. S. (2019). Circular integration of processes industries and economies. *Renewable and Sustainable Energy Reviews, 107*, 507–515. https://doi.org/10.1016/j.rser.2019.03.039

Zimmer, T., Rudi, A., Müller, A.-K., Fröhling, M., & Schultmann, F. (2017). Modeling the impact of competing utilization paths on biomass-to-liquid (BtL) supply chains. *Applied Energy, 208*, 954–971. https://doi.org/10.1016/j.apenergy.2017.09.056

The Path to Introduce Nanofibrillated Cellulose (NFC) to the Chilean Market

Alex Berg and Felipe Guzmán

Abstract Chemical and mechanical degradation of cellulose pulp with minimal impact on cellulose fibers results in an interesting material known as nanofibrillated cellulose (NFC). As a glucose polymer with a linear structure, NFC can be easily modified by chemical means, which leads to the possibility of modifying its compatibility with a variety of polymeric matrices. Then, due to its remarkable mechanical properties, it has great potential a mechanical reinforcer for multiple materials. In UDT (Technology Development Unit, Universidad de Concepción), we have extensively studied different strategies to obtain, modify, and use NFC as mechanical reinforcer of thermoplastic and thermosetting polymers. Thu, in the following chapter, we explore and discuss our study of the potential of NFC as reinforcing additive for important products in the Chilean industry.

1 Introduction: Towards a Sustainable Bioeconomy

Today, Chile has more than 16 million hectares of forest, 86% of which corresponds to natural forests and 14% to plantations (Raga, 2009). The most prominent planted species include *Pinus radiata* (56% of the plantations), *Eucalyptus globulus,* and *Eucalyptus nitens* (World Bank, 2020). In fact, these three species cover 92% of the total planted surface, owing to the number of cycles and high yields attainable with these within Chilean growing conditions. Indeed, the average growth rate for *Pinus radiata* is around 25 m^3/ha/year, while for *Eucalyptus sp.* it is around 40 m^3/ha/year. Such production levels have facilitated a continuous increase in forestry exports over the last few years. In 2018, exports revenue of forestry products accounted for US $6,84 billion, an increase of 27% compared to 2015 (INFOR, 2021). These figures place Chile among the top 20 leading exporters of forest products in the world. The

A. Berg (✉) · F. Guzmán
Unidad de Desarrollo Tecnológico, Universidad de Concepción. Avenida Cordillera 2634, Parque Industrial Coronel, 419996 Coronel, Chile
e-mail: a.berg@udt.cl

F. Guzmán
e-mail: felipe.guzman@sgs.com

A. Ch. Braun et al. (eds.), *Eco-Industrial Development as an Industrial Strategy*,
Green Energy and Technology, https://doi.org/10.1007/978-3-031-73576-9_13

main subsectors in forest exports are pulp and sawn timber, accounting for more than 60% of exports. Other important export products include woodchips, plywood, several types of paper (paperboard, newspaper, corrugated paper, wrapping paper, and printing and writing paper), tissue (toilet paper, paper towels, paper napkins, and facial paper), and sanitary products (diapers and sanitary pads).

In this chapter, we analyze the path our research team has taken thus far in seeking to introduce nanofibrillated cellulose, a rather interesting and versatile product that takes advantage of these byproducts, into the Chilean market. We discuss the possibilities of this product and narrow them down to promising markets. Next, we describe promising methods we have used to adapt the inherent characteristics of nanofibrillated cellulose, improve its compatibility with different products, and enhance its performance in specific applications.

2 Nanofibrillated Cellulose Production: Possibilities and Our Experience so Far

2.1 *The State of the Art*

A wide variety of biobased products have been developed in order to address increasing demand for products made from renewable, non-petroleum-based resources. However, several economic barriers hinder their entry to the market, as most well-established products derived from fossil fuels are rather inexpensive compared to their natural alternatives. As a result, the transition to a green economy is a slow and experimental process. A tentative strategy may be to develop highly functional products capable of outperforming their competitors, yet this outcome is currently quite challenging to achieve while maintaining a cost-effective production scheme. Another possibility is beginning market penetration with a conservative approach, by creating additives capable of improving commercially available products, increasing their efficacy and thus reducing their overall usage (Maksymiv et al., 2021). In that context, nanocellulose comes to mind as an ideal candidate. Owing to the hierarchical structure and semi-crystalline nature of cellulose, nanoparticles can be produced from this naturally occurring polymer using mechanical or chemical deconstruction processes. When cellulose particles are disintegrated to sizes with at least one dimension in nanoscale (1–100 nm), the resulting material is referred to as nanocellulose (Trache et al., 2016). Mechanical disaggregation is often carried out by exposing a suspension of cellulose fibers in a liquid medium (usually water) to the action of high shearing forces, which can be accomplished by refining, grinding, steam explosion, and/or homogenization processes. During such treatment, strongly entangled networks of nanofibrils are created that have both crystalline and amorphous domains, which are often called nanofibrillated cellulose (NFC). These suspensions maintain several inherent characteristics of cellulose, such as renewability, biodegradability, and non-toxicity. However, they stand out as exceptional materials owing to the very high mechanical properties that only manifest in such a format, as covalent bond breakage enhances hydroxyl group accessibility which leads to increased hydrogen bond formation between glucose units. Indeed, when cellulose is fibrillated to the nanoscale the fibers display high surface area-to-volume ratio, high Young's modulus and tensile strength, low coefficient of thermal expansion, as well as an exceptional hydrogen-bonding capacity, which make NFC such a high-performing and versatile material (Maksymiv et al., 2021; Nechyporchuk et al., 2016).

NFCs have demonstrated excellent functionality in a variety of applications, which has enabled them to reach industrial-scale production in countries such as Japan and Sweden. However, even in these cases, NFC production is somewhat limited and does not live up to its potential. We attribute this to two main factors:

(1) First, NFCs are primarily manufactured using highly energy-demanding methods. This is due to the high specific surface of NFCs, as fibrillation entails high hydrogen bond formation, leading to hard fiber agglomerations that become difficult to disperse.
(2) Second, the fibrillation of cellulose must be carried out in a liquid medium, with irreversible fiber agglomeration (called hornification) occurring after drying. Therefore, NFCs mostly exist as aqueous dispersions with low solid contents. As a result, these products require large storage areas and have high transportation costs.

In order to address the first limitation, that is, to reduce the excessive energy consumption involved in the mechanical disaggregation of cellulose, several pretreatments tailored to partially fibrillate the polymeric matrix and/or to reduce inter-fiber interaction have been proposed, tested, and evaluated. We can classify these into four main techniques (Nechyporchuk et al., 2016; Yi et al., 2020):

- **Cellulose hydrolysis** to partially fibrillate cellulose before the mechanical treatment, thus reducing energy consumption during that step
- **Chemical modification of cellulose**, such as using coupling agents (lauric acid, malic acid, etc.) to reduce the hydroxyl content on the cellulose surface, thereby reducing the inter-cellulose and fiber interaction and facilitating the mechanical separation of the fibers
- **Functionalization of hydroxyl groups on the cellulose surface**, such as using TEMPO to oxidize surface hydroxyl groups; cationization of cellulose fibers through the introduction of positive and negative charges; nucleus addition to introduce specific groups or components to promote the swelling of cellulose in water; or increasing the fiber charge to make fibers repel each other and reduce inter-fiber cohesion
- **Bacterial/enzymatic pretreatment** to start bond breakage by the action of specific enzymes

Although these pretreatments considerably reduce energy consumption during the fibrillation step, all of them have limitations. The first three methods tend to alter some properties of NFC in varying degrees, depending on the operation conditions, and are considered to be better-suited as strategies to adapt the product for specific applications that require modified behavior. The fourth method, on the other hand, must be carried out under rather constrained operation conditions to ensure optimum performance, since it leaves the hydroxyl groups unaltered, maintaining their original properties. It can even be used in concert with other pretreatment methods in order to obtain highly functional NFC variants, while further reducing energy demands during fibrillation (Yi et al., 2020).

In spite of the variety of options that have been developed to address the first limitations, attempts to overcome the second limitation have been unsuccessful until now, mainly because removing NFC from a liquid medium inevitably leads to the mentioned irreversible fiber hornification that prevents them from being redispersed for later usage, regardless of the chosen drying technique.

On another note, several post-treatments have been tested to enhance product performance by chemically modifying NFC's surface composition, thus making it possible to improve NFC compatibility with hydrophobic matrices without altering its inherent properties. Such techniques increase the range of materials with which it can form composites, expanding its potential applications even further. Today, such applications include construction (for making structural composites), automotive (for manufacturing parts based on micropatterns), electronics (as a membrane for electroacoustic devices), pharmaceutical (for biomedical applications), and cosmetic usages, among others (Nechyporchuk et al., 2016).

Considering such varied possibilities, as well as the expected challenges of introducing a new and relatively unknown product into the Chilean market, our research at the Technical Development Unit (UDT) of Universidad de Concepción aimed to tailor NFC for applications as an additive meant to enhance the mechanical properties of thermoplastic and thermostable resins, as these products belong to well-established markets that would benefit from innovation.

2.2 *Our Experience so Far*

We have been studying NFC production and modification for the last six years. Over this time, we have managed to accumulate considerable knowledge and experience regarding the effects of several variables on the quality and properties of the resulting products. In this section, we describe both the advantages and limitations of different processing configurations.

2.2.1 Evaluation of Different Processing Settings: Advantages and Limitations

Materials

Bleached Kraft Eucalyptus pulp, obtained from CMPC S. A.

Methodology

Cellulose Fibrillation

We employed various combinations of three different operations units during this step:

- Hammer mill (Peerless H20AA): A pretreatment performed in all the experiments, its main objective was to reduce average particle size of the cellulose to favor subsequent processes.
- Refiner (Moorim P&P, Laboratory Plant 12-1C): A commercial refiner with 12″ concentric discs capable of fibrillating the pulp to the micro scale, reaching the minimum size required for processing in the next stage. It operates optimally with suspensions of 1–2% consistency, maintaining a distance between the discs of less than 0.05 mm and a rotation speed of 3,000. Samples obtained after 15, 30, and 45 cycles through the refiner were characterized. This step was also performed in all the experiments, but the number of cycles was varied to compare its influence on fibrillation.
- Homogenization (GYB High-Pressure Homogenizer, GYB1200-10S-N, 1200-10S-N): Fiber suspensions obtained from a 30-cycle refining process were taken to a high-pressure homogenizer (200 bar), where they were recirculated multiple times. This unit enabled further reductions in the particles' size, reaching nanoscale and thus increasing their aspect ratio and viscosity, leading to a product with gel-like behavior. Samples of different fibrillation degrees were taken after 15, 30, and 45 cycles. Homogenization was not performed in all the experiments.

Film Preparation

We prepared films using the samples obtained at different stages of the refining and homogenization processes. All samples were diluted to 0.5% w/w and mixed in an IKA T50 digital Ultraturrax for 1 min at 5,000 rpm. Afterwards, we sonicated the samples for 2 min, using aClean-120HD ultrasonic cleaner. Finally, we poured the required measure for a 20 g/m^2 sample into a petri dish, and dried it at 50 °C until reaching constant weight.

Transmittance

We measured the transmittance of the samples obtained at different stages of the refining and homogenization processes, using a (brand model) spectrophotometer. To that end, we diluted the samples to 0.1% w/w and stirred them in an IKA T50 digital Ultraturrax for 1 min at 5,000 rpm. We measured the transmittance of each sample at 800 λ on triplicate.

Viscosity

We measured cinematic viscosity of the NFC samples in a Fungilab SMART SMAL 201,102 Brookfield viscometer. We diluted the samples to 0.5% w/w, mixed them in an IKA T50 digital Ultraturrax for 4 min at 10,000 rpm, and kept them under

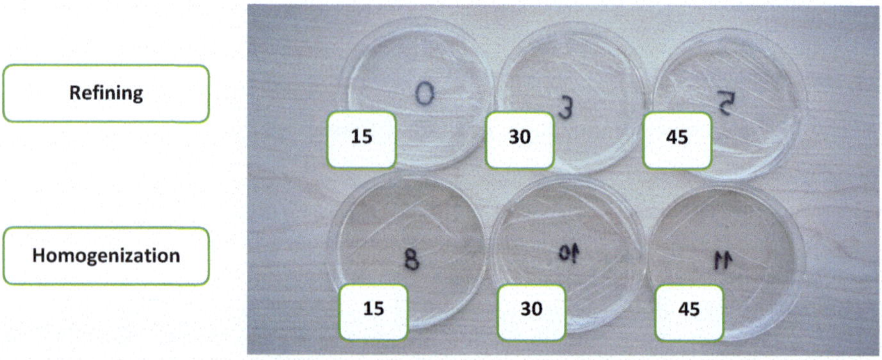

Fig. 1 Films prepared from NFC samples obtained after 15, 30, and 45 refining cycles (top), and from samples additionally processed by 15, 30, and 45 homogenization cyces (below)

a controlled environment at 25 °C and 50% moisture for 24 h. We measured this property at 25 °C and speeds of 30, 60, and 100 rpm, using the appropriate Spindle, depending on sample behavior.

Results

Film Preparation

In Fig. 1, we present a picture of films prepared with the samples from different refining and homogenization stages.

We observe stark differences in transparency between the refining and homogenization samples. This suggests a correlation between transparency and the degree of fibrillation, which is reasonable, since smaller molecules should have a lower capacity to retain light. However, when looking for differences between samples obtained from the same process, i.e. comparing variances in the number of processing cycles in the same unit only, this phenomenon is not as evident. Thus, we surmise that light transmittance should be measured and analyzed in order to properly establish this hypothesis.

Transmittance

We carried out transmittance tests in order to fully understand the optical differences between the films prepared from different samples of aqueous NFC. The results are presented in Table 1:

The results shown in Table 1 are consistent with the hypothesis proposed in the previous section: that the degree of fibrillation has a close relationship with transparency. We may attribute this behavior not only to the reduction of particle size, but also to a possible reduction of non-cellulosic content (residual lignin and

Table 1 Transmittance of samples obtained at different stages of the refining and homogenization processes

Sample	Transmittance
Refiner, 15 cycles	19.2
Refiner, 30 cycles	24.5
Refiner, 45 cycles	30.1
Homogenizer, 15 cycles	39.2
Homogenizer, 30 cycles	47.2
Homogenizer, 45 cycles	56.2

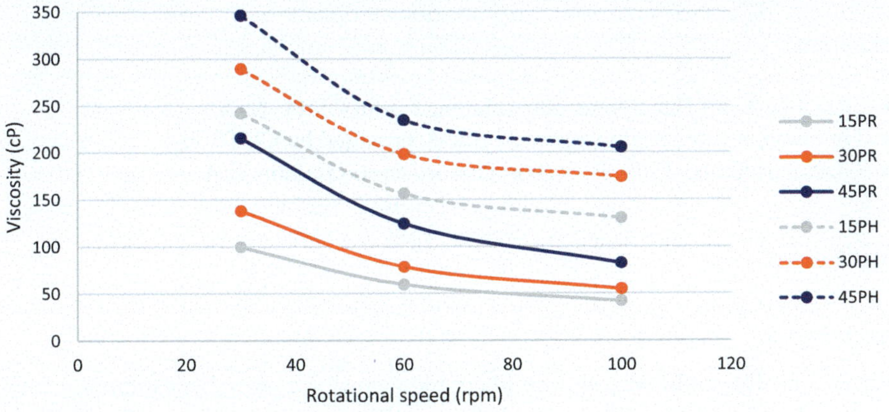

Fig. 2 Viscosity of aqueous NFC samples obtained at different stages of the refining and homogenization processes

hemicelluloses that remain in the samples) due to the effect of the shearing forces applied during fibrillation.

Viscosity

In Fig. 2, we present the results of the cinematic viscosity of aqueous NFC samples obtained from refining and homogenization.

In Fig. 2, we observe that viscosity increases with fibrillation and that such differences tend to decrease as the rotational velocity of the Spindle increases. This behavior is consistent with the key concepts that describe the properties of NFC. That is, as molecular size decreases, intermolecular interactions increase as more hydroxyl groups become available and steric hindrance is reduced, thereby elevating the reinforcing capabilities of NFC and resulting in suspensions with gel-like properties.

2.2.2 Chemical and Biological Modification: Reliable Pathways to Improve Performance?

The literature describes a variety of processes that have been tested to reduce energy consumption during the mechanical fibrillation stage, and/or to alter NFC properties to improve their performance in specific applications. Our research so far has focused on minimizing production costs and increasing NFC compatibility with hydrophobic polymeric matrices. To that end, both enzymatic and chemical pretreatments and post-treatments have been tested and evaluated.

Materials

- Bleached Kraft Eucalyptus pulp, obtained from CMPC S. A.
- Quimizyme B (endoglucanase enzymes), provided by CHT Ltda.
- Glacial acetic acid (99.9% w/w), obtained from Oxiquim S.A.

Methodology

Acetylation

We chemically modified cellulose fibers by means of a partial substitution of the hydroxyl groups of the cellulose chain by acetyl groups via esterification with acetic acid. To that end, we suspended the fibers in glacial acetic acid (99.99% concentration) under a 1/40 ratio (cellulose/acid). We heated the mixture to 120 °C for 2 h, either before or after fibrillation. Then we evaluated the resulting products with the same tests used to characterize aqueous cellulose fibers.

Enzymatic Treatment

In addition, we tested an enzymatic treatment in order to promote mechanical fibrillation and achieve a better aspect ratio of the fibrils at a lower cost. We evaluated the effectiveness of the enzymatic treatment at different stages of the NFC production process, as follows.

We performed these experiments using acetic acid as a fibrillation medium and executing the maximum number of cycles (45) in both the refining and homogenization processes. For the enzymatic treatment, we adjusted the fiber suspension to 2.5% consistency. Next, we added the enzyme Quimizyme B to the suspension in a 0.5% concentration with respect to the solid mass of fibers. The suspension was continuously stirred at 48 °C (enzyme activation temperature) for 60 min. Afterwards, we deactivated the enzyme by increasing the temperature above 70 °C. Finally, we diluted the fiber suspension to adapt it for the next processing stage.

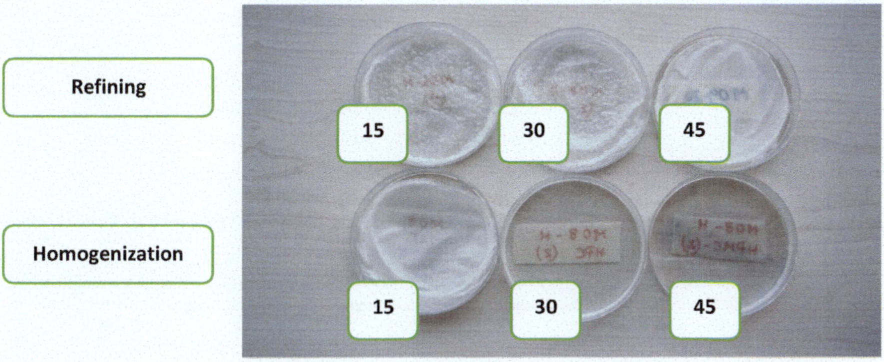

Fig. 3 Films prepared from NFC samples acetylated post-fibrillation, obtained after 15, 30, and 45 refining cycles (top), and after 15, 30, and 45 homogenization cycles (below)

Fig. 4 Films prepared from acetylated (top) and aqueous (bottom) NFC samples, obtained after 15, 30, and 45 refining cycles

Results

Film Preparation

In Fig. 3., we present an image of the films prepared with the acetylated samples, obtained at different stages of the refining and homogenization processes. Note that only the samples that were heated up post-fibrillation are included, as those that were heated prior to fibrillation displayed rather poor transparency, regardless of the degree of fibrillation.

In order to facilitate comparison, in Figs. 4 and 5 we also present images of the films prepared from both the aqueous NFC samples and the acetylated samples, obtained from the refining and homogenization processes, respectively.

In Fig. 4, we can see that the differences between acetylated and aqueous samples obtained from the refining process are markedly apparent, irrespective

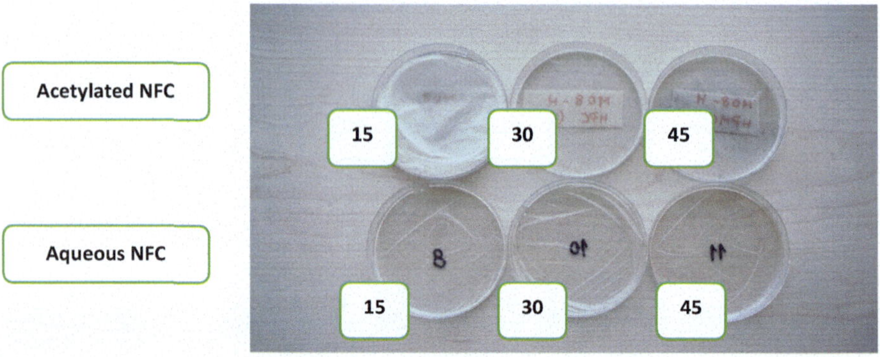

Fig. 5 Films prepared from acetylated (top) and aqueous (bottom) NFC samples, obtained after 15, 30, and 45 homogenization cycles

Table 2 Tested NFC processing scheme configurations

Processing scheme	Code
Refining–Homogenization	RH
Enzymatic treatment–Refining–Homogenization	ERH
Refining–Enzymatic treatment–Homogenization	REH
Acetylation–Refining–Homogenization	ARH
Refining–Acetylation–Homogenization	RAH
Refining–Homogenization–Acetylation	RHA

of the number of refining cycles. In Fig. 5, we can observe that such differences decrease with homogenization. This suggests that acetylated NFC has lower light-transmittance capacity, making it require further fibrillation to achieve transparency. This hypothesis should be supported with transmittance measurements.

Transmittance

In order to support the observations made about films prepared with acetylated NFC samples, we carried out transmittance tests on the prepared films (Table 3).

The results shown in Table 2 confirm the hypothesis we made in the previous section, i.e. that acetylated NFC has lower light transmittance than the aqueous version, which also increases with fibrillation. This entails that higher fibrillation is required to achieve the same transparency. However, other properties should be analyzed individually.

Viscosity

In Fig. 6, we present the results of our measurements of cinematic viscosity for the acetylated NFC samples obtained from refining and homogenization.

Table 3 Transmittance of aqueous and acetylated NFC samples obtained at different stages of the refining and homogenization processes

	Transmittance	
Sample	Aqueous NFC	Acetylated NFC
Refiner, 15 cycles	19.2	13.6
Refiner, 30 cycles	24.5	18.2
Refiner, 45 cycles	30.1	25.8
Homogenizer, 15 cycles	39.2	30.5
Homogenizer, 30 cycles	47.2	38.2
Homogenizer, 45 cycles	56.2	46.2

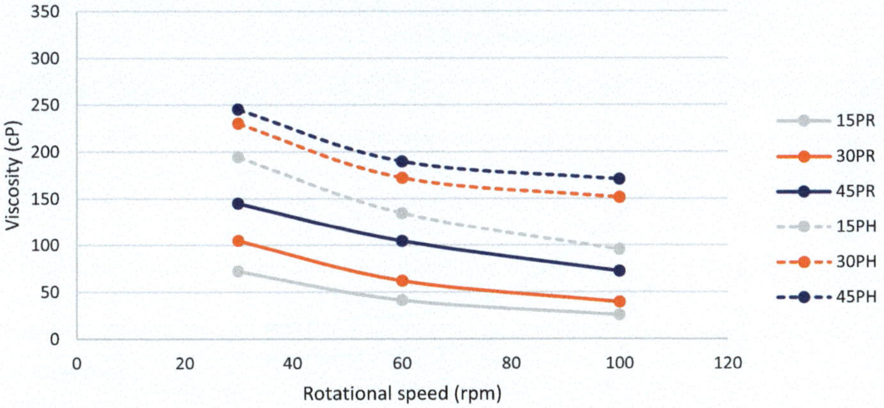

Fig. 6 Viscosity of acetylated NFC samples obtained at different stages of the refining and homogenization processes

To facilitate comparison, in Figs. 7 and 8 we present the viscosity of the aqueous NFC samples and the acetylated samples, obtained from the refining and homogenization processes, respectively.

In Fig. 6 we can observe that, similarly to the case of aqueous NFC, the viscosity of the acetylated NFC samples increases as the degree of fibrillation augments. We also attribute this to the rise of available hydroxyl groups, which interact with each other via hydrogen bonds. On the other hand, Fig. 7 and 8 show that the viscosity of the acetylated samples tends to be lower than the aqueous samples fibrillated under the same conditions (number of processing cycles). This is consistent with our previous observations, as the esterification reaction substitutes a fraction of the hydroxyl groups, limiting hydrogen bonds and thus lowering the number of intermolecular interactions.

To test the effectiveness of enzymatic treatments and, at the same time, to evaluate the effects of acetylation before and after fibrillation (beyond the optical properties), we additionally tested several production configurations, which are summarized in

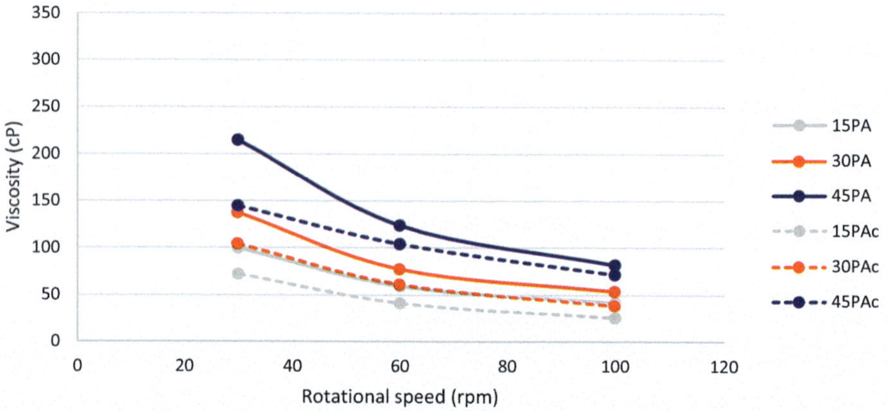

Fig. 7 Viscosity of aqueous and acetylated NFC samples obtained at different stages of the refining process

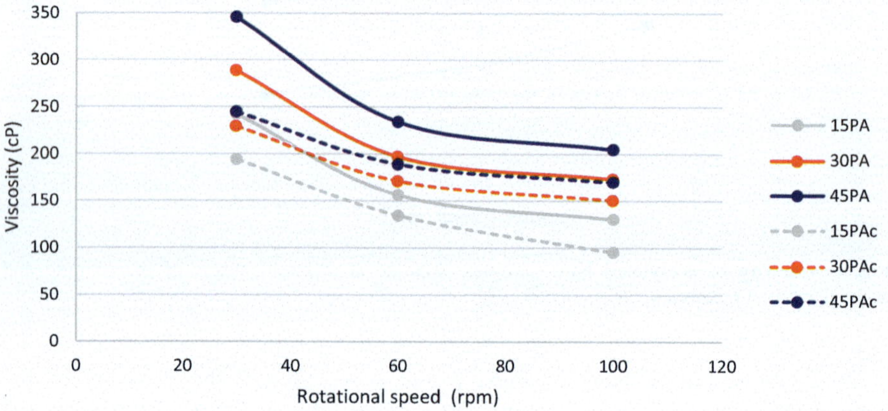

Fig. 8 Viscosity of aqueous acetylated NFC samples obtained at different stages of the homogenization process

Table 2. In Fig. 9, we present the viscosity of the resulting product of each of these production schemes.

Figure 9 reveals that both acetylation and enzymatic treatment enhance viscosity and therefore the intermolecular interactions within NFC. In both cases, we note that maximum viscosity is reached when the studied treatment is applied between the refining and homogenization step. We attribute this behavior to the fact that refining partially opens the cellulose fibers, providing better access to the hydroxyl groups (for acetylation) and ester bonds (for enzymatic treatment), while the true fibrillation to the nanoscale occurs during homogenization. This also explains why these configurations do not work as post-treatments, as bonding cleavage is more effective during homogenization, and these steps only serve to ease this process.

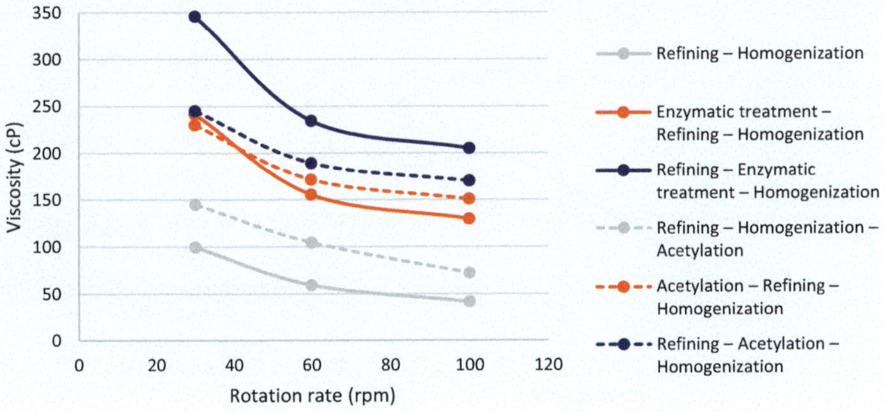

Fig. 9 Viscosity of NFC samples derived from different processing schemes

3 Nanofibrilated Cellulose Applications: Adapting to the Current Needs of the Chilean Market

3.1 Overview

Our goal is to develop NFC-based reinforcing additives for wood adhesives and plastics, capable of improving product performance to reduce consumption or increase performance.

3.2 Our Experience so Far

3.2.1 Urea Formaldehyde Adhesives

Our first target market is the wood adhesive industry. The objective of this investigation is to devise an NFC-based additive capable of homogeneous dispersion in urea–formaldehyde matrices, while reinforcing their adhesive performance to reduce adhesive consumption in particleboard manufacture.

Materials

- Commercial urea–formaldehyde (UF) adhesive, provided by Resinas del Bío-Bío S.A.
- Maple wood sheets (for bonding testing)

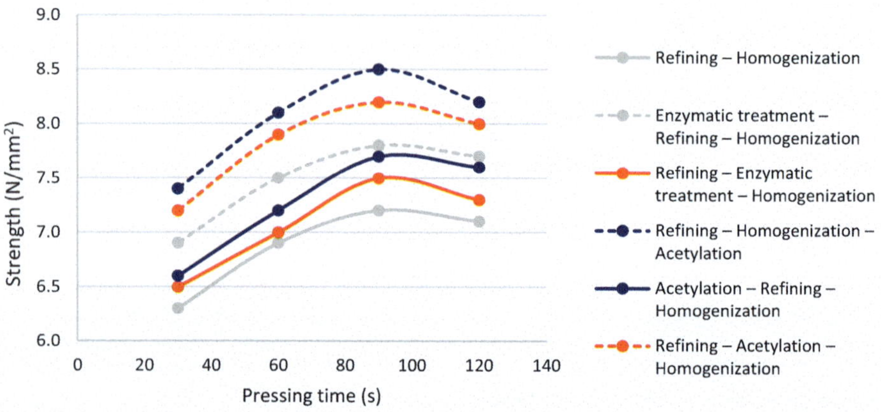

Fig. 10 Adhesive bonding strength of NFC samples derived from different processing schemes

- Aqueous and acetylated NFC samples, produced at UDT using the methodology described in Sect. 2.2.1.2

Methodology

We mixed 25 g samples of the UF resin with 1% w/w of NFC with respect to the solid content of the adhesive. The mixture was then stirred with an IKA T50 digital Ultraturrax for 2 min at 10,000 rpm.

Afterwards, we tested the adhesive performance of the mixture in a standard Adhesive Bonding Evaluation System (ABES), using maple wood sheets of 2.00 × 11.70 × 0.07 cm dimension. We smeared an area of 2.00 cm × 0.50 cm at the top of these wooden sheets with 0.0200 ± 0.0002 g of the NFC-reinforced adhesive. The tests were performed at 160 °C for 30, 60, 90, and 10 s. Every condition was measured three times.

Results

In Fig. 10 the adhesive bonding strength of the six NFC samples are presented.

In Fig. 10, we can observe that adhesive bonding strength may be improved by both acetylation and enzymatic pretreatment. Normally, one would expect samples reinforced with additives with higher viscosities to perform better, however this is not the case. Indeed, the highest performance is observed in acetylated samples, which can be attributed to a higher compatibility of the acetyl groups with the polarity of the UF matrix. In fact, we can see that the enzymatically treated sample with the highest viscosity displays the lowest performance among the modified samples, which supports the hypothesis that this processing scheme affords the optimal fibril disaggregation and therefore has the highest amount of available hydroxyl groups (which are highly hydrophilic). Such considerations demonstrate that acetylation is indeed a reliable path to improve NFC compatibility with urea–formaldehyde suspensions and has become the starting point of an extraneous development process. Afterwards, the research focus at UDT shifted to finding the processing conditions that generate a product with an adequate degree of substitution, acetyl group distribution, and degree of fibrillation for this application. Today, this product has been successfully optimized for commercial application.

3.2.2 Polylactic Acid

In recent years, our research group has focused on improving bioplastics performance—polylactic acid (PLA) in particular—to make these either commercially attractive within their target markets or to enhance their properties to access alternative niche markets.

PLA is a biodegradable, bio-based polymer that has several interesting properties: a high modulus of elasticity, high strength, thermoplasticity, transparency, and biocompatibility, even compared to other biopolymers such as thermoplastic starch. However, its brittle behavior and low crystallization kinetics limit its potential to a few applications. To overcome these drawbacks, much research has focused on modifying the polymer via branching chain reactions, by mixing it with other biopolymers, and through the addition of fillers. Several authors have shown that plasticizers such as polyethylene glycol (PEG) or citrate esters improve the crystallization kinetics. Among these, PEG is one of the most efficient at plasticizing bioplastics due to its good miscibility. Thus, our research focused on incorporating a variety of NFCs in PEG, which was then added to PLA.

Materials

- PEG 400, obtained from Reutter S.A.
- Six samples of acetylated NFC with varied degrees of substitution (0.07, 0.24, 0.47, 0.65, 0.75, and 1.06)
- Commercial 4032D polylactic acid (mainly used for injection), provided by Shanghai Huiang Industrial Co., Ltd.

Methodology

Preparing NFC-PLA Composites

First, we studied the effects of the dosage of NFC with respect to PLA. To that end, we prepared NFC-PLA composites using PEG as the plasticizer. We determined that 7.5% (w/w) of PEG with respect to the PLA was an adequate dosage and therefore

Table 4 Quantity of each component used in the prepared composites

NFC (% w/w)	PLA mass (g)	PEG mass (g)	NFC dry mass (g)
0.5%	45	3.38	0.23
1%	45	3.38	0.45
2%	45	3.38	0.90
3%	45	3.38	1.35

used this in each mixture. In Table 4, we present the composition of the prepared NFC-PLA composite samples.

We added the NFC samples to PEG using an electric stirrer over a hot plate kept at 60 °C, in order to work over the melting point of PEG 1500 (~58 °C) and to partially evaporate the water contained in the acetylated NFCs.

The mixtures were kept at this temperature for 10 min until reaching the consistency of a uniform paste. Next, the product was placed in a petri dish previously covered with aluminum foil. The mixtures were then placed in an oven at 60 °C to eliminate any trace of humidity during the night, removed the next day, and placed in a desiccator.

The plasticized NFCs were ground to the point of homogeneity. Before mixing, the PLA was oven-dried at 60 °C to minimize moisture content. The plasticized NFCs were slowly added to the PLA in a torque rheometer operated at 170 °C and 70 rpm for 10 min, focusing on maintaining torque levels over 0 at all times. After incorporating all of the reinforced plasticizer, the machine was kept running for at least 1.5 min to ensure total mixing; most cases required a minimum of 5 min.

Each mixture was subsequently ground to reduce the particles to a suitable format for injection molding.

Mechanical Properties

First, preparation of test samples suitable for standard evaluation was required. To that end, a HAAKE mini-jet injector with a molding piston system was used. Seven type IV test samples (detailed in Fig. 11) of each mixture were manufactured in accordance with the ASTM D638 standard. The piston was preheated at 175 °C for 5 min. Samples of 2.5 g were inserted into a mold maintained at 100 °C. The injection was carried out at 400 bar for 30 s.

The evaluation of mechanical properties was carried out using a smart Tens 005 universal machine in accordance with the ASTM D638 standard: "Tests to determine the tensile properties of reinforced and unreinforced plastics and is applied to the types of tubular specimens and crowbars (Types I-V) and rods."

Results

The elastic modulus of the produced NFC-PLA composites is presented in Fig. 12.

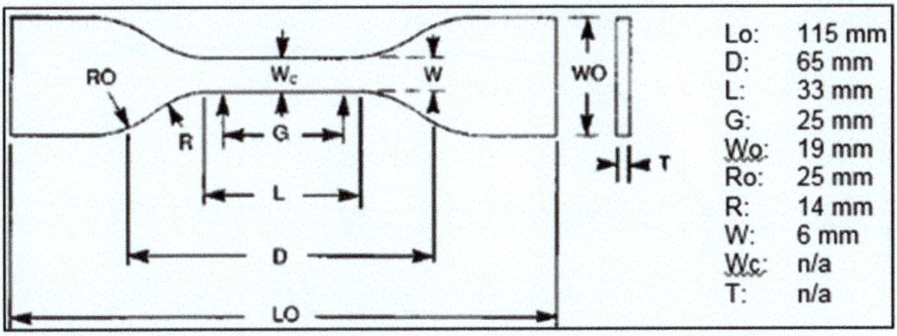

Fig. 11 Type IV test samples, in accordance with ASTM D638 standard

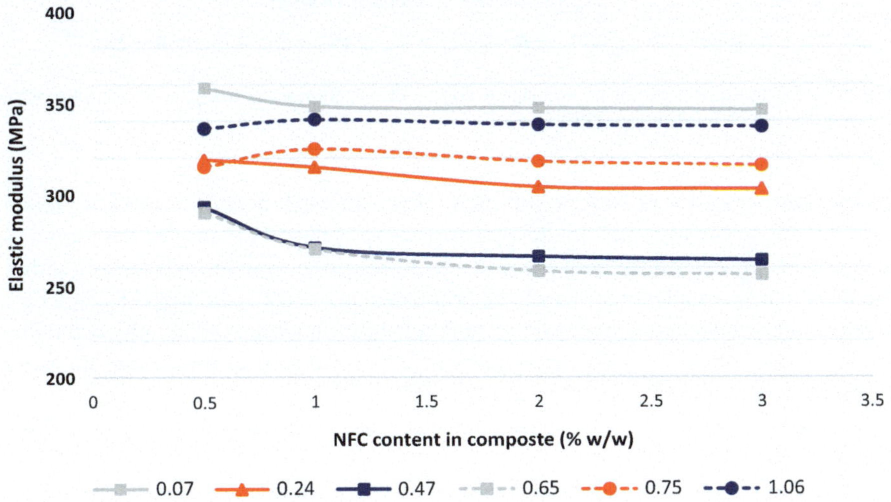

Fig. 12 Elastic modulus of acetylated NFC-PLA composites, using NFC samples with varying degrees of substitution (DS)

The modulus of elasticity, also known as Young's modulus, is the measure of the toughness and stiffness of polymeric material, expressed in MPa. The higher the value (modulus), the stiffer the material. Conversely, materials with low values are less stiff and easier to bend under load.

In Fig. 12, we observe that the addition of acetylated NFC can reduce the modulus of elasticity between 6.9% and 44.5% with respect to PLA + PEG400, i.e. the additive helps to lower the stiffness of the original material, making it more flexible. The results show that the sharpest decrease in the elasticity modulus occurs when NFC with a degree of substitution (DS) of 0.65 is added. We also notice that NFCs with a DS of 0.07 and 1.06 display a similar behavior, suggesting that, after a certain point, the benefits of acetylation are offset. We also considered the possibility of

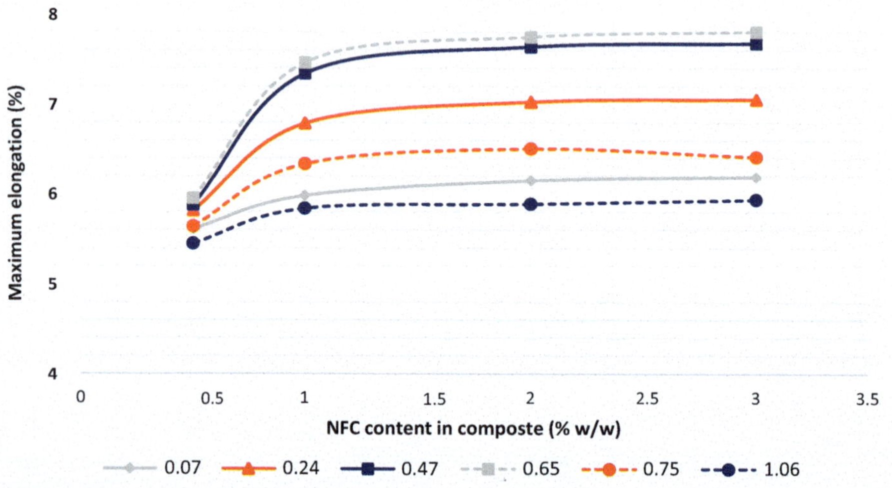

Fig. 13 Maximum elongation of acetylated NFC– PLA composites, using NFC samples with varying degrees of substitution (DS)

adding more than 1% of NFC to the PLA. However, such quantities did not yield a notable effect. We attributed the low performance of the NFC samples with low DS to low NFC compatibility with the plasticizer, resulting in poor distribution within the PLA matrix. On the other hand, the performance of NFC with high DS may be a consequence of excessive hydroxyl groups substitution within certain glucose units, leading to partial cellulose acetate formation, which reduces the reinforcing capacity of the additives.

In Fig. 13, we present the maximum elongation tests for the acetylated NFC samples.

Maximum elongation measures the increase in length that a material experiences before breaking when subjected to a tensile stress. The maximum elongation is expressed as a percentage (%) with respect to the initial length. In Fig. 13 we observe that the addition of acetylated NFC to PLA has a positive effect on the maximum elongation of the composite material. The samples with a DS between 0.24 and 0.65 yield the highest performance, reaching its peak with DS 0.65. Such behavior can be attributed to the same phenomena described to explain the optimal reduction in the Young's modulus. It is worth mentioning that, in both cases, the optimum value is reached with the 3% w/w addition of acetylated NFC, with a DS of 0.65.

As with the case of urea–formaldehyde adhesives, these results are a promising starting point to develop acetylated NFCs even further. In the meantime, we improved our production process and optimized our operating conditions. This has allowed our research team to obtain a modified NFC with better acetyl groups distribution, leading to a reduction in cellulose acetate formation. Our new samples have enabled us to shift the optimum DS value to a higher degree, and to enhance the reported reinforcing effect.

4 Conclusions

1. If fibrillation is carried out in a succession of refining and homogenization steps, it is possible to reduce cellulose particle size to the nanoscale. This allows for a product with gel-like behavior and high transparency.
2. Enzymatic treatments can increase the degree of fibrillation of nanofibrillated cellulose (NFC). Maximum performance is attained if the NFC is partially fibrillated beforehand.
3. Acetylation is a suitable method for increasing NFC compatibility with matrices of different polarities. Partial fibrillation before this chemical treatment results in a product with better acetyl group distribution.
4. Acetylated NFC is a promising candidate to increase the bonding strength of urea–formaldehyde adhesives.
5. Acetylated NFC can be inserted into plastic matrices by solubilizing it in compatible plasticizers such as PEG. In this way, it is possible to reduce the stiffness and enhance the flexibility of such products.

References

INFOR. (2021). *Statistical Bulletin N°180: Chilean statistical yearbook of forestry 2021*. Santiago, Chile: Intistituo Forestal INFOR.

Maksymiv, Y., Yakubiv, V., Pylypiv, N., Hryhoruk, I., Piatnychuk, I., & Popadynets, N. (2021). Strategic Challenges for Sustainable Governance of the Bioeconomy: Preventing Conflict between SDGs. *Sustainability, 13*, 8308. https://doi.org/10.3390/su13158308

Nechyporchuk, O., Belgacem, M. N., & Bras, J. (2016). Production of cellulose nanofibrils: A review of recent advances. *Industrial Crops and Products, 93*, 2–25. https://doi.org/10.1016/j.indcrop.2016.02.016

Raga, F. (2009). The Chilean Forestry Sector and associated risks. *Trébol*, 10–19.

Trache, D., Tarchoun, A. F., Derradji, M., Hamidon, T. S., Masruchin, N., Brosse, N., & Hussin, M. H. (2016). Nanocellulose: From fundamentals to advanced applications. *Frontiers in Chemistry, 8*, 392. https://doi.org/10.3389/fchem.2020.00392

World Bank. (2020). *Chile forest note*. World Bank.

Yi, T., Zhao, H., Mo, Q., Pan, D., Liu, Y., Huang, L., Xu, H., Hu, B., & Song, H. (2020). From cellulose to cellulose nanofibrils—A comprehensive review of the preparation and modification of cellulose nanofibrils. *Materials, 13*, 5062. https://doi.org/10.3390/ma13225062

Techno-Economic Analysis of Mobile Production Plants for Bioenergy Carriers

Tobias Zimmer, Simon Glöser-Chahoud, and Frank Schultmann

Abstract In Chile, forestry and agriculture produce significant amounts of lignocellulosic byproducts, such as forest residues or cereal straw. These biomass resources are currently used to a limited extent for bioenergy, but a significant share is disposed of without utilization, for instance via stubble burning. A major challenge in the recovery of biomass residues is the low density of the material, which results in small truckloads and high transportation costs. Mobile or containerized plants can be located directly at the harvesting site, where they can produce densified bioenergy carriers that can be easily transported. This study presents an economic assessment of mobile plants for the pre-treatment of forest residues in Chile. Two non-stationary configurations, fully mobile and satellite plants, are compared with large, conventional facilities. The results indicate that mobile plants cannot match the economies of scale of the largest commercial plants, but their modular design is advantageous at smaller capacities. In Chile, they would be most suitable for more remote locations and for decentralized utilization concepts.

1 Introduction

There is an increasing interest in intensifying the use of biomass resources for the production of electricity, heat, and biofuels. Biomass is generally considered as a sustainable and renewable source of energy that mitigates climate change and promotes the development of rural areas. Despite these benefits, the intensified use of biomass resources is hindered by technical and economic challenges. Biomass feedstock is frequently characterized by high moisture content, low energy density, and varying chemical composition. Due to these properties, raw biomass is considered

T. Zimmer (✉) · F. Schultmann
Karlsruhe Institute of Technology (KIT), Institute for Industrial Production (IIP), Karlsruhe, Germany
e-mail: tobias.zimmer@outlook.de

T. Zimmer · S. Glöser-Chahoud
TU Bergakademie Freiberg (TU BAF), Chair of Corporate Sustainability and Environmental Management, Freiberg, Germany

A. Ch. Braun et al. (eds.), *Eco-Industrial Development as an Industrial Strategy*, Green Energy and Technology, https://doi.org/10.1007/978-3-031-73576-9_14

a low-quality fuel in comparison with fossil fuels. In addition, raw materials accrue over wide regions, which increases collection, transportation, and handling costs. Feedstock recovery is frequently restricted by seasonality and weather conditions, while storage capabilities are limited by biological degradation.

The use of pre-treatment technologies has gained interest in recent years, as it has the potential to overcome both the technical and logistical hurdles associated with bioenergy production. During pre-treatment, the physical properties of the feedstock are altered by mechanical or thermal processes. Currently, the most important pre-treatment technology is pelletization. Pellets have a higher density and lower moisture content than raw biomass and exhibit superior properties with respect to handling and combustion. In addition, thermal processes such as torrefaction and pyrolysis can be used to produce bio-based energy carriers with increased energy density and properties similar to those of coal and oil. Due to these favorable properties, pre-treatment technologies are suitable for integration into different biomass value chains, providing heat, electricity, or chemicals.

Some of these value chains could also be viable and promising in Chile. At the moment, Chile is highly dependent on imported fossil fuels. Oil and natural gas account for 40% and 24% of the country's primary energy demand. With 98% of the oil and 90% of the gas supplied by external sources, more than 60% of the primary energy is imported. This dependency on foreign fuels has led to problems in the past, for instance during the Argentina gas crisis in 2004. Tapping into domestic biomass resources could reduce this dependency and contribute to a more stable energy supply. Since the Chilean electricity net shows a high degree of fragmentation, high-quality bio-based energy carriers suitable for transportation and storage are particularly promising. In addition, major cities in south-central Chile suffer from high levels of particulate matter (PM 10 and PM 2.5) due to the combustion of moist firewood for heating (Reyes et al., 2019). With regard to these benefits and the large biomass potential throughout Chile, the question arises why a large-scale production of pellets and biofuels has not been put into practice yet. The total capacity of wood pellets in Chile is less than 150,000 tons per year while biofuels are not produced at all (Cremonez et al., 2015).

The fairly slow development in the area of advanced biofuels has economic as well as technological reasons. With regard to biofuels, Chile has no mandatory quotas comparable to those in the European Union (García et al., 2011). Another economic obstacle apart from the role of the state is the uncertain market demand. The primary customers of pellets in the European Union and the United States are households and small-scale consumers. Currently, Chilean households are still reluctant to acquire high-quality pellet stoves, which are considered expensive (Goh et al., 2013). The slower development of biofuel production capacities in Chile in comparison with other countries in Latin America can be explained by the biomass resources available. In Chile, the climate does not favor the cultivation of sugarcane, which is the main feedstock for ethanol production in Brazil and Argentina. This is not to say that the production of biofuels is not a viable option in Chile, but biofuels would have to be produced on the basis of wood or agricultural residues (Cremonez et al., 2015).

With respect to pre-treatment technologies, the general conditions in Chile are promising and challenging at the same time. Since household demand for pellets is highly uncertain, torrefaction and pyrolysis could provide access to new customers such as power plants or CHP plants. Torrefied pellets and bio-oil are suitable for co-firing in large coal- and gas-fired plants that are more likely to invest in advanced combustion technologies. Fast pyrolysis has also been suggested as an intermediate processing step in the production of biofuels from agricultural and forest residues. As Chile does not grow large amounts of sugarcane, wood and straw could be promising alternative raw materials if upgraded by pyrolysis. However, little assistance in terms of incentives or supporting legislation can be expected from the government. Thus, the prospects of success will depend almost solely on cost efficiency of the respective value chain.

In this subchapter, the case of mobile plants for the production of densified bioenergy carriers will be evaluated by means of a techno-economic analysis. First, the structure of a biomass supply chain with decentralized pre-treatment will be explained. Second, technologies for the energetic densification of biomass and concepts for a mobile or containerized production will be presented. Finally, a brief case study based on an economic assessment will illustrate the potential of mobile plants to utilize residues from Chilean forest plantations.

2 Decentralized Pre-treatment and Densified Bioenergy Carriers

2.1 Biomass Value Chains

In general, biomass supply chains are characterized by a high degree of complexity, which results from the vast number of possible configurations. Reasons for this include the different geographical conditions of the raw material as well as different options for handling during storage, transshipment and transport activities. In addition, different technological alternatives for biomass processing add to the diversity of supply chain arrangements (Gold & Seuring, 2011). According to Nunes et al. (2020), a comprehensive approach for planning and optimization, which covers all steps along the value chain, is essential to achieve a cost-effective utilization of biomass resources. Figure 1 shows the steps required to process and deliver biomass and illustrates the central issue regarding the use of mobile pre-treatment plants.

According to Sahoo et al. (2019), a supply chain design with decentralized energy densification not only reduces costs but also minimizes risks. In the section below, the different parts of a supply chain are examined in more detail in order to identify the potential benefits of decentralized and mobile densification processes. Raw material collection plays a central role in the supply chain as a result of its considerable volatility and the resulting significant costs. Albashabsheh and Heier Stamm (2019) concluded in their analysis that raw material can account for over 40% of total costs.

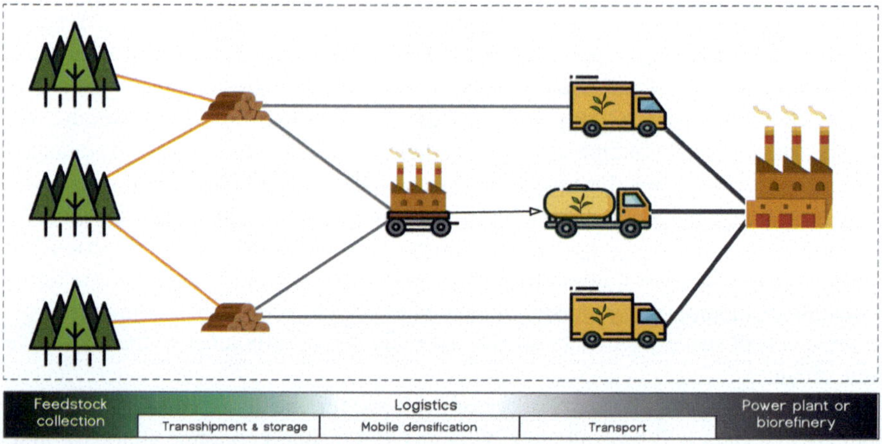

Fig. 1 Biomass supply chain with mobile densification

Although this result must be seen in the context of the selected region, in this case the United States, it does indicate the importance of this processing step.

In general, a distinction must be made between coupled and decoupled extraction (Mirkouei et al., 2017). While in a coupled extraction the residues are collected at the same time timber and other merchantable wood products are harvested, in decoupled extraction the residues are left behind. In order to optimally coordinate the deployed resources, the local conditions on a harvesting site must be taken into account. Depending on the available space to move the equipment and the slope of the terrain, a variety of challenges arise. For example, on steep terrain, the simultaneous harvesting of timber and residues is often the only cost-efficient approach. On the other hand, especially steep terrain can limit the available space of the machinery, making simultaneous harvesting impossible (Spinelli et al., 2019) Therefore, careful consideration of the geographic conditions is an essential step in the planning and evaluation of biomass collection processes.

The way in which the biomass feedstock is collected also has far-reaching implications for the following logistical processes:

- For immediate subsequent processing, all necessary equipment must be available at the harvesting site. In addition to the requirement of sufficient storage space, this makes coordinating the machinery more challenging. If the residues are collected after the primary product is harvested, space for intermediate storage must be created (Spinelli et al., 2019).
- In general, forest residues can be processed directly at the harvesting site. Alternatively, consolidation or transshipment points of various sizes can be created. However, the transport of completely untreated residues, without any chipping, bundling, or baling, is usually not economically viable. Exceptions here are very small catchment areas with transport distances of less than 20 km (Mirkouei et al., 2017).

- Forest residues have a very low bulk density of 50–150 kg/m3, which leads to a situation where the volume and not the weight limits the amount of feedstock that can be loaded onto a typical truck. As a result, the transportation of forest residues without previous processing is rather inefficient in terms of the energy content per truckload (Zamora-Cristales et al., 2017).

To evaluate and compare the available options for biomass transportation, the characteristics of the road network must be taken into account. According to Mirkouei et al. (2016), three types of roads can be distinguished in a biomass supply chain:

- Country roads and highways
- Unpaved gravel roads within the forest area
- Unpaved clay roads within the forest area

The resulting restrictions with respect to transport speed and the different requirements regarding the versatility of the transport vehicles have already been analyzed extensively. In order to increase capacity while maintaining the same degree of maneuverability, the use of an tractor unit with two trailers is a common approach. Zamora-Cristales et al. (2015) were able to demonstrate that transportation with two trailers can be more efficient due to the increased loading capacity, provided that sufficient turning space is available at a central collection point. However, under certain circumstances–such as narrow dirt roads or steep terrain–a hook-up point is required where trailers can be temporarily stored for sequential loading. This process leads to increased handling time, which makes the creation of a hook-up point only viable for long transport distances. In the case study that was carried out by Zamora-Cristales et al. (2015) in the US state of Oregon, this situation only occurred for distances of more than 110 km.

According to Rentizelas et al. (2009), raw material storage in biomass supply chains can be divided into the following categories:

- Closed depots with integrated hot air drying
- Covered, open depots without drying
- Outdoor storage with plastic covering

In many studies, however, storage is considered negligible, as it represents only a minor part of the total costs (Albashabsheh & Heier Stamm, 2019). Nonetheless, storage cannot be completely disregarded: if the feedstock remains at the harvesting site for an excessive period of time without precautions, this can lead to substantial material losses. Inappropriate storage processes also cannot only impair the quality of the raw material, but also result in health risks. Since biomass is usually characterized by high humidity, incorrect storage brings the risk of fungal development with harmful effects on the respiratory system of workers.

2.2 *Processes for Biomass Densification*

A fundamental problem of biomass residues is their relatively low quality compared to fossil energy carriers. According to Kaltschmitt et al. (2016), this can be explained by the chemical and mechanical properties of the raw material, such as the ash content or bulk density. In this context, especially the high water content of up to 50% is worth mentioning, as it has a decisive impact on the energy content and the efficiency of transport processes. However, the properties of biomass residues can be improved through mechanical or thermo-chemical pre-treatment (Fig. 2).

2.2.1 Mechanical Densification Processes

The most common processes for mechanical densification of biomass are pelletizing and briquetting. As both processes are based on the same principle, the final products are classified by their diameter, which is less than 25 mm for pellets and more than 25 mm for briquettes (Kaltschmitt et al., 2016). In both cases, the treatment results in the following advantages:

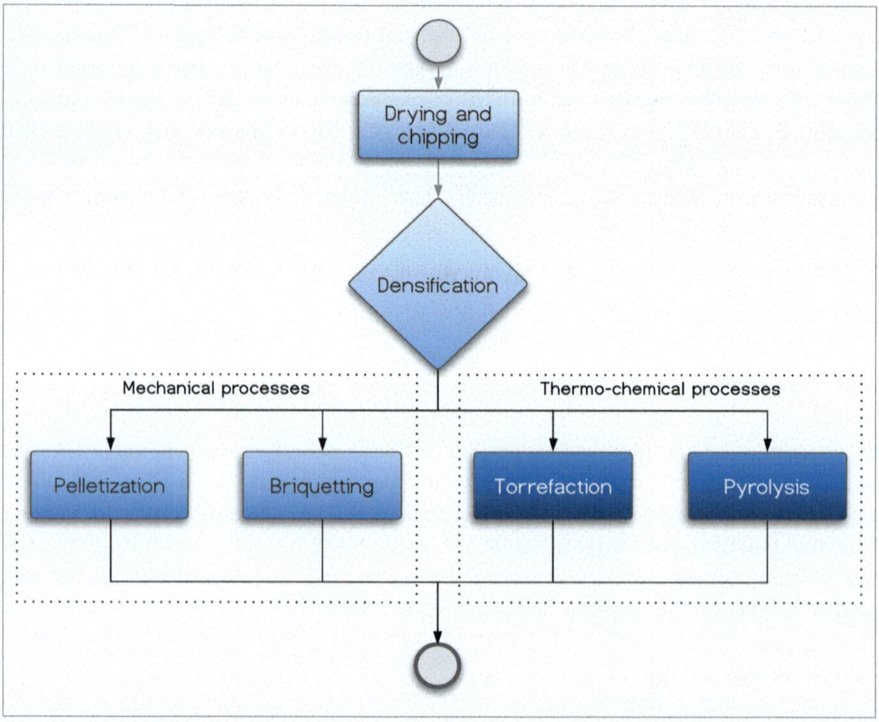

Fig. 2 Mechanical and thermo-chemical processes for biomass densification

- Increased efficiency of transportation and handling processes due to higher energy density
- Improved flow and feeding properties, and reduced dust formation
- Lower water content and thus improved storability
- High level of fuel homogeneity

The general procedure for producing mechanically densified energy carriers can be summed up as follows: First, if forest residues are used as feedstock, the material is shredded into wood chips using a chipper. The wood chips are then dried to achieve the required water content of less than 12% and the dry biomass is milled into a powder. The ground material is then conditioned with superheated steam and pressed into pellets or briquettes. Finally, the pellets are cooled and fine particles are removed by sieving.

The key difference between the technologies comes from their flexibility: briquetting presses are able to press raw materials with a higher water content and larger particles into valuable briquettes. In addition, the infrastructure requirements and the energy demand are lower (Tumuluru et al., 2011). However, this has the effect that briquettes–compared to pellets–have a lower energy density.

2.2.2 Thermo-Chemical Densification Processes

According to Sharma et al. (2019), thermo-chemical processes in general can be categorized into three groups, depending on the oxygen supply:

- Combustion–full oxygen supply
- Gasification–partial oxygen supply
- Pyrolysis–strongly reduced or no oxygen supply

Of these three technologies, only pyrolysis leads to solid and liquid products that can be transported with a truck or kept in storage. As this sub-chapter focuses on logistical aspects, especially transportation costs, only the two pyrolysis processes–fast pyrolysis and torrefaction–are considered in the following analysis while combustion and gasification are excluded.

In general, all pyrolysis processes can be divided into two steps. First, the water content is reduced by heating up the feedstock. When the temperature is increased further, the second process of pyrolytic decomposition begins. Here, the basic molecular components of the biomass–hemicellulose and cellulose–are broken down. The lignin, which is responsible for the stability of woody biomass, is transformed into coke at temperatures of 230 °C and above. At temperatures above 400 °C, the production of methane and methanol reaches its peak (Kaltschmitt et al., 2016).

Thus, pyrolysis processes offer the possibility of densifying biomass by converting it into a mix of gaseous, liquid and solid components. The proportion of the end products, however, depends on the selected process parameters. Therefore, pyrolysis processes are often classified based on process temperature and residence time. With regard to temperature, a distinction is made between low (450–550 °C), medium

(600–800 °C), and high (900–1050 °C) temperatures (Sharma et al., 2019). In terms of residence time, it is possible to distinguish between fast (residence time <3 s), intermediate (10–30 s), and slow (from 10 min up to several hours/days) pyrolysis (Kaltschmitt et al., 2016).

It can generally be observed that a reduction in residence time results in a higher proportion of gases within the end product mix and a corresponding reduction in the proportion of solids. Furthermore, higher process temperatures lead to an increased proportion of gases and a reduction in the liquid components. The resulting gas is generally not suitable for transport and is therefore fed into the heating process as fuel (Chen et al., 2018).

In order to implement a bioenergy value chain with decentralized, thermo-chemical pre-treatment, the processes of torrefaction and fast pyrolysis were identified as advantageous in various studies (Brown et al., 2013).

Torrefaction

Torrefaction is a form of mild pyrolysis and is characterized by the slow heating of biomass in an oxygen-deficient atmosphere to 225–300 °C, which leads to the decomposition of hemicellulose (Prins et al., 2006). The residence time varies between ten minutes and two hours depending on the design of the process. This results in a mass loss of 30–50% with a remaining energy content of 75–90% (Christoforou & Fokaides, 2019).

Torrefaction improves the quality of the fuel with regard to different properties:

- Increased heating value
- Reduction of the water content to below 2%
- Hydrophobizing and outdoor storability
- Improved grindability

The efficiency of the process can be increased by reaching an autothermal state. In this state, the energy content of the gases produced corresponds to the energy required for the heating processes, which means that the reaction can be continued without an external energy supply (Bergman, 2005). As torrefied biomass is brittle and not easy to handle, torrefaction is often combined with pelletization to obtain torrefied pellets as the main product. This combination of both processes results in a superior product, as torrefaction improves the negative properties of the pellets or briquettes and vice versa. In this context, the improved grindability and the reduced water content resulting from torrefaction significantly increase the storability. On the other hand, the mechanical compression from pelletization leads to a reduction in dust formation and additional densification in terms of volume (Christoforou & Fokaides, 2019).

Fast Pyrolysis

The process of fast pyrolysis takes place at temperatures between 450–600 °C. In contrast to torrefaction, fast pyrolysis breaks down a major part of the structure of wood by an increased decomposition of cellulose and lignin. Its aim is to achieve rapid heat and mass transfer, which maximizes the yield of the liquid product, which

Table 1 Characteristics of woody residues and derived bioenergy carriers

Property	Wood	Regular wood pellets	Torrefied wood pellets	Fast pyrolysis bio-oil
Water content [m%]	35	7–10	1–5	15–30
Energy content [MJ/kg]	11	16	20	18
Bulk density [kg/m3]	550	650	850	1200
Energy density [GJ/m3]	5.8	10.5	18.4	23.8

is called bio-oil. As a result, the residence time is reduced to less than three seconds (Kaltschmitt et al., 2016). As the obtained pyrolysis oil has a significantly lower calorific value than petroleum products, it is often refined into marketable products in a subsequent step, for instance via hydroprocessing. In the context of decentralized pre-treatment and energetic densification, so-called bio-slurry is a major alternative to bio-oil as an intermediate energy carrier. Here, bio-oil and biochar are mixed to form a suspension, whereby up to 90% of the energy content of the pyrolysis products is retained and can be transported (Dahmen et al., 2012).

The properties of untreated wood are compared to the bioenergy carriers obtained from pelletization, torrefaction (with pelletization) and fast pyrolysis in Table 1. The comparatively low water content and the suitability for storage that comes with hydrophobization are decisive factors in favor of torrefaction. Bio-oil, the main product of fast pyrolysis, offers the highest density in terms of energy per cubic meter. However, as a liquid energy carrier, it requires particular equipment for storage and transportation, such as storage tanks and tanker trucks. Mechanical densification processes are primarily recognized for their simplicity and robustness. Keeping in mind that in a decentralized setting, for instance on the harvesting site of a forest plantation where the supporting infrastructure is very limited, a less sophisticated process is often more suitable. The choice between pelletizing and briquetting must be made with regard to the given raw materials and the space available for the pre-treatment process. Here, a trade-off between the higher bulk and energy density of pelletizing and the process robustness of briquetting has to be considered.

2.3 *Mobile Biomass Processing*

Mobile and semi-mobile plants are specialized plants that can be moved from one location to another. Lei et al. (2016) define mobile plants as plants that can be relocated in order to satisfy the demand for specific services from customers located in the vicinity of the plant. A key aspect here is that mobile plants offer their services only when they are physically located in a particular place. There is usually no value-adding activity during transportation to the plant, and a certain setup time to start operations at a new location is usually required. In most cases, the level of activity is proportional to the length of time a mobile plant stays at a specific location. A key advantage of mobile or semi-mobile plants compared to stationary plants is that they can be used very flexibly, as they can be deployed at any location. For a given planning period, this means that the location of the mobile plant can be changed several times (Güden & Süral, 2014). The mobility of mobile and semi-mobile plants can vary greatly. The specified time spent at a site and the required dismantling and installation time are a decisive criterion here. Different plants require very different times to be made ready for transport or to be installed at a new location.

The main motivation for mobile and semi-mobile plants is to minimize the total cost of transportation and operation, while maximizing the coverage of customer demand. An important aspect here is the trade-off between the reduced transportation costs and less operating hours resulting from more frequent relocations. Transportation costs play an important role for mobile plants with regard to several aspects. In various models, therefore, not only the transportation costs of the mobile plant, but also the transportation costs of the customers (who have to move to the plant) are taken into account. This type of cost should be included, for example, if there are disproportionately large transportation costs for customers, materials, or similar. It can therefore be advantageous to accept higher transportation costs for a mobile plant if cost savings can be achieved elsewhere as a result. In other applications of mobile facilities, cost is naturally not the first concern. Examples include disaster response and mobile healthcare facilities (Doerner et al., 2007). Here, the focus is primarily on covering the existing demand as much as possible with a limited set of resources. Table 2 gives examples of scenarios addressed in the literature, where mobile systems are used to produce goods or provide services.

With regard to the design of the plant, a distinction can be made between three levels of mobility:

1. Mobile plants, where the production equipment is permanently mounted on a truck. This design results in a stand-alone unit that can freely move from one location to another without additional support from trucks or cranes.
2. Semi-mobile plants, where the production equipment is installed in one or more containers. Such plants cannot move on their own, but the containerized design greatly facilitates the relocation to another location.

Table 2 Examples of mobile systems from different areas

References	Application
Delmelle et al. (2014)	Mobile classrooms
Güden and Süral (2014)	Mobile concrete mixers
Zhen et al. (2014)	Ambulances and mobile healthcare
Sharifzadeh et al. (2015)	Mobile biofuel production
Lei et al. (2016)	Mobile radio-telephone stations
Mirkouei et al. (2016)	Mobile biofuel production
Wishon and Villalobos (2016)	Mobile supermarkets
Xie et al. (2016)	Mobile maintenance of locomotives

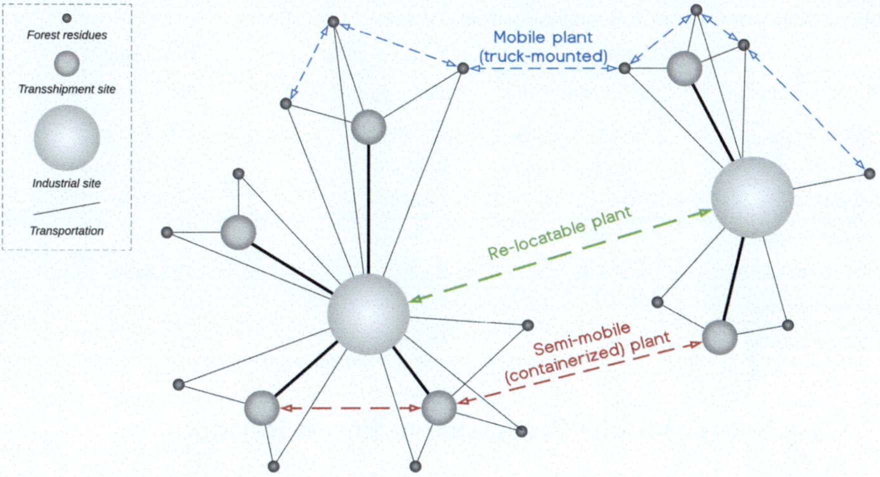

Fig. 3 Relocation of non-stationary plants of different capacities

3. Relocatable plants that can be disassembled into individual components for transport.

In general, it can be expected that smaller and more mobile plants have lower setup times and therefore change their location more often in order to reduce transportation costs. In contrast, containerized or re-locatable plants stay at a single location for a longer period of several weeks or months (Fig. 3).

In order to use a non-stationary plant in the most efficient way, the time lost during transport and setup must compared to the reduction in transportation costs or other benefits of a more flexible production process. As a result, the setup time required to start operations at a new location is a key parameter for all non-stationary plants.

However, in the case of bioenergy, the mobile plants known today are mostly demonstration units that are not operated commercially. Thus, it has not been possible so far to measure actual setup or relocation times of mobile systems in practice. Techno-economic assessments addressing the benefits of mobile plants therefore have to rely on educated assumptions with respect to the setup time.

The study of Polagye et al. (2007) covers mobile, semi-mobile, and relocatable plants and presents estimates for throughput and setup times of all three systems. As this study has been widely cited in the literature, the plant characteristics shown in Table 3 are used for the economic analysis that follows below.

Table 3 Characteristics of non-stationary plants (Polagye et al., 2007)

	Mobile	Transportable	Re-locatable
Mobility setup	Semitrailer	Multiple containers	Industrial site
Capacity	10 t/d	100 t/d	500 t/d
Location	Logging deck	In forest	Outside forest
Setup time	4 h	4 days	2 months
Transport speed	40 km/h	40 km/h	80 km/h
Catchment radius	0.12 km	5 km	70 km
Moves per year	23	3	1
Site separation	2 km	50 km	80 km

3 Case Study: Mobile Processing of Forest Residues in Chile

Mobile or containerized plants have the potential to reduce costs and emissions resulting from biomass transportation and to enable more decentralized, local bioenergy concepts. As a result, these technologies could be beneficially applied to the Chilean forestry and bioenergy sector. In Chile, forest residues are already utilized to a great extent, mostly to produce heat and power for sawmills and cellulose plants. However, transportation distances are often long and can reach up to 150 km for large biomass power plants like the Valdivia plant of Arauco. For a large share of the biomass feedstock used in Chilean power plants, the cost of the delivered fuel corresponds more or less to the transportation cost of the biomass.

In the following section, the potential of mobile facilities to produce bioenergy carriers from forest residues is analyzed in the example of the Bío-Bío region in Chile. The assessment aims to answer two questions: First, can biomass supply costs be reduced by densifying processes such as pelletization, torrefaction, and pyrolysis? Second, is there an advantage in carrying out these processes directly at the harvesting site using mobile facilities? For this purpose, an economic model was created to analyze the optimal number of relocations and to compare stationary and mobile concepts for different densification processes. The different supply chain options considered by this model are shown in Fig. 4 in the form of a flowchart.

Based on GIS data, forest roads and potential plant locations were identified. Next, the economic impact of using mobile energy densification technologies directly at the harvesting site (Option C) or at a consolidation point along the gravel roads (Option B) was evaluated. The results were compared to the reference case where no mobile process was used (Option A). Literature estimates were used to model the plants in terms of their cost, production capacity and setup time.

To model the road network, especially the roads within a forest plantation, an example forest area was analyzed using GIS software. Figure 5 shows an overview image of the approximately 75 km^2 large area.

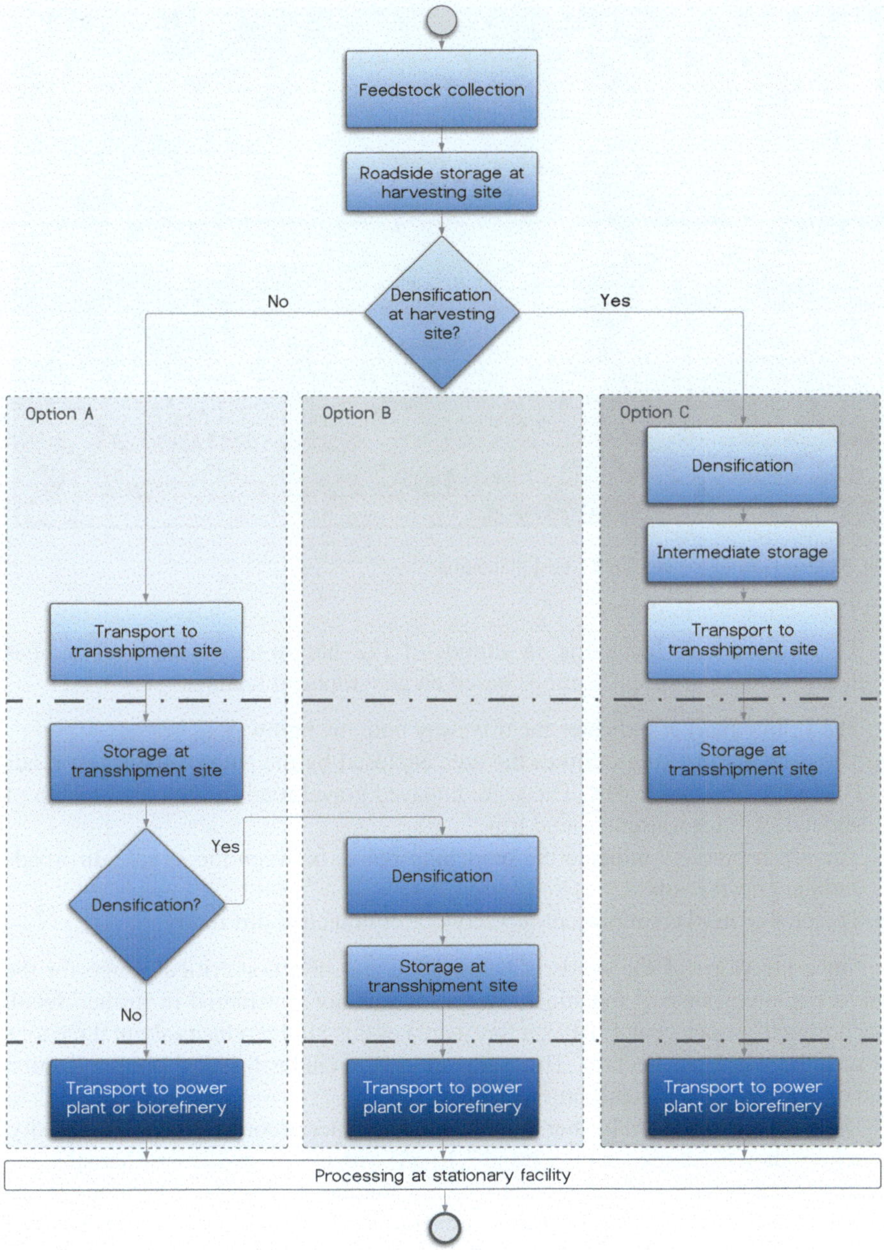

Fig. 4 Supply chain options considered for the economic evaluation

Fig. 5 Overview of the analyzed forest plantations

The entire area is located at an altitude of 122–886 m above sea level and has been divided into different sections based on geographical features:

- The yellow marker indicates the driveway onto the highway.
- The dark red marking outlines the area enclosed by the path of the forest roads P-32-Q, P-334 and P-418. These are unpaved gravel roads with a total length of about 43 km for a single round trip.
- The green markers indicate the branching points between the simple dirt roads and the gravel roads.
- The orange markers represent the network of branched dirt roads.

Since the slope of the roads is below 15% and in a non-critical range for the harvesting and transport machines, steepness was not considered in further detail. According to Acuña et al. (2017), a forest area of this kind produces about three tons of residual wood per hectare. This value already includes losses during collection and due to terrain that is difficult to access.

Unpaved dirt roads are further divided into two types according to their tortuosity, which are shown enlarged in Figs. 6 and 7. Roads with low tortuosity are characterized by their straight path and a comparatively low number of branches. Tortuous roads, on the other hand, have significantly more complications, resulting in the formation of a network of roads. Based on the modelled network of forest roads, transportation distances were then calculated for plant locations within or near forest plantations. These distances were used in the economic assessment to estimate transportation costs and to identify the optimal number of relocations for non-stationary plants.

Two cases representing a higher and lower capital investment for non-stationary plants were considered in the economic assessment. In the reference case, the capital

Fig. 6 Example of forest roads with low tortuosity

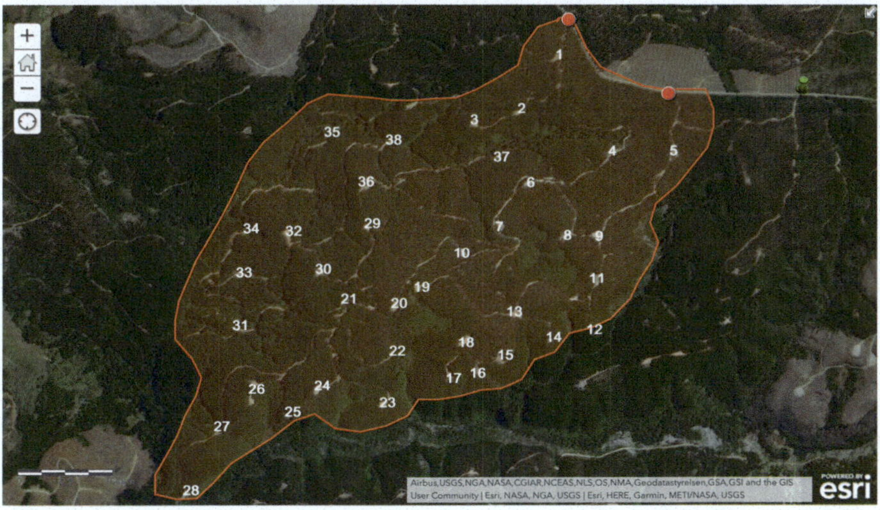

Fig. 7 Example of forest roads with high tortuosity

investment of a containerized satellite plant is 13% lower in comparison with a stationary plant of the same capacity due to reduced construction expenses. This estimate fits very well with the announcement of the Drax group to invest 40 MUSD in three satellite plants. Under the same assumptions, the investment for a mobile facility would be 30% lower, as no yard improvements or service facilities would have to be installed for such a stand-alone unit. The development case represents an optimistic, future scenario, whereas non-stationary plants have successfully entered the market and a larger number of such plants are being used commercially. It can be assumed that the cost per plant could be further reduced in that case through the containerized, standardized design that requires less planning and engineering effort. Results for stationary and non-stationary pellet plants in both cases are shown in Table 4. It can be seen that mobile and satellite plants can drastically reduce transportation distances as well as their resulting costs and emissions.

For non-stationary plants, the question arises as to how often they should change their location. The more frequently the location is changed, the more transport costs can be reduced. However, production cannot take place during the transport and setup time, so that costs are incurred for each relocation in proportion to the required time. Figures 8 and 9 show the development of setup and transportation costs with an increasing number of relocations. It can be seen that 7 relocations per year are optimal for a mobile plant and 3 relocations per year for a semi-mobile plant.

The comparison of costs in Table 4 indicates that the cost of mobile and satellite plants tends to be higher than the cost per ton of a large, stationary plant that can take full advantage of economies of scale. However, the result of this comparison depends very much on the capacity assumed for the stationary plant. It should therefore be critically examined whether very large capacities are realistic with regard to the development status of the technology or the regional availability of biomass feedstock.

Table 4 Comparison of stationary, mobile, and satellite pellet plants

		Stationary	Reference case		Development case	
			Mobile	Satellite	Mobile	Satellite
Capacity	Tons/year	200,000	5,000	40,000	5,000	40,000
Total capital investment	MEUR	48.7	2.59	13.81	2.07	10.11
Optimal number of relocations		–	7	3	8	4
Transport distance	km	37	0.8	5.2	0.8	4.4
Transport emissions	CO2eq/ton feedstock	4.84	0.11	0.67	0.10	0.57
Capital cost	EUR/ton product	39.65	84.39	56.21	67.47	41.14
Transport cost	EUR/ton product	14.88	0.34	2.07	0.30	1.77
Setup cost	EUR/ton product	–	0.34	2.07	0.30	1.77
Total cost	EUR/ton product	54.53	85.07	60.35	68.07	44.67

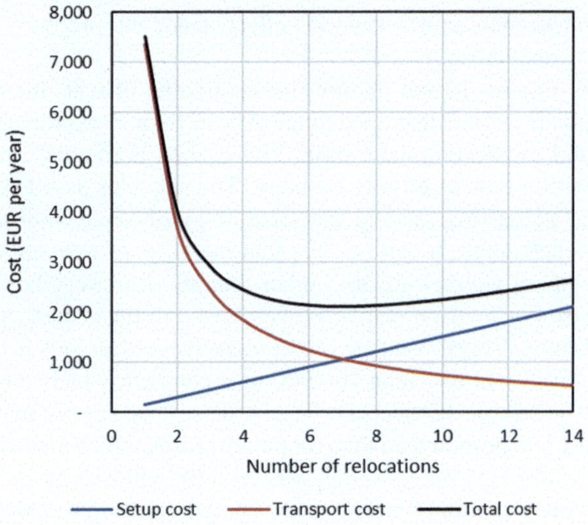

Fig. 8 Optimal relocations of a mobile pellet plant (5,000 tons/year)

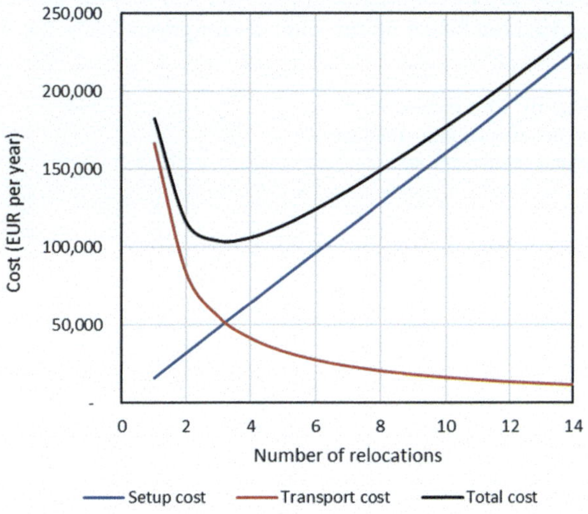

Fig. 9 Optimal relocations of a satellite pellet plant (40,000 tons/year)

In this subchapter, the three processes pelletization, torrefaction and pyrolysis were considered as options for the decentral densification of biomass. A look at current commercial projects for all three processes reveals major differences regarding the capacities that have been implemented. While pellet plants can reach capacities of more than 400,000 tons per year, the realistic dimension of a fast pyrolysis plant is currently more in the range of 70,000 tons per year. With regard to the situation in Chile, the question arises whether a pellet plant of more than 100,000 tons of output per year would be feasible in practice. On the one hand, large plants, especially biomass power plants and pulp mills, are quite common. However, on the other, the sector for wood pellets stands out precisely because of the large number of smaller suppliers.

Figures 10 and 11 show the economies of scale for stationary pellet and pyrolysis plants and the discussed range of costs for non-stationary plants between the reference and the development case. The dashed horizontal line marks the costs attainable with a typical project capacity. This capacity is 400,000 tons per year in the case of pelletizing, and in the case of pyrolysis 70,000 tons per year. In the case of pelletization, it can be seen that the costs of non-stationary plants with the current state of technology are higher than those of very large pellet plants. However, the costs of satellite plants are very close to these costs and may even be lower in the future. A big advantage is that such costs are already achieved at a considerably lower capacity of less than 100,000 tons per year, which can be a big advantage for more decentralized concepts or in remote regions. For fast pyrolysis, the economies of scale are less pronounced than for pelletization, which means that the non-stationary plants are closer to the costs that are currently achieved by commercial projects. As a result, a semi-mobile concept could be even more promising here than for pelletization. However, it must be taken into account that currently no semi-mobile pyrolysis plants are in commercial operation and that the corresponding cost estimation was carried out based on the same assumptions as for a containerized pellet plant.

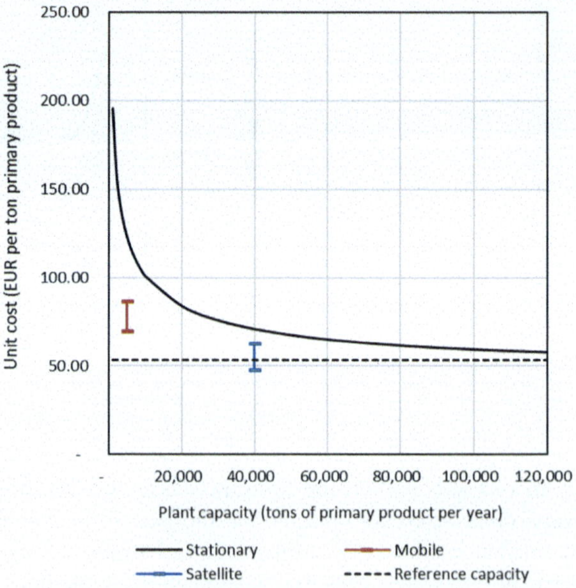

Fig. 10 Cost ranges of stationary and non-stationary plants (pelletization)

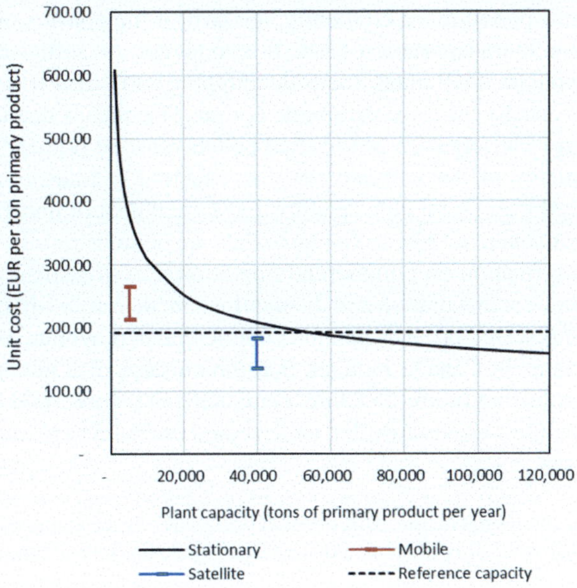

Fig. 11 Cost ranges of stationary and non-stationary plants (fast pyrolysis)

4 Discussion

As biomass feedstock is characterized by a low bulk density, efficient transportation is a challenge that can prevent the utilization of biomass resources. Decentralized pretreatment and densification using mobile or containerized plants can reduce costs and emissions resulting from biomass transportation. In the case study, a brief economic assessment was carried out to compare plant and transportation costs and to illustrate the potential of non-stationary plants. Beyond that, however, there are numerous more qualitative factors to consider for such mobile concepts that would be very difficult to reflect in a cost assessment. For instance, the containerized, standardized design of the plants could facilitate maintenance and operation in comparison with a dedicated larger plant. In addition, as the total investment is considerably lower, smaller operators such as sawmills, farmers, or cooperatives could be potential owners or operators of containerized plants. In contrast, it is obvious that bioenergy projects above a certain magnitude would be only implemented by large companies from the forestry or energy sector. There are, however, also further advantages to larger, stationary plants that have not been considered here. One example is process integration of the thermochemical processes torrefaction and pyrolysis, for instance with a CHP plant. Also, direct subsequent processing and upgrading of the products, e.g. by hydroprocessing of pyrolysis oil, can be accomplished more easily at an integrated site where the bio-oil can be pumped through pipelines.

In most cases, lower transport costs alone are not a sufficient incentive for processes like torrefaction or pyrolysis. Although their densified products can be transported more efficiently, the cost of the conversion plant outweighs the reduction in transportation costs. If wood chips are sufficient as a fuel, for instance for a biomass CHP plant, roadside chipping and direct transportation to the power plant is usually the most economic option. This means that the use of high-quality bioenergy carriers such as torrefied pellets must be primarily motivated by the increased quality of the fuel and not the reduction in transportation costs. In general, non-stationary plant concepts involve a great deal of uncertainty. Most of the mobile or containerized plants known to date are demonstration plants, and there is no reliable information on commercial operation. Key logistical parameters such as the setup time for installation and dismantling are therefore still unknown. However, a number of research projects and the commercial deployment of the first containerized pellet plants in Canada indicate that the concept of a non-stationary plant design has a promising future. In Chile, containerized plants could be particularly interesting for medium-sized sawmills, which could use them to produce bioenergy carriers on their own.

Acknowledgements This research was funded by the German Federal Ministry of Education and Research (BMBF), grant number 031B0056A (SeMoBioEnergy).

References

Acuña, E., Cancino, J., Rubilar, R., & Silva, L. (2017). Volume, physical characteristics and costs of harvest residue utilization of Pinus radiata as an energy source. *Custos E Agronegócio, 13*(1), 442–463.

Albashabsheh, N. T., & Heier Stamm, J. L. (2019). Optimization of lignocellulosic biomass-to-biofuel supply chains with mobile pelleting. *Transportation Research Part e: Logistics and Transportation Review, 122*, 545–562. https://doi.org/10.1016/j.tre.2018.12.015

Bergman, P. (2005). *Combined torrefaction and pelletisation: The TOP process. ECN-C: 2005–073.* Netherlands Energy Research Foundation.

Brown, D., Rowe, A., & Wild, P. (2013). A techno-economic analysis of using mobile distributed pyrolysis facilities to deliver a forest residue resource. *Bioresource Technology, 150*, 367–376. https://doi.org/10.1016/j.biortech.2013.10.018

Chen, X., Zhang, H., Xiao, R. (2018). Mobile autothermal pyrolysis system for local biomass conversion: process simulation and techno-economic analysis. *Energy & Fuels, 32*(4), 4178–4188. https://doi.org/10.1021/acs.energyfuels.7b03172

Christoforou, E., & Fokaides, P. A. (2019). Advances in solid biofuels. *Springer International Publishing*. https://doi.org/10.1007/978-3-030-00862-8

Cremonez, P. A., Feroldi, M., Feiden, A., Gustavo Teleken, J., José Gris, D., Dieter, J., de Rossi, E., & Antonelli, J. (2015). Current scenario and prospects of use of liquid biofuels in South America. *Renewable and Sustainable Energy Reviews, 43*, 352–362. https://doi.org/10.1016/j.rser.2014.11.064

Dahmen, N., Henrich, E., Dinjus, E., & Weirich, F. (2012). The bioliq® bioslurry gasification process for the production of biosynfuels, organic chemicals, and energy. *Energy, Sustainability and Society, 2*(1), 3. https://doi.org/10.1186/2192-0567-2-3

Delmelle, E. M., Thill, J.-C., Peeters, D., & Thomas, I. (2014). A multi-period capacitated school location problem with modular equipment and closest assignment considerations. *Journal of Geographical Systems, 16*(3), 263–286. https://doi.org/10.1007/s10109-013-0195-2

Doerner, K., Focke, A., & Gutjahr, W. J. (2007). Multicriteria tour planning for mobile healthcare facilities in a developing country. *European Journal of Operational Research, 179*(3), 1078–1096. https://doi.org/10.1016/j.ejor.2005.10.067

García, A. E., Carmona, R. J., Lienqueo, M. E., & Salazar, O. (2011). The current status of liquid biofuels in Chile. In *5th Dubrovnik Conference on Sustainable Development of Energy, Water & Environment Systems, 36*(4), 2077–2084. https://doi.org/10.1016/j.energy.2010.06.005

Goh, C. S., Junginger, M., Cocchi, M., Marchal, D., Thrän, D., Hennig, C., Heinimö, J., Nikolaisen, L., Schouwenberg, P.-P., Bradley, D., Hess, R., Jacobson, J., Ovard, L., & Deutmeyer, M. (2013). Wood pellet market and trade: A global perspective. *Biofuels, Bioproducts and Biorefining, 7*(1), 24–42. https://doi.org/10.1002/bbb.1366

Gold, S., & Seuring, S. (2011). *Supply chain and logistics issues of bio-energy production.* http://ac.els-cdn.com/S0959652610003240/1-s2.0-S0959652610003240-main.pdf?_tid=e115a1cae0ad3c2fbbea64d7e551d995&acdnat=1334907434_d9ea52de7915ce34a4b6bde677f76751

Güden, H., & Süral, H. (2014). Locating mobile facilities in railway construction management. *Omega, 45*, 71–79. https://doi.org/10.1016/j.omega.2014.01.001

Kaltschmitt, M., Hartmann, H., & Hofbauer, H. (2016). *Energie aus Biomasse.* Springer. https://doi.org/10.1007/978-3-662-47438-9

Lei, C., Lin, W.-H., & Miao, L. (2016). A two-stage robust optimization approach for the mobile facility fleet sizing and routing problem under uncertainty. *Computers & Operations Research, 67*, 75–89. https://doi.org/10.1016/j.cor.2015.09.007

Mirkouei, A., Haapala, K. R., Sessions, J., & Murthy, G. S. (2017). A review and future directions in techno-economic modeling and optimization of upstream forest biomass to bio-oil supply chains. *Renewable and Sustainable Energy Reviews, 67*, 15–35. https://doi.org/10.1016/j.rser.2016.08.053

Mirkouei, A., Mirzaie, P., Haapala, K. R., Sessions, J., & Murthy, G. S. (2016). Reducing the cost and environmental impact of integrated fixed and mobile bio-oil refinery supply chains. *Journal of Cleaner Production, 113*, 495–507. https://doi.org/10.1016/j.jclepro.2015.11.023

Nunes, L., Causer, T. P., & Ciolkosz, D. (2020). Biomass for energy: A review on supply chain management models. *Renewable and Sustainable Energy Reviews, 120*, 109658. https://doi.org/10.1016/j.rser.2019.109658

Polagye, B., Hodgson, K., & Malte, P. (2007). An economic analysis of bio-energy options using thinnings from overstocked forests. *Biomass and Bioenergy, 31*(2–3), 105–125. https://doi.org/10.1016/j.biombioe.2006.02.005

Prins, M. J., Ptasinski, K. J., & Janssen, Frans J. J. G. (2006). Torrefaction of wood: Part 2. Analysis of products. *Journal of Analytical and Applied Pyrolysis, 77*(1), 35–40. https://doi.org/10.1016/j.jaap.2006.01.001

Rentizelas, A. A., Tolis, A. J., & Tatsiopoulos, I. P. (2009). *Logistics issues of biomass: The storage problem and the multi-biomass supply chain.* http://ac.els-cdn.com/S1364032108000142/1-s2.0-S1364032108000142-main.pdf?_tid=daa776813618ff157e5e45d3a89869c3&acdnat=1337589265_ae15af007b230bb11537453a1e450592

Reyes, R., Schueftan, A., Ruiz, C., & González, A. D. (2019). Controlling air pollution in a context of high energy poverty levels in southern Chile: Clean air but colder houses? *Energy Policy, 124*, 301–311. https://doi.org/10.1016/j.enpol.2018.10.022

Sahoo, K., Bilek, E., Bergman, R., & Mani, S. (2019). Techno-economic analysis of producing solid biofuels and biochar from forest residues using portable systems. *Applied Energy, 235*, 578–590. https://doi.org/10.1016/j.apenergy.2018.10.076

Sharifzadeh, M., Garcia, M. C., & Shah, N. (2015). Supply chain network design and operation: Systematic decision-making for centralized, distributed, and mobile biofuel production using mixed integer linear programming (MILP) under uncertainty. *Biomass and Bioenergy, 81*, 401–414. https://doi.org/10.1016/j.biombioe.2015.07.026

Sharma, A., Jakhete, A., Sharma, A., Joshi, J. B., & Pareek, V. (2019). Lowering greenhouse gas (GHG) emissions: Techno-economic analysis of biomass conversion to biofuels and value-added chemicals. *Greenhouse Gases: Science and Technology, 9*(3), 454–473.

Spinelli, R., Visser, R., Björheden, R., & Röser, D. (2019). Recovering energy biomass in conventional forest operations: A review of integrated harvesting systems. *Current Forestry Reports, 5*(2), 90–100.

Tumuluru, J. S., Wright, C. T., Hess, J. R., & Kenney, K. L. (2011). A review of biomass densification systems to develop uniform feedstock commodities for bioenergy application. *Biofuels, Bioproducts and Biorefining, 5*(6), 683–707. https://doi.org/10.1002/bbb.324

Wishon, C., & Villalobos, R. J. (2016). Alleviating food disparities with mobile retailers: Dissecting the problem from an OR perspective. *Computers & Industrial Engineering, 91*, 154–164. https://doi.org/10.1016/j.cie.2015.11.014

Xie, S., Chen, X., Wang, Z., Ouyang, Y., Somani, K., & Huang, J. (2016). Integrated planning for multiple types of locomotive work facilities under location, routing, and inventory considerations. *Interfaces, 46*(5), 391–408. https://doi.org/10.1287/inte.2016.0857

Zamora-Cristales, R., Sessions, J., Boston, K., & Murphy, G. (2015). Economic optimization of forest biomass processing and transport in the Pacific Northwest USA. *Forest Science, 61*(2), 220–234. https://doi.org/10.5849/forsci.13-158

Zamora-Cristales, R., Sessions, J., & Marrs, G. (2017). Economic implications of grinding, transporting, and pretreating fresh versus aged forest residues for biofuel production. *Canadian Journal of Forest Research, 47*(2), 269–276. https://doi.org/10.1139/cjfr-2016-0221

Zhen, L., Wang, K., Hu, H., & Chang, D. (2014). A simulation optimization framework for ambulance deployment and relocation problems. *Computers & Industrial Engineering, 72*, 12–23. https://doi.org/10.1016/j.cie.2014.03.008

Production of Alternative Fuels and Chemicals from Green Hydrogen and Various Carbon Sources

Simon Glöser-Chahoud, Paul Heinzmann, Uwe Langenmayr, and Frank Schultmann

Abstract Chemical synthesis offers the possibility to produce different base chemicals and synthetic fuels from various carbon sources. In particular, the formation of synthesis gas from green hydrogen produced through electrolysis using renewable energy and CO_2 stemming from sources such as flue gases or direct air capture (DAC) are currently being broadly discussed as a promising source for the production of synthetic fuels. However, the required electrolyzer capacity for the production of synthetic fuels is linked to high capital expenditure and renewable energy from wind and solar power is subject to volatility. These present challenges to the economic viability of e-fuel production, particularly in areas with comparatively high levelized cost of electricity (LCOE). In this contribution, we provide a technological overview of different production pathways of electricity-based synthetic fuels (e-fuels) and perform a techno-economic evaluation of e-fuel production. Costs below €2 /l for e-fuels are only achievable in areas where renewable energy is available at full load over longer periods of time, and where overall costs of renewable electricity are below €0.03/kWh. While these conditions are not met in Germany, Chile has high potential for renewable electricity and could become a key global player in the production of e-fuels and green chemicals.

1 Introduction

With the "Green Deal," the European Commission has set ambitious targets to reach climate neutrality by 2050 (Fetting, 2020). This will require a range of measures including strong investments in alternative energy supply and in the ramp-up of a

S. Glöser-Chahoud (✉)
Technical University of Freiberg (TU BAF), Chair of Corporate Sustainability and Environmental Management, Freiberg, Germany
e-mail: simon.gloeser-chahoud@bwl.tu-freiberg.de

P. Heinzmann · U. Langenmayr · F. Schultmann
Karlsruhe Institute of Technology (KIT), Institute for Industrial Production (IIP), Karlsruhe, Germany
e-mail: frank.schultmann@kit.edu

A. Ch. Braun et al. (eds.), *Eco-Industrial Development as an Industrial Strategy*, Green Energy and Technology, https://doi.org/10.1007/978-3-031-73576-9_15

hydrogen infrastructure and supply system. Different EU member states but also non-EU countries including Chile have developed individual hydrogen strategies (Cheng & Lee, 2022). The economics of green hydrogen production through electrolysis based on renewable electrical power (mainly solar and wind energy) depend strongly on local conditions affecting the potential for renewable power generation. Chile is among the countries with the highest renewable potential for both wind (Satymov et al., 2022) and solar energy (Zimmer et al., 2022). This makes Chile an attractive location for green hydrogen production and further processing e.g. through chemical synthesis to achieve additional value creation and to enable better transportation in conventional oil tankers and infrastructure (Stoll et al., 2023). Chile could become an important global supplier of green hydrogen or hydrogen-based alternative fuels and green chemicals. German industry partners Siemens Energy AG and Porsche AG, together with further partners from the energy sector, are already running a pilot plant for the production of green methanol in Punta Arenas in Chile's Patagonia region, for which a scale-up to industrially relevant capacities is planned in the coming years (Ziegler, 2023).

In addition, Chile has high biomass potential stemming from agricultural and wood residues (see for example the previous case study in PART VI about mobile production plants for bioenergy carriers) that could supplement green hydrogen processing. Recent techno-economic assessments of alternative fuel production for instance have shown that the best economic performance is achieved when biomass gasification is combined with additional green hydrogen and subsequent chemical synthesis (Dietrich et al., 2018; Weyand et al., 2023). This combination, which is also referred to as power and biomass-to-liquids (PBtL), increases the carbon efficiency of the process and enables stoichiometric conditioning of synthesis gas to achieve the highest yields of synthetic fuels and chemicals (Dossow et al., 2021). As green hydrogen processing, chemical synthesis, and related infrastructure are of high relevance in the German and European strategy towards carbon neutrality, this area of research and development could become a future field of German-Chilean cooperation, both in science and industrial cooperation.

The following article provides an overview of green hydrogen-based value chains mainly regarding the production of liquid fuels (e-fuels) and synthetic base chemicals considering both technological challenges and economic implications. After a technical description of the different processing routes, we provide an economic analysis of the cost structure of e-fuel production. Subsequently, we discuss the production potential in Chile and Germany and conclude that Chile has the best conditions to become a strategic player on the global market for green hydrogen and its derivatives. Global aspirations to reduce greenhouse gas (GHG) emissions and to make both the transportation sector and industrial processes more sustainable will foster the demand for green hydrogen and synthetic fuels in the coming years.

2 Green Hydrogen Production and Processing Routes for Alternative Fuels Through Chemical Synthesis

The following subsections provide a brief overview of the different technologies utilized in the production of electricity-based alternative fuels through chemical synthesis. As illustrated in Fig. 1, the value chain of classical e-fuels begins with the production of green hydrogen through electrolysis and the separation of CO_2 from a point source or from air. Next, synthesis gas is formed which is the basis for the subsequent chemical synthesis to produce hydrocarbons. Finally, further treatment and refining may be necessary to achieve the intended fuel property and quality. The various processing steps are briefly described hereafter, beginning with electrolysis technologies.

Synthesis gas from renewable resources may also be created through direct gasification of conditioned biomass or through gasification of biomass-based pyrolysis oil and gas. As the stoichiometric ratio between carbon monoxide and hydrogen in the synthesis gas stemming from biomass gasification is not optimal due to the comparatively low hydrogen content, the addition of green hydrogen from electrolysis to the synthesis gas may strongly increase the carbon efficiency of the overall process (Weyand et al., 2023). This combination of the PtL (power-to-liquids) and BtL (biomass-to-liquids) processes is referred to as PBtL (power and biomass-to-liquids). The different processing steps for the production of synthetic fuels and chemicals are described in greater detail in the following sections.

2.1 *Hydrogen Production Through Electrolysis*

Electrolysis enables the splitting of water into hydrogen and oxygen using electrical energy. When powered by renewable sources such as wind energy or solar power, the resulting hydrogen is considered carbon neutral and is known as green hydrogen. There are three major technologies of electrolyzers that have achieved

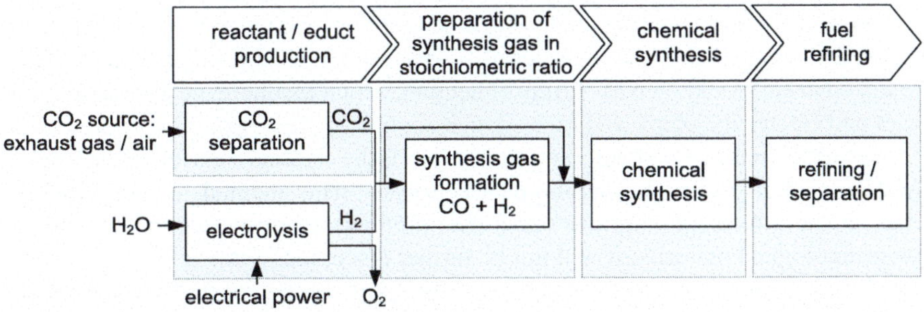

Fig. 1 General elements of the processing route to produce e-fuels and further chemicals from green hydrogen and carbon dioxide (Wachsmuth et al., 2022)

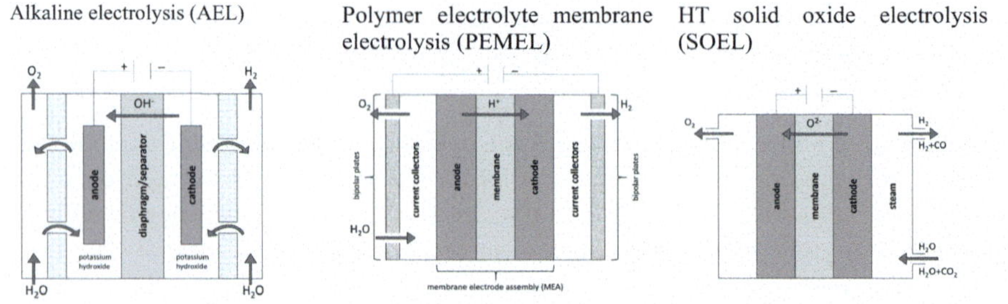

Fig. 2 Basic principles of major electrolyzer technologies. While AEL can be seen as a fully commercialized technology, PEMEL is about to reach TRL 9 with various demonstration projects under construction or in planning. HT SOEL is still under development with first small-scale technology demonstrations and an estimated TRL of around 6 (Wachsmuth et al., 2022)

sufficiently advanced technology readiness levels for use in commercial settings or, at the very least, within research initiatives at the demonstration stage (see Fig. 2). The existing systems can be further distinguished regarding low-temperature (LT) and high-temperature (HT) systems according to their process temperatures (Pitschak et al., 2017; Töpler & Lehmann, 2017).

Alkaline electrolysis (AEL) is considered the most developed technology, boasting the longest history of research and development. This system typically requires an electrolyte, such as potassium hydroxide (KOH) or sodium hydroxide (NaOH), which, when dissolved in water, results in the creation of hydroxide ions. These ions then travel through the diaphragm (or separator), and hydrogen is generated from OH- anions by the addition of electrons at the cathode (Tremel, 2018).

Polymer electrolyte membrane electrolysis (PEMEL) ranks as the second technology on the verge of widespread commercialization. Like AEL, PEMEL is a low-temperature (LT) electrolysis process that primarily uses desalinated water and electricity. The fundamental architecture of a PEMEL unit consists of a cell featuring an anode and a cathode, each separated by a solid membrane that is permeable to protons. Unlike AEL, this membrane acts not only as a separator but also fulfills the role of the electrolyte. The electrodes are typically crafted from carbon-based precious metals, and the entire cell is encased within bipolar plates. These plates are designed with small channels to precisely control the flow of water and gases. They are linked to entities known as current collectors, which facilitate the transfer of both gases and electricity (Zapf, 2017).

Another technology in membrane electrolysis is the alkaline electrolyte membrane (AEM) or anion exchange membrane electrolysis. This method, unlike PEMEL, relies on the movement of anions rather than protons. However, due to its relatively low technological maturity, it will not be further discussed in this text.

Solid oxide electrolysis (SOEL) represents a high-temperature (HT) electrolysis process that operates at significantly higher temperatures and theoretically has the capability to produce synthesis gas in a single step through a process known as co-electrolysis (detailed in the following section). This feature is particularly relevant for subsequent chemical synthesis, as it may eliminate the need for a separate synthesis gas formation step, thus enhancing the overall efficiency of the process. The HT SOEL unit also comprises a solid electrolyte that divides the two half-cells into an anode and a cathode (made of ceramic). Typically, the electrolyte is made of yttrium-stabilized zirconium oxide. This separator blocks gas flow while allowing the passage of oxygen ions (Pitschak et al., 2017). As illustrated in Fig. 2, water in the form of steam is fed into the cell on the side of the cathode. This steam is then split into hydrogen and oxygen ions. The oxygen ions pass the electrolyte and lose electrons to form molecular oxygen on the anode side. On the cathode side, the molecular hydrogen leaves the electrolysis cell.

The high-temperature solid oxide electrolysis (HT SOEL) boasts notable efficiency advantages primarily due to favorable thermodynamic conditions. Unlike low-temperature (LT) electrolysis technologies that convert liquid water into gaseous hydrogen and oxygen, resulting in increased entropy and respective efficiency losses, HT SOEL employs gaseous water and operates at higher pressures, enhancing its efficiency significantly. However, in terms of technological readiness level (TRL), HT SOEL is less advanced compared to the other two technologies mentioned, necessitating further advancements for dependable industrial-scale operation. Additionally, HT SOEL's operational flexibility is somewhat inferior to that of PEMEL, which could pose challenges when utilizing fluctuating renewable energy sources. Given that HT SOEL requires substantial heat, pairing it with a concentrated solar power plant (CSP) or integrating it with an exothermal synthesis process appears most promising for future industrial applications. Such integration could optimize heat use, further elevating the system's overall efficiency. Main technical parameters from various literature sources for the respective electrolyzer technologies are listed in Table 1. The techno-economic assessment in the next section builds upon PEM electrolyzers, as this technology seems most promising in the nearby future, due also to its flexibility regarding load management (Pitschak et al., 2017; Zapf, 2017).

2.2 *Formation of Synthesis Gas*

Synthesis gas, or syngas, consists of a combination of carbon monoxide (CO) and hydrogen (H_2), with their proportions tailored to suit the specific needs of the synthesis process. For processes that use CO_2 as a raw material, it is necessary to incorporate a step that converts CO_2 to CO when direct utilization of CO_2 is not feasible due to the design and reaction mechanisms of the synthesis reactor. A variety of methods can be employed for the conversion of CO_2 to CO, including both catalytic and electrochemical technologies. The most relevant technologies are sketched in Fig. 3 and are briefly described in the following.

The simplest and most mature processing route is through methanation (this synthesis process is directly possible from CO_2 and H_2) with subsequent reforming

Table 1 Main technical parameters of the different electrolyzer technologies (Wachsmuth et al., 2022)

Type of electrolysis-technology	Charge carrier	Feedstock	Operating temp. in °C	Pressure in bar	Stack lifetime in h	Stack efficiency % HHV	System eff. % HHV
AEL	OH^-	Electricity/Water	60–80	1–30	60,000–90,000	60–84	60–70
PEMEL	H^+	Electricity/Water	50–80	1–50	20,000–60,000	64–84	65–75
(Co-)SOEL	O^{2-}	Electricity/Steam/ CO_2	650–1000	1	3,500–20,000	89–98[a]	80–90[b]

[a]Excluding the provision of external thermal energy, [b]Including external thermal energy

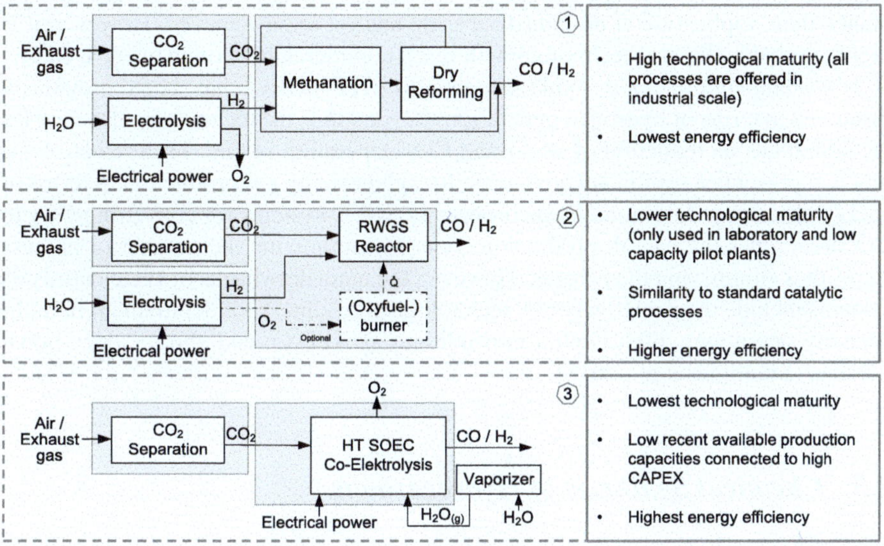

Fig. 3 Major technologies and processing routes for the production of synthesis gas: (1) methanation and subsequent reforming; (2) reverse water–gas shift reactor (RWGS); (3) co-electrolysis of H_2O and CO_2

to synthesis gas, which is an established process in natural gas valorization. Dry reforming can be seen as a modified steam reforming, in which water is substituted with CO_2 to produce a synthesis gas with a low H_2/CO-ratio according to the following reaction (Eq. 1).

$$CH_4 + CO_2 \rightarrow 2CO + 2H_2 \quad \Delta H_R^0 = 247,3kJ/mol \tag{1}$$

The process takes place at elevated temperatures higher than 800 °C to obtain technically feasible conversion levels. As the dry reforming reaction is endothermal, thermal energy has to be added, usually through the operation of a fired bundle tube reactor (Wittich et al., 2020). While this technology has high maturity, it is disadvantaged by low efficiency and the comparatively extensive process with two separate reactors (methanation and reforming).

An alternative that offers higher efficiency, yet is less advanced in development, is the reverse water–gas shift (RWGS) reaction. This process directly converts CO_2 and H_2 into synthesis gas and water (Eq. 2).

$$CO_2 + H_2 \rightarrow CO + H_2O \quad \Delta H_R^0 = 41,5kJ/mol \tag{2}$$

Due to the stability of CO_2, a significant amount of energy is required to reduce it to CO. This endothermic reaction typically occurs at temperatures ranging from 700 °C to 1,000 °C. Although RWGS reactors are operational in various demonstration plants, their application at an industrial scale has yet to be achieved. However, it is highly probable that this technology will reach commercialization in the near future.

High-temperature solid oxide electrolysis (HT SOEC) has been previously discussed in terms of hydrogen generation. Beyond this, the technology also enables the simultaneous reduction of water and CO_2. Operating in this HT co-electrolysis mode, it presents a significant advantage for synthesizing gas production in terms of energy efficiency. Additionally, incorporating SOEC systems for converting CO_2 to CO streamlines the overall production chain by minimizing the number of process steps, thus simplifying the process. However, the main drawbacks of HT-electrolysis systems include their relatively low technological readiness for deployment in high-capacity operations, high capital expenditures (CAPEX), and shorter operational lifespan (Smolinka et al., 2018).

2.3 Chemical Synthesis of Hydrocarbons

As illustrated in Fig. 4, there are various technologies for the chemical synthesis of hydrocarbons. While ammonia might become an important hydrogen carrier in future, the Haber–Bosch synthesis to produce ammonia is not further discussed here, as this contribution only focuses on the synthesis of hydrocarbons. Major chemical synthesis processes were developed in Germany in the early twentieth century. Today, these are mature technologies applied and optimized based on fossil resources, mainly natural gas as a source for synthesis gas production or coal gasification, which has been an important resource for chemical synthesis in the past, while today coal gasification for chemical synthesis is only applied in China. The different synthesis processes mainly differ in operating conditions, catalyst materials, and available reactor concepts. While all synthesis processes can be supplied by synthesis gas (the usual process route in fossil-based plants), DME and methanol synthesis as well as methanation can be operated directly with a gaseous mixture of CO_2 and H_2 as educt. This can be advantageous as the step of synthesis gas formation e.g. through an RWGS reactor (cf. Fig. 3) is not mandatory.

To generate liquid fuels, which are applicable in widely used combustion engines, the two processing routes via Fischer–Tropsch synthesis (FTS) for diesel and jet fuel formation and methanol synthesis, which can be further processed to gasoline fuel via the methanol-to-gasoline route (MTG), are the most relevant processes. This is also the case for the production of green chemicals where the paraffins and waxes formed in the FTS synthesis and methanol as a starting point for the formation of a whole range of further base chemicals (e.g. ethylene, propylene, and aromatics) are key raw materials for chemical industries. Therefore, this contribution focuses on FTS and methanol synthesis, while the additional synthesis processes shown in Fig. 4 are not further discussed here. Methanation for example has the advantage of a highly selective and less complex synthesis, however gaseous methane needs to be transported as CNG or LNG (compressed or liquefied natural gas) and therefore provides only a limited advantage compared to green hydrogen from electrolysis.

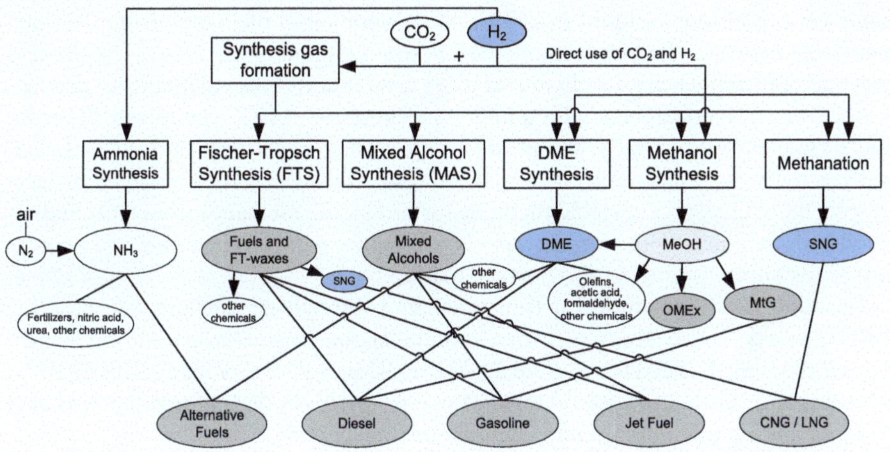

Fig. 4 Chemical synthesis pathways and subsequent products/fuels. Abbreviations: DME: dimethyl ether; SNG: synthetic natural gas; MtG: methanol-to-gasoline; OMEx: oxymethylene ethers; CNG: compressed natural gas; LNG: liquefied natural gas (Wachsmuth et al., 2022)

2.3.1 Methanol Synthesis

Methanol serves as a crucial base chemical in the production of a wide range of chemical products, including formaldehyde and acetic acid. In recent debates surrounding e-fuels and chemical energy carriers, the synthesis of methanol has occupied a central position. This prominence is due to the well-established synthesis processes available and the ability to directly utilize CO_2 as a feedstock, making methanol synthesis a pivotal element in the transition towards more sustainable energy and chemical production methods. The reaction of CO_2 and CO with hydrogen to form methanol is performed according to the following reaction equations (Eqs. 3 and 4):

$$CO_2 + 3H_2 \rightarrow CH_3OH + H_2O \quad \Delta H_R^0 = -49,31kJ/mol \tag{3}$$

$$CO + 2H_2 \rightarrow CH_3OH \quad \Delta H_R^0 = -90,64kJ/mol \tag{4}$$

Both reactions are exothermic and decrease the volume of gas, meaning that in methanol synthesis, lower temperatures and higher pressures result in a higher yield of methanol at chemical equilibrium. Catalysts are employed to ensure a quick conversion of CO and CO_2 under optimal reaction conditions. Commercial processes for synthesizing methanol from synthesis gas operate at temperatures ranging from 220 to 290 °C and pressures between 50 to 125 bar. Copper-zinc-aluminum mixtures are the conventionally used catalyst materials (Aresta et al., 2019).

In conventional processes for methanol synthesis, maintaining a CO_2 concentration of 2–10% in the synthesis gas is found to enhance the yield while minimizing the creation of by-products like higher alcohols and ketones. Within the framework of power-to-liquid strategies, the generation of methanol using CO_2 and renewable hydrogen (power-to-methanol) is receiving increased focus. The process concepts applicable here closely mirror those used in conventional methanol production. However, feed streams with a high concentration of CO_2 can hinder the reaction's kinetics. Additionally, the production of water negatively impacts the catalyst's activity, as water molecules take up space on the catalyst's active surface. Despite these challenges, viable methods for producing methanol from CO_2 and H_2 are already in place, with ongoing efforts to enhance catalyst materials and process design for improved efficiency (Bertau, 2014).

Gasoline can be synthesized from methanol through the methanol-to-gasoline (MtG) process. The resulting fuel shares similar characteristics with traditional gasoline, eliminating the need for modifications to existing vehicles or the fuel distribution infrastructure. In this process, methanol is converted via the intermediate product DME according to the following reaction equation (Eq. 5):

$$nCH_3OH \rightarrow \frac{n}{2}CH_3OCH_3 + \frac{n}{2}H_2O \rightarrow (CH_2)_n + nH_2O \qquad (5)$$

The term $(CH_2)_n$ represents hydrocarbons with n carbon atoms. Due to the steric hindrance of the micropores of the catalyst, the size of the hydrocarbons produced is limited to ten carbon atoms. A lower reactor temperature favors a higher gasoline yield, but the octane number also decreases, as well as the production ratio of the largest and undesirable component durene (1, 2, 4, 5-tetramethylbenzene, $C_{10}H_{14}$). MtG processes are usually carried out at temperatures between 300 °C and 450 °C. The pressure level significantly affects the composition of the product. Lower pressures result in the production of more olefins, while higher pressures favor the formation of aromatic compounds (Arnold et al., 2019; Bertau et al., 2014).

The development and initial large-scale implementation of methanol-to-gasoline (MtG) technology were undertaken by the petroleum company Mobil (today Exxon Mobil) in response to the surge in oil prices during the oil crisis. Mobil commenced operations of a commercial plant in New Zealand in 1985. However, following the decline in oil prices after the crisis ended, the plant was sold twelve years later and repurposed for methanol production. Despite this, various companies continue to offer licensed MtG and gas-to-gasoline (GtG) processes. Since 2009, a facility with a capacity of 100,000 tonnes per annum of gasoline has been operated in the Chinese province of Shanxi, which utilizes coal gasification for the provision of synthesis gas (Bertau, 2014; Hofbauer & Rauch, 2017).

2.3.2 Fischer–Tropsch Synthesis

The origins of Fischer–Tropsch synthesis (FTS) trace back to the research conducted by Franz Fischer and Hans Tropsch at the Kaiser Wilhelm Institute in Mühlheim/Ruhr, Germany, during the 1920s. The large-scale implementation of FTS plants, aimed at producing fuels and chemical industry products, commenced in Germany in 1935 through the efforts of Ruhrchemie (König, 2016).

Fischer–Tropsch synthesis is a process catalyzed by heterogeneous catalysts, where synthesis gas is transformed into a range of hydrocarbons with varying chain lengths and structures, along with different by-products. The primary goal of the FTS process is to produce hydrocarbons that can be used as fuels, chemical industry feedstocks, lubricants, and waxes. The hydrocarbons generated, which include oxygenated compounds, vary in their chemical and physical characteristics. Alkanes and alkenes are the main outputs of FTS but the process also produces alcohols, carbonyls, and carboxylic acids among its range of products. In general, FTS can be described by the following reaction equation (Ail & Dasappa, 2016):

$$CO + 2H_2 \rightarrow (-CH_2-) + H_2O \quad \Delta H_R^{250^\circ C} = -158,5kJ/mol$$

In this process, the CH_2 generated can be considered a fundamental building block for constructing hydrocarbons with varying chain lengths and functional groups. The reactions occurring at the catalyst's active sites are intricate and not fully comprehended. The literature contains various proposed reaction schemes, which are the subject of ongoing debate. There is no consensus on a single mechanism that can fully account for the diverse range of products formed; hence, it's believed that multiple mechanisms overlap to explain the production of the wide range of products (Eilers, 2018).

Fischer–Tropsch synthesis processes are categorized into two main types based on the reaction conditions, with reaction temperature being the key differentiator. Higher temperatures tend to promote the creation of short-chain molecules and methane, while lower temperatures favor the production of waxy, long-chain hydrocarbons. High-temperature Fischer–Tropsch synthesis (HT-FTS) operates within the 300 °C to 350 °C range, utilizing exclusively iron catalysts due to the significant increase in unwanted methane production observed with cobalt catalysts at elevated temperatures. HT-FTS results in a high output of hydrocarbons with chain lengths comparable to conventional gasoline and short-chain olefins. Conversely, low-temperature Fischer–Tropsch synthesis (LT-FTS) processes generate a greater yield of n-alkanes with longer chain lengths and waxy substances, operating at temperatures between 200 and 250 °C. Both iron and cobalt catalysts are suitable for use in LT-FTS (Hofbauer & Rauch, 2017).

Fischer–Tropsch synthesis (FTS) processes can be conducted in various types of reactors, including fixed bed, fluidized bed, and bubble column reactors. Fixed bed reactors for FTS are typically configured as tube bundle reactors. Given that FTS is a process that releases a considerable amount of heat, effective heat management is crucial to maintain appropriate reaction conditions and prevent catalyst degradation through overheating or coking. An innovative variant of fixed bed reactors are microchannel reactors. While these have not yet been adopted in commercial settings, they are the subject of extensive research due to their potential for achieving high yields and selectivity (Hofbauer & Rauch, 2017). Fluidized bed reactors are applied due to their relatively homogeneous temperature distribution, but they are only suitable for use in LT-FTS systems (Eilers, 2018). Bubble column reactors are characterized by their relatively uniform mixing and the heat storage capacity of the suspension, which facilitates easy temperature regulation. These reactors are capable of handling high capacities. However, the challenge lies in the abrasion between catalyst particles and the reactor walls, necessitating a sophisticated method for separating the fine particles from the product stream. Scaling up bubble column reactors presents difficulties due to the intricate fluid dynamics involved (Ail & Dasappa, 2016).

2.4 *Potential Carbon Sources*

Besides green hydrogen, a carbon source is required for the production of synthetic hydrocarbons. Classical e-fuels, or renewable fuels of non-biological origin (RFNBOs) as defined in EU legislation (e.g. the EU Renewable Energy Directive), build upon CO_2 as the carbon source. To be considered an RFNBO in the EU, the energy content of the fuel has to be fully based on renewable electrical power; hence, the carbon source should not have any caloric value. However, as previously discussed, synthesis gas can be produced from various hydrocarbons including biomass. In the case of biomass gasification, the stochiometric ratio between hydrogen and carbon monoxide in the synthesis gas is not optimal, resulting in low carbon efficiency. Only around 25% of the carbon content of the biomass ends up in the synthetic fuel, while up to 75% percent is emitted in the form of CO_2 (Weyand et al., 2023). With an admixture of green hydrogen to the synthesis gas, a carbon efficiency of almost 100% can be achieved. Recent techno-economic assessments have also shown that the cost efficiency of so-called PBtL (power and biomass-to-liquid) processes is even higher than that of PtL (power-to-liquid) or BtL (biomass-to-liquid) processes (Weyand et al., 2023). As Chile has both very good conditions for renewable energy generation as well as a high biomass potential, the concept of PBtL plants could also be an attractive alternative to single biomass combustion or gasification. The CO_2 for classical e-fuel production may originate from a point source such as an industrial facility with process-related CO_2 emissions (e.g. a cement plant with high CO_2 emissions from the calcination process) or from direct air capture (DAC). The major advantage of a point source is the comparatively high concentration of CO_2 in the exhaust gases, while DAC provides the possibility to collect CO_2 from the air without spatial limitations. This is of particular relevance if the production site of the PtL plant is in remote areas with optimal conditions for renewable power generation but without proximity to a potential point source of CO_2. Both concepts of carbon capture are briefly described in the following subsections.

2.4.1 Carbon Capture Technologies for CO_2 Point Sources (CO_2 Scrubbing)

The two main methods used for CO_2 scrubbing are adsorption and absorption to separate the CO_2 in the flue gas from other chemical elements and compounds, e.g. nitrogen, hydrogen, sulphur dioxide, or nitrogen oxides.

The adsorption process involves the attachment of gaseous compounds to the surface of a solid material, known as the adsorbent. Conversely, in the absorption process, the gas is captured by being dissolved or chemically reacted with a liquid absorbent (Jansen et al., 2015). Chemical absorption is considered the leading technology in post-combustion facilities due to the low pressure of the flue gas, which is more conducive to chemical absorption than to physical adsorption. Aqueous alkanolamines in particular such as monoethanolamine (MEA) are well developed and closest to a commercial usage. For MEA, the thermal energy demand is around 3.8–4.0 GJ/tonne CO_2 (Chen et al., 2020). The energy required for these processes substantially affects the energy penalty, with chemical absorption processes applied for carbon capture in conventional combustion plants experiencing efficiency losses in the range of 4–14%. In addition to the demand for thermal energy, the need for mechanical energy is also crucial during the regeneration process. In particular, high pressure can facilitate the regeneration process but necessitates an additional supply of energy (Goto et al., 2013).

Both adsorption and absorption processes require extra energy, typically in the form of thermal and mechanical energy such as heat and pressure. Processes that rely on temperature changes are known as temperature swing adsorption, while those dependent on pressure changes are called pressure swing adsorption or vacuum swing adsorption. In addition to adsorption and absorption, membrane technology can be employed to separate CO_2 from flue gas or other gas mixtures by leveraging material permeability. This process allows the targeted substance to pass through the membrane while blocking other components. Membrane technology primarily relies on pressure differentials and does not necessitate thermal energy. However, despite the initial use of polymeric membranes for CO_2 separation from flue gases, membrane durability remains a challenge. The presence of other components in the flue gas can deteriorate the membrane's permeability or selectivity, significantly reducing its efficiency, sometimes even irreversibly (Kárászová et al., 2020).

2.4.2 Direct Air Capture (DAC)

Direct air capture (DAC) offers an alternative to point-source CO_2 capture by extracting CO_2 directly from the ambient air. This method's advantage lies in its decentralized implementation, making it particularly compatible with decentralized energy sources like renewables. By storing the captured CO_2 permanently, DAC can achieve a negative carbon balance. However, its main drawback is the significant demand for thermal and electrical energy. The regeneration process requires approximately 1,420–2,250 kWh of heat per tonne of CO_2 captured, and the operation of pumps, compressors, and other electrical components consumes an additional 366–764 kWh of electricity per tonne of CO_2 (Fasihi et al., 2019). There are two general approaches of DAC based on either absorption or adsorption as briefly described in the following sections.

The high-temperature aqueous solution approach consists in general of two cycles. In the first cycle, CO_2 is absorbed from the ambient air into a sodium hydroxide (NaOH) sorbent at room temperature and ambient pressure, creating sodium carbonate (Na_2CO_3). The second cycle regenerates the absorbent by mixing it with calcium hydroxide ($Ca(OH)_2$). In this process, calcium carbonate ($CaCO_3$) emerges and NaOH is separated, which can be reused in the first cycle. Using a temperature of around 900 °C, CO_2 is released from the $CaCO_3$. The emerging CaO of this process can be hydrated and recycled (Fasihi et al., 2019).

The low-temperature solid sorbent approach uses chemical adsorption to separate CO_2 from the ambient air. The first step includes the saturation of the sorbent under room temperature and ambient pressure. After the adsorption process, the vessel is closed and the remaining air removed. The CO_2 in the sorbent is removed by increasing the temperature inside the vessel to around 80–110 °C depending on the sorbent used. After the regeneration process is finished, the adsorption process restarts. One main advantage of this approach is the possible utilization of the low-gradient heat of other processes.

Viebahn et al. (2019) found that direct air capture (DAC) technology has achieved a technology readiness level (TRL) of 5 for the high-temperature (HT) aqueous sorbent method and TRL 6 for the low-temperature (LT) solid sorbent method. While the technical risks associated with these approaches are relatively low, the economic risks are high, primarily due to the substantial energy consumption and competition from point source carbon capture technologies. Based on their analysis, Viebahn et al. (2019) concluded that reaching a TRL of 9 would require several development and upscaling steps aimed at cost reduction. Furthermore, they highlighted additional uncertainties such as the future market potential, the short timeframe for achieving carbon neutrality by 2050, the need to expand production capacities, and the lack of public and political discussion on the topic as potential barriers to widespread adoption.

3 Techno-Economic Assessment and Cost Structure

A key factor for the potential future realization of large-scale e-fuel production facilities is their cost structure and the related profitability of synthetic fuel production and commercialization. There have been various studies and publications in recent years analyzing the potential costs of producing e-fuels at locations with high renewable energy potential and importing these fuels to Germany via shipping in a manner currently performed with crude oil. A comparison of the estimated cost structure for the production of FTS diesel in the most relevant recent studies for Germany is provided in Fig. 5.

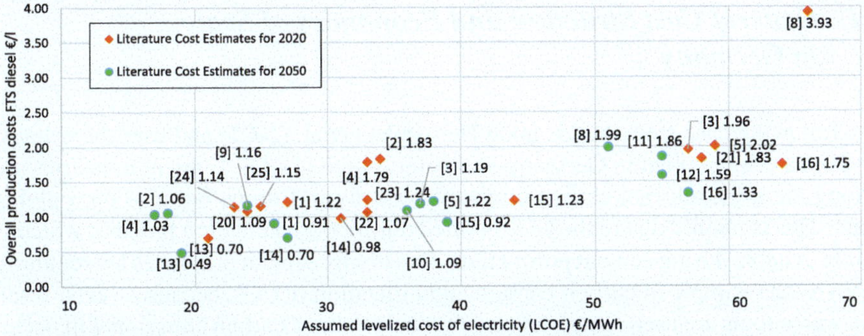

Source specification regarding location and technology assumptions for different scenarios	[13]: Prognos/Fraunhofer Umsicht/DBFZ 2018; Import optimistic from North Africa to Germany (2% WACC); SOEC; 2030/2050
[1]: Agora/Frontier Economics 2018; Geothermal Energy / Hydropower in Island; Alkaline/PEM	[14]: Prognos/Fraunhofer Umsicht/DBFZ 2018; Import optimistic from North Africa to Germany (7% WACC); SOEC; 2030/2050
[2]: Agora/Frontier Economics 2018; PV in North Africa; Alkaline/PEM	[15]: Prognos/Fraunhofer Umsicht/DBFZ 2018; Import realistic from North Africa to Germany (2% WACC); SOEC; 2030/2050
[3]: Agora/Frontier Economics 2018; PV/Onshore in Northafrica; Alkaline/PEM	[16]: Prognos/Fraunhofer Umsicht/DBFZ 2018; Import realistic from North Africa to Germany (7% WACC); SOEC; 2030/2050
[4]: Agora/Frontier Economics 2018; PV in the Middle East; Alkaline/PEM	[17]: EA NRW/Enervis 2019; PV in North Africa; Electrolyzer not specified
[5]: Agora/Frontier Economics 2018; PV/Onshore in the Middle East; Alkaline/PEM	[18]: EA NRW/Enervis 2019; Wind onshore in North Africa; Electrolyzer not specified
[6]: Leopoldina/Acatech 2017; PV Dubai; Electrolyzer not specified	[19]: EA NRW/Enervis 2019; Wind offshore in North Africa; Electrolyzer not specified
[7]: Leopoldina/Acatech 2017; Wind Marocco; Electrolyzer not specified	[20]: Tremel 2018; Wind onshore Marocco; PEM; Preise Methanol in Dieseläq. 2025-2030
[8]: UBA/Öko-Institut/KIT/Infras 2016; Turkey / Egypt; Electrolyzer not specified	[21]: Tremel 2018; Wind offshore Chile; PEM; Preise Methanol in Dieseläq. 2025-2030
[9]: FVV 2018; PV-Wind-Hybrid MENA; Alkaline	[22]: Tremel 2018; Hydropower in Scandinavia point source of CO2; PEM; Preise Methanol in Dieseläq. 2025-2030
[10]: Dena/EWI 2018; Algeria; Alkaline	[23]: Tremel 2018; Hydropower in Scandinavia DAC; PEM; Preise Methanol in Dieseläq. 2025-2030
[11]: FVV/LBST 2016; in North Africa; Alkaline/PEM	[24]: Tremel 2018; PV in Dubai; PEM; Preise Methanol in Dieseläq. 2025-2030
[12]: FVV/LBST 2016; in North Africa; SOEC	[25]: Tremel 2018; Wind onshore/PV in Australia; PEM; Preise Methanol in Dieseläq. 2025-2030

Fig. 5 Results of a broad literature analysis regarding the cost structure of importing e-fuels to Germany. The figure represents different scenarios from the respective studies (Ausfelder et al., 2017; Bründlinger et al., 2018; Hobohm et al., 2018; Kasten et al., 2016; Kramer, 2018; Schindler, 2019; Tremel, 2018; Verkehrswende, 2019)

The major cost drivers for the subsequent fuel are the costs for renewable electricity provision. Further major differences between respective studies occur in the assumed investment volume for electrolyzers and the full-load hours of hydrogen production as well as the capacity load of the final FTS plant. Due to the expected reduction of prices for electrolyzers, mainly driven by the ramp-up of electrolyzer production capacities and respective economies of scale, e-fuel production costs are expected to significantly decline in the future (see green spots in comparison to red spots in Fig. 5).

Figure 5 provides a broad overview of the cost structure of e-fuel production regarding different locations and technologies from various literature sources. To illustrate the dependency of fuel production costs on the levelized cost of electricity (LCOE) provision, we plotted the resulting production costs of the synthetic fuels over the literature based LCOE of different renewable sources at different locations. While Fig. 5 provides a general overview from different literature sources, a more detailed analysis of specific production costs based on the results of a recent research project funded by Germany's federal state of Baden-Württemberg (Toedter et al., 2023) is provided in the next section.

3.1 General Cost Structure and Economies of Scale in Germany

The following case study builds upon Heinzmann et al. (2021) and uses the exhaust gases of a cement plant in Germany as a point source of CO_2. Cement industries are among the most relevant carbon emitters, not only in Europe but also on a global scale. This is mainly due to the CO_2 emitted during the calcination process, which is barely evitable during cement production. This is why cement industries are regularly given as an example of carbon capture and utilization (CCU), because—even when renewable fuels are used—high levels of process-related carbon emission will occur. The process conception regarding CO_2 scrubbing, H_2 electrolysis, syngas formation, chemical synthesis, and subsequent fuel refining is summarized in Fig. 6. The figure also includes the current technology readiness levels (TRLs) of the respective processing steps. As already mentioned in the technical background sections, the downstream processing steps of chemical synthesis and fuel refining are mature processes that have been realized in various fossil-based FTS plants, while the formation of synthesis gas from CO_2 und H_2 in a reverse water–gas shift reactor (RWGS) has not yet reached full technological maturity. However, it is expected that the first large-scale plants using RWGS reactor technology will be realized in the coming years (Heinzmann et al., 2021). Major drivers for the overall production costs of e-fuels are depreciations and capital costs of the investments for the facility (CAPEX), mainly driven by electrolyzer costs, the costs for renewable electricity (levelized cost of electricity, LCOE), and the full-load hours of the overall production plant. Further operational costs arise for maintenance, insurance, and taxes. In order to gain detailed mass and energy balances of the process sketched in Fig. 6 and to get an understanding of the dimensions of relevant plant components, a process simulation of the process was performed using AspenPlus (Heinzmann et al., 2021).

An assumed production capacity of 100,000 tons of fuel per year (at a utilization rate of 91%) corresponds to the use of CO_2 emissions from an average German cement plant. For the techno-economic assessment, several assumptions regarding

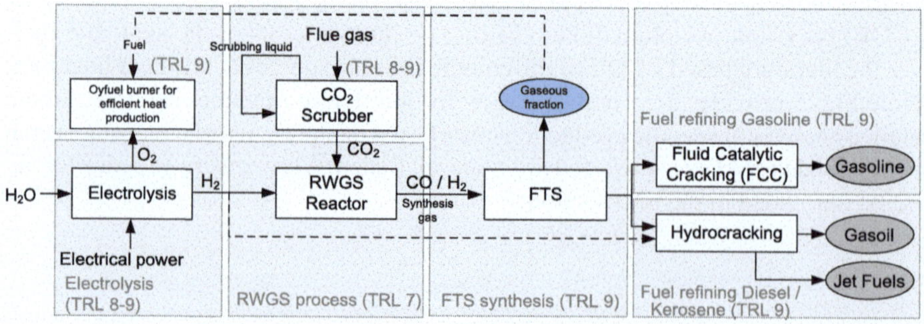

Fig. 6 Basic structure of a Fisher-Tropsch synthesis plant with CO_2 scrubbing from exhaust gases, e.g. from cement production. TRLs are based on (Wachsmuth et al., 2022)

major cost components are necessary. As these assumptions are linked to uncertainty, we developed different cost scenarios based on the data provided in Table 2. The respective scenarios mainly refer to the investment costs (CAPEX) for the electrolyzers and related costs for capital provision (weighted average costs of capital) as well as maintenance and insurance costs. In all scenarios, a reduction of CAPEX over time was assumed as it is expectable that current costs for electrolyzer capacities will decrease in the course of a ramp-up of production capacities and related economies of scale due to mass production.

As summarized in Fig. 7, the CAPEX for the electrolyzers dominates overall investments in all scenarios. However, the investment volume is expected to decline over time and strongly varies between the different scenarios. This is a direct result of the uncertainties regarding current investment costs for electrolyzers as indicated in Table 2. The operational expenditure (OPEX) is dominated by the electricity provision costs as illustrated in Fig. 8. However, the capital costs resulting in annuities for the high investments in Fig. 7 also contribute heavily to the overall production cost. As clearly shown in Fig. 8, based on the current cost structure and assuming a levelized cost of electricity of €51 /MWh (which is relatively low regarding current electricity costs in Germany), overall production costs between €2 and €2.50 per liter of fuel can be expected.

This cost level might decline slightly over time due to decreasing CAPEX for electrolyzers, leading to a reduction of depreciations and annuities. However, due to the dominating costs for electrical power and Germany not being the best location to produce cheap renewable electricity, a significant reduction in overall production

Table 2 Basic parameters for the cost estimation (Heinzmann et al., 2021; Toedter et al., 2023)

Parameter		Conservative	Base case	Optimistic
Depreciation period		20 years		
Weighted average costs of capital (WACC)		10%	7%	5%
Annual costs for maintenance of the chemical plant		8%	6%	4%
CAPEX of electrolyzers [€/kW_{el}]	2022	€800 /kW_{el}	€1000 /kW_{el}	€1200 /kW_{el}
	2030	€490 /kW_{el}	€810 /kW_{el}	€1120 /kW_{el}
	2050	€210 /kW_{el}	€510 /kW_{el}	€800 /kW_{el}
Annual costs for maintenance of electrolyzers	2022	3.5%		
	2030	3.3%		
	2050	3.9%		
Taxes and insurance		2%	1.5%	1%
Plant overhead		2%	1.5%	1%
Levelized cost of electricity (LCOE)		€51 /MWh		
Annual full load hours		8000 h/a		

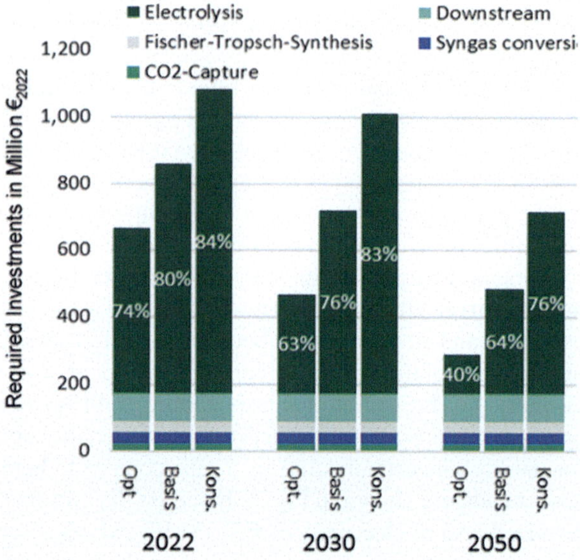

Fig. 7 CAPEX for the regarded synthesis plant

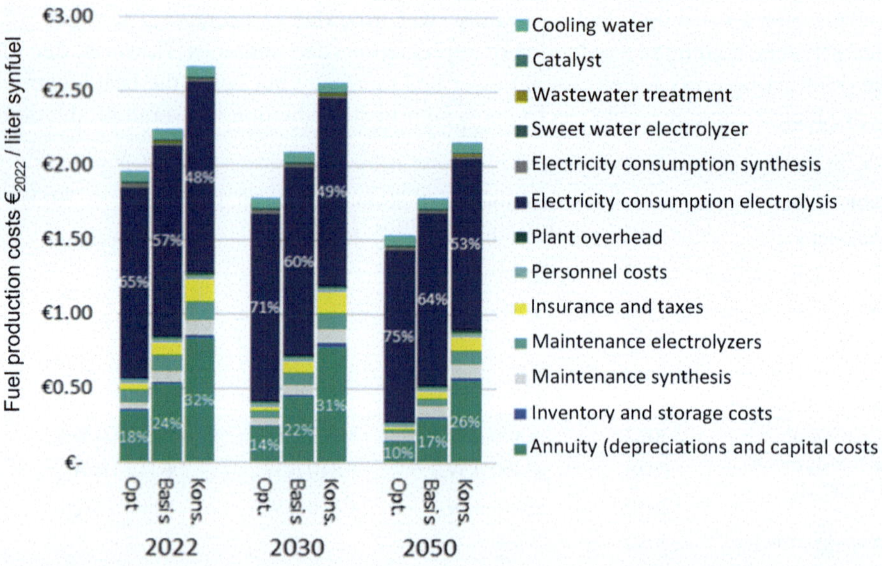

Fig. 8 Total production costs under different scenarios

costs for e-fuels can only be achieved in countries where renewable electricity can be provided at significantly lower costs than in Germany. Generally, recent publications about e-fuel production in Germany have shown limited capacity potential for production sites in Germany that still meet the requirements defined by the European Renewable Energy Directive (Langenmayr & Ruppert, 2023a, 2023b). As pointed out

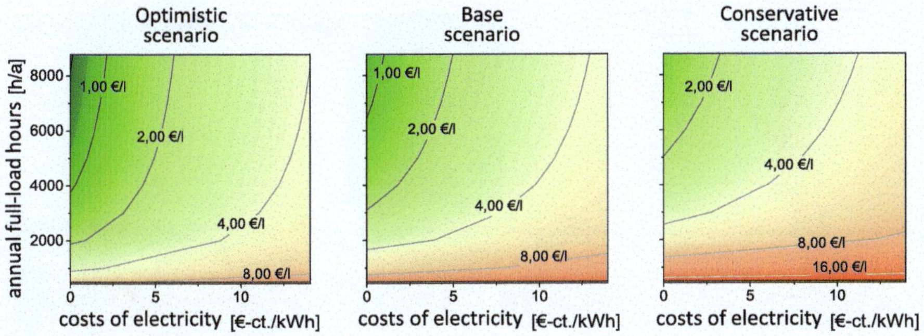

Fig. 9 Relationship between production costs, full load hours of the synthesis plant, and provision costs of electrical power

by Heinzmann et al. (2021) and as illustrated in Fig. 9, a prerequisite for achieving production costs of FTS fuels significantly below €2 per liter is the availability of cheap renewable electricity for as long as possible (full-load hours of the electrolyzers). This could lead to an additional need for electricity storage e.g. through batteries for peak shaving of renewable electricity production and for a continuous supply of the electrolyzers. Also, hydrogen storage to enable a continuous supply of the less flexible syngas production and synthesis process will be necessary for large-scale operations. This requires additional location-specific plant optimization as suggested for example by Langenmayr & Ruppert (2023a, 2023b).

However, the most important aspect to achieve cost levels for fuels of around €1 per liter are full-load hours for electrolyzers of above 6000 h/a and LCOE of below 3ct/kWh (see Fig. 9). These conditions can only be achieved at locations with very high potential for renewable electricity production, in the best case using a combination of solar power and wind energy. This brings Chile into play as one of the most promising locations for the production of renewable synthetic fuels.

3.2 *Chile as One of the Most Promising Locations for the Production of Alternative Fuels and Green Chemicals*

As pointed out in Fig. 9, the availability of low-cost renewable energy is a prerequisite to produce e-fuels at theoretically competitive prices. This requires either high solar irradiation for electricity production from photovoltaics and concentrated solar power (CSP), or high and continuous wind speed for electricity production through wind turbines. With its various geographic and climatic regions, Chile provides outstanding locations for both photovoltaics or CSP and perfect conditions for wind energy.

Northern Chile, including the Atacama Desert, receives some of the highest solar irradiation levels in the world. This region is renowned for its clear skies and

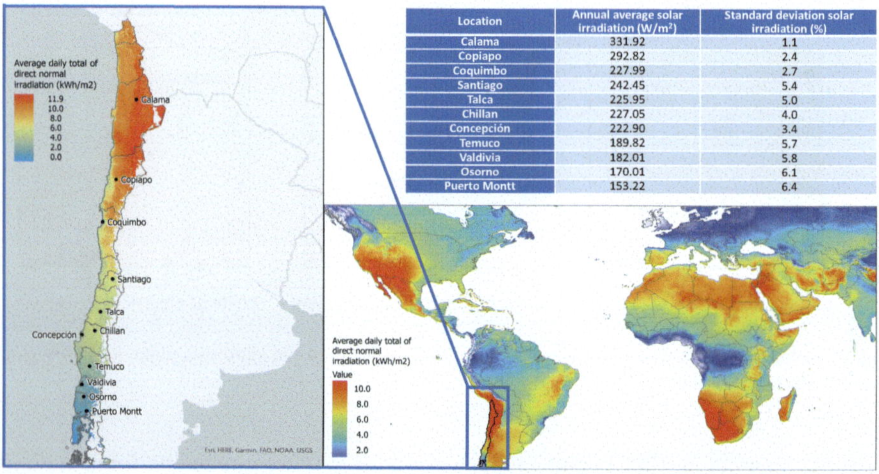

Location	Annual average solar irradiation (W/m²)	Standard deviation solar irradiation (%)
Calama	331.92	1.1
Copiapo	292.82	2.4
Coquimbo	227.99	2.7
Santiago	242.45	5.4
Talca	225.95	5.0
Chillan	227.05	4.0
Concepción	222.90	3.4
Temuco	189.82	5.7
Valdivia	182.01	5.8
Osorno	170.01	6.1
Puerto Montt	153.22	6.4

Fig. 10 Solar radiation intensity in Chile. The Atacama Desert in particular shows perfect conditions for solar energy from photovoltaics or even CSP (Zimmer et al., 2022)

minimal cloud cover throughout the year, providing optimal conditions for solar energy generation (cf. Fig. 10).

The Atacama Desert further offers vast expanses of uninhabited and arid land, which is suitable for setting up large-scale solar power installations without the land use conflicts that might occur in more densely populated areas. Figure 10 summarizes the solar irradiation of different locations in Chile und underlines the outstanding potential of the Atacama Desert, especially the region between Copiapo and Calama. While this region is highly suitable for solar energy provision, sweet water for supplying the electrolyzers could be a problem in this dry area. However, sea water desalination is an option for providing sweet water that has been addressed and analyzed in literature, not only in Northern Chile but also for other dry regions with high solar power potential such as North Africa or the Middle East (Delpisheh et al., 2021).

Besides these desert areas, Chile's varied geography includes coastal regions and mountains, offering opportunities for diverse combinations of wind and solar energy. Patagonia is especially renowned for its strong and steady winds, particularly in the southern regions of Chile (Fig. 11). The area's geographic and atmospheric conditions contribute to consistent wind patterns that are ideal for wind energy production. The region offers vast, sparsely populated areas that are suitable for the installation of large wind farms. The availability of such spaces minimizes conflicts over land use and allows for the scalability of wind energy projects. The consistent wind conditions in Patagonia lead to high capacity factors for wind turbines, meaning that turbines can produce electricity at or near their maximum capacity for a significant portion of the time. This efficiency makes wind projects more economically viable in the region. Recent publications assume LCOEs of around \$25 /MWh in Patagonia and full load hours of over 6000 h/a for wind turbines. Regarding the cost structure in

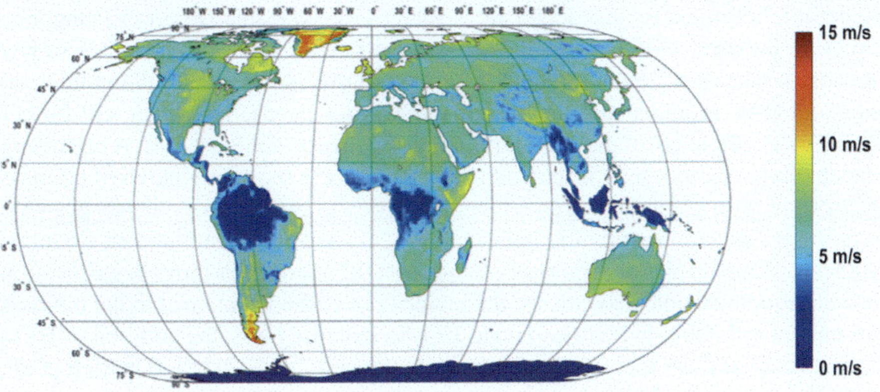

Fig. 11 Global wind map showing the highest potential for wind energy in Patagonia and Greenland (Satymov et al., 2022)

Fig. 9, this is in a range where fuel costs can be significantly reduced as compared to Germany and could reach the range of around 1 €/l as also indicated by the literature analysis in Fig. 5. Patagonia's exceptional wind resources, combined with supportive government policies and a growing focus on renewable energy, position it as a key region for wind energy development in Chile and South America. This is also the main reason for Siemens Energy and Porsche AG running their Haru Oni pilot plant for the production of green methanol in Punta Arenas in Patagonia. However, while the potential is significant, the development of wind energy in Patagonia must consider environmental impacts and work closely with local communities to ensure sustainable and responsible growth of the sector. Particularly if large amounts of the produced energy are exported in the form of hydrogen or e-fuels, consideration of sustainability aspects and cooperation with local communities is inevitable and a key factor for the long-term success of Chile becoming a global player in e-fuel production.

Chile has demonstrated a strong commitment to renewable energy through supportive government policies, regulatory frameworks, and incentives aimed at increasing the share of renewables in its energy mix. This political backing further enhances Chile's attractiveness for investments in renewable energy projects. These factors combined with the unique geological, climate, and weather-related conditions make Chile a highly favorable location for the development and expansion of e-fuels and green chemicals production, aligning with global trends towards sustainable and renewable energy sources.

Last but not least, the extensive biomass potential of Chile could be a further advantage in the production of synthetic fuels. As described in Sect. 2.4, beside sweet water provision for the electrolyzers, the carbon source for the formation of synthesis gas could also become a challenge in the production of e-fuels. Point sources such as cement plants from which CO_2 can be recovered from the flue gases and directly used in chemical synthesis (concept of carbon capture and utilization) are definitely limited in abandoned areas such as the Atacama Desert or Patagonia. Direct air capture (DAC) is a widely discussed possibility to provide CO_2 in any location, however, the technological maturity of DAC is not yet at full commercial scale and DAC requires additional energy (heat and electricity) which will further decrease the efficiency of the overall production process (see Sect. 2.4). A promising concept that is increasingly discussed in the literature is the combination of biomass gasification with green hydrogen from electrolysis to produce a synthesis gas with an optimum stoichiometric ratio between CO/CO_2 and H_2 for subsequent chemical synthesis. While in classical biomass-to-liquids (BtL) processes through gasification and subsequent chemical synthesis, the majority of carbon contained in the biomass will not enter the fuel but will end up as CO_2 in exhaust gases, the addition of H_2 to the synthesis gas can increase the carbon efficiency to up to 100% (Dietrich et al., 2018; Weyand et al., 2023). Hence, power and biomass-to-liquids (PBtL) combines the processes of power-to-liquid and biomass-to-liquid, utilizing both renewable energy (electricity) and biomass as feedstocks to produce liquid fuels. As Chile's extensive agricultural and forestry sectors produce vast amounts of residues that could be utilized for bioenergy production, PBtL processes could also contribute the production of renewable synthetic fuels and green chemicals in Chile. However, PBtL processes currently exist only at the technical demonstration phase. Large-scale industrial implementation will require additional research and will have to consider precise logistics planning (transportation and storage) for the required biomass.

4 Discussion and Conclusions

This contribution shows the general potential of renewable synthetic fuel production in Chile and Germany from a techno-economic perspective. As demonstrated here, the overall production costs mainly depend on the levelized cost of renewable electricity and the availability of power to enable high levels of full-load hours for the electrolyzers. This is of key relevance to compensate for the high CAPEX of the electrolyzers and to achieve sufficient return on invest. In Germany, large-scale production of PtL will only be possible at limited levels as renewable electricity is needed in all sectors (especially households, industries, and transportation) and will neither be available at sufficiently low prices, nor will a high level of full-load hours for the electrolyzers be achievable. This makes the economics of e-fuel production in Germany at a large scale questionable. In contrast, Chile has outstanding potential for renewable energies, particularly regarding solar and wind energy. The region of Patagonia in particular has some of the highest potential for wind energy in the world, offering opportunities for large-scale wind energy projects. One of the world's highest solar irradiance levels is found in the Atacama Desert, making it an ideal location for photovoltaic (PV) and concentrated solar power (CSP) plants. The low levelized cost of electricity and the high availability of renewable electricity makes Chile one of the most promising locations for the production of green hydrogen and synthetic fuels. The high potential for renewable electricity production is supplemented by a high biomass potential from agriculture and the forestry sector, which could serve as an additional carbon source through biomass gasification and PBtL processes (power and biomass-to-liquids).

While the current discussion about chemical synthesis is dominated by the idea to produce synthetic energy carriers (synfuels), one should keep in mind that both FTS waxes and methanol can form the basis for the production of a broad range of green chemicals. Methanol in particular can be transformed into numerous other chemicals and products, making it incredibly versatile for industrial purposes. It is used as a feedstock for the synthesis of formaldehyde, acetic acid, and various other chemicals that serve as intermediates in the production of plastics, paints, explosives, and pharmaceuticals (Bertau, 2014). Methanol can for example be converted into olefins and aromatics through processes like methanol-to-olefins (MTO) and methanol-to-aromatics (MTA). These compounds are crucial for the petrochemical industry, serving as the basis for manufacturing polymers, fibers, and numerous other petrochemical products. Figure 12 summarizes the basic value chain of methanol that can serve as a source of a huge range of chemical products including ethylene and propylene, which are among the most important raw materials for plastics production. Methanol's role as a base chemical in the chemical industry is underscored by its versatility, adaptability to green and renewable production methods, and its critical position in the supply chain of numerous essential chemicals and products. Its importance is likely to grow as the focus on sustainability and renewable energy sources increases.

When regarding the production of synthetic fuels, it is important to point out that e-fuels should not be considered a competitor to electric mobility but rather as a necessary add-on for those areas of the transportation sector that cannot be electrified through batteries and where fuel cells and green hydrogen are also only limited options. The big advantage of e-fuels is that the existing fleet of combustion engines

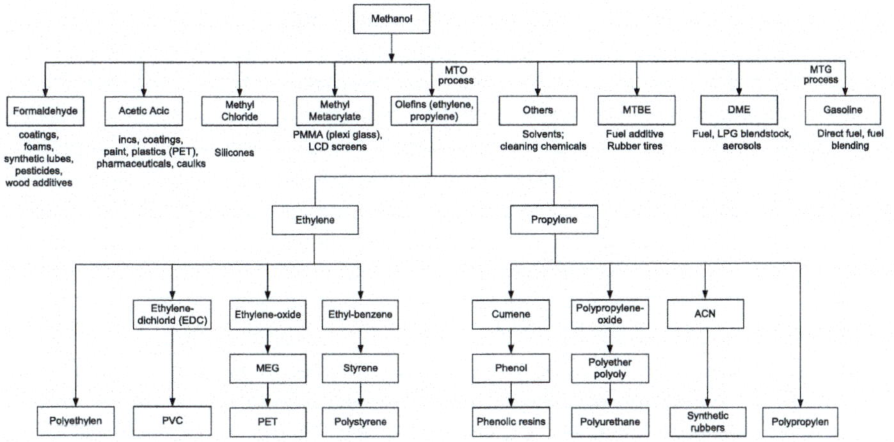

Fig. 12 Methanol value chain in chemical industries. The potential transformation of methanol to olefins (MTO process) in particular opens a wide range of applications for plastics production

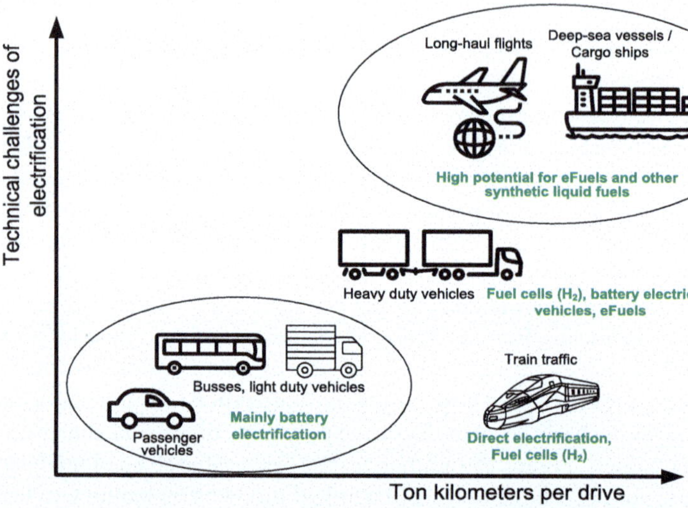

Fig. 13 Main transportation modes and their suitability for electrification. For long-haul aircraft and shipping in particular, there is currently no climate-neutral alternative except for alternative fuels

and respective infrastructure for fossil fuels can be further used without additional investments or modification. As illustrated in Fig. 13, the most relevant fields of application for e-fuels will be aviation, shipping, and long-haul heavy road traffic, while passenger cars and light-duty vehicles show better potential for electrification through batteries.

As a result, both electric mobility and e-fuels are inevitable for a sustainable transition within the transportation sector. This becomes even more obvious when regarding the required resources (especially land use) for the production of e-fuels. Germany, as an exemplary consumer, uses around 2300 PJ (LHV) of liquid fuels in the traffic sector (UBA, 2024). The overall efficiency of synfuel production from green hydrogen will not exceed 50% due to thermodynamic limits. The main losses occur during electrolysis (see Table 1), but the chemical synthesis and CO_2 separation also require additional energy and cause further losses in efficiency. Assuming an average wind turbine power of 5 MW peak and a yearly power generation equivalent to 6,000 full-load hours per year (which is already quite optimistic), a total of 40,000–50,000 wind turbines would have to be installed in Patagonia only to meet the current level of fuel consumption by Germany. This is by far more than all the wind turbines currently installed in Germany, reported to be around 30,000 (BWE, 2024). These figures clearly show the necessity to significantly reduce overall consumption of liquid and gaseous fuels and to only utilize these fuels in areas where electrification is not possible. As demonstrated here, Chile's huge potential for renewable energy production could make the country a key global player in the business of renewable synthetic fuels and green chemicals.

References

Ail, S. S., & Dasappa, S. (2016). Biomass to liquid transportation fuel via Fischer Tropsch synthesis–Technology review and current scenario. *Renewable and Sustainable Energy Reviews, 58*, 267–286. https://doi.org/10.1016/j.rser.2015.12.143

Aresta, M., Karimi, I. A., & Kawi, S. (2019). *An economy based on carbon dioxide and water: Potential of large scale carbon dioxide utilization.* Springer.

Arnold, U., Haltenort, P., Herrera Delgado, K., Niethammer, B., & Sauer, J. (2019). Die Rolle von Dimethylether (DME) als Schlüsselbaustein synthetischer Kraftstoffe aus erneuerbaren Rohstoffen. SpringerIn W. Maus (Ed.), *Zukünftige Kraftstoffe* (Vol. 92, pp. 532–561). Berlin Heidelberg.

Ausfelder, F., Drake, F. D., Erlach, B., Fischedick, M., Henning, H. M., Kost, C. et al. (2017): Sektorkopplung–Untersuchungen und Überlegungen zur Entwicklung Eines Integrierten Energiesystems (Schriftenreihe Energiesysteme der Zukunft). Hg. v. Nationale Akademie der Wissenschaften Leopoldina. acatech – Deutsche Akademie der Technikwissenschaften. Berlin

Bertau, M., Offermanns, H., Plass, L., Schmidt, F., & Wernicke, H.-J. (2014). *Methanol: The basic chemical and energy feedstock of the future: Asinger's vision today.* Springer.

BWE. (2024). *Windenergie in Deutschland - Zahlen und Fakten.* Bundesverband Windenergie.

Bründlinger, T., König, J. E., Frank, O., Gründig, D., Jugel, C., Kraft, P., Krieger, O., Mischinger, S., Prein, P., & Seidl, H. (2018). Dena-leitstudie integrierte energiewende: Impulse für die gestaltung des energiesystems bis 2050. Deutsche Energie-Agentur GmbH (dena), ewi Energy Research & Scenarios, Berlin/Köln

Chen, H., Cheng, M., Liu, L., Li, Y., Li, Z., & Cai, N. (2020). Coal-fired chemical looping combustion coupled with a high-efficiency annular carbon stripper. *International Journal of Greenhouse Gas Control, 93*, 102889. https://doi.org/10.1016/j.ijggc.2019.102889

Cheng, W., & Lee, S. (2022). How green are the national hydrogen strategies? *Sustainability, 14*, 1930. https://doi.org/10.3390/su14031930

Delpisheh, M., Haghghi, M. A., Athari, H., & Mehrpooya, M. (2021). Desalinated water and hydrogen generation from seawater via a desalination unit and a low temperature electrolysis using a novel solar-based setup. *International Journal of Hydrogen Energy, 46*, 7211–7229. https://doi.org/10.1016/j.ijhydene.2020.11.215

Dietrich, R.-U., Albrecht, F. G., Maier, S., König, D. H., Estelmann, S., Adelung, S., Bealu, Z., & Seitz, A. (2018). Cost calculations for three different approaches of biofuel production using biomass, electricity and CO2. *Biomass and Bioenergy, 111*, 165–173. https://doi.org/10.1016/j.biombioe.2017.07.006

Dossow, M., Dieterich, V., Hanel, A., Spliethoff, H., & Fendt, S. (2021). Improving carbon efficiency for an advanced Biomass-to-Liquid process using hydrogen and oxygen from electrolysis. *Renewable and Sustainable Energy Reviews, 152*, 111670. https://doi.org/10.1016/j.rser.2021.111670

Fasihi, M., Efimova, O., & Breyer, C. (2019). Techno-economic assessment of CO2 direct air capture plants. *Journal of Cleaner Production, 224*, 957–980. https://doi.org/10.1016/j.jclepro.2019.03.086

Fetting, C. (2020). The European green deal. ESDN report 53.

Goto, K., Yogo, K., & Higashii, T. (2013). A review of efficiency penalty in a coal-fired power plant with post-combustion CO2 capture. *Applied Energy, 111*, 710–720. https://doi.org/10.1016/j.apenergy.2013.05.020

Heinzmann, P., Glöser-Chahoud, S., Dahmen, N., Langenmayr, U., & Schultmann, F. (2021). Techno-ökonomische Bewertung der Produktion regenerativer synthetischer Kraftstoffe. Working Paper Series in Production and Energy. Karlsruhe.

Hobohm, J., der Maur, A., auf, Dambeck, H., Kemmler, A., Koziel, S., Kreidelmeyer, S., Piégsa, A., Wendring, P., Meyer, B., & Apfelbacher, A. (2018). „Status und Perspektiven flüssiger Energieträger in der Energiewende ". Prognos AG Berlin.

Hofbauer, H., & Rauch, R. (2017). Biomass to Liquid (BtL). In *Encyclopedia of sustainability science and technology*, (pp. 1–18). https://doi.org/10.1007/978-1-4939-2493-6_986-1

Jansen, D., Gazzani, M., Manzolini, G., van Dijk, E., & Carbo, M. (2015). Pre-combustion CO2 capture. *International Journal of Greenhouse Gas Control, 40*, 167–187. https://doi.org/10.1016/j.ijggc.2015.05.028

Kárászová, M., Zach, B., Petrusová, Z., Červenka, V., Bobák, M., Šyc, M., & Izák, P. (2020). Post-combustion carbon capture by membrane separation, review. *Separation and Purification Technology, 238*, 116448. https://doi.org/10.1016/j.seppur.2019.116448

Kasten, P., Mottschall, M., Köppel, W., Degünther, C., Schmied, M., & Wüthrich, P. (2016). *Erarbeitung einer fachlichen Strategie zur Energieversorgung des Verkehrs bis zum Jahr 2050.*

Kramer, U. (2018). *Defossilisierung des Transportsektors: Optionen und Voraussetzungen in Deutschland*. FVV.

Langenmayr, U., & Ruppert, M. (2023a). Renewable origin, additionality, temporal and geographical correlation–eFuels production in Germany under the RED II regime. *Energy Policy, 183*, 113830. https://doi.org/10.1016/j.enpol.2023.113830

Langenmayr, U., & Ruppert, M. (2023). Calculation of synthetic energy carrier production costs with high temporal and geographical resolution. Working Paper Series in Production and Energy. KIT, Karlsruhe.

Pitschak, B., Mergel, J., & Müller, M. (2017). Elektrolyse-Verfahren. In Töpler, J., & Lehmann, J. (eds) *Wasserstoff und Brennstoffzelle: Technologien und Marktperspektiven*, (pp. 207–227, 2nd edn.) Springer Vieweg.

Satymov, R., Bogdanov, D., & Breyer, C. (2022). Global-local analysis of cost-optimal onshore wind turbine configurations considering wind classes and hub heights. *Energy, 256*, 124629. https://doi.org/10.1016/j.energy.2022.124629

Schindler, J. (2019). Gestehungskosten von PtX-Produkten im Vergleich Zwischen Deutschland und Nordafrika. EnergieAgentur. NRW, Wuppertal.

Smolinka, T., Wiebe, N., Sterchele, P., Palzer, A., Lehner, F., Jansen, M., Kiemel, S., Miehe, R., Wahren, S., & Zimmermann, F. (2018). Industrialisierung der Wasserelektrolyse in Deutschland: Chancen und Herausforderungen für nachhaltigen Wasserstoff für Verkehr, Strom und Wärme. NOW GmbH, Berlin.

Stoll, T., Nemati, E., Kulzer, A. C., & Leistner, P. (2023). Greenhouse gas and cost analysis of current state e-fuels. *MTZ Worldw, 84*, 74–79. https://doi.org/10.1007/s38313-023-1522-8

Töpler, J., & Lehmann, J. (Eds.). (2017). *Wasserstoff und Brennstoffzelle: Technologien und Marktperspektiven* (2nd ed.). Springer Vieweg.

Tremel, A. (2018). Electricity-based Fuels. SpringerLink Bücher.

Toedter, O., Heinzmann, P., Glöser-Chahoud, S., Schultmann, F., Langenmayr, U., Ruppert, M., Fichtner, W., Dahmen, N., Sauer, J., et al. (2023). *Ergebnisbericht reFuels–Kraftstoffe neu denken*. KIT.

Verkehrswende, A. (2019). *Die zukünftigen Kosten strombasierter synthetischer Brennstoffe.*

Viebahn, P., Scholz, A., & Zelt, O. (2019). The potential role of direct air capture in the German energy research program—Results of a multi-dimensional analysis. *Energies, 12*, 3443. https://doi.org/10.3390/en12183443

Wachsmuth, J., Kleinschmitt, C., Eckstein, J., Wietschel, M., Glöser-Chahoud, S., Heinzmann, P., Lengenmayr, U., Pflegler, D., Kunze, R., & Meier, J. (2022). Technical assistance to assess the potential of renewable liquid and gaseous transport fuels of non-biological origin (RFNBOs): Final report. Publications Office of the European Union, Luxembourg.

Weyand, J., Habermeyer, F., & Dietrich, R.-U. (2023). Process design analysis of a hybrid power-and-biomass-to-liquid process–An approach combining life cycle and techno-economic assessment. *Fuel, 342*, 127763. https://doi.org/10.1016/j.fuel.2023.127763

Wittich, K., Krämer, M., Bottke, N., & Schunk, S. A. (2020). Catalytic dry reforming of methane: insights from model systems. *ChemCatChem, 12*, 2130–2147. https://doi.org/10.1002/cctc.201902142

Zapf, M. (2017). Stromspeicher und Power-to-Gas im deutschen Energiesystem: Rahmenbedingungen, Bedarf und Einsatzmöglichkeiten. Springer Fachmedien.

Ziegler, M. (2023). Sektorübergreifende Sicht und Bewertung. *MTZ Motortech Z, 84*, 58–59. https://doi.org/10.1007/s35146-023-1484-z

Zimmer, T., Rudi, A., Glöser-Chahoud, S., & Schultmann, F. (2022). Techno-economic analysis of intermediate pyrolysis with solar drying: A Chilean case study. *Energies, 15*, 2272. https://doi.org/10.3390/en15062272

Eco-Industrial Utilization of Subsoil Resources—Introduction

Multi-use of Geothermal Systems: Perspectives for Renewable Baseload Energy and Green Mining

Valentin Goldberg, Fabian Nitschke, Diego Morata, Eva Schill, and Thomas Kohl

Abstract The challenges of climate change necessitate a shift from fossil fuels to renewable energy sources, with a particular need for sustainable base load-capable energies. Geothermal energy emerges as a promising solution, offering ecological and economic benefits with a growing installed capacity globally. In 2021, geothermal electricity capacity reached 16 GWe, and direct use of geothermal heat achieved 108 GWt. In countries like Chile and Germany, hydrothermal systems can provide significant energy outputs and moreover contribute to supplying critical raw materials essential for the energy transition, such as lithium. Geothermal energy production harnesses the Earth's internal heat via deep wells, producing naturally occurring hot fluids. These fluids, enriched with elements like lithium, present opportunities for multi-use approaches producing both energy and raw materials. This dual approach can enhance the sustainability of the mining sector, the economic feasibility of geothermal projects, and the diversification of global value chains. This study highlights the potential of hydrothermal geothermal systems in Chile and Germany to address global energy challenges. By leveraging geothermal energy for both power generation and raw material extraction, these systems can provide a resilient, sustainable solution to the energy and material demands of the future.

V. Goldberg (✉) · F. Nitschke · T. Kohl
Institute of Applied Geosciences, Chair for Geothermal Energy and Reservoir Technology, Karlsruhe Institute of Technology, Adenauerring 20b, 76131 Karlsruhe, Germany
e-mail: valentin.goldberg@kit.edu

V. Goldberg · D. Morata
Department of Geology and Andean Geothermal Center of Excellence (CEGA), Facultad de Ciencias Físicas y Matemáticas, Universidad de Chile, Santiago, Chile

E. Schill
Institute for Nuclear Waste Disposal, Karlsruhe Institute of Technology, Hermann-Von-Helmholtz-Platz 1, 76344 Eggenstein-Leopoldshafen, Germany

Institute of Applied Geosciences, Technical University Darmstadt, Schnittspahnstraße 9, 64287 Darmstadt, Germany

A. Ch. Braun et al. (eds.), *Eco-Industrial Development as an Industrial Strategy*, Green Energy and Technology, https://doi.org/10.1007/978-3-031-73576-9_16

1 Introduction

Climate change and the resulting necessity for transitions in energy and mobility are creating enormous challenges and will continue to do so for upcoming generations. Fossil CO_2-emitting energy carriers must be replaced by renewable alternatives. In particular, base load-capable energy sources such as coal and gas will leave a large gap in the energy landscape, a gap that must be filled with sustainable base load-capable energies. Furthermore, the newly emerging technologies create a large and diverse demand for high-tech materials. The electrification of the mobility sector in particular presents a significant challenge for global supply chains of critical raw materials.

On a global scale, geothermal energy demonstrates potential for ecological, economical provision of base load heat and energy with an installed capacity that is continuously growing. In 2021 the worldwide installed renewable electricity capacity of the geothermal sector was 16 GW_e, a growth of 30% within a five-year period. The installed direct use of geothermal heat demonstrated even higher growth of 50% in the same period, reaching 108 GW_t in 2020 (Huttrer et al., 2020; Lund & Toth, 2021). Nonetheless, despite the significant progress of the past years, only a negligible amount of the global geothermal potential is currently tapped. In countries with strong volcanic activity such as Chile, the sheer amount of subsurface energy is obvious. But even countries without these natural resources, such as Germany, have abundant geothermal potential (Agemar et al., 2014a, 2014b; Aravena et al., 2016; Morata et al., 2020).

Concerning deep geothermal energy, production from hydrothermal systems is the most promising approach in both example countries of Chile and Germany. With this type of use, naturally occurring deep fluids are accessed by deep production wells and are brought to the surface with large flow rates of up to several hundred liters per second for powering electricity-generating turbines or directly supplying heating grids or industrial installations. The cooled water is reinjected into the subsurface after energy generation. Due to the high temperatures and pressures in the reservoirs, water–rock interactions can lead to a complex chemistry of the deep fluids. Very often, these fluids also demonstrate an enrichment of various critical elements needed in the energy transition process. The occurrence of lithium (Li) in geothermal brines in particular and its potential for extraction has been widely discussed of late. Lithium concentrations of up to 200 mg/L, in combination with the high flow rates typical for geothermal production, result in large circulated quantities of this commodity over the course of a year (Goldberg et al., 2021, 2022). Multi-use of hydrothermal systems for energy and raw material production would have several benefits. For established Li-producing countries such as Chile, such an approach could help make the mining sector more sustainable. For countries fully dependent on Li imports, the new approach could support at least partial independence from overseas imports, leading to a more resilient raw material strategy. For this reason, it is of great interest to drive this development forward on a bi-national level to reduce the negative environmental impact globally.

2 Fundamentals of Hydrothermal Geothermal Energy Production

The Earth's interior offers nearly endless energy sources in terms of heat. The mean temperature in the Earth's subsurface is 5,000 °C while it is 14 °C at the surface, a difference that causes a natural flow of heat from the Earth's interior to the surface. The mean temperature increase in the crust is 3 °C per 100 m depth. A minor part of this heat (30%) is coming from the Earth's mantle or core, while the majority is produced directly in the Earth's crust by the decay of naturally occurring radioisotopes such as uranium (^{238}U, ^{235}U), thorium (^{238}U, ^{235}U), and potassium (^{40}K) (Stober & Bucher, 2012).

Water or steam circulating in deep aquifers and structures is a natural heat carrier transporting the Earth's heat to the surface. The use of natural outflowing hydrothermal resources by mankind for bathing, heating, or medical uses dates back at least as far as the Paleolithic era, and is further not reserved for humans only (Fig. 1).

Hydrothermal energy production targets thermal fluids at depths greater than 400 m. A typical set-up includes at least one production well pumping the water to the surface facility for power production or heat extraction. To maintain the water balance, the cooled water is reinjected into the same reservoir formation at a sufficient interval to avoid premature cooling of the water at the production well by a so-called thermal breakthrough (Fig. 3).

Hydrothermal reservoirs can be characterized by their heat source, the structure of their fluid pathways, and their water chemistry, and can be further classified into different play types (Moeck, 2014). In Chile, most geothermal systems are dominated by the Quaternary volcanism of the Andean volcanic arc, including more than 300 geothermal systems and over 200 potentially active volcanoes all along the Chilean Andes (Aravena et al., 2016). The volcanic fields are assigned to the magmatic geothermal plays, receiving their heat from a deep, intrusive heat source (Fig. 2a). The fluids are typically of meteoric origin, seeping into the subsurface and reaching the vicinity of the heat source. Along the site-specific heat source and

Fig. 1 A Japanese macaque's direct use of geothermal heat at the Nagano volcanic springs (Asteiner, CC BY-SA 3.0 https://creativecommons.org/licenses/by-sa/3.0, via Wikimedia Commons)

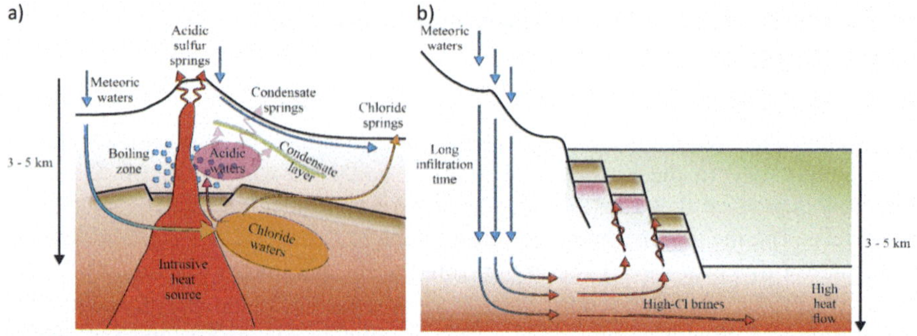

Fig. 2 Schematic recharge and discharge system of a volcanic geothermal setting (**a**), and schematic recharge and discharge system of extensional geothermal setting (**b**). Based on (Moeck, 2014)

geological structure, different hydrothermal reservoirs can form within one volcanic system. Typically, a deep, rather chloride-rich fluid is heated up reaching temperatures of approximately 300 °C and feeding different subsystems. A direct ascent in the periphery of a volcano can form chloride-rich hot springs. Fluids can further rise along the heat source, becoming continuously enriched with volcanic gases and forming acidic sulfur springs at the summit of a volcano. Structurally driven pressure relief can also initiate boiling and degassing, causing a phase separation. Condensation of volcanic gases can form a shallower, acidic condensate layer and condensate springs (Henley & Ellis, 1983; Moeck, 2014).

In Germany, three larger geothermal regions exist: the Upper Rhine Graben (URG), the North German Basin (NGB), and the Bavarian Molasse Basin (BMB). In comparison to the systems in Chile, the regions are not driven by a magmatic heat source but by extensional regimes or orogenesis. The different regions are further detailed and explained in the chapter "Challenges and Opportunities for Lithium Extraction from Geothermal Systems in Germany – Part 2: Potential and Production Scenarios in Germany." (Goldberg et al., 2024a, 2024b) The rift system of the URG has caused subsidence and sedimentary infilling (Fig. 2b). This led to an extensive fault system partially connecting several fluid reservoirs, with occasional contributions of surface waters as well (Sanjuan et al., 2016). For the deep reservoir fluids, a salinity of up to 120 g/L and the corresponding fluid chemistry and isotope values generally indicate a very complex genesis from different fluids as well as several evaporation cycles, mineral precipitation, and dissolution events.

3 Perspectives for Chile and Germany

As of 2021, forty-two deep geothermal power plants were in operation in Germany. Of these, thirty plants have produced direct heat, two have produced only electricity, and ten produced in a cascade process firstly electricity and then downstream heat

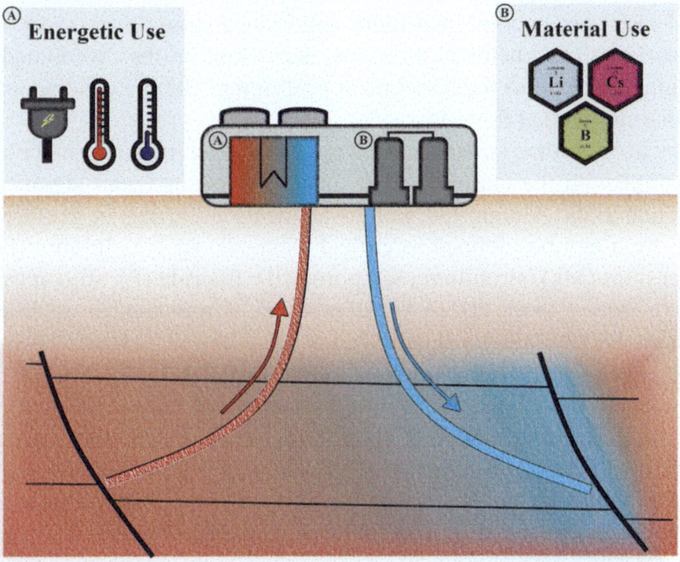

Fig. 3 Multi-use concept for a deep hydrothermal doublet system. An ideal process would aim for a cascade use in energy. Primary electricity production using the initially very high temperatures (>100 °C) is followed by direct heating use. The new multi-use concept would then further extract valuable raw materials before reinjecting the fluids back into the subsurface

(Agemar et al., 2014a, 2014b). In total, 350 MW_t and 47 MW_e have been installed in these facilities. Although there is already a significant geothermal sector in Germany, it is still far behind its potential considering a market potential for heat alone of more than 70 GW_t installed capacity, potentially able to cover 25% of Germany's heat demand (Bracke et al., 2022). In Chile, only one geothermal power plant is currently operational. The first geothermal power plant in South America, Cerro Pabellon is located in northern Chile and began operations in late 2017 with an installed capacity of 48 MW_e (Cappetti et al., 2020; Morata et al., 2020). During 2022, this installed capacity was expanded by an additional 33 MWel, reaching 81 MW_e in total (Morata et al., 2020). However, this extension of use only scratches the surface of the enormous power generation potential of 1300–3800 MW_{el} solely for already explored areas (Ministerio de Energia Gobierno de Chile, 2018), equivalent to 13% of Chile's total installed energy capacity (Garcia, 2021).

In both countries, then, there is significant and proven geothermal potential for baseload energy that would be able to cover large quantities of the demand for energy but remains mostly untapped. One key aspect is local acceptance. Another major hurdle in both countries is the investment costs, which, in combination with the finding risk, presents an economic risk for investors.

Another treasure that could improve local acceptance and reduce the economic risk also lies within geothermal energy which has so far remained almost completely undiscovered or unused. Due to the high temperatures and pressures prevailing in geothermal reservoirs, geothermal fluids can induce strong water–rock interaction in the subsurface. The chemical reactions, depending on the given conditions and reservoir minerals, lead to the enrichment of different chemical elements dissolved in the geothermal fluids (Table 1).

The large fluid volume streams constantly circulating in geothermal power plants in combination with the enrichment of specific elements entail significant potential in raw materials. Among the various elements occurring in geothermal fluids worldwide are a number defined as critical elements, such as lithium (Li), cesium (Cs), magnesium (Mg), strontium (Sr), boron (B), fluoride (F), silicon (Si), or antimony (Sb) (Goldberg et al., 2021). For Li production from geothermal brines, Direct Li Extraction (DLE) approaches are already advanced in part and have overcome small-scale prototype tests (Stringfellow & Dobson, 2021).

A new concept of advanced multi-use hydrothermal systems (Fig. 3) producing electricity, heat, and raw materials could bring several national and international benefits. In mining countries such as Chile, the new approach powered by geothermal energy could help make the local mining sector more sustainable. Due to the water's inherent energy, external energy consumption would be minimized. By using DLE,

Table 1 Examples of the chemical composition of geothermal fluids circulating within geothermal plants in Germany (Bruchsal) and Chile (Cerro Pabellon) (Eggeling et al., 2018; Giudetti et al., 2020)

	Cerro Pabellon (CP_1)	Bruchsal (GB2)
Downhole temperature [°C]	255	134
pH	6.89	5.2
Na [mg/L]	5,182	36,800
K [mg/L]	871	3,220
Ca [mg/L]	543	7,250
Mg [mg/L]	0.17	340
Cl [mg/L]	10,066	72,500
SO_4 [mg/L]	22	303
F [mg/L]	10.9	4
Br [mg/L]	10.4	304
SiO_2 [mg/L]	531	116
B [mg/L]	273	44
HCO_3 [mg/L]	34	363
Li [mg/L]	59	155
Rb [mg/L]	9.5	27.6
Cs [mg/L]	32	<0,01
Sr [mg/L]	13.4	380
Sb [mg/L]	9.5	0.097
TDS [mg/L]	16,769	~130,000
Data Source	(Giudetti et al., 2020)	(Eggeling et al., 2018)

further consumption of large amounts of water could be avoided—a typical topic of conflict in mining. Moreover, the newly tapped resources could open up larger global market shares for Chile's lithium sector. And for a country like Germany, which currently imports most of its lithium, this micro-invasive mining approach opens up the perspective for a renaissance of the local mining sector.

With the studies we have conducted, we want to outline the manifold potential available to address the significant energy challenges of our time. Hydrothermal systems are already deriving base load energy decentrally, renewably, and economically. Additional, multi-use raw material concepts could also make mining sectors more environmentally friendly and provide a new source of high-tech materials. Our aim is to use these two countries, Chile and Germany, to demonstrate the enormous global potential for an ecological, resilient supply of energy and raw materials.

A first description of different extraction technologies is presented in the chapter, "Challenges and Opportunities for Lithium Extraction from Geothermal Systems in Germany–Part 1: A Literature Review of Existing Extraction Technologies," by Goldberg et al. (2024a) outlining different challenges for the implementation of DLE in geothermal cycles. The chapter "Challenges and Opportunities for Lithium Extraction from Geothermal Systems in Germany–Part 2: Potential and Production Scenarios in Germany" by Goldberg et al. (2024b) shows the possible contribution of geothermal Li to the national battery sector. The last chapter of this book part, "Raw Material Potential of Geothermal Systems in Chile," from Peréz et al. (2024) transfers these results to the hydrothermal systems in Chile and demonstrates the significant potential for the Chilean mining sector.

References

Agemar, T., Alten, J., Ganz, B., Kuder, J., Kühne, K., Schuhmacher, S., & Schulz, R. (2014a). The geothermal information system for Germany-GeotIS. *ZDGG, 165*(2), 129–144.

Agemar, T., Weber, J., & Schulz, R. (2014b). Deep geothermal energy production in Germany. *Energies, 7*(7), 4397–4416.

Aravena, D., Muñoz, M., Morata, D., Lahsen, A., Parada, M. Á., & Dobson, P. (2016). Assessment of high enthalpy geothermal resources and promising areas of Chile. *Geothermics, 59*, 1–13. https://doi.org/10.1016/j.geothermics.2015.09.001

Bracke, R., Huenges, E., Acksel, D., Amman, F., Bremer, J., Bruhn, D., Budt, M., Bussman, G., Görke, J.-U., Grün, G., Hahn, F., Hanßke, A., Kohl, T., Kolditz, O., Regenspurg, S., Reinsch, T., Rink, K., Sass, I., Schill, E., & Will, H. (2022). Roadmap for deep geothermal energy for Germany. In *Fraunhofer research institution for energy infrastructure and geothermal energy (IEG)*, (Vol. 40). https://doi.org/10.24406/publica-248

Cappetti, G., Giorgi, N., Arias, A., Volpi, G., Rivera, G., Cei, M., Fedeli, M., Marzio, G. Di, Pasti, M., Massei, S., & Baccarin, F. (2020). The cerro pabellón geothermal project (Chile): From surface exploration to energy production. In *Proceedings World Geothermal Congress 2020* (pp. 1–9), Reykjavik.

Eggeling, L., Herr, K., Goldberg, V., Siefert, D., Köhler, J., Kölbel, T. & Reith, S. (2018). *Schlussbericht zum Verbundprojekt ANEMONA. Anlagenmonitoring als Schlüsseltechnologie für den erfolgreichen Betrieb von Geothermiekraftwerken in Deutschland* (F. und I. EnBW Energie Baden-Württemberg AG (ed.)).

Garcia, N. (2021). *Matriz energética y eléctrica en Chile*. https://obtienearchivo.bcn.cl/obtienearchivo?id=repositorio/10221/32492/1/BCN_Matriz_energetica_electrica_en_Chile.pdf

Giudetti, G., Tempesti, L., Power, E. G., & Pisano, V. A. (2020). First geochemical data from cerro pabellón geothermal project (Apacheta Region, Chile). *Proceedings World Geothermal Congress 2020*, *8*(2010), 1–10.

Goldberg, V., Nitschke, F., & Kluge, T. (2022). Herausforderungen und Chancen für die Lithiumgewinnung aus geothermalen Systemen in Deutschland–Teil 2: Potenziale und Produktionsszenarien in Deutschland. *Grundwasser*. https://doi.org/10.1007/s00767-022-00523-4

Goldberg, V., Kluge, T., & Nitschke, F. (2024a). Challenges and opportunities for lithium extraction from geothermal systems in Germany-Part 1: A literature review of existing extraction technologies. In A. C. Braun, G. Espinosa, D. Tröger, & T. Hirth (Eds.), *Eco-industrial development as an industrial strategy-contributions from a German-Chilean research partnership*. Springer Nature.

Goldberg, V., Nitschke, F., & Kluge, T. (2024b). Challenges and opportunities for lithium extraction from geothermal systems in Germany Part 2: Potential and production scenarios in Germany. In A. C. Braun, G. Espinosa, D. Tröger, & T. Hirth (Eds.), *Eco-industrial development as an industrial strategy-contributions from a German-Chilean research partnership*. Springer Nature.

Goldberg, V., Winter, D., Nitschke, F., Rath, M., Held, S., Spitzmüller, L., Budach, I., Pavez, M., Morata, D., Koschkowski, J., & Kohl, T. (2021). The potential of raw material extraction from thermal brines–Successful milestones of the BrineMine project. *OIL GAS*, *1*, 26–33. https://doi.org/10.19225/210306

Henley, R. W., & Ellis, A. J. (1983). Geothermal systems ancient and modern: A geochemical review. *Earth Science Reviews, 19*(1), 1–50. https://doi.org/10.1016/0012-8252(83)90075-2

Huttrer, G. W., Rica, C., & Salvador, E. (2020). *Geothermal Power Generation in the World 2015–2020 Update Report*. 1–17.

Lund, J. W., & Toth, A. N. (2021). Direct utilization of geothermal energy 2020 worldwide review. *Geothermics, 90*, 101915. https://doi.org/10.1016/j.geothermics.2020.101915

Ministerio de Energia Gobierno de Chile. (2018). Mesa de geotermia. https://energia.gob.cl/sitios-relacionados/mesa-de-geotermia

Moeck, I. S. (2014). Catalog of geothermal play types based on geologic controls. *Renewable and Sustainable Energy Reviews, 37*, 867–882. https://doi.org/10.1016/j.rser.2014.05.032

Morata, D., Aravena, D., Lahsen, A., Muñoz, M., & Valdenegro, P. (2020). Chile Up-Date: The first South American geothermal power plant after one century of exploration. In *Proceedings World Geothermal Congress 2020*.

Pérez, I., Goldberg, V., Nitschke, F., Morata, D., Koschikowski, J., & Kohl, T. (2024). Raw Material Potential of Geothermal Systems in Chile. In A. C. Braun, G. Espinosa, D. Tröger, & T. Hirth (Eds.), *Eco-industrial development as an industrial strategy-contributions from a German-Chilean research partnership*. Springer Nature.

Sanjuan, B., Millot, R., Innocent, C., Dezayes, C., Scheiber, J., & Brach, M. (2016). Major geochemical characteristics of geothermal brines from the Upper Rhine Graben granitic basement with constraints on temperature and circulation. *Chemical Geology, 428*, 27–47. https://doi.org/10.1016/j.chemgeo.2016.02.021

Stober, I., & Bucher, K. (2012). *Geothermie*. Springer Verlag Berlin.

Stringfellow, W. T., & Dobson, P. F. (2021). Technology for the recovery of lithium from geothermal brines. *Energies, 14*(20), 6805. https://doi.org/10.3390/en14206805

Eco-Industrial Utilization of Subsoil Resources—Case Studies from Chile and Germany

Challenges and Opportunities for Lithium Extraction from Geothermal Systems in Germany: Part 1: A Literature Review of Existing Extraction Technologies

Valentin Goldberg, Tobias Kluge, and Fabian Nitschke

Abstract This study assesses the status of currently available extraction technologies for lithium recovery from geothermal waters based on recent scientific studies and identifies potential technical challenges. Available technologies are evaluated in terms of their applicability and integrability in geothermal energy production. Current research projects have validated various extraction methods at laboratory and prototype scale. Scaling up to an industrial process has not yet been performed. Accordingly, information regarding continuous operation as well as site-specific challenges and actual economic viability is lacking. The amount of recoverable lithium depends primarily on the concentration of lithium dissolved in the water, the extraction efficiency and rate, and the amount of extractant used. The interaction of these factors determines the process technology and the size of the extraction infrastructure. Depending on the chosen Li extraction method, the physicochemical properties of the water (pH, Eh, T, p, etc.) are altered during extraction, which can increase scaling and corrosion potential. The current status indicates an early to medium technology readiness level while reaching lithium extraction efficiencies of 50–90% in laboratory experiments. Under the disproportionately higher challenges

The following chapter is a translation of the German journal article "Herausforderungen und Chancen für die Lithiumgewinnung aus geothermalen Systemen in Deutschland Teil 2: Potenziale und Produktionsszenarien in Deutschland" (https://doi.org/10.1007/s00767-022-00523-4). It was originally published under an Open Access license in the peer-reviewed journal Grundwasser (Goldberg et al., 2022a). The translation was conducted by the original authors of the initial study.

V. Goldberg (✉) · F. Nitschke
Institute of Applied Geosciences, Chair for Geothermal Energy and Reservoir Technology, Karlsruhe Institute of Technology, Adenauerring 20b, 76131 Karlsruhe, Germany
e-mail: valentin.goldberg@kit.edu

F. Nitschke
e-mail: fabian.nitschke@kit.edu

T. Kluge
Institute of Applied Geosciences, Chair for Geochemistry and Economic Geology, Karlsruhe Institute of Technology, Adenauerring 20b, 76131 Karlsruhe, Germany
e-mail: tobias.kluge@kit.edu

A. Ch. Braun et al. (eds.), *Eco-Industrial Development as an Industrial Strategy*, Green Energy and Technology, https://doi.org/10.1007/978-3-031-73576-9_17

of ongoing operation in a geothermal plant, extraction efficiencies at the bottom of this range are considered realistic.

Keywords Geothermal fluids · Raw material extraction · Direct lithium extraction · Technology overview · Maturity assessment

Header Technologies for lithium extraction from thermal waters

1 Introduction

The current energy transition from fossil fuels to renewable energy sources requires the availability of a variety of strategic raw materials. One of the key metals defined as a critical raw material by the European Commission is lithium (Europäische_ Kommission, 2020). Until the beginning of the twenty-first century, the main applications for this alkali metal were in the ceramics and glass industries, as a lubricant, or in aluminum alloys. Today, most lithium is used for the production of lithium-ion batteries (Liu et al., 2019; Schmidt, 2017). Due to their higher energy density, low discharge rates, long life cycles, and fast charging compared to nickel-metal hydrid or lead-acid batteries (Kavanagh et al., 2018; Lee et al., 2011; Schmidt, 2017), these batteries first found application in cameras or mobile phones. Today, the main growth market is electric cars. Large electrical storage capacities are needed to compete with the achievable ranges of vehicles with internal combustion engines powered by fossil fuels. This requirement results in correspondingly large batteries (Lee et al., 2011; Srinivasan et al., 2008) with a proportion of pure lithium in an electric car battery of around 10–14 kg (Xu et al., 2020).

Due to high demand and the increase in market price, new lithium sources are being evaluated globally, also to find more environmentally friendly alternatives to conventional mining or extraction from the brines of salt lakes. Currently, lithium extraction from geothermal fluids is being discussed as a new option, which could also make lithium production possible in Germany. This literature review evaluates different technologies proposed for the extraction of lithium from geothermal fluids. Published experimental results are discussed and used to explicate the basic principles of extraction processes and to highlight the technical challenges for integration into a geothermal power plant process. In the second part of this study (Goldberg et al., 2022b), the lithium market is examined in more detail and the potential contribution of geothermal lithium production in Germany is estimated.

1.1 Geothermal Fluids in Germany

To evaluate the applicability of existing lithium extraction technologies to geothermal fluids in Germany, the fluid chemistry of relevant geothermal regions is described below (Table 1).

In Germany, energy from deep geothermal fluids is obtained mainly from three major geological structures: the Upper Rhine Graben (URG), the Bavarian Molasse Basin (BMB), and the North German Basin (NGB). These three regions have low to moderate geothermal reservoir temperatures (120–250 °C) and fall under the category of non-magmatic geothermal systems (Moeck, 2014; Sanjuan et al., 2022).

The Malm aquifer in the Bavarian Molasse Basin, especially around Munich, represents one of the horizons with the strongest expansion of geothermal energy production in Germany. However, the fluids there demonstrate no significant lithium concentrations and generally yield only a very low total dissolved solid load (approx. 2–6 g/L) (Mayrhofer et al., 2014).

The North German Basin is a sedimentary basin on a crystalline basement. Its reservoir fluids, however, have fundamentally different chemistry compared to the Molasse Basin (Moeck, 2014; Sanjuan et al., 2022). For example, at the Großschönebeck geothermal site solution contents >300 g/L are reached. The main ions of the fluids are Ca, Na, and Cl, which account for 70–98% of the total salinity (Regenspurg et al., 2010, 2016). In addition, the fluids contain high amounts of non-condensable gases in a gas–water volume ratio of 1:1–1.6:1 (Regenspurg et al., 2010). The main components of the gas phase are N_2 and CH_4 and in smaller amounts He, CO_2, H_2, Ar, and heavy hydrocarbons (Regenspurg et al., 2010).

Table 1 Fluid chemistry of geothermal fluids from different geothermal regions in Germany

	Unit	Central Bavarian Molasse Basin	Upper Rhine Graben	North German Basin (example: Groß Schönebeck)
pH	–	6.4–7.5	4.96–7.82	7.05
Na^+	mg/L	120–140	827–105,000	33,600
K^+	mg/L	16–32	70.9–4,030	2,560
Ca^{2+}	mg/L	35–410	112–11,700	45,400
Mg^{2+}	mg/L	12–58	3.96–1,930	233
Li^{2+}	mg/L	–	4.5–210	212
SiO_2	mg/L	–	10.7–201	93.7
Cl^-	mg/L	120–900	1,480–167,000	138,384
SO_4^-	mg/L	3.3–16	140–2,730	94.4
HCO_3^-	mg/L	190–330	124–1,027	–
Reference:	–	Mayrhofer et al. (2014)	Sanjuan et al. (2016)	Regenspurg et al. (2016)

The geothermally utilized reservoir fluids in the Upper Rhine Graben exhibit salinities of up to 120 g/L. These Na-Cl fluids have gas–water ratios of up to 1.6:1, making them similar to the fluids in the North German Basin (Eggeling et al., 2018; Sanjuan et al., 2016). A gas fraction of 0.13 wt% was measured at the site of Soultz-sous-Forêts, and about 0.5 wt% at the Bruchsal site (Eggeling et al., 2018; Sanjuan et al., 2016). The main gas component is typically CO_2 with 80–90 vol.%. In addition, N_2 (6–10 vol.%) and CH_4 (2–7 vol.%) were measured in the fluid. At the Landau site, nitrogen is the main component with about 48 vol.% whereas the share of CO_2 is 44 vol.%.

1.2 Geothermal Fluids as a Lithium Resource

Lithium occurs in relatively high concentrations in the geothermal fluids of the URG and NGB, though at levels not comparable to the highest concentrations of the salars in South America (Table 2). The lithium concentration, reservoir size, and flow rate considered economically sufficient for extraction depends on the level of technology readiness level, the extraction process, and the world market price. Seawater contains only 0.18 mg/L of lithium (Stoffyn-Egli & MacKenzie, 1984). Moderate lithium concentrations >1 mg/L are found in Europe in a few select regions. In addition to the URG and NGB, those regions include Cornwall in northern England, the Massif Central in France, the Paris Basin, and the western Apennines (Sanjuan et al., 2022). Little has been published on the concentration of lithium in geothermal fluids from other European regions. As a result, the potential in other European regions cannot be assessed conclusively.

In Germany, the highest Li concentrations and flow rates are reached mainly in reservoirs developed by deep drilling (Table 2). The highest concentrations approach the lower limit of the concentration range of salars in Argentina, where lithium is already being commercially produced. With an effective technological approach, these fluids could be an environmentally sound and economically viable resource in Germany.

1.3 Water-Solids Interactions in Geothermal Power Plants

In a geothermal power plant circuit, the water–gas-solids equilibrium is affected in different ways, which can lead to uncontrolled mineral precipitation. This process is referred to as "scaling," and the mineral precipitates are termed "scales." An important driver of scaling is the temperature reduction required for energy production. Mineral phases such as amorphous SiO_2, $BaSO_4$, $CaSO_4$, NaCl, and KCl show a decrease in solubility in water with decreasing temperatures (Appelo, 2015; Hörbrand et al., 2018). One group of minerals that pose a particular challenge, especially in high-temperature systems, are amorphous silicate phases (Goldberg et al., 2021;

Table 2 Lithium concentrations of different fluids in Germany and South America. Lithium concentrations and extraction conditions of fluids from geothermal, hydrocarbon and hot springs in Germany and salars in Chile and Argentina are compared

Region	Extraction conditions	Li concentrations	References
Upper Rhine Graben Insheim, Landau, Bruchsal (D) Soultz-sous-Forêts, Rittershoffen (F)	Geothermal wells: Temperatures: $\leq$ 165 °C Flow rates: $\leq$ 80 L/s Depth: 2500–5000 m	160–190 mg/L	Dilger et al. (2021), Sanjuan et al. (2016), Uhde (2021)
North German Basin Groß-Schönebeck	Research well: Temperature: 150 °C Flow rate: < 15 L/s Depth: $\leq$ 4240 m	212 mg/L	Dilger et al. (2021), Regenspurg et al. (2016)
North German Basin Altmark	Gas field formation fluid: Depth: $\leq$ 3500 m	50–375 mg/L	Lüders et al. (2010)
Southwest German Molasse Basin	Hydrocarbon wells: Temperature: 50–100 °C Depth: 1400–2500 m	0–150 mg/L	Stober (2014)
Taunus Mountains Bad Nauheim	Hot spring: Temperature: $\leq$ 33 °C Flow rate: $\leq$ 9 L/s Depth: 180 m	$\leq$ 12 mg/L	Kirnbauer, (2008), Loges et al. (2012)
Chile Salar de Atacama	Li extraction projects: Flow rates: 600–1600 L/s	600–5000 mg/L	Garcés and Alvarez (2020)
Argentina Salar de Cauchari, Salar de Olaroz, Salar Hombre Muerto, Salar Rincon	Li extraction projects	340–900 mg/L	Xu et al. (2021)

Gunnarsson & Arnórsson, 2005; Spitzmüller et al., 2021). In a fluid containing around 2.6 mol of dissolved NaCl (about 150 g/L), on the order of salinities in the URG, approximately 200 mg/L Si are soluble at 150 °C, 110 mg/L at 100 °C, and about 40 mg/L at 25 °C (Chen & Marshall, 1982; Marshall & Warakomski, 1980). In the case of barite ($BaSO_4$), solubility in water containing 3 mol of NaCl (TDS of approximately 180 g/L) changes from approx.. 100 mg/L at 150 °C to approx.. 77 mg/L at 95 °C and to approx.. 44 mg at 50 °C (Monnin, 1999; Templeton, 1960). Mineral phases such as $CaCO_3$ or $CaMg[CO_3]_2$, on the other hand, demonstrate retrograde solubility behavior, i.e. improved solubility with decreasing temperature.

The solubility of CO_2 also increases in the presence of decreasing temperatures (Appelo, 2015; Hörbrand et al., 2018). However, CO_2 solubility deteriorates significantly with decreasing system pressure, which is of significance for the (surface) depressurization of water. In water at 150 °C with 2 mol NaCl (about 120 g/L), ca. 42 g CO_2 (4.2 wt%) is soluble at a pressure of 300 bar in a geothermal reservoir. At the same pressure and salinity, the solubility changes to approx. 40 g/L at 60 °C. When this water is depressurized at 150 °C to 100 bar, the solubility drops to approx. 21 g/L and at 10 bar to approx. 1.5 g/L. At 10 bar and a temperature of 60

°C, comparable to conditions in a geothermal power plant after energy production, 5 g/L (0.5 wt.%) of CO_2 is soluble (Duan & Sun, 2003; Duan et al., 2006). The outgassing of CO_2 dissolved in water as carbonic acid leads to an increase in pH and a shift in the speciation of dissolved inorganic carbon (DIC). Pure water can dissolve 0.88 g/L $CaCO_3$ at a CO_2 partial pressure of 100 bar at 100 °C. When the CO_2 partial pressure is reduced to 10 bar at the same temperature, the solubility decreases to 0.51 g/L (Miller, 1952). However, at the same CO_2 partial pressure of 10 bar, 0.99 g/L $CaCO_3$ is soluble again at 60 °C due to the retrograde solubility of CO_2 and $CaCO_3$ (Miller, 1952). Consequently, water–mineral interactions must be considered not only site-specifically, but also section-specifically (with regards to the pumping and reinjection of the fluid and processing during extraction) in a geothermal system.

Equally complex is the dissolution behavior of SiO_2 as a function of pH and dissolved ions in water. In pure water at pH 7, approximately 61 mg/L of Si is soluble as SiO_2 at 25 °C (Marshall & Warakomski, 1980). As pH increases, the solubility of amorphous SiO_2 increases steadily to approx. 260 mg/L at a pH of 10.4 (Marsh et al., 1975). In the presence of Ca and pH > 10, the amount of aquatic Si species is greatly reduced due to the precipitation of calcium silicate hydrate (CSH) phases (Greenberg & Chang, 1965; Spitzmüller et al., 2021). In the presence of Ca, the solubility at 25 °C and pH 10.5 is only approx. 41 mg/L, and at pH 11.4 it is as low as 7.1 mg/L (Greenberg & Chang, 1965). In addition, the change in redox potential and pH can result in the formation of BaSr sulfates and polymetallic sulfides. In this process, ^{226}Ra and ^{228}Ra can be incorporated into sulfates and ^{210}Pb into sulfides (Nitschke et al., 2014; Regenspurg et al., 2015a).

Those processes described are challenging in particular due to the prevailing flow rates in geothermal power plants in the URG of up to 80 L/s. Even relatively low mineral precipitation per liter can cause larger amounts of solids to accumulate over time and lead to damage to the aboveground facility, reduction of the well diameter or even loss of a well, as well as clogging of pathways in the injection area of a reservoir. Therefore, scales present one of the major challenges in geothermal energy production regardless of reservoir type (Eggeling et al., 2018; Gunnarsson & Arnórsson, 2005; Hörbrand et al., 2018; Nitschke et al., 2014; Regenspurg et al., 2015a, 2015b; Scheiber et al., 2019b). Nevertheless, scales do not constitute an uncontrollable constraint on geothermal energy production. With appropriate measures (addition of inhibitors, pressure maintenance), scaling can be controlled, though these interventions alway requires site-specific evaluation and are subject to ongoing research (Eggeling et al., 2018; EVA-M, 2021; Scheiber et al., 2019a). Accordingly, uncontrolled precipitation in geothermal or extraction systems is also regarded as a major challenge for the planned extraction of valuable elements (Goldberg et al., 2021; Spitzmüller et al., 2021).

2 Methods

2.1 Standardized Technology Assessment—Technology Readiness Level

The Technology Readiness Level (TRL) scale was used to quantify the status of different technologies. This scale was developed by NASA in the 1970s to systematically classify the technological maturity of developments and make the status of different technologies comparable (Mankins, 1995, 2009). The TRL scale continues to be used in the European Union's Horizon 2020 program to define funding targets. For this purpose, the European Commission has tested, confirmed, and adapted the applicability of the TRL scale to energy issues and explicitly to geothermal energy. This adapted scale deviates slightly from the original version in some respects (De Rose et al., 2017). The different levels of maturity and their definitions are provided in Table 3.

The (financial) expenditure required to reach the individual stages is by no means linear and moreover technology-specific (Mankins, 1995, 2009; Straub, 2015). In the first three TRL stages, the costs and effort are low to moderate compared to any later system application. (Mankins, 2009; van der Spek et al., 2017). The development and validation of initial system components for TRL 4 represent a significant first leap that can be many times the investment required to reach TRL 3 (Mankins, 2009). Moderate to high costs must be assumed for reaching TRL 5, similar to those for reaching TRL 4 or a multiple thereof (Mankins, 2009; van der Spek et al., 2017). Reaching an initial demonstrator of final size and capability (TRL 6/7) usually represents an escalation of expenditure and constitutes a significant portion of the total development cost. Due to the significantly more complex level of detail, this stage brings a large number of unforeseeable technical issues that require additional adjustments. This level of development usually cannot be achieved by regular research facilities alone (Mankins, 2009; van der Spek et al., 2017). Completing and implementing a first production facility of its kind in the target environment (TRL 8) represents the end of system development with an estimated effort equivalent to five to ten times the cost of all previous stages (Mankins, 2009; van der Spek et al., 2017). TRL 9 describes the successful operation of the new technology. Costs for reaching TRL 9 can still be high, depending on the technology, but are generally much lower than for reaching TRL 8 (Mankins, 2009; van der Spek et al., 2017).

On the basis of the TRL scale, different lithium extraction technologies were evaluated and compared using a standard case (a flow rate of 80 L/s and a lithium concentration of 200 mg/L). These values correspond to a best-case assumption for known geothermal reservoirs in Germany.

Table 3 Definition of technology readiness levels according to Mankins (1995, 2009), adaptated to geothermal systems (De Rose et al., 2017)

TRL	Requirements
TRL 1	**Basic principles observed and documented** (De Rose et al., 2017; Mankins, 1995, 2009) Geothermal-specific: theoretical concepts, applications, and barriers identified (De Rose et al., 2017)
TRL 2	**Formulation of technological concept and/or application** (Mankins, 1995, 2009) Consideration of interfaces and commercial offerings (De Rose et al., 2017) Geothermal-specific: first simulation of the model (De Rose et al., 2017)
TRL 3	**Basic function proven by experimentation or analysis/proof of concept** (De Rose et al., 2017; Mankins, 1995, 2009) Geothermal-specific: simulation concept validated (De Rose et al., 2017)
TRL 4	**Validation of individual components in a laboratory environment** (Mankins, 1995, 2009) Small-scale demo with auxiliary systems tested in laboratory (De Rose et al., 2017) Geothermal-specific: prototype ready for testing phase (De Rose et al., 2017)
TRL 5	**Validation of individual components in a relevant environment** (Mankins, 1995, 2009) Large-scale demo with auxiliary systems completed, concretized economic consideration (De Rose et al., 2017) Geothermal-specific: Prototype tested and additional functions integrated (De Rose et al., 2017)
TRL 6	**Model or prototype tested in a relevant environment** (Mankins, 1995, 2009) Technological pilot plant in relevant environment, manufacturing strategy defined (De Rose et al., 2017) Geothermal-specific: Full-size prototype tested (De Rose et al., 2017)
TRL 7	**System prototype tested in the target environment** (Mankins, 1995, 2009) Demo in operating environment, manufacturing approach demonstrated (De Rose et al., 2017) Geothermal-specific: Full-size prototype installed on-site and connected (De Rose et al., 2017)
TRL 8	**Actual system completed and confirmed by testing and demonstration in the target environment** (Mankins, 1995, 2009) Geothermal-specific: Technology has reached the final stage and technically successful application is possible; economic and financial problems have been solved; marketing measures are underway (De Rose et al., 2017)
TRL 9	**Actual system confirmed through successful operation in the target environment** (Mankins, 1995, 2009) System is fully operational concerning production, profitability, and the market (De Rose et al., 2017)

3 Extraction Technologies

When raw materials are extracted from fluids, the extraction technique should alter the chemical state of the brine in the power plant and reservoir as little as possible while achieving a high extraction efficiency. Any change in fluid composition should neither adversely affect the surface installations nor the downhole area around the injection well. The criteria used in this study to evaluate the extraction technologies were ion selectivity, the ratio of material cost to lithium yield, the chemical properties of the required material (cycle stability), and the required energy input. The goal of our comparison is to identify which technology aspects present particular challenges for use in geothermal power plants. In addition, we provide a derived technology maturity level based exclusively on scientifically peer-reviewed studies. In general, it should be noted that direct comparison and evaluation of extraction technologies are hampered by inconsistent data (Battistel et al., 2020; Liu et al., 2019; Stringfellow & Dobson, 2021).

3.1 Liquid–Liquid Extraction

In the case of liquid–liquid extraction, dissolved substances of a liquid starting medium are transferred into a liquid solvent (hence the term solvent extraction) (Liu et al., 2019). The basic prerequisite is that the starting medium does not mix with the solvent but exchanges the substance to be extracted on contact. Organic solvents can take up relevant amounts of lithium chloride and also show selectivity toward other cations (Liu et al., 2019; Stringfellow & Dobson, 2021). Promising approaches for recovering lithium from high-salinity fluids include the use of tributyl phosphates (TBP) diluted with methyl isobutyl ketone or kerosene (Liu et al., 2019; Nguyen & Lee, 2018; Shi et al., 2018; Xiang et al., 2016, 2017; Yu et al., 2019), the use of crown ethers (Liu et al., 2019; Swain, 2016; Zhang et al., 2021), or the use of ionic liquids (Liu et al., 2019; Park et al., 2014; Shi et al., 2017).

For extraction by TBP, a co-extraction reagent, e.g. $FeCl_3$, is used in addition to the solvents. A complex is formed from the $FeCl_3$ in chloride-rich fluids together with the solvents, which binds the lithium via a cation exchange process (Fig. 1). For this process, the two liquids and the coextraction reagent have to reach chemical equilibrium. Subsequently, the phases are separated again and the lithium is recovered from the solvent complex utilizing highly concentrated acid. Finally, the solvent and coextraction reagent are regenerated using caustic solutions (Liu et al., 2019; Nguyen & Lee, 2018; Shi et al., 2018; Xiang et al., 2016).

The function of crown ethers is based on the "Hard and Soft Acids and Bases" concept (HSAB/Pearson concept) (Pearson, 1963; Swain, 2016). According to this concept, strong acids preferentially bind to strong bases and weak acids to weak bases (Pearson, 1963). In crown ethers, oxygen acts as a donor atom, forming a strong base and showing very good reactivity with dissolved alkali metal cations such as lithium, which in turn represent a strong acid (Liu et al., 2019; Swain, 2016; Zhang et al., 2021). The cations are thus bound in the center of the crown ether structure (Fig. 1; Boulatov et al., 1999; Itoh et al., 1991). After the uptake of the lithium, the organic ether phase can be separated from the water phase, such as via centrifugation, and the lithium can be redissolved from the ethers with acid (Itoh et al., 1991; Zhang et al., 2021). Competing ions such as Na, Ka, Mg, or Ca can be incorporated into

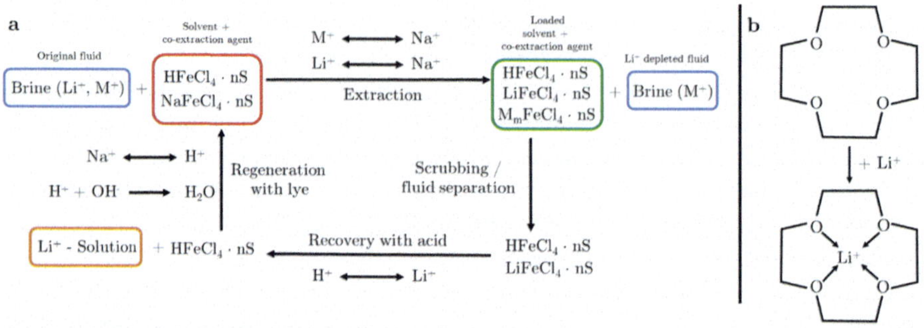

Fig. 1 Illustration of a liquid–liquid extraction cycle, adapted after D. Shi et al. (2018). Shown are the four typical extraction steps using a solvent (nL) and a co-extraction reagent, in our example tributyl phosphates (TBP) and FeC14, respectively. The extraction in the case example is performed from a brine containing lithium and other interfering cations (M+). Right: Principle of incorporation of lithium into a crown ether structure according to Swain (2016)

the ether structure at the same scale as lithium (Itoh et al., 1991; Zhang et al., 2021). Therefore, the recovery of Li from the ethers proceeds in several steps. Na and K can be dissolved from the ethers at a pH of 7–8, since Li is released only at a lower pH. Ca and Mg remain in the organic phase (Itoh et al., 1991).

Ionic liquids are salts with melting temperatures below 100 °C that consist of organic cations and organic or inorganic anions (Park et al., 2014; Shi et al., 2017). For targeted raw material extraction, they can serve directly as extractants or be used in combination with other organic solvents as co-extractants (Liu et al., 2019; Shi et al., 2017; Stringfellow & Dobson, 2021; Swain, 2017). This approach also requires the ionic liquids to first be separated from the lithium-bearing fluid and subsequently treated with acid for lithium extraction (Shi et al., 2017).

During the application of liquid–liquid extraction, up to 90% of the lithium contained was extracted from aqueous solutions in the laboratory (Liu et al., 2019; Yu et al., 2019). In extraction using TBP, an extraction efficiency of 90% was achieved after five extraction passes. For single contact, the rate was approximately 40% (Yu et al., 2019). For the most part, the solvents discussed can be regenerated and reused after lithium extraction. This requires large amounts of acid to mobilize the lithium, as well as mostly caustic soda for regeneration (Liu et al., 2019; Shi et al., 2017; Yu et al., 2019). Accordingly, material consumption must be included in the economic analysis, and the residual acids and caustic solutions need to be considered in terms of their ecological footprint. In addition, small amounts of solvent are lost during contact with the carrier fluid as well as during subsequent reprocessing (Liu et al., 2019; Shi et al., 2017). In addition, the corrosiveness of acids and bases puts high demands on the material of reaction vessels and piping in an extraction plant. Another challenge is the volumetric flow rates for liquid–liquid extraction. Depending on the solvent, geothermal fluid-to-solvent ratios vary from 1:1 to 1:2.5 (Flexer et al., 2018; Garrett, 2004; Shi et al., 2017; Yu et al., 2019; Zhang et al., 2021). This requires a correspondingly high solvent input for the high flow rates in geothermal power plants, but the solvents require sufficient contact time with the source fluid of at least 10–30 min (Shi et al., 2017; Yu et al., 2019). A continuous flow system with flow rates of 80 L/s would require a reaction volume of 48–144 m^3 to achieve appropriate retention times for the geothermal fluid. For extraction, the volume of the solvent must also be provided. Assuming a 1:1 ratio of solvent to geothermal fluid, between 96 m^3 and 288 m^3 would be required for retention times of 10–30 min. Separation of the phases again requires a reaction time of the same order of magnitude but can be shortened by the use of centrifuges (Shi et al., 2018; Swain, 2017; Yu et al., 2019). The dissolved ions in the redissolution water can also pose a challenge as they can negatively affect Li recovery and redissolution (Yu et al., 2019).

Generally, liquid–liquid technologies offer high potential and have achieved very good extraction efficiencies in the laboratory. The series of experiments most closely corresponding to the reality of a geothermal power plant was conducted in the laboratory at room temperature with a geothermal fluid. The fluid was concentrated to about 350 g/L total salinity before the experiments and contained 700 mg/L Li (Yu et al., 2019). As a result, the basic function was experimentally demonstrated and tested in a laboratory setting corresponding to a TRL of 3–4 (De Rose et al., 2017; Mankins, 1995, 2009). However, for technical application at the scale of a geothermal power plant, different parameters such as reaction time, required flow rate, and chemical use should be optimized to enable economic application. Solutions also need to be developed for the maintenance of pressure and temperature required in a geothermal cycle.

3.2 *Selective Extraction by Sorption*

Different approaches and materials are being explored for the use of inorganic sorbents for selective ion separation, such as titanium oxides, manganese oxides, aluminum hydroxides, or zeolites. The current focus of research projects is to optimize the performance of sorbents concerning their loading capacity and reusability (Liu et al., 2019; Stringfellow & Dobson, 2021). The basic principle is based on ion exchange and physical adsorption and absorption. In the processes, the dissolved raw material is bound to a solid that is added to or passed through the fluid. After contact between the sorbent and the raw material to be recovered, the latter must be redissolved from the loaded sorbent. Depending on the application, the solid is first filtered off for this purpose, or the flow cell in which it is located is treated directly.

Manganese oxides with spinel structure demonstrate a very good efficiency for lithium extraction (Liu et al., 2019; Slunitschek et al., 2021; Stringfellow & Dobson, 2021; Weng et al., 2020). The manganese oxides are synthesized with the addition of lithium, which is incorporated into the crystal structure. After synthesis, the lithium is removed from the structure and a gap remains, creating a (lithium) ion sieve (Liu et al., 2019; Weng et al., 2020). Lithium is bound via redox reactions and ion exchange on and in the crystal lattice (Fig. 2; Liu et al., 2019; Weng et al., 2020). During ion exchange, H^+ cations from the crystal lattice are exchanged with Li^+ cations from the aqueous solution. The release of H^+ into the water lowers the pH, so buffering to high pH has a positive effect on the loading capacity and required reaction time (Liu et al., 2019; Weng et al., 2020). In the redox reaction, Mn^{3+} is assumed to disproportionate and decompose into equal parts of Mn^{2+} and Mn^{4+}. This allows for the movement of electrons in the crystal lattice (Fig. 2) and thus incorporation of Li^+. However, it also leads to dissolution of the sorbent and the release of Mn^{2+} into the fluid (Feng et al., 1992; Liu et al., 2019). After contact between the sorbent and the lithium-rich fluid, the lithium can be dissolved from the sorbent via acid treatment (Liu et al., 2019; Ryu et al., 2019; Stringfellow & Dobson, 2021; Weng et al., 2020). The loading capacities for seawater or brine range from 20–60 mg/g for manganese sorbents (Liu et al., 2019; Ryu et al., 2016; Weng et al., 2020; Zandevakili et al., 2014). Capacities >30 mg/g were obtained at pH values between 10 and 13 (Liu et al., 2019). High initial lithium levels promote a more efficient extraction (Ryu et al., 2016). Thus, with the appropriate amounts of sorbent, up to 90% of lithium could be extracted from different fluids in a laboratory setting (Zandevakili et al., 2014). The highest extraction efficiencies (90%) and loading capacities (62 mg/g) were obtained in experiments at 30 °C using 100 mg of sorbent for 100 mL of fluid and a reaction time of 120 h. The brine of a salt lake with approx. 230 g/L salinity and approx. 17 mg/L Li was used for this purpose (Zandevakili et al., 2014).

Titanium oxide-based sorbents are very similar to manganese oxide sorbents regarding their reaction mechanisms, but show higher stability over multiple loading and unloading cycles and are less selective (Liu et al., 2019; Stringfellow & Dobson, 2021). For an optimal pH range of 8–13, their loading capacities are in the range of 20–30 mg/g (Liu et al., 2019).

Lithium extraction with aluminum hydroxides is based on the incorporation of lithium into octahedral gaps of layered aluminum hydroxide minerals (e.g. gibbsite) with the formation of lithium aluminum double hydroxide chloride (LADH-CL)

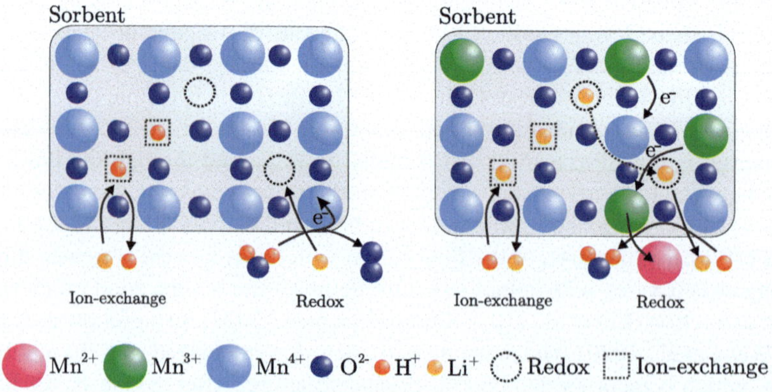

Fig. 2 Principle of lithium extraction with manganese oxides according to Feng et al. (1992). Left: Incorporation of lithium into the structure of manganese oxides. Incorporation takes place via two processes: ion exchange and redox reaction. Right: Redissolution of lithium from the structure by treatment with acid

(Isupov et al., 1999; Paranthaman et al., 2017; Stringfellow & Dobson, 2021; Wu et al., 2019). At the same time, chloride is bound between each layer, which facilitates lithium chloride production during redissolution (Jiang et al., 2020; Wu et al., 2019). Compared to the above sorbents, the loading is much lower, reaching less than 8 mg/g (Isupov et al., 1999; Jiang et al., 2020; Stringfellow & Dobson, 2021). No acid is required for desorption, which minimizes the loss of sorbent. However, this requires large amounts of water at a water:sorbent ratio of 100:1 (Isupov et al., 1999). In a sensitivity analysis (Jiang et al., 2020), an optimal pH of 7 was determined, as well as a loading time of 60 min for 50% loading and 600 min for reaching equilibrium. Considering these parameters, up to 90% of lithium could be recovered from brines in the laboratory (Isupov et al., 1999; Wu et al., 2019).

Another method is the extraction of lithium using natural and artificially synthesized zeolites. These achieve loading capacities of 5 mg/L via incorporation by ion exchange and are still in the early stages of development for use in thermal fluids (Stringfellow & Dobson, 2021; Wísniewska et al., 2018).

Direct lithium extraction using sorbents is among the best-investigated methods. Of those sorbents, manganese and titanium sorbents achieve the highest loading capacities (Liu et al., 2019; Stringfellow & Dobson, 2021). Nevertheless, the material input at loadings of max. 40 mg/g sorbent is not insignificant if fluids with lithium concentrations of up to 200 mg/L are to be treated at flow rates of 80 L/s. To achieve the sometimes long reaction times of 24–120 h (Liu et al., 2019; Ryu et al., 2019; Zandevakili et al., 2014) for full loading, very large reaction vessels are required for continuous volume flow. Even if the extraction is not part of the geothermal energy production unit, it must be possible to integrate appropriate reaction vessels, retention times, etc. into the thermal water loop. Another specific challenge for this type of sorbent is that they are degraded during desorption, which further increases material consumption. The consequences of introducing manganese or titanium into the geothermal reservoir must also be considered. In this context, dissolution rates vary between 1 and 7% during the first use of a freshly synthesized sorbent and are about 1% per cycle thereafter (Ryu et al., 2016; Zandevakili et al., 2014). For the optimal extraction range with high pH values, bases are necessary to buffer the system. These buffers further increase the material input. In basic pH ranges, there is also the risk of carbonate and CSH scales that could reduce the effectiveness of the sorbent. In addition, the incorporation of competing ions, such as Na, into the crystal lattice may also occur (Zandevakili et al., 2014). Material consumption also increases due to the acid required to redissolve the lithium.

LADH-CL require bases (or more rarely acids) to buffer the system if the optimum pH range of 7 is to be achieved. Disadvantages are the low loading of maximum 8 mg/g and the high water requirement. According to Jiang et al. (2020), to achieve an extraction rate of 50%, a reaction time of approximately 60 min is required. The reaction volume must be able to hold the flow rate of at least 60 min, which would correspond to 288 m^3 at 80 L/s. Thus, for fluids with a concentration of 200 mg/L lithium, 57,600 g of lithium would circulate. To extract 50% of this within one hour, a total capacity of 8 mg/g would require about 7.2 t of aluminum hydroxide sorbent and 100 times this amount, i.e. 720 t of water, for redissolution. Even though this water can be used in the next cycle after treatment, the volume is still a key parameter for the dimensions of the plant. The high volume of water also results in a low lithium content in the extracted solution, which requires concentration and water recovery before further processing.

In summary, the different sorbents and their functions are very well investigated in the laboratory setting and achieve efficient results. Experiments with geothermal fluids have demonstrated the high potential of the sorbents. The technological readiness level of manganese and titanium sorbents is assumed to be between 3 and 4 based on the available literature, as individual components and the basic function have already been tested in experiments. According to reports of successful field tests (Rettenmaier et al., 2021; Stringfellow & Dobson, 2021), the technology readiness level of aluminum hydroxides is estimated to be higher. For the usage of titanium or manganese sorbents with real fluids in a geothermal power plant, stability needs to be improved. In the case of aluminum hydroxide, the loading capacity is the focus of development. In general, for sorbents, the kinetics where lithium uptake occurs will need to be improved. Since retention times significantly affect the size of the reaction vessels required, given the high flow rates in geothermal power plants, rapid partial loading of the sorbents is preferable to high-rate (or even complete) but significantly slower loading. Furthermore, long retention times introduce new problems of temperature maintenance and kinetic reactions.

3.3 Electrochemical Methods

The basic principle of the electrochemical extraction of lithium is based on the selective attraction of positively charged lithium cations to a working electrode under the application of a voltage. Anions and interfering elements are bound to a counter electrode (Battistel et al., 2020; Calvo, 2019; Liu et al., 2019). Similarly to sorbents, manganese and titanium oxides can be used as working electrodes (Battistel et al., 2020; Calvo, 2019; Liu et al., 2019; Stringfellow & Dobson, 2021). The major advantage of the electrochemical approach is the rapid binding of lithium and redissolution without the use of chemicals (Battistel et al., 2020). The electrochemical methods show good selectivity even at low initial lithium levels starting at 7 mg/L, but are limited to a maximum concentration of about 350 mg/L lithium (Battistel et al., 2020; Palagonia et al., 2017). The most commonly discussed methods are the electrochemical ion pump and electrodialysis methods. Both techniques use a working electrode that can bind lithium. The maximum loading coefficients are 30–40 mg/g (Battistel et al., 2020; Calvo, 2019; Liu et al., 2019; Stringfellow & Dobson, 2021).

In the case of ion pumps, Li is bound by applying a voltage to the working electrode and chloride is bound to the counter electrode, which is made of nickel or silver, for example (Fig. 3; Battistel et al., 2020; Liu et al., 2019; Romero et al., 2021; Zhao et al., 2019). In the next step, the fluid is flushed from the chamber and replaced with a recovery solution (seawater or brine). In a third step, the bound lithium is released by voltage reversal (Battistel et al., 2020; Zhao et al., 2019).

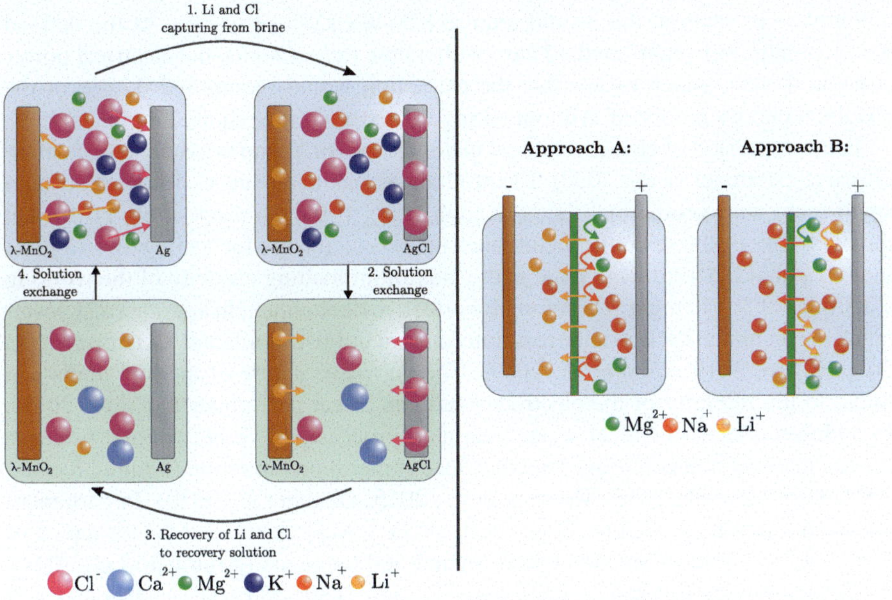

Fig. 3 Left: Schematic representation of an electrochemical ion pump using manganese oxide as a working electrode for lithium. Adapted from Battistel et al. (2020). Right: Schematic representation of two different approaches for lithium enrichment using electrodialysis. Approach A is based on a membrane that is permeable for lithium but largely blocks other ions. Approach B uses a lithium-blocking membrane. Adapted from Liu et al. (2019)

Electrodialysis combines electrochemical and membrane-based methods (Fig. 3) and additionally uses ion-selective membranes (Li et al., 2019a; Liu et al., 2019; Mroczek et al., 2015; Stringfellow & Dobson, 2021). These membranes selectively separate lithium cations and chloride anions when a voltage is applied, thus allowing for enrichment (Mroczek et al., 2015). In a second step, a voltage reversal can be used to deliver the recovered raw material back to a carrier solution (Battistel et al., 2020; Li et al., 2019a; Liu et al., 2019). The technology has been tested for low-salinity geothermal fluids in the laboratory (Mroczek et al., 2015). For higher-salinity fluids, multiple bipolar membranes have been connected in sequence in laboratory tests, allowing for the incremental removal of ions from the fluid phase (Calvo, 2019; Li et al., 2019a).

Since electrochemical extraction methods are based on the same mechanisms as in lithium-ion batteries, the processes are well understood (Liu et al., 2019). Loading times are generally lower (<20 min) than for inorganic sorbentia (Battistel et al., 2020). No chemicals are needed for redissolution or regeneration as applying a corresponding voltage is sufficient. Depending on the method, the required energy varies from 1 to 60 Wh/mol, without considering pump energy (Battistel et al., 2020; Liu et al., 2019). In a geothermal power plant with a flow rate of 80 L/s and 200 mg/L lithium, 8300 mol of lithium (about 0.31 t/h LCE) circulate in one hour. For a complete extraction, this would require 8.3–498 kWh of energy. At the defined geothermal power plant standard case with a flow rate of 80 l/s, the electrical power is about 4 MW, which means that the extraction would require 0.2–12.4% of the electrical energy produced, without taking into account the required pump energy.

In a comparison of electrochemical methods, the ion pump is assumed to be more effective (Battistel et al., 2020). Electrodialysis has been successfully tested on a small scale with geothermal fluids (Kanoh et al., 1993; Mroczek et al., 2015), but the durability of the selective membranes poses a challenge for scaling up to a plant scale (Li et al., 2019a). One issue is the maximum loading capacity of the working electrodes of 30–40 mg/g. Therefore, for effective deployment in a geothermal power plant, the technical design of the reaction cell is of utmost significance (Battistel et al., 2020; Calvo, 2019; Zhao et al., 2019). The corrosive nature of thermal fluids is a threat to the integrity of the electrodes and can affect performance (Calvo, 2019). In addition, electrochemical water treatment processes have been found to cause pH fluctuations that can trigger uncontrolled mineral precipitation (Arulrajan et al., 2021; Dykstra et al., 2017; Mroczek et al., 2015; Obata et al., 2020). In particular, the accumulation of divalent cations such as Ca^{2+}, Mg^{2+}, Fe^{2+}, and the formation of carbonate precipitates can affect performance and longevity (Arulrajan et al., 2021). When membranes are used in electrodialysis, there is an additional risk that they will become clogged by precipitates (Li et al., 2019a; Zhang et al., 2018).

Despite many unresolved issues, electrochemical methods are considered to have high potential for commercial use at a plant scale, mainly due to short loading time and the low need for chemicals (Battistel et al., 2020; Calvo, 2019; Liu et al., 2019; Zhao et al., 2019). Based on initial successful laboratory tests, the technology readiness level is estimated to be between 3 and 4.

3.4 Membrane Technologies

Like electrodialysis, pure membrane technologies are based on the use of lithium-selective membranes. Selective separation can be achieved by ion size, surface charge, or chemical and physical properties (Stringfellow & Dobson, 2021). There is a wide variation in membrane technologies, where the membrane itself can cause separation or be the carrier for solvent or sorbents (Li et al., 2019a).

Nanofiltration, one example of membrane technology, enables selective separation of multivalent from monovalent ions (Li et al., 2019a, 2019b; Sun et al., 2015). Nanofiltration treatment is particularly useful for magnesium-rich fluids, as Mg competes with Li in many methods due to similar ionic radii (Schmidt, 2017; Sun et al., 2021). Separation occurs in a pressure-driven process via pore size as well as membrane charge (Li et al., 2019a, 2019b; Sun et al., 2015). Nanofiltration systems are already available on an industrial scale and are also used for water treatment. The technique has been tested for lithium extraction from complex fluids (Li et al., 2019a, 2019b; Sun et al., 2015). In the context of lithium removal, nanofiltration is always used in combination with a pre- or post-treatment to separate the interfering ions since other monovalent cations such as Na or K also pass through the membrane (Li et al., 2019a, 2019b; Somrani et al., 2013; Sun et al., 2015).

Another approach is to combine membrane distillation and crystallizers. Membrane distillation is a temperature-driven process that facilitates contactless concentration to the crystallization limit via a water-repellent membrane (Li et al., 2019a; Quist-Jensen et al., 2016a, 2016b). The temperature-driven process enables net energy-neutral treatment of water for geothermal systems. In this process, a warm input solution is conveyed past a hydrophobic membrane with a cold permeate stream on the other side, creating a vapor pressure gradient along the membrane. The water can pass through the membrane as vapor and condenses in the permeate stream. The removal of water molecules in the form of the separated vapor results in the relative enrichment of solids in the input fluid stream (Quist-Jensen et al., 2016a; Winter et al., 2017). The permeate water stream could even be used to recover fresh water as a co-product (Li et al., 2019a; Macdonio, 2015). Similarly to nanofiltration, membrane distillation only allows for the enrichment of mineral phases up to the saturation limit. Large-scale plants for water treatment using membrane distillation have already been tested in continuous operation with seawater (34 g/L salinity) (Winter et al., 2017) and at a prototype scale at the Insheim geothermal power plant in the URG (Goldberg et al., 2021). The economic viability of water extraction by membrane distillation has been demonstrated for high-salinity fluids (about 100 g/L) from shale gas extraction (Tavakkoli et al., 2017). A separate process or crystallizer must be used for actual Li extraction (Li et al., 2019a; Quist-Jensen et al., 2016a, 2016b).

In the supported liquid membrane method, membranes are loaded with solvents used for liquid–liquid extraction (Li et al., 2019a; Sharma et al., 2016). These membranes can then be permeated like other membrane systems (Ma et al., 2000). This preparation combines the advantages of liquid–liquid treatment with significantly reduced solvent and space requirements. Under laboratory conditions, this approach was tested for lithium extraction with low-salinity solutions and achieved extraction rates of over 90% in recirculating processes within 120 min at pH values of 9.5 and 12.5, respectively (Ma et al., 2000; Sharma et al., 2016).

Membranes can also be equipped with ion sieves, which facilitates application at plant scale (Li et al., 2019a). This combines the large surface area and selectivity of the sorbents with the advantages of the membrane, which keeps the sorbent stationary and ensures low energy consumption (Li et al., 2019a). Membranes loaded with the sorbents were able to achieve loadings of 30 mg/g. Most of the loading in this approach occurs within the first 60 min of contact time, showing comparatively fast kinetics (Li et al., 2019a; Sun et al., 2016).

Membrane processes can be used for various applications in the recovery of raw materials from thermal fluids. Nanofiltration and membrane distillation are effective methods for concentrating dissolved components in the fluids and can be very well integrated into a geothermal cycle via their pressure- or temperature-driven process. Concentration can also bring low-salinity fluids to lithium concentrations that enable the extraction of raw material. Membrane technologies for direct lithium separation, combining the sorbent or solvent with membranes, are still at a developmental stage.

Selective extraction has been successfully performed in laboratory testing, but not yet expanded to a plant scale (Li et al., 2019a, 2019b; Ma et al., 2000; Sharma et al., 2016; Sun et al., 2016). The technology readiness level ranges from 3 to 4. The combinations have improved the performance of each extraction method in some areas, but challenges such as lithium redissolution with acids or sorbents or solvent degradation remain. In addition, there are new membrane-specific challenges such as clogging of the membranes by mineral precipitates in the pores. One exception is membrane distillation, which shows a lower propensity for clogging. Other challenges include the complexity of membrane production and the high costs associated with it (Li et al., 2019a).

4 Current Extraction Projects

Research is being conducted worldwide on techniques for direct lithium extraction from salar brines or thermal fluids with high lithium concentrations (Table 4). Examples include work by Livent in Catamarca (Argentina) and by several companies in Qinghai (China) (Grant, 2020). Pilot work by Livent (formerly FMC) dates back to the 1990s (Grant, 2020). Commercial production with direct lithium extraction using sorption technology (sorbents not specified) has existed since the 2000s in Catamarca (Argentina) (Grant, 2020). The first attempts to apply this technology to thermal waters were conducted in the Salton Sea region (Imperial Valley, USA). Following initial laboratory studies of lithium extraction by precipitation of lithium aluminate complexes (Bertold & Baker, 1976) and lithium carbonate (Palmer et al., 1975), thermal waters there were taken into consideration as an important resource for lithium (Bertold & Baker, 1976). However, what was actually implemented was the extraction of other elements such as K and Ca, Mn and Fe (1965–1990, various companies) or Zn (CalEnergy around the year 2000) (Stringfellow & Dobson, 2021). SimbolMaterials completed the setup for commercial lithium production from geothermal fluids in the 2010s (Stringfellow & Dobson, 2021). However, due to the volatility of the market, companies there were under severe financial pressure, and most projects had to be abandoned prematurely. Increasing demand and rising market prices for lithium (Schmidt, 2017) have brought in new players who say they expect to start production in the Salton Sea geothermal field beyond the pilot stage as early as 2024 (Energy Source Minerals, 2022). Several projects dealing with direct lithium extraction from thermal fluids are also underway in Europe (Table 4).

5 Discussion and Conclusion

From a technical standpoint, direct extraction of lithium from thermal fluids is not trivial, yet it has been successfully achieved in the laboratory on multiple occasions. The next hurdle is to implement the technologies in field demonstration units and

Table 4 Current research projects on and commercial approaches regarding direct lithium extraction from geothermal fluids. Information about the (industrial) projects is not taken from scientifically peer-reviewed publications due to lack of availability but sourced from company communications that cannot be independently verified and evaluated

	Project	Partner or Company	Funding	Extraction Method	Details	References
In Europe						
URG	Unlimited	EnBW, KIT, Universität Göttingen, Bestec, Hydrosion	BMWK, €2.7 million	Selective sorbents, mainly manganese oxides	Prototype tests According to the project consortium, up to 20,000 car batteries could be produced per year from the lithium theoretically extracted at the Bruchsal geothermal power plant (URG)	Unlimited (2022)

(continued)

Table 4 (continued)

	Project	Partner or Company	Funding	Extraction Method	Details	References
URG	EuGeLi	Eramet Ideas, BASF, BRGM, Chimie Paris Tech, EIfER, Electricité de Strasbourg, IFPEN, VITO, Vrije Universitet Brussel	EIT Raw Materials	Unspecified sorbents	In a pilot project at the Rittershoffen geothermal power plant (URG), sorbent application was adapted to the temperature and fluid chemistry of the URG. Lithium could be extracted at operating pressure and temperature. Recovery through low-salinity eluate	Eramet (2022a, 2022b), Rettenmaier et al. (2021)

(continued)

Table 4 (continued)

	Project	Partner or Company	Funding	Extraction Method	Details	References
Germany	Li + fluids	BGR, Fraunhofer IEG, Fraunhofer UMSICHT	BMWK		Evaluation of the potential of lithium extraction in Germany and assessment of the extraction methods known so far. Lithium release rates from the deep rock will be determined together with a utility analysis of co-production in geothermal operation	Stechern (2022)

(continued)

Table 4 (continued)

	Project	Partner or Company	Funding	Extraction Method	Details	References
URG	Vulcan Energy Resources	Australian-German company	Commercial	Aluminum hydroxide with pretreatment to remove competing ions	A demonstrator is to start direct lithium extraction at the Landau or Insheim (URG) sites. Objective: Pure lithium chloride concentrate. Commercial lithium production is planned for 2024. Vulcan Energy Resources holds several exploration licenses in the URG and has signed several offtake agreements for the lithium to be produced with Umicore, LG, Renault Group, Stellantis, and Volkswagen. Committed to deliver 179,000–225,000 t of LiOH (276,000–347,000 t of Li_2CO_3) between 2025 and 2031	Wedin and Harrison (2021), Wedin (2022), Vulcan Energy Resources (2021a, 2021b, 2021c, 2021d, 2021e, 2021f)

(continued)

Table 4 (continued)

	Project	Partner or Company	Funding	Extraction Method	Details	References
France	Geolith (Sustainable Lithium Solutions)	French company; partners: Mines Paris Tech, ADEME	Commercial	Sorbents fixed to microfibers	Geolith promises economic production from a lithium concentration of 50 mg/L. A mobile test plant has been available since 2021	Geolith (2022)
France	Adionics	French company	Commercial	"Thermal Swing Salt Absorption" (variant of liquid–liquid extraction)	"Thermal Swing Salt Absorption" technique developed and patented. The aim is the direct extraction of LiCl from salars and geothermal fluids	Adionics (2022)
Salton Sea (USA)						
	Energy Source Minerals	US-American company	Commercial	Sorbents	Current projects at John L. Featherstone geothermal power plant, projected capacity of 20,000 t LiOH (31,000 t LCE) per year	Energy Source Minerals (2022)

(continued)

Table 4 (continued)

	Project	Partner or Company	Funding	Extraction Method	Details	References
	Controlled Thermal Resources	US-American company	Commercial	Sorbents, ion exchange	Current projects at Hell's Kitchen geothermal power plant, projected capacity of 20,000 t LiOH (31,000 t LCE) per year	Controlled Thermal Resources (2022)
	Berkshire Hathaway Renewables	US-American company	Commercial, \$6 million California Energy Commission, \$14.9 million DOE	Ion exchange	The plan is to build a demonstrator for LiOH extraction by spring 2022 and start commercial production by 2024	Lithium Valley (2022)

test them in the operational environment. A comparison of the technologies (Table 5) shows that all methods have hurdles to overcome when integrated into a power plant process, but also present individual advantages. The main challenges for achieving technological readiness can be summarized in four main points:

- **Retention time/space requirements**: Retention time is a decisive parameter for integration into a power plant process, as it has a significant influence on the size of the reaction infrastructure. Electrochemical methods and liquid–liquid extraction stand out favorably, as high extraction efficiencies of 90% are achieved in short periods (< 60 min) with these methods. In contrast, the different sorbents can only achieve low extraction rates (e.g. 50%) at similar periods (≈60 min). Contact durations of several hours to days are required to fully load the sorbents. These periods can be shortened, such as by combination with membrane processes. Large space requirements may also be arise from the quantity of extraction or redissolving substances needed. This quantity is significantly influenced by the specific loading capacity Q of the substances used:

$$Q = \frac{(c_0 - c_e) * V_{fluid}}{m_{sub}}$$

  Q = loading capacity [mg/g]
  c_0 = initial lithium concentration [mg/L]
  c_e = lithium concentration after extraction [mg/L]
  V_{fluid} = processed volume of thermal fluid [L]
  m_{sub} = mass of extraction substance used [g]

  Liquid-liquid extraction solvents are required for effective extraction at a ratio of one to two and a half times the thermal fluid flow, which is a substantial space requirement and thus a disadvantage. For lithium aluminum double hydroxide chloride (LADH-CL), a large stationary water volume is required, amounting to several 100-1000 tons including processing equipment.

  In particular, the large reaction volumes are a challenge for power plants where the thermal water loop is kept under pressure to avoid scaling. Depending on the extraction method, an enormous pressure content product can arise due to the pressure level and volume flow. This can result in significant demands on the extraction infrastructure and technical implementation.
- **Material decomposition/requirements**: The decomposition of extraction substances and the chemicals needed during operation are important factors in the economic implementation of a method. Electrochemical methods and extraction using aluminum hydroxide sorbents are the only methods that do not require additional chemicals for extraction in laboratory tests (Battistel et al., 2020; Isupov et al., 1999). All other solvents and sorbents require acids and bases for lithium redissolution and regeneration and for setting an ideal pH range during extraction. In addition, partial dissolution of the extraction substances was also observed in the sorbents.

Table 5 Different technologies for direct lithium extraction from thermal fluids. The table gives an overview of the advantages and disadvantages of the technologies, operating parameters during the extraction process, and an estimation of the technology readiness level (TRL) based on the literature review

Method	Technologies	Advantages	Disadvantages	Extraction parameters	TRL
Liquid–liquid extraction	Tributyl phosphates as solvent Crown ether as solvent Ionic liquids as solvent or for co-extraction	Short loading time Extraction agent recyclable	High consumption of chemicals for redissolution and regeneration Large space requirement for the required fluid flows High costs (for crown ethers) Possible corrosion due to acids and bases used for redissolution, regeneration, and buffering	Extraction efficiency: 40–90% Loading time: 10–50 min Solvent-brine ratio: 1:1–1:2.5	3–4
Inorganic sorbents	Manganese and titanium sorbents (ion screens)	High loading capacities	Sorbents are degraded by use and regeneration (approx. 1% per cycle) pH change can lead to scaling Long loading time Corrosion may occur due to acids and bases used for redissolution, regeneration, and buffering	Extraction efficiency: 40–95% Loading capacity: 20–60 mg/g Ideal pH value during extraction: 10–13 Loading time: 60 min –120 h	3–4

(continued)

Table 5 (continued)

Method	Technologies	Advantages	Disadvantages	Extraction parameters	TRL
	Aluminum hydroxide	Low chemical consumption Good recyclability	Low loading capacity Large stationary volume of water required for redissolution	Extraction efficiency: 50–90% Loading capacity: <8 mg/g Ideal pH value during extraction: 7 Loading time: 60 min–10 h Water requirement for stripping: ~ 100 times the amount of sorbents	5
Electrochemical methods	Ion pump Electrodialysis	Short loading time No chemical consumption Processes related to those in a battery Works efficiently even at low lithium levels	Additional energy demand pH effects of the process can lead to scaling Can only be used up to a lithium content of 350 mg/L	Loading capacity of the working electrode: 30–40 mg/g pH value fluctuations due to method used: ±2 Loading time: <20 min Additional power demand: 1–60 Wh/mol lithium	3–4
Membrane technologies	Nanofiltration/ membrane distillation	Energy-efficient Can be well integrated into geothermal cycles Known for water treatment and used industrially Additional possibility of water extraction	Integrity of the membranes is compromised by scaling Without additional processes, lithium is only relatively enriched and not selectively extracted	–	3–4

(continued)

Table 5 (continued)

Method	Technologies	Advantages	Disadvantages	Extraction parameters	TRL
	Supported Liquid Membrane	–	Very high pH values required for optimum conditions pH value change can lead to scaling Possible corrosion due to the use of acids and bases for redissolution, regeneration, and buffering	Extraction rate: 90% Loading time: 120 min pH value: 9.5–12.5	3–4
	Combination with ion sieves	High loading capacities Faster kinetics than sorbents alone	Sorbent is degraded by lithium recovery pH value change can lead to scaling Long loading time Possible corrosion due to acids and bases used for redissolution, regeneration, and buffering	Loading capacity: 30 mg/g Loading time: <60 min Other parameters analogous to manganese & titanium sorbentia: Extraction efficiency: 40–95% Ideal pH value for extraction: 10–13	

– **Scaling potential**: The scaling potential (risk of uncontrolled mineral precipitation) is significantly influenced by pH value, temperature, and the duration of interaction with the extraction materials. In the technologies described with long retention times (>60 min), kinetic reactions can occur that would not be observed with a short retention time (Setiawan et al., 2019). The risk of precipitation may be increased by potential cooling during the contact time with the sorbent. Adjusting pH to improve extraction conditions alters fluid chemistry and can induce scaling. In electrochemical processes, the application of voltage can cause local pH changes of ±2 that can induce scaling (Arulrajan et al., 2021; Dykstra et al., 2017; Mroczek et al., 2015; Obata et al., 2020). The sorbents as well as the supported-liquid membranes have an ideal pH range that is only reached by buffering. The best efficiencies have been experimentally demonstrated at pH 9–13 for manganese and titanium oxide (Liu et al., 2019) and pH 7 for LADH-CL.

At high pH (>10), different mineral phases, such as CSH phases, supersaturate and can lead to precipitation (Goldberg et al., 2021; Greenberg & Chang, 1965; Spitzmüller et al., 2021).

- **Energy consumption**: Energy consumption influences the economic viability of combined energy and raw material production. With an electricity-generating power plant, the amount of electricity that can be fed into the grid is reduced, but is available in a CO_2-neutral and reliable manner. In the case of a thermal-only power plant, the electricity would have to be additionally purchased. All methods involve additional energy consumption by pumps, stirring vessels, centrifuges, or control technologies. In the case of the electrochemical methods, additional energy is needed for extraction.

Comparison of the different technologies shows that selective extraction of lithium from geothermal fluids is technologically possible and has been validated in the laboratory. However, transfer to a prototype or plant scale is no trivial matter. Extraction efficiencies of 90% were achieved in idealized laboratory tests at room temperature, steady-state operation, and mostly only with long retention times. However, long retention times are very difficult to achieve in a geothermal power plant due to large volume flows. For implementation, lower extraction efficiencies in support of faster kinetics are therefore preferable, or else working on a partial stream, as otherwise a significantly higher material input and space requirement must be planned for. Achieving 50% efficiency requires a 60-min retention time for various methods (Jiang et al., 2020; Yu et al., 2019), which presents a challenge with respect to flow rates of 80 L/s or higher. This is particularly true since the thermal fluids are immediately returned to the subsurface during energy production and thus normally have a short surface retention time (Brasser et al., 2014; Schilling et al., 2022).

Extraction is complicated by the corrosion and scaling potential of the fluids, which is a known issue from geothermal energy production. However, this can be controlled with a targeted treatment strategy (Eggeling et al., 2018; Nitschke et al., 2014; Scheiber et al., 2019a, 2019b). Corrosion and scaling issues increase as a result of fluid concentration processes, cooling during the reaction time, or pH changes to improve extraction efficiency. How this affects lithium extraction at a production scale in detail, however, can only be evaluated in a long-term in-situ plant test (TRL 7–8). Spot prototype tests provide only limited information. Grant (2020), for example, highlights that a simple transfer of findings from one plant to other sites is not possible without limitations due to differences in the thermal fluid composition. The efficiency of extraction in ongoing operation thus remains speculative due to the lack of large-scale demonstration cases. The extraction efficiencies achieved in the laboratory were 50–90% under ideal conditions with very long retention times in some cases. Since the demands on material, technology, and volume capacity of the reactors are disproportionately higher in ongoing plant operation with real fluids and high flow rates, comparably high efficiencies cannot be expected. Therefore, with the current status of technology, extraction efficiencies during ongoing operation are more plausibly at the lower end of the range seen in laboratory experiments, rather than the frequently discussed, more optimistic scenarios.

Acknowledgements The authors would like to thank the Helmholtz Association for research funding under the Geoenergy subtopic in the MTET (Materials and Technologies for the Energy Transition) program of the research field "Energy". The BMBF (Federal Ministry of Education and Research) Client II is thanked for funding within the BrineMine project (grant number 033R190B). Special thanks are also due to the chair of Geothermal Energy and Reservoir Technology, headed by Professor Dr. Thomas Kohl, and the chair of Geochemistry and Economic Geology, headed by Professor Dr. Jochen Kolb, at the Institute of Applied Geosciences at KIT for making this work possible and for the lively scientific exchange. Furthermore, this work grew through many technical discussions, for which we would like to especially thank Dr. Joachim Koschikowski, Dr. Daniel Winter, Gerrit Fuelling, and Klemens Slunitschek. We thank the anonymous reviewers and the editors for their constructive suggestions for improvement.

References

Adionics. (2022). Adionics, https://www.adionics.com/technology/. Access date: 31 January 2022

Appelo, C. A. J. (2015). Principles, caveats and improvements in databases for calculating hydrogeochemical reactions in saline waters from 0 to 200°C and 1 to 1000atm. *Applied Geochemistry, 55*, 62–71. https://doi.org/10.1016/j.apgeochem.2014.11.007

Arulrajan, A. C., Dykstra, J. E., Van Der Wal, A., & Porada, S. (2021). Unravelling pH changes in electrochemical desalination with capacitive deionization. *Environmental Science and Technology, 55*, 14165–14172. https://doi.org/10.1021/acs.est.1c04479

Battistel, A., Palagonia, M. S., Brogioli, D., La Mantia, F., & Trócoli, R. (2020). Electrochemical methods for lithium recovery: A comprehensive and critical review. *Advance Materials, 32*. https://doi.org/10.1002/adma.201905440

Bertold, C.E. & Baker, D.H. (1976) Lithium recovery from geothermal fluids. In Lithium resources and requirements by the year 2000. Geological Survey Professional Paper (Vol. **1005**, pp. 61–66).

Boulatov, R., Du, B., Meyers, E. A., & Shore, S. G. (1999). Two novel lithium-15-crown-5 complexes: An extended LiCl chain stabilized by crown ether and a dimeric complex stabilized by hydrogen bonding with water. *Inorganic Chemistry, 38*, 4554–4558. https://doi.org/10.1021/ic9904790

Brasser, T., Feige, S., Frieling, G., Herbert, H.-J., Heinen, C., Strack, C. & Vieten, C. (2014) Geosys Systemanalyse der geothermalen Energieerzeugung Teil B Ausführliche Ergebnisdokumentation.

Calvo, E. J. (2019). Electrochemical methods for sustainable recovery of lithium from natural brines and battery recycling. *Current Opinion in Electrochemistry, 15*, 102–108. https://doi.org/10.1016/j.coelec.2019.04.010

Chen, C. T. A., & Marshall, W. L. (1982). Amorphous silica solubilities IV. Behavior in pure water and aqueous sodium chloride, sodium sulfate, magnesium chloride, and magnesium sulfate solutions up to 350°C. *Geochimica Et Cosmochimica Acta, 46*, 279–287.

Controlled Thermal Resources. (2022). No Title. Retrieved July 15, 2022, from https://www.cthermal.com/projects

Dilger, G., Deinhardt, A., Reimer, D., Link, K., Zwicklhuber, K., & Götzl, G. (2021) Tiefe Geothermie-Projekte in Deutschland 2021/2022. Geotherm. Energie, Fachzeitschrift für Geotherm. Forsch. und Anwendung Deutschland, Österreich und der Schweiz, 30(100)

Duan, Z., & Sun, R. (2003). An improved model calculating CO 2 solubility in pure water and aqueous NaCl solutions from 273 to 533 K and from 0 to 2000 bar. *Chemical Geology, 193*, 257–271.

Duan, Z., Sun, R., Zhu, C., & Chou, I. M. (2006). An improved model for the calculation of CO2 solubility in aqueous solutions containing Na+, K+, Ca2+, Mg2+, Cl-, and SO42-. *Marine Chemistry, 98*, 131–139. https://doi.org/10.1016/j.marchem.2005.09.001

Dykstra, J. E., Keesman, K. J., Biesheuvel, P. M., & van der Wal, A. (2017). Theory of pH changes in water desalination by capacitive deionization. *Water Research, 119*, 178–186. https://doi.org/10.1016/j.watres.2017.04.039

Eggeling, L., Herr, K., Goldberg, V., Siefert, D., Köhler, J., Kölbel, T., & Reith, S. (2018). Schlussbericht zum Verbundprojekt ANEMONA. Anlagenmonitoring als Schlüsseltechnologie für den erfolgreichen Betrieb von Geothermiekraftwerken in Deutschland.

Energy Source Minerals. (2022). Energy Source Minerals. Retrieved January 31, 2022, from, https://www.esminerals.com/project

Eramet: Eramet (2022a). Retrieved January 31, 2022, from https://www.eramet.com/en/eramet-and-electricite-de-strasbourg-announce-success-first-pilot-test-extract-lithium-geothermal

Eramet: Eramet (2022b). Retrieved January 31, 2022, from https://www.eramet.com/en/activities/innovate-design/eugeli-project

Europäische_Kommission. (2020). Widerstandsfähigkeit der EU bei kritischen Rohstoffen: Einen Pfad hin zu größerer Sicherheit und Nachhaltigkeit abstecken. Mitteilung der Kommission an das Eur. Parlam. den Rat, den Eur. Wirtschafts- und Sozialausschuss und den Ausschuss der Reg. 27.

EVA-M. (2021). Verbundvorhaben EvA-M: Einsatz von Ausfällungsinhibitoren im Molassebecken ' Begleitendes Monitoring und Experimente in unterschiedlichen Skalen, Teilvorhaben A: Inhibitorevaluierung unter Anlagen- und Reservoir-Bedingungen, Anlagenmonitoring und Modellie.

Feng, Q., Miyai, Y., Kanoh, H., & Ooi, K. (1992). Li+Extraction/insertion with spinel-type lithium manganese oxides. Characterization of redox-type and ion-exchange-type sites. *Langmuir, 8*, 1861–1867.

Flexer, V., Baspineiro, C. F., & Galli, C. I. (2018). Lithium recovery from brines: A vital raw material for green energies with a potential environmental impact in its mining and processing. *Science of the Total Environment, 639*, 1188–1204. https://doi.org/10.1016/j.scitotenv.2018.05.223

Garrett, D. E. (2004). *Handbook of lithium and natural calcium chloride* (Vol. 235). Elsevier. https://doi.org/10.1016/B978-012276152-2/50037-2

Garcés, I., & Alvarez, G. (2020). Water mining and extractivism of the Salar de Atacama, Chile. *WIT Transactions on Ecology and the Environment, 245*, 189–199. https://doi.org/10.2495/EID200181

Geolith. (2022). Geolith. Retrieved January 31, 2022, from https://en.geolith.fr/copie-de-technologies.

Goldberg, V., Kluge, T., & Nitschke, F. (2022a). Herausforderungen und Chancen für die Lithiumgewinnung aus geothermalen Systemen in Deutschland Teil 1: Literaturvergleich bestehender Extraktionstechnologien. *Grundwasser*. https://doi.org/10.1007/s00767-022-00522-5

Goldberg, V., Nitschke, F., & Kluge, T. (2022b). Herausforderungen und Chancen für die Lithiumgewinnung aus geothermalen Systemen in Deutschland Teil 2: Potenziale und Produktionsszenarien in Deutschland. Grundwasser – Zeitschrift der Fachsektion Hydrogeologie. https://doi.org/10.1007/s00767-022-00523-4

Goldberg, V., Winter, D., Nitschke, F., Rath, M., Held, S., Spitzmüller, L., Budach, I., Pavez, M., Morata, D., Koschkowski, J., & Kohl, T. (2021). The potential of raw material extraction from thermal brines–Successful milestones of the BrineMine project. *Oil Gas, 1*, 26–33. https://doi.org/10.19225/210306

Grant, A. (2020). From Catamarca to Qinghai: The commercial scale direct lithium extraction operations. Jade Cove Partners (pp. 1–10)

Greenberg, S. A., & Chang, T. N. (1965). Investigation of the colloidal hydrated calcium silicates. II. Solubility relationships in the calcium oxide-silica-water system at 25°. *Journal of Physical Chemistry, 69*, 182–188. https://doi.org/10.1021/j100885a027

Gunnarsson, I., & Arnórsson, S. (2005). Impact of silica scaling on the efficiency of heat extraction from high-temperature geothermal fluids. *Geothermics, 34*, 320–329.

Hörbrand, T., Baumann, T., & Moog, H. C. (2018). Validation of hydrogeochemical databases for problems in deep geothermal energy. *Geothermal Energy, 6*. https://doi.org/10.1186/s40517-018-0106-3

Isupov, V. P., Kotsupalo, N. P., Nemudry, A. P., & Menzeres, L. T. (1999). Aluminium hydroxide as selective sorbent of lithium salts from brines and technical solutions. *Studies in Surface Science and Catalysis, 120 A*, 621–652. https://doi.org/10.1016/s0167-2991(99)80567-9

Itoh, T., Billah, M., Honjo, T., & Terada, K. (1991). Separation and determination of a trace amount of lithium as its thenoyltrifluoroacetone complex with 12-crown-4 by means of synergic extraction and flame photometry. *Analytical Sciences, 7*, 47–50. https://doi.org/10.2116/analsci.7.Supple_47

Jiang, H., Yang, Y., Sun, S., & Yu, J. (2020). Adsorption of lithium ions on lithium-aluminum hydroxides: Equilibrium and kinetics. *Canadian Journal of Chemical Engineering, 98*, 544–555. https://doi.org/10.1002/cjce.23640

Kanoh, H., Ooi, K., Miyai, Y., & Katoh, S. (1993). Electrochemical recovery of lithium ions in the aqueous phase. *Separation Science and Technology, 28*, 643–651. https://doi.org/10.1080/01496399308019512

Kavanagh, L., Keohane, J., Garcia Cabellos, G., Lloyd, A., & Cleary, J. (2018). Global lithium sources-Industrial use and future in the electric vehicle industry: A review. *Resources, 7*, 57. https://doi.org/10.3390/resources7030057

Kirnbauer, T. (2008). Hydrothermale Bildungen des Thermalwassersystems von Bad Nauheim (Wetterau) und dessen Alter. *Jahrbücher Des Nassau. Vereins Für Naturkd., 158*, 39–96.

Lee, J.-S., Tai Kim, S., Cao, R., Choi, N.-S., Liu, M., Lee, K. T., & Cho, J. (2011). Metal-air batteries with high energy density: Li-Air versus Zn-Air. *Advanced Energy Materials, 1*, 34–50. https://doi.org/10.1002/aenm.201000010

Li, X., Mo, Y., Qing, W., Shao, S., Tang, C. Y., & Li, J. (2019a). Membrane-based technologies for lithium recovery from water lithium resources: A review. *Journal of Membrane Science, 591*, 117317. https://doi.org/10.1016/j.memsci.2019.117317

Li, Y., Zhao, Y. J., Wang, H., & Wang, M. (2019b). The application of nanofiltration membrane for recovering lithium from salt lake brine. *Desalination, 468*, 114081. https://doi.org/10.1016/j.desal.2019.114081

Lithium Valley. (2022). Lithium Valley. Retrieved January 31, 2022, from https://www.energy.ca.gov/data-reports/california-power-generation-and-power-sources/geothermal-energy/lithium-valley

Liu, G., Zhao, Z., & Ghahreman, A. (2019). Novel approaches for lithium extraction from salt-lake brines: A review. *Hydrometallurgy, 187*, 81–100. https://doi.org/10.1016/j.hydromet.2019.05.005

Loges, A., Wagner, T., Kirnbauer, T., Göb, S., Bau, M., Berner, Z., & Markl, G. (2012). Source and origin of active and fossil thermal spring systems, northern Upper Rhine Graben, Germany. *Applied Geochemistry, 27*, 1153–1169. https://doi.org/10.1016/j.apgeochem.2012.02.024

Lüders, V., Plessen, B., Romer, R. L., Weise, S. M., Banks, D. A., Hoth, P., Dulski, P., & Schettler, G. (2010). Chemistry and isotopic composition of Rotliegend and Upper Carboniferous formation waters from the North German Basin. *Chemical Geology, 276*, 198–208. https://doi.org/10.1016/j.chemgeo.2010.06.006

Ma, P., Chen, X. D., & Hossain, M. M. (2000). Lithium extraction from a multicomponent mixture using supported liquid membranes. *Separation Science and Technology, 35*, 2513–2533. https://doi.org/10.1081/SS-100102353

Macdonio, F. (2015). Encyclopedia of membranes-Membrane distillation (MD) (pp. 1–9). Springer.

Mankins, J. C. (1995) Technology readiness levels: A white paper. Retrieved January 31, 2002, from http://www.hq.nasa.gov/office/codeq/trl/trl

Mankins, J. C. (2009). Technology readiness assessments: A retrospective. *Acta Astronautica, 65*, 1216–1223. https://doi.org/10.1016/j.actaastro.2009.03.058

Marsh, A. R., Klein, G., & Vermeulen, T. (1975). Polymerisation kinetics and equilibria of silicic acid in aqueous systems. Lawrence Berkeley Lab. Rep.

Marshall, W. L., & Warakomski, J. M. (1980). Amorphous silica solubilities-II. Effect of aqueous salt solutions at 25°C. *Geochemica Cosmochem., 44*, 915–924.

Mayrhofer, C., Niessner, R., & Baumann, T. (2014). Hydrochemistry and hydrogen sulfide generating processes in the Malm aquifer, Bavarian Molasse Basin, Germany. *Hydrogeology Journal, 22*, 151–162. https://doi.org/10.1007/s10040-013-1064-2

Miller, J. P. (1952). A Portion of the system calcium carbonate-carbon dioxide-water, with geological implications. *American Journal of Science, 250*, 161–203.

Moeck, I. S. (2014). Catalog of geothermal play types based on geologic controls. *Renewable and Sustainable Energy Reviews, 37*, 867–882. https://doi.org/10.1016/j.rser.2014.05.032

Monnin, C. (1999). A thermodynamic model for the solubility of barite and celestite in electrolyte solutions and seawater to 200°C and to 1 kbar. *Chemical Geology, 153*, 187–209.

Mroczek, E., Dedual, G., Graham, D., & Bacon, L. (2015). Lithium extraction from wairakei geothermal fluid using electrodialysis. In *Proceeding of World Geothermal Congress 2015* (Vol. 6).

Nguyen, T. H., & Lee, M. S. (2018). A review on the separation of lithium ion from leach liquors of primary and secondary resources by solvent extraction with commercial extractants. *Processes, 6*, 1–15. https://doi.org/10.3390/pr6050055

Nitschke, F., Scheiber, J., Kramar, U., & Neumann, T. (2014). Formation of alternating layered Ba-Sr-sulfate and Pb-sulfide scaling in the geothermal plant of Soultz-sous-Forêts. *Journal Mineral Geochemistry, 191*, 145–156. https://doi.org/10.1127/0077-7757/2014/0253

Obata, K., Van De Krol, R., Schwarze, M., Schomäcker, R., & Abdi, F. F. (2020). In situ observation of pH change during water splitting in neutral pH conditions: Impact of natural convection driven by buoyancy effects. *Energy & Environmental Science, 13*, 5104–5116. https://doi.org/10.1039/d0ee01760d

Palagonia, M. S., Brogioli, D., & Mantia, F. L. (2017). Influence of hydrodynamics on the lithium recovery efficiency in an electrochemical ion pumping separation process. *Journal of the Electrochemical Society, 164*, E586–E595. https://doi.org/10.1149/2.1531714jes

Palmer, T. D., Howard, J. H., & Lande, D. P. (1975) *Geothermal development of the Salton trough*, California and Mexico, UCRL-51775. Livermore

Paranthaman, M. P., Li, L., Luo, J., Hoke, T., Ucar, H., Moyer, B. A., & Harrison, S. (2017). Recovery of lithium from geothermal brine with lithium-aluminum layered double hydroxide chloride sorbents. *Environmental Science and Technology, 51*, 13481–13486. https://doi.org/10.1021/acs.est.7b03464

Park, J., Jung, Y., Kusumah, P., Lee, J., Kwon, K., & Lee, C. K. (2014). Application of ionic liquids in hydrometallurgy. *International Journal of Molecular Sciences, 15*, 15320–15343. https://doi.org/10.3390/ijms150915320

Pearson, R. G. (1963). Hard and soft acids and bases. *Journal of the American Ceramic Society, 85*, 7.

Quist-Jensen, C. A., Ali, A., Mondal, S., Macedonio, F., & Drioli, E. (2016a). A study of membrane distillation and crystallization for lithium recovery from high-concentrated aqueous solutions. *Journal of Membrane Science, 505*, 167–173. https://doi.org/10.1016/j.memsci.2016.01.033

Quist-Jensen, C. A., Macedonio, F., & Drioli, E. (2016b). Membrane crystallization for salts recovery from brine—An experimental and theoretical analysis. *Desalination and Water Treatment, 57*, 7593–7603. https://doi.org/10.1080/19443994.2015.1030110

Regenspurg, S., Feldbusch, E., Byrne, J., Deon, F., Driba, D. L., Henninges, J., Kappler, A., Naumann, R., Reinsch, T., & Schubert, C. (2015a). Mineral precipitation during production of geothermal fluid from a Permian Rotliegend reservoir. *Geothermics, 54*, 122–135. https://doi.org/10.1016/j.geothermics.2015.01.003

Regenspurg, S., Milsch, H., & Schaper, J. (2015b). Copper in geothermal brine: origin, reactions, risks, and chances. *World Geothermal Congress, 2015*, 5.

Regenspurg, S., Feldbusch, E., Norden, B., & Tichomirowa, M. (2016). Fluid-rock interactions in a geothermal Rotliegend/Permo-Carboniferous reservoir (North German Basin). *Applied Geochemistry, 69*, 12–27. https://doi.org/10.1016/j.apgeochem.2016.03.010

Regenspurg, S., Wiersberg, T., Brandt, W., Huenges, E., Saadat, A., Schmidt, K., & Zimmermann, G. (2010). Supplement 3. *Chemie der Erde, 70*, 3–12. https://doi.org/10.1016/j.chemer.2010.05.002

Rettenmaier, D., Millot, R., Blandine, G., Sanjuan, B., Elodie, J., Sophie, L., Clement, B. & Zorn, R. (2021). Rohstoff Lithium aus Geothermie–und aus Europa (EuGeLi - European Geothermal Lithium). In: Der Geothermie Kongress 2021. Essen

Romero, V. C. E., Llano, K., & Calvo, E. J. (2021). Electrochemical extraction of lithium by ion insertion from natural brine using a flow-by reactor: Possibilities and limitations. *Electrochemistry Communications, 125*, 106980. https://doi.org/10.1016/j.elecom.2021.106980

De Rose, A., Buna, M., Strazza, C., Olivieri, N., Stevens, T., Peeters, L., & Tawil-Jarnault, D. (2017). Technology readiness level: Guidance principles for renewable energy technologies. European Union, Luxembourg

Ryu, T., Haldorai, Y., Rengaraj, A., Shin, J., Hong, H. J., Lee, G. W., Han, Y. K., Huh, Y. S., & Chung, K. S. (2016). Recovery of lithium ions from seawater using a continuous flow adsorption column packed with granulated chitosan-lithium manganese oxide. *Industrial and Engineering Chemistry Research, 55*, 7218–7225. https://doi.org/10.1021/acs.iecr.6b01632

Ryu, T., Shin, J., Ghoreishian, S. M., Chung, K. S., & Huh, Y. S. (2019). Recovery of lithium in seawater using a titanium intercalated lithium manganese oxide composite. *Hydrometallurgy, 184*, 22–28. https://doi.org/10.1016/j.hydromet.2018.12.012

Sanjuan, B., Gourcerol, B., Millot, R., Rettenmaier, D., Jeandel, E., & Rombaut, A. (2022). Lithium-rich geothermal brines in Europe: An up-date about geochemical characteristics and implications for potential Li resources. *Geothermics, 101*. https://doi.org/10.1016/j.geothermics.2022.102385

Sanjuan, B., Millot, R., Innocent, C., Dezayes, C., Scheiber, J., & Brach, M. (2016). Major geochemical characteristics of geothermal brines from the Upper Rhine Graben granitic basement with constraints on temperature and circulation. *Chemical Geology, 428*, 27–47. https://doi.org/10.1016/j.chemgeo.2016.02.021

Scheiber, J., Hettkamp, T., Schindler, M., Wahl, G. & Baumgärtner, J. (2019a). Schlussbericht Verbundvorhaben Sulfidinhibition und Deponierbarkeit (SUBITO): Verbesserung der Wirtschaftlichkeit von Geothermieanlagen durch Inhibition natürlich belasteter Ablagerungen.

Scheiber, J., Seibt, A., Jähnichen, S., Degering, D. & Mouchot, J. (2019b). Combined application of inhibitors for scale and corrosion mitigation: Lessons learned. In *European Geothermal Congress*, (pp. 11–14).

Schilling, F., Stober, I., Müller, B. & Kiefer, E. (2022). Fragen und Antworten zur Tiefen Geothermie. Karlsruhe.

Schmidt, M. (2017). *Rohstoffrisikobewertung-Lithium: Rohstoffinformationen 33*.

Setiawan, F. A., Rahayuningsih, E., Petrus, H. T. B. M., Nurpratama, M. I., & Perdana, I. (2019). Kinetics of silica precipitation in geothermal brine with seeds addition: Minimizing silica scaling in a cold re-injection system. *Geothermal Energy, 7*, 22. https://doi.org/10.1186/s40517-019-0138-3

Sharma, A. D., Patil, N. D., Patwardhan, A. W., Moorthy, R. K., & Ghosh, P. K. (2016). Synergistic interplay between D2EHPA and TBP towards the extraction of lithium using hollow fiber supported liquid membrane. *Separation Science and Technology, 51*, 2242–2254. https://doi.org/10.1080/01496395.2016.1202280

Shi, C., Jing, Y., Xiao, J., Wang, X., Yao, Y., & Jia, Y. (2017). Solvent extraction of lithium from aqueous solution using non-fluorinated functionalized ionic liquids as extraction agents. *Separation and Purification Technology, 172*, 473–479. https://doi.org/10.1016/j.seppur.2016.08.034

Shi, D., Zhang, L., Peng, X., Li, L., Song, F., Nie, F., Ji, L., & Zhang, Y. (2018). Extraction of lithium from salt lake brine containing boron using multistage centrifuge extractors. *Desalination, 441*, 44–51. https://doi.org/10.1016/j.desal.2018.04.029

Slunitschek, K., Kolb, J., & Eiche, E. (2021). Lithiumextraktion aus geothermischen Solen: Moderne Adsorptionsmethode zur Rohstoffgewinnung. Wiley Analytical Magazine. Retrieved January 31, 2022, from https://analyticalscience.wiley.com/do/, https://doi.org/10.1002/was.00050231

Somrani, A., Hamzaoui, A. H., & Pontie, M. (2013). Study on lithium separation from salt lake brines by nanofiltration (NF) and low pressure reverse osmosis (LPRO). *Desalination, 317*, 184–192. https://doi.org/10.1016/j.desal.2013.03.009

van der Spek, M., Ramirez, A., & Faaij, A. (2017). Challenges and uncertainties of ex ante techno-economic analysis of low TRL CO2 capture technology: Lessons from a case study of an NGCC with exhaust gas recycle and electric swing adsorption. *Applied Energy, 208*, 920–934. https://doi.org/10.1016/j.apenergy.2017.09.058

Spitzmüller, L., Goldberg, V., Held, S., Grimmer, J. C., Winter, D., Genovese, M., Koschikowski, J., & Kohl, T. (2021). Selective silica removal in geothermal fluids: Implications for applications for geothermal power plant operation and mineral extraction. *Geothermics, 95*, 102141. https://doi.org/10.1016/j.geothermics.2021.102141

Srinivasan, V., Hafemeister, D., Levi, B., Levine, M. & Schwartz, P. (2008) Batteries for vehicular applications. In *AIP Conference Proceedings* (pp. 283–296). AIP

Stechern, A. (2022) BGR koordiniert Projekt: Forschungsverbund untersucht Lithium-Potenziale in Deutschland. Retrieved July 15, 2022, from https://www.bgr.bund.de/DE/Gemeinsames/Oeffentlichkeitsarbeit/Pressemitteilungen/BGR/bgr-2021-10-26_lithium-potenziale-deutschland.html;jsessionid=31A00B00135E288DDE576D5CC4D2F455.1_cid331?nn=1544712

Stober, I. (2014). Hydrochemical properties of deep carbonate aquifers in the SW German Molasse basin. *Geothermal Energy, 2*. https://doi.org/10.1186/s40517-014-0013-1

Stoffyn-Egli, P., & MacKenzie, F. T. (1984). Mass balance of dissolved lithium in the oceans. *Geochemica et Cosmochimica Acta, 48*, 859–872.

Straub, J. (2015). In search of technology readiness level (TRL) 10. *Aerospace Science and Technology, 46*, 312–320. https://doi.org/10.1016/j.ast.2015.07.007

Stringfellow, W. T., & Dobson, P. F. (2021). Technology for the recovery of lithium from geothermal brines. *Energies, 14*, 6805. https://doi.org/10.3390/en14206805

Sun, D., Meng, M., Yin, Y., Zhu, Y., Li, H., & Yan, Y. (2016). Highly selective, regenerated ion-sieve microfiltration porous membrane for targeted separation of Li+. *Journal of Porous Materials, 23*, 1411–1419. https://doi.org/10.1007/s10934-016-0201-4

Sun, S. Y., Cai, L. J., Nie, X. Y., Song, X., & Yu, J. G. (2015). Separation of magnesium and lithium from brine using a Desal nanofiltration membrane. *Journal of Water Process Engineering, 7*, 210–217. https://doi.org/10.1016/j.jwpe.2015.06.012

Sun, Y., Wang, Q., Wang, Y., Yun, R., & Xiang, X. (2021). Recent advances in magnesium/lithium separation and lithium extraction technologies from salt lake brine. *Separation and Purification Technology, 256*, 117807. https://doi.org/10.1016/j.seppur.2020.117807

Swain, B. (2016). Separation and purification of lithium by solvent extraction and supported liquid membrane, analysis of their mechanism: A review. *Journal of Chemical Technology and Biotechnology, 91*, 2549–2562. https://doi.org/10.1002/jctb.4976

Swain, B. (2017). Recovery and recycling of lithium: A review. *Separation and Purification Technology, 172*, 388–403. https://doi.org/10.1016/j.seppur.2016.08.031

Tavakkoli, S., Lokare, O. R., Vidic, R. D., & Khanna, V. (2017). A techno-economic assessment of membrane distillation for treatment of Marcellus shale produced water. *Desalination, 416*, 24–34. https://doi.org/10.1016/j.desal.2017.04.014

Templeton, C. C. (1960). Solubility of barium sulfate in sodium chloride solutions from 25°C to 95°. *Chemical Engineering Science Data, 5*, 514–516.

Uhde, J. (2021). Welcome to the geothermal power plant insheim. In: 9th European Geothermal Workshop

Unlimited: Unlimited. Retrieved January 31, 2022, from https://www.geothermal-lithium.org/

Vulcan Energy Resources. (2021a). Vulcan Energy Resources ASX Announcement. Positive Pre-Feasibility Study, Highlights–Plant Design. Retrieved January 31, 2022, from https://vul.live.irmau.com/site/PDF/6a9c79d2-5c99-40d7-ba32-72cd6ea3a6f6/PositivePreFeasibilityStudy

Vulcan Energy Resources. (2021b). Diversified mix of offtakers sharing our decarbonisation ambitions. Retrieved January 31, 2022, from https://www.linkedin.com/posts/vulcan-energy-resources_negotiating-offtake-agreements-and-collaborating-activity-6874247684241293312-XIfh

Vulcan Energy Resources. (2021c). Binding lithium offtake agreement with Volkswagen Group. Retrieved January 31, 2022, from https://vul.live.irmau.com/site/PDF/4a8714c5-51c8-4cd5-8ae8-2e684c200107/BindingOfftakeAgreementSignedwithVolkswagenGroup

Vulcan Energy Resources. (2021d). Binding lithium offtake agreement with Renault. Retrieved January 31, 2022, from https://vul.live.irmau.com/site/PDF/346fc118-1b04-4c8c-98e0-50cc3534ff14/BindingOfftakeAgreementSignedwithRenault

Vulcan Energy Resources. (2021e). Quarterly Activities Report-March 2021. (2021e). Retrieved January 31, 2022, from https://vul.live.irmau.com/site/PDF/88d6ee23-6598-4127-b206-fac8ae8a74df/QuarterlyActivitiesReport

Vulcan Energy Resources. (2021f). Vulcan Energy signs lithium supply agreement with Stellantis. Retrieved January 31, 2022, from https://vul.live.irmau.com/site/PDF/99900137-8ade-467e-b794-74528c328bb7/VulcansignslithiumsupplyagreementwithStellantis

Wedin, F. (2022). Vulcan energy resources: Corporate presentation Q1 2022. Retrieved January 31, 2022, from https://www.v-er.eu

Wedin, F., & Harrison, S. (2021). Direct Lithium Extraction (DLE)-Technical Update. Vulcan Energy Resources Ltd.

Weng, D., Duan, H., Hou, Y., Huo, J., Chen, L., Zhang, F., & Wang, J. (2020). Introduction of manganese based lithium-ion Sieve-A review. *Progress in Natural Science: Materials International, 30*, 139–152. https://doi.org/10.1016/j.pnsc.2020.01.017

Winter, D., Koschikowski, J., Gross, F., Maucher, D., Düver, D., Jositz, M., Mann, T., & Hagedorn, A. (2017). Comparative analysis of full-scale membrane distillation contactors-methods and modules. *Journal of Membrane Science, 524*, 758–771. https://doi.org/10.1016/j.memsci.2016.11.080

Wísniewska, M., Fijałkowska, G., Ostolska, I., Franus, W., Nosal-Wiercinska, A., Tomasewska, B., Goscianska, J., & Wójcik, G. (2018). Investigations of the possibility of lithium acquisition from geothermal water using natural and synthetic zeolites applying poly(acrylic acid). *Journal of Cleaner Production, 195*, 821–830.

Wu, L., Li, L., Evans, S. F., Eskander, T. A., Moyer, B. A., Hu, Z., Antonick, P. J., Harrison, S., Paranthaman, M. P., Riman, R., & Navrotsky, A. (2019). Lithium aluminum-layered double hydroxide chlorides (LDH): Formation enthalpies and energetics for lithium ion capture. *Journal of the American Ceramic Society, 102*, 2398–2404. https://doi.org/10.1111/jace.16150

Xiang, W., Liang, S., Zhou, Z., Qin, W., & Fei, W. (2016). Extraction of lithium from salt lake brine containing borate anion and high concentration of magnesium. *Hydrometallurgy, 166*, 9–15. https://doi.org/10.1016/j.hydromet.2016.08.005

Xiang, W., Liang, S., Zhou, Z., Qin, W., & Fei, W. (2017). Lithium recovery from salt lake brine by counter-current extraction using tributyl phosphate/FeCl3 in methyl isobutyl ketone. *Hydrometallurgy, 171*, 27–32. https://doi.org/10.1016/j.hydromet.2017.04.007

Xu, C., Dai, Q., Gaines, L., Hu, M., Tukker, A., & Steubing, B. (2020). Future material demand for automotive lithium-based batteries. *Communications Materials, 1*, 10. https://doi.org/10.1038/s43246-020-00095-x

Xu, P., Hong, J., Qian, X., Xu, Z., Xia, H., Tao, X., Xu, Z., & Ni, Q. Q. (2021). Materials for lithium recovery from salt lake brine. *Journal of Materials Science, 56*(1), 16–63. https://doi.org/10.1007/s10853-020-05019-1

Yu, X., Fan, X., Guo, Y., & Deng, T. (2019). Recovery of lithium from underground brine by multistage centrifugal extraction using tri-isobutyl phosphate. *Separation and Purification Technology, 211*, 790–798. https://doi.org/10.1016/j.seppur.2018.10.054

Zandevakili, S., Ranjbar, M., & Ehteshamzadeh, M. (2014). Recovery of lithium from urmia lake by a nanostructure MnO2ion sieve. *Hydrometallurgy, 149*, 148–152. https://doi.org/10.1016/j.hydromet.2014.08.004

Zhang, P., Pabstmann, A., Gray, S., & Duke, M. (2018). Silica fouling during direct contact membrane distillation of coal seam gas brine with high sodium bicarbonate and low hardness. *Desalination, 444*, 107–117. https://doi.org/10.1016/j.desal.2018.07.015

Zhang, Z., Jia, Y., Liu, B., Sun, H., Jing, Y., Zhang, Q., Shao, F., Qi, M., & Yao, Y. (2021). Study on behavior of lithium ion in solvent extraction and isotope separation. *Journal of Molecular Liquids, 324*, 114709. https://doi.org/10.1016/j.molliq.2020.114709

Zhao, X., Yang, H., Wang, Y., & Sha, Z. (2019). Review on the electrochemical extraction of lithium from seawater/brine. *Journal of Electroanalytical Chemistry, 850*, 113389. https://doi.org/10.1016/j.jelechem.2019.113389

Challenges and Opportunities for Lithium Extraction from Geothermal Systems in Germany: Part 2: Potential and Production Scenarios in Germany

Valentin Goldberg, Fabian Nitschke, and Tobias Kluge

Abstract This study provides background information on the potential for lithium production from geothermal fluids in Germany. Growing demand and dependence on poorly diversified overseas sources point toward the high strategic importance of domestic resources within Germany and the EU. Potentially lower CO_2 emissions and reduced land use during lithium production are additional aspects that need to be considered in comparison to traditional mining. Based on the technology comparison for direct lithium extraction from geothermal fluids and the current state of geothermal energy production in Germany and the French part of the Upper Rhine Graben, different scenarios for the extractable amount of lithium carbonate were calculated. In the most optimistic scenario, taking into account all currently active wells, a maximum production of 7200 t/a of lithium carbonate equivalent can be expected. This could cover 5–19% of the annual demand of planned battery cell production in Germany. Key parameters for the process design are the usable volume fraction of the geothermal fluid and extraction efficiency. Considerable uncertainties still remain in assessing the scale and sustainability of cultivating this resource. To exploit the great potential of this technology, these key issues need to be addressed.

The following chapter is a translation of the German journal article "Herausforderungen und Chancen für die Lithiumgewinnung aus geothermalen Systemen in Deutschland Teil 2: Potenziale und Produktionsszenarien in Deutschland" (https://doi.org/10.1007/s00767-022-00523-4). It was originally published under an Open Access license in the peer-reviewed journal Grundwasser (Goldberg et al. 2022b). The translation was conducted by the original authors of the initial study.

V. Goldberg (✉) · F. Nitschke
Institute of Applied Geosciences, Chair for Geothermal Energy and Reservoir Technology, Karlsruhe Institute of Technology, Adenauerring 20b, 76131 Karlsruhe, Germany
e-mail: valentin.goldberg@kit.edu

F. Nitschke
e-mail: fabian.nitschke@kit.edu

T. Kluge
Institute of Applied Geosciences, Chair for Geochemistry & Economic Geology, Karlsruhe Institute of Technology, Adenauerring 20b, 76131 Karlsruhe, Germany
e-mail: tobias.kluge@kit.edu

A. Ch. Braun et al. (eds.), *Eco-Industrial Development as an Industrial Strategy*, Green Energy and Technology, https://doi.org/10.1007/978-3-031-73576-9_18

Keywords Lithium market · Geothermal fluids · Potential study · Production scenarios · Potential in Germany · Resource assessment

Header German potential for geothermal lithium

1 Introduction

The most important field of application for lithium today is the production of lithium-ion batteries (Liu et al., 2019; Schmidt, 2017). The market with the largest growth potential is the use of lithium in electric vehicles, which can be competitive with combustion engine vehicles due to the low discharge rates of lithium-ion batteries, their long life, and their comparably short charging times. To achieve a sufficiently long range, a large storage capacity in the region of 60–100 kWh is needed, resulting in battery sizes of 600–800 kg and volumes of 0.4–0.6 m^3 (Lee et al., 2011; Srinivasan et al., 2008; Xu et al., 2020). The amount of pure lithium required for such a storage capacity amounts to 10–14 kg (Xu et al., 2020).

Due to the development of electromobility, a great increase in lithium demand can be expected. (Adams, 2020; Jones et al., 2021; Martin et al., 2017; Meng et al., 2021; Schmidt, 2017). Alternatives to conventional mining or extraction from the brines of salars in South America are being evaluated around the globe to find new ways to meet this demand. One important focus has been placed on environmentally friendlier alternatives to make the automotive industry more sustainable in at least one part of its value chain. A much-discussed approach is lithium extraction from geothermal fluids. The successful realization of domestic lithium production from geothermal fluids could make Germany less dependent on global markets for raw materials, offer an opportunity to buffer global price fluctuations with regional supply chains, and cushion against global supply shortages. Additionally, the combined material and energetic utilization of a geothermal reservoir could lead to a symbiosis with positive economic effects for the geothermal sector, all while rendering the extraction of raw materials more sustainable. The combination of renewable energy with raw materials production requiring only small amounts of land and having a minimal environmental impact represents a great opportunity for a modern and future-oriented use of resources in Europe and Germany.

There are several technical approaches worth considering for extracting lithium from geothermal waters, which were compared in the first part of this study (Goldberg et al., 2022). Lithium extraction from brines is possible and has been demonstrated in the laboratory and with small-scale prototype systems (Battistel et al., 2020; Liu et al., 2019; Ma et al., 2000; Mroczek et al., 2015; Sharma et al., 2016; Wu et al., 2019; Yu et al., 2019; Zandevakili et al., 2014). However, the integration of this extraction into the process of a geothermal power plant results in several challenges that need to be examined in long-term testing procedures. In particular, the chemistry of the geothermal fluids and the high flow rates, elevated temperatures, and pressures in geothermal cycles result in specific requirements for extraction agents and system components. Furthermore, extraction efficiencies determined in the laboratory need to be continuously tested and adjusted during system operation. It is uncertain whether the same efficiencies can be reached on an industrial scale as under laboratory conditions. In view of the current state of the art, no exact prediction can be made about the costs of extraction, which will depend to a great extent on the exact technology used. Nevertheless, geothermal reservoirs can potentially contribute to cushioning rising lithium prices and ensuring a domestic supply of raw materials. The contribution of lithium production from brines to the regional market at the current state of technology will be further explored in this study.

1.1 The Market and Projections

The development of annual global lithium production (Fig. 1), measured in lithium carbonate equivalent (LCE), indicates a strong upward trend in the last twenty-five years from approx. 40,000 t in 1995 to 420,000 t in 2020. By the end of 2021, a production of 520,000 t is expected. Although 2008 showed a clear decline in production due to the global economic crisis, in 2010 the market had already recovered and reached record highs (Trading Economics, 2022; German Lithium Participation, 2021; Hohmann, 2021). The political decision to transition towards electromobility, as seen in major markets such as Europe and China, has resulted in a stronger increase in growth since 2015 than in prior decades. In recent years, electromobility has expanded to such an extent that even the optimistic projections of the 2017 Electric Vehicle Scenario as well as those made in 2019 were significantly exceeded in 2021 (Martin et al., 2017; Meng et al., 2021). If growth remains unchanged, a doubling of lithium demand is predicted by 2025, reaching 1,000,000 t LCE (Adams, 2020; DERA, 2021a; Jones et al., 2021). A comparison of predicted demand with the maximum global lithium production capacity of 640,000 t LCE in an optimistic scenario (Schmidt, 2017) indicates a risk of a global lithium deficit of 360,000 t LCE in 2025. For 2030, a deficit of up to 1,000,000 t LCE is predicted (Schmidt, 2021).

Global lithium production is mainly concentrated in only three countries and four companies (Fig. 2). The largest lithium producer by far is Australia, where lithium is mined in the form of the mineral spodumene and mostly exported to China (DERA, 2021a; Schmidt, 2017, 2021). Chile and Argentina are the second- and third-largest producers, respectively (DERA, 2021a). Lithium is obtained there via evaporation and precipitation from the brines of salars. This small number of lithium-producing countries is seen as problematic for European imports, especially since there is virtually no European lithium production. While there is only a medium risk associated with the number of countries and the weighted risk due to the safety and stability of the producing countries (DERA, 2021b), it should be noted that geological availability is not synonymous with free availability for global industries. Of the concentrates extracted from material mined in Australia, for instance, 95%

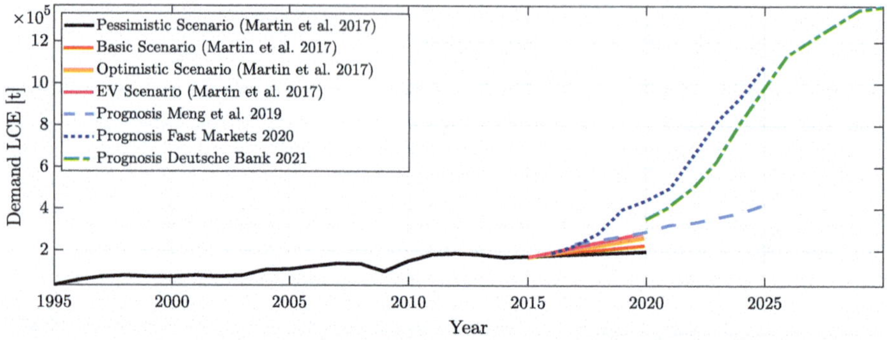

Fig. 1 Global annual lithium production as lithium carbonate equivalents (LCE) since 2015 and predictions of different scenarios until 2030 (Data origin: Adams 2020, Jones et al., 2021, Martin et al., 2017, Meng et al., 2021). In 2020, 420,000 t of LCE were produced, significantly exceeding earlier projections from 2017

will go to China due to long-term supply contracts and is therefore unavailable to Europe or Germany as a raw material (DERA, 2021a; Schmidt, 2021).

The planned expansion of battery cell production beginning in 2022 at nine production sites in Germany is projected to produce a total capacity of 55 GWh in the medium term and reach up to 215 GWh at its completion. This increase in battery production would result in an additional lithium demand of 7,000–28,000 t annually (equaling 37,000–149,000 t LCE/a) (DERA, 2021a). To meet this demand, Germany would need 3–15% of the predicted global demand in LCE in 2025 without an existing domestic production at its disposal.

Changes in supply and demand also have strong effects on the development of lithium prices (Fig. 3). Production costs vary depending on the production type. While the mining costs of lithium ores are rather low compared to the market price for lithium (US$250–400/t), the processing costs to reach battery quality are higher

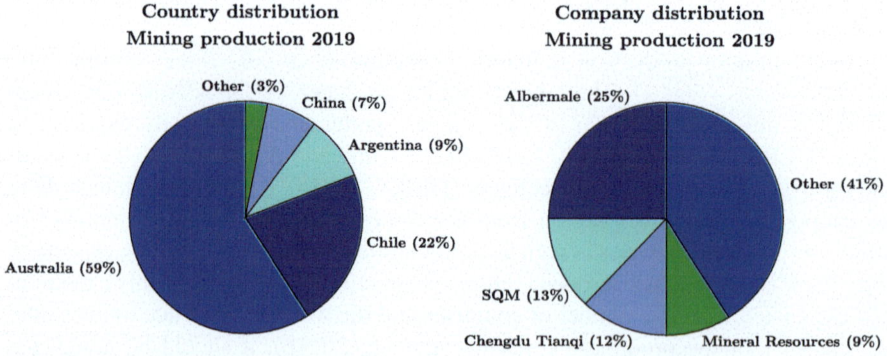

Fig. 2 Country and company concentrations of lithium mining production in 2019 (Data origin: DERA 2021a)

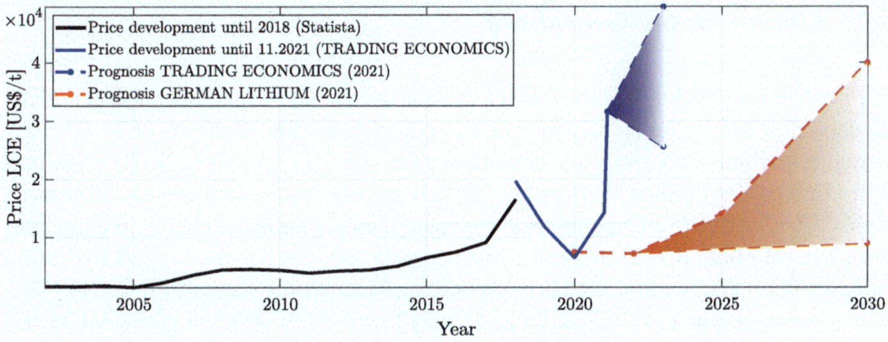

Fig. 3 Price development for LCE from 2000 to 2021 and forecasts up to 2030 (Data origin: Hohmann 2021, Trading Economics 2022, German Lithium Participation 2021

than those of extracting lithium carbonate from brine (Schmidt, 2017). Production costs for one ton of lithium carbonate in battery quality from the world's largest lithium mine (Greenbushes, Australia) are around US$4,500, higher than the costs of producing lithium carbonate from the salt lakes of the Atacama Desert, which are currently US$2,500–3,000 (Schmidt, 2017).

Since 2002, prices for lithium carbonate have increased from US$1,590/t to more than US$30,000/t in 2021 (Fig. 3). The expansion of production in Australia in particular is the result of the strong price increase between 2016 and 2018. It led to a temporary oversupply which, in combination with delayed demand, caused prices to fall in 2020 (DERA, 2021a). However, lithium prices recovered quickly and reached a new record in the fourth quarter of 2021. Different scenarios predict further growth in the lithium market (Adams, 2020; Jones et al., 2021; Martin et al., 2017; Meng et al., 2021; Schmidt, 2017). According to a study from the German Mineral Resources Agency (DERA) (Schmidt, 2017), an annual growth in demand between 7.3% and 12.8% is expected. The conservative model for the minimum case (7.3%) would result in a 23% surplus of supply (109,000 t LCE) in 2025, making a rise in the lithium carbonate price unrealistic. In 2017, a supply deficit of 19% (121,000 t LCE) was estimated for the maximum case of a 12.8% growth in demand, causing a corresponding increase in prices (Schmidt, 2017). With recent predictions indicating an even larger deficit (36% or 360,000 t LCE), continued price increases are very likely. Forecasts differ on price developments (Fig. 3): In 2020, a development between US$8,000 and US$14,000/t LCE was expected for 2025 due to the temporarily low prices (German Lithium Participation, 2021). On the other hand, studies from 2021 predict prices in 2023 to be between US$25,600 and US$49,600/t LCE (Trading Economics, 2022) and between US$9,000 and US$40,000/t LCE in 2030 (German Lithium Participation, 2021), expecting a considerably greater increase when taking the current values into account.

1.2 Lithium in the Geosphere

The percentage of lithium in the Earth's crust amounts to $1.8 \cdot 10^{-3}\%$ by weight (Hans Wedepohl, 1995). Elements with similar percentages are niobium ($1.9 \cdot 10^{-3}\%$), nitrogen, gallium ($1.5 \cdot 10^{-3}\%$), scandium ($1.6 \cdot 10^{-3}\%$), cobalt ($2.4 \cdot 10^{-3}\%$), and lead ($1.5 \cdot 10^{-3}\%$) (Hans Wedepohl, 1995). Like all other alkali metals, lithium is highly reactive due to its free valence electron. It therefore never occurs in elemental form but only bound with minerals. Furthermore, lithium has the highest hydration enthalpy of all alkali metals and is hygroscopic. The melting point of lithium is 180.5 °C, its boiling point 1342 °C, and it has a very low density of 0.534 g/cm^3 (at 25 °C) (Lide, 2005). A look at the electrochemical properties of lithium reveals very positive properties for applications in electrical engineering. Its electrochemical potential of 3.05 V is the highest of all metals. (In comparison, sodium is 2.71 V and zinc 0.76 V.) The ionic radius of lithium is 60 pm (Li^+), allowing it to substitute ions of similar ionic radius in metals like magnesium or aluminum (Kavanagh et al., 2018).

1.3 Potential Hydrothermal Plays for Lithium Extraction in Germany

Sites for deep geothermal energy in Germany are generally located within one of three major geological structures: the active intercontinental rift of the Upper Rhine Graben (URG), the Bavarian Molasse Basin (BMB), and the North German Basin (NGB). The latter two are sedimentary basins on top of a crystalline basement. All three major structures are non-magmatic play systems with low to medium temperatures of 120–250 °C (Moeck, 2014; Sanjuan et al., 2022).

In 2021, a total of forty-two sites were operational in Germany producing thermal water from depths of >400 m (Fig. 4, geothermal production data from GeotIS; Agemar et al., 2014a; Agemar et al., 2014b). Thirty of these sites are used solely for heat generation, three geothermal sites generate electricity, and nine sites are used for both heat and electricity production. In total, they provide approximately 350 MW of heating capacity and 47 MW of electrical capacity. The lithium concentration in deep waters in Germany ranges widely, from <2 mg/L up to 400 mg/L (Lüders et al., 2010; Sanjuan et al., 2022; Stober, 2014). For the higher concentrations within this range, extraction at a reservoir of sufficient size appears profitable. It is important to stress that water composition can vary widely according to the specific geological conditions of the site and even in different parts of the same geological structure or at a short geographic distances. The concept of lithium extraction from deep geothermal fluids proposes accessing the resource via deep drilling analogously to geothermal production of heat and electricity. Current efforts focus on the shared use of existing thermal water production.

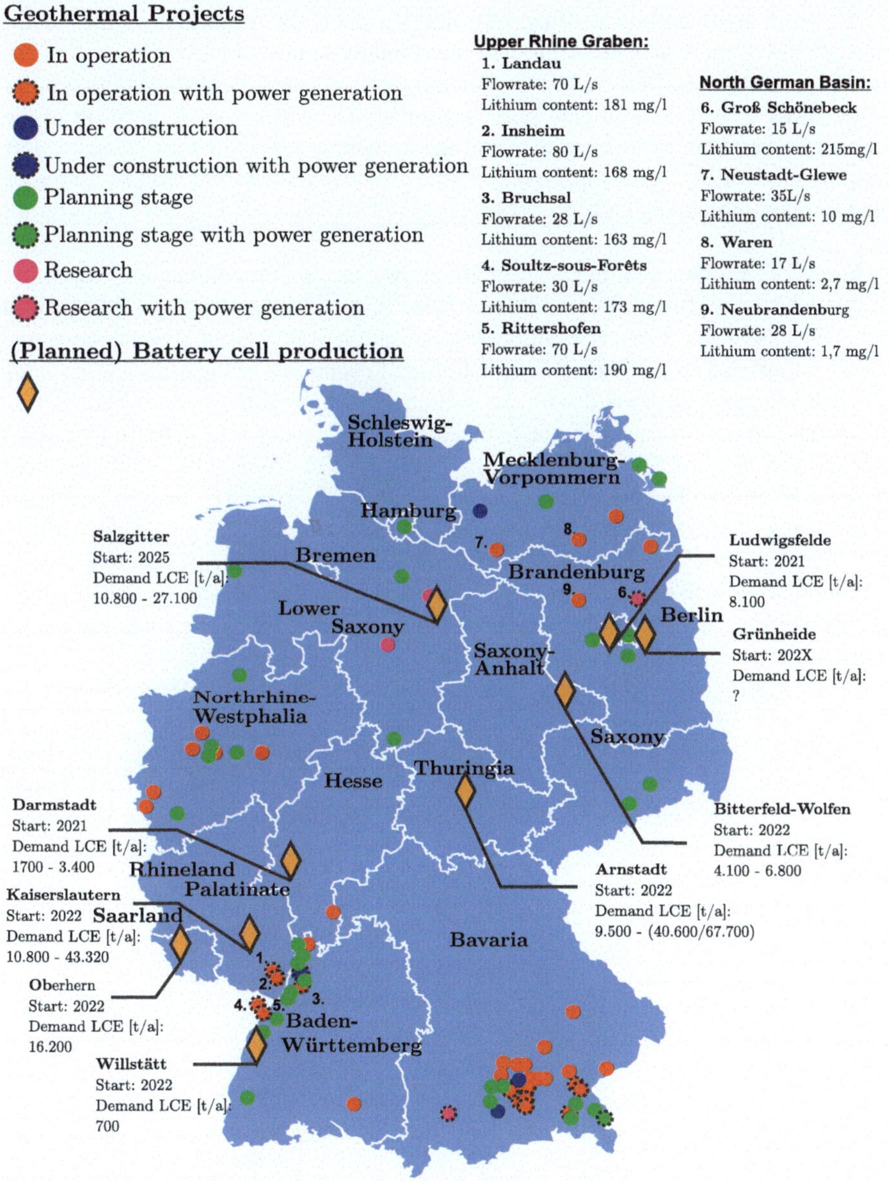

Fig. 4 Location of geothermal sites and of planned battery cell production in Germany. The map shows the lithium content and flow rate of the geothermal sites potentially suited for raw material extraction and indicates the forecasted lithium demand for battery cell production. The demand was calculated based on the planned annual capacities. (Data sources: DERA (2021a), Agemar et al. (2014a) (GeotIS), (2014b))

Although hydrocarbon drilling into the Muschelkalk formation of the southwestern BMB geological structure has occasionally indicated high lithium concentrations (>150 mg/L; Stober, 2014), geothermal fluids used from the Malm aquifer in the BMB show no significant potential for lithium extraction. As a result of the interaction between meteoric water and the carbonate reservoir rock, these waters are only moderately saline (approx. 2–6 g/L). They contain 4–900 Nmol/L of gases, the most common being CO_2 (6–90%), N_2 (5–73%), and CH_4 (3–55%) (Mayrhofer et al., 2014).

Fluids from the sedimentary NGB, however, are fundamentally different compared to the BMB. For example, brine from the geothermal site at Groß Schönebeck (NGB) contains a lithium concentration of >300 mg/L, presumably due to the evaporation of sea water or the solution of evaporites in the sedimentary units of the Rotliegend, combined with the subsequent interaction of water and rock in the metal-rich permo-carboniferous vulcanites (Regenspurg et al., 2016). Dominant ions are Ca, Na, and Cl, accounting for 70–98% of the total salinity (Regenspurg et al., 2010, 2016). The gas phase consists mainly of N_2 and CH_4, as well as smaller percentages of He, CO_2, H_2, Ar, and heavy hydrocarbons. The gas–water volume ratio ranges from 1:1 to 1.6:1 (Regenspurg et al., 2010). The anaerobic environment in combination with the high salinity and the absence of sulfides potentially enables metals like Fe, Cu, or Zn to be present as chloride complexes, increasing their solubility in the reservoir and eventually leading to high concentrations in the thermal fluid and scaling issues during geothermal energy generation. (Regenspurg et al., 2015).

The thermal water reservoirs in the URG are influenced by the rift system, which caused subsidence and sedimentary infilling. This led to an extensive fault system that partially connects several fluid reservoirs, including reservoirs in the Paleozoic bedrock and those in the overlying Mesozoic-Cenozoic rock, with occasional contributions of surface waters as well (Sanjuan et al., 2016a, 2016b). For the deep reservoir fluids, a salinity of up to 120 g/L and the corresponding fluid chemistry and isotope values generally indicate a very complex genesis from different fluids as well as several evaporation, mineral precipitation, and dissolution events. Dilution by meteoric waters further influences the fluid evolution. Interaction of these highly saline waters with sedimentary units of Buntsandstein and crystalline bedrock at high temperatures (225 ± 25 °C) leads to an accumulation of metallic ions (Sanjuan et al., 2016a, 2016b). The related brines are characterized as Na-Cl fluids and are dominated by high gas contents. CO_2 typically constitutes the main component of the gas phase (80–90 Vol%). Secondary gas components of the fluid are N_2 (6–10 Vol%) and CH_4 (2–7%). The geothermal site at Landau is an exception with approx. 48% N_2 and 44% CO_2. The gas–water ratio under standard conditions of 1.6:1 is very similar to that of the North German Basin (Eggeling et al., 2018; Sanjuan et al., 2016a, 2016b).

Differences between the geological plays in Germany and the variable distribution of their lithium concentration raises new questions regarding the exploration, resource evaluation, and exploitation of geothermal reservoirs for combined material and energy utilization. The origin of the lithium in deep fluid reservoirs remains a major open question. Furthermore, consistent fluid genesis models for the different plays and geothermal sites are yet to be developed and are still subject to active research. High lithium concentrations cannot be explained by high temperatures or reservoir depths alone, which becomes apparent when comparing the Muschelkalk and Malm aquifers of the Molasse Basin (Stober, 2014). High lithium concentrations are instead the result of the interplay of seawater evaporation, fluid mixing, and rock-water interaction, particularly at elevated temperature and high chloride concentration (Regenspurg et al., 2015, 2016; Sanjuan et al., 2016a, 2022).

Looking at the properties of the reservoir rock can provide helpful clues with respect to lithium in geothermal brines. Experiments on the fluid-rock interaction show that specific types of reservoir rocks in the URG or in the Rotliegend sandstone of the NGB have a higher potential for lithium emission into the formation water (Drüppel et al., 2020; Regenspurg et al., 2015). Experiments on the fluid interaction with granite confirm a significant correlation between the lithium solution potential and the salinity of a fluid (Drüppel et al., 2020).

Three major mechanisms are expected to influence the genesis of fluids containing high lithium concentration in the URG and NGB:

- Initial lithium accumulation during the evaporation of seawater. This fluid remains in the respective layer after the initial phase and is potentially diluted by other fluids. Lithium accumulation has been observed during the experimental evaporation of seawater (e.g. Babel and Schreiber, 2014).
- Subsequent interaction of the fluid with the surrounding minerals in the reservoir at increased temperatures, e.g. the crystalline bedrock or lower Triassic sandstone (Sanjuan et al., 2022). The interaction with the granite bedrock leads to an alteration of layered silicate, especially muscovite, biotite, and chlorite, and to a release of lithium into the fluid (Drüppel et al., 2020). This lithium-rich fluid then remains in the respective layers and could in a later stage be diluted by other fluids.
- In a more complex formation scenario, deep thermal brines integrate fluids from different depths and reservoirs through mixing via fluid circulation through fault zones (Burisch et al., 2018). End-members are seawater, highly saline waters from halite dissolution, and a meteoric component. If the saline fluids interact with the reservoir rock over a long period of time (multiple 100 ka) and at increased temperatures, lithium and other minerals with affinity to chloride can accumulate. It remains unknown which other types of reservoir rocks and sedimentary layers apart from granite, Triassic Buntsandstein, Rotliegend, or Zechstein can contribute to significant lithium concentrations by interaction with the initial brine.

2 Methods

Two site-specific scenarios were considered during the calculation of a first estimate of the maximum amount of lithium that could theoretically be extracted from the geothermal power plants operational in Germany as of December 2021. The scenarios were calculated only for geothermal power plants with a lithium concentration of > 1 mg/L (see Fig. 4). The lithium concentrations $c\left[\frac{mg}{L}\right]$ were taken from literature sources (Naumann, 2000; Regenspurg et al., 2015; Sanjuan et al., 2016a, 2016b; Schallenberg, 1999) and the flow rates $Q\left[\frac{L}{s}\right]$ of the respective heat and power generating plants derive from information provided by the German Geothermal Association (Dilger et al., 2021) and other literature data (Egert et al., 2020; Maurer et al., 2020).

Merely multiplying the lithium concentration of the water with the flow rate of a geothermal plant led to an overestimation of the raw material production potential. The following factors must be considered for a more realistic estimate of the extractable amount of lithium from geothermal reservoirs:

- **System availability of the geothermal power plant** ($A = Availability[\%]$): The most reliable geothermal plants generate power 80–95% of the time (Faltlhauser, 2016; Uhde, 2021). However, for some sites dealing with scaling and corrosion problems due to a complex fluid composition, a much lower availability must be expected. For a starting approximation, we orient ourselves by the availability of power-generating plants and assume an optimistic value of 90% availability.
- **Exploitable partial flow rate** $q_Q[\%]$: Some facilities in the URG with large extraction potential have flow rates of 70–80 L/s (Dilger et al., 2021). The extraction procedure (see accompanying article by Goldberg et al. 2022a) requires sufficient time for efficient extraction as well as large amounts of extractants in the form of sorbents, solvents, or water. The necessary volume and infrastructure must be reconciled with the requirements of a geothermal plant, potentially allowing for extraction from only a portion of the total flow (Goldberg et al., 2022a). This constraint cannot be quantified without a large-scale demonstrator of element extraction. Therefore, we base our calculations on 100% of the flow rate and note that a lower fraction will likely be used in practice.
- **Lithium concentration over time**: The geothermal utilization of deep fluids in Germany is based on doublet systems (or similar configurations). Tracer tests show that parts of the fluid return from the injection well into the production system during the monitoring period (Egert et al., 2020; Sanjuan, et al., 2016a, 2016b). As a consequence, lithium concentration will diminish over time when extracting lithium. Without detailed information a priori and because of the high dependency on site conditions, this factor is difficult to calculate. We therefore assumed a constant lithium concentration for a first approximation but also note here that changes in the lithium concentration over time are likely, depending on hydrogeological conditions in the sub-surface.
- **Resource extraction efficiency** ε **[%]**: In laboratory experiments, up to 90% lithium extraction from waters has often been achieved. Upscaling and integration of extraction processes into the operations of a power plant are likely to result in lower efficiencies due to the following reasons: competing ions in the selective extraction process, insufficient retention times of the geothermal fluid in the extraction system (e.g. due to high flow rates), high temperatures unsuitable

for certain techniques, and more. For this reason, two different extraction efficiencies—50% and 90%—were used for the calculations (Table 1). This mirrors the range of extraction efficiencies achieved in the laboratory using different extraction methods (Goldberg et al., 2022a).

Based on flow rates, the lithium concentration of the related fluid, and the reducing factors, we calculated the site-specific amount of lithium that could theoretically be extracted per year and converted it into lithium carbonate (Li_2CO_3), as ($m = Lithiummass(LCE)\left[\frac{t}{a}\right]$):

$$m = Q \cdot A \cdot c \cdot q_Q \cdot \varepsilon \tag{1}$$

$m = Lithium\ mass\left[\frac{t}{a}\right]$.

$Q = Flowrate\left[\frac{L}{s}\right]$: (site-specific).

$A = Availability$: Assumption 90% (329 days).

$c = Lithiumconcentration\left[\frac{mg}{L}\right]$: (site-specific).

$q_Q = Exploitablepartialflowrate$: Assumption 100%

$\varepsilon = Extractionefficiency$: Assumption 50 and 90%

Equation (1) was used to calculate an extraction scenario of 50% and another of 90% for all power plants listed in Table 1. This was then combined with different price

Table 1 Lithium concentration in the fluid, volumetric flow produced, and theoretically extractable amount of lithium for geothermal power plants in the Upper Rhine Graben and the North German Basin. Lithium concentrations are from Sanjuan, et al. (2016a), Regenspurg et al. (2015), Schallenberg (1999) and Naumann (2000). Volumetric flows are taken from the work of Dilger et al. (2021), Egert et al. (2020), and Maurer et al. (2020)

Geothermal power plant	Li-conc. fluid [mg/L]	Volumetric Flow Q [L/s]	Extractable Amount of Li at efficiencies: 50%/90% LCE [t/a]	Market value worst-case: 50% extraction at US$9,000 /t LCE (German Lithium Participation, 2021)	Market value best-case: 90% extraction at US$50,000 /t LCE (German Lithium Participation, 2021)
Bruchsal	163	28	345/620	$3,105,000	$31,000,000
Insheim	168	80	1015/1826	$9,135,000	$91,300,000
Landau	181	70	957/1722	$8,613,000	$88,600,000
Soultz sous Forêts	173	30	392/705	$3,528,000	$35,250,000
Rittershoffen	190	70	1004/1807	$9,036,000	$90,350,000
Groß Schönebeck	215	15	243/438	$2,187,000	$21,900,000
Neustadt-Glewe	10	35	26/48	$234,000	$2,400,000
Waren	2.7	17	3/6	$27,000	$300,000
Neubrandenburg	1.8	28	4/7	$36,000	$350,000
SUM	-	**373**	**3989/7229**	**$35,901,000**	**$361,450,000**

predictions, creating a worst-case scenario with the lowest prediction (US$9,000/t LCE) and the 50% extraction scenario, as well as a best-case scenario with the highest price prediction (US$50,000/t LCE) and the 90% extraction scenario. It should be noted, however, that even the worst-case scenario is based on very optimistic assumptions of a constant lithium concentration in the geothermal reservoir during the production period, 90% availability of the production fluid, and utilization of the entire flow rate.

3 Results

3.1 *Potential of Existing Power Plants for Lithium Extraction*

The high lithium content in some geothermal fluids and the large volume of their respective reservoirs represent a promising resource (Fig. 5). The combined fluid volume produced by currently operational geothermal wells in the German parts of the URG and the NGB known for a high Li concentration amounts to around 270 L/s. When the French sites in the URG at Soultz-sous-Forêts and Rittershoffen are also taken into consideration, the total amount reaches 370 L/s of lithium-rich thermal fluids (Table 1). These values form the basis for the calculation of the extraction scenarios according to Eq. (1).

For the geothermal power plant at Insheim, Eq. (1) results in a maximum quantity of 1000–1800 t/a LCE depending on the extraction efficiency. Similar quantities are found for the plants at Rittershoffen (1000–1800 t/a) and Landau (950–1720 t/a). Due

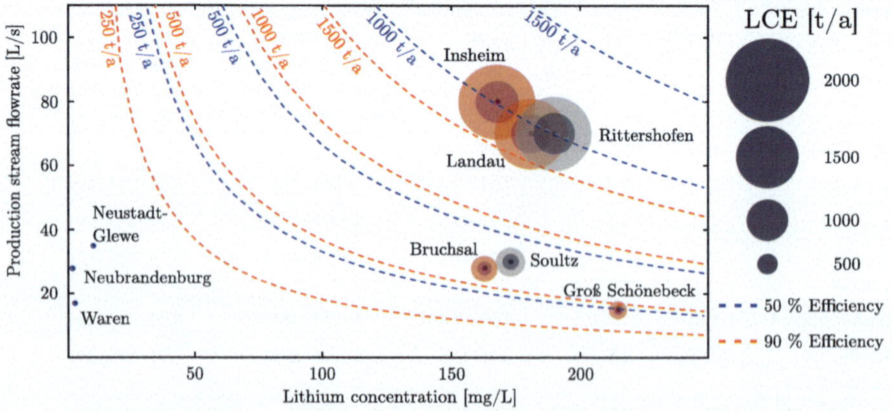

Fig. 5 Theoretically extractable amount of LCE (t/a) from the geothermal systems in the Upper Rhine Graben and the North German Basin for extraction efficiencies of 50% (blue) and 90% (red). Further assumptions for determining potential are a constant lithium concentration over the production period, fluid content usability of 100%, and the availability of geothermal production 90% of the time

to lower flow rates, the sites at Soultz-sous-Forêts and Bruchsal allow for a potential lithium extraction of only 390–700 t/a LCE and 350 to 620 t/a LCE, respectively. At the Groß Schönebeck site, which contains the highest lithium concentration (215 mg/L), the theoretical flow rate of only 15 L/s results in 240–440 t/a LCE. The three other sites in North Germany (Neustadt-Glewe, Waren, and Neubrandenburg) have a lithium concentration between only 1 and 10 mg/L, resulting in much smaller quantities of <50 t/a LCE, respectively.

In a 50% efficiency scenario, the total production volume of all sites together, including those in in the French URG, amounts to approx. 4000 t LCE per year (assuming 90% availability of geothermal power plants and 100% utilization of flow rates). This would equate to 3–11% of the total demand for battery cell production planned in Germany. Depending on price predictions (US$9000, US$25,600, or US$50,000/t LCE), the production volume would have a market volume of US$36, 102, or 200 million. With global production reaching 420,000 t LCE in 2020 (Fig. 1), the total amount of extractable LCE from the geothermal plants listed above could add around 1% (2% with 90% efficiency) to the world market, close to the amount of Europe's entire current production (Kavanagh et al., 2018).

4 Discussion

The calculated scenarios represent production volumes that would be globally measurable and make a relevant contribution to lithium production in Germany and Europe. It should be noted that German industry is transitioning from buying batteries elsewhere towards having its own battery cell production, which entails increased demand for lithium as a raw material. The planned expansion of domestic cell production is confronted with nonexistent lithium extraction and processing, resulting in complete dependency on the global lithium market. The time frames for developing lithium extraction from geothermal fluids are relatively long due to the long lead times for the construction of geothermal power plants. With exploration and approval procedures and the subsequent construction phase, it can take five to eight years until a power plant begins operations (Geothermie Uhde, 2021; Unterhaching, 2021). Nonetheless, this development is crucial to tap into additional reservoirs of lithium-rich geothermal fluids.

Our results show that high flow rates are essential for extracting sufficiently large amounts of lithium. As existing sites only have a limited ability to extend their flow rates due to the reservoir hydraulics, new wells and geothermal power plants are the most decisive factors to ensure relevant lithium production from geothermal fluids in Germany.

4.1 Extraction Scenarios

At the current state of research and technology, estimates are subject to large degrees of uncertainty and simplified assumptions. Important factors for the potential amount of extractable lithium are the availability of the geothermal fluid flow and the percentage of lithium that can actually be extracted from it. The operational availability of geothermal systems was assumed to be 90%. However, an availability of 90% is a very high estimate corresponding to electricity-generating geothermal systems (Faltlhauser, 2016; Uhde, 2021). Heat-generating geothermal plants (for instance in the NGB) often have a significantly reduced availability of 30–60% due to the seasonal nature of demand for heating (Sandrock et al., 2020). As the availability of the geothermal fluid flow equals the resource availability and linearly affects the calculations, the full load hours of the heat-generating plants limit the lithium production. This reduces the expected lithium production by one- to two-thirds if it is not adjusted for cases of combined heat and power generation.

Extraction efficiency has so far been derived only from laboratory tests and small-scale demonstrators. On a laboratory scale, multiple methods were successfully validated and mostly yielded high efficiencies of in part >90%. Successful upscaling to an industrial process has yet to occur. Available documentation accounts for a Technology Readiness Level (TRL) of around 3–5 for most lithium extraction techniques (Battistel et al., 2020; Liu et al., 2019; Ma et al., 2000; Mroczek et al., 2015; Sharma et al., 2016; Wu et al., 2019; Yu et al., 2019; Zandevakili et al., 2014). The nine-level scale of the Technology Readiness Level was developed by NASA to classify the maturity level of new technologies in development (Mankins, 1995, 2009). In this scale, TRL 4–5 describes the initial prototype testing. The communication from some central players in the URG includes reports of on-site demonstrators, indicating a TRL of 5 (Eramet, 2021; Vulcan Energy Resources, 2021). It should be noted that the next step towards a demonstrator that reaches the dimensions and efficiency of an industrial installation (TRL 6/7) represents a much greater obstacle than all earlier stages (Mankins, 2009; Straub, 2015; van der Spek et al., 2017). Scaling an installation requires a much higher level of detail than previous prototypes, often causing complications that increase costs and other expenditures (Mankins, 2009; Straub, 2015; van der Spek et al., 2017). The costs of this stage are usually between the same amount and up to more than twice as much as for reaching all previous Readiness Levels (Mankins, 2009; van der Spek et al., 2017). The subsequent construction and implementation of the first real industrial installation (TRL 8) then requires five to ten times as much financial investment as all previous steps combined, depending on the technology (Mankins, 2009; van der Spek et al., 2017). As the advanced TRL stages have not yet been reached to our current knowledge, it is not possible to determine the extraction efficiency actually achievable during future continuous extraction operations, nor to estimate the exact costs.

4.2 Reservoir and Resource Evaluation

One major factor for investment in the extraction of valuable elements from geothermal fluids is the available reservoir volume in combination with the concentration of the target element. In contrast to conventional mining, schematized practices for resource evaluation were rarely applied until now. An estimation of the resource size of lithium in geothermal fluids was carried out by Vulcan Energy. In the Ortenau license area, a resource size of 13.2 million tons of LCE was estimated (Vulcan Energy Resources, 2019). However, the estimation was graded with the highest degree of insecurity ("inferred resource") available for a resource. A classification of the Australian Joint Ore Reserves Committee Code (JORC) from 2012 was used for the evaluation. A more precise estimation of the resource size of a deep geothermal reservoir is challenging, as many parameters specific to each reservoir must be determined which are in part still being evaluated in ongoing research projects:

- Volume of the fluid reservoir
- Porosity
- Natural and artificially modified permeability in the aquifer
- Possible aquitards and filled fractures reducing or impeding permeability
- Fractures and fault zones connecting to aquifers below or above, allowing for an exchange between reservoirs in different stratigraphic units
- Details of natural fluid circulation in the geothermal reservoir
- Connections of the power plant's geothermal circulation to the fluid circulation in the reservoir (What percentage of the geothermal reservoir is connected to the deep wells and its fluid circulation? Do the production well and the injection well form a short circuit, and if so, to what extent?)
- Origin of the lithium and potential additional accumulation in the reservoir; depletion rate during extraction

To operate a lithium extraction installation that is economically viable, the behavior of the fluid chemistry over time is of crucial importance, followed by the resource size. In the case of a hydraulic short circuit between the production well and the injection well that resulted in an extraction of injected water volume after just a few weeks, the lithium concentration would quickly drop (in the course of weeks or months) to a point where production would become unprofitable. A well-established connection of the production well to the deep geothermal reservoir with only a minor percentage of injected water contributing to the production is therefore essential. At the Soultz-sous-Forêts site, tracer tests showed that after ninety days, 25% of the injected water is extracted again. Furthermore, the highest tracer concentration was observed after only thirteen days (Egert et al., 2020; Sanjuan et al., 2016a, 2016b). In contrast, at the Rittershofen geothermal site, the tracer return after twenty-five days was only around 0.2% (Sanjuan et al., 2016a, 2016b). If completely lithium-free water was pumped back into the reservoir at the injection well (equivalent to an extraction efficiency of 100%), the stated tracer return would correspond to the dilution of the lithium concentration in the given time frame. This dilution would lead to a continuous decrease of the extractable lithium over time and would negatively impact the amount of lithium that could be produced. In addition, the system parameters would have to be continuously adapted to the changing lithium concentrations, negatively affecting the extraction performance. The wide range of 0.2–25% return of the injected fluid during the tracer tests demonstrates how site-specific this factor is and that the reduction in fluid lithium concentrations over time strongly depends on the system design and hydrogeologic parameters. For instance, at Soultz-sous-Forêts, the system was designed as an EGS (Enhanced/Engineered Geothermal System), a design that necessitates a high fluid return rate. Consequently, an estimation of the potential for lithium extraction cannot simply be transferred from one site to another without taking site-specific changes in hydrogeology, fluid chemistry, and characteristics of the geothermal production site (T, p, flow rate, injection T,..) into account.

In this context, the aspect of a possible "recharge" of lithium as a cause of water–rock interactions remains undetermined. Due to the current lack of operational in-situ lithium extraction, no empirical knowledge is available. Estimations can only be based on laboratory experiments and simulations and remain simplified as they cannot fully reflect the conditions in an aquifer. A study by Drüppel et al. (2020) investigated the leaching behavior of granite when exposed to fluids at 70 °C and at 200 °C with 2 molar NaCl solution. At 70 °C, lithium is released to the lithium-free fluid amongst other things from the alteration of layered silicates like muscovite, biotite, and chlorite. In experiments at 200 °C with granite and monzonite, the leaching solution had a lithium concentration of 1–2 mg/L at the end of a thirty-six-day test period. This result shows that a lithium recharge from the reservoir is possible under high temperatures. For a more precise estimate of the magnitude of this effect in an exploited thermal water system, flow paths and interaction time with reservoir rock as well as the rock-water ratio must be taken into account. This approach may allow for an estimation of the scale of lithium concentration in a fluid that would be injected lithium-free and left in the thermal water aquifer for a specific time until it is extracted again.

In general, further detailed research and models are needed regarding the behavior of water that is depleted of lithium and potentially other ions hindering an extraction and injected back into a geothermal reservoir. Selective extraction will disturb the chemical equilibrium between the geothermal fluid and the minerals in the reservoir. After re-injection of the lithium-poor and chemically altered fluid after selective extraction, the fluid attains a new chemical equilibrium by water–rock interaction that could cause mineral precipitation or dissolution in the geothermal reservoir and therefore affect production sustainability.

Another relevant aspect is the potential presence of radionuclides in deep, high-salinity thermal fluids (Eggeling et al., 2018; Nitschke et al., 2014; Regenspurg et al., 2014; Scheiber et al., 2019). Lithium extraction attempts to very selectively separate lithium from the rest of the fluid, making enrichment of radionuclides unlikely due to their very different physical properties (e.g. significantly larger ionic radii of radionuclides). However, changes in the element concentration of the fluid, temperature reduction, or changes in Eh–pH conditions can enhance the formation of scales. Among minerals occurring during scaling, BaSr-sulfates and polymetallic sulfides can incorporate radionuclides. In the sulfates, ^{226}Ra and ^{228}Ra and in the sulfides ^{210}Pb are incorporated (Nitschke et al., 2014). Due to the lack of direct contact, the resulting radioactivity normally does not pose a hazard to humans or the environment during plant operations. Over time, however, the mineral precipitates can accumulate on pipe walls or filters, which must be separately disposed of after removal. Pipes from sites in the URG have been classified as "metal waste contaminated by hazardous substances" (AVV-170409), which could be cleaned of radionuclides by melting and thus be recycled (Eggeling et al., 2018). Radionuclides are not a direct obstacle to extraction, but the costly disposal of mineral precipitates with possible radionuclide contents must be included in a holistic economic consideration.

5 Conclusions

Global lithium demand will continue to grow at a strong pace. The main driver for this is the increasing need for lithium-ion high-performance energy storage, for the expansion of e-mobility in particular. In this context, the German lithium market will change significantly due to the large-scale battery production in planning for at least nine production sites. The current import of ready-manufactured batteries will have to be replaced by the import of raw materials. The proposed battery production would meet up to 15% of global demand in 2025, with a planned use of 37,000–149,000 t LCE for battery cell production in Germany. Rising demand could create a global lithium deficit of 360,000 t LCE (36% of global demand) in that same year, which may have a particular impact on new players on the market due to existing long-term contracts with companies and countries already producing batteries (such as China).

Until now, lithium production has been poorly diversified, with more than 80% of all lithium being produced in Australia and Chile. Lithium extraction and processing in Germany (or at least in Europe) is advisable from a geostrategic standpoint, especially since the processing of mineral ores is mostly conducted in China. In addition, extraction has been associated with environmental impacts, which are magnified by the long transport distances to battery cell manufacturing facilities. Domestic production from geothermal fluids could set new standards in terms of minimizing the CO_2 footprint and the extent of environmental impacts.

In Germany, lithium occurs in high concentrations in deep geothermal fluids. Concentrations at relevant sites in the Upper Rhine Graben range between 160 and 190 mg/L. In the North German Basin, concentrations in the Groß Schönebeck well reach up to 215 mg/L. In fluids of the Rotliegend, concentrations of close to 400 mg/L were detected in isolated cases. Access to these deposits has so far been achieved only by deep drilling. The co-production of lithium at currently operational geothermal energy production sites has the potential to make the first contribution to covering Germany's growing demand for lithium. In Germany and France, there are nine actively productive deep wells at lithium-relevant locations (the URG and NDB). Cumulatively, a volume flow of about 370 l/s is produced at these sites. From this total flow, a maximum of 7200 t/a LCE could be produced in a very optimistic forecast scenario (extraction efficiency 90%, availability 90%, constant lithium concentration in the geothermal fluid, and a usable fluid fraction of 100%). The value of this simplified estimate corresponds to about 5–19% of the annual demand of planned battery cell production in Germany. This is a purely theoretical, highly simplified potential estimate. Nevertheless, it shows the potential for responsible, independent raw material production in Germany, which could play an important role especially in the case of an impending lithium shortage.

The adaptability of the specific extraction technology to the characteristics of the geothermal fluid and power plant operation and the question of potential changes in lithium concentration due to re-circulation of injected fluid are critical for the actual feasibility. Should the breakthrough to an industrial process succeed, the following site-specific techno-economic key questions, which could not be taken into account in the very optimistic consideration carried out here, must still be clarified for implementation:

- How large is the geothermal fluid reservoir and how sustainably can it be managed?
- How does the lithium concentration behave over time during production?
- Based on fluid chemistry, which extraction method is best suited?
- Can extraction efficiency and surface fluid retention time be optimized at the site? Can such design be integrated into the operating plant process?
- How large are the volume flows resulting from the process engineering treatments? Can these be handled? What material input is necessary at what price?
- What infrastructures need to be built?
- Do plant operating parameters (temperature, pressure, full load hours) need to be adjusted for extraction?
- What are the energy requirements for selective lithium extraction?
- How does the scaling potential in all plant components develop through coupling with lithium extraction?
- What influence does the extraction (possibly with pretreatment) have on the chemical composition of the fluid? How does this affect the reservoir and the existing plant components?
- Which substances are added to the fluid (inhibitors, extraction media, etc.) and removed (raw material, co-precipitate, scaling)? Is this compatible with legal requirements?
- What substances and quantities are generated for waste disposal? What are the related costs?

Due to the complexity of the topic, the exceptional importance of future raw material supply, as well as increasing interest on the part of society, scientific support should continue to be provided for this process. More cooperation between researchers and the industrial sector would be highly constructive for achieving efficient technological progress.

Acknowledgements The authors would like to thank the Helmholtz Association for research funding within the Geoenergy subtopic in the MTET (Materials and Technologies for the Energy Transition) program of the Energy research field. Thank you to the BMBF (Federal Ministry of Education and Research) Client II for funding the BrineMine project (Grant Number 033R190B). Special thanks are also due to the chairs of Geothermal Energy and Reservoir Technology and Geochemistry and Economic Geology at the Institute of Applied Geosciences at the KIT, headed by Prof. Thomas Kohl and Prof. Jochen Kolb, for making this work possible, and for the lively scientific exchange. Furthermore, this work grew through many technical discussions, for which we would like to especially thank Dr. Joachim Koschikowski, Dr. Daniel Winter, Gerrit Fuelling, Klemens Slunitschek, Rebekka Reich, and Michael Schmidt. Last but not least, we would like to thank the anonymous reviewers and the editors for their constructive suggestions for improvement.

References

Adams, W. (2020). *Why there's still time to avoid a lithium supply crunch*. https://insights.fastmarkets.com/why-there-is-still-time-to-avoid-a-lithium-supply-crunch-final/

Agemar, T., Alten, J., Ganz, B., Kuder, J., Kühne, K., Schuhmacher, S., & Schulz, R. (2014a). The Geothermal Information System for Germany-GeotIS. *ZDGG, 165*(2), 129–144.

Agemar, T., Weber, J., & Schulz, R. (2014b). Deep geothermal energy production in Germany. *Energies, 7*(7), 4397–4416.

Babel, M., Schreiber, B. C. (2014). Geochemistry of evaporites and evolution of seawater. In *Treatise on geochemistry*, Bd. 9, S. 483–560. Elsevier, Amsterdam

Battistel, A., Palagonia, M. S., Brogioli, D., La Mantia, F., & Trócoli, R. (2020). Electrochemical methods for lithium recovery: A comprehensive and critical review. *Advanced Materials, 32*(23). https://doi.org/10.1002/adma.201905440

Burisch, M., Walter, B. F., Gerdes, A., Lanz, M., & Markl, G. (2018). Late-stage anhydrite-gypsum-siderite-dolomite-calcite assemblages record the transition from a deep to a shallow hydrothermal system in the Schwarzwald mining district, SW Germany. *Geochimica Et Cosmochimica Acta, 223*, 259–278. https://doi.org/10.1016/j.gca.2017.12.002

DERA. (2021a). *Batterierohstoffe für die Elektromobilität.-DERA Themenheft*.

DERA. (2021b). *DERA-Rohstoffliste 2021-DERA Rohstoffinformationen 49*. DERA-Deutsche Rohstoffagentur in der Bundesanstalt für Geowissenschaften und Rohstoffe.

Dilger, G., Deinhardt, A., Reimer, D., Link, K., Zwicklhuber, K., & Götzl, G. (2021). Tiefe Geothermie-Projekte in Deutschland 2021/2022. In *Geothermische Energie, Fachzeitschrift Für Geothermische Forschung Und Anwendung in Deutschland, Österreich Und Der Schweiz, 30. Jahrgang, Heft Nr. 100*.

Drüppel, K., Stober, I., Grimmer, J. C., & Mertz-Kraus, R. (2020). Experimental alteration of granitic rocks: Implications for the evolution of geothermal brines in the Upper Rhine Graben, Germany. *Geothermics, 88*(May), 101903. https://doi.org/10.1016/j.geothermics.2020.101903

Egert, R., Gholami Korzani, M., Held, S., & Kohl, T. (2020). Implications on large-scale flow of the fractured EGS reservoir Soultz inferred from hydraulic data and tracer experiments. *Geothermics, 84*(November 2019). https://doi.org/10.1016/j.geothermics.2019.101749

Eggeling, L., Herr, K., Goldberg, V., Siefert, D., Köhler, J., Kölbel, T., & Reith, S. (2018). *Schlussbericht zum Verbundprojekt ANEMONA. Anlagenmonitoring als Schlüsseltechnologie für den erfolgreichen Betrieb von Geothermiekraftwerken in Deutschland* (F. und I. EnBW Energie Baden-Württemberg AG (ed.)).

Eramet. (2021). *Eramet*. https://www.eramet.com/en/eramet-and-electricite-de-strasbourg-announce-success-first-pilot-test-extract-lithium-geothermal

Faltlhauser, M. (2016). Zahlen und Fakten zur Stromversorgung in Deutschland 2016. *Wirtschaftsbeirat Bayern*, 1–24.

Geothermie Unterhaching. (2021). *Meilensteine: Geothermie Unterhaching*. https://www.geothermie-unterhaching.de/cms/geothermie/web.nsf/id/pa_historie.html

German Lithium Participation. (2021). *Reale Preisentwicklung und prognostizierter Preistrend für battery grade Lithiumcarbonat*. https://germanlithium.com/preisentwicklung-lithium/

Goldberg, V., Kluge, T., & Nitschke, F. (2022a). Herausforderungen und Chancen für die Lithiumgewinnung aus geothermalen Systemen in Deutschland—Teil 1: Literaturvergleich bestehender Extraktionstechnologien. *Grundwasser*. https://doi.org/10.1007/s00767-022-00522-5

Goldberg, V., Nitschke, F., Kluge, T. (2022b). Herausforderungen und Chancen für die Lithiumgewinnung aus geothermalen Systemen in Deutschland—Teil 2: Potenziale und Produktionsszenarien in Deutschland. *Grundwasser—Zeitschrift der Fachsektion Hydrogeologie*. https://doi.org/10.1007/s00767-022-00523-4

Hans Wedepohl, K. (1995). The composition of the continental crust. *Geochimica Et Cosmochimica Acta, 59*(7), 1217–1232. https://doi.org/10.1016/0016-7037(95)00038-2

Hohmann, M. (2021). *Durchschnittlicher Preis von Lithiumcarbonat weltweit in den Jahren von 2002 bis 2018 (in US-Dollar je Tonne)* (Statista GmbH (ed.)). https://de.statista.com/statistik/daten/studie/979746/umfrage/durchschnittlicher-preis-von-lithium-weltweit/

Jones, D., Blanchard, C., Templeman, L., & Fitzpatrick, L. (2021). *Is lithium a ' green ' alternative?* https://www.dbresearch.com/PROD/RPS_ENPROD/PROD0000000000518496/Q%26A_series%3A_Is_lithium_a_'green'_alternative%3F.pdf?undefined&realload=NfunTjMdzVdyuNFxqpz3sEw9wumftGRGj5ktGV2dFcWO~6dsE0AWj62WZeRSLfc8

Kavanagh, L., Keohane, J., Garcia Cabellos, G., Lloyd, A., & Cleary, J. (2018). Global lithium sources—industrial use and future in the electric vehicle industry: A review. *Resources, 7*(3), 57. https://doi.org/10.3390/resources7030057

Lee, J.-S., Tai Kim, S., Cao, R., Choi, N.-S., Liu, M., Lee, K. T., & Cho, J. (2011). Metal-air batteries with high energy density: Li-Air versus Zn-Air. *Advanced Energy Materials, 1*(1), 34–50. https://doi.org/10.1002/aenm.201000010

Lide, D. R. (2005). *CRC handbook of chemistry and physics (Internet V)*. CRC Press.

Liu, G., Zhao, Z., & Ghahreman, A. (2019). Novel approaches for lithium extraction from salt-lake brines: A review. *Hydrometallurgy, 187*(May), 81–100. https://doi.org/10.1016/j.hydromet.2019.05.005

Lüders, V., Plessen, B., Romer, R. L., Weise, S. M., Banks, D. A., Hoth, P., Dulski, P., & Schettler, G. (2010). Chemistry and isotopic composition of Rotliegend and Upper Carboniferous formation waters from the North German Basin. *Chemical Geology, 276*(3–4), 198–208. https://doi.org/10.1016/j.chemgeo.2010.06.006

Ma, P., Chen, X. D., & Hossain, M. M. (2000). Lithium extraction from a multicomponent mixture using supported liquid membranes. *Separation Science and Technology, 35*(15), 2513–2533. https://doi.org/10.1081/SS-100102353

Mankins, J. C. (1995). *Technology readiness levels: A white paper*. October, from http://www.hq.nasa.gov/office/codeq/trl/trl.

Mankins, J. C. (2009). Technology readiness assessments: A retrospective. *Acta Astronautica, 65*(9–10), 1216–1223. https://doi.org/10.1016/j.actaastro.2009.03.058

Martin, G., Rentsch, L., Höck, M., & Bertau, M. (2017). Lithium market research–global supply, future demand and price development. *Energy Storage Materials, 6*, 171–179. https://doi.org/10.1016/j.ensm.2016.11.004

Maurer, V., Gaucher, E., Grunberg, M., Koepke, R., Pestourie, R., & Cuenot, N. (2020). Seismicity induced during the development of the Rittershoffen geothermal field, France. *Geothermal Energy, 8*(1). https://doi.org/10.1186/s40517-020-0155-2

Mayrhofer, C., Niessner, R., & Baumann, T. (2014). Hydrochemistry and hydrogen sulfide generating processes in the Malm aquifer, Bavarian Molasse Basin, Germany. *Hydrogeology Journal, 22*(1), 151–162. https://doi.org/10.1007/s10040-013-1064-2

Meng, F., McNeice, J., Zadeh, S. S., & Ghahreman, A. (2021). Review of lithium production and recovery from minerals, brines, and lithium-ion batteries. *Mineral Processing and Extractive Metallurgy Review, 42*(2), 123–141. https://doi.org/10.1080/08827508.2019.1668387

Moeck, I. S. (2014). Catalog of geothermal play types based on geologic controls. *Renewable and Sustainable Energy Reviews, 37*, 867–882. https://doi.org/10.1016/j.rser.2014.05.032

Mroczek, E., Dedual, G., Graham, D., & Bacon, L. (2015). Lithium Extraction from Wairakei Geothermal fluid using electrodialysis. In *Proceedings of the World Geothermal Congress 2015, April*, 6.

Naumann, D. (2000). Salinare Tiefenwässer in Norddeutschland-Salinare Tiefenwässer in Norddeutschland Gas- und isotopengeochemische Untersuchungen. *Scientific Technical Report STR00/21*.

Nitschke, F., Scheiber, J., Kramar, U., & Neumann, T. (2014). Formation of alternating layered Ba-Sr-sulfate and Pb-sulfide scaling in the geothermal plant of Soultz-sous-Forêts. *Journal of Mineralogy and Geochemistry, 191*(2), 145–156. https://doi.org/10.1127/0077-7757/2014/0253

Regenspurg, S., Dilling, J., Mielcarek, J., Korte, F., & Schkade, U. K. (2014). Naturally occurring radionuclides and their geochemical interactions at a geothermal site in the North German Basin. *Environmental Earth Sciences, 72*(10), 4131–4140. https://doi.org/10.1007/s12665-014-3306-6

Regenspurg, S., Feldbusch, E., Norden, B., & Tichomirowa, M. (2016). Fluid-rock interactions in a geothermal Rotliegend/Permo-Carboniferous reservoir (North German Basin). *Applied Geochemistry, 69*, 12–27. https://doi.org/10.1016/j.apgeochem.2016.03.010

Regenspurg, S., Milsch, H., & Schaper, J. (2015). Copper in geothermal brine: Origin , reactions, risks, and chances. *World Geothermal Congress 2015, April*, 5.

Regenspurg, S., Wiersberg, T., Brandt, W., Huenges, E., Saadat, A., Schmidt, K., & Zimmermann, G. (2010). Supplement 3. *Chemie der Erde, 70*(SUPPL. 3), 3–12. https://doi.org/10.1016/j.chemer.2010.05.002

Sandrock, M., Maaß, C., Weiselder, S., Westholm, H., Schulz, W., Löschan, G., Bausch, C., Kreuter, H., Reyer, D., Mangold, D., Riegger, M., & Köhler, C. (2020). Kommunaler Klimaschutz durch Verbesserung der Effizienz in der Fernwärmeversorgung mittels Nutzung von Niedertemperaturwärme- quellen am Beispiel tiefengeothermischer Ressourcen-Abschlussbericht 31/2020. *Umweltbundesamt, Climate Ch*(February), 357. https://www.environment.gov.za/sites/default/files/reports/environmentoutlook_chapter11.pdf

Sanjuan, B., Gourcerol, B., Millot, R., Rettenmaier, D., Jeandel, E., & Rombaut, A. (2022). Lithium-rich geothermal brines in Europe: An up-date about geochemical characteristics and implications for potential Li resources. *Geothermics, 101*(July 2021). https://doi.org/10.1016/j.geothermics.2022.102385

Sanjuan, B., Millot, R., Innocent, C., Dezayes, C., Scheiber, J., & Brach, M. (2016a). Major geochemical characteristics of geothermal brines from the Upper Rhine Graben granitic basement with constraints on temperature and circulation. *Chemical Geology, 428*, 27–47. https://doi.org/10.1016/j.chemgeo.2016.02.021

Sanjuan, B., Scheiber, J., Gal, F., Touzelet, S., Genter, A., & Villadangos, G. (2016). Inter-well chemical tracer testing at the Rittershoffen geothermal site (Alsace, France). *European Geothermal Congress*, 19–24.

Schallenberg, K. (1999). Das Geothermische Heizwerk in Neustadt-Glewe im Betriebsjahr 1996. In *Scientific Technical Report/Geoforschungszentrum Potsdam* (Vol. 04).

Scheiber, J., Hettkamp, T., Schindler, M., Wahl, G. & Baumgärtner, J. (2019). *Schlussbericht Verbundvorhaben Sulfidinhibition und Deponierbarkeit (SUBITO): Verbesserung der Wirtschaftlichkeit von Geothermieanlagen durch Inhibition natürlich belasteter Ablagerungen* (Issue November). https://doi.org/10.2314/KXP:1691178594

Schmidt, M. (2017). *Rohstoffrisikobewertung-Lithium: Rohstoffinformationen 33*.

Schmidt, M. (2021). Raw material requirements for lithium-ion batteries and hydrogen electrolysis-Supply risks for Europe as an industrial location? *ULM Electrochemical Talks*. https://uect.de/program/overview.html

Sharma, A. D., Patil, N. D., Patwardhan, A. W., Moorthy, R. K., & Ghosh, P. K. (2016). Synergistic interplay between D2EHPA and TBP towards the extraction of lithium using hollow fiber supported liquid membrane. *Separation Science and Technology (Philadelphia), 51*(13), 2242–2254. https://doi.org/10.1080/01496395.2016.1202280

Srinivasan, V., Hafemeister, D., Levi, B., Levine, M., & Schwartz, P. (2008). Batteries for Vehicular Applications. *AIP Conference Proceedings, 1044*, 283–296. https://doi.org/10.1063/1.2993726

Stober, I. (2014). Hydrochemical properties of deep carbonate aquifers in the SW German Molasse basin. *Geothermal Energy, 2*(1). https://doi.org/10.1186/s40517-014-0013-1

Straub, J. (2015). In search of technology readiness level (TRL) 10. *Aerospace Science and Technology, 46*, 312–320. https://doi.org/10.1016/j.ast.2015.07.007

Trading Economics. (2022). *Lithium.* https://tradingeconomics.com/commodity/lithium

Uhde, J. (2021). Welcome to the Geothermal Power Plant Insheim. In *9th European Geothermal Workshop.*

van der Spek, M., Ramirez, A., & Faaij, A. (2017). Challenges and uncertainties of ex ante techno-economic analysis of low TRL CO2 capture technology: Lessons from a case study of an NGCC with exhaust gas recycle and electric swing adsorption. *Applied Energy, 208*(September), 920–934. https://doi.org/10.1016/j.apenergy.2017.09.058

Vulcan Energy Resources. (2019). *Vulcan energy resources ASX announcement (4 December 2019), Largest JORC Lithium Resource in EuropeVulcan Zero Carbon Lithium™ Project.* https://www.investi.com.au/api/announcements/vul/269c4f08-99d.pdf. Zugegriffen: 31. Jan. 2022.

Vulcan Energy Resources. (2021). *Quarterly Activities Report-March 2021.* https://vul.live.irmau.com/site/PDF/88d6ee23-6598-4127-b206-fac8ae8a74df/QuarterlyActivitiesReport

Wu, L., Li, L., Evans, S. F., Eskander, T. A., Moyer, B. A., Hu, Z., Antonick, P. J., Harrison, S., Paranthaman, M. P., Riman, R., & Navrotsky, A. (2019). Lithium aluminum-layered double hydroxide chlorides (LDH): Formation enthalpies and energetics for lithium ion capture. *Journal of the American Ceramic Society, 102*(5), 2398–2404. https://doi.org/10.1111/jace.16150

Xu, C., Dai, Q., Gaines, L., Hu, M., Tukker, A., & Steubing, B. (2020). Future material demand for automotive lithium-based batteries. *Communications Materials, 1*(1), 10. https://doi.org/10.1038/s43246-020-00095-x

Yu, X., Fan, X., Guo, Y., & Deng, T. (2019). Recovery of lithium from underground brine by multistage centrifugal extraction using tri-isobutyl phosphate. *Separation and Purification Technology, 211*(July 2018), 790–798. https://doi.org/10.1016/j.seppur.2018.10.054

Zandevakili, S., Ranjbar, M., & Ehteshamzadeh, M. (2014). Recovery of lithium from Urmia Lake by a nanostructure MnO2ion sieve. *Hydrometallurgy, 149*, 148–152. https://doi.org/10.1016/j.hydromet.2014.08.004

Raw Material Potential of Geothermal Systems in Chile

Ignacio Pérez, Valentin Goldberg, Fabian Nitschke, Diego Morata, Joachim Koschikowski, and Thomas Kohl

Abstract This study analyzes the economic potential of raw materials that can be produced from the thermal waters of Chile. Beyond the widely known energetic potential of geothermal systems, a combined raw material production could add economic value to geothermal projects and enable micro-invasive mining processes using renewable energy. Raw materials such as lithium are key to Chile's development and could also be an important incentive for new geothermal energy projects. Chile is a country known for its large mineral deposits and its geothermal potential. However, the focus of raw material production is on hard-rock mining and evaporation ponds in the salars, which can consume high amounts of energy, water, and space. The potential of a green mining approach in terms of geothermal energy and raw material

The following chapter is a translation of the German journal article "Herausforderungen und Chancen für die Lithiumgewinnung aus geothermalen Systemen in Deutschland Teil 2: Potenziale und Produktionsszenarien in Deutschland" (https://doi.org/10.1007/s00767-022-00523-4). It was originally published under an Open Access license in the peer-reviewed journal Grundwasser (Goldberg et al. 2022b). The translation was conducted by the original authors of the initial study.

I. Pérez (✉) · V. Goldberg · D. Morata
Department of Geology and Andean Geothermal Center of Excellence (CEGA), Facultad de Ciencias Físicas y Matemáticas, Universidad de Chile, Santiago, Chile
e-mail: ignacio.perez.1@ug.uchile.cl

V. Goldberg
e-mail: valentin.goldberg@kit.edu

D. Morata
e-mail: dmorata@ing.uchile.cl

I. Pérez · V. Goldberg · F. Nitschke · T. Kohl
Institute of Applied Geosciences, Division Geothermal Energy and Reservoir Technology, Karlsruhe Institute of Technology (KIT), Karlsruhe, Germany
e-mail: fabian.nitschke@kit.edu

T. Kohl
e-mail: thomas.kohl@kit.edu

I. Pérez · J. Koschikowski
Fraunhofer Institute for Solar Energy Systems ISE, Freiburg, Germany
e-mail: joachim.koschikowski@ise.fraunhofer.de

A. Ch. Braun et al. (eds.), *Eco-Industrial Development as an Industrial Strategy*, Green Energy and Technology, https://doi.org/10.1007/978-3-031-73576-9_19

production is yet untapped in Chile while receiving high industrial and scientific interest in countries such as the United States (US) or Germany. In this study, the potential of seven raw materials present in thermal springs in Chile is analyzed, based on a compilation of more than 800 samples from twenty different studies throughout the country. The quantifications of potentially producible amounts of commodities are then compared to the current production in Chile and the economic potential is outlined on the basis of current market prices.

1 Introduction

Beyond their energy potential, geothermal waters may also contain high concentrations of strategic raw materials in dissolved form. Strategic raw materials are those that are of great economic importance to the industry, the country, or a particular geographic area. This importance may be due to the economic value of the material, growing demand for future technologies, the scarcity of other resources, or the desire to become independent from global imports. The mining industry is transforming to the use of renewable energy and sustainable raw material extraction (Brantes, 2016). In the case of Chile, for example, lithium is an important factor in the country's industry and has a critical role to play in the global energy transition (Flexer et al., 2018). Therefore, long-term production must be ensured to maintain a strong position for Chile against global competitors. Lithium mining is associated with high water consumption, which is particularly critical since Chile is facing drought issues across much of its territory. Mining companies are therefore striving more and more for efficient water use and recovery, with measures such as seawater desalination plants (Campero & Harris, 2019). In this regard, direct raw material extraction from geothermal fluids could complement these approaches.

Many communities across the country have problems maintaining a steady supply of drinking water. Furthermore, the available water often contains thermal water components that render it unfit for drinking due to high amounts of toxic elements such as arsenic or antimony (Pincetti-Zúniga et al., 2022). In these locations, extraction of these elements could have the additional advantage of providing fresh water as a byproduct.

This report evaluates the potential of raw material extraction from thermal brines in Chile and compares it with current reports on conventional mining from the National Geology and Mining Service (SERNAGEOMIN) and the Chilean Copper Commission (COCHILCO). In addition, various studies have been incorporated from other countries that already have experience in extracting minerals from geothermal brines or tapping into the economic potential of raw materials, such as Mexico (Mercado & Hurtado, 1992), Italy (Allegrini et al., 1992), the Netherlands (Biagini, 2022), and the US (Neupane & Wendt, 2017).

2 Geothermal Waters and the Chilean Raw Material Market

In 2020, the installed capacity of geothermal energy worldwide was around 16,000 MW_{el}, with the US being the world leader with around 3,700 MW_{el} (Huttrer, 2020). With ninety active volcanoes on its territory (www.sernageomin.cl, 2023) and its position in the Pacific Ring of Fire, similarly to most of the geothermal plants in the US (Neupane & Wendt, 2017), Chile possesses a technically exploitable potential of up to 3,800 MW_{el} in its explored areas alone (Mesa de Geotermia report, page 5, 2018). In 2017, the first geothermal plant in South America, Cerro Pabellón, was inaugurated in Chile with a total capacity of 48 MW_{el} and later extended by 33 additional MW_{el}, reaching 81 MW_{el} in total (Capetti et al., 2020). When compared to the total installed energy capacity in Chile, 28,495 MW_{el} (Garcia, 2021*)*, it can be deduced that the explored areas of territory alone could supply 13% of the national demand while only 0.3% is developed with Cerro Pabellón. Thus, the necessity of continuing the exploration, investigation, and subsequent use of these resources in an optimal way becomes more and more apparent.

Extracting raw materials from the brines already used for power generation could be one way to maximize the economic feasibility of geothermal projects. On the other hand, it would also help making the Chilean mining sector more sustainable and/or provide drinking water to communities.

The typically high flow rates in geothermal plants (e.g. 560 L/s at Cerro Pabellón, Capetti et al., 2020) in combination with the elevated concentrations of various elements result in the circulation of large amounts of commodities. In comparison to conventional mining, the raw materials dissolved in thermal brines come along with the fluids' inherent energy in terms of heat that could be used for powering the extraction process. These factors could facilitate economic competitiveness over established extraction approaches.

In order to initiate this kind of mining in Chile, a hydrochemical study of the thermal brines and a general geological and hydrological study of each geothermal system (to estimate recharge, lithology, access to the reservoir, etc.) are necessary for future research and investment. For this study, seven potential raw materials that occur in the thermal waters of Chile were examined (Fig. 1). These were selected due to their elevated concentration, strategic factors for the country, or on the basis of whether their extraction could enhance the efficiency of geothermal energy production.

2.1 *Potential Target Elements in Chile*

Boron

Boron resources of Chile are located in the evaporitic salt flats in the north of the country, genetically related to the volcanic activity of the Upper Tertiary–Quaternary (Chong et al., 2000). The borate ulexite and boric acid are produced in the regions of

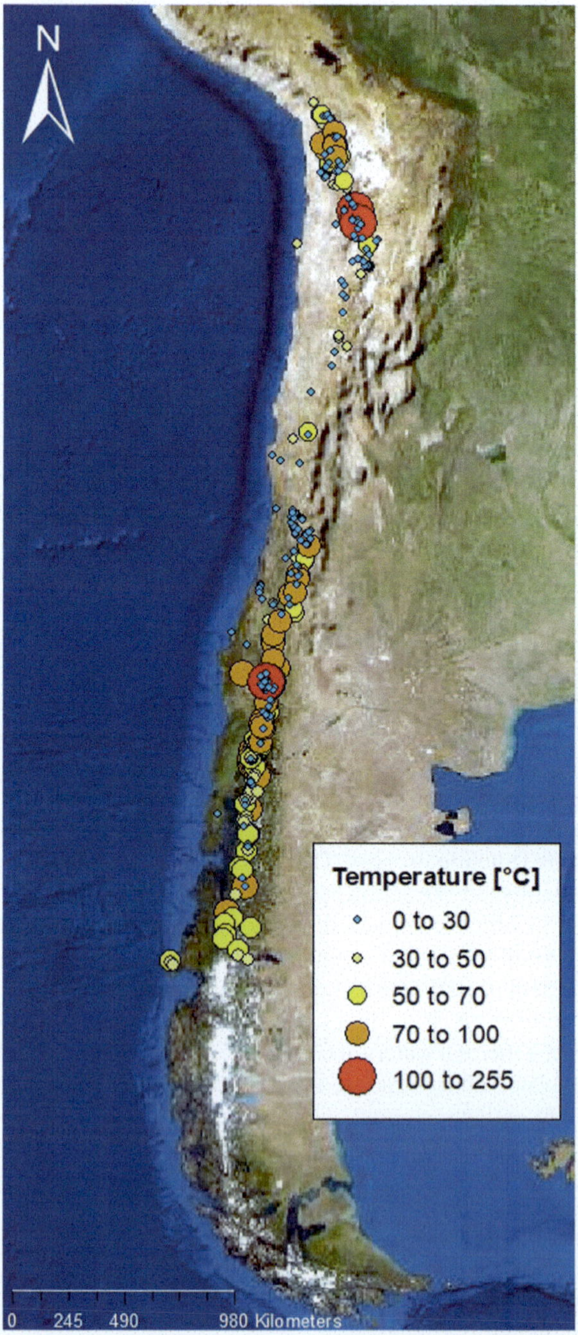

Fig. 1 Distribution of hot springs in Chile and their respective measured temperatures. The main thermal springs used in this report were selected due to their economic potential and are mostly located in the country's geothermal fields

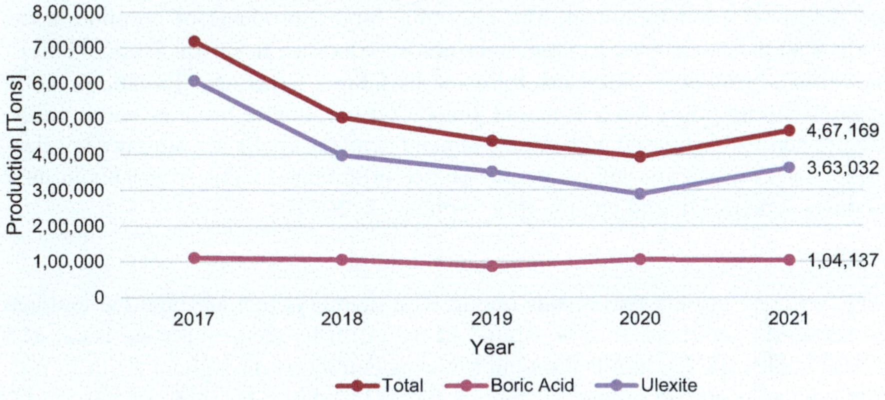

Fig. 2 Graph showing the production of boron compounds in the last five years in Chile. The production of 363,032 tons of ulexite in 2021 implied an increase of 26% compared to 2019, while the production of boric acid decreased by 2% from 105,470 to 104,137 tons (Servicio Nacional de Geología y Minería, 2021)

Arica y Parinacota and Antofagasta through the exploitation and processing of the saline crusts and brines of the Andean and pre-Andean salt flats (Servicio Nacional de Geología y Minería, 2021). The main applications of these boron compounds are in the manufacture of detergents, glass, fiberglass, wood preservatives, fire retardants, and agricultural micronutrients. Ulexite production is mainly from Andean salt flats and boric acid comes mainly from the processing of ulexite (Servicio Nacional de Geología y Minería, 2021). Chile's Quiroborax is the third-largest boric acid producer in the world, with an installed capacity of 100,000 tons per year (Fig. 2). The company extracts ulexite from the Salar de Surire.

Due to China having only low-grade boron reserves and the anticipated increase in demand for boron in that country, exports to China are expected to remain stable over the coming years (USGS Mineral Commodity Summaries, 2023). The importance of boron is established by its wide variety of applications in products. Although the fiberglass sector occupies 50% of boron uses, the remaining uses include the manufacture of soaps and detergents, insulation such as textiles, enamels, and ceramics, fuel for planes and rockets, nuclear applications, optics, etc. (Comisión Chilena del Cobre, 2013).

Bromine

Bromine is produced from brines after the separation of sodium chloride and potash, and is found in seawater, salt lakes, and thermal brines. Seawater contains bromine at about 65 mg/L, but bromine is found in much higher concentrations (2.500 to 10.000 mg/L) in inland seas and continental brines (Bromine Science and Environmental Forum (BSEF) (2023)). The leading global applications of bromine are the production of brominated flame retardants as well as industrial uses. Bromine compounds are also used in a variety of other applications, including drilling fluids and industrial water treatment. The US is the largest producer of bromine in the world with the Smackover Formation brines in Arkansas, from which about half of the world's bromine is extracted, followed by China, Israel, and Jordan. Recently bromine selling prices have risen and global consumption of bromine and bromine compounds also increased with strong demand from the appliance, automotive, electronic, and construction industries, which use brominated flame retardants in their products (USGS Mineral Commodity Summaries, 2023).

Cesium

Cesium is one of only two metals (along with mercury) that are liquid at ambient temperature conditions. It is not found in nature in its elemental state because it is easily oxidized. Cesium is found in low concentrations throughout Earth's crust. Pollucite, a cesium aluminum silicate, is found in lithium-bearing granites worldwide, with most of the world's reserves located at Bernic Lake in Canada (USGS Mineral Commodity Summaries, 2023). Cesium occurs in brines in Chile and China, but these have not been commercially exploited (Angulo, 2018). The petroleum industry consumes more than half of the cesium produced annually. The primary application is in cesium formate muds used for high-pressure and temperature drilling for oil and gas production and exploration. Cesium is also used in GPS, internet and cellular telephone transmissions, and aircraft guidance systems (USGS Mineral Commodity Summaries, 2023). World production is estimated to be more than 45 tons per year and world consumption is estimated to be less than 50 tons per year, with nearly half of all consumption occurring in the US (Angulo, 2018). Because cesium metal is not traded in large-scale public markets, no market price is available. In 2020, one company offered 25 g of cesium formate for US$41,4 (USGS Mineral Commodity Summaries, 2023), which is the price used for analysis in this report.

Lithium

Lithium with its application in rechargeable lithium-ion-batteries (LIB) is one of the key elements for the ongoing energy transition and electrification. It can be found in pegmatites, brines, geothermal fields, oil wells, and seawater (Comisión Chilena del Cobre, 2013).

In the Andes, lithium has its origin in magmatic processes, where it is remobilized as a consequence of erosion and rock leaching by thermal waters and enriched in the salt flats at the borders between Argentina, Bolivia, and Chile (Meixner et al., 2020). The brines and salt flats of northern Chile are well known for their important lithium reserves, with Chile being the second-largest producer of lithium compounds in the world (30,200 metric tons annually) after Australia (55,000 metric tons annually). The salt flats bear various good properties for efficient Li extraction, such as low Mg/Li ratios. High ratios require more surface area of evaporation ponds to crystallize as much magnesium salts as possible prior to Li extraction. For this reason, the Salar de Atacama is considered the most favorable reservoir, as it has a low Mg/Li ratio in its brines (Comision Chilena del Cobre, 2013).

Chile also possesses more than 65% of lithium reserves from brines and 51% of lithium reserves worldwide (Servicio Nacional de Geología y Minería, 2021).

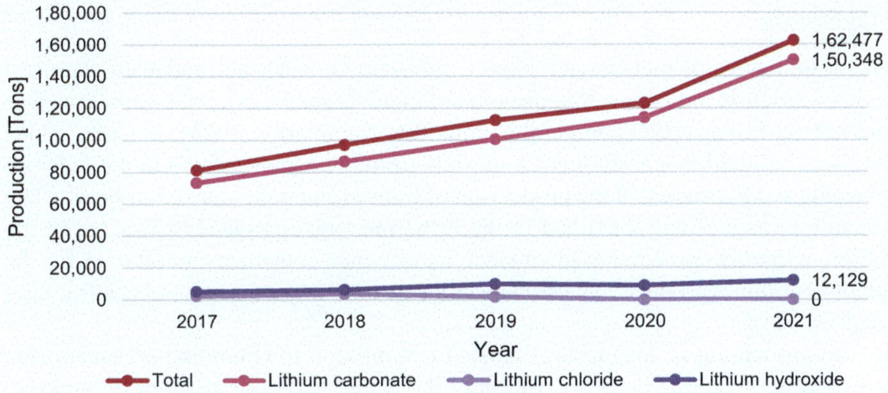

Fig. 3 Graph showing the production of lithium compounds in the last five years in Chile. Production grew by 32% from 2020 to 2021, increasing from 123,290 to 162,477 tons (Servicio Nacional de Geología y Minería, 2021)

In Chile, SQM is the main company in charge of lithium production in the Salar de Atacama and the Salar del Carmen, which together represent a production capacity of 120,000 tons of lithium carbonate per year, with an expected production of 180,000 tons by 2022 (Annual report SQM, 2021).

The lithium compounds produced in Chile (Fig. 3) and their uses are:

- **Lithium carbonate**: Used mainly in the manufacture of rechargeable batteries for electric vehicles (80% of global lithium end-use, according to USGS, 2023)
- **Lithium chloride**: Mainly used in arc welding and the manufacture of aluminum heat exchangers (Servicio Nacional de Geología y Minería, 2021)
- **Lithium hydroxide**: Used in the production of lubricating greases that can be applied in extreme conditions of temperature and load. It is also used in the manufacture of ceramics and colorants, in the purification of gases, as an absorber of carbon dioxide in spacecraft and submarines, and as a medium for heat transfer and as a storage of battery electrolytes (Servicio Nacional de Geología y Minería, 2021)
- **Lithium sulfate**: Used in medicine for the treatment of bipolar disorder and in the battery industry as a precursor to producing lithium hydroxide (Servicio Nacional de Geología y Minería, 2021)

Lithium carbonate (71%) is the main product in demand over lithium hydroxide (24%). It is expected that in the future, lithium hydroxide will be the main product of lithium demand (57% by 2030), which is primarily explained by an increase in the manufacture of NMC batteries (nickel-lithium-cobalt, manganese) (Gonzalez, 2021).

Magnesium

Magnesium is a light metal used for applications for human health and in the chemical and construction industries. Magnesium compounds can be extracted from seawater and natural brines (USGS Mineral Commodity Summaries, 2023). In Chile, SQM produces residual magnesium chloride salts from the brines located in the Salar de Atacama as a byproduct of the production of lithium and potassium compounds. The remaining magnesium contained in the brine must be completely extracted in the lithium carbonate processing plant since its presence constitutes a pollutant for the quality of lithium products. Magnesium chloride is mainly used for road conditioning (SQM, 2023).

The production of magnesia is carried out through the pumping of waters rich in magnesium and the decarbonization of the waters with limestone or dolomite to precipitate the hydroxide. Magnesium oxide (MgO) is the raw material for various magnesium chemical products, and an important component in refractory bricks for steel and copper smelting furnaces. In Chile, there are potential magnesium resources in various salt flats (Comision Chilena del Cobre, 2013).

Potassium

Potassium is the seventh-most abundant element in the Earth's crust. Most of the potassium resources are found in stratified saline deposits, which contain high-grade deposits and large reserves at a low cost of exploitation. Important potassium production also takes place from the evaporation of surface brines present in salt flats and saline lakes. In Chile, potassium chloride and potassium sulfate are produced from the processing of chloride brines with high potassium content (22 g/l) in Salar de Atacama, as seen in Fig. 4 (Servicio Nacional de Geología y Minería, 2021).

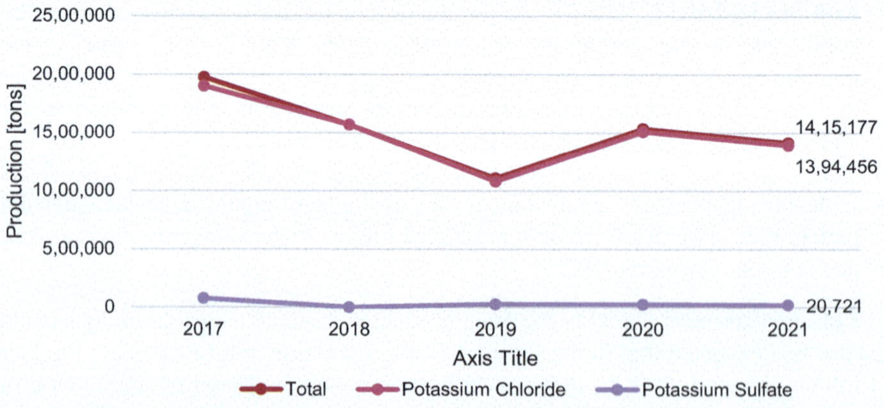

Fig. 4 Graph showing the production of potassium compounds in the last five years in Chile. Production of potassium chloride and sulfate in 2021 equaled 1,415,177 tons, a decrease of 8% compared to 2020 (Servicio Nacional de Gegología y Minería, 2021)

The potassium compounds are potassium chloride and potassium sulfate, mining-industrial products that are used mainly as potassium fertilizers. Chile's SQM is the only company in the world to produce the three types of potash fertilizers in existence. No substitutes exist for potassium as an essential plant nutrient. Furthermore, it is an essential nutritional requirement for animals and humans as well (USGS Mineral Commodity Summaries, 2023).

One of the most important projections is for the increasing relevance of potassium batteries and sodium batteries, which are presented as promising alternatives to lithium-ion batteries. These batteries would be more sustainable, affordable, and safer than their lithium-ion counterparts, though they have a lower energy density than lithium-ion batteries (Lu et al., 2021).

Silica

Silicon is one of the most abundant elements in the Earth's crust. Due to the increased solubility of silica with increasing temperature, it is typically enriched in geothermal brines due to water–rock interactions (Fournier & Rowe, 1966; Henley, 1983; Iler, 1979).

In Chilean silica production (Fig. 5), the main siliceous resources are quartz and siliceous sands (Servicio Nacional de Geología y Minería, 2021). The main consumers of silica have been producers of aluminum alloys and the chemical industry, which uses silica in the manufacture of silicones. The semiconductor and solar energy industries, which manufacture chips for computers and photovoltaic cells from high-purity silicon, are also consumers of silicon metal (USGS Mineral Commodity Summaries, 2023).

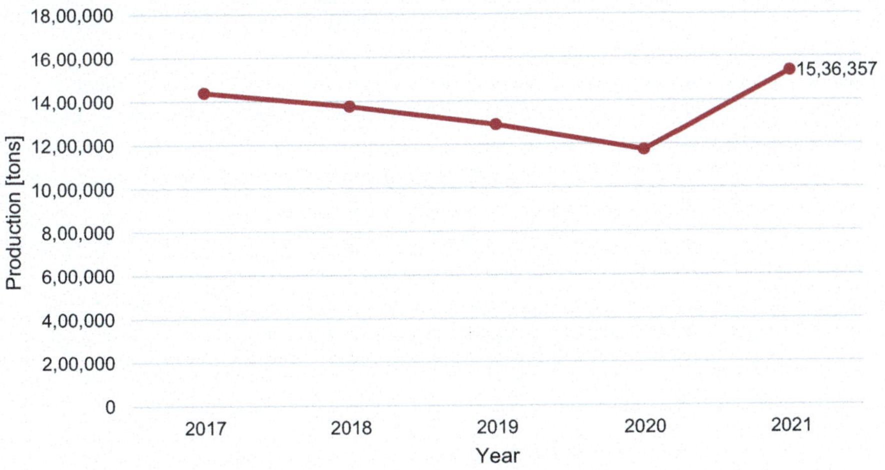

Fig. 5 Graph showing the production of silica compounds in the last five years in Chile. Production values have remained stable between 1,1Mt and 1,5Mt (Servicio Nacional de Geología y Minería, 2021)

High silica concentration in geothermal brines is a common problem affecting the efficiency of geothermal energy production as well as in water treatment processes (Gunnarsson & Arnórsson, 2005; Cob et al., 2014; Mi & Elimelech, 2013; Badruzzaman et al., 2011; Zhang et al., 2018). During the cooling and concentration of fluids, silica tends to precipitate, generating scaling problems. Therefore, beyond silica's value as a raw material, selective silica extraction is economically beneficial in different industries. To maintain a zero-waste strategy and enhance economic potential, an after-use of the potentially extracted silica is advisable.

3 Methodology

For this report, twenty studies investigating the hydrochemistry of hot springs in Chile were compiled based on the data collected by Goldberg et al. (2021), which provide more than 800 hydrochemical samples with major and trace elements throughout Chile. The highest concentrations of each constituent of interest were selected to produce a first impression of the potential economic recovery they may have in Chile. Using those concentrations, a theoretical scenario for potential recovery was calculated, assuming flow rates and availabilities from geothermal power plants.

For the calculation of the economic potential (EP), a simplified extraction scenario was used based on Goldberg et al. (2022):

$$\boldsymbol{EP = c_{max} \cdot Q_a \cdot \varepsilon \cdot A \cdot P} \quad (1)$$

- C_{max} = Maximum concentration of each element taken from the compiled database
- Q_a = Annual brine circulation based on the flow rate: theoretical scenario based on the combined flow rate of all wells of Cerro Pabellón Geothermal Plant (560 L/s, Cappetti et al. 2021)
- ε = Extraction efficiency: conservative assumption of 50%, based on literature data for direct lithium extraction (Goldberg et al., 2022)
- A = Availability of a geothermal plant: assumption of 90%, equals 36 days of maintenance per year
- P = Market price of the specific commodity: based on the accumulated element concentration in the brine for each element, the producible amount of the commonly traded compound was used and the associated price (Table 1)

Table 1 Results of the different scenario calculations. The table gives the traded compounds and related annual extraction rates, market prices, economic values, and the location with the highest potential. Differences between the economic value and the product of producible quantity and price are the result of the rounding of the results

Element	Compound	Potential extraction rate [t/y]	Market price [US$/t]	Resulting value [US$/y]	Location of the highest concentration
Boron	Boric acid (H_3BO_3)	46,000	430 (USGS, 2023)	21,800,000	Puchuldiza-Tuja
Bromine	–	1,200	3,500 (USGS, 2023)	4,700,000	El Tatio
Cesium	Cesium formate ($CHCsO_2$)	340	1,656,000 (USGS, 2023)	615,000,000	Cerro Pabellón
Lithium	Lithium carbonate (Li_2CO_3)	2,600	68,000 (Trading Economics, February 2022)	195,000,000	Cerro Pabellón
Magnesium	Magnesium chloride ($MgCl_2$)	7,400	545 (ChemAnalyst, February 2023)	4,400,000	Irruputuncu Volcano
Potassium	Potassium chloride (KCl)	17,000	660 (ChemAnalyst, February 2023)	12,200,000	Cerro Pabellón
Silicon	Silica (SiO_2)	6,600	300 (Dolley, 2018)	2,100,000	Tolhuaca

4 Results

The samples used for this analysis are mainly from northern Chile: Cerro Pabellón, Puchuldiza-Tuja, El Tatio, and Irruputuncu Volcano, while only one sample is from Tolhuaca Volcano in southern Chile. These locations represent the main geothermal fields of Chile. The samples have temperatures ranging from 36 °C (Irruputuncu Volcano) to 255 °C (Cerro Pabellón) and are from both surface and deep wells.

4.1 *Occurrence and Raw Material Potential*

Boron

Boron was measured in 519 samples, taken mainly in the north of Chile and ranging between 0.002 mg/L and 1,020 mg/L. The mean value of these samples is 41 mg/L, with the highest concentrations measured in the Puchuldiza-Tuja geothermal field (Tassi et al., 2010). With a concentration of 1,020 mg/L, a theoretical mass flow of 100,000 [t/year] is obtained and with an extractable amount of approximately

46,000 [t/year], a potential recovery of US$20,000,000 per year could be obtained in the form of boric acid.

Bromine

In the case of bromine there are 267 data points, sourced from the central part of the country to the north and ranging between 0.003 mg/L and 154.5 mg/L. The mean value of these is 5.24 mg/L, with the highest concentration located in El Tatio (Nicolau et al., 2014).

With the concentration of 155 mg/L, a theoretical bromine mass flow of 2,700 [t/year] is obtained. The extractable amount of approximately 1200 [t/year] lead to a potential recovery of approximately US$4,300,000 per year.

Cesium

Regarding cesium, there are 111 data points, located mainly in the north of the country and ranging between 0.0005 mg/L and 32 mg/L. The mean value is 4.6 mg/L, with the highest concentrations measured in the data from the Cerro Pabellón well (Giudetti & Tempesti, 2020).

With a concentration of 32 mg/L, a theoretical cesium formate mass flow of 750 [t/year] is obtained, with an extractable amount of approximately 340 [t/year] and a potential recovery of approximately US$560,000,000 per year. These numbers have to be seen as sheer hypothetical potential and result from the high prices of the small traded units in the available data (USGS Mineral Commodity Summaries, 2023). These prices are not realistic for large-scale production, since the business-to-business traded cesium prices that reflect the realistic values are not publicly available.

Lithium

Lithium was measured in 538 data points, located mainly from the center of the country to the north, and ranging between 0.0006 mg/L and 62 mg/L. The mean value is 6 mg/L, with the highest concentrations measured in the data from the Cerro Pabellón well (Giudetti & Tempesti, 2020).

With the maximum concentration (62 mg/L in Cerro Pabellón) and the assumed flow rate, we obtain a theoretical lithium carbonate mass flow of 5,800 [t/year], with an extractable amount of 2,600 [t/year]. The potential recovery is approximately US$178,000,000 per year at this location alone.

Magnesium

Magnesium is present in 575 data points, located from the southern part of the country to the north and ranging between 0.001 mg/L and 561 mg/L. The mean value is 20.5 mg/L, with the highest concentration found in the Irruputuncu volcano in northern Chile (Risacher et al., 2011).

With the concentration of 561 mg/L, we obtain a theoretical magnesium mass flow of 16,500 [t/year]. With an extractable rate of approximately 7.400 [t/year], this amounts to a potential recovery of approximately US$4,000,000 per year solely at this location in the form of magnesium chloride.

Potassium

In the case of potassium, there are 617 data points, located mainly from the center of the country to the extreme north and ranging between 0.1 and 1,125 mg/L. The mean value is 75.5 mg/L, with the highest concentrations measured in the data from the Cerro Pabellón well (Giudetti & Tempesti, 2020).

With this concentration of 1,125 mg/L, we obtain a theoretical potassium mass flow of 37,700 [t/year], with an extractable amount of approximately 17,000 [t/year] and a potential recovery of US$11,200,000 per year in the form of potassium chloride.

Silica

Silica is measured in 442 data points, located from the southern part of the country to the north and ranging between 1.16 and 832 mg/L. The mean value is 120.7 mg/L, with the highest concentrations measured in the Tolhuaca geothermal field (Transmark Renewables).

With the concentration of 832 mg/L, we obtain a theoretical silica mass flow of 14,700 [t/year], with an extractable amount of approximately 6,600 [t/year] and a potential recovery of approximately US$2,000,000 per year at this location alone.

5 Discussion and Conclusion

In addition to Chile's great potential to produce energy from geothermal resources, this study demonstrates that there is significant economic potential for the raw materials dissolved in the country's geothermal waters, which, if extracted in the right way, would have several benefits as mentioned below. A number of raw materials are circulated in market-relevant annual amounts at the Cerro Pabellón geothermal power plant (Table 2). Further potential is revealed by the concentrations of different elements occurring in other thermal systems in combination with theoretical flow rates based on Cerro Pabellón.

Table 2 Comparison between approximate potential production and the current national production of Chile

Element/ Compound	Location	Potential production [t/year]	Total national production [t/year]	Percentage of national production [%]
Lithium	Cerro Pabellón	2,600	160,000	1,6
Boron	Puchuldiza-Tuja	46,000	470,000	10
Potassium	Cerro Pabellón	17,000	1,400,000	1.2
Silica	Tolhuaca	6,600	1,500,000	0.5
Magnesium	Irruputuncu	7,400	No data	–
Cesium	Cerro Pabellón	340	No data	–
Bromine	El Tatio	1,200	No data	–

In Table 2, the potential production amounts from geothermal brines are compared with the existing conventional production of each raw material in Chile in 2021 (Servicio Nacional de Geología y Minería, 2021).

The potential to produce an additional 1.6% of Chile's current annual lithium production represents a significant value considering that the calculation is based on the fluids circulated by only one geothermal power plant. In this specific case (Cerro Pabellón plant), this amount is currently produced to surface and reinjected back into the well without any recovery of the raw material.

The amount of dissolved raw materials in the thermal waters can vary over orders of magnitudes. However, the high flow rates in a geothermal plant would compensate for low concentration, resulting in high economic values. In terms of processing, the fact that the raw materials are already dissolved in the water constitutes a significant advantage, since it is not necessary to process large volumes of rock to access them, instead using practically the infrastructure of the geothermal plant.

The extraction of lithium and other minerals from salt flats has a number of disadvantages, namely that the process is dependent on the evaporation rate of the area and that Chile's salt flats are unique and delicate systems with flora and fauna (Garcés & Álvarez, 2012, 2020; Muñoz-Pardo et al., 2015). Green mining from a geothermal source would be less environmentally invasive and would not be dependent on climatic conditions, thus ensuring continuous extraction.

The great potential to obtain boron from geothermal sources (amounting to 10% of the country's current total production) is particularly promising because China, one of Chile's main trading partners, expects to see increased demand for this mineral (USGS Mineral Commodity Summaries, 2023). Accordingly, boron imports from Chile should remain constant or even grow over the next several years, making it necessary to diversify and increase the country's production of this mineral.

The demand for lithium compounds are expected to maintain their long-standing upward trend, mainly due to the use in the automotive industry and battery storage (Comisión Chilena del Cobre, 2020). Also, the current production costs of conventional lithium compounds are estimated to be 4,100 to 5,750 US$/t, while the expected cost of producing lithium compounds from geothermal brines is 4,000 US$/t (Warren, 2021). It can also be compared to the potential recovery of lithium compounds at Cerro Pabellón (US$178,000,000/year), with the initial cost for the field and power plant. In the United States, US$2,500 of costs per installed kW are estimated (The U.S. Department of Energy), resulting in US$203,000,000 for 81 MW_{el} as installed in Cerro Pabellón (without considering mineral extraction costs and other expenses). A comparison of these values demonstrates how the extraction of raw materials would be tremendously beneficial to make investment in geothermal projects more attractive.

Regarding the high prices for cesium, it is worth noting that the world's total consumption of this element could be reached with production at only one geothermal well, meaning that new technologies requiring cesium could be supported. However, cesium is usually found together with ions such as Na^+, K^+, Ca^{2+}, and Mg^{2+}, which hinders the efficiency and selectivity of cesium recovery (Chen et al., 2022). Therefore, the main challenge is to find the appropriate technologies and methods to recover an element that has a promising future on the market.

Another objective of this study is to add value to geothermal projects and thus encourage investment. In addition to the extraction of raw materials, the efficiency of the processes in a geothermal plant can be improved with the selective extraction of scaling minerals. Extraction of silicon may not be lucrative from a raw material point of view, but it can improve the efficiency of geothermal energy production and facilitate the extraction of associated raw materials without the risk of uncontrolled mineral precipitation (scaling). Scaling occurs when minerals precipitate from the liquid and deposit on the surface of the geothermal wells and equipment due to changes of the physico-chemical conditions (such as pH, pressure, and temperature). The use of the silica as a raw material could perhaps compensate for some of the costs of the necessary descaling treatment.

From a social point of view, geothermal projects have less impact on the surface environment than conventional mining. The salt flats of northern Chile interact with aquifer reservoirs where water is a scarce commodity and communities depend on its use. Alternative methods with less water consumption are thus highly advisable. In addition, the extraction of toxic minerals such as arsenic from thermal spring waters (Bundschuh and Maity, 2015) could help supply drinking water to local communities, along with renewable energy generation and heating systems.

This study demonstrates the manifold potential of the geothermal systems of Chile beyond their already known and proven energy potential (Aravena et al., 2016). To unleash this potential it is necessary to research and explore the reservoirs and fluid origins, in addition to developing suitable extraction technologies. The recovery of raw materials from geothermal fluids could significantly increase the economic feasibility of sustainable geothermal energy production as well as create a more sustainable mining approach, not only in Chile but also far beyond its borders.

Acknowledgements The authors would like to thank the Helmholtz Association for research funding within the "Geoenergy" subtopic of the MTET (Materials and Technologies for the Energy Transition) program of the research field "Energy." The BMBF (Federal Ministry of Education and Research) Client II is thanked for funding within the project BrineMine (Grant Number 033R190B). This research is a contribution to the Anid-Fondap projects 15090013, 15200001, and ACE210005 Centro de Excelencia en Geotermia de los Andes (CEGA).

References

Allegrini, G., Luccioli, F., & Trivella, A. (1992). Industrial uses of geothermal fluids at Larderello (Vol. 21, Issue 5).

Angulo, M. (June 14, 2018). Mineral Resource of the month: Cesium. U.S. Geological Survey. EARTH Magazine.

Aravena, D., Muñoz, M., Morata, D., Lahsen, A., Parada, M. Á., & Dobson, P. (2016). Assessment of high enthalpy geothermal resources and promising areas of Chile. *Geothermics, 59*, 1–13. https://doi.org/10.1016/j.geothermics.2015.09.001

Badruzzaman, M., Subramani, A., DeCarolis, J., Pearce, W., & Jacangelo, J. G. (2011). Impacts of silica on the sustainable productivity of reverse osmosis membranes treating low-salinity brackish groundwater. *Desalination, 279*(1–3), 210–218. https://doi.org/10.1016/j.desal.2011.06.013

Biagini, M. (2022). Geothermal mining: The potential for environmentally friendly extraction of valuable components from geothermal brines in the Netherlands. B.sc Minor Project.

Brantes, R. (2016). Proyección del consumo de energía eléctrica en la minería del cobre 2016–2027. *BSEF, International Bromine Council.*

Bromine Science and Environmental Forum (BSEF), International Bromine Council. (2023). https://www.bsef.com/about-bromine/what-is-bromine/production/

Bundschuh, J., & Maity, J. P. (2015). Geothermal arsenic: Occurrence, mobility and environmental implications. *Renewable and Sustainable Energy Reviews, 42*, 1214–1222.

Campero, C., & Harris, L. M. (2019). The legal geographies of water claims: Seawater desalination in mining regions in Chile. *Water (Switzerland), 11*(5). https://doi.org/10.3390/w11050886

Cappetti, G., Giorgi, N., Arias, A., Volpi, G., Rivera, G., Cei, M., Fedeli, M., Marzio, G. di, Pasti, M., Massei, S., & Baccarin, F. (2020). *The Cerro Pabellón Geothermal Project (Chile): From Surface Exploration to Energy Production.*

Chen, S., Dong, Y., Wang, H., Sun, J., Wang, J., Zhang, S., & Dong, H. (2022). Highly efficient and selective cesium recovery from natural brine resources using mesoporous Prussian blue analogs synthesized by ionic liquid-assisted strategy. *Resources, Conservation and Recycling, 186*. https://doi.org/10.1016/j.resconrec.2022.106542

Chong, G., Pueyo, J. J., & Demergasso, C. (2000). Los yacimientos de boratos de Chile. *Revista Geológica De Chile, 27*(1), 99–119. https://doi.org/10.4067/S0716-02082000000100007

Cob, S., Hofs, B., Maffezzoni, C., Adamus, J., Siegers, W. G., Cornelissen, E. R., Genceli Güner, F. E., & Witkamp, G. J. (2014). Silica removal to prevent silica scaling in reverse osmosis membranes. *Desalination, 344*, 137–143. https://doi.org/10.1016/j.desal.2014.03.020

Comisión Chilena del Cobre. (2013). Monitoreo de los minerales industriales de Chile. Análisis de los recursos salinos. Santiago, Chile.

Comision Chilena del Cobre. (2020). Estadísticas de la minería no metálica 2005–2020. https://www.cochilco.cl/Paginas/Estadisticas/Bases%20de%20Datos/Estad%C3%ADsticas-de-la-Miner%C3%ADa-No-Met%C3%A1lica.aspx

Dolley, T. (2018). 2018 US Geological Survey Minerals Yearbook–Silica.

Flexer, V., Baspineiro, C., & Galli, C. (2018). Lithium recovery from brines: A vital raw material for green energies with a potential environmental impact in its mining and processing. *Science of the Total Environment, 639*, 1188–1204. https://doi.org/10.1016/j.scitotenv.2018.05.223

Fournier, R. O., & Rowe, J. J. (1966). Estimation of underground temperatures from the silica content of water from hot springs and wet-steam wells. *American Journal of Science, 264*, 685–697.

Garcés, I., & Alvarez, G. (2012). Implicancias del cambio climático sobre la biodiversidad del Salar de Aguas Calientes I, Chile.

Garcés, I., & Álvarez, G. (2020). Water mining and extractivism of the Salar de Atacama, Chile. *WIT Transactions on Ecology and the Environment, 245*, 189–199.

García, N. (2021). Matriz energética y eléctrica en Chile. Biblioteca del Congreso Nacional de Chile.

Giudetti, G., Tempesti, L. (2020). First geochemical data from Cerro Pabellon Geothermal Project (Apacheta region, Chile). In *Proceedings World Geothermal Congress*, Reykjavik, Iceland.

Goldberg, V., Winter, D., Nitschke, F., Rath, M., Held, S., Spitzmüller, L., Budach, I., Pavez, M., Morata, D., Koschikowski, J., Kohl, T., Goldberg, V., Winter, D., Nitschkea, F., Rath, M., Held, S., Spitzmüller, L., Budach, I., Pavez, M., & Kohl, T. (2021). The potential of raw material extraction from thermal brines-Successful milestones of the BrineMine project. https://doi.org/10.19225/210306

Goldberg, V., Kluge, T., & Nitschke, F. (2022). Herausforderungen und Chancen für die Lithiumgewinnung aus geothermalen Systemen in Deutschland–Teil 1: Literaturvergleich bestehender Extraktionstechnologien. *Grundwasser*. https://doi.org/10.1007/s00767-022-00522-5

Gonzalez, S. (2020). Oferta y demanda de litio hacia el 2030. Comision chilena del cobre, Santiago, Chile.

Gunnarsson, I., & Arnórsson, S. (2005). Impact of silica scaling on the efficiency of heat extraction from high-temperature geothermal fluids. *Geothermics, 34*(3), 320–329. https://doi.org/10.1016/j.geothermics.2005.02.002

Henley, R. W. (1983). PH and silica scaling control in geothermal field development. *Geothermics, 12*(4), 307–321.

Huttrer, G. W. (2020, April). Geothermal power generation in the world 2015–2020 update report. In *Proceedings World Geothermal Congress* (Vol. 2020, p. 17).

Iler, R. K. (1979). *The Chemistry of silica: Solubility, polymerization, colloid and surface properties, and biochemistry* (p. 866). Wiley.

Lu, Y., Saroja, A. P. V. K., Wei, R., & Xu, Y. (2021). Engineering metal selenides for sodium-and potassium-ion batteries. In *Cell reports physical science* (Vol. 2, Issue 9). Cell Press. https://doi.org/10.1016/j.xcrp.2021.100555

Meixner, A., Sarchi, C., Lucassen, F., Becchio, R., Caffe, P. J., Lindsay, J., Rosner, M., & Kasemann, S. A. (2020). Lithium concentrations and isotope signatures of Palaeozoic basement rocks and Cenozoic volcanic rocks from the Central Andean arc and back-arc. *Mineralium Deposita, 55*(6), 1071–1084. https://doi.org/10.1007/s00126-019-00915-2

Mercado, S., & Hurtado, R. (1992). *Potash extraction from Cerro Prieto Geothermal Brine* (Vol. 21, Issue 5).

Mesa de Geotermia. (2018). Rol de la geotermia en el desarrollo de la matriz eléctrica chilena. Ministerio de Energía, Chile. http://www.minenergia.cl/mesa-geotermia

Mi, B., & Elimelech, M. (2013). Silica scaling and scaling reversibility in forward osmosis. *Desalination, 312*, 75–81. https://doi.org/10.1016/j.desal.2012.08.034

Muñoz-Pardo, J. F., Ortiz-Astete, C. A., Mardones-Pérez, L., & de Vidts-Sabelle, P. (2015). Funcionamiento hidrogeológico del acuífero del núcleo del salar de Atacama, Chile. *Tecnología Y Ciencias Del Agua, 19*(3), 69–81.

Neupane, G., & Wendt, D. S. (2017). Assessment of mineral resources in geothermal brines in the US rare earth elements view project climate change view project assessment of mineral resources in geothermal brines in the US. http://gdr.openei.org/files/194/GEOTHERM_ALL.xls

Nicolau, C., Reich, M., & Lynne, B. (2014). Physico-chemical and environmental controls on siliceous sinter formation at the high-altitude El Tatio geothermal field, Chile. *Journal of Volcanology and Geothermal Research, 282*, 60–76. https://doi.org/10.1016/j.jvolgeores.2014.06.012

Pincetti-Zúniga, G. P., Richards, L. A., Daniele, L., Boyce, A. J., & Polya, D. A. (2022). Hydrochemical characterization, spatial distribution, and geochemical controls on arsenic and boron in waters from arid Arica and Parinacota, Northern Chile. *Science of the Total Environment, 806*. https://doi.org/10.1016/j.scitotenv.2021.150206

Risacher, F., Fritz, B., & Hauser, A. (2011). Origin of components in Chilean thermal waters. *Journal of South American Earth Sciences, 31*(1), 153–170. https://doi.org/10.1016/j.jsames.2010.07.002

Servicio Nacional de Geología y Minería. (2021). Anuario de la Mineria chilena. SERNAGEOMIN, Santiago, Chile.

Sociedad Química y Minera de Chile S.A. (2021). *Annual Report 2021*. https://www.sqm.com

Sociedad Química y Minera de Chile (SQM). (2023). https://www.sqm.com/productos

Tassi, F., Aguilera, F., Darrah, T., Vaselli, O., Capaccioni, B., Poreda, R. J., & Delgado Huertas, A. (2010). Fluid geochemistry of hydrothermal systems in the Arica-Parinacota, Tarapacá and

Antofagasta regions (northern Chile). *Journal of Volcanology and Geothermal Research, 192*(1–2), 1–15. https://doi.org/10.1016/j.jvolgeores.2010.02.006
Trading Economics. (2023). Lithium Summary. Tradingeconomics.com.
US Geological Survey. (2023). Mineral commodity summaries, 2021. *Government Printing Office*. https://doi.org/10.3133/mcs2023
Warren, P. (2021). Techno-economic analysis of lithium extraction from geothermal brines (No. NREL/TP-5700-79178). National Renewable Energy Lab. (NREL), Golden, CO (United States).
Zhang, P., Pabstmann, A., Gray, S., & Duke, M. (2018). Silica fouling during direct contact membrane distillation of coal seam gas brine with high sodium bicarbonate and low hardness. *Desalination, 444*, 107–117. https://doi.org/10.1016/j.desal.2018.07.015

Urban Eco-Industrial Development in South America—Introduction

Symbiosis Between Urban and Industrial Metabolism

Lorenzo Reyes-Chamorro

Abstract The chapter delves into the intricate relationship between urban and industrial metabolism, shedding light on the symbiotic interactions that occur between these two systems. This chapter aims at introducing the innate connections between urban processes and industrial processes. These connections have been widely studied in literature and are usually compared with human and biological phenomena. In that sense, in this chapter the concept of symbiosis between urban and industrial metabolism is discussed, as a way to contextualise and localise the contribution of the solutions presented in the following chapters. Overall, the chapter highlights the significance of understanding and optimizing the flows within urban systems for a more sustainable future, and thus places the empirical studies in the book section on urban eco-industrial development in a wider context.

1 Symbiosis Between Urban and Industrial Metabolism

Throughout history, researchers from different disciplines such as sociology, ecology, and biology have frequently compared human systems with biological organisms. With this comparison, metabolism—defined as "the chemical processes in living things that change food, etc. into energy and materials for growth" (Oxford University Press, 2023)—can be understood in the human system as the process of using supplies to perform any actions necessary for the correct operation of the system, and may produce residues as an eventual result (Kennedy et al., 2007). Usually, the concept of urban metabolism (UM) is applicable when the system in question is an entire city. Indeed, the UM concept can be traced back to Marx (Pincetl et al., 2012) but it wasn't until 1965 (Wolman, 1965) when the idea of metabolism in cities began to gather attention. Figure 1 shows a simplified diagram of a city in which the main flows are identified.

L. Reyes-Chamorro (✉)
Universidad Austral de Chile, General Lagos, 2086 Valdivia, Chile
e-mail: lorenzo.reyes@uach.cl

A. Ch. Braun et al. (eds.), *Eco-Industrial Development as an Industrial Strategy*,
Green Energy and Technology, https://doi.org/10.1007/978-3-031-73576-9_20

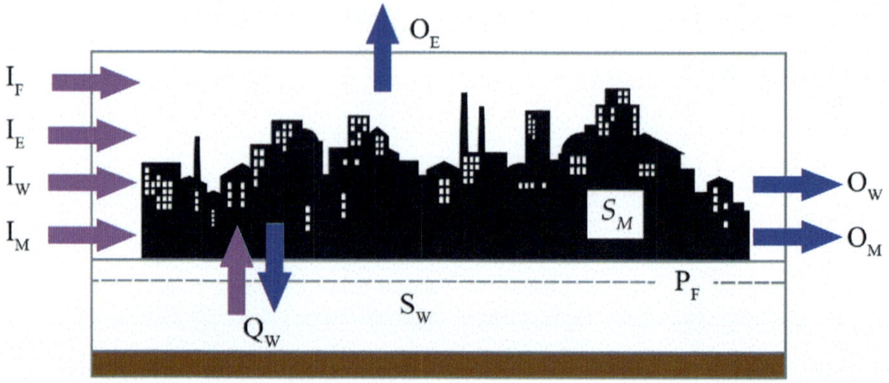

Fig. 1 A simplified diagram of urban metabolism (Derrible et al, 2021), where inputs (I), outputs (O), internal flows (Q), and storage (S) are classified by production (P) of water (W), energy (E), material (M), and food (F)

One of the earliest applied studies to use this concept to quantify the flows of materials, water, and energy was one for Brussels, Belgium (Duvigneaud & Denaeyer-De Smet, 1977) as presented in Fig. 2.

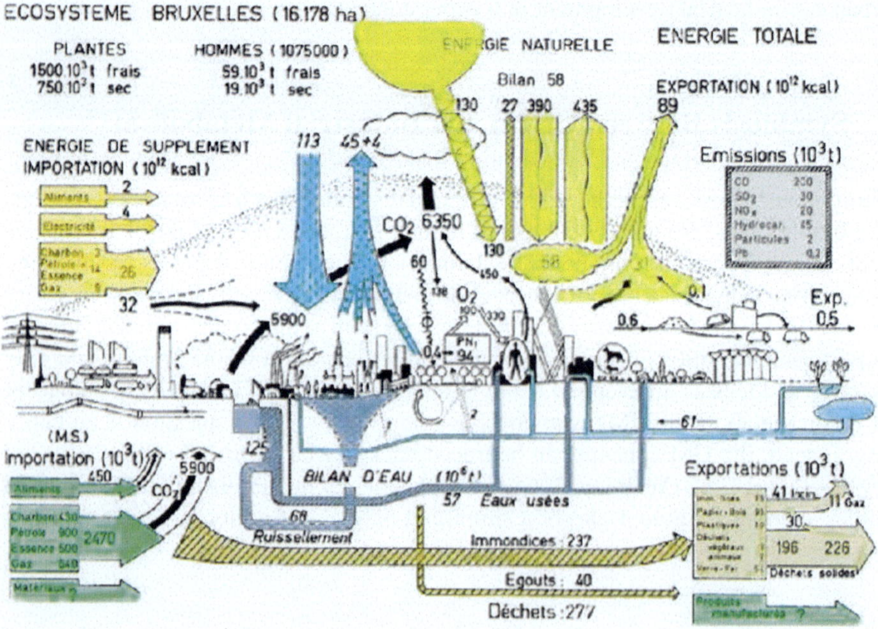

Fig. 2 The urban metabolism of Brussels. Duvigneaud and Denaeyer-De Smet (1977) (colored in Federico Cinquepalmi's "Towards (R)evolving Cities," 2021)

A particular subsystem inside the city is that of industry, for which the concept of industrial metabolism (IM) can similarly be defined. Specifically, IM can be understood as the set of processes that convert raw material into refined materials by means of energy and labor, eventually producing waste or residues (Ayres & Simonis, 1994). Since the magnitude of the flows in industrial activity is relatively large with respect to the total of the city, it has been natural to analyze IM separately from UM (Fischer-Kowalski, 2002). Indeed, when both are compared, UM usually refers only to the processes related to human activities (e.g., transportation, food, commerce, and so on).

It is thus clear that UM and IM are strongly interconnected and interdependent. Following the biological analogy, one can speak of *symbiosis between UM and IM* (Hess, 2010). In this respect, processes such as reuse or recycling become crucial when evaluating the interdependence between both (Ferrão & Fernández, 2013). Since UM and IM are linked in reality anyway, this approach can help develop mechanisms and tools to minimize negative ecological impacts (Dong et al., 2014; Kim et al., 2018; Shah et al., 2020). For example, concepts of eco-industrial development can be used as part of urban and regional planning (Gibbs & Deutz, 2005). Indeed, the concept of circular metabolism is often used when looking at minimizing residues and unnecessary inputs as shown in Fig. 3.

This minimization, usually referred to in the literature as *efficiency (*Conke & Ferreira, 2015; Gan et al., 2023; Zhang & Yang, 2007*)*, has been the focus of persistent research in the last decades. Yet focusing only on efficiency could lead to inequity (Dijst et al., 2018) and thus well-being should instead be the focus, where both efficiency and equity must be taken into account concurrently as shown in Fig. 4 (Broto et al., 2012; McCulligh & Fregoso, 2019). Recent literature has consequently raised awareness of the need for a multi-disciplinary view of UM and IM (Broto et al., 2012; Molina-Prieto et al., 2019; Tröger et al., 2023).

In this context, when studying the symbiosis between UM and IM apart from technological aspects related to the conversion of inputs to outputs, for which a very large body of literature already exists (e.g., McManus & Gibbs, 2008; Ohnishi et al., 2017; Yang & Lay, 2004), the following elements are to be considered.

1. **Economies of scale**: Processes for reuse or recycling are usually difficult to put into practice on a small scale. Therefore, when designing new methods for taking advantage of residues from industry or from citizens, it is key to analyze how these can quickly be scaled up.
2. **Spatial scale**: Cities are geographically constrained systems. Thus, land use needs to be part of the variables under consideration when evaluating well-being.
3. **Social acceptance**: Although the minimization of residues such as contaminants or greenhouse gas emissions is the objective, citizens could potentially oppose projects that optimize well-being, such as the well-known NIMBY (Hermansson, 2007) conflicts. These challenges to acceptance are therefore to be identified and proactively tackled.
4. **Implementation mechanisms**: Regulation is key for the development of new alternatives for the reuse or recycling of residues. It is consequently important

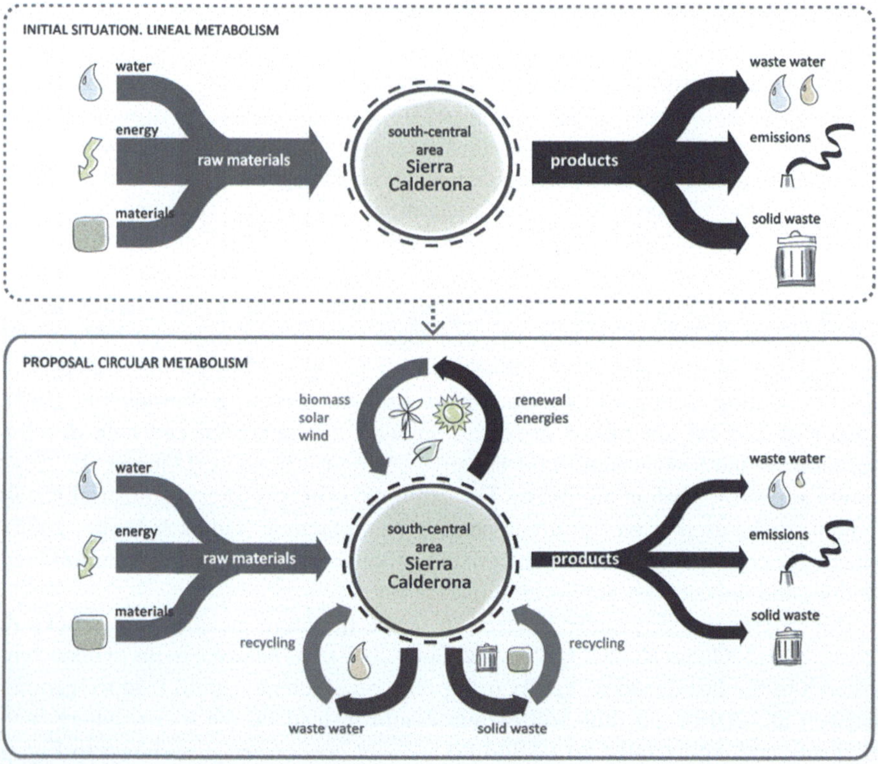

Fig. 3 Linear and circular metabolism, taken from a study on the particular case of Sierra Calderona (Galan & Perrotti, 2019)

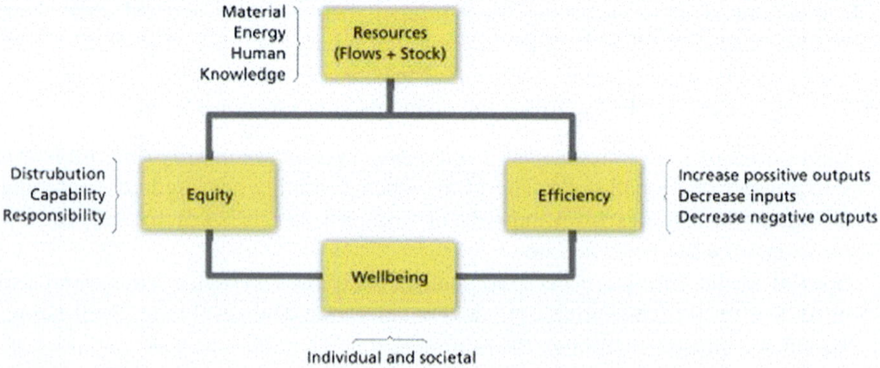

Fig. 4 Efficiency and equity in urban and industrial metabolism

to identify whether new mechanisms or modifications of existing ones would be needed to deploy a new technology.

5. **Centralized versus decentralized**: When regulations are centralized at the national level (as with most Latin American countries, for example), the implementation of new strategies could be difficult. In that respect, close work with local authorities is always recommended in the implementation of new conversion projects.

The above list represents the main aspects where attention is needed, according to the author, and is not intended to be an exhaustive analysis. Indeed, other elements could be added to the list depending on political, economic, social, technical, legal, and/or ecological aspects (Perera, 2017).

In this part of the book, four applied experiences of UM and IM symbiosis are presented, one in each of the following chapters.

First, the use of urban residues such as cigarette butts, milk packages, and industrial residues such as cellulose are analyzed as possible materials for the construction of pavements when incorporated into asphalt mixtures. These considered materials represent typical residues for the city of Valdivia, Chile. The chapter by Andrés-Valeri et al. (2024) presents an experimental analysis of the different pavement structures that result from each mixture.

Second, the example of using treated urban wastewater in the mining industry of Antofagasta, Chile is presented. The chapter by Chacana et al. (2024) first contextualizes the situation by presenting the current situation in the zone. Next, it shows the potential for the community of wastewater recycling and the current implementation of a new treatment system that incorporates UM and IM symbiosis.

Third, the possibility of taking advantage of large-scale mining processes in urban mining is studied using the Chilean example. The chapter by Vega-Garcia et al. (2024) dissects the entirety of the mining process and analyzes in detail what sub-processes can be applied to urban mining, especially in the case of electric and electronic waste.

Fourth, Tröger et al. (2024) study urban and industrial symbiosis to identify the challenges in its implementation, using several Latin American cities as case studies. This more general vision focuses not only on the technological aspect of converting waste PET bottles into construction panels, but also analyzes the impact on local communities by incorporating informal waste-pickers in the process, a symbiosis taking place mostly "under the radar" (Tröger et al., 2023).

These four chapters are depicted in Fig. 5, with the inputs and outputs of each experience also identified.

Finally, the last two chapters of this topic provide examples of the application of the UM and IM symbiosis concept using two particular case studies.

The first of these, discusses the need to design urban areas in an adaptive way (Flores et al., 2024). In this chapter, which focuses on the city of Padre Las Casas, Chile, the maximization of the use of local resources is considered as a way to improve urban resilience to natural disasters.

Vallebuona and Schmidt (2024) present the design and implementation of the ICELab, a laboratory conceived for scientific work in rough weather conditions and

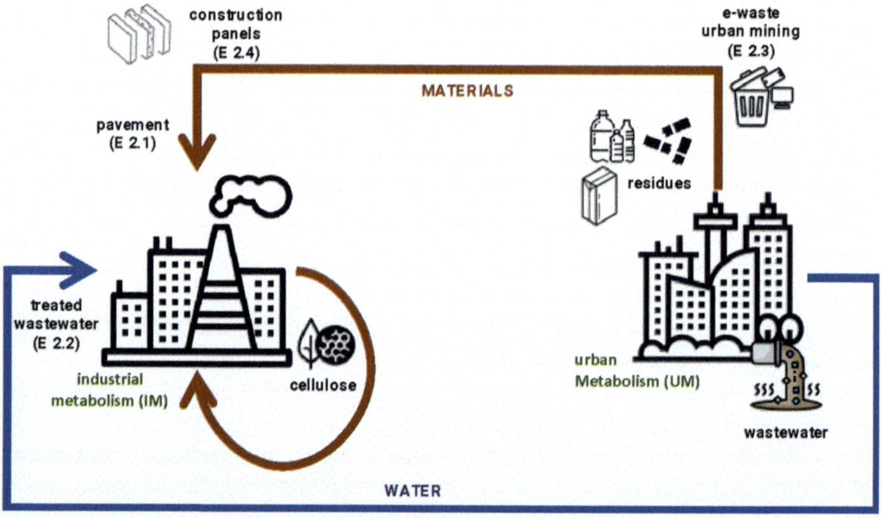

Fig. 5 Specific experiences presented in Part E showing the symbiosis between urban and industrial metabolism

in areas with restricted access to large urban zones. The façade of the laboratory has been built using untreated natural wood from industrial residues, treats and reuses its wastewater, and uses local renewable resources such as wind to supply its electricity. In a sense, the ICELab thus replicates various aspects at the nexus of urban and industrial metabolism, albeit in small towns in otherwise peripheral rural areas.

References

Andrés-Valeri, V. C., Raposeiras, A. C., Movilla-Qesada, D., Muñoz-Caceres, O., & Lagos-Varas, M. (2024). Sustainable pavement materials: Using urban and industrial residues in porous asphalt mixtures. In A. C. Braun, G. Espinosa, D. Tröger, & T. Hirth (Eds.), *Eco-industrial development as an industrial strategy - contributions from a German-Chilean research partnership*. Springer Nature.

Ayres, R. U., & Simonis, U. E. (1994). Industrial metabolism: Restructuring for sustainable development.

Broto, V. C., Allen, A., & Rapoport, E. (2012). Interdisciplinary perspectives on urban metabolism. *Journal of Industrial Ecology, 16*(6), 851–861. https://doi.org/10.1111/j.1530-9290.2012.00556.x

Chacana, J., Herrera, S., Cruz, C., & Guerra, P. (2024). Water, cities, and industries in an arid zone: The case study of Antofagasta, Chile. In A. C. Braun, G. Espinosa, D. Tröger, & T. Hirth, (Eds.), *Eco-industrial development as an industrial strategy-contributions from a German-Chilean research partnership*. Springer Nature.

Conke, L. S., & Ferreira, T. L. (2015). Urban metabolism: Measuring the city's contribution to sustainable development. *Environmental Pollution, 202*, 146–152. https://doi.org/10.1016/j.envpol.2015.03.027

Derrible, S., Cheah, L., Arora, M., & Yeow, L. W. (2021). Urban metabolism. In W. Shi, M. F. Goodchild, M. Batty, M. P. Kwan, & A. Zhang (Eds.), Urban Informatics (pp. 5–114). The Urban Book Series. Springer. https://doi.org/10.1007/978-981-15-8983-6_7

Dijst, M., Worrell, E., Böcker, L., Brunner, P., Davoudi, S., Geertman, S., Harmsen, R., Helbich, M., Holtslag, A. A., Kwan, M., Lenz, B., Lyons, G., Mokhtarian, P. L., Newman, P., Perrels, A., Ribeiro, A. P., Rosales, J., Thomson, G., Urge-Vorsatz, D., & Zeyringer, M. (2018). Exploring urban metabolism—Towards an interdisciplinary perspective. *Resources, Conservation and Recycling, 132*, 190–203. https://doi.org/10.1016/j.resconrec.2017.09.014

Duvigneaud, P., & Denaeyer-De Smet, S. (1977). L'écosystème urbs. L'écosystème urbaine Bruxellois, Productivité biologique en Belgique.

Dong, H., Ohnishi, S., Fujita, T., Geng, Y., Fujii, M., & Dong, L. (2014). Achieving carbon emission reduction through industrial & urban symbiosis: A case of Kawasaki. *Energy, 64*, 277–286. https://doi.org/10.1016/j.energy.2013.11.005

Ferrão, P., & Fernández, J. E. (2013). Sustainable urban metabolism. *The MIT Press*. https://doi.org/10.7551/mitpress/8617.001.0001

Fischer-Kowalski, M. (2002). Exploring the history of industrial metabolism. In R. U. Ayres & L. W. Ayres (Eds.), *A handbook of industrial ecology* (pp. 16–26). Edward Elgar.

Flores, A., Agurtoo, L., Ibieta, P., & Valderrama, C. (2024). Adaptive urban design for resilience in the Chilean City of Padre Las Casas. In A. C. Braun, G. Espinosa, D. Tröger, & T. Hirth, (Eds.), *Eco-industrial development as an industrial strategy-contributions from a German-Chilean research partnership*. Springer Nature.

Galan, J., & Perrotti, D. (2019). Incorporating metabolic thinking into regional planning: The case of the sierra calderona strategic plan. *Urban Planning, 4*(1), 152–171. https://doi.org/10.17645/up.v4i1.1549

Gan, L., Wan, X., Ma, Y., & Lev, B. (2023). Efficiency evaluation for urban industrial metabolism through the methodologies of emergy analysis and dynamic network stochastic block model. *Sustainable Cities and Society, 90*, 104396. https://doi.org/10.1016/j.scs.2023.104396

Gibbs, D., & Deutz, P. (2005). Implementing industrial ecology? Planning for eco-industrial parks in the USA. *Geoforum, 36*, 452–464. https://doi.org/10.1016/j.geoforum.2004.07.009

Hermansson, H. (2007). The ethics of NIMBY conflicts. *Ethic Theory and Moral Practice, 10*, 23–34. https://doi.org/10.1007/s10677-006-9038-2

Hess, G. (2010). The ecosystem: Model or metaphor? *Journal of Industrial Ecology, 14*, 270–285. https://doi.org/10.1111/j.1530-9290.2010.00226.x

Kennedy, C., Cuddihy, J., & Engel-Yan, J. (2007). The Changing metabolism of cities. *Journal of Industrial Ecology, 11*, 43–59. https://doi.org/10.1162/jie.2007.1107

Kim, H.-W., Dong, L., Choi, A. E. S., Fujii, M., Fujita, T., & Park, H. (2018). Co-benefit potential of industrial and urban symbiosis using waste heat from industrial park in Ulsan, Korea. *Resources, Conservation and Recycling, 135*, 225–234. https://doi.org/10.1016/j.resconrec.2017.09.027

McCulligh, C., & Fregoso, G. (2019). Defiance from Down River: Deflection and dispute in the urban-industrial metabolism of pollution in Guadalajara. *Sustainability, 11*(22), 6294. https://doi.org/10.3390/su11226294

McManus, P., & Gibbs, D. (2008). Industrial ecosystems? The use of tropes in the literature of industrial ecology and eco-industrial parks. *Progress in Human Geography, 32*, 525–540. https://doi.org/10.1177/0309132507088118

Molina-Prieto, L, Suárez-Serrano, M., & Villa-Camacho, M. (2019). Bucle multidisciplinar para la sustentabilidad urbana. Revista de Arquitectura (Bogotá), *21*(2), 76–88. https://doi.org/10.14718/RevArq.2019.21.2.2048

Oxford University Press. (2023). Metabolism noun-Definition, pictures, pronunciation and usage notes. https://www.oxfordlearnersdictionaries.com/us/definition/english/metabolism?q=metabolism

Ohnishi, S., Dong, H., Geng, Y., Fujii, M., & Fujita, T. (2017). A comprehensive evaluation on industrial & urban symbiosis by combining MFA, carbon footprint and emergy methods—Case of Kawasaki, Japan. *Ecological Indicators, 73*, 513–524. https://doi.org/10.1016/j.ecolind.2016.10.016

Perera, R. (2017). *The PESTLE analysis*, Nerdynaut.

Pincetl, S., Bunje, P., & Holmes, T. (2012). An expanded urban metabolism method: Toward a systems approach for assessing urban energy processes and causes. *Landscape and Urban Planning, 107*(3), 193–202. https://doi.org/10.1016/j.landurbplan.2012.06.006

Shah, I. H., Dong, L., & Park, H. (2020). Tracking urban sustainability transition: An eco-efficiency analysis on eco-industrial development in Ulsan, Korea. *Journal of Cleaner Production, 262*, 121286. https://doi.org/10.1016/j.jclepro.2020.121286

Tröger, D., Araneda, A. A. B., Busnelli, R., Yajnes, M., Williams, F., & Braun, A. C. (2023). Exploring eco-industrial development in the global south: Recognizing informal waste-picking as urban-industrial symbiosis? *Cleaner Waste Systems, 5*, 100096. https://doi.org/10.1016/j.clwas.2023.100096

Tröger, D., Becerra Araneda, A. A., Busnelli, R., & Braun, A. C. (2024). Enhancement of the urban habitat through novel construction products from solid industrial residues: Opportunities and challenges for urban-industrial symbiosis by including informal waste-pickers. In A. C. Braun, G. Espinosa, D. Tröger, & T. Hirth (Eds.), *Eco-industrial development as an industrial strategy-contributions from a German-Chilean research partnership*. Springer Nature.

Vallebuona, R., & Schmidt, S. (2024). ICELab, a Prototype module for environmental scientists. A joint project of the Karlsruhe Institute of Technology (KIT), Germany and the Universidad de Concepción (UdeC), Chile. In A. C. Braun, G. Espinosa, D. Tröger, & T. Hirth, (Eds.), *Eco-industrial development as an industrial strategy-contributions from a German-Chilean research partnership*. Springer Nature.

Vega-Garcia, D., Ramirez, A., Gutierrez, L., Bustos, C., & Salinas, L. (2024). Opportunities for the sustainable development of urban mining: lessons learned from the large-scale mining industry. In A. C. Braun, G. Espinosa, D. Tröger, & T. Hirth (Eds.), *Eco-industrial development as an industrial strategy-contributions from a German-Chilean research partnership*. Springer Nature.

Wolman, A. (1965). The metabolism of cities. *Scientific American, 213*(3), 179–190.

Yang, P. P., & Lay, O. B. (2004). Applying ecosystem concepts to the planning of industrial areas: A case study of Singapore's Jurong Island. *Journal of Cleaner Production, 12*, 1011–1023. https://doi.org/10.1016/j.jclepro.2004.02.028

Zhang, Y., & Yang, Z. (2007). Eco-efficiency of urban material metabolism: A case study in Shenzhen, China. *Acta Ecologica Sinica, 27*(8), 3124–3131. https://doi.org/10.1016/S1872-2032(07)60067-5

Urban Eco-Industrial Development in South America—Case Studies from Chile and Germany

Sustainable Pavement Materials: Using Urban and Industrial Residues in Porous Asphalt Mixtures

Valerio C. Andrés-Valeri, Aitor C. Raposeiras, Diana Movilla-Quesada, Osvaldo Muñoz-Caceres, and Manuel Lagos-Varas

Abstract Urban and industrial residues are generally difficult to recycle or reuse. Although combustion soot, cellulose ash, cigarette butts, and Tetra Brik Aseptic© containers are abundantly available in Chile, the country's recycling industry is struggling to find an affordable techno-economical solution. In this chapter, we will show experimental results and an analysis of particle loss, air void, and the stability of porous asphalt mixtures manufactured for different types of pavement, and show that all of them can be incorporated into porous asphalt mixtures without sacrificing mechanical and functional capabilities, even increasing their mechanical strength and durability.

V. C. Andrés-Valeri (✉)
Departamento de Ingeniería Civil, Universidad de Las Palmas de Gran Canaria, Las Palmas de Gran Canaria, España
e-mail: valerio.andres@ulpgc.es

A. C. Raposeiras
Departamento de Ingeniería Mecánica, Escuela Politécnica Superior de Zamora, Universidad de Salamanca, Zamora, España
e-mail: araposeiras@usal.es

D. Movilla-Quesada
Departamento de Construcción y Agronomía, Escuela Politécnica Superior de Zamora, Universidad de Salamanca, Zamora, España
e-mail: dmovilla@usal.es

O. Muñoz-Caceres · M. Lagos-Varas
Gi2V Research Group, Institute of Civil Engineering, Faculty of Engineering Sciences, Universidad Austral de Chile, Valdivia, Chile
e-mail: osvaldo.munoz@uach.cl

M. Lagos-Varas
e-mail: manuel.lagos@uach.cl

A. Ch. Braun et al. (eds.), *Eco-Industrial Development as an Industrial Strategy*, Green Energy and Technology, https://doi.org/10.1007/978-3-031-73576-9_21

1 Introduction

Transportation networks are physically realized through transportation infrastructures that connect different points in order to allow the movement of people and goods from one point to another in a cycle that brings economic, cultural, and social development. One of the most important transportation infrastructures is roads and pavement. They have been used for hundreds of years to allow the movement of wheeled traffic in better conditions than those provided by the natural landscape in terms of travel time, safety, and amenities. Although technical and technological solutions have changed over time, nowadays roads maintain the same objective. Yet, they have been adapted in order to meet new requirements in terms of the bearing capacity, regularity, and safety required by motorized vehicles. Furthermore, nowadays road infrastructures are much more complex than those developed hundreds of years ago, allowing higher travel speeds and better travel conditions.

Road infrastructures are built using different types of pavement that can be classified according to their mechanical response to the traffic loads: flexible pavement, rigid pavement, and interlocking pavement. Interlocking pavement is composed of discrete prefabricated elements, also called pavers, that are interlocked with one another to form a matrix of elements that resists traffic loads through frictional forces among the pavers. On the other hand, flexible and rigid pavement is made by using a mineral aggregate skeleton coated with a binder, which can be of several different types. While a cementitious binder is used for rigid pavement, giving rise to so-called concrete pavement, in flexible pavement the binder used is a bituminous binder leading to what is called asphalt pavement. Asphalt and concrete pavements are normally made of different layers of asphalt and concrete mixtures, respectively, placed over a compacted and eventually stabilized layer of soil. Even if they are quite similar in composition, their mechanical performance is vastly different due to the properties of the binders used in asphalt and concrete mixtures. While concrete pavement resists vertical loads directly and without important deformations before cracking, in rigid performance, asphalt pavement partially transmits the vertical loads to inferior layers resisting through a visco-elasto-plastic model. In general, asphalt pavement is much more common than concrete pavement in road infrastructures, mainly due to the lower costs and increased benefits of this material in terms of comfort.

Asphalt mixtures can be defined as specific combinations of the materials they are composed of: bituminous binders, coarse aggregates, fine aggregates, and filler or very fine aggregates. Eventually, other elements and substances can be used as additives and additions in order to provide some specific characteristics. Even if asphalt mixtures can be classified according to different criteria, normally they are catalogued according to their gradation in dense-graded asphalt mixtures, open-graded asphalt mixtures, and stone-mastic asphalt mixtures.

1.1 Porous Asphalt

Porous asphalt (PA) mixtures are a special type of open-graded asphalt mixture, also called porous friction course (PFC) or open-graded friction course (OGFC) in the USA (Alvarez et al., 2011). These mixtures are designed for maintaining high air void content, normally higher than 18–20%, while providing the necessary bearing capacity and durability in resisting the traffic loads applied during a reasonable service life (usually 6–12 years). While the granular skeleton provides the mechanical strength, the interconnected air voids provide paths for water infiltration (Andrés Valeri et al., 2018). The high air void content in porous asphalt mixtures enables rainwater infiltration, pollutant filtering and increased skidding resistance and road safety, as well as a reduction in hydroplaning, water splash, and glare effects in wet weather conditions. In fact, spray and splash have been proven to decrease in ranges of 90–95% (Nicholls & Daines, 1997; Nichols, 1999), and visibility loss was reduced by 40–50% in relation to conventional pavement surfaces (Rungruangvirojn & Kanitpong, 2010), reducing the number of wet weather crashes and accidents to between 50 and 100% (Dale, 2012; Rand, 2004; Takahashi, 2013). Different studies showed that the use of porous asphalt mixtures also has great benefits in terms of traffic, enabling higher vehicle speeds and traffic capacities (Cooley et al., 2009), and also reducing fuel consumption (Bennert et al., 2005).

Despite the benefits of porous mixtures, there are still big challenges that have to be addressed for their wider application. For instance, the durability of PA mixtures is usually in the range of 6–12 years (Fay & Akin, 2014; Hernandez-Saenz et al., 2016; Mallick et al., 2000; Root, 2009), which is slightly shorter than the service life of other type of asphalt mixtures (20 years) (Smith et al., 1998). The successive application of vertical stresses can cause excessive plastic deformations (Alvarez et al., 2011), producing pavement failure. Shear stresses, produced by vehicle movements and maneuvers, may result in raveling damage. Raveling is the phenomenon that occurs when aggregate particles at the pavement surface are broken off (Wurst & Putman, 2013), leading to windshield damage, dangerous driving conditions (Molenaar et al., 2010), and a consequent reduction in the service life of the pavement. The environmental conditions that affect porous asphalt mixtures on-site can significantly modify their mechanical behavior over time. The freeze and thaw cycles in cool climates can produce microcracks due to the freezing of the retained water inside the porous matrix, reducing the mechanical strength of the materials (Kevern et al., 2010; Sol-Sánchez et al., 2015). On the other hand, bituminous binders can be degraded by the water contact affected by oxidation mechanisms (Hernandez Noguera et al., 2014), this effect being much more critical in PA mixtures than in other asphalt mixtures due to the internal pore structure that allows binder oxidation throughout the entire pavement layer (Molenaar et al., 2010). Other environmental conditions such as temperature and solar radiation contribute to the aging of PA mixtures, making them brittle and increasing their raveling potential (Kuang et al., 2014; Mansour & Putman, 2013).

To ensure the good mechanical response of PA mixtures, it is necessary to obtain an adequate thickness of the binder film on the aggregates to ensure mixture durability and to prevent raveling damage. With this goal, stabilizing additives are normally incorporated for stabilizing mixtures with high binder contents and to prevent binder draindown (Wurst & Putman, 2013). Different studies have pointed out the suitability of using different kinds of fibers for this purpose, highlighting cellulose fibers (Wu et al., 2006) or mineral fibers (Cooley Jr. et al., 2000). The addition of cellulose fibers to PA mixtures, in a range of 0.2–0.5% by weight of mixture, helps to increase the binder content thus preventing draindown problems, providing significant improvements in the long-term raveling resistance by sacrificing some infiltration capacity (Alvarez et al., 2011; Cooley et al., 2009; Chen et al., 2012, 2013; Lyons & Putman, 2013). The particle losses measured by the Cantabro test can be decreased by between 5 and 20% in absolute values depending on the binder type used, also under aged conditions. According to the literature reviewed, cellulose fibers have also shown a positive effect on the indirect tensile strength (ITS), where values of 15–20% higher in PA mixtures with fibers were observed. However, the effect of cellulose fibers in the mixtures' water sensitivity are questionable, with improvements not being observed in the literature.

Traditionally, the addition of mineral filler to PA mixtures has been used to improve their mechanical strength and durability, reducing draindown problems and increasing the mastic stiffness and the aggregate coat. The main road agencies recommend the use of limestone filler, hydrated lime, or ordinary Portland cement. However, whereas some authors pointed out the better performance of hydrated lime as a filler material in PA mixtures (Aman et al., 2014), additions in the range of 2–3% by mass of aggregates were suggested by some road agencies (Alvarez et al., 2006). The absorption of carboxylic compounds of the hydrated lime by the aggregate particles enhances the bonding between aggregates and binder, resulting in stronger mixtures. In fact, the inclusion of hydrated lime in PA mixtures increases the raveling resistance, the indirect tensile strength, the indirect tensile strength ratio, and the resistance to oxidation in relation to PA mixtures with limestone powder (Ameri & Esfahani, 2008). Recently conducted research that includes hydrated lime as a replacement for mineral filler concluded that the indirect tensile strength in wet conditions was increased by 220% by only replacing 50% of the mineral filler with hydrated lime (Mora Aldana & Zamora Castillo, 2019). Additionally, other studies have pointed out that replacing mineral filler with cellulose ash increases the adhesiveness between binder and aggregates (Movilla-Quesada et al., 2018), while preliminary results with combustion soot in densely graded asphalt mixtures showed improved stability, fluency, and indirect tensile strength test results.

The composition of asphalt mixtures suggests the suitability of the potential reuse of industrial byproducts and urban residues for replacing the needed raw materials in total or partially. A great deal of research has been done that supports the effectiveness of some residues in replacing conventionally used mineral aggregates, fibers, filler, and bituminous binders, producing even better asphalt mixtures than those developed with virgin materials.

1.2 *Research Objective*

One of the main limits for the wider application of PA mixtures is its reduced service life in relation to dense graded asphalt mixtures, mainly due to the open structure, which leads to a higher susceptibility to degradation mechanisms due to the effect of water or other environmental factors. This limitation normally led to the necessity of using high-quality materials that can provide improved characteristics to mitigate these problems, finally resulting in increased production costs. The main goal of the research project that is the basis for developing the research presented in this chapter was to reduce the production costs and increase the service life of PA mixtures by incorporating 4 different residues and byproducts locally available in Chile. Two of the residues were studied as replacement for mineral filler (combustion soot and cellulose ash), and the other two were incorporated as a stabilizing additive (Tetra Pak© containers and cigarette butts) for increasing bitumen content, and hence the durability and service life of the mixtures.

2 Materials and Methods

2.1 *Materials*

2.1.1 Bituminous Binder

In this research, Chilean bituminous binder CA-24 was used for producing reference and experimental PA mixtures. Dosages used were 5.0 and 6.0% by weight of mixture. The main characteristics of the used bituminous binder are shown in Table 1.

Table 1 Main characteristics of Chilean CA-24 bituminous binder

Characteristic	Value
Absolute viscosity at 60 °C and 300 mm/Hg (posies)	3,600
Penetration at 25 °C, 100 g, 5 s (1/10 mm)	54
Density at 25 °C and 5 cm/min	+150
% of xylol	Negative
Solubility in trichlorethylene (%)	+99
Flame point (°C)	+232
Softening point	52.6
Penetration index	−0.1

2.1.2 Mineral Aggregates

The mineral skeleton was built up with limestone mineral aggregates locally available in the Los Ríos region ($\rho = 2.97$ T/m^3). The gradation used was adjusted to the PA12 gradation band defined by the Chilean Road Manual (MOP, 2018). Two different gradations, called gradation 1 (G1) and gradation 2 (G2), were used. Gradation G1 was adjusted with the Bailey method (Vavrick et al., 2002) in order to maximize the air void content, while G2 was a gradation adjusted to the center of a PA-12 gradation band. Gradation G1 was selected according to the Chilean requirements for PA-12 mixtures, and designed for maximizing the air void content in the mixture. The Bailey method is a systematic approach to blending aggregates that provides aggregate interlock as the backbone of the structure. The method provides a set of tools that allows a better understanding of the relationship between aggregate gradation and air voids in compacted mixtures (Vavrick et al., 2002). After applying this method to the range of gradations that can be used for PA-12 mixtures according to the Chilean requirements, the gradation with higher air void content was selected. In this gradation the finer part of the aggregates was reduced and the dosages of the coarse aggregates were optimized. The final gradations selected according to these criteria are shown in Fig. 1.

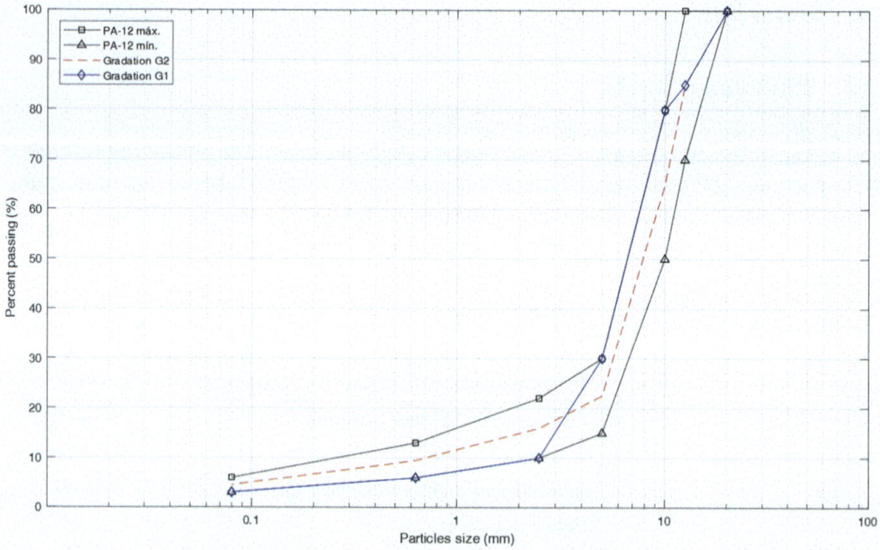

Fig. 1 Gradations used: Gradation G1 adjusted by using the Bailey method for PA12 gradation band, and Gradation G2, adjusted to the center of the PA12 Gradation band

2.1.3 Recycled Materials

Four different residues were used as a replacement or an addition in experimental PA mixtures. Residues were selected on the basis of local availability, composition, characteristics and experience in looking for their compatibility with asphalt mixtures. The residues that were used in this research (Fig. 2) were:

1. cellulose ash (CA, $\rho = 2.48$ g/cm^3),
2. combustion soot (CS, $\rho = 1.31$ g/cm^3),
3. recycled Tetra Brik Aseptic© packages (TBA), and
4. cigarette butts (CB).

These residues were used as raw material in experimental PA mixtures. While CA and CS were used as filler replacement, TBA and CB were used as stabilizing additives.

Combustion soot is a particulate material, with a nominal particle size of around 100–500 nm, rich in carbon, calcite, iron oxide, and other aromatic compounds obtained as a byproduct of the biomass combustion (Casanova et al., 2019; Kim

Fig. 2 Used residues: **a** combustion soot (CS), **b** cellulose ash (CA), **c** recycled cigarette butts (CB) and **d** recycled Tetra Brik Aseptic © (TBA) containers

et al., 2011). This material is a common residue in some industrial activities and in residential heating systems in Chile, were near 95% of the houses use biomass combustion for heating (Schueftan et al., 2016).

Cellulose ash, similarly to CS, is obtained as a byproduct of biomass combustion and is mainly composed of silicon oxide. This material was first used in asphalt mixtures in the 1950s (Mistry et. al., 2019). Since then, various research work has been done to demonstrate the benefits of this material in asphalt mixtures (Movilla-Quesada et al., 2015), showing improvements in the strength and water sensitivity of warm asphalt mixtures and lowering the particle losses produced by the vehicles' movement over the pavement.

Cigarette butts are a non-degradable form of cellulose acetate fibers, similar to natural cotton (Marinello et al., 2020), estimated to comprise nearly 875,000 tons of this residue annually worldwide by 2025 (Novotny et al., 2009). Some authors previously studied some of the uses of this residue in asphalt mixtures, highlighting the work developed by Mohajerani et al. (2017), who investigated the possibility of incorporating it in an encapsulated form during the mixing of asphalt mixtures with relatively good results.

Tetra Brik Aseptic© (TBA) packaging has been widely used worldwide to commercialize liquid foods such as milk or fruit juices, resulting in an annual production of nearly 30 billion containers (close to 20 g each), and is one of the most common residues in recycling centers. This kind of container is mainly composed of cellulose (63%), low-density polyethylene (30%) and aluminum (7%) (Andrés-Valeri et al., 2018), making it difficult to recycle due to the need to separate these different compounds. The composition of TBA packages raises the possibility of reusing them as a stabilizing additive for asphalt mixtures to substitute for conventionally used virgin cellulose or mineral fibers.

2.2 Experimental Matrix

Materials used as filler replacement were used for replacing 100% of the filler quantity defined by the gradation used. For obtaining the quantity of filler that should be incorporated into the mixture for replacing mineral filler, the specific methodology shown in the IRAM 1542 standard (Instituto argentino de racionalización de materiales, 1993) was used. The method described in this standard defines two parameters: the critical concentration (C_c) and the volumetric concentration (C_v). Based on these parameters, the number of alternative fillers to be incorporated into the mixture is suggested not to surpass the 1.00 value of the C_v/C_c ratio in order to maintain the viscous performance of the mixture. While C_c depends on the weight of the filler, the apparent volume of the material in the mixture and the actual volume through a sedimentation analysis of the dispersion particles of the filler at rest in a continuous medium (paraffin) and Eq. 1, C_v can be obtained through Eq. 2 based on filler and binder characteristics.

$$C_c = \frac{M_f}{V \times \rho_f} \tag{1}$$

$$C_v = \frac{\frac{M_f}{\rho_f}}{\frac{M_f}{\rho_f} + \frac{M_b}{\rho_b}} \tag{2}$$

- C_c: critical concentration
- C_v: volumetric concentration
- V: sediment volume (g/cm^3)
- M_f: filler weight in grams
- M_b: binder weight in grams
- ρ_f: filler density (g/cm^3)
- ρ_b: binder density (g/cm^3).

In this research, 4 different C_v/C_c ratios were used, based on the IRAM1542 and previous research: 0.50, 0.75, 1.00 and 1.25. Measured C_c for combustion soot was 0.45, and for cellulose ash it was 0.22. On the other hand, materials that were used as stabilizing additives (TBA and CB) were dosed in percentages of between 0.3 and 0.5% by weight of mixture. These values are considered adequate according to the state-of-the-art and preliminarily developed tests at the starting point of the research. Residues were initially washed off and air dried, and finally shredded in order to obtain homogeneous materials that can be adequately mixed in asphalt mixtures. Experimental PA mixtures developed in this research are shown in Table 2.

2.3 *Laboratory Tests*

All PA mixtures developed were tested in order to obtain their mechanical and hydraulic characteristics. The laboratory tests developed within the framework of this research are summarized in Table 3.

2.3.1 Particle Loss

The raveling resistance was assessed by measuring the particle loss through the Cantabro test (EN 12697-17), performed at 25 °C. This test measures the Particle Loss (*PL*) of asphalt specimens after being placed in the Los Angeles Drum for 10 min at 30 rpm by using Eq. 3.

$$PL(\%) = 100 \times \frac{(W_1 - W_2)}{W_1} \tag{3}$$

Table 2 Experimental matrix

Mix	% binder (b/w of mix)	Gradation	Fiber type	% fiber (b/w of mix)	Filler type	Filler C_v/C_c
R1	6.0	G1	–	–	Minerals	1.00
M1–F1	6.0	G1	TBA	0.3	Minerals	1.00
M2–F1	6.0	G1	TBA	0.4	Minerals	1.00
M3–F1	6.0	G1	TBA	0.5	Minerals	1.00
M1–F2	6.0	G1	CB	0.3	Minerals	1.00
M2–F2	6.0	G1	CB	0.4	Minerals	1.00
M3–F2	6.0	G	CB	0.5	Minerals	1.00
R2	5.0	G2	–	–	Minerals	1.00
M4–R1	5.0	G2	–	–	Soot	0.50
M5–R1	5.0	G2	–	–	Soot	0.75
M6–R1	5.0	G2	–	–	Soot	1.00
M7–R1	5.0	G2	–	–	Soot	1.25
M4–R2	5.0	G2	–	–	Cellulose ash	0.50
M5–R2	5.0	G2	–	–	Cellulose ash	0.75
M6–R2	5.0	G2	–	–	Cellulose ash	1.00
M7–R2	5.0	G2	–	–	Cellulose ash	1.25

Table 3 Laboratory tests performed

Parameter	Method/standard	Comments
Bulk density	EN 12697-5	–
Total air voids (AV)	EN 12697-8	–
Particle loss (Cantabro test)	EN 12697-17	Aged specimens according to Spanish standard NLT 362/92
Marshall's stability test	EN 12697-34	–

where w_1 and w_2 are the initial and final weights of the specimens in grams. The test was performed on dry, newly-built samples and in conditioned specimens according to Spanish standard NLT 362/92 in order to obtain an indicator of the water effect in the strength of the mixtures. The conditioning method consisted of submerging the specimens in water at 60 °C for 24 h and conditioning them for another 24 h at an air temperature of 25 °C. It was recommended that the *PL* should not exceed 35% in these conditions when using polymer modified bituminous binders. Some details about the test are shown in Fig. 3.

Particle losses can be interpreted as an indicator of the raveling damage, as it is a measurement of the particles that are broken up from asphalt samples after being tested by the particle loss test, so the obtained value is an estimate of the damage that can affect pavement materials due to the frictional forces produced by vehicles tires.

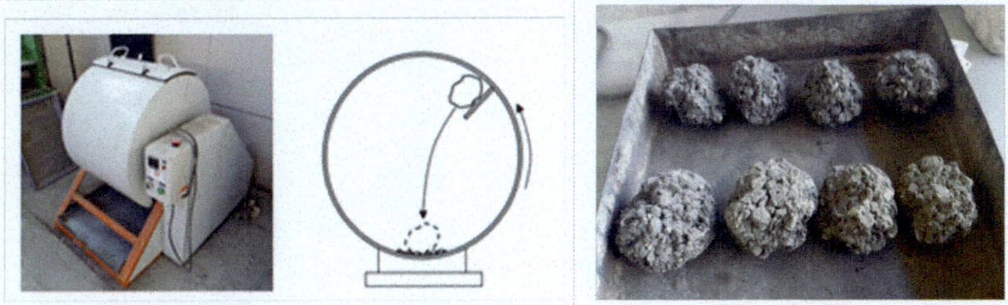

(a) (b)

Fig. 3 **a** Los Angeles drum device and **b** samples after being tested

Fig. 4 Marshall stability test device

2.3.2 Marshall Stability Test

The Marshall stability test (EN 12697-34) allows one to obtain the Marshall coefficient by measuring the maximum diametral compressive strength and deformation that the mixture can resist, called stability and fluency, respectively. Prior to carrying out the test, the samples should be conditioned in a water bath at 60 °C for 40 min according to EN 12697-34. This test provides information about the bearing capacity of the mixture and the level of deformation that the mixture can support previous to the cracking. In Fig. 4, the laboratory device used for performing this test is shown.

2.3.3 Bulk Density and Total Air Voids

In order to obtain the air void content of the mixture it is necessary to measure bulk density and particle density. The air voids in the mixture were later obtained as the relationship between the particle density and the bulk density of these particles. A

Fig. 5 Cylindrical samples used in the research

minimum of 20% air voids for ensuring the infiltration capacity of the pavement structure under rainy conditions is recommended for PA mixtures.

All tests were performed on cylindrical specimens of 102 mm in diameter (Fig. 5) according to the standards reported above, and the results of the tests provided for each mixture are the average of the results obtained in 4 different samples.

3 Results and Discussion

The results for all laboratory tests are obtained as the average of 4 measurements. As shown in Fig. 5, the fiber addition to porous asphalt mixtures generally reduces air voids in relation to reference mixture (R1). However, all the mixtures with fiber addition (M1F1 to M3F2) showed air voids over the limit defined by the main international standards to ensure good hydrological capacity (i.e. 20%). Comparing mixtures with TBA and CB fibers, it can be observed that both fibers produce similar reductions in the void content, but it is slightly higher with cigarette butts, however.

Despite the similar capacity of mixtures with fibers, TBA recycled fibers show much better performance in relation to cigarette butt fibers. Even if both fibers serve as stabilizing fibers for PA mixtures, the composition of the chemical compounds retained in used cigarette butts seems to reduce almost all mechanical strength parameters, producing, in general, weaker materials. Especially interesting are the results related to particle losses in Fig. 6, where the addition of TBA fibers shows the improvement of the performance of PA mixtures, reducing particle losses up to 50% in dry conditions and 40% in wet conditions in relation to reference mixtures, indicating lower raveling damage and hence, reducing pavement disaggregation and the surface particle breakage typical of PA mixtures.

On the contrary, mixtures with cigarette butts show, in general, poor particle loss results except for the M1–F2 mixture (6% binder and 0.3% cigarette butts), which shows similar resistance to fragmentation than the reference mixture. Similar conclusions can be obtained by analyzing the stability and fluency results in Fig. 7, where mixtures with TBA fibers were shown to be up to 25% stronger and up to 25% less deformable than reference mixtures. Also in this test, cigarette butt fibers were shown to produce, in general, weaker mixtures, except for the M1–F2 mixture, which

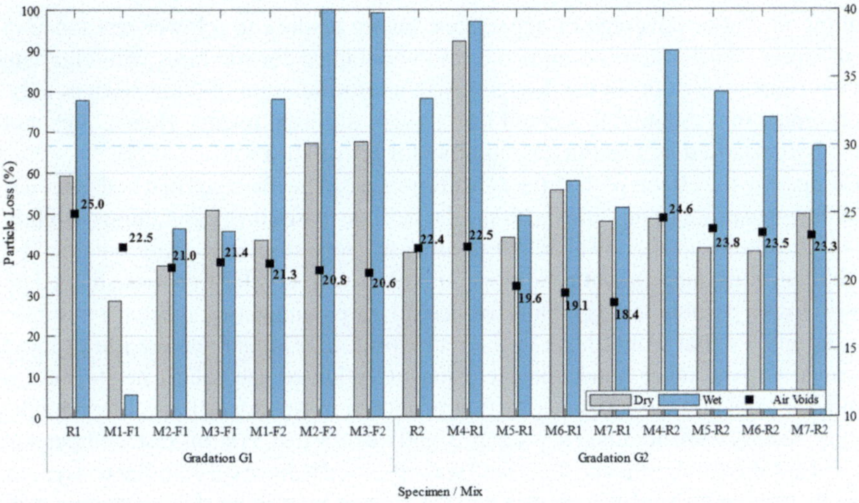

Fig. 6 Void ratio and particle loss results

showed great performance, increasing stability by 70% and reducing the fluency by 30% in relation to reference mixtures.

Analyzing the different dosages, the results obtained for TBA recycled fibers suggest incorporating it at a dosage of between 0.2 and 0.4% of the weight of the mixture in order to achieve the best mechanical performance. The results obtained for the M1–F2 mixture, which is the mixture with the lowest content of CB fibers, suggests that the poor results obtained with this material can be produced by an overdosage of these fibers, which may excessively absorb the bituminous binder,

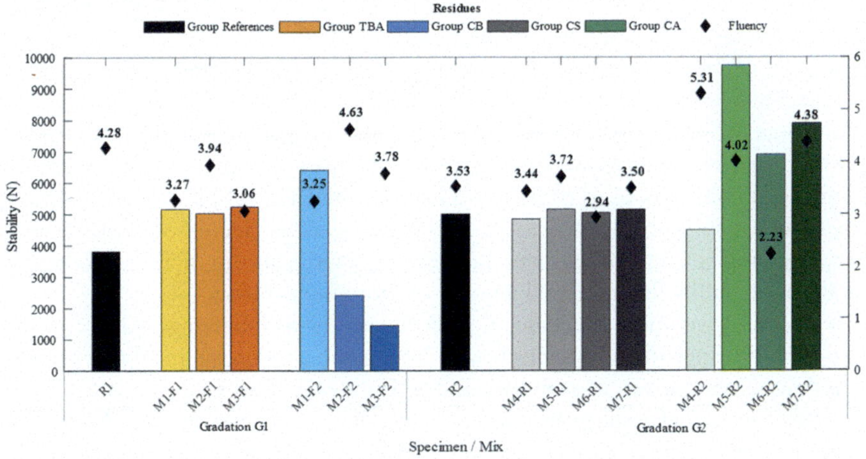

Fig. 7 Marshall stability test results

resulting in a poor coverage of aggregates, finally leading to a lower resistance in the mixtures. Furthermore, even if the results obtained for mixtures with cigarette butt fibers are very poor for the dosages used, a reformulation of these dosages to be incorporated into the mixtures could lead to very different results. Hence, it cannot be totally discarded as a stabilizing additive for PA mixtures.

Comparing the effects of the filler substitution, it can be observed that both residues were shown to produce mixtures with similar particle losses in dry conditions, slightly better for cellulose ash, however, but in all cases very similar to those obtained by the reference mixture. In wet conditions, mixtures with combustion soot showed up to 60% lower particle losses than reference mixtures, indicating a better resistance to water effects that would lead, in the real world, to a longer service life of these mixtures. On the other hand, the inclusion of cellulose ash seems not to have a significant effect on the resistance to fragmentation in wet conditions. Interestingly, inverted results were obtained in a stability and fluency test. For this test, combustion soot did not show a significant effect, and was observed to have very similar test results for both parameters to those observed in the reference mixture. However, mixtures with cellulose ash were shown to improve stability up to 95% while fluency is slightly increased by 15% in relation to reference mixtures. Analyzing the dosages used, it can be observed that for both recycled filler materials generally, the concentration to be used, according to IRAM152, ranges between 0.75 and 1 of C_v/C_c relationships.

4 Conclusions

The inclusion of some common residues in cities and industries in the south of Chile and around the world demonstrated a possible practice for reducing the storage of these residues and their potential impact on the environment. The residues used in this research and included as fibers or filler in PA mixtures was shown to produce materials similar to those produced with conventionally used raw materials, even increasing some mechanical parameters, and consequently leading to stronger and more durable materials.

TBA fibers showed satisfactory results as stabilizing additives in PA mixtures. Important improvements in the strength, deformability and raveling resistance in dry and wet conditions were observed. The good mechanical performance shown by these fibers is probably due to their composition that increases adhesiveness between aggregates and bituminous binders together with the increased bonding capacity provided by the increased amount of bituminous binder.

CB fibers showed poor mechanical performance as a stabilizing additive in PA mixtures in the range of the dosages tested. In general, a reduction in all mechanical and hydraulic parameters was observed. The only exception to this was the M1–F2 mixture. The good results observed with the M1–F2 suggest the possibility that the poor results obtained in this research with CB were due to an overdosage of this material, which would lead to poor coverage of mineral aggregate particles, resulting in weak mechanical performance.

The replacement of mineral filler with combustion soot showed that it was a suitable replacement for mineral filler in PA mixtures. Even if this material did not significantly reduce the particle losses in dry conditions, it was able to improve raveling resistance in wet conditions if the void content was reduced to values under 20%, however.

Cellulose ash was shown to provide similar raveling resistance to PA mixtures than mineral filler in both dry and wet conditions. Additionally, impressive results were observed in Marshall's stability test, where the stability was increased in practically all mixtures, highlighting the suitability of incorporating CA as a filler replacement in PA mixtures.

Further research is needed for extending the conclusions of this research to other types of asphalt mixtures and including other mechanical and functional pavement parameters in order to achieve a better understanding of the potential benefits and drawbacks of the residues used in pavement performance. Considering the promising results obtained in this research, other residues produced in cities and industries can also be considered for inclusion in asphalt mixtures in order to create a catalogue that can encourage practitioners to extend their application in asphalt pavement construction.

Acknowledgements The authors would like to thank the INNOVING 2030 program of the Faculty of Engineering Sciences at the Austral University of Chile and to the FONDECYT 11190909 project for funding this research. The authors would also like to thank BIOFOREST S.A. and BITUMIX S.A. for supporting this research and providing the raw materials used to develop the laboratory work.

References

Alvarez, A. E., Martin, A. E., Estakhri, C. K., Button, J. W., Glover, C. J., & Jung, S. H. (2006). *Synthesis of current practice on the design, construction, and maintenance of porous friction courses* (No. FHWA/TX-06/0-5262-1). https://static.tti.tamu.edu/tti.tamu.edu/documents/0-5262-1.pdf

Alvarez, A. E., Martin, A. E., & Estakhri, C. (2011). A review of mix design and evaluation research for permeable friction course mixtures. *Construction and Building Materials, 25*(3), 1159–1166. https://doi.org/10.1016/j.conbuildmat.2010.09.038

Aman, M. Y., Shahadan, Z., & Noh, M. Z. M. (2014). A comparative study of anti-stripping additives in porous asphalt mixtures. *Jurnal Teknologi, 70*(7). https://doi.org/10.11113/jt.v70.3594

Ameri, M., & Esfahani, M. A. (2008). Evaluation and performance of hydrated lime and limestone powder in porous asphalt. *Road Materials and Pavement Design, 9*(4), 651–664. https://doi.org/10.3166/rmpd.9

Andrés-Valeri, V. C., Rodriguez-Torres, J., Calzada-Perez, M. A., & Rodriguez-Hernandez, J. (2018). Exploratory study of porous asphalt mixtures with additions of reclaimed tetra pak material. *Construction and Building Materials, 160*, 233–239. https://doi.org/10.1016/j.conbuildmat.2017.11.067

Bennert, T., Fee, F., Sheehy, E., Jumikis, A., & Sauber, R. (2005). Comparison of thin-lift hot-mix asphalt surface course mixes in New Jersey. *Transportation Research Record, 1929*(1), 59–68. https://doi.org/10.3141/1929-08

Casanova, F., Raposeiras, A., & Movilla-Quesada, D. (2019). *Estudio experimental de la sensibilidad al agua y rigidez de mezclas asfálticas fabricadas con hollín como filler de aportación.* [Bachelor degree dissertation, Austral University of Chile]. http://cybertesis.uach.cl/tesis/uach/2018/bmfcis718e/doc/bmfcis718e.pdf.

Chen, J. S., Sun, Y. C., Liao, M. C., & Huang, C. C. (2012). Effect of binder types on engineering properties and performance of porous asphalt concrete. *Transportation Research Record, 2293*(1), 55–62. https://doi.org/10.3141/2293-07

Chen, J. S., Sun, Y. C., Liao, M. C., Huang, C. C., & Tsou, K. W. (2013). Evaluation of permeable friction course mixes with various binders and additives. *Journal of Materials in Civil Engineering, 25*(5), 573–579. https://doi.org/10.1061/(ASCE)MT.1943-5533.0000631

Cooley Jr, L. A., Brown, E. R., & Watson, D. E. (2000). Evaluation of open-graded friction course mixtures containing cellulose fibers. *Transportation Research Record, 1723*(1), 19–25. http://www.interfibe.com/pdf/NCATStudy.pdf

Cooley, L. A., Brumfield, J. W., Mallick, R. B., Mogawer, W. S., Partl, M., Poulikakos, L., & Hicks, G. (2009). *Construction and maintenance practices for permeable friction courses* (Vol. 640). Transportation Research Board. https://nap.nationalacademies.org/read/14310

Dale, R. (2012). *Permeable friction courses (PFC) in Texas.* Technical report, Presentation for the Annual School for County Commissioners Courts, College Station, TX. https://www.ltrc.lsu.edu/ltc_11/pdf/Update%20on%20Permeable%20Friction%20Courses%20(PFC)%20in%20Texas.pdf

Fay, L., & Akin, M. (2014). Snow and ice control on porous and permeable pavements: literature review and state of the practice. *Transportation Research Board 93rd Annual Meeting Transportation Research Board* (14-2839). https://trid.trb.org/view/1288499

Noguera, J. A. H., Quintana, H. A. R., & Gómez, W. D. F. (2014). The influence of water on the oxidation of asphalt cements. *Construction and Building Materials, 71*, 451–455. https://doi.org/10.1016/j.conbuildmat.2014.08.064

Hernandez-Saenz, M. A., Caro, S., Arámbula-Mercado, E., & Martin, A. E. (2016). Mix design, performance and maintenance of permeable friction courses (PFC) in the United States: State of the art. *Construction and Building Materials, 111*, 358–367. https://doi.org/10.1016/j.conbuildmat.2016.02.053

Instituto argentino de racionalización de materiales. (1993). *IRAM 1542/92: Método de determinación de la relación crítica ("concentración critica") y de la densidad.* https://es.scribd.com/document/447074865/Norma-IRAM-1542

Kevern, J. T., Wang, K., & Schaefer, V. R. (2010). Effect of coarse aggregate on the freeze-thaw durability of pervious concrete. *Journal of Materials in Civil Engineering, 22*(5), 469–475. https://doi.org/10.1061/(ASCE)MT.1943-5533.0000049

Kim, K. H., Sekiguchi, K., Kudo, S., & Sakamoto, K. (2011). Characteristics of atmospheric elemental carbon (char and soot) in ultrafine and fine particles in a roadside environment, Japan. *Aerosol and Air Quality Research, 11*(1), 1–12. https://doi.org/10.4209/aaqr.2010.07.0061

Kuang, D., Yu, J., Feng, Z., Li, R., Chen, H., Guan, Y., & Zhang, Z. (2014). Performance evaluation and preventive measures for aging of different bitumens. *Construction and Building Materials, 66*, 209–213. https://doi.org/10.1016/j.conbuildmat.2014.04.016

Lyons, K. R., & Putman, B. J. (2013). Laboratory evaluation of stabilizing methods for porous asphalt mixtures. *Construction and Building Materials, 49*, 772–780. https://doi.org/10.1016/j.conbuildmat.2013.08.076

Mallick, R. B., & Cooley L. A. J. (2000). Design, construction, and performance of new-generation open-graded friction courses. *Proceedings of the Association of Asphalt Paving Technologists, 69*, 391–423. https://rosap.ntl.bts.gov/view/dot/14062

Mansour, T. N., & Putman, B. J. (2013). Influence of aggregate gradation on the performance properties of porous asphalt mixtures. *Journal of Materials in Civil Engineering, 25*(2), 281–288. https://doi.org/10.1061/(ASCE)MT.1943-5533.0000602

Marinello, S., Lolli, F., Gamberini, R., & Rimini, B. (2020). A second life for cigarette butts? A review of recycling solutions. *Journal of Hazardous Materials, 384*, 121245. https://doi.org/10.1016/j.jhazmat.2019.121245

Ministerio de Obras Publicas de Chile (MOP). (2018). *Manual de Carreteras. Vol. 5. Especificaciones técnicas generales de construcción.* https://portal.ondac.com/601/articles-59864_doc_pdf.pdf

Mistry, R., Karmakar, S., & Kumar Roy, T. (2019). Experimental evaluation of rice husk ash and fly ash as alternative fillers in hot-mix asphalt. *Road Materials and Pavement Design, 20*(4), 979–990. https://doi.org/10.1080/14680629.2017.1422791

Mohajerani, A., Tanriverdi, Y., Nguyen, B. T., Wong, K. K., Dissanayake, H. N., Johnson, L., & Rezaei, A. (2017). Physico-mechanical properties of asphalt concrete incorporated with encapsulated cigarette butts. *Construction and Building Materials, 153*, 69–80. https://doi.org/10.1016/j.conbuildmat.2017.07.091

Molenaar, A. A. A., Hagos, E. T., & Van de Ven, M. F. C. (2010). Effects of aging on the mechanical characteristics of bituminous binders in PAC. *Journal of Materials in Civil Engineering, 22*(8), 779–787. https://doi.org/10.1061/(ASCE)MT.1943-5533.0000021

Mora-Aldana, E. M., & Zamora-Castillo, F. R. (2019). Influencia de la cal hidratada en mezclas asfálticas drenantes. https://repository.ucatolica.edu.co/handle/10983/23895

Movilla-Quesada, D., Muñoz, O., Raposeiras, A. C., & Castro-Fresno, D. (2018). Thermal suspectability analysis of the reuse of fly ash from cellulose industry as contribution filler in bituminous mixtures. *Construction and Building Materials, 160*, 268–277. https://doi.org/10.1016/j.conbuildmat.2017.11.046

Movilla-Quesada, D., Raposeiras, A. C., Castro-Fresno, D., & Peña-Mansilla, D. (2015). Experimental study on stiffness development of asphalt mixture containing cement and $Ca\ (OH)_2$ as contribution filler. *Materials & Design, 74*, 157–163. https://doi.org/10.1016/j.matdes.2015.02.026

Nicholls, J. C., & Daines, M. E. (1997). Spray suppression by porous asphalt. *TRL Staff Paper PA2092/92*. https://trid.trb.org/view/475042

Nicholls, J. C. (1999). *Review of UK porous asphalt trials* (Vol. 264). Thomas Telford. https://trid.trb.org/view/483244

Novotny, T. E., Lum, K., Smith, E., Wang, V., & Barnes, R. (2009). Cigarettes butts and the case for an environmental policy on hazardous cigarette waste. *International Journal of Environmental Research and Public Health, 6*(5), 1691–1705. https://doi.org/10.3390/ijerph6051691

Rand, D. A. (2004). Permeable friction courses TxDOT experiences. In *SEAUPG 2004 Conference, Baton Rouge, LA*. http://www.seaupg.org/PDF/2004/Wed%20DRand.pdf

Root, R. E. (2009). *Investigation of the use of open-graded friction courses in Wisconsin*. Wisconsin Highway Research Program. http://digital.library.wisc.edu/1793/53403

Rungruangvirojn, P., & Kanitpong, K. (2010). Measurement of visibility loss due to splash and spray: Porous, SMA and conventional asphalt pavements. *International Journal of Pavement Engineering, 11*(6), 499–510. https://doi.org/10.1080/10298430903578945

Schueftan, A., Sommerhoff, J., & González, A. D. (2016). Bosques energía sociedad nº5. Demanda de leña y políticas de energía en el centro-sur de Chile. https://bibliotecadigital.infor.cl/handle/20.500.12220/27301

Smith, K., Von Quintus, H., Killingsworth, B., Barton, R., & Kobia, K. (1998). Review of life-cycle costing analysis procedures. *Final Report prepared by: Brent Rauhut Engineering, ADI Group, and submitted by ERES Consultants for Ontario Ministry of Transportation.*

Sol-Sánchez, M., Moreno-Navarro, F., García-Travé, G., & Rubio-Gámez, M. C. (2015). Laboratory study of the long-term climatic deterioration of asphalt mixtures. *Construction and Building Materials, 88*, 32–40. https://doi.org/10.1016/j.conbuildmat.2015.03.090

Takahashi, S. (2013). Comprehensive study on the porous asphalt effects on expressways in Japan: Based on field data analysis in the last decade. *Road Materials and Pavement Design, 14*(2), 239–255. https://doi.org/10.1080/14680629.2013.779298

Vavrick, W. R., Huber, G., Pine, W. J., & Bailey, R. (2002). Method summarized for gradation selection in hot-mix asphalt design. *Transportation Research Circular E-C044, TRB, National Research Council, Washington, DC.* https://trb.org/publications/circulars/ec044.pdf

Wu, S. P., Gang, L. I. U., Mo, L. T., Zheng, C. H. E. N., & Ye, Q. S. (2006). Effect of fiber types on relevant properties of porous asphalt. *Transactions of Nonferrous Metals Society of China, 16*, s791–s795. https://doi.org/10.1016/S1003-6326(06)60327-0

Wurst, J. E., III., & Putman, B. J. (2013). Laboratory evaluation of warm-mix open graded friction course mixtures. *Journal of Materials in Civil Engineering, 25*(3), 403–410. https://doi.org/10.1061/(ASCE)MT.1943-5533.0000611

Water, Cities, and Industries in an Arid Zone: The Case Study of Antofagasta, Chile

Jaime Chacana-Olivares, Sebastián Herrera-León, Constanza Cruz, and Paula Guerra

Abstract The Antofagasta region is located within the Atacama Desert of northern Chile. This region is facing a severe scarcity of water, with water availability below the threshold considered necessary for sustainable development. Nonetheless, with its population of about 700,000 inhabitants, Antofagasta is the second-largest contributor to the Chilean economy with some of the world's largest copper mines and a growing lithium industry. This chapter presents the technological response that has been implemented at the regional level to address the water scarcity challenge, highlighting the installed capacities of seawater desalination plants as sources of water for the industrial and urban sectors as well as the growing implementation of wastewater treatment for reuse.

1 Introduction

The Atacama Desert is located in northern Chile and southern Peru, spanning more than 165,000 square kilometers and stretching almost 1,000 km between latitudes 17° S to 28° S. It is widely recognized as not only one of the driest regions on Earth but also

J. Chacana-Olivares (✉) · S. Herrera-León · C. Cruz
Department of Chemical and Environmental Engineering, Universidad Católica del Norte, Antofagasta, Chile
e-mail: jaime.chacana@ucn.cl

S. Herrera-León
e-mail: sebastian.herrera@ucn.cl

C. Cruz
e-mail: constanza.cruz@ucn.cl

C. Cruz
Centro de Investigación Tecnológica del Agua en el Desierto-CEITSAZA, Universidad Católica del Norte, Antofagasta, Chile

P. Guerra
Department of Chemical and Environmental Engineering, Universidad Técnica Federico Santa María, Santiago, Chile
e-mail: paula.guerra@usm.cl

A. Ch. Braun et al. (eds.), *Eco-Industrial Development as an Industrial Strategy*, Green Energy and Technology, https://doi.org/10.1007/978-3-031-73576-9_22

a hyper-arid one, characterized by extremely low levels of precipitation and scarce vegetation (Gómez-Silva et al., 2008, 2022; Houston & Hartley, 2003; McKay et al., 2003; Tapia et al., 2018). The Atacama Desert boasts distinctive landscapes, including salt flats, volcanic cones, and sand dunes. The region's desert climate is marked by limited rainfall and considerable day-to-night temperature variations. Thanks to the Humboldt Current, the coastal areas feature a more moderate climate, whereas the Andean highlands are subject to colder temperatures and sporadic snowfall.

In Chile, the Atacama Desert is characterized by significant economic activities such as mining, tourism, and agriculture, all of which heavily depend on the availability of water resources. Nevertheless, this region is facing a severe water scarcity problem, particularly in the region of Antofagasta where the annual availability of water from rainfall runoff is 52 cubic meters per inhabitant, significantly lower than the national average and below the threshold of 2,000 cubic meters per inhabitant considered necessary for sustainable development worldwide (The World Bank, 2011). The water scarcity crisis in the Antofagasta region has resulted in a decline in agricultural production and an increase in conflicts between mining companies and local communities that challenge the supply of drinking water (Ruffino et al., 2022).

The issue of water scarcity in the Antofagasta region is a complex and multidimensional problem that necessitates a comprehensive approach addressing the needs of both human and natural systems. This situation presents a remarkable case study due to the extreme scenario of water scarcity in the region and the responses from industry and cities in tackling this challenge. In this context, seawater desalination plants have been established in the Antofagasta region to address the increasing demand for water resources from the industrial and urban sectors. In addition, wastewater recycling plants have been installed as a complementary strategy to tackle the pressing challenge of water scarcity, with the goal of ensuring long-term regional water security. In the following sections, we will provide an overview of the Antofagasta region and highlight its current and future implementation of seawater desalination plants and wastewater treatment and recycling processes.

2 Antofagasta Region

The Antofagasta region spans approximately 126,049 square kilometers, making it Chile's second-largest region. Extending from the Pacific coastline to the Andes Mountains, the longest length of the region stretches about 350 km inland and reaches an elevation of 6,739 m above sea level at its highest point. Rainfall in the Antofagasta region primarily occurs during the summer months, originating from the Andes Mountains (Clarke, 2006). At elevations exceeding 3,000 m above sea level, annual precipitation levels average around 150 mm per year.

From an administrative perspective, the Antofagasta region is divided into three provinces (Antofagasta, El Loa, and Tocopilla) and nine communes (Antofagasta, Mejillones, Sierra Gorda, Taltal, Calama, Ollagüe, San Pedro de Atacama, Tocopilla, and María Elena). With a population of approximately 700,000 people, most inhabitants reside in two urban areas: the regional capital, Antofagasta city, and Calama, often referred to as the mining capital of Chile.

The economy of Antofagasta is predominantly driven by metallic and non-metallic mining, especially the extraction and processing of copper ores. Some of the world's largest copper mines, such as Escondida and Chuquicamata, are located in the region. According to data from the Central Bank of Chile, the gross domestic product (GDP) of the Antofagasta region amounted in 2021 to 27,633,000 million Chilean pesos, accounting for 11% of Chile's national GDP and making it the second-largest contributor to the national economy after the country's capital, Santiago de Chile (Central Bank of Chile, 2021). This region contributes 51% of the country's mining GDP and delivers 55% of national copper exports and 100% of Chile's lithium production, which corresponds to 26% of global production (Central Bank of Chile, 2021; National Geology & Mining Service, 2020; Prieto et al., 2022).

The most prominent economic sector in Antofagasta is mining, constituting 71% of the region's total GDP, generating numerous job opportunities, and stimulating economic expansion. The second-largest contributor to the region's economy in 2021 was the financial and business services sector (4.5% of regional GDP), followed by the construction sector (3.7%). This data highlights the significance of mining as the primary economic activity in the Antofagasta region.

In addition, the primary consumers of water resources in the Antofagasta region are associated with mining, agriculture, urban use, and energy production. The copper mines need considerable amounts of water for extraction and processing, making mining the largest consumer of water in the area. Although the region's arid conditions limit agricultural activities, water is still required for irrigation in certain areas where crops are cultivated, such as the Loa River basin. As previously mentioned, a significant portion of the population resides in urban centers like Antofagasta and Calama cities; therefore, water resources are essential to fulfill the daily needs of households and businesses such as drinking water and sanitation. Furthermore, water resources play a critical role in energy production, particularly in thermal power plants that rely on water for cooling purposes.

The hydrography of the Antofagasta region is predominantly characterized by endorheism, meaning that most basins in this region do not drain into the sea. One exception is the Loa River, Chile's longest river, which drains into the sea in the northern part of Antofagasta region. The region comprises ten hydrographic basins in total, with the Loa River basin being the most significant contributor of conventional water resources. Other notable basins in the region include the Salar de Atacama basin and those located in the high-altitude Andes Mountain areas (General Directorate of Water, 2016a).

In terms of the production of water for human consumption, Aguas Antofagasta S.A. holds a concession for providing sanitary services in the urban sector of the region (General Directorate of Water, 2016b). This encompasses the production and distribution of potable water, as well as the collection and disposal of sewage. The company operates two supply systems: the Northern System and the Southern System. Expanded in 2014, the Northern System meets the water requirements of

Table 1 Water consumption by sector in the Antofagasta region

Sector	Water consumption (l/s)		
	2015	2030	2040
Urban	1,242	1,428	1,574
Agriculture	1,465	1,341	1,344
Mining	4,789	3,156	2,409
Industrial	940	1,548	1,892
Total	8,437	7,474	7,220

the cities of Antofagasta, Calama, Tocopilla, Mejillones, and the towns of Sierra Gorda and Baquedano. In contrast, the Southern System serves the water needs of Taltal commune. Considering the water scarcity suffered by this area, the Antofagasta region demonstrates a notable achievement in maintaining a 100% coverage rate for the production and distribution of potable water. A substantial portion of this water supply is obtained from the Loa River basin, playing a critical role in providing water resources for the region.

Table 1 provides a comprehensive analysis of water consumption across various sectors in the Antofagasta region (General Directorate of Water, 2017). It is relevant to emphasize that the water usage of the mining sector includes the use of desalinated water for operational activities. A thorough examination of the desalination plants operating or under evaluation within the region is presented in the following section. Furthermore, it is noteworthy that approximately 90% of water consumption in the urban sector takes place in the cities of Antofagasta and Calama, demonstrating how the population is concentrated in those two urban areas. Consequently, the wastewater treatment plants operating in those cities play a key role in maintaining high water quality standards and achieving a higher level of water reuse throughout the region. Further details regarding these plants are presented upon in the subsequent sections.

3 The Current and Future Situation of Water Treatment in Antofagasta

3.1 Seawater Desalination Plants to Alleviate Water Scarcity in Antofagasta Region

Table 2 lists the desalination plants in operation and pending approval (under evaluation) in Antofagasta region. As this information shows, there are currently more than twelve desalination plants operating at an industrial scale in Antofagasta, four desalination projects in various stages of evaluation, and one desalination plant under construction. The desalination plants already in operation produce a total of 5,852 l/s of desalinated water, with future projects slated to increase this capacity by 129.6% to reach the total of 7,588 l/s. It is worth mentioning that all of the desalination plants summarized in Table 2 are based on reverse osmosis (RO) technology for the treatment of seawater, and a general description of them based on antecedents given by Ramírez and Ramírez (2022) is presented in this section. Main water treatment plants and water distribution systems are presented in Fig. 1.

As can be observed in Table 2, most of the currently operational desalination plants work to satisfy the water demands of the mining sector, which is 4,768 l/s of desalinated water, while the demand for water from the urban sector is around 414 l/s of desalinated water. It is relevant to mention that Antofagasta city hosts the largest desalination system operating in South America, the El Coloso desalination plant, which supplies industrial water to the largest copper mine in the world, Escondida mine. After its expansion, this desalination plant will be capable of producing a maximum of 3,025 l/s of desalinated water, a significant volume in comparison to other desalination plants in the region. The cost of the first El Coloso desalination plant was US$160 million, while its expansion is estimated to cost US$3.43 billion.

The Spence desalination plant, which produces 1,000 l/s of desalinated water, was built and launched in two stages. The first stage provides 800 l/s of desalinated water to supply the requirements of the Minera Spence Primary Minerals project. The second stage produces an additional 800 l/s of desalinated water to supply other

Table 2 Desalination plants in operation or pending approval in Antofagasta region (Ramírez & Ramírez, 2022)

Desalination plant	Designed capacity [l/s]	Sector	City	Status
El Coloso	525	Mining (copper)	Antofagasta	In operation
Escondida EWS (El Coloso expansion)	2,500	Mining (copper)	Antofagasta	In operation
Spence	1,600	Mining (copper)	Mejillones	In operation
Sierra Gorda	63	Mining (copper)	Sierra Gorda	In operation
Antucoya	48	Mining (copper)	Maria Elena	In operation
Distrito Centinela	32.5	Mining (copper)	Sierra Gorda	In operation
Distrito Centinela 2	140	Mining (copper)	Sierra Gorda	Under evaluation
RT Sulfuros (Distrito Norte)	840 (with a maximum extension to 1,956)	Mining (copper)	Tocopilla	Under evaluation
Desaladora Norte	850	Urban	Antofagasta	In operation
Expansion Desaladora Norte	634	Urban	Antofagasta	Under construction
Tocopilla	100	Urban	Tocopilla	In operation
Michilla	75	Urban	Sierra Gorda	In operation
Moly-cop	4.4	Industrial	Mejillones	In operation
Angamos	55	Energy	Mejillones	In operation
Expansion Angamos	158	Energy	Mejillones	Under evaluation

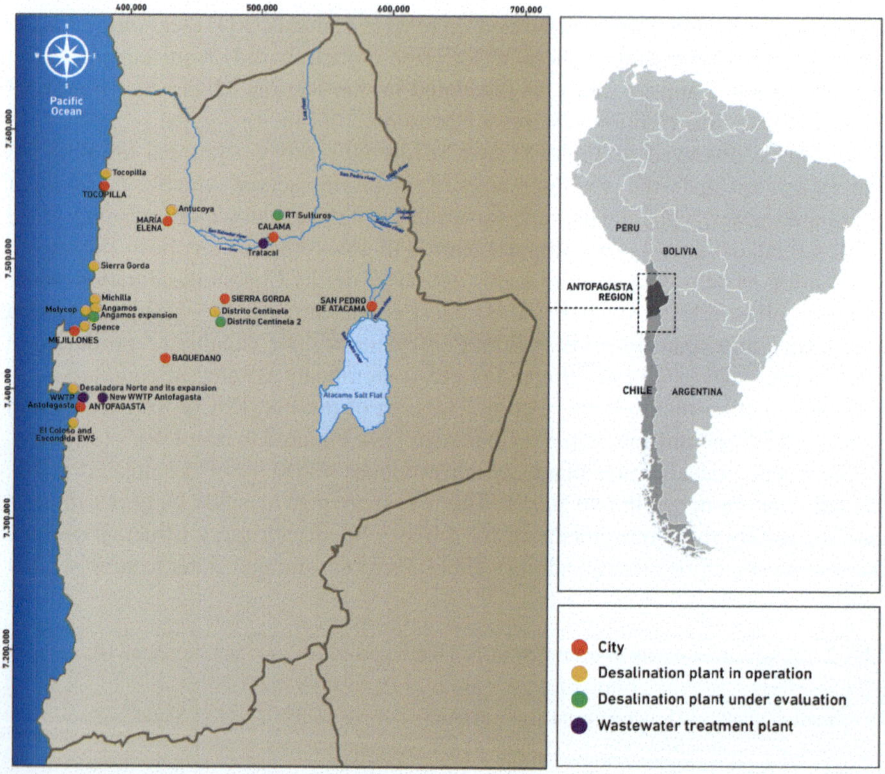

Fig. 1 Main water and wastewater treatment facilities in the Antofagasta region, including locations of main cities and industries (own elaboration)

initiatives of Minera Spence, BHP Billiton, and/or third party users. The Sierra Gorda desalination plant treats water from a thermoelectric plant in a mining project that processes copper-molybdenum sulfide ores. The process at Sierra Gorda mine uses direct seawater, and the desalination plant treats only a fraction of the flow (60 l/s) for different purposes, mainly for human consumption.

Antofagasta Minerals company has three desalination projects in Antofagasta region: two in operation and one under evaluation. All projects are connected to the same water distribution system which aims to satisfy the water demands of the company's various mining projects. The Antucoya desalination plant produces 48 l/s of desalinated water and was installed with the objective of meeting the water demands of a mining project and human consumption. The Distrito Centinela desalination plant, located near Esperanza mine, produces 32.5 l/s of desalinated water mainly used for human consumption and copper concentrate washing. It is relevant to mention that the Esperanza mine uses seawater without desalination in its production process to obtain copper concentrate, making it the first company in the world to incorporate this modern and sustainable initiative on a large scale. Of the total water used, only a fraction is desalinated. The Distrito Centinela 2 desalination plant will provide 140 l/s of desalinated water, also mainly for human consumption, and is part of a new mining project from the company. Another desalination plant for the mining sector is RT Sulfuros, which will supply water required by a new mining project for processing copper sulfide ores. In the first stage of the mining project, desalination plants will produce 630 l/s with their output expected to increase to 1,950 l/s of desalinated water. The first desalination plant owned by the Chilean government, the project will involve a total investment estimated at US$1 billion.

In the Antofagasta region, 70% of the population lives in coastal cities. In this context, there are currently four desalination plants operating in the region to supply water for the urban sector. The Desaladora Norte plant began operations in 2003, with a plant capacity of 600 l/s of desalinated water. As a result of increasing population growth in Antofagasta city and the need to provide security to the drinking water supply system, the desalination plant was expanded, increasing its capacity by 250 l/s of desalinated water for a total production of 850 l/s. Currently, the plant's infrastructure supplies 85% of the demand of Antofagasta city and 100% of Mejillones city, with another expansion under construction. The expansion will increase production by 53%, enabling the plant to supply 100% of the current and future demand of Antofagasta and Mejillones cities. In addition, the Tocopilla desalination plant has been supplying 100% of the demand of the city's 23,000 inhabitants since 2020, with production levels of 75 l/s desalinated water and the potential to expand to 100 l/s.

The Michilla desalination plant was donated to the municipality of Mejillones city by the Antofagasta Minerals company to meet the water needs of a fishing village of ca. 250 inhabitants. The Moly-cop desalination plant supplies industrial desalinated water for the manufacturing and cooling process of steel balls. The Angamos plant has a production capacity of 55 l/s of desalinated water, with ultrafiltration membrane pretreatment and two RO steps to obtain high-quality water for power generation at the Angamos plant. In addition, a second-phase expansion to 158 l/s is under evaluation, which would enable the plant to supply water to other users.

One of the most significant technical challenges that these desalination plants face is the recurrence of harmful algal blooms events along the Antofagasta coast (Herrera-León et al., 2019). This phenomenon can cause irreversible damage to the surface of RO membranes, thus affecting the operation of desalination plants. Moreover, algal bloom events can cause health problems among the population as they can alter the physical, chemical, and biological characteristics of water organisms (Wang et al., 2018). In addition, RO is an electricity-intensive process that can generate a large amount of greenhouse gas emissions when that electricity is derived predominantly from the combustion of fossil fuels (Gude, 2016; Shemer & Semiat, 2017). The amount of electricity required annually for the operation of a desalination plant using RO technology is a function of its capacity and the unitary electrical energy consumption, which is variable and depends on different factors including plant performance, capacity, and fed water quality (Gude, 2016). According to Shahabi et al. (2017), the RO process requires 3.1 kW/h of electricity to produce one cubic meter of desalinated water from seawater.

Additionally, the RO process creates large volumes of waste brine that are commonly discharged into the ocean (Jones et al., 2019; Missimer & Maliva, 2018). Waste brine is mainly characterized by a high salinity concentration that is about 1.5–2 times higher than seawater concentration, depending on the intake water salinity and recovery rate of the RO process (Sola et al., 2020). Despite the high salinity, it may also contain chemical reagents commonly used in the pre-treatment, membrane, and filter-cleaning process that are discharged into the waste brine, such as antiscalants, antifouling and antifoaming agents, coagulants, biocides, cleaning solutions, as well as adjustors of pH and hardness (Grossowicz et al., 2020; Jones et al., 2019). This situation may negatively impact marine ecosystems if waste brine is not properly managed (Sola et al., 2020). For instance, antiscalants may induce local eutrophication by introducing phosphorus (e.g., polyphosphates), while coagulants may enhance water turbidity and coloration (Petersen et al., 2018). Other potential environmental damages include reduction of the abundance and diversity of benthic communities, pH fluctuations, and an increase of heavy metals where waste brine is disposed (Grossowicz et al., 2020; Petersen et al., 2018; Sola et al., 2020; Zarzo, 2018).

3.2 *Wastewater Treatment and Recycling as an Alternative to Address Water Scarcity in Antofagasta*

The Antofagasta region has made significant progress in the coverage of wastewater treatment, following the national trend. This is attributed to the implementation of stricter discharge regulations and increased investment and implementation of wastewater treatment plants (WWTP), particularly in the last two decades. According to the Superintendence of Sanitary Services (SISS) (2018), a total volume of 42.9 million cubic meters of wastewater was treated in the region by a total of nine WWTP, providing treatment coverage to 610,943 inhabitants (SISS, 2020) (Fig. 1). Table 3 presents the existing WWTP in the Antofagasta region, along with the technologies used and the receiving body or user of the treated water.

Out of the total treated wastewater, approximately 69.5% was discharged into the sea through submarine outfalls after preliminary treatment, while 24.4% was discharged into rivers and 8.2% was recycled for irrigation and/or industrial use after secondary treatment (Ministry of Public Works, 2020). Although the percentage of recycled treated wastewater is higher than the national average—1.3% of the total treated wastewater—the current situation reflects a significant loss of water resources that could be recycled, potentially increasing the net availability of water in the region. Given the water scarcity scenario in the northern part of the country, it is imperative to move beyond wastewater treatment for purposes of environmental sanitation only.

The cities of Antofagasta and Calama, the most populous in the Antofagasta region, account for approximately 95% of the regional population and have WWTP managed by the Concessionaire of Sanitary Services S.A. (ECONSSA) and Tratacal

Table 3 Wastewater treatment plants located in the Antofagasta region, 2019 (SISS, 2020)

Company	WWTP	Technology	Destination
Aguas Antofagasta S.A.	Submarine outfall—Mejillones	Submarine outfall	Sea
Aguas Antofagasta S.A.	Submarine outfall—Taltal	Submarine outfall	Sea
Aguas Antofagasta S.A.	Submarine outfall—Tocopilla	Submarine outfall	Sea
Aguas Antofagasta S.A.	WWTP—Sierra Gorda	Activated sludge	Irrigation
Aguas Antofagasta S.A.	WWTP—Baquedano	Activated sludge	Irrigation
ECONSSA S.A.	Submarine outfall—Gran Antofagasta	Submarine outfall	Sea
ECONSSA S.A.	WWTP—Lodos Antofagasta	Activated sludge	Third part
ECONSSA S.A.	WWTP—San Pedro Coloso	Worm filter	Irrigation
Tratacal S.A.	WWTP—Calama	Activated sludge	Water course

S.A., respectively. Antofagasta city operates a preliminary treatment plant located along the coastline in the central-northern sector. The city's geographical layout allows for gravitational runoff in the east–west direction, but not in the north–south direction (ECONSSA, 2013). Up to 150 l/s, which is approximately 10% of the wastewater treated at the facility, is directed to an activated sludge process to produce water of sufficient quality for recycling. According to ECONSSA (2021), while the treated waters from Antofagasta were initially used for agricultural irrigation, they are currently sold to various customers located in the industrial neighborhood of La Negra, which is situated 20 km east of Antofagasta city, predominantly to industries involved in the production of copper and zinc products.

The remaining 90% of the wastewater from the municipality is conveyed to the preliminary treatment plant, where the effluent is discharged around 1,000 m offshore at a depth of 30 m through an underwater outfall with a design capacity of 1,500 l/s. It is important to highlight that the primary goal of the pre-treatment process is the removal of coarse and fine solids, sand, and fats, as the regulatory standards for liquid waste discharge into marine waters are less strict compared to discharge into continental surface waters and for other purposes. In Calama, the second-most populous city in the region, around 165,000 residents are served by Tratacal S.A., a company that operates a gravity-based collection system with an average slope from northeast to southwest, leading to an activated sludge process with aerobic digestion of sludge. Finally, the treated water is discharged into Quetena creek, upstream from the San Salvador River, which eventually converges with the Loa River (Centro de Ecología Aplicada, 2020).

One project to highlight on the regional level is entitled New Wastewater Treatment Plant for Reuse in Antofagasta, which was submitted by ECONSSA for environmental evaluation and approved in December 2020. This project is expected to significantly impact the future capacity for urban wastewater recycling in the region, with a new plant to be constructed in the industrial sector of Salar del Carmen-La Negra, away from the urban center of Antofagasta. Conduits and elevations will be established to transfer the pre-treated water from the current location of the plant to the new site. Currently, the company is preparing a tender for the construction of this new treatment plant (Bnaméricas, 2023).

The proposed treatment system utilizes activated sludge technology, which employs oxidation to break down and stabilize the organic matter in the effluent. The plant will have three modules, each with the capacity to treat 300 l/s, totaling a combined capacity of 900 l/s. Each module will have two aerated biological reactors, and sedimentation tanks and a contact chamber will be used for disinfection. A lime dosing system will be incorporated to ensure sludge stabilization, which will be processed in gravity thickeners and a dewatering stage to reduce its moisture content.

It is important to note that the objective of this project is to provide added value to the treated wastewater, allowing it to be used as input for industries located in the La Negra and Mantos Blancos clusters, without foreseeing the discharge of treated effluent into any receiving body. Based on the current amount of wastewater generated and the design capacity of this plant alone, the rate of recycled, treated wastewater in the Antofagasta region will exceed 60%, which will significantly reduce the demand for and pressure on fresh water resources, positioning the region at the forefront of such initiatives at a national and Latin American level.

This project is strategic for advancing towards sustainable development in the region, as it will help reduce marine pollution caused by the outfall and contribute to the development of industrial production in the region (Inter-American Development Bank, 2022). Moreover, this project will provide a more cost-effective alternative to seawater desalination, which entails high energy consumption and costs, resulting in a significant carbon footprint. Furthermore, wastewater recycling as an alternative is attractive in the long term as it reduces investment costs for disposal and water supply (Council for Infrastructure Policies, 2018; ECONSSA, 2013).

The arid climate characteristics of the Antofagasta region, coupled with its growing population and production demand, have required water sources to be increased and diversified. Desalination solves water quality problems and meets the demand for drinking water, and private initiatives from mining companies meet the demand for fresh water in production processes, which reduces the onus to not stress water sources. However, the reuse of wastewater remains an opportunity and challenge to meet the growing production demand. Chilean laws on wastewater discharge allow for the reuse of treated wastewater from sewage treatment plants in production processes, which opens up a great opportunity to reduce costs and make desalinated water available for human consumption.

4 Conclusions

Antofagasta region is facing a severe water scarcity scenario with an annual availability of water from rainfall runoff significantly below the threshold considered necessary for sustainable development worldwide. Seawater desalination plants have been established in the Antofagasta region to address the increasing demand for water from the industrial and urban sectors. There are currently more than twelve desalination plants operating at an industrial scale in Antofagasta, with a total capacity of 5,852 l/s of desalinated water. Another important strategy to tackle water scarcity in the Antofagasta region is the treatment and recycling of wastewater by WWTPs. However, this strategy still constitutes a complementary measure, as out of the total volume of 42.1 million cubic meters of treated wastewater produced in the region by a total of nine WWTPs, barely 8.2% is recycled for irrigation and/or industrial use. There are still many opportunities to increase recycling, with benefits such as the reduction of seawater desalination with its significantly higher carbon footprint, the reduction of brine discharged into the ocean and its associated environmental liabilities, and the reduction of costs, especially for the agricultural sector.

Acknowledgements The authors acknowledge the support of the FSEQ210038 and FOVI210079 projects of the National Agency of Investigation and Development (ANID), Chile.

References

Bnaméricas. (2023). *Nueva planta de tratamiento de aguas servidas para reúso en Antofagasta* [New wastewater treatment plant for reuse in Antofagasta]. https://www.bnamericas.com/es/perfil-proyecto/nueva-planta-de-tratamiento-de-agua-servidas-ptas-para-reuso-de-antofagasta

Central Bank of Chile. (2021). *Statistics Database – National Accounts*. https://si3.bcentral.cl/Siete/ES/Siete/Cuadro/CAP_CCNN/MN_CCNN76/CCNN2018_PIB_II_ACT_N/637899748373607863?cbFechaInicio=2021&cbFechaTermino=2021&cbFrecuencia=ANNUAL&cbCalculo=NONE&cbFechaBase=

Centro de Ecología Aplicada. (2020). *Diagnóstico del caudal ambiental del río Loa, región de Antofagasta* [Diagnosis of the environmental flow of the Loa river, Antofagasta region]. Gobierno Regional de Antofagasta, Chile. https://mma.gob.cl/antofa-doc/2020_07_GOA002_INF_V1_InfFinal.pdf

Council for Infrastructure Policies. (2018). *Reúso de agua servida transformaría a Antofagasta en un gran parque* [Reuse of wastewater would transform Antofagasta into a great park]. http://www.infraestructurapublica.cl/reuso-agua-servida-transformaria-antofagasta-gran-parque/

Clarke, J. D. (2006). Antiquity of aridity in the Chilean Atacama Desert. *Geomorphology, 73*(1–2), 101–114. https://doi.org/10.1016/j.geomorph.2005.06.008

ECONSSA Chile S.A. (2013). *Sistema de Disposición Final de las Aguas Servidas de Antofagasta* [Final disposal system for Sewage in Antofagasta]. https://www.econssachile.cl/pdf/docs/estudios/GHD_WEB.pdf

ECONSSA Chile S.A. (2021). *Planta de reúso de AS Salar del Carmen* [Salar del Carmen Activated Sludge Reuse Plant]. https://www.econssachile.cl/proyectos/planta-de-reuso-de-as-salar-del-carmen

General Directorate of Water. (2016a). *Atlas del Agua, Chile* [Water Atlas, Chile]. Santiago, Chile. https://snia.mop.gob.cl/sad/Atlas2016parte1.pdf

General Directorate of Water. (2016b). *Plan Estratégico para la Gestión de los Recursos Hídricos Región de Antofagasta* [Strategic Plan for the management of water resources Antofagasta region]. Santiago, Chile. https://snia.mop.gob.cl/repositoriodga/handle/20.500.13000/4382

General Directorate of Water. (2017). *Estimación de la demanda actual, proyecciones futuras y caracterización de la calidad de los recursos hídricos en Chile* [Estimation of current demand, future projections and characterization of the quality of water resources in Chile]. Ministerio de Obras Públicas, Santiago, Chile. https://dga.mop.gob.cl/Estudios/01%20Informe/Informe%20Final%20Vol%20II.pdf

Gómez-Silva, B., & Batista-García, R. A. (2022). The Atacama Desert: A biodiversity hotspot and not just a mineral-rich region. *Frontiers in Microbiology, 13*, 812842. https://doi.org/10.3389/fmicb.2022.812842

Gómez-Silva, B., Rainey, F. A., Warren-Rhodes, K. A., McKay, C. P., & Navarro-González, R. (2008). Atacama Desert soil microbiology. In P. Dion & C.S. Nautiyal (Eds.), *Microbiology of extreme soils*. Soil biology (Vol. 13). Springer. https://doi.org/10.1007/978-3-540-74231-9_6

Grossowicz, M., Ofir, E., Shabtay, A., Wood, J., Biton, E., Belkin, N., Frid, O., Sisma-Ventura, G., Kress, N., Berman-Frank, I., & Gal, G. (2020). Modeling the effects of brine outflow from desalination plants on coastal food-webs of the Levantine basin (eastern Mediterranean Sea). *Desalination, 496*, 114757. https://doi.org/10.1016/j.desal.2020.114757

Gude, V. G. (2016). Desalination and sustainability–an appraisal and current perspective. *Water Research, 89*, 87–106. https://doi.org/10.1016/j.watres.2015.11.012

Herrera-Leon, S., Cruz, C., Kraslawski, A., & Cisternas, L. A. (2019). Current situation and major challenges of desalination in Chile. *Desalination and Water Treatment, 171*, 93–104. https://doi.org/10.5004/dwt.2019.24863

Houston, J., & Hartley, A. J. (2003). The Central Andean west-slope rainshadow and its potential contribution to the origin of hyperaridity in the Atacama Desert. *International Journal of Climatology: A Journal of the Royal Meteorological Society, 23*(12), 1453–1464. https://doi.org/10.1002/joc.938

Inter-American Development Bank. (2022). *El reúso de agua residual tratada en America Latina y el Caribe: 10 estudios de caso* [The reuse of treated wastewater in Latin America and the Caribbean: 10 case studies]. Water and Sanitation Division, Inter-American Development Bank. https://doi.org/10.18235/0004515

Jones, E., Qadir, M., van Vliet, M. T., Smakhtin, V., & Kang, S. M. (2019). The state of desalination and brine production: A global outlook. *Science of the Total Environment, 657*, 1343–1356. https://doi.org/10.1016/j.scitotenv.2018.12.076

McKay, C. P., Friedmann, E. I., Gómez-Silva, B., Cáceres-Villanueva, L., Andersen, D. T., & Landheim, R. (2003). Temperature and moisture conditions for life in the extreme arid region of the Atacama Desert: Four years of observations including the El Niño of 1997–1998. *Astrobiology, 3*, 393–406. https://doi.org/10.1089/153110703769016460

Ministry of Public Works. (2020). *First report. National Water Board*, Chile. https://www.infraestructurapublica.cl/wp-content/uploads/2020/03/Mesa_Nacional_del_Agua_2020_Primer_Informe_Enero.pdf

Missimer, T. M., & Maliva, R. G. (2018). Environmental issues in seawater reverse osmosis desalination: Intakes and outfalls. *Desalination, 434*, 198–215. https://doi.org/10.1016/j.desal.2017.07.012

National Geology and Mining Service. (2020). *Anuario de la minería de Chile 2019 [Chilean Mining Yearbook 2019]*. Ministry of Mining.

Petersen, K. L., Frank, H., Paytan, A., & Bar-Zeev, E. (2018). Impacts of seawater desalination on coastal environments. In V.G. Gude (Ed.), *Sustainable desalination handbook* (1st ed., pp. 437–463). Butterworth-Heinemann.

Prieto, M., Calderón-Seguel, M., Fragkou, M. C., & Fuster, R. (2022). The (not-so-free) Chilean water model. The case of the Antofagasta Region, Atacama Desert, Chile. *The Extractive Industries and Society, 11*, 101081. https://doi.org/10.1016/j.exis.2022.101081

Ramírez, J., & Ramírez L. (2022). *Catastro de plantas desaladoras y sistemas de implusión de agua de mar en Chile 2022–2023* [Cadaster of desalination plants and seawater impulsion systems in Chile 2022–2023]. Minería Chilena, Chile.

Ruffino, B., Campo, G., Crutchik, D., Reyes, A., & Zanetti, M. (2022). Drinking water supply in the region of Antofagasta (Chile): A challenge between past, present and future. *International Journal of Environmental Research and Public Health, 19*(21), 14406. https://doi.org/10.3390/ijerph192114406

Shahabi, M. P., McHugh, A., Anda, M., & Ho, G. (2017). A framework for planning sustainable seawater desalination water supply. *Science of the Total Environment, 575*, 826–835. https://doi.org/10.1016/j.scitotenv.2016.09.136

Shemer, H., & Semiat, R. (2017). Sustainable RO desalination–Energy demand and environmental impact. *Desalination, 424*, 10–16. https://doi.org/10.1016/j.desal.2017.09.021

Sola, I., Fernández-Torquemada, Y., Forcada, A., Valle, C., del Pilar-Ruso, Y., González-Correa, J. M., & Sánchez-Lizaso, J. L. (2020). Sustainable desalination: Long-term monitoring of brine discharge in the marine environment. *Marine Pollution Bulletin, 161*, 111813. https://doi.org/10.1016/j.marpolbul.2020.111813

Superintendency of Sanitary Services (SISS). (2018). *Informe de Gestión del Sector Sanitario—2018* [Health sector management report—2018], Chile. https://www.siss.gob.cl/586/articles-17722_recurso_1.pdf

Superintendency of Sanitary Services (SISS). (2020). *Informe de Gestión del Sector Sanitario—2020*, [Health sector management report—2020], Chile. https://www.siss.gob.cl/586/articles-19431_recurso_1.pdf

Tapia, J., González, R., Townley, B., Oliveros, V., Álvarez, F., Aguilar, G., Menzies, A., & Calderón, M. (2018). Geology and geochemistry of the Atacama Desert. *Antonie Van Leeuwenhoek, 111*, 1273–1291. https://doi.org/10.1007/s10482-018-1024-x

The World Bank. (2011). *Chile: Diagnóstico de la gestión de los recursos hídricos* [Chile: Diagnosis of water resources management]. Department of Environment and Sustainable Development, Region for Latin America and the Caribbean.

Wang, K., Saththasivam, J., Yiming, W., Loganathan, K., & Liu, Z. (2018). Fast and efficient separation of seawater algae using a low-fouling micro/nano-composite membrane. *Desalination, 433*, 108–112. https://doi.org/10.1016/j.desal.2018.01.032

Zarzo, D. (2018). Beneficial uses and valorization of reverse osmosis brines. In *Emerging technologies for sustainable desalination handbook* (pp. 365–397). Butterworth-Heinemann, United Kindgom.

Opportunities for the Sustainable Development of Urban Mining: Lessons Learned from the Large-Scale Mining Industry

Dennis Vega-Garcia, Andrés Ramirez-Madrid, Leopoldo Gutierrez, Carlos Bustos, and Luis Salinas

Abstract Interest in urban mining has been growing in recent years, especially in waste from electric and electronic equipment (WEEE). The interest in WEEE is not only due to the concentration at which valuable key elements are found, but also the growing waste streams that are being generated by humankind. On the other hand, standard large-scale mining is facing new operating challenges because the number of elements of interest in ores is continuously decreasing. Being a very well-established industry, large-scale mining has vast know-how that can be transferred to urban mining. This chapter will present a comparative review of the stages that are similar to urban and large-scale mining (i.e., mining, pre-treatment and concentration, and chemical metallurgy), showing the advantages and disadvantages of chemical and physical processing routes for both urban and conventional mining. The most relevant findings in this chapter are (1) the importance of the size reduction and concentration processes to reduce the potentially hazardous waste produced in the recovery of elements, and (2) the lack of foresight that conventional mining has had with its operating problems when the quality of the raw materials decreased.

D. Vega-Garcia (✉)
Sustainable Minerals Institute International Centre of Excellence Chile (SMI-ICE-Chile), The University of Queensland, Santiago, Chile
e-mail: d.vega@smiicechile.cl

A. Ramirez-Madrid · L. Gutierrez · C. Bustos
Department of Metallurgical Engineering, Universidad de Concepción, Concepción, Chile

A. Ramirez-Madrid · L. Gutierrez
Water Research Center for Agriculture and Mining (CRHIAM), Universidad de Concepción, Concepción, Chile

L. Salinas
Department of Earth Science and Engineering, Imperial College London, London, UK

A. Ch. Braun et al. (eds.), *Eco-Industrial Development as an Industrial Strategy*, Green Energy and Technology, https://doi.org/10.1007/978-3-031-73576-9_23

1 Introduction

The extractive industry of non-renewable natural resources has been facing new challenges over the last several decades. These challenges range from opportunities in social policies to environmental regulations worldwide. This has emerged from the need to conceive, implement, and generate lower impacts on the environment and its surroundings. This approach has brought society to a new level of consciousness with respect to production processes that are more focused on reducing the amount of waste, using available natural resources more effectively, and encouraging responsible consumption (Marimuthu et al., 2021). The main goal of these actions is to reduce the carbon footprint and the anthropogenic effects of industrial processes. The mining industry has not been exempted from seeking more sustainable alternatives of production.

In the mining industry, and in the metal mining industry in particular, these challenges are being added to the gradual reduction of the mineral being exploited (Estrada et al., 2020; Kelm et al., 2013; Uribe et al., 2016), the decrease in the content of the elements of interest in the ore deposits, and the need to exploit deposits with the presence of minerals that are more difficult to process (Castillo et al., 2020; Ramirez et al., 2020a, 2020b; Rebolledo et al., 2017). Likewise, this industry is facing the need to increase the processing capacity of their plants because they are required to supply larger demands of metals. However, it is also facing the challenge of using mining resources more effectively, in particular water (Castro & Laskowski, 2011; Laskowski et al., 2019; Ramos et al., 2013).

Urban mining, broadly speaking, involves all of the recovery processes and reuse of materials that are considered waste, and the reject products, or products that have completed their life cycle in terms of human use. In this sense, urban mining can be seen as recycling systems and the systematic reuse of materials produced as residues with the objective of reducing raw material consumption (Pascuas Rengifo et al., 2018). However, these days urban mining is considered to be a recycling process of technological residues. This relates to the increase in demand for electronic products and their reduced life cycle, either for planned obsolescence or for rapid advances in technology.

When electronic waste (residue) is recycled, a large number of key elements can be recovered and returned to the productive system. Ceramics, polymers, metals, and rare earths can be recovered, reprocessed, and used in the production of new technologies or other applications, thus reducing dependence on the exploitation of mining resources. In this respect, the mining industry has a great deal of experience with methods and processes that can be adapted to urban mining. To compare urban mining and the mining industry, we can start by describing the mining industry. In this case, we will focus on the large copper mining industry, which has the greatest share of participation in the Chilean economy.

Latin America is a strategic area for the production of raw materials, producing large amounts of concentrates and metals. This is mainly because of the large availability of these non-renewable natural resources in the area, of which copper is one of the most-produced commodities. Metallic copper can be obtained via two different types of production: hydrometallurgy and pyrometallurgy. The selected process for a given application depends on the type of mineral and the feasibility of its being leached. In general terms, oxides and hydroxides of copper are treated in hydrometallurgical processes, while sulfurs are processed in pyrometallurgy processes. Both options start with the mining operation, which involves all the activities from the mineral extraction from the deposits until the delivery of the ore to the processing plants. The initial focus of the mining operation is to first exploit the richer areas (higher grades) of the ore deposits, which are normally easier to process. The mining operations are as selective as possible because the main goal is to optimize the amount of minerals being removed and transported. In the processing plants, ore minerals are sequentially crushed and milled until a particle size in which the desired degree of liberation of the mineral is achieved. The degree of liberation indicates how much the target component is liberated from the non-valuable components. Crushers and mills are operated in open and/or closed circuits with classifiers to ensure the correct particle size product goes on to the next stage. Crushing circuits normally include screens as classifiers in dry processes, while mills work in a circuit with hydrocyclones.

Both means of obtaining copper consider various processes and can be very extensive to describe in detail. Therefore, this chapter briefly presents the similarities and know-how that can be taken from the large-scale mining industry and incorporated into urban mining. The know-how is presented along with potential processing alternatives that can be applied in urban mining. This is aligned with the principles of the circular economy, which are focused on optimizing and recycling residues that can be returned to the productive system and, thus, reducing the need to exploit more resources.

Hydrometallurgical Route

The processing of copper oxides and hydroxides starts with the comminution process, which is the process of particle size reduction. The average particle size for the hydrometallurgy process (e.g., leaching) is between 10 and 20 mm. These minerals are agglomerated using water and sulfuric acid and then deposited into large leaching heaps that range from 3 to 8 m in height. In this operation, leaching heaps are irrigated from the top with acidic solutions by aspersion or trickling. The solution goes through the interstitial space inside the heaps, thereby extracting copper from the ore. In the lower part of the heaps, the solution containing copper (pregnant leach solution or PLS) is collected and transported to the next stage, solvent extraction, or SX. In the SX process the copper ions are removed from the PLS by bringing it into contact with an organic extractant, resulting in an organic phase with a high concentration of copper ions. Afterward, the copper ions present in the organic phase are re-extracted into a solution richer in copper ions. Finally, the solution is carried out to the last two stages, electrowinning and electrorefining, in which cathodes of copper with 99.99% purity are obtained (see Fig. 1).

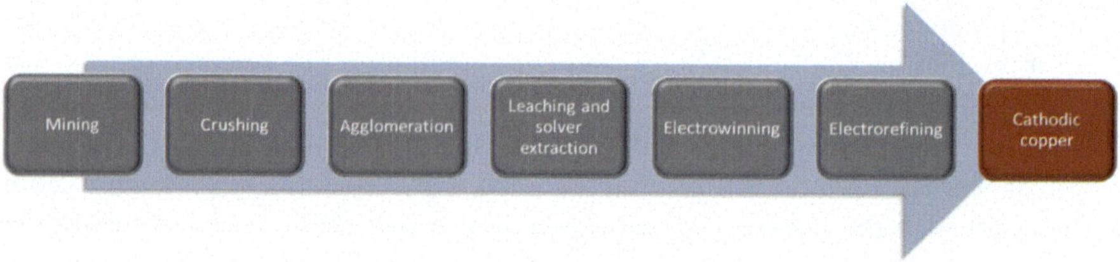

Fig. 1 Schematic of hydrometallurgical line to copper obtention from oxide minerals

Pyrometallurgical Route

The processing of sulfide minerals includes more unit operations than those for oxide minerals. Since the urban mining and big mining industry have more points of comparison in common when minerals are processed in pyrometallurgical processes, more lessons can be taken from this route.

Once the mineral is delivered to the processing plant, the reduction in the particle size takes place in the comminution process, which consists in a series of unit operations such as the crushing, milling, and classification of particles. Minerals are normally processed in the crushing units in dry processes and exit them with a particle size below 1 cm. The product of the crushing units is mixed with water to form a slurry that is then fed to the milling units. The product of the milling process has a particle size below 150 microns so that it assures the adequate liberation of the mineral of interest. The slurry containing the liberated minerals is fed into the concentration stage in which two products are obtained. One product contains the valuable mineral and the other the gangue or non-valuable material, which are called concentrate and tailings, respectively. Before being disposed of in tailing dumps, tailings are thickened to remove as much water as possible. On the other hand, concentrates are thickened and filtered to reduce the water content to its minimum. Filtered concentrates are then taken to pyrometallurgical processes.

Pyrometallurgical processes start with a drying process to remove the superficial water. Afterwards, the concentrate enters the fusion process, where a series of chemical reactions of oxidation of some oxides and reduction of others take place. The product of the fusion process is a copper alloy that is then taken to the refining process, after which electrolytic copper is obtained (99.99% of Cu), as seen in Fig. 2.

2 From Conventional Mining to Urban Mining

In this section, a comparison of the stages of processing will be made. The whole process has been divided in three sections, which are common to both conventional and urban mining. The stages compared are mining, pre-treatment and concentration, and chemical metallurgy. Although, the outcomes of this comparison could be applied to other countries, the focus of this chapter is on Chile.

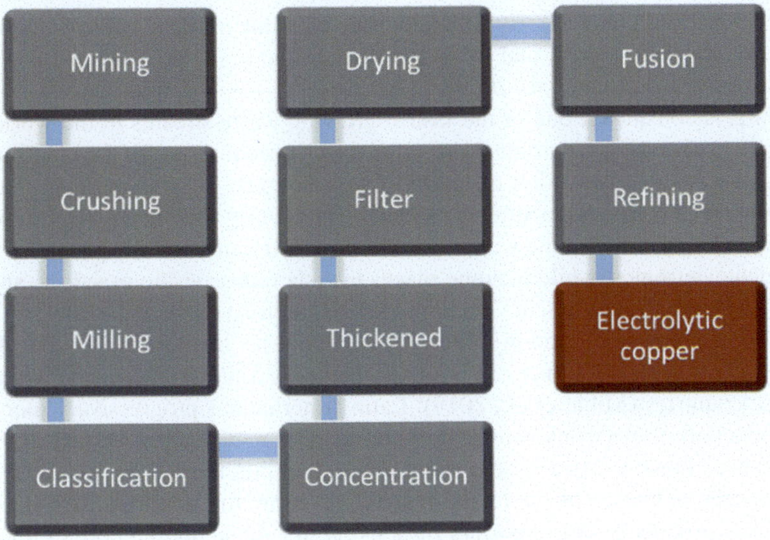

Fig. 2 Schematic of the pyrometallurgical route to obtain copper from sulfide minerals

2.1 *Mining Stage*

Mining plays a predominant role in Latin America and particularly in Chile, as the country is one of the world's largest copper producers. However, since economic benefits are obtained from the exploitation of non-renewable resources, the direct and indirect contribution to the internal economy must be publicly reported yearly. In fact, this activity contributes approximately 11% of the gross domestic product (GDP), 50% of the exports, and 26% of the foreign investments in the country (Consejo Minero, 2019).

The mineral grade has decreased dramatically over time, impacting the costs of excavation and crushing every year, and increasing energy consumption (Fernando et al., 2019). Besides, considering the environmental impact that arises when dealing with large volumes of waste rock, it can be argued that operations must become more versatile in order to overcome the affected natural and economic conditions.

A good starting point for the industry is to tackle these issues by paving a road to sustainability. That is, using resources without compromising their potential use or availability for future generations (Mudd, 2010). In the same vein, the concept of urban mining is a branch that has been gaining in popularity over the last few years and is directly related to an efficient use of resources. Therefore, this concept has not only become necessary, but essential when the possibility of facilitating the lives of generations to come is considered.

The key assumption of urban mining is that waste from electrical and electronic equipment (WEEE or e-waste) can be recovered in an analogous manner to conventional mining, producing secondary raw material from waste under the 3R concept (reduce, reuse, and recycle). By following this path, the intrinsic value of industrial products can be recovered, and these can be returned to the production chain at an even lower cost (Xavier et al., 2019).

WEEE is considered one of the fastest growing waste streams in the world and is becoming an emerging problem due to adverse consequences for the environment and human health (Awasthi et al., 2018). Yet, important efforts are being made to overcome WEEE disposal. Strategies vary from country to country, which mainly depend on the disposal of the waste. For example, African countries generally reuse discarded electronics, while in some places in Asia rather unsafe practices, such as landfilling and incineration, are performed (Duzgoren-Aydin et al., 2007).

Electronic waste production in developing countries is lower than in developed countries. The same difference can be observed in waste management and regulations in those countries (Mihai et al., 2019). Latin America has progressively increased the production of electronic waste, but there is poor management and treatment of the residues, however. In a more specific case, in Chile there is only one law that promotes the reduction of waste and fosters recycling: the so-called REP law. The law makes producers and importers responsible for financing the correct management of the waste generated by the products that are marketed on the national market, whether they are imported or of national manufacture. Therefore, there are companies dedicated only to disassembling and exporting WEEE (Araújo et al., 2012; Steubing et al., 2010). This has allowed them to disassemble and export residues with commercial value and return them to the value chain following the principles of the circular economy.

Electronic waste can be seen as a metallic deposit that has already been processed in which many key elements are present, but in concentrations similar to those in ore deposits. In this sense, from the experience of large-scale mining, it can be said that having a large group of elements present in the waste makes an extraction project available. Yet, it is essential to have collection areas and storage for WEEE that would work as in high-grade mining.

It can be suggested that this branch of the circular economy has a competitive advantage with respect to the exploitation of traditional mineral resources (Arora et al., 2021; Zeng et al., 2021). Substantially more metals of interest can be extracted from WEEE compared to traditional primary exploitation. At the same time, lower operating-costs are commonly combined with a drastic reduction in the negative impacts on the ecosystem, if done formally. It is, therefore, necessary to continue in the search for a better understanding of the current and future extraction mechanisms of metals from WEEE and the identification of the limitations that have historically hampered the massive development of this industry.

Historically, large-scale mining has developed due to its ability to specialize its exploitation of specific deposits with clearly defined objectives and products of interest, leaving aside both complex products and products with no economic interest. WEEE contains many key elements that are sometimes present in easy-to-recover systems and others in complex systems. Even though large-scale mining has had inconvenient strategies that have provided very enriching lessons, the possibility of processing complex areas has not yet been studied. Therefore, the mining industry has found low yields and done little development in this area. This is, nowadays, a research gap that has not been explored in Latin America. Yet, it appears to be an important opportunity for urban mining.

Selective exploitation is a strategy to maintain, or even increase, the production of elements of interest in deposits that have complex behaviors in conventional mining (Knissel et al., 1996). Current technological developments for conventional mining (either due to their high clay content, very low grades and, in some cases, with the presence of contaminants penalized in the commercialization) can be applied to urban mining. However, there are still some challenges that have already been identified in urban mining:

1. It requires that collection areas are dispersed throughout the operation site where there are dismantling areas in which the components of interest are clearly separated for each process.
2. It needs adequate transportation of the material to the specialized industries that can process each valuable material, following the environmental regulations of a given territory.
3. It should take advantage of the existing industrial infrastructures for the exploitation of resources and promote new specialized industries.

While urban mining is a process that has been around for a few decades, it is rather new when compared to conventional mining. Urban mining has clear challenges regarding its capacity for processing and pollutant emissions, which are a relatively recent matter. By comparing the complex products and challenges of large-scale mining with the number of polluting products associated with WEEE (e.g., heavy metals, polluting gases, or liquids), finding viable alternatives for storage and processing could be seen as an interesting R&D opportunity. Therefore, it is possible to promote the appropriate development of these processes in Latin America thanks to the increasing environmental awareness in the community and regulation makers.

2.2 Pre-treatment and Concentration

In large-scale mining, the quality of the pre-treatment and concentration processes determines the quantity of valuable material that will be taken to the following stages. Special interest is given to those minerals from which concentrates can be obtained, such as sulfide minerals. The mineral particle size that is input into the concentrator stage is one of the most important parameters. This is because the particle size is in direct relationship to the degree of liberation of the mineral of interest. The smaller the particle size, the higher the degree of liberation. The particle size also influences the hydrodynamics in the forthcoming stages of the concentration of minerals. Among the most frequently used methods of mineral concentration one can find are hydraulic washing, magnetic separation, and flotation. In hydraulic washing, the separation of particles is achieved by the difference in their specific gravity, magnetic separation uses the magnetic properties of the particles to separate them from the undesired material, and flotation uses the hydrophobicity of the mineral to separate it from the hydrophilic minerals in a three-phase system (air-water-mineral).

The need to increase the efficiency of comminution processes (processes of particle size reduction) in the mining industry led to the development of energy-saving technologies. This is the case for impact crushers, autogenous (AG) and semi-autogenous (SAG) mills, and high-pressure grinding rolls (HPGR). Even though these technologies have shown high performance in capacity and energy efficiency, it is always necessary to have ball mills and hydrocyclones for an adequate particle size before the concentration process. In this line, the know-how from comminution can be taken to urban mining as a direct implementation of this technology, e.g., reducing the particle size until the desired degree of liberation is reached using classifiers to recycle particles larger than that. However, the know-how can also be applied by adapting them accordingly, e.g., concentrating the desired WEEE element into the intermediate stages of comminution given the difference in density between metals and plastics.

In the recycling of WEEE, the dismantling and selection of pieces are done manually in stages, with basically no technology whatsoever involved. This has resulted in high labor demand and working hours. Therefore, the processing capacity is limited and costly, which is the main drawback for the large-scale implementation of WEEE recycling and processing. Large-scale mining has demonstrated that high processing capacity does not necessarily promote high efficiency, but it allows cost reductions. This is mainly because it focuses on having a high-quality product in the concentration processes in which a selective separation is carried out. To allow this, a pre-defined particle size in the comminution stage is key.

The know-how in the management of large volumes of ores and concentration of minerals by chemical and physical properties are part of what can be contributed to the processing of WEEE. It is necessary, after a basic dismantling, to implement comminution processes similar to those in the mining industry (crushing and milling) for an adequate liberation of particles. A correct liberation of particles will ease the mechanical separation of components. This means reducing the particle sizes down to the point where differences in density, hydrophobicity, magnetic and electric properties are large enough that they can be used as a driving force to separate valuable materials.

Large-scale mining has started focusing on having polymetallic-oriented processes, which are basically processes designed to obtain as many by-products as possible. This also should be the case for WEEE processing. Having complex systems (i.e., with many valuable components) as raw material is a challenge to think about technical, economical, and environmentally sustainable routes to obtain many valuable products.

It is important to mention that the dismantling process can be highly contaminant. Some components are difficult to handle such as volatile compounds, lead, mercury, barium, cadmium, and antimony, among others. They also contain organic and inorganic compounds that are dangerous for human health, such as cathode ray tubes (CTR), batteries, liquid crystal displays (LCD), capacitors, among others (Kaya, 2016; Singh et al., 2016).

In a correct recycling process, it is key to have an idea of the kinds of components present in WEEE for a better selection of the processes to be used. It is necessary to deeply understand the properties of the materials being processed. This includes densities, and magnetic, conductive, and surface properties. For instance, gravimetric concentration systems are very effective for materials with high density differences, which allows polymeric minerals to be separated from ceramics (Alter, 2005; Kaya, 2016; Saitoh et al., 1976; Shent et al., 1999).

The mining industry has several challenges nowadays. Processing large quantities of minerals that are mainly disposed of (the fraction of valuable material is small) requires high investment and energy. This generates tailings, which have other problems: large negative visual impact in the area of disposal and potential risks over time (during and post operation). This has taught us that mining, besides being a highly profitable business, requires new perspectives to give added value to its residues, especially because they are non-renewable resources. Urban mining and, in general, the circular economy, are working toward that in adding value to residues. This is a learning that contributes to a new responsible way of producing and making the most out of the available resources.

2.3 *Chemical Metallurgy*

In this section, the most commonly used metallurgical approaches for the extraction of metals from WEEE will be described with a special emphasis on hydrometallurgical processes.

Metamorphosis of Waste

It is important to consider the direct relationship that exists between metallurgy and the treatment of WEEE. Each technique bases its procedure on combinations of physical, chemical, thermal, and metallurgical processes. Each process will have advantages and disadvantages depending on the product to be extracted, environmental consequences, and operational costs. Therefore, a specific processing route must be chosen to maintain a balance between pros and cons.

Generally, to face the problems of each technique, it is recommended that the metal recovery process be carried out in several stages. In other words, that means a subsequent series of stages of physical/mechanical methods, separation and concentration methods, and metallurgical treatments. All of this is done in order to develop a combined process that supports the best performance (Chauhan et al., 2018). The recovery of metals from urban mining, in most cases, requires a previous stage before grinding so that later, the hydrometallurgical and pyrometallurgical processes can be carried out in the most efficient manner.

Metallurgical Processes

A variety of alternatives are available to link conventional metallurgical processes with the recovery of metals after segregation, size reduction, and the potential physical separation of metallic and non-metallic fractions. For instance, an advantage of using pyrometallurgical processes is the simultaneous disassembly that is likely to occur when introducing different types of WEEE to a furnace. However, there are some economic disadvantages that would be important to consider before choosing this option. The most conventional processing routes for WEEE are summarized in Fig. 3.

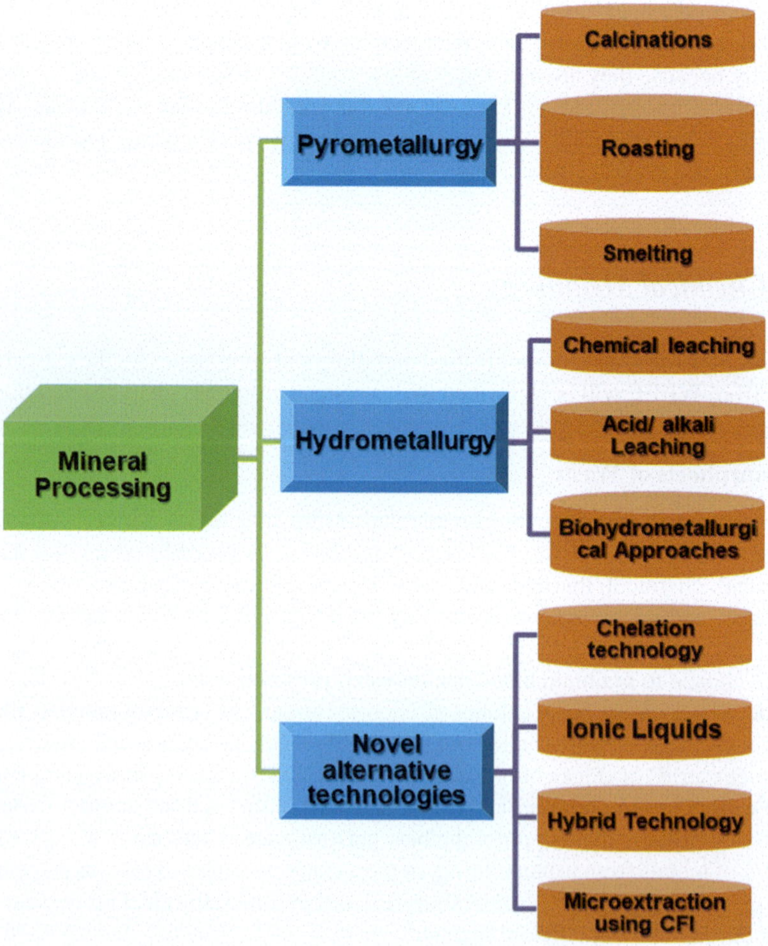

Fig. 3 Metallurgical approaches for the extraction of metals from WEEE

Pyrometallurgical Route

Pyrometallurgy is a traditional approach to metal recovery. Between its successive stages, it typically involves incineration, arc furnace smelting, blast furnace or Cu smelting, and high temperature roasting (Chauhan et al., 2018). Still, while this alternative has generated a good recovery of base metals, it has some environmental impacts that prevent it from being considered as a future alternative when seeking sustainability. Among them:

- A massive contribution of energy is required to maintain the temperature at which pyrometallurgical processes are carried out.
- There is considerable emission of dust and gases (among them CO and CO_2 that are commonly used as reducing agents) that can directly affect the ecosystem.
- Matter is very difficult to transport.
- High recoveries are required for the process to be economically viable.
- Enormous quantities of mixed oxides are generated, which causes substantial losses of precious metals.

There are very few smelting plants that are available for the formal processing of metal components from WEEE worldwide. The Aurubius smelter in Germany (Aurubius, 2021) and the Ronnskar smelter in Sweden (New Boliden, 2021) are two examples of registered smelters that are available today for the formal processing of metal components worldwide.

However, it would not be technically nor economically appropriate to abandon this way of obtaining elements in both traditional and urban mining. Large-scale mining has had technological developments that make it relevant to review some of them. For instance, before the high temperature processes, the concentration of elements of interest should be maximized. In the case of mining, concentrates with copper content higher than 25% are sold, which means that a significant part is not necessarily of interest. The most common copper ore is chalcopyrite, so assuming that the concentrate is pure chalcopyrite, the highest possible concentration of copper is 34.6% (Fairthorne et al., 1997). While the mineralogical and processing conditions are not always favorable for maximizing the concentrate grade, it can be said that large-scale mining has progressed in relation to the presence of elements that have negative effects in both pyrometallurgical processes and in the environment. For instance, sulfuric acid is obtained as a by-product in exhaust gas scrubbing (sulfuric acid is used to leach copper oxide minerals, among other applications). Also, pyrite, which is found in concentrates, is present in the exothermal reaction that allows for a reduction in the amount of fuel needed in the fusion process.

On this basis, in the processing of WEEE, more research in the development of cleaner and safer technologies is required, which is the focus of many researchers worldwide, especially considering that WEEE is composed of many materials that can be harmful contaminants.

Hydrometallurgical Routes

Hydrometallurgical routes are the most commonly used branches to recover precious metals and other metals of interest in the treatment of electronic waste, particularly from printed circuit boards, cell phones, and electronic devices. What makes this technique advantageous over pyrometallurgy is the high selectivity and its lower expected environmental impacts on the ecosystem. Some of the advantages of this route (when compared to pyrometallurgy) include:

- Much lower emissions of harmful gases.
- Lower operational costs.
- Lower energy requirements.

One of the disadvantages of this route is that it can be slower than pyrometallurgical processes. Hydrometallurgical methods consist mainly of oxidative leaching for the extraction, purification, and recovery of metals (Iannicelli-Zubiani et al., 2017). One of the key parameters to achieve a good recovery of the metal of interest is an appropriate selection of the leaching reagents, which will vary according to the extraction requirements of each process. The leaching reagents most typically used can be divided into three large groups and can be seen in Fig. 4.

Different leaching processes are found in large-scale mining, such as heap leaching, dump leaching in an agitation leach tank, and pressure leaching, which are selected according to the specific compounds of interest. It seems unlikely to extend the industrial application of the available leaching technologies to processing copper oxides in WEEE processing. However, techniques and technologies for other elements could be more applicable. For instance, in gold mining, concentration processes are done before leaching in agitation tanks, which significatively increases the recovery rate of gold.

In WEEE processing, a concentration process should be done after dismantling and selection to take only the parts of interest to the leaching stage, avoiding the contamination of parts without elements of interest. Unlike technologies with slow kinetics (i.e., heap leaching), autoclave and closed reactors can be used for enriched solutions to selectively concentrate elements of interest. Techniques such as adsorption with activated carbon, ion-exchange in resin, cementation processes, or selective precipitation can be used. It is important to highlight that although large-scale mining does not use multi-step leaching, there are experiences that can be brought to urban mining. Paradigms of processing scale and selectivity must be broken based on an adequate selection of technologies from the beginning of the processing.

3 Conclusions

This chapter showed how physical and chemical processes involved in large-scale mining could be applied to process waste from electrical and electronic equipment (WEEE) in urban mining.

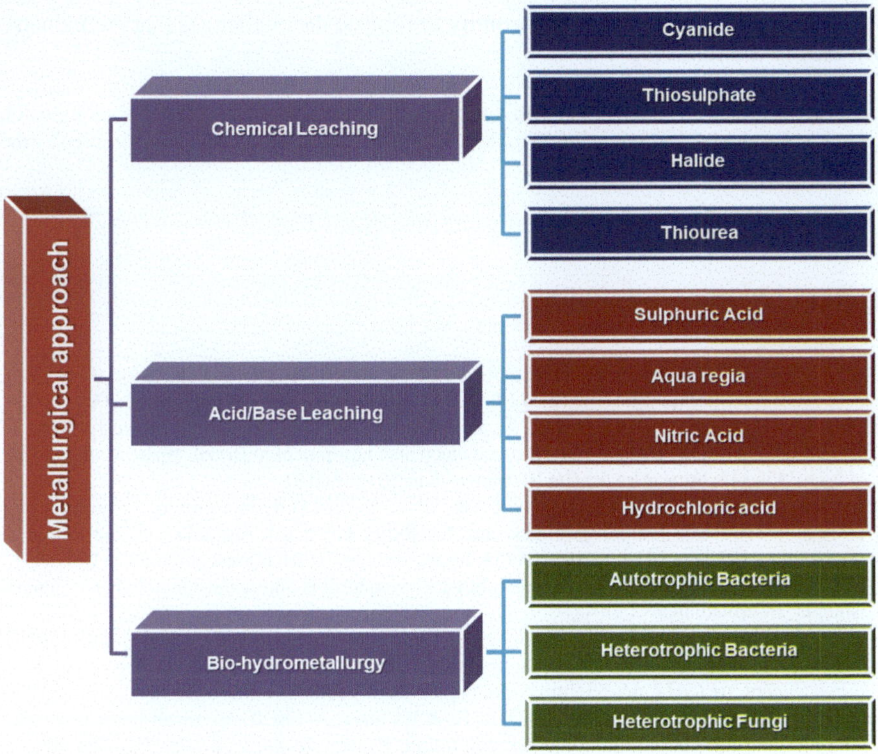

Fig. 4 Classification of the leaching reagents used for the treatment of WEEE

It described the advantages and disadvantages of chemical and physical processing routes for both urban and conventional mining. The economic and environmental potential of WEEE was also shown. The economic potential is justified because the concentration of elements of interest in WEEE is higher than those in ores for conventional mining. The environmental impact is related to the reduction of potential harmful pollutants from WEEE.

In the near future, urban and conventional mining are going to be linked in aspects such as innovation and technology development. The challenge is to find sustainable processes at a large scale that allow us to recover minerals of interest more efficiently. As in large-scale mining, in urban mining it is necessary to identify areas for the application of logistics, metallurgy, and chemical processes to address variations in the raw materials.

Although the hydrometallurgy and pyrometallurgy processes that are currently being proposed for recovery of elements from WEEE are claimed to be environmentally friendly, it is necessary to do more research focused on improving size reduction and concentration processes.

Another lesson that can be taken from large-scale mining is the lack of foresight for operating problems such as the reduction in grade or quality of ores. Therefore, with urban mining there is an opportunity to foresee the challenges that the industry will face in the future.

Acknowledgements The authors wish to acknowledge the financial support of the Water Research Centre for Agriculture and Mining (CRHIAM) of the Universidad de Concepción sponsored by the ANID/FONDAP/15130015 project.

References

Alter, H. (2005). The recovery of plastics from waste with reference to froth flotation. *Resources, Conservation and Recycling, 43*(2), 119–132. https://doi.org/10.1016/j.resconrec.2004.05.003

Araújo, M. G., Magrini, A., Mahler, C. F., & Bilitewski, B. (2012). A model for estimation of potential generation of waste electrical and electronic equipment in Brazil. *Waste Management, 32*(2), 335–342. https://doi.org/10.1016/j.wasman.2011.09.020

Arora, M., Raspall, F., Des, D., Fearnley, L., & Silva, A. (2021). Urban mining in buildings for a circular economy: Planning, process and feasibility prospects. *Resources, Conservation & Recycling,* 174(September 2020), 105754. https://doi.org/10.1016/j.resconrec.2021.105754

Aurubius. (2021). Smelter Intermediates. https://www.aurubis.com/en/products/other-products/smelter_intermediates

Awasthi, A. K., Cucchiella, F., D'Adamo, I., Li, J., Rosa, P., Terzi, S., Wei, G., & Zeng, X. (2018). Modelling the correlations of e-waste quantity with economic increase. *Science of the Total Environment, 613–614*, 46–53. https://doi.org/10.1016/j.scitotenv.2017.08.288

Castillo, I., Gutierrez, L., Hernandez, V., Diaz, E., & Ramirez, A. (2020). Hemicelluloses monosaccharides and their effect on molybdenite flotation. *Powder Technology, 373*, 758–764. https://doi.org/10.1016/j.powtec.2020.07.032

Castro, S., & Laskowski, J. S. (2011). Froth flotation in saline water. *KONA Powder and Particle Journal, 29*(29), 4–15. https://doi.org/10.14356/kona.2011005

Consejo Minero. (2019). Cifras actualizadas de la minería. Consejo Minero, 4. https://consejominero.cl

Chauhan, G., Jadhao, P. R., Pant, K. K., & Nigam, K. D. P. (2018). Novel technologies and conventional processes for recovery of metals from waste electrical and electronic equipment: Challenges & opportunities—A review. *Journal of Environmental Chemical Engineering, 6*(1), 1288–1304. https://doi.org/10.1016/j.jece.2018.01.032

Duzgoren-Aydin, S. C., Aydin, N. S., & Wong, M. H. (2007). Evidence of excessive releases of metals from primitive e-waste processing in Guiyu China. *Environmental Pollution, 148*(1), 62–72. https://doi.org/10.1016/j.envpol.2006.11.006

Estrada, D., Echeverry, L., Ramirez, A., & Gutierrez, L. (2020). Molybdenite flotation in the presence of a polyacrylamide of low anionicity subjected to different conditions of mechanical shearing. *Minerals, 10*(10), 1–12. https://doi.org/10.3390/min10100895

Fairthorne, G., Fornasiero, D., & Ralston, J. (1997). Effect of oxidation on the collectorless flotation of chalcopyrite. *International Journal of Mineral Processing, 49*(1), 31–48.

Fernando, W. A. M., Ilankoon, I. M. S. K., Krishnan, S. G., & Chokshi, V. (2019). Los efectos de la forma y estructura del empaque en la distribución de líquidos de los sistemas de lixiviación en pilas: Adición de PCB como partículas no minerales. *Hydrometallurgy, 187*(May), 149–157. https://doi.org/10.1016/j.hydromet.2019.05.010

Iannicelli-Zubiani, E. M., Cristiani, C., Dotelli, G., & Gallo Stampino, P. (2017). Recovery of valuable metals from electronic scraps by clays and organo-clays: Study on bi-ionic model solutions. *Waste Management, 60*, 582–590. https://doi.org/10.1016/j.wasman.2016.07.035

Kaya, M. (2016). Recovery of metals and nonmetals from electronic waste by physical and chemical recycling processes. In *Waste management* (Vol. 57, pp. 64–90). https://doi.org/10.1016/j.wasman.2016.08.004

Kelm, U., Helle, S., Jerez, O., & Pincheira, M. (2013). What are copper clays? Geometallurgical implications. In U. Kelm, S. Helle, O. Jerez, & M. Pincheira (Eds.), *Proceedings of Copper* (pp. 257–266).

Knissel, W., Hilger, J., & Fahlbusch, M. (1996) Significance of selective mineral exploitation for economical and environmentally beneficial underground ore mining, *International Journal of Rock Mechanics and Mining Sciences and Geomechanics* Abstracts 33, 326A. https://doi.org/10.1142/S0950609895000163

Laskowski, J. S., Castro, S., & Gutierrez, L. (2019). Flotation in seawater. *Mining, Metallurgy and Exploration, 36*(1), 89–98. https://doi.org/10.1007/s42461-018-0018-6

Marimuthu, R., Sankaranarayanan, B., Ali, S. M., Jabbour, A. B. L. de S., & Karuppiah, K. (2021). Assessment of key socio-economic and environmental challenges in the mining industry: Implications for resource policies in emerging economies. *Sustainable Production and Consumption*, 27, 814–830. https://doi.org/10.1016/j.spc.2021.02.005

Mihai, F. C., Gnoni, M. G., Meidiana, C., Ezeah, C., & Elia, V. (2019). Waste electrical and electronic equipment (WEEE): Flows, quantities, and management—A global scenario. *Electronic Waste Management and Treatment Technology*, 1–34. https://doi.org/10.1016/B978-0-12-816190-6.00001-7

Mudd, G. M. (2010). The Environmental sustainability of mining in Australia: Key mega-trends and looming constraints. *Resources Policy, 35*(2), 98–115. https://doi.org/10.1016/j.resourpol.2009.12.001

New Boliden. (2021). Boliden Rönnskär - Boliden. https://www.boliden.com/operations/smelters/boliden-ronnskar

Pascuas Rengifo, Y., Correa Cruz, L., & Marlés Betancourt, C. (2018). Residuos electrónicos: análisis de las implicaciones socioambientales y alternativas frente al metabolismo urbano—Dialnet. *Ciencia, Docencia y Tecnologia, 26*(56), 242–252. https://dialnet.unirioja.es/servlet/articulo?codigo=6610213

Ramirez, A., Gutierrez, L., & Laskowski, J. S. (2020a). Use of "oily bubbles" and dispersants in flotation of molybdenite in fresh and seawater. *Minerals Engineering, 148*. https://doi.org/10.1016/j.mineng.2020.106197

Ramirez, A., Gutierrez, L., Vega-Garcia, D., & Reyes-Bozo, L. (2020b). The depressing effect of kaolinite on molybdenite flotation in seawater. *Minerals, 10*(6), 1–14. https://doi.org/10.3390/min10060578

Ramos, O., Castro, S., & Laskowski, J. S. (2013). Copper-molybdenum ores flotation in sea water: Floatability and frothability. *Minerals Engineering, 53*, 108–112. https://doi.org/10.1016/j.mineng.2013.07.009

Rebolledo, E., Laskowski, J. S., Gutierrez, L., & Castro, S. (2017). Use of dispersants in flotation of molybdenite in seawater. *Minerals Engineering, 100*, 71–74. https://doi.org/10.1016/j.mineng.2016.10.004

Saitoh, K., Nagano, I., & Izumi, S. (1976). New separation technique for waste plastics. *Resource Recovery and Conservation, 2*(2), 127–145. https://doi.org/10.1016/0304-3967(76)90004-4

Shent, H., Pugh, R. J., & Forssberg, E. (1999). A review of plastics waste recycling and the flotation of plastics. *Resources, Conservation and Recycling, 25*(2), 85–109. https://doi.org/10.1016/S0921-3449(98)00017-2

Singh, N., Li, J., & Zeng, X. (2016). Global responses for recycling waste CRTs in e-waste. *Waste Management, 57*, 187–197. https://doi.org/10.1016/j.wasman.2016.03.013

Steubing, B., Böni, H., Schluep, M., Silva, U., & Ludwig, C. (2010). Assessing computer waste generation in Chile using material flow analysis. *Waste Management, 30*(3), 473–482. https://doi.org/10.1016/j.wasman.2009.09.007

Uribe, L., Gutierrez, L., & Jerez, O. (2016). The depressing effect of clay minerals on the floatability of chalcopyrite. *Mineral Processing and Extractive Metallurgy Review, 37*(4), 227–235. https://doi.org/10.1080/08827508.2016.1168419

Xavier, L. H., Giese, E. C., Ribeiro-Duthie, A. C., & Lins, F. A. F. (2019). Sustainability and the circular economy: A theoretical approach focused on e-waste urban mining. *Resources Policy*, October 2017, 101467. https://doi.org/10.1016/j.resourpol.2019.101467

Zeng, X., Xiao, T., Xu, G., Albalghiti, E., Shan, G., & Li, J. (2021). Comparing the costs and benefits of virgin and urban mining. *Journal of Management Science and Engineering*. https://doi.org/10.1016/j.jmse.2021.05.002

Enhancement of the Urban Habitat Through Novel Construction Products from Solid Industrial Residues: Opportunities and Challenges for Urban-Industrial Symbiosis by Including Informal Waste-Pickers

Danny Tröger, Abraham Alexis Becerra Araneda, Roberto Busnelli, and Andreas Ch. Braun

Abstract Rural migration and rapid urban growth have led to socio-environmental problems in many Latin American cities, of which waste accumulation, housing shortages, and unemployment are three of the most obvious. Conventional waste management strategies and scholarly discourses on the circular economy pay little attention to the practice of informal waste collection, and there is a lack of literature on industrial ecology in Latin America. The example of Santiago de Chile is used to highlight the problems of conventional management of industrial residues and household solid waste. Using examples from several Latin American countries with a focus on Buenos Aires, Argentina, a strategy for an urban-industrial symbiosis is developed that addresses SWM. This eco-industrial development strategy is based simultaneously on the involvement of informal waste pickers and the development of waste-based construction products to improve the urban habitat by providing sustainable construction products. The technical aspect is covered by international guidelines for the design of new sustainable products. Finally, we compile the perspectives of the stakeholders involved and their options for implementing this strategy.

D. Tröger (✉) · A. Ch. Braun
Kassel Institute for Sustainability, University of Kassel, Kassel, Germany
e-mail: danny.troeger@uni-kassel.de

A. A. Becerra Araneda · R. Busnelli
Instituto de Arquitectura y Urbanismo, Escuela de Hábitat y Sostenibilidad, Universidad Nacional de General San Martín, Buenos Aires, Argentina

A. Ch. Braun et al. (eds.), *Eco-Industrial Development as an Industrial Strategy*, Green Energy and Technology, https://doi.org/10.1007/978-3-031-73576-9_24

1 Socio-Environmental Problems of Latin American Metropolitan Areas

Within the last decades, a rural exodus and rapid population growth have often overwhelmed urban planning in Latin American cities. Meanwhile, industrial advancement has created more artificial materials that are not naturally degraded in the environment or degrade too slowly. As a result, large amounts of industrial residues have accumulated in the cities, whether in commercial situations or in private households, creating increased needs for waste management. Yet waste management systems are already overburdened with the removal of household and industrial waste from districts planned by the municipalities, not to mention from informal settlements. In Latin America, waste management is focused on bringing waste to legal landfills and decreasing the occurrence of illegal dumps (Margallo et al., 2019). Adequate waste treatment or recycling is rare. Instead, the incineration of both industrial and household waste in open fires is common. This generates and releases toxic compounds, especially when plastics are involved (Verma et al., 2016). In any case, the focus on removing waste from the streets or other visible locations alone will be insufficient to achieve sustainable development. Rather, increasing the volume of waste being recovered (reused, upcycled, or recycled) will be necessary to reduce the consumption of primary resources as well as lower climate-relevant emissions (Nelles et al., 2016). On the other hand, the widespread role of informal waste picking and recycling in Latin America is largely ignored (or at least hardly noticed) by society (Medina, 2007; Villalba, 2020).

In this chapter, we present ideas on how to mitigate the problems mentioned above, compiling those ideas into an eco-industrial development strategy. The core idea is to explore the potential of solid waste for the production of construction materials and products from the solid industrial residues that would otherwise end up in legal or illegal landfills. The metropolitan areas of Latin America present enough manpower and recoverable waste (supply) as well as a need for housing (demand), with self-constructed, low-quality houses being widespread (Masotti et al., 2011). Subsequently, we also discuss which general conclusions can be drawn from this strategy for the framework of eco-industrial development.

1.1 Eco-Industrial and Urban Development

From the perspective of eco-industrial development (EID), the described waste problem is a matter of material flow(s) which can be differentiated as follows (see also introductory chapter to this book part): On the one hand, there is the industrial metabolism, i.e., all commercially manufactured products including their byproducts and residues of the production processes, and on the other hand, there is the urban metabolism, i.e., the materials discarded by private households.

The general idea of EID is to minimize the need for primary raw materials and the generation of unwanted materials (waste/solid industrial residues) that end up in landfills or in illegal incineration. In terms of industrial ecology, that means establishing and improving the link between the industrial and urban metabolisms, which is also called urban-industrial symbiosis. In fact, EID strategies have already been successfully used as an urban planning tool to achieve environmental goals, as demonstrated by the following examples.

In the metropolitan area of Ulsan, South Korea, industrial waste generation was reduced by 35% at the regional level within fifteen years through EID promotion and urban-industrial symbiosis (Shah et al., 2020). In Japan, industrial symbiosis was successfully established through utilization of byproducts and eco-sharing utilities, while industrial and urban symbiosis resulted in the reduction of waste for incineration or landfilling through traditional and new recycling industries, among other measures. Legislative framework in Japan contributed to the implementation of these symbioses, as well as government subsidies (van Berkel et al, 2009). For the city of Kawasaki, Japan (Geng et al., 2010) estimated "that recycling mixed paper, mixed plastics, and organic wastes and utilizing the recycled materials in industrial production will potentially reduce about 69 kt CO_2(e) emissions and 8 kt incineration ashes to be landfilled in 2015." In China, environmental efficiency increased significantly in some cities from 2003 to 2012 in response to industrial ecological transformation (Song et al., 2016).

So far, there is a lack of literature on EID in Latin America. The studies mentioned are from countries whose political and social contexts are different from those in Latin America. In addition, the simple transfer of circular economy solutions to other countries is fraught with pitfalls (Iyamu et al., 2020). In particular, the social aspects will get the short end of the stick (Gall et al., 2020; Vanhuyse et al., 2021).

1.2 Industrial Ecology in the Construction Sector

Only a few studies have been conducted on the application of eco-industrial strategies to the links between urban metabolism and the construction industry. These studies mainly address resource conservation through savings in packaging materials and byproducts, efficient logistics, and shorter transport routes, or the energy efficiency and durability of the new housing (Hossain et al., 2020). Gonzalez-Garcia and Dias (2019) focus on enhancing environmental impact through means of urban symbiosis on the scale of cities, which they calculate though life cycle assessment. They conclude that there is much potential to improve the environmental impact of cities using sustainable construction materials and more energy-efficient housing. Building materials and products made of recycled or secondary raw materials, also called non-traditional or non-conventional building materials, hardly play a role in the applications for which circular economy and urban-industrial symbioses approaches are discussed (Winans et al., 2017). One of the few examples discussed in academic literature is related to the use of agri-food waste for producing new building materials (Giordano et al., 2020).

Basically, the literature on industrial ecology in the construction sector revolves around solutions within the industrial metabolism, but not around engaging and connecting it to the urban metabolism. In fact, these are circular economy solutions and little attention is paid to the social dimension, such organizational, urban, and territorial planning aspects and development issues.

1.3 Linking the Urban and Industrial Metabolisms: General Suggestions and Open Questions

Among the barriers to implementing eco-industrial strategies, which have been well studied in the relevant literature (Cohen-Rosenthal, 2017; Gibbs, 2009; Gibbs & Deutz, 2005), we consider the following aspects as particularly relevant challenges for the development of EID strategies in Latin America:

- Estimating the actual contribution of eco-industrial strategies to sustainable development (in contrast and comparison to conventional economic approaches) is methodologically difficult and is in fact rarely undertaken, especially for the social dimension (Walker et al., 2021; Gall et al., 2018).
- Developing the interface between eco-efficiency in industries and local communities is an important success factor for the long-term sustainability of eco-industrial development, yet this remains a challenging task for public administration (Yedla & Park, 2017).
- The marketability of "green" products must be promoted by eliminating biases against products made from recycled materials (Yedla & Park, 2017).
- Trust between the actors involved is crucial for successful cooperation, but is actually severely limited due to the institutional logic of action of each respective actor (Gibbs, 2003).
- Conflicting worldviews can easily lead to differing risk perception around industrial development (Masuda & Garvin, 2006).
- The development and creation of business models is the key to success, but highly case-specific (Fraccascia, 2018; Gall et al., 2018).
- The data on the flows and exchanges of matter and energy between actors (industry, society, and environment) is unreliable, making it difficult to propose urban and industrial metabolism strategies (López de Munain et al., 2021).

We also see another challenge, which the following quote from Brown and McGranahan (2016) brought to our attention:

> Social problems can be exacerbated in the context of promoting green economy and sustainability if informal activities are ignored. The poorest sellers and traders, of all people, would not be able to comply with stricter environmental and safety standards, thus driving their activity further into illegality.

We fully agree with this argument. However, the literature on EID thus far hardly deals with this problem, not even that which develops links between the industrial and urban metabolisms. In this line of argumentation, we see one central and challenging question:

How can the informal recovery sector be included into the formal cycles of waste reduction and the transformation of waste into certified sustainable construction products?

From our own observations (such as participating in meetings of consortia exchanging green economy strategies and respective lobby events), we have the impression that Latin American institutions searching for circular economy solutions often hire European consulting companies. Those, in turn, usually present circular economy solutions to the problem of industrial residues, which are technically highly sophisticated but would in our opinion fail to address the social dimension of sustainability at the very least, if not the ecologic and economic dimensions as well (Kovacic et al., 2019; Nelles et al., 2016).

As efforts have recently been made in Chile to promote the circular economy, including the enactment of a general circular economy law (Law Nº 20.920) and the development and publication of national roadmaps for the circular economy in general and specifically for the construction sector (Sala & Zegers, 2021), we will focus our discussion on the example of Santiago de Chile.

We start with a rough description of waste management in Santiago de Chile (Sect. 2), followed by a compilation of examples of successful inclusion of informal waste pickers from different Latin American countries, Argentina in particular (Sect. 3). We also present technical aspects on building materials/products from secondary raw materials and link this with social and institutional issues. In Sect. 4, we provide some food for thought and impetus on how the experiences from Buenos Aires could be implemented in Chile, addressing several stakeholders. In the last section we will draw general conclusions and discuss the relevance of our findings on the eco-industrial framework.

2 Household Waste Management (Urban Metabolism) in Santiago de Chile

The study site is the city of Santiago de Chile (from now on "Santiago"), the capital of the Metropolitan Region, with about eight million inhabitants representing nearly 47% of the national population (INE, 2018). In 2019, Santiago generated about 45% of the total municipal solid waste in Chile (Chile, 2021). The amount of household solid waste (HSW) produced in Santiago from 1995 to 2019 increased from about 0.8 to about 1.3 (kg/capita*day) (Chile, 2021). In 2006, 46% of the total waste produced in Santiago fell into the HSW category (Bräutigam et al., 2012).

The stakeholders in the HSW management (HSWM) of Santiago are defined by three hierarchical levels: national policies for standards and global goals; regional government for controlling the development of landfills; and local policies developed by municipalities that manage the collection, transport, and disposal of waste (Law Nº 18,695; Chile, 2006). The HSWM goals are to safeguard health by preventing the spread of disease, to protect the environment, and to conserve resources and energy. The most relevant stakeholders of HSWM are environmental directives, municipalities, waste-producing companies, landfills, private recycling plants, non-governmental organizations (NGOs), the National Movement of Waste Pickers of Chile, and informal or formal social actors such as waste-pickers (*recicladores de base*), whether individual or gathered in cooperatives.

Each of Santiago's thirty-seven municipalities defines the actors of its HSWM and the annual budget for related services (Aguilera et al., 2003). In 2015, the end disposal into landfills was dominated mainly by an oligopoly of three large private companies: Urbaser Danner Group, a Spanish-American company; Consorcio Santa Marta, a Chilean holding company; and Veolia, a French company (Valenzuela-Levi, 2020). Several municipalities contract private actors to collect and transport waste (Reyes, 2004; Valenzuela-Levi, 2020). The contracts between the municipality and these private actors are usually long-term and require exclusivity about the final destination of the HSW (Reyes, 2004; Valenzuela-Levi, 2020). For instance, the contracts between the Urbaser Danner Group and twenty-two municipalities of Santiago are for a period of more than thirty years, from 1995 to 2027 (Pizarro & Jara, 2015).

The recycling route in the HSWM of Santiago is performed by informal and formal actors in the following sequence: (1) waste-pickers collect waste from the streets and dumps and perform a first sorting; (2) private companies and charities transport the waste sorted by waste-pickers; and (3) medium to large private industries recycle the materials recuperated by waste-pickers. In 2018, Chile had implemented HSW recycling services in 54.8% of its municipalities (Valenzuela-Levi, 2019). However, the average amount of waste recycled by private recycling plants in 2019 was only 1.7% (Valenzuela-Levi, 2019, 2021). Other data from the Chilean Ministry of the Environment show that the recovery of HSW[1] reached almost 1% in 2018 (Chile, 2020). Although the municipalities have defined their HSWM plan, very few have introduced a segregated HSW collection system (González Martínez, 2011). In 2008, for example, 30% of the waste collected in the municipality of Nunoa for recycling was in fact not recyclable (González Martínez, 2011). All these facts promote a low recycling rate[2] that has been maintained over the years, as shown in Fig. 1 with data from 2007 (Barton, 2016). These figures show that around 84% of the total HSW produced in Santiago is disposed of in landfills and lacks any treatment (mechanical,

[1] Recovery refers to the set of actions aimed at recovering a waste, one or more of the materials that compose it, and/or their calorific value. Includes preparation for reuse, recycling, and energy recovery (Chile, 2020).

[2] To increase the recycling rate, in 2021 Decree No. 12 of the Ministry of the Environment established goals for the collection and recovery of waste. This decree applies to companies, not micro-enterprises as defined in Law No. 20,416 (CONAMA, 2010), that introduce more than 300 kg of containers per year into the market.

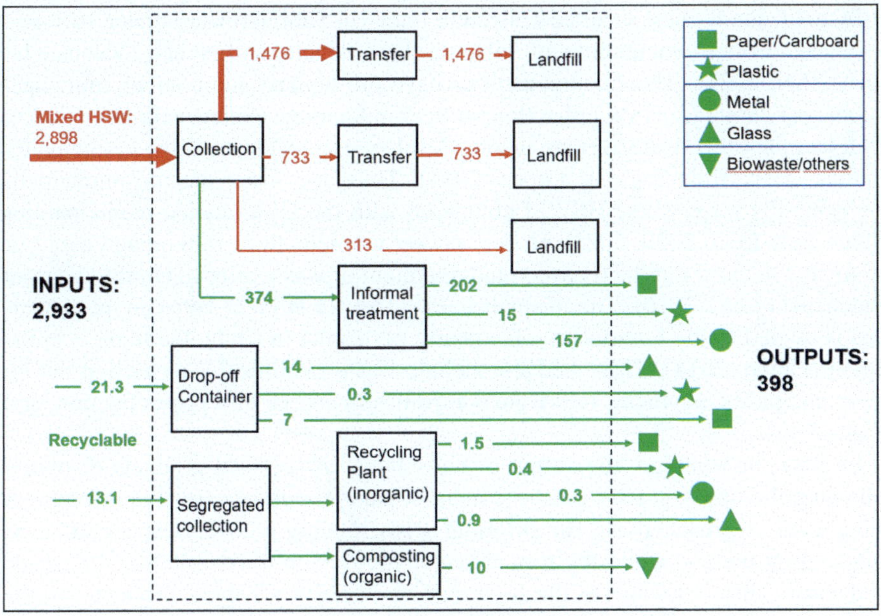

Fig. 1 Mass flow of inputs and outputs in the waste management system in Santiago de Chile in 2007. The numbers in the figure indicate 1 kt/year. Red arrows indicate disposed, unrecovered, mixed household solid waste (HSW) and green arrows indicate recovered HSW. Symbols correspond to the specific type of recovered material. Adapted from Barton & Kopfmüller (2016)

biological, or thermal) prior to its disposal, due to deficient laws and policies and the low tax on landfilling, among other reasons (González Martínez, 2011). Chile has the lowest recycling rate amongst countries of the Organization for Economic Cooperation and Development (OECD) (OECD, 2020), where the average recycling rate is 35%.

Figure 1 is based on material flow analysis (MFA), which is a good starting point for developing urban and industrial metabolism strategies. According to Ayres (1994), industrial metabolism is an integrated collection of processes that convert raw materials, energy, and labor into finished products plus wastes. Hence, Fig. 1 is missing the energy/emission flows for collection, recovery, and disposal processes of HSW, the MFA on waste-based products that arise from the recycled volume, and the MFA related to labor. The latter is related to the diversity of actors that show interdependency and cooperation in the HSMW. Korhonen (2004) reports that this diversity of actors is needed to comply with the principles that drive the development of natural ecosystems, which are roundput, diversity, locality, and gradual change. Since industrial ecosystems would function as an analog of natural ecosystems (Erkman, 1997), the principles of the latter allow for the development of industrial ecology strategies, as shown by several authors (Ayres, 1994; Jensen et al., 2011; Korhonen, 2004).

In 2017, the highest separate collection rates[3] in Santiago were under 10% and belonged to the municipalities of Vitacura, Providencia, and Nunoa (Valenzuela-Levi, 2020, 2021). These municipalities have the highest autonomous municipal income per capita for HSW collection services in Santiago (Valenzuela-Levi, 2020). Such income reflects the average income level among residents of the municipality (Palma, 2014; Ruiz-Tagle & López, 2014). Thus, the lower-income municipalities generally have worse HSWM compared with the high-income municipalities (Valenzuela-Levi, 2020, 2021). Hence, poorer municipalities in Santiago avoid or are unable to carry out the recovery and treatment of waste prior to landfill disposal (Valenzuela-Levi, 2019). The inequality in budget for HSWM between municipalities is related to the high-income inequality per capita in Chile, the highest in the region (Palma, 2014). These budgets include waste treatment costs carried out by the municipality, payments to private companies in charge of waste collection, and disposal costs.

In 2007, in Santiago, the output of mass flows of recycled products increased following the order: glass < plastic < metal < paper/cardboard (Fig. 1), where < is being used as greater than. The informal sector mainly recovers paper/cardboard followed by metals, while the formal sector focuses on glass followed by paper/cardboard. This is related to the high current demand for these products by the Chilean industry and due to the higher prices for these types of waste compared to the others (Table 1). This table summarizes the composition of HSW in Santiago and the prices of secondary raw materials, i.e., the raw materials that arise from waste recovery. Although some municipalities such as La Pintana in Santiago produce a high volume of composting (Valenzuela-Levi, 2020), this activity is still scarce in Santiago considering the high proportion of organic food in HSW (Fig. 1 and Table 1).

Recycled ferrous metal in Chile is normally used by local industry but non-ferrous metals are exported due to local technological limitations (Florisbela & Astorga, 2006). On the other hand, it could be argued that the volume recycled also adjusts for the infrastructure available for recycling. However, a 2020 report (InvestChile, 2021) notes that only 50% of installed capacity for plastic recycling is used. Hence, only 8.5% of the 990,000 tons of plastic consumed per year in Chile is recycled (InvestChile, 2021). Low quality of recycled plastic due to poor source separation and low density discourage plastic collection. The main form of plastic collected is PET bottles due to its high commercial value and ease of recycling (Florisbela & Astorga, 2006). As approximately 80% of the plastic recycling infrastructure is located in Santiago (InvestChile, 2021), it is difficult to promote decentralized recovery policies at the national level; Chile is the longest country in the world at more than 4,200 km from north to south. The most widely recovered plastic products are of polyethylene (PE), polypropylene (PP), and polyethylene terephthalate (PET) (InvestChile, 2021). Finally, several types of waste are not treated in Santiago, among which textiles, packaging materials, and various plastics stand out with more than 10% presence in

[3] Separate collection rate corresponds to the weight of waste collected for recovery vs. the weight of waste collected for disposal at a landfill (Valenzuela-Levi, 2019).

Table 1 Characteristics of HSW in Santiago and of secondary raw materials in Chile

Related to HSW[a]		Related to secondary raw material[b]	
Material	Composition (%)	Material	Price (US$/Mg)
Organic food	53.6		
Yard waste	1.8		
Cardboard	3.2	Cardboard	74
Paper	12.2	White paper	160
		Newspaper	66
		Magazine paper	76
Plastics	10.2	PET bottles	76
Tetra pak	0.8		
Hygienic articles	6.5		
Glass	3.2	Glass	50
Metals	1.1	Ferrous metals	100–120
		Aluminum	900
		Copper	4000
Leather	0.2		
Wood	0.5		
Textiles	3.2		
Dust/Ash	1.2		

[a]Oddou (2008), [b]González Martínez (2011)

the city's HSW (Table 1). As this waste has little or no sales market in Chile, it is collected and sent to landfills without pretreatment.

In summary, the freedom of the municipalities of Santiago to choose HSWM actors has favored the privatization of services and better efficiency in the richest municipalities (Orellana, 2009). Campaigns to improve waste management and the inclusion of actors are carried out by sectors at the municipal level (EIU, 2017). Data shows that solely investing more money into privatized, oligopolistic HSWM is insufficient to increase the efficiency of waste recovery in Santiago (Valenzuela-Levi, 2020, 2021). Several authors agree that the inclusion of informal waste-pickers increases the rate of waste removed and recycled in countries of the so-called Global South, which includes Chile according to evaluation criteria (Guibrunet et al., 2017; Medina, 2007; Navarrete & Navarrete, 2018; Vergara et al., 2015).

3 Informal Waste-Picking Activity in Countries of the Global South

Waste-pickers are important social actors in Latin America and Asia, that brings together 2% of the population (Medina, 2000). The term refers to people who live from the sale of recovered waste found in city streets, garbage dumps, and municipal garbage trucks (Scheinberg et al., 2011). These individuals are often at the bottom of the recycling chain, lack the capacity to process and store recovered materials, and are exploited by other actors (Wilson et al., 2006). In 2010, informal waste-picker activity recovered approximately 60% of the waste recycled in Chile. Of the total waste generated in Santiago in 2007 12.7% was recycled by informal waste-pickers vs. the 0.85% recycled by private recycling plants (Barton & Kopfmüller, 2016); see Fig. 1. In 2011 the informal recycling rate was 14% and was still higher than the rate recycled by private plants (CONAMA, 2010). The incorporation of waste-pickers into Chile's HSWM would also reduce national CO_{2eq} emissions by around 3.5%, as reported by a study of Life Cycle Assessment (LCA) (Rojas et al., 2018). Despite these facts, in 2013 around 60,000 people worked as informal waste-pickers. Such informality persists despite the implementation of Law 20,920 (2016) in 2016 (Extended Producer Responsibility, REP) (Giovanini & Huybrechts, 2017), which formally recognizes waste-picking activity in Chile. It should be noted that the service offered by waste-pickers is not paid for by the municipalities (EIU, 2017). These actors live from the sale of what they recover, which fluctuates depending on a number of factors (such as weather, lack of transportation, competition between private waste-pickers collecting in the same area) (EIU, 2017).

HSWM actors in the countries of the Global South are different from those of industrialized nations or countries of the Global North (Medina, 2007). As seen in the Chilean context, recycling in the Global South is closely tied to informal activities (Medina, 2007). Hence, direct transfer of HSWM practices from the Global North to the Global South in order to improve the HSWM efficiency of the latter may not be appropriate. The inclusion of waste-pickers in the countries of the Global South is a win–win solution, as was reported (Wilson et al., 2012) after an analysis of benchmark indicators from twenty cities on six continents. However, as long as the work of waste-pickers is kept informal and these individuals have low power in comparison to other actors, this activity and its contribution to HSWM will continue to be ignored and undervalued by city planners and by society.

In addition to the problems and shortcomings inherent to informal waste-picking activity, these individuals suffer stigmatization and marginalization from society, as summarized by Medina in his book on the world's scavengers (Medina, 2007). Negative perceptions of waste-picker endure and are sometimes aggravated due to the lack of knowledge about the problems waste-pickers face and the benefits they generate (EIU, 2017). Table 2 summarizes perceptions and facts around the role of waste-picking activity in HSWM. It has been argued that informal waste-pickers are inefficient in separating and recovering waste and hence have a deleterious effect on HSWM efficiency. However, the data collected from Global South countries (Medina,

Table 2 Perceptions and facts about the informal activity of waste-pickers (Medina, 2007)

Perceptions	Facts
Recent phenomenon	Create jobs that help to reduce local poverty
Marginal activity	Municipalities can save money and minimize expenditures on facilities
Disorganized activity	Supply inexpensive materials to industry, thus improving competitiveness
Static activity	Conserve natural resources and help clean up the urban environment
Has minimal economic impact	By recovering materials from the waste stream, dumps and landfills can extend their lifespan
A nuisance that must be eliminated	
The poorest of the poor	Increasing volume of secondary raw materials to decrease pollution and save water and energy
No place in modern waste management systems	

2007; Rojas et al., 2018) and Fig. 1 show that the participation of waste-pickers increases HSWM efficiency and decreases the global environmental impact.

3.1 Formalization of Informal Waste-Pickers in Chile

The inclusion of informal waste-pickers by a formalization plan can help to reduce social stigma and empower this actor in their role in HSWM (Medina, 2007). In Chile, Law Nº 20,267 (2008) regulates the procedure to formalize recyclers active in HSWM. However, of the total number of waste-pickers in Chile, only 982—less than less than 2% of the total—are certified according to the (Chile Valora, 2022) database (see profile: P-3830-9611-011-V01 Reciclador(a) de base). On the other hand, it has been shown (Giovanini & Huybrechts, 2017) that the procedure to organize individual waste-pickers into cooperatives is still unknown and bureaucratic. A study from the municipality of Santiago (EIU, 2017) reports that 82.1% of the waste-pickers there are disorganized and that 85.7% estimate they receive little help from institutions and people. The process of organization begins with registration with the Chilean Ministry of the Environment, where the cooperative must belong to the Registry of Emissions and Transfers of Pollutants (RETC). Later, the cooperative must make agreements with NGOs or private corporations before seeking support from municipalities (Giovanini & Huybrechts, 2017). These authors report that certain NGOs supported by private companies have run into conflicts in their work with groups of waste-pickers in Santiago. These NGOs impose their environmental policies (top-down approach) instead of maintaining a collaborative approach and listening to the demands of the waste-pickers.

In Santiago, the freedom to define the HSWM process within a context of inequity has promoted different policies of integration or exclusion of waste-pickers (Navarrete & Navarrete, 2018). According to Navarrete and Navarrete (2018), policies can be classified into five schools of thought (approaches), as shown in Table 3. They conclude that the "co-production" approach is appropriate to promote policies that include waste-pickers in the HSWM of Santiago. The same conclusion has been reported in other countries of the Global South (Dias, 2016; Fergutz et al., 2011; Fidelis & Colmenero, 2018; Hartmann, 2018; Wilson et al., 2006).

The formalization and organization of waste-pickers' work does not ensure inclusion in HSWM when they are in competition with large, private actors, nor does it guarantee their job security. In Santiago, formal waste-picker organizations must compete with the large companies that dominate the HSWM market (Giovanini & Huybrechts, 2017), hence their vulnerability and the imbalance of power is maintained. For some authors, this has suggested that the current inclusion of waste-pickers in Santiago is more a situation of "organized informality" (Giovanini, 2014). Furthermore, waste-picking activity generally involves poor working conditions, a lack of social and health insurance, and job instability (Giovanini & Huybrechts, 2017; Carenzo, 2020). Security measures are scarce in these informal jobs. The work of waste-picking reduces life expectancy, increases the mortality rate, and

Table 3 Approaches (school of thought) on the incorporation of informal waste-picking activity in municipal waste management

Approach	Situation of waste-picking activity	Policy implication
Dualist	Worker lacks economic growth and poverty is maintained. Marginal survival activity in the absence of formal work. Counter-cyclical to economic growth	No support. Repression and creation of formal jobs to reduce its number
Voluntarist	Worker as entrepreneur, but poverty is perpetuated (low productivity and efficiency due to low skills and undercapitalization). Counter-cyclical to economic growth	No support. Approach of carrot (pro-formal) and stick (repression of informal)
Eructuralist	Worker as source of cheap recyclable materials. Enterprises consume those materials in order to reduce input costs and thus increase profit. Capitalist exploitation. Pro-cyclical to economic growth	Soft support. Promote associations to reinforce power (negotiate better prices and working conditions)
Legalist	Worker connected with formal industry. Positive role by improving the competitiveness of local industries. Counter-cyclical to economic growth (the prices of imported raw material increase during economic crisis, so increase demand for cheaper substitutes)	Neglect support, as government intervention drawn back. Legalize activity
Co-production	Worker as micro-entrepreneurs in need of governmental support. Counter-cyclical (survival activity) and pro-cyclical (microentrepreneurs)	Strong support

Adapted from Navarrete and Navarrete (2018)

increases the incidence of infectious diseases (Medina, 2000; Wilson et al., 2006). In the municipality of Santiago, 70.3% of waste-pickers use human-powered vehicles to transport waste and 51.8% must collect waste at home (EIU, 2017). These situations and working conditions generate multiple diseases that (Meneses, 2012) summarizes in his work. In addition, although the waste collected by waste-pickers comes from non-hazardous HSW, certain waste can emit substances that are harmful to health during its degradation (Meneses, 2012) and is a relevant source of vector-borne diseases.

3.2 *Experiences with Inclusion of Waste-Pickers*

A review of the academic literature suggests that countries in the Global South have varying degrees of interest in including the informal sector and the waste-picker actor in HSWM. Figure 2 summarizes the number of papers from six Latin American countries that include keywords related to informal waste-pickers in the waste management system. Three scientific databases were used (Google Scholar, Scopus and Science Direct) for the literature review. According to the total amount of works from each country analyzed, interest in the subject increases in the following order: Paraguay < Ecuador < Chile < Bolivia = Peru < Colombia < Argentina < Mexico < Brazil. Brazil is the most studied territory in the world on the topic of waste-pickers, according to Scopus, contributing 69% of the works in Fig. 2. Scopus also shows that interest in the subject has increased sharply since 2003 and that the oldest work on waste-pickers and informality focused on Calí, Colombia (Birkbeck, 1978). Due to this interest and the socio-political similarity of those countries with Chile, results that encourage the inclusion of waste-pickers from Brazil, Colombia and Argentina are summarized below. According to an inclusive recycling policy report (EIU, 2017), Buenos Aires (Argentina), Sao Paulo (Brazil) and Bogota (Colombia) lead Latin America and the Caribbean. This ranking arises from indicators that group three dimensions (regulatory, organizational, and market) around the inclusion of the waste-picker.

3.2.1 Brazil

Indicators are developed that evaluate the environmental impact of different waste managements in order to classify their performance (Deus et al., 2020). Such performance increases when the waste-pickers are included (Deus et al., 2020). A method to evaluate the performance of waste-picker cooperatives in their operational activities related to the recycling chain was proposed (Fidelis & Colmenero, 2018). This method improves the efficiency of cooperatives in waste recovery, favors their socio-productive inclusion in the waste management process, and reduces their marginalization in society (Fidelis & Colmenero, 2018). These indicators (Deus et al., 2020) and method (Fidelis & Colmenero, 2018) require reliable, transparent, and easily

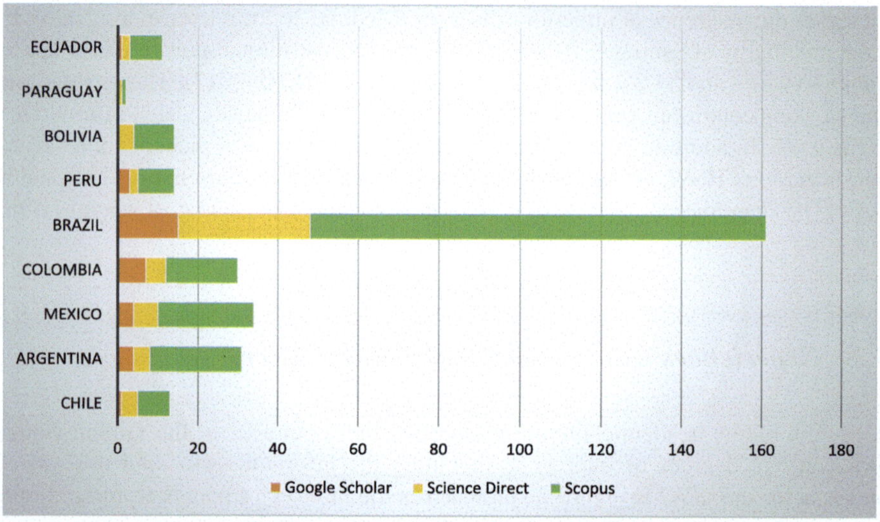

Fig. 2 Number of works issued until 2021 reported in three scientific databases: Google Scholar, Scopus and Science Direct. Works that feature the following keywords in their abstracts: "waste," "management," "picker or informal" and "Chile, Peru, Mexico, Brazil, Colombia or Argentina" are quantified

accessible information on the inputs-outputs of the activities carried out by the HSWM actors. In general, such information is of both low quality and limited availability (Nogueira et al., 2020).

3.2.2 Colombia

In order to register and manage the amount of material recovered by waste-pickers, the Bogota Recyclers Association (ARB) in Colombia has implemented the Recyclers and Recycling Registration Platform. In 2017, this platform reported 14,000 tons of material recovered for recycling that arose from the link between 9000 waste-pickers, 2000 warehouses, and 350 vehicles for the transport of waste (CEPAL & CEMPRE, 2020). The recovered material comes from homes and large generators. Such information allows for an analysis of possibilities and strategies to integrate formal and informal actors into HSWM using a strength, weakness, opportunity, and threat (SWOT) analysis, as studied by Pardo and Piña (2016) for Bogota.

3.2.3 Argentina

The HSWM actors in Buenos Aires, Argentina, are administered by the Provincial Organization for Sustainable Development (OPDS). OPDS classifies waste generators into small or large based on whether they produce a waste volume lower or higher than 1 Mg/day, respectively. Large generators have to: (1) report the quantity of urban solid waste produced per month and how much is potentially recoverable by OPDS resolutions No. 317/20, 85/14, 139/13, 137/13, and 317/20; and (2) pay for the transportation and final disposal of waste in a public landfill called Coordinación Ecológica Área Metropolitana Sociedad del Estado (CEAMSE), in accordance with National Law No. 14273. The fraction of recyclable waste is treated at a "sustainable destination," which are plants for the separation, conditioning, and/or recovery of recyclable waste (dry materials and organic fractions of solid urban waste). OPDS encourages waste-pickers organized in cooperatives to become sustainable destinations, which resulted in more than 70 such destinations being created in Buenos Aires (Giorgio & Girotto, 2012). Those waste-pickers receive a salary for their work and become an actor between the waste-producing industry and the recycling industry. However, a lack of empowerment of cooperatives vis-à-vis private actors still remains in Argentina's HSWM (Carenzo, 2020).

In Buenos Aires, a project that brought together various actors with the El Amanecer de los Cartoneros cooperative[4] developed software that quantifies and records the volume of recovered materials (Carenzo, 2020). This software classifies the type of waste, the collection area, the types of materials arising after separation and treatment, and the sales of the recovered materials. This digital tool has improved the collection efficiency of the waste-pickers by offering communication channels between them. In addition, the cooperative can now define collection targets in order to improve income. It should be noted that this open-source software is being distributed among other cooperatives in Argentina.

The project mentioned above is promoted by the Network of Technical Assistance to the Popular Economy (RITEP) (RITEP, 2021). RITEP is a platform for projects and proposals that seeks to include waste-pickers in the Argentine production chain and to value their knowledge and innovations and their role in HSWM. As suggested by some authors (Carenzo, 2020; Cozzens & Sutz, 2014; Fressoli et al., 2014; Heeks et al., 2014), the political, citizen, business, and scientific-technical communities must recognize and value the "grassroots innovations" of waste-pickers. The latter must also accept that the scientific sector can favor and improve the management of their productive units.

[4] The project brought together this cooperative of 3600 waste-pickers, the Technical Assistance Network for the Popular Economy (RATEP), the Argentine Federation of Cartoneros, Carreros and Recicladores (FACYR), the Social Actors Demand Council Program (PROCODAS), the National University of Tres de Febrero (UNTREF), and the Clementina cooperative with its WOW software project.

3.3 *Empowering Waste-Picking Activity*

3.3.1 Cooperation with the Academy

The Argentinean Institute of Architecture and Urbanism (IA) of the National University of San Martín (UNSAM) studies how to increase the social value and empowerment of waste-pickers in HSWM and in the production chain. The academy is a relevant actor to empower and include waste-pickers in HSWM (Medina, 2007). IA-UNSAM works together with private companies to develop waste-based products for buildings that require simple equipment for their manufacture (Becerra & Busnelli, 2021). The development of waste-based products also plays a fundamental role in the fulfillment of at least nine of the seventeen Sustainable Development Goals of the Agenda 2030 (Opoku, 2019). IA-UNSAM follows the principles of industrial ecology as a development strategy and not only purely technical- or market-oriented approaches. For instance, an Atlas of Solid Wastes is also in development to convert wastes into new inputs for construction and thus help to create a market of reuse- or recycling-based products (Busnelli, 2018). The new products are expected to be manufactured by cooperatives from recovered waste, which will then be distributed in the local market to build social housing. Hence, the IA-UNSAM research seeks to reduce the amount of waste, create jobs, and build affordable housing to mitigate the current lack.

The lack of affordable housing is an issue in both Argentina and Chile (Borsdorf et al., 2016). Furthermore, there is a connection between the development of settlements in Santiago and informal economic activity (Hillier et al., 2000). These settlements have been located on the outskirts of the city since the mid-1980s (Borsdorf et al., 2016). In general, social housing in Chile is lacking in material quality (Mora et al., 2014; Vergara et al., 2019), especially in insulation. The primary source of heating is firewood and kerosene with 26.2% of total energy consumption going towards heating (Becerra et al., 2018). Such options are used by households of low socio-economic status as they are the cheapest, but also release the highest level of pollutants in terms of emissions (Jones, 1999). In this context, the abovementioned IA-UNSAM research would also contribute to increasing the efficiency of HSWM, thus reducing the environmental impact by circularity as waste-pickers would introduce recovered materials into the value chain; additionally, the material quality of social housing would be improved, and indoor comfort preserved.

The IA-UNSAM research also encompasses the training of waste-pickers' organizations on technical concepts of design and the processing of products using recovered waste. The aim is to produce secondary raw materials by increasing the knowledge and technology of waste-pickers on waste treatment. The limited technology of various waste-pickers' organizations makes it difficult to jump from mere collection and primary separation to deep separation and waste processing. In this context, the Argentinean Reciclando Sueños cooperative together with the Universidad Nacional de Quilmes, Argentina, and Red Latinoamericana de Catadores y Recicladores (Red LACRE) have developed machinery to classify, and process recovered materials

(Carenzo, 2014). The machinery is manufactured by reusing recovered objects or is adapted from other machinery to crush, dry, wash, and press recovered material (Carenzo, 2020). Hence, the recovered material is both raw material for manufacturing machinery and raw material that "feeds" this machinery, i.e., an example of circularity in the waste treatment itself. The selling price of recovered materials increased by up to 200% with the use of this machinery developed by the cooperative (Carenzo, 2014).

3.3.2 Development of Waste-Based Materials and Products

After improving waste processing, the next technological step in the empowerment of cooperatives may be education on waste-based product design following reuse and recycling routes. Reuse is preferred over recycling. Recycling requires more technology than reuse because it involves transforming a product (of a specific shape and organoleptic properties) into a secondary raw material with a suitable form (granulate, scale) for further processing. The waste transformation involved in a recycling route requires additional equipment/facilities and inputs (energy, matter) compared to the reuse route. As a result, the environmental benefits of recycling are less than for reuse for different types of products (textile (Sandin & Peters, 2018), metal (Ng & Chau, 2015), and plastic (Ross & Evans, 2003)).

Figure 3 shows the manufacturing route of a panel for home insulation made from reused PET bottle waste. This panel is being studied by IA-UNSAM and is based on the insulation system developed by the 3C Construcciones company (3C, 2021). The PET bottle-based insulation panel can be adapted to different climatic regions and was developed for the construction of social housing (Becerra & Busnelli, 2021). The panel is comprised of a network of stacks of PET bottles within a wooden framework. To build a panel with the dimensions 3 m × 3 m, 2880 recycled bottles with a total mass of approximately 96 kg are used. This means that more than 10,000 discarded PET bottles are reused in the construction of a room of 9 m square. The design of the insulation panel follows the principles of two guidelines for the development of sustainable products: the principles of green engineering (Anastas & Zimmerman, 2013) and the material passport (BAMB, 2019), as described in Table 4. Both guidelines are useful tools to include in further strategies for urban and industrial symbiosis for Santiago. Both guidelines correspond to applications of products and processes of industrial ecology because they increase the efficiency of products and processes and make them compatible with nature. On the one hand, the principles of green engineering couple conventional engineering processes with natural processes. On the other hand, the material passport centralizes information on materials/products useful for the application of green engineering in the design of a material/product. Therefore, these guidelines are connected and have symbiosis.

The materials passport and the principles of green engineering are relevant guidelines across the world, as shown by a study on global business models for circular economy issued for the Chilean state program Construye2025 (Sala & Zegers, 2021). That study identified 119 cases of a circular economy model in construction, which

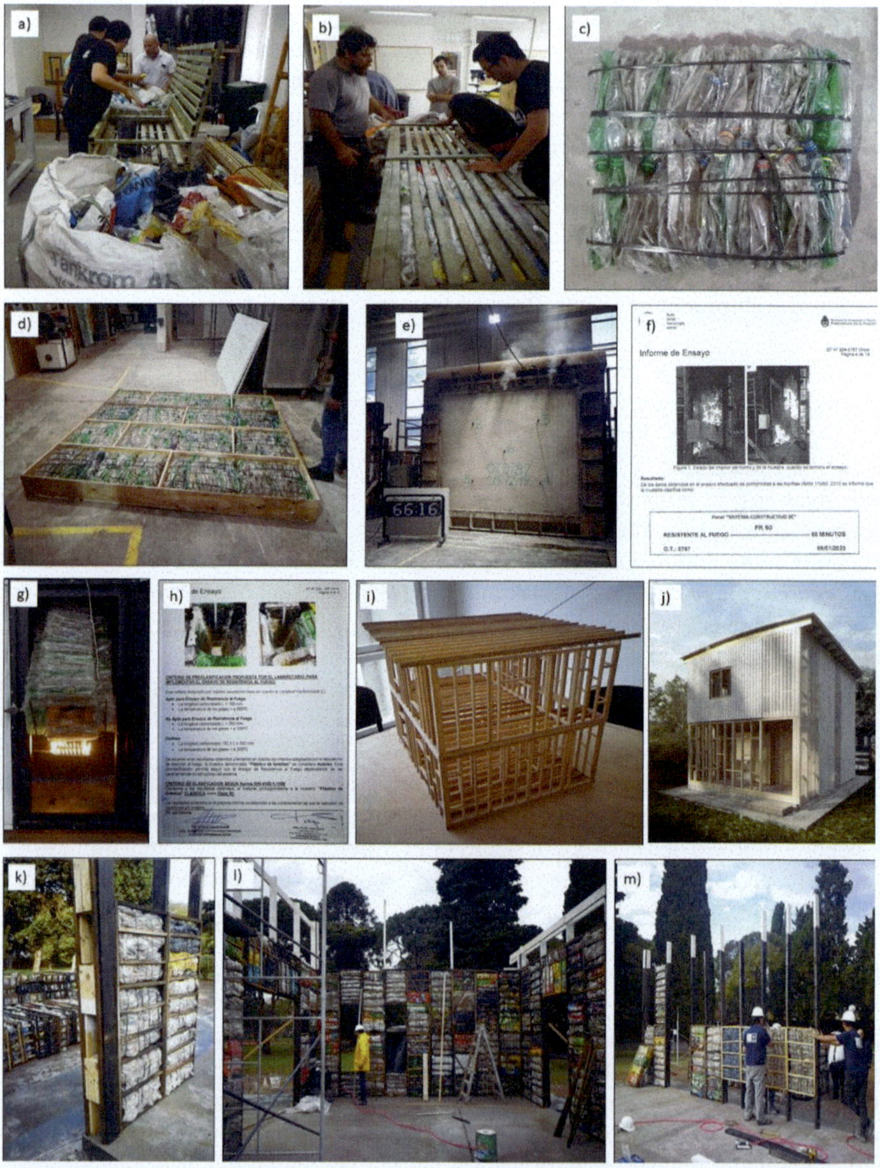

Fig. 3 Fabrication (**a–b**) and characterization (**e–h**) processes of the insulation system based on waste PET bottles. A house prototype using the insulation system (**i**, **j**) and other waste-based products. Construction of the house (**k–m**)

Table 4 Guidelines of green engineering principles and material passports fulfilled during the design of the insulation panel based on bottle waste. Letters in parentheses refer to the images in Fig. 3

Green engineering	**Minimize complexity and diversity**. The insulation panel consists of compacted bottles inside a wooden frame without adhesives (d). This eases disassembly and helps with separation after the product lifespan
	Non-hazardous inputs and outputs. PET and wood in this insulation panel would have no harmful effects on human health during the system lifespan
	Design for a future life. PET bottles could become new products after ending their lifespan in the insulation panel. The degradation level is lowered as PET bottles are not exposed to direct UV radiation from the sun
	Minimize resource use and promote renewability instead of depletion, as the insulation panels use raw materials from reused PET bottle waste
	Maximize efficiency by increasing the quantity of reused PET bottles without loss of technical performance (fire resistance, thermal insulation, etc.). The characterization process is relevant here for fulfilling Argentinean standards (e–i)
	Integrate local materials and energy as the product uses local waste and energy during the fabrication which comes from local labor and local PET waste
	Minimize both energy and material use during separation of materials and components. The PET bottle stacks can easily be disassembled due to the absence of adhesives and thus recovered for future applications
Material passport	Information on the material origin: PET from urban wastes from the San Martín district and locally sourced pine wood
	Technical data about performance (e–i), which is relevant for the health of the environment and home occupants
	Knowledge about the chemical composition of bottle waste and the wood frame
	The lifespan of the insulation panel is defined
	Clear job opportunities due to self-construction of the panel (a, b)
	Standardization in the fabrication as each step is well defined
	Technical data to probe this insulation panel is competitive compared with traditional options

were grouped into ten approaches. The two main approaches are associated with treating waste as a resource (49.6% of the total) and circular design (35.3% of the total). The first approach is related to the materials passport and the second with the principles of green engineering. Construye2025 is using these results to define its Roadmap Circular Economy (RCE) to promote inclusive policies with various actors around the circular economy and thus around HSWM processes. In sum, the materials passport and the green engineering principles should be implemented as soon as possible in product development practices, especially those developed by recyclers.

In the Chilean roadmap for a circular economy in construction (Construye, 2025, 2020), the role of the waste-picker as a relevant actor is poorly discussed. The RCE slogan of Construye2025 is for "…a country that manages its resources efficiently, positively impacting the social, environmental and economic spheres, and with the participation of all stakeholders" (Construye, 2025, 2020). However, the main sector included in this roadmap is the private sector with 53% of total participation, while civil society has a participation of only 2%. According to the official RCE document, this poor rate of participation occurred "…despite having made a broad call for participation through social networks and other media, which may be explained due to the technical and specific nature of the subject." According to a report on recycling in Latin America and the Caribbean (EIU, 2017), the waste-pickers of Santiago were summoned to elaborate upon or update legal instruments. However, their proposals and demands were not always incorporated. It seems that in addition to the focus on the privatization of HSWM in Santiago, neither waste-pickers nor citizens are considered relevant actors in the elaboration of policies on HSWM or circular economy.

Principles of green engineering

Anastas and Zimmerman (2013) proposed twelve principles of green engineering focused on achieving sustainability in the design of new products. Some principles demonstrate synergy with others, meaning that a balance between them allows for the creation of better designs. The two fundamental concepts that material designers must connect are life cycle and inherence, i.e., that a design stems from the nature of the material and not dependent on something external. Table 3 shows that more than 50% of these principles are fulfilled with the development of the PET bottle-based insulation system.

The principles of green engineering promote reuse and recycling for the development of materials, products, and systems. Current challenges stem from difficulties in the reuse and recycling of materials/products that are increasingly complex and mixed, which translates into a more complex waste stream to separate and treat. For example, many plastic products contain several chemical additives, some of them harmful, which present a problem when it comes to deciding the end of the product's life. Functional and aesthetic demands can increase the difficulty of designing with less complexity and diversity of small materials. In today's homes it is common to find products (doors, windows, furniture) composed of different materials with different natures (metal, ceramic, wood, concrete, etc.) and connected by different types of glue (usually non-recyclable and complex to remove). This leitmotif of "the different" brings hidden costs for our environment: problematic disassembly and complicated identification and separation of individual materials for recycling or reuse. The latter becomes even more complex when the characterization of individual materials is weak or unknown.

Characterization is the process by which the characteristics and properties of materials/products are investigated, and constitutes a key step in their certification. This process is based on specific standards established for a market. In Chile, certification in the construction sector is based on the Chilean standard (NCh, 2022). Asdrubali et al. (2015) summarized the characterization of more than twenty insulation products for buildings made from waste materials or natural sources (such as plants and fruits). These authors observed that several products are still in a prototype stage and lack information about their technical and environmental performance, meaning that characterization is still incomplete. They conclude that this poor characterization becomes a barrier to achieving certification and acceptance in the market. Therefore, the guidelines for collecting data on the characteristics of a material/product throughout its life cycle are relevant. At this point, another guideline on the design of reuse- and recycle-based products could facilitate the process, namely that of the material passport.

Material Passports

Within the framework of the EU's Horizon 2020 program, the BAMB (Buildings As Material Banks) project (BAMB, 2019) promotes a circular economy in construction by developing a document known as the material passport. This passport summarizes sets of material characteristics in order to value their current use and seek recycling, recovery, or reusing opportunities. BAMB designed the material passport as a repository that stores, links, and provides information on building materials throughout their life cycle. These passports detail information on chemical, physical, biological, and process aspects for different hierarchical levels: materials, components, products, and systems. For instance, the passport connects details about material toxicity, resistance of the product in various environments, inputs/outputs during fabrication, and reuse or recycling opportunities. This passport is also useful for weighing the reuse and recycling value of products depending on how they are linked to a building. In sum, the material passport is a centralized and detailed source about the techno-environmental and socio-economic aspects of building products or systems. This centralization is a major current challenge because it requires multidisciplinary cooperation, as data sets connect aspects of knowledge with beneficial or harmful synergies.

The material passport should be promoted from the birth of new products based on reused or recycled sources. This document could increase the trust and eventual acceptance of these waste-based products among the construction sector and society, especially as they are non-traditional options. The certification of non-traditional building products requires greater efforts compared to traditional options because the former sometimes lack applicable standards. For example, the fire resistance of the PET bottle-based insulation panel (Fig. 3) was characterized by an Argentinean standard (IRAM, 11950, 2010) that does not include bottle-shaped geometries for sampling. As another standard on fire resistance was lacking, the PET bottles were cut and thus adapted to the shape requirements of the standard (Fig. 3g). Non-traditional products may also encounter skepticism from stakeholders in certification because these options lack accumulated construction experience compared to traditional options. These drawbacks show that certification based solely on the technical aspect and on experience is insufficient to weigh the environmental and social benefits promoted by these non-traditional options. The PET bottle-based panel complies with technical standards and also demonstrates advantages in other aspects: environmental benefits in terms of soil recovery, emissions reduction, and the reuse of waste bottles; social benefits in creating jobs in regions with low technological development and improving quality of life; and finally economic benefits as well.

The principles of green engineering and the material passport become very useful tools for creating guidelines for product design and weighing environment/social/technical benefits. However, their application also requires advancements in other tools to assess environmental and social issues, including the creation of databases for Life Cycle Analysis (LCA), which are still under development in the countries of the Global South, and the development of Environmental Product Declarations for construction (ISO, 14025, 2006, EN 15804, 2012), which are rare in Latin American countries.

In sum, teaching the principles of green engineering and material passports to stakeholders involved in HSWM would encourage the development of waste-based products with a techno-socio-environmental-economic approach. This would in turn promote the incorporation of recovered materials in the production chain, which increases the HSWM efficiency in Santiago by promoting recycling but not disposal. New products based on such guidelines are expected to be competitive due to global interest in environmental and social issues. If cooperatives develop these products, they could increase their social value (acceptance by society), revenues, and participation in the waste management and production chains. Also, this empowerment would be greater if it is supported by a Chilean product certification that weighs the social/environmental benefits of the waste-based products.

4 Inclusion of Waste-Pickers for Enhancing Urban-Industrial Symbiosis in Santiago: The Action Map

With our proposals, we address different groups and stakeholders. For better clarity, we have arranged our proposals regarding the different stakeholders in Fig. 4 with an illustration around the waste-picker as the central actor.

Table 5 summarizes the barriers and challenges that hinder the inclusion of waste-pickers in the current HSWM in Santiago. This table also summarizes proposals or measures to include waste-pickers, which are in concordance with the literature discussed in this chapter. The measures are also expected to encourage participation and synergies among other HSWM actors.

Figure 4 summarizes the map of actions that recyclers can take to encourage the inclusion of waste-pickers in the HSWM process. This map of actions should be supported with the governmental measures summarized in Table 5. Figure 4 also includes measures from other actors to support the empowerment of waste-pickers. These follow experiences from other Latin American countries (see Sect. 3.2) and are summarized in Table 6.

Figure 4 identifies the interaction between waste-pickers with each actor in the HSWM context. In order to maximize improvements to HSWM efficiency, some waste-picker actions could follow a hierarchical order in time. For example, the interaction with academia and other institutions could take place prior to interaction with the government in order to increase the chances of acquiring funds and improving the final benefits. Therefore, waste-pickers organized in cooperatives should define

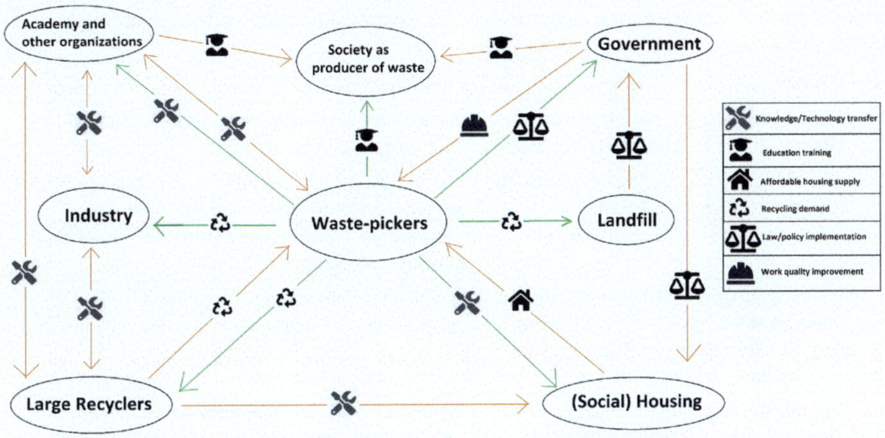

Fig. 4 Scheme of the activity between main actors in the management of household solid waste in Santiago. The arrows indicate the interactions between actors and the direction of the action. The roles of waste-pickers (green arrows) and other actors (orange arrows) are detailed in Table 5. The symbols and their meanings are listed in the upper-right box

a work plan to prioritize the timeframe of their action plan. The context of each cooperative will depend on the integration policies promoted by each municipality, as shown in Table 3. The action plan of Fig. 4 is expected to become a useful tool for formulating policies for the inclusion of waste-pickers in Santiago.

5 Conclusions and Outlook

5.1 A Change in Perspective Creates Potential for Urban-Industrial Symbiosis

To summarize the last sections, we will now briefly explain how our proposals address the challenges for EID strategies listed in the introduction.

In general, it is difficult to estimate the share of recycled material from household waste in a given area, which is made even more difficult when a significant proportion of recycling takes place informally. This makes a realistic material flow analysis difficult, which would provide a useful basis for calculation from an economic and business management point of view. If the informal waste-picking were formalized and the production of building materials from these secondary raw materials promoted and advertised, it would be possible to estimate or calculate the amount of material saved from landfilling based on the quantity of the manufactured construction products.

Regarding the challenges of the interface between the local community, public administration, and further stakeholders, which requires trust building as it is often

Table 5 Barriers and corrective measures e.g., by public administration to include recyclers in HSWM in Santiago

N°	Barrier	Challenge
1	HSWM has a high level of privatization (Reyes, 2004; Valenzuela-Levi, 2020)	Reduce the level of privatization in HSWM by including social actors
2	The recycling service performed by the recycler is not necessarily paid (Giovanini & Huybrechts, 2017)	Include payment methods for the waste-picker' services and supervise such payment. See example of "sustainable destination" actor from (Giorgio & Girotto, 2012)
3	Bureaucratic process to set up contracts between waste-pickers' organizations and municipalities (Giovanini & Huybrechts, 2017)	Reduce the bureaucracy to contract waste-pickers' organizations as much as possible
4	Stigmatization of waste-pickers amongst citizens (Medina, 2007)	Make the work of waste-pickers visible among citizens through campaigns to publicize their achievements
5	Municipal funds for HSWM are unequal as they depend on local revenues and taxes (Valenzuela-Levi, 2019, 2020)	Encourage policies that increase the equity of funds for HSWM among municipalities
6	Waste-picker organizations have less power than private companies (Navarrete & Navarrete, 2018; Giovanini & Huybrechts, 2017)	Promote interactions between HSWM actors and improve waste-pickers' training on waste separation and treatment
7	Lack of infrastructure for waste-pickers' work (Navarrete & Navarrete, 2018; Giovanini & Huybrechts, 2017)	More competitive funds to improve the quality of the work of waste-pickers and to promote cooperation with other HSWM actors
8	Lack of trust between HSWM actors (Giovanini & Huybrechts, 2017)	Increase transparency of policies and promotion of platforms that share data on recovered waste streams. See examples from Colombia, Argentina, and Brazil
9	Exclusive and long-term contracts with private enterprises/companies make it difficult to include other actors (Valenzuela-Levi, 2020)	Reduce level of exclusivity and duration of contracts with private companies
10	Poor participation of social actors in the development of HSWM and circular economy policies (Sala & Zegers, 2021; EIU, 2017)	Include the opinions of citizens and value the knowledge developed by recyclers. See examples on "gross innovations" (Cozzens & Sutz, 2014; Fressoli et al., 2014; Heeks et al., 2014)
11	Little disclosure of the real contribution of the informal sector to HSWM. Few studies show comparative data between informal and formal private/economic sector (Barton & Kopfmüller, 2016)	Report more frequently on the volume recovered by waste-pickers compared to other HSWM actors

(continued)

Table 5 (continued)

N°	Barrier	Challenge
12	Lack of cooperation between the "bottom" (the waste-pickers) and the top (from technology experts, etc.) (Carenzo, 2020)	Promote policies that encourage a collaborative approach between waste-pickers and experts (Carenzo, 2020; Cozzens & Sutz, 2014; Fressoli et al., 2014; Heeks et al., 2014)
13	The recycler as an actor at the beginning of the recycling chain (mere collection and basic sorting), unable to generate knowledge or transform waste into products	Value the knowledge developed by waste-pickers to transform residues through reuse and recycling using low-tech methods. See examples (Carenzo, 2020; Cozzens & Sutz, 2014; Fressoli et al., 2014; Heeks et al., 2014; Smith et al., 2017)

characterized by conflicting worldviews, we suggest a shift in the perspective on the marginalized praxis of waste-picking, as suggested by Medina (2007), Carenzo (2020) and Becerra et al. (2018). Waste-picking activity represents an adaptive response to scarcity (Medina, 2007). The perception and acknowledgement of this activity as a contribution to a cleaner urban environment is a necessary condition for exploiting the potential of our approach to improving the linkage between urban and industrial metabolisms (Tröger et al., 2023). For just transitions to circular economy, the recognition and inclusion of informal waste-picking is mandatory (Purvis et al., 2023; Valencia et al., 2023). The formalization of this activity makes it possible to use the existing potential for enhancing household waste management (urban metabolism). It also allows the involved people to develop new products, where we see great potential for construction products, as these require large quantities of materials. The application of the principles of green engineering as well as material passporting for developing waste-based construction products is technically low-threshold, helpful, and conducive for cooperatives. At the same time this might address the value-based purchasing of potential buyers. We see a further success factor in the support from public academic institutions like universities, such as through providing consultation, pre-tests for certification, and accompaniment of the certification process. With this strategy, a considerable amount of material could find its way back from the urban to the industrial metabolism instead of ending up in the landfill.

5.2 *Impetus for Future Directions in Eco-Industrial Development*

In this article we illuminated a widespread blind spot in conventional municipal waste management and in the theory of eco-industrial development. Models of waste management are often adopted from industrialized countries. At the same time, the actual social practice of informal waste collection in countries of the Global South is hardly noticed by the institutions. Our proposal, based on an urban-industrial ecology

Table 6 Measures to include waste-pickers and to improve the efficiency of municipal solid waste management in Santiago by a framework of eco-industrial development

From…	To…	Measure
Government	Society	Reward the separation at origin and promote education campaigns
Recyclers	Waste-picker	Promote recovery by Chilean law and use 50% of the available infrastructure for plastic recycling (InvestChile, 2021)
Government	Waste-picker	Improve the safety equipment of waste-pickers to improve the quality of their work
Academy	Industry, recycler	Advise on innovation to develop new waste-based products and improve the efficiency of the recycling process
Academy	Society	Promote education campaigns on waste separation at its source
Landfill	Government	Discourage landfill disposal by increasing the tax per ton on landfilled waste, and promote laws and policies to pretreat waste
Government	Housing	Promote projects that involve the construction of affordable housing using reused/recycled materials
Housing	Waste-picker	Special housing prices for the informal sector in order to promote the improvement of quality of life and work environment
Academy	Waste-picker	Develop low-cost technology that is easy to implement and operate to collect, separate, and treat waste. Bottom-up and top-down approaches must work together to increase synergies between waste-pickers and experts
Recyclers	Housing	Fabrication of waste-based products with affordable prices for building housing
Industry Waste-picker	Recycler	Sharing knowledge of waste streams to promote an industrial ecology framework
	Industry, recycler	Improve waste treatment efficiency to greater acceptance and sale prices
	Academy	Encourage mutual learning relationships to improve waste recovery processes
	Citizen	Education campaigns to improve separation at origin (less final disposal and improvement of recycling efficiency)
	Government	Apply to funds with other institutions/the academy to improve facilities and technology
	Landfill	Improve the efficiency of waste treatment to reduce final disposal
	Housing	Development of certified waste-based products for housing construction

perspective, forms a sharp contrast and is contesting common perspectives on household waste management, sophisticated circular economy solutions, and construction product innovation. The framework of urban-industrial ecology allows us to see the social praxis independent from conventional perspectives on urban development and therefore realize the existing potential (Tröger et al., 2023). Instead of transferring the business models of circular economy and the corresponding technical solutions from Europe to Latin America, we propose the creation of novel construction products made of solid industrial residues locally generated by formalized waste-pickers and cooperatives and the creation of business models that correspond to the social praxis. To establish an urban-industrial symbiosis in Latin American cities, this enhanced understanding of industrial ecology might be useful to illuminate the full existing potential for sustainable development.

References

3C. (2021). 3C: construir relaciones de valor, no sólo casas con ecoladrillos. Retrieved December 19, 2022, from https://www.lavoz.com.ar/ciudadanos/3c-construir-relaciones-de-valor-no-solo-casas-con-ecoladrillos

Aguilera, J. P., Salazar, R., Sepúlveda, M. A., Conejeros, M. S., & Malfanti, I. S. (2003). Residuos sólidos domiciliarios: Logística, una herramienta moderna para enfrentar este antiguo problema. *Revista Ingeniería Industrial, 5*(1), 77–87.

Anastas, P. T., & Zimmerman, J. B. (2013). Design through the 12 principles of green engineering. *Environmental Science and Technology, 37*(5), 94A-101A.

Asdrubali, F., D' Alessandro, F., & Schiavoni, S. (2015). A review of unconventional sustainable building insulation materials. *Sustainable Materials and Technologies, 4*, 1–17.

Ayres, R. (1994). Industrial metabolism: Theory and practice. In B. Allenby & D. Richards (Eds.), *The greening of industrial ecosystems* (1st ed., pp. 23–37). National Academy of Engineering.

BAMB. (2019). Materials passports – Best practice innovative solutions for a transition to a circular economy in the built environment. Retrieved December 19, 2022, from https://www.bamb2020.eu/wp-content/uploads/2019/02/BAMB_MaterialsPassports_BestPractice.pdf

Barton, J. R., & Kopfmüller, J. (Eds.). (2016). *Santiago 2030. Escenarios para la planificación estratégica.* RIL editores - Instituto de Estudios Urbanos y Territoriales UC, Santiago de Chile

Becerra, A., & Busnelli, R. (2021). Caracterización de nuevos aislantes para envolventes arquitectónicas con materiales no tradicionales. *Revista Ingeniería De Obras Civiles, 11*(1), 14–24.

Becerra, M., Jerez, A., & Valenzuela, M. (2018). Life quality disparity: Analysis of indoor comfort gaps for Chilean households. *Energy Policy, 121*, 190–201.

Birkbeck, C. (1978). Self-employed proletarians in an informal factory: The case of Cali's garbage dump. *World Development, 6*(9–10), 1173–1185.

Borsdorf, A., Hidalgo, R., & Vidal-Koppmann, S. (2016). Social segregation and gated communities in Santiago de Chile and Buenos Aires. A Comparison. *Habitat International, 54*, 18–27.

Bräutigam, K. R., Gonzalez, T., Szanto, M., Seifert, H., & Vogdt, J. (2012). Municipal solid waste management in Santiago de Chile: Challenges and perspectives towards sustainability. In D. Heinrichs, K. Krellenberg, B. Hansjürgens, & F. Martínez (Eds.), *Risk habitat megacity* (pp. p279-301). Springer.

Brown, D., & McGranahan, G. (2016). The urban informal economy, local inclusion and achieving a global green transformation. *Habitat International, 53*, 97–105.

Busnelli, R. (2018). Atlas de residuos sólidos industriales del partido de Gral. San Martín Nuevos modelos de gestión para la construcción de viviendas. *Paper presented at the Congreso Internacional Vivienda y Ciudad: Debate en torno a la Nueva Agenda Urbana*, Córdoba, Argentina, June 2018

Carenzo, S. (2020). Contesting informality through innovation "from below": epistemic and political challenges in a waste pickers cooperative from Buenos Aires (Argentina). *Tapuya: Latin American Science, Technology and Society, 3*(1):441–471

Carenzo, S. (2014). Lo que (no) cuentan las máquinas: La experiencia sociotécnica como herramienta económica (y política) en una cooperativa de "cartoneros" del gran buenos aires. *Antipod Rev Antropol Arqueol, 18*, 109–135.

Chile Valora. (2022). Comisión Sistema Nacional Certificación de Competencias Laborales, Registro de personas certificadas. Retrieved December 19, 2022, from https://certificacion.chilevalora.cl/ChileValora-publica/candidatosList.html?limpiarFiltros

Chile. (2006). Ministerio del Interior (2006). Ley N°18.695: Ley organica constitucional de municipalidades. Retrieved December 19, 2022, from https://www.bcn.cl/leychile/navegar?idNorma=30077

Chile. (2020). Ministerio del Medio Ambiente. (2020). Resumen del Estado del Medio Ambiente para la Ciudadanía 2020. Retrieved December 19, 2022, from https://sinia.mma.gob.cl/wp-content/uploads/2020/12/Resumen-Ejecutivo-IEMA-2020.pdf

Chile. (2021). Ministerio del Medio Ambiente (2021) Sexto Reporte del Estado del Medio Ambiente 2021. Retrieved December 19, 2022, from https://sinia.mma.gob.cl/wp-content/uploads/2022/06/REMA2021.pdf

Cohen-Rosenthal, E. (2017). What is eco-industrial development? In E. Cohen-Rosenthal & J. Musnikow (Eds.), *Eco-industrial strategies, unleashing synergy between economic development and the environment* (1st ed., pp. p1-16). Routledge.

Comisión Económica para América Latina y el Caribe (CEPAL), Ministerio del Medio Ambiente/ Compromiso Empresarial para el Reciclaje (CEMPRE). (2020). Encuesta a municipios sobre gestión de residuos sólidos domiciliarios 2019, Documentos de Proyectos (LC/TS.2020/137). CEPAL, Santiago de Chile.

Consejo Nacional de Medio Ambiente (CONAMA). (2010). First Report of Solid Waste Management in Chile; CONAMA: Santiago de Chile.

Construye2025. (2020). Hoja de Ruta RCD Economía Circular En Construcción 2035. Available via DIALOG. Retrieved December 19, 2022, from https://construye2025.cl/rcd/wp-content/uploads/2020/08/HDR-PAGINA_RCD_200825.pdf

Cozzens, S., & Sutz, J. (2014). Innovation in informal settings: Reflections and proposals for a research agenda. *Innovation and Development, 4*(1), 5–31.

Deus, R., Mele, F. D., & Bezerra, B. S. (2020). A municipal solid waste indicator for environmental impact: Assessment and identification of best management practices. *Journal of Cleaner Production, 242*, 118433.

Dias, S. M. (2016). Waste pickers and cities. *Environment and Urbanization, 28*(2), 375–390.

Erkman, S. (1997). Industrial ecology: An historical view. *Journal of Cleaner Production, 5*(1–2), 1–10.

Fergutz, O., Dias, S., & Mitlin, D. (2011). Developing urban waste management in Brazil with waste picker organizations. *Environment and Urbanization, 23*(2), 597–608.

Fidelis, R., & Colmenero, J. C. (2018). Evaluating the performance of recycling cooperatives in their operational activities in the recycling chain. *Resources, Conservation and Recycling, 130*, 152–163.

Florisbela, A., & Astorga, A. (2006). La integración del sector informal en la gestión de los residuos sólidos urbanos, enfoque al sector de la chatarra (The integration of the informal sector in the MSW management, focus to the scrap sector). Deutsche Gesellschaft für Technische Zusammenarbeit GmbH, Santiago deh Chile.

Fraccascia, L. (2018). Industrial symbiosis and urban areas: A systematic literature review and future research directions. *Procedia Environmental Science, Engineering and Management, 5*(2), 73–83.

Fressoli, M., Arond, E., Abrol, D., Smith, A., Ely, A., & Dias, R. (2014). When grassroots innovation movements encounter mainstream institutions: Implications for models of inclusive innovation. *Innovation and Development, 4*(2), 277–292.

Gall, M., Wiener, M., Oliveira, C. C., Lang, R. W., & Hansen, E. G. (2020). Building a circular plastics economy with informal waste pickers: Recyclate quality, business model, and societal impacts. *Resources, Conservation and Recycling, 156*, 104685. https://doi.org/10.1016/j.resconrec.2020.104685

Geng, Y., Tsuyoshi, F., & Chen, X. (2010). Evaluation of innovative municipal solid waste management 755 through urban symbiosis: A case study of Kawasaki. *Journal of Cleaner Production, 18*, 993–1000.

Gibbs, D. (2003). Trust and networking in inter-firm relations: The case of eco-industrial development. *Local Economy, 18*(3), 222–236.

Gibbs, D. (2009). Eco-industrial parks and industrial ecology: Strategic niche or mainstream development? In *The social embeddedness of industrial ecology*. Edward Elgar, Cheltenham.

Gibbs, D., & Deutz, P. (2005). Implementing IE? Planning for eco-industrial parks in the USA. *Geoforum, 36*, 452–464.

Giordano, R., Montacchini, E., & Tedesco, S. (2020). All You Can't Eat: Research and experiences from agri-food waste to new building products in a circular economy perspective. In R. Salomone, A. Cecchin, P. Deutz, A. Raggi, & L. Cutaia (Eds.), *Industrial symbiosis for the circular economy*. Strategies for Sustainability (1sd edn, pp. 149–164). Springer.

Giorgio, S., & Girotto, L. (2012). El Organismo Provincial para el Desarrollo Sostenible (OPDS) y los municipios en el marco de la ley de radicación industrial: Delegación y descentralización en la regionalización provincial. Retrieved December 19, 2022, from https://repositoriosdigitales.mincyt.gob.ar/vufind/Record/SEDICI_e74b6fdbb78e28bbaaa3fa34ff809e15

Giovannini, M. (2014). De la Economía Popular a la Economía Social y Solidaria: el Caso de los Recicladores de Base en Santiago de Chile, Euricse Working Papers 73|14.

Giovannini, M., & Huybrechts, B. (2017). How inclusive is inclusive recycling? Recyclers' perspectives on a cross-sector partnership in Santiago de Chile. *Local Environment, 22*(12), 1497–1509.

González-García, S., & Dias, A. C. (2019). Integrating lifecycle assessment and urban metabolism at city level: Comparison between Spanish cities. *Journal of Industrial Ecology, 23*(5), 1062–1076.

González Martínez, T. M. (2011). Analysis of different municipal solid waste management systems in Santiago de Chile. Universität Stuttgart.

Guibrunet, L., Calvet, M. S., & Broto, V. C. (2017). Flows, system boundaries and the politics of urban metabolism: Waste management in Mexico City and Santiago de Chile. *Geoforum, 85*, 353–367.

Hartmann, C. (2018). Waste picker livelihoods and inclusive neoliberal municipal solid waste management policies: The case of the La Chureca garbage dump site in Managua, Nicaragua. *Waste Management, 71*, 565–577.

Heeks, R., Foster, C., & Nugroho, Y. (2014). New models of inclusive innovation for development. *Innovation and Development, 4*(2), 175–185.

Hillier, B., Greene, M., & Desyllas, J. (2000). Self-generated Neighbourhoods: The role of urban form in the consolidation of informal settlements. *Urban Design International, 5*, 61–96.

Hossain, M. U., Ng, S. T., Antwi-Afari, P., & Amor, B. (2020). Circular economy and the construction industry: Existing trends, challenges and prospective framework for sustainable construction. *Renewable and Sustainable Energy Reviews, 130*, 109948.

INE. (2018). Memoria del Censo 2017 Instituto Nacional de Estadísticas. Retrieved December 19, 2022, from http://www.censo2017.cl/memoria/descargas/memoria/libro_memoria_censal_2017_final.pdf

InvestChile. (2021). Infraestructura para el Reciclaje: Envases y Embalajes Plásticos. Retrieved December 19, 2022, from https://fch.cl/wp-content/uploads/2021/07/e-Book-PactoPlastico-esp.pdf

IRAM 11950. (2010). Métodos de ensayos de resistencia al fuego de los elementos constructivos.

ISO 14025. (2006). Environmental labels and declarations—Type III environmental declarations—Principles and procedures.

Iyamu, H. O., Anda, M., & Ho, G. (2020). A review of municipal solid waste management in the BRIC and high-income countries: A thematic framework for low-income countries. *Habitat International, 95*, 102097.

Jensen, P. D., Basson, L., & Leach, M. A. (2011). *Reinterpreting Industrial Ecology. Journal of Industrial Ecology, 15*(5), 680–692.

Jones, A. P. (1999). Indoor air quality and health. *Atmospheric Environment, 33*, 4535–4564.

Korhonen, J. (2004). Industrial ecology in the strategic sustainable development model: Strategic applications of industrial ecology. *Journal of Cleaner Production, 12*, 809–823.

Kovacic, Z., Strand, R., & Völker, T. (2019). *The circular economy in Europe: Critical perspectives on policies and imaginaries (Routledge explorations in sustainability and governance)*. Routledge.

Law N°18.695: Fija normas especiales para las empresas de menor tamaño, Ministerio de Economía, Fomento y Reconstrucción de Chile (2010). Retrieved December 19, 2022, from https://www.diariooficial.interior.gob.cl/publicaciones/2021/03/16/42906/01/1910141.pdf

Law N° 20267: Crea el sistema nacional de certificación de competencias laborales y perfecciona el estatuto de capacitación y empleo (2008). Retrieved December 19, 2022, from https://www.bcn.cl/leychile/navegar?idNorma=272829&idParte=8052184&idVersion=

Law N°20.920: Establece Marco para la gestión de residuos, la Responsabilidad Extendida del Productor y fomento al reciclaje, Ministerio del Medio Ambiente de Chile (2016). Retrieved December 19, 2022, from https://www.bcn.cl/leychile/navegar?idNorma=1090894

López de Munain, D., Castelo, B., & Ruggerio, C. A. (2021). Social metabolism and material flow analysis applied to waste management: A study case of Autonomous City of Buenos Aires, Argentina. In: *Waste management (New York, N.Y.)* 126, S. 843–852. https://doi.org/10.1016/j.wasman.2021.04.014

Margallo, M., Ziegler-Rodriguez, K., Vázquez-Rowe, A. R., Irabien, Á., & Kahhat, R. (2019). Enhancing waste management strategies in Latin America under a holistic environmental assessment perspective: A review for policy support. *Science of the Total Environment, 689*, 1255–1275.

Masotti, C., Ferrante, A., Boiardi, L., & Fabbri, C. (2011). Self help retrofitting technologies for low cost housing construction. El caso de Vila Novo Ouro Preto, Brasil. *Procedia Engineering, 21*, 625–632.

Masuda, J. R., & Garvin, T. (2006). Place, culture, and the social amplification of risk. *Risk Analysis, 26*(2), 437–454.

Medina, M. (2000). Scavenger cooperatives in Asia and Latin America. *Resources, Conservation and Recycling, 31*, 51–69.

Medina, M. (2007). *The world's scavengers: Salvaging for sustainable consumption and production (globalization and the environment)*. AltaMira Press.

Meneses, E. (2012). Centro de valoración de residuos domiciliarios para su reciclaje. Retrieved December 19, 2022, from https://repositorio.uchile.cl/handle/2250/112567

Mora, P., Sabatini, F., Fulgueiras, M., & Innocenti, D. (2014). Disyuntivas en la política habitacional chilena. Available via DIALOG. Retrieved December 19, 2022, from http://politicaspublicas.uc.cl/wp-content/uploads/2015/02/estudio.pdf

Navarrete, P., & Navarrete, N. (2018). Unleashing waste-pickers' potential: Supporting recycling cooperatives in Santiago de Chile. *World Development, 101*, 293–310.

Nelles, M., Grünes, J., & Morscheck, G. (2016). Waste management in Germany—Development to a sustainable circular economy? *Procedia Environmental Sciences, 35*, 6–14.

Ng, W. Y., & Chau, C. K. (2015). New life of the building materials- recycle, reuse and recovery. *Energy Procedia, 75*, 2884–2891.

Nogueira, Z. J. L., Jacobsen Leopoldino, C., & Harue Yamane, L. (2020). Waste pickers organizations and municipal selective waste collection: Sustainability indicators. *Waste Management, 118*, 219–231.

Norma Chilena (NCh). (2022). Listado de normas técnicas obligatorias. Retrieved December 19, 2022, from http://normastecnicas.minvu.cl

Oddou, M. (2008). Caracterización por comunas en la estación de transferencia de Quilicura (Characterization by municipalities at the Quilicura Transfer Station). KDM-Grupo Urbaser Kiasa, Santiago de Chile.

OECD. (2020). Environment at a Glance Indicators, OECD Publishing, Paris. Retrieved December 19, 2022, from https://www.oecd.org/environment/environment-at-a-glance/Circular-economy-waste-and-materials-May-2020.pdf

Opoku, A. (2019). Biodiversity and the built environment: Implications for the sustainable development goals (SDGs). *Resources, Conservation and Recycling, 141*, 1–7.

Orellana, A. (2009). La gobernabilidad metropolitana de Santiago: La dispar relación de poder de los municipios. *EURE (Santiago), 35*(104), 101–120.

Palma, J. G. (2014). Has the Income Share of the Middle and Upper-middle Been Stable around the "50/50 Rule", or Has it Converged towards that Level? The "Palma Ratio" Revisited. *Development and Change, 45*(6), 1416–1448.

Pardo, M. C., & Piña, W. A. (2016). Solid waste management in Bogota´: The role of recycling associations as investigated through SWOT analysis. *Environment, Development and Sustainability, 19*, 1067–1086.

Pizarro, G., & Jara, M. (2015). El mapa de Chile del millonario negocio de la basura. Retrieved December 19, 2022, from https://ciperchile.cl/2015/05/26/el-mapa-de-chile-del-millonario-negocio-de-la-basura/

Purvis, B., Celebi, D., & Pansera, M. (2023). A framework for a responsible circular economy. *Journal of Cleaner Production, 400*, 136679. https://doi.org/10.1016/j.jclepro.2023.136679

Reyes, S. (2004). Santiago: la difícil sustentabilidad de una ciudad neoliberal. In C. de Mattos, M.E. Ducci, A. Rodríguez, & G. Yáñez (Eds.), *Santiago en la Globalización ¿Una Nueva Ciudad?* (1st edn., pp. 189–218). Eure Libros.

RITEP (2021). Recuperadores urbanos. Retrieved December 19, 2022, from https://intercambiotecnico.ritep.org/recuperadores-urbanos/

Rojas, A., Yabar, H., Mizunoya, T., & Higano, Y. (2018). The potential benefits of introducing informal recyclers and organic waste recovery to a current waste management system: The case study of Santiago de Chile. *Resources, 7*, 18–40.

Ross, S., & Evans, D. (2003). The environmental effect of reusing and recycling a plastic-based packaging system. *Journal of Cleaner Production, 11*(2003), 561–571.

Ruiz-Tagle, J., & López, E. (2014). El estudio de la segregación residencial en Santiago de Chile: Revisión crítica de algunos problemas metodológicos y conceptuales. *EURE (Santiago), 40*(119), 25–48.

Sala, B. H., & Zegers, C. C. (2021). Portafolio de modelos de negocio en economía circular para la construcción: Informe final de la consultoría. Iniciativa de la Hoja de Ruta RCD y Economía Circular en Construcción. Available via DIALOG. Retrieved December 19, 2022, from https://construye2025.cl/wp-content/uploads/2022/07/Portafolio-de-modelos%E2%80%A8-de-negocio-en-economia-circular-para-la-construccion-Informe-final-de-la-consultoria.pdf

Sandin, G., & Peters, G. M. (2018). Environmental impact of textile reuse and recycling - A review. *Journal of Cleaner Production, 184*, 353–365.

Scheinberg, A., Spies, S., Simpson, M. H., & Mol, A. P. J. (2011). Assessing urban recycling in low- and middle-income countries: Building on modernised. *Habitat International, 35*, 188–198.

Shah, I.-H., Dong, L., & Park, H.-S. (2020). Tracking urban sustainability transition: An eco-efficiency analysis on eco-industrial development in Ulsan, Korea. *Journal of Cleaner Production, 262*, 121286. https://doi.org/10.1016/j.jclepro.2020.121286

Smith, A., Fressoli, M., Abrol, D., Arond, E., & Ely, A. (2017). *Grassroots innovation movements*. Taylor & Francis.

Song, M., Guan, Y., Wang, J., & Zhao, J. (2016). Evaluation of urban industrial ecological transformation in China. *Clean Technologies and Environmental, 18*, 2649–2662.

The Economist Intelligence Unit (EIU). (2017). Avances y desafíos para el reciclaje inclusivo: Evaluación de 12 ciudades de América Latina y el Caribe. EIU, Nueva York.

Tröger, D., Becerra, A. A. A., Busnelli, R., Yajnes, M., Williams, F., & Braun, A. C. (2023). Exploring eco-industrial development in the global south: Recognizing informal waste-picking as urban-industrial symbiosis? *Cleaner Waste Systems, 5*, 100096. https://doi.org/10.1016/j.clwas.2023.100096

UNE-EN 15804:2012+A2. (2020). Sustainability of construction works—Environmental product declarations—Core rules for the product category of construction products.

Valencia, M., Solíz, M. F., & Yépez, M. (2023). Waste picking as social provisioning: The case for a fair transition to a circular economy. *Journal of Cleaner Production, 398*, 136646. https://doi.org/10.1016/j.jclepro.2023.136646

Valenzuela-Levi, N. (2019). Factors influencing municipal recycling in the Global South: The case of Chile. *Resources, Conservation and Recycling, 150*, 104441. https://doi.org/10.1016/j.resconrec.2019.104441

Valenzuela-Levi, N. (2020). Waste Political Settlements in Colombia and Chile: Power, Inequality and Informality in Recycling. *Dev Change, 51*(4), 1098–1122.

Valenzuela-Levi, N. (2021). Poor performance in municipal recycling: The case of Chile. *Waste Management, 133*, 49–58.

Van Berkel, R., Fujita, T., Hashimoto, S., & Geng, Y. (2009). Industrial and urban symbiosis in Japan: Analysis of the eco-town program 1997–2006. *Journal of Environmental Management, 90*, 1544–1556.

Vanhuyse, F., Fejzić, E., Ddiba, D., & Henrysson, M. (2021). The lack of social impact considerations in transitioning towards urban circular economies: A scoping review. *Sustainable Cities and Society, 75*, 103394. https://doi.org/10.1016/j.scs.2021.10339

Vergara, L. M., Gruis, V., & Flier, K. (2019). Understanding housing management by low-income homeowners: Technical, organisational and sociocultural challenges in Chilean Condominium housing. *Buildings, 9*(3), 65–86.

Vergara, S. E., Damgaard, A., & Gomez, D. (2015). The efficiency of informality: Quantifying greenhouse gas reductions from informal recycling in Bogotá, Colombia. *Journal of Industrial Ecology, 20*(1), 107–119.

Verma, R., Vinoda, K. S., Papireddy, M., & Gowda, A. N. S. (2016). Toxic pollutants from plastic waste- a review. *Procedia Environmental Sciences, 35*, 701–708.

Villalba, L. (2020). Recent evolution of the informal recycling sector in Argentina within the 'popular economy': Measuring its impact through a case study in Tandil (Buenos Aires). *Waste Management & Research, 38*(9), 1037–1046.

Walker, A. M., Vermeulen, W. J., Simboli, A., & Raggi, A. (2021). Sustainability assessment in circular inter-firm networks: An integrated framework of IE and circular supply chain management approaches. *Journal of Cleaner Production, 286*, 125457.

Wilson, D. C., Rodic, L., Scheinberg, A., Velis, C. A., & Alabaster, G. (2012). Comparative analysis of solid waste management in 20 cities. *Waste Management & Research, 30*(3), 237–254.

Wilson, D. C., Velis, C., & Cheeseman, C. (2006). Role of informal sector recycling in waste management in developing countries. *Habitat International, 30*, 797–808.

Winans, K., Kendall, A., & Deng, H. (2017). The history and current applications of the circular economy concept. *Renewable and Sustainable Energy Reviews, 68*, 825–833.

Yedla, S., & Park, H.-s. (2017). Eco-industrial networking for sustainable development: Review of issues and development strategies. *Clean Technologies and Environmental Policy, 19*(2), 391–402.

Adaptive Urban Design for Resilience in the Chilean City of Padre Las Casas

Alejandro Flores Lizama, Leonardo N. A. Agurto Venegas, Paulina Ibieta Illanes, and Cristian Valderrama

Abstract This article addresses problems related to urban growth in areas considered vulnerable to natural risks, in particular the risks of flooding and waterlogging in areas of the city of Padre Las Casas, Chile. These risks arose due to a lack of urban planning in the face of major urban growth and development over the last thirty years, with the city experiencing a growth rate of 29.5% between 2002 and 2017 (CCHC. Cámara Regional Temuco, 2019). On this basis, we seek to understand the importance of considering preexisting natural elements in urban planning in order to decrease the natural threats that may be present, understanding the urban and rural environment as something unpredictable and in constant flux. Finally, we research urban resilience and adaptability strategies in the face of natural risks and the possible vulnerabilities of a system in relation to flooding and waterlogging threats, in order to define a methodology that permits intervention and restoration of natural elements in that location with the objective of mitigating these threats and ensuring the adaptation and transformation capacity of the most at-risk sectors.

1 Introduction

Chilean urban planning has been characterized by its severity, consisting of strict land use rules and construction codes. Historically, this has led to a paradoxical outcome: cities that grow expansively without any apparent order, the deterioration of preexisting urban spaces, and the aggravation of urban inefficiency and quality of life problems (Sabatini & Soler, 1995). On this basis, the phenomenon of urban

A. Flores Lizama · L. N. A. Agurto Venegas (✉) · P. Ibieta Illanes · C. Valderrama
Institute of Architecture and Urbanism, Faculty of Architecture and Arts, Universidad Austral de Valdivia, Ernst Kasper Building, Helena Haverbeck street. Campus Isla Teja, Valdivia, Chile
e-mail: leonardo.agurto@uach.cl

P. Ibieta Illanes
e-mail: maria.ibieta@uach.cl

C. Valderrama
e-mail: cvalderramaa@uach.cl

A. Ch. Braun et al. (eds.), *Eco-Industrial Development as an Industrial Strategy*, Green Energy and Technology, https://doi.org/10.1007/978-3-031-73576-9_25

growth in Chilean cities is characterized by the ineffectiveness of urban planning, which has been overwhelmed by the accelerated growth of cities.

In the face of this situation, various factors that appear as causes of this paradoxical result. The main problem among these is the lack of housing arising from migrations into urban centers, generating a need to expand the city and giving rise to various forms of independent management that are alien to urban planning. In this sense, direct public management by the state is associated with various social housing policies and programs that have facilitated low-density design solutions in urban peripheries. Meanwhile, indirect public management, associated with housing subsidies, favored the profitability of housing projects for the private sector and private companies, facilitating the appearance and expansion of urban land markets and boosting expansion towards the periphery in search of cheap land and economically profitable projects.

1.1 The Case of Padre Las Casas

"The city of Padre Las Casas is a relatively new town, located in the Araucanía region at [the] south of Chile. It became administratively independent from Temuco and declared a different city in 1995" (Municipio Padre Las Casas, 2019). In this sense, this town used to be a neighborhood of the city of Temuco founded in 1899, then called Villa Alegre as shown in Fig. 1.

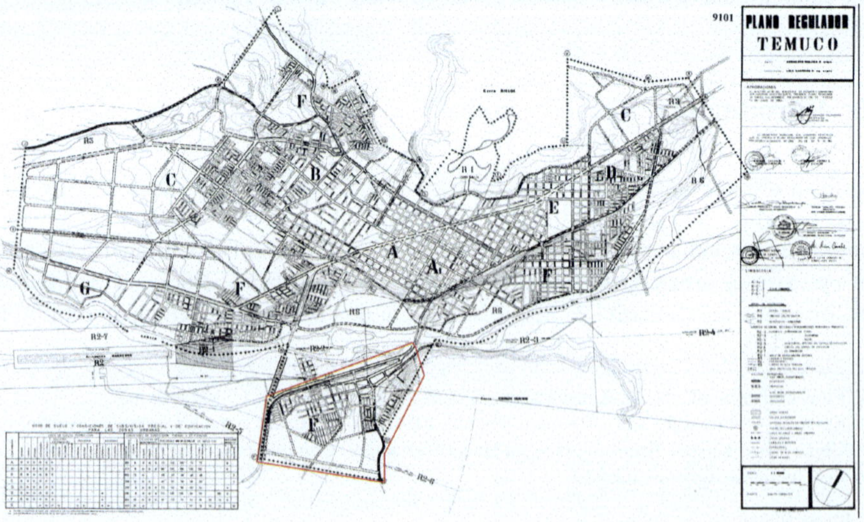

Fig. 1 Regulatory map of Temuco–Padre Las Casas, 1983. Red indicates the Padre Las Casas area, then known as Villa Alegre. *Source* Municipalidad de Padre Las Casas

Accordingly, during recent years this neighborhood saw explosive growth beyond extant urban planning regulations, leading to accelerated expansion beginning in the late 1980s and 1990s as shown in Fig. 2, finally leading to it being recognized as an independent town apart from Temuco. "Via this expansion, the city of Padre Las Casas occupied ecosystems needed for the long-term subsistence of the city, including forests and wetlands" (Laboratorio cambio social, 2020), which stretched from the Cautín River basin to the north down to the Vilcún River basin to the south (Fig. 3).

The real estate development that resulted from this exponential urban growth, the aforementioned public policies, and a lack of consideration for preexisting natural elements have led to wetlands being filled in and a general deterioration of the preexisting natural environment, with the construction of roads and housing being prioritized over the conservation of these natural elements. This has led to segregation between the neighborhoods at the south of the city, separated by spaces corresponding to former wetlands that now have high risks of waterlogging and flooding that also affect the residential sectors, as illustrated in Fig. 4. "This generates accessibility problems, flooding risks, temperature rises due to vegetation loss, decreased

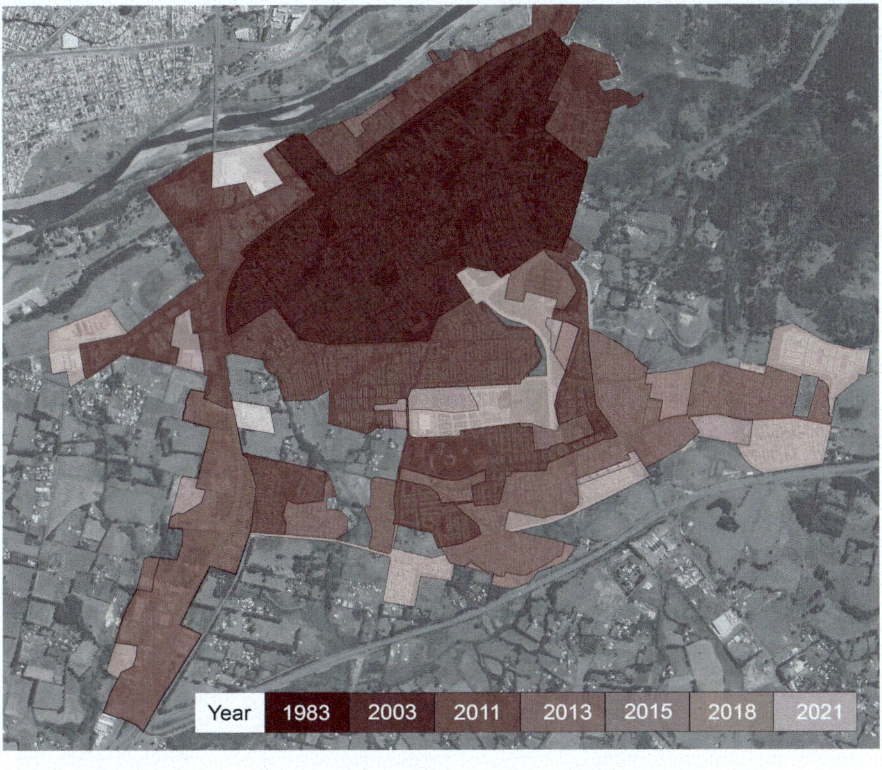

Fig. 2 The urban growth of Padre Las Casas in recent decades. *Source* Author

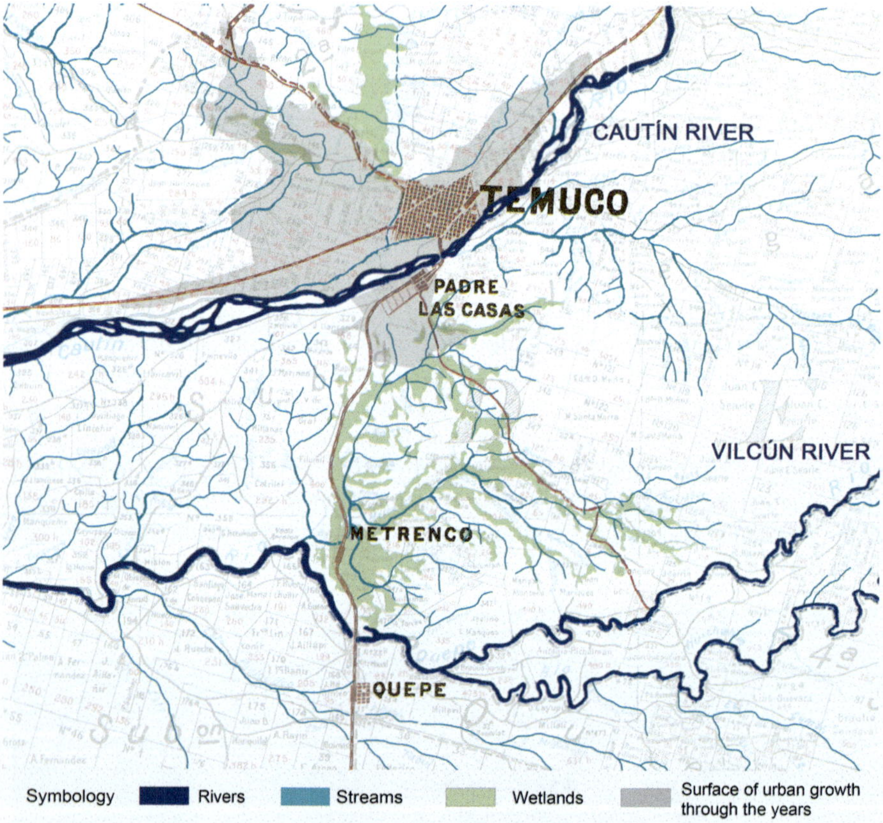

Fig. 3 Location of the Temuco-Padre Las Casas urban area, in relation to the natural environment. *Source* Author

soil humidity, and also restricts community mobility systems" (Laboratorio cambio social, 2020).

2 Preliminary Study

The form of growth in the main intermediate cities of Chile has historically led to urban expansions that do not consider natural elements in their context, which includes a dismissal of natural environmental risks. This has occurred amid a lack of restrictions for occupying zones near floodplains, wetlands, rivers, and other natural areas, an indication that urban planning regulations have been inefficient in many sectors of the Chilean territory. There are current efforts to improve this via new regulations such as the urban wetlands law #21,202, aimed at regulating and protecting wetlands ecosystems within cities.

Fig. 4 Cartography problems of the city of Padre Las Casas. *Source* Author

In this sense, Arenas et al. (2010) stated that the urban sprawl of cities towards areas exposed to various types of natural hazards has presented high costs for their population, the state, and the productive sector. The same authors mention that two constituent factors of natural risk faced by urban settlements can be identified, namely the hazards and vulnerabilities, as shown in Fig. 5. In this equation, hazard is defined as "an external risk factor, represented by the potential occurrence of a naturally originating event" (Arenas et al., 2010). This threat can be evaluated by gathering background information and evidence on the ground in order to form an idea about the level of danger of a phenomenon that occurs with a certain regularity and intensity. Vulnerability, meanwhile, is "considered to be the internal risk factor of a subject, object or system exposed to a hazard, corresponding to its disposition to being damaged" (Ayala-Carcedo & Olcina, 2002, as cited in Arenas et al., 2010). In this sense, vulnerability is composed of physical, economic, environmental, and social factors within a given urban system in the face of a natural threat.

On this basis, one response type to natural hazards comes from the concept of resilience, which, when understood from a socio-ecological perspective as defined by Chelleri along with other authors, is related to the adaptation capacity of an urban system in the face of a natural risk event. In this sense, this perspective considers the generic and emerging properties of complex adaptive systems, which are able to

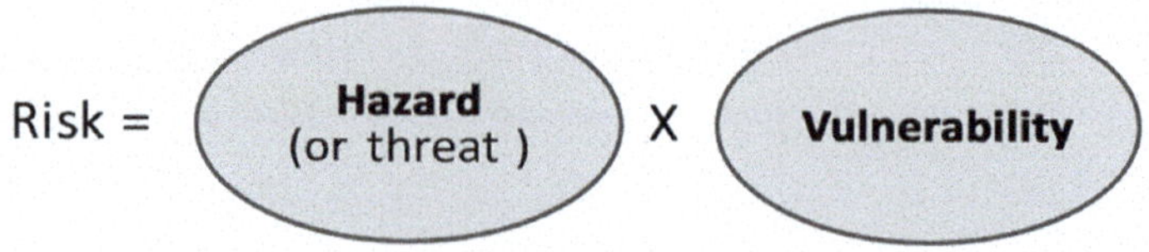

Fig. 5 Risk is the function of hazard times vulnerability. *Source* Arenas, Lagos and Hidalgo (2010)

adapt, transform, and learn while navigating unpredictable evolutionary trajectories (Chelleri et al., 2015).

These authors also indicate that urban resilience is not the exact opposite of vulnerability. The paradigm of "greater resilience means lower vulnerability" is the easiest simplification of the concept of resilience. By considering it to be a rule, focusing only on strategies that decrease vulnerability and therefore natural risk, we deny the possibility of change and adaptation. This way of understanding resilience is consequently a key basis for the development of this article as it is the first approach to the concept of urban adaptability that corresponds to the capacity of a system to resist and recover from the impact of a given natural hazard, ensuring adaptation and transformation capacity over time. The conceptual perspective of resilience has been used to generate various theoretical models that evaluate the ability of an urban system to respond to a natural hazard event. A notable conceptualization by Ahern (2011) differentiates two focuses from the perspective of sustainability, via which international urban planners have addressed these risks.

According to Ahern (2011), the first of these is a fail-safe model, which arose during the modern era of the 20th century and corresponds to a determinist idea of nature, science, and ecology. In this sense, from an urban planning perspective there is also a static and deterministic vision where urban sustainability appears as a stable element that is durable over time; a fail-safe condition which, once achieved, can persist for generations.

We can thus conclude that this vision sought to generate answers that would completely suppress an urban system's vulnerabilities to a given hazard, eliminating the possibility that this response could fail, and without generating any plans or alternatives in case failure occurred. However, in subsequent years an alternative conception appeared that sought to understand urban and natural environments as inherently variable elements, uncertain and inclined to unexpected changes.

This presents a paradox: How can a static conception of the landscape be sustainable in a context of disruption and unpredictable change arising from natural and urban surroundings? Considering this question, the resilience capacity of settlements conceived via the landscape and urban planning requires identification of these stochastic processes and the disruptions that a given landscape or urban locale is inclined to face in order to generate a "safe-to-fail" model.

In this sense, this model seeks to strengthen the adaptability of an urban system, where resilience mainly appears via planning strategies rather than as a normative framework, since to be effective it must be directly guided by the physical, environmental, social, and economic factors of a given place, which can help identify urban system vulnerabilities(Acuti and Bellucci, 2019). According to this, we can conclude that resilience strategies should be applied with consideration of various aforementioned urban system factors, on the basis of understanding the ecological system associated with the changing natural environment that contains the urban structure associated with the form of the city, serving as a basis for the socioeconomic system and governance institutions that finally shape the urban system. The multilayered systems mindset of DNV GL's Systems and Urban Resilience Framework (SURF), as depicted in Fig. 6, represents a model that views urban areas as systems with unique profiles of mutually interconnected ecological, structural, socioeconomic, and governance systems.

Based on this safe-to-fail model, Ahern (2011) identified five urban planning and design strategies to build urban resilience capacity:

- **Multifunctionality**: A way to generate sustainable systems in limited spaces within compact cities, via the combination and relation of various functions that are close to each other.
- **Redundancy and modulation**: The existence of multiple urban system components that play a similar function or support each other, so that if one fails due to a natural risk, others will fill this need in an emergency.

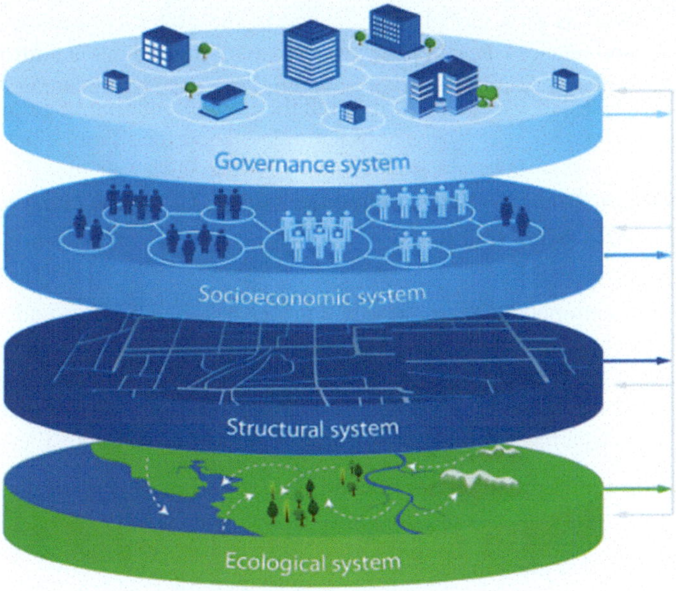

Fig. 6 The multilayered systems mindset in DNV GL's Systems and Urban Resilience Framework (SURF) model. *Source* Christiansen, AF & Libby, B. 2017. Changing Cities in a Changing Climate. DNV GL Group Technology and Research, Position Paper 1–2017

- **Socio- and biodiversity**: Natural elements or users that can respond variously in the face of a given disruptive event. More diversity will provide a greater possibility for different responses to address a given natural hazard.
- **Muti-scaled networks and connectivity**: Networks are the systems that support functions within an urban environment. When this is understood as a system that carries out various necessary functions for the city, connectivity becomes the critical element relating these functions between their various scales.
- **Planning and adaptive design**: Adaptive planning considers a lack of knowledge about disturbances and changes as an opportunity to "learn by doing." If we intend to carry out innovative and adaptable urban design, we must recognize an inherent potential for failure, generating design proposals that are safe in case this happens ("safe-to-fail").

On these grounds, and in continuation with previous statements, we must implement these resilience strategies across urban layers to strengthen resilience and adaptability at various urban system scales.

In this sense, Chelleri et al. (2015) also approach the challenges of resilience from different scales and their mutual implications, mainly considering a time scale and a space scale. The former establishes that we must recognize environmental change possibilities over time, meaning that adopting a deterministic focus could open or close the window for future development patterns and compromise system adaptability. The second challenge, related to urban metabolism, refers to inefficient and unsustainable resource use on a local scale due to consumption and pollution/waste generation, as well as on a global scale related to the global connections between people, goods, and services. Resilience must also involve strategies at multiple scales, without compromising resources and/or generating damage to one area over others.

2.1 *Approximation to Water Management: Urban Resilience to Hydrological Hazards*

As previously mentioned, to design resilient urban systems we must understand the natural and urban environments and their inherent possibility for change. In this sense, considering existing problems in Padre Las Casas as presented earlier, we must understand the natural hydrological cycle and the impacts produced on it by anthropic effects as a basis for adaptability and increasing system resilience.

Understanding the Natural Water Cycle

Mahaut (2009) summarizes the different stages of the hydrological cycle, showing the importance of this phenomenon for the earth's ecological system:

> The sun is the driving force in this cycle, providing the energy needed for water to evaporate. Water from the oceans, seas, and lakes evaporates into the air, condensing into fine suspended droplets and forming clouds. Solar energy heats the atmosphere, creating pressures and depressions that displace these huge masses of water in the form of clouds. The water

> molecules have weight: a depression or cooling changes the phase of the water in the clouds and the liquid water falls due to gravity as precipitation. Then, still by gravity, the water runs off, burrows into the soil or infiltrates it, shaping all of Earth's landscapes. Vegetation absorbs part of the water from the surface or from the soil for its growth. This absorbed water is returned to the atmosphere by transpiration of water through the pores of the plant (Mahaut, 2009, p. 57).

Understanding and Mapping the Impacts of Anthropic Activities on the Natural Water Cycle

Human activities involving the water cycle can be summarized in four main categories: climate change, resource extraction, surface modifications, and pollution. In this sense, climate change can generally affect water cycles via acceleration while natural resource extraction can decrease ecological biodiversity, affecting water absorption by vegetation and the evapotranspiration process.

The main impacts that can lead to flooding and waterlogging hazards, as shown in Fig. 7, is human activity that makes surfaces impermeable (roofs, paving, plugging of bare soils, etc.), reducing rainwater filtration in the soil towards the phreatic layers and considerably increasing runoff and flood risks in the area, thus affecting the most densely populated zones.

The contamination of water by household or industrial wastewater discharge and runoff (which is loaded with particles such as dust, sand, heavy metals, etc.) qualitatively transforms the water cycle. Quantitative or qualitative imbalances can appear in nature as soon as humans interrupt the natural water cycle.

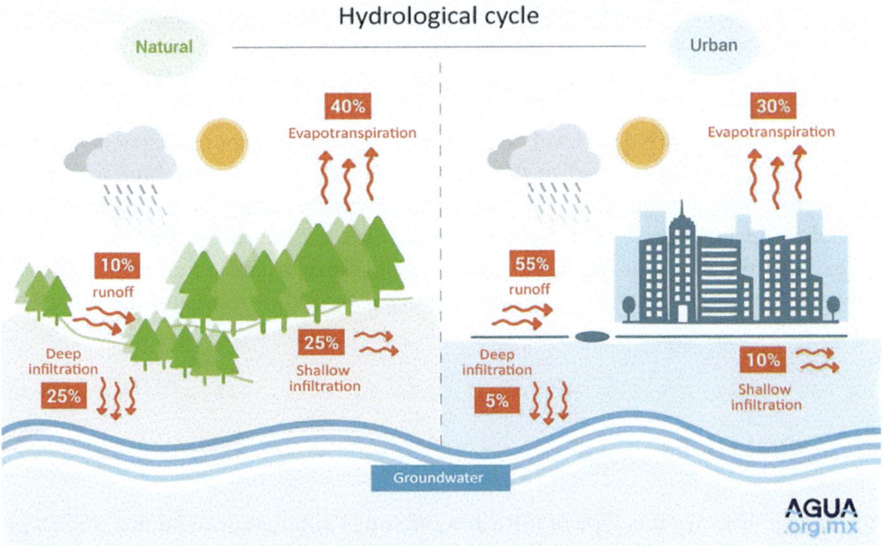

Fig. 7 The water cycle. *Source* Dr. José Arturo Gleason/ITAC

2.2 Sustainable Water Resource Management Strategies

Following up on the previous points, one existing strategy that can help restore the preexisting ecosystem of Padre Las Casas is a blue-green infrastructure system, with the aim of protecting natural water cycles by improving the adaptability and resilience of the city against flooding and waterlogging risks.

According to Ghofrani et al. (2017) a blue-green infrastructure (BGI) "is an interconnected network of natural and designed landscape components, including water bodies and green and open spaces, which provide multiple functions such as: (i) water storage for irrigation and industry use, (ii) flood control, (iii) wetland areas for wildlife habitat or water purification, and many others."

- **Blue network components**: These are any body of water such as rivers, canals, lakes, wetlands, etc. that exist in a city or urban system, and which can contribute to system resilience against hydrological hazards via sustainable water use.
- **Green network components**: "This, in turn, can be considered to be the hybrid infrastructure of green spaces including forests, wetlands, parks, roofs and green walls which together can contribute to ecosystem resilience by contributing benefits to human settlements" (Demuzere et al., 2014).

As shown in Fig. 8, Demuzere et al. indicate the physical, psychological, and social benefits produced by the implementation of green urban infrastructures, directly associated with climate change mitigation and the adaptation capacity of urban systems against natural hazards. In line with these points and with the objectives of the present article, the main direct benefits produced by green infrastructure relevant to water management are the reduction of flooding problems by decreasing runoff and drought in the city, as well as the improvement of water quality via purification. "Assessment of bioretention cells has shown a reduction in peak flows by at least 96.5% for small to medium-sized storm events" (Hunt et al., 2008 as cited in Demuzere et al., 2014).

The social and individual-level contributions related to response capacity against natural hazards are connected to the proximity of people to these blue and green ecosystems. The aforementioned uses also provide a range of recreational and psychological benefits, which create opportunities for education and the creation of ties with the community about natural environments and potential hazards as shown in Fig. 9. This increases the adaptability potential of residents and the urban system.

2.3 Conclusions

Design responses for this type of infrastructure must aim to reduce human impact on the water cycle in order to decrease the vulnerabilities of urban systems and ensure their adaptability capacity. In this sense, the basic principles for decreasing flooding and/or waterlogging risk are as follows.

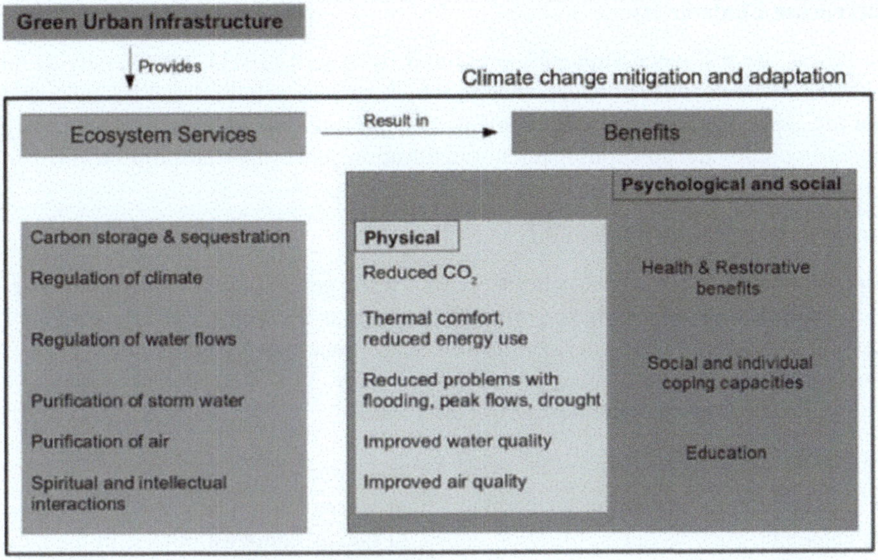

Fig. 8 Green urban infrastructure services and benefits within a climate change mitigation and adaptation framework. *Source* Demuzere et al. (2014)

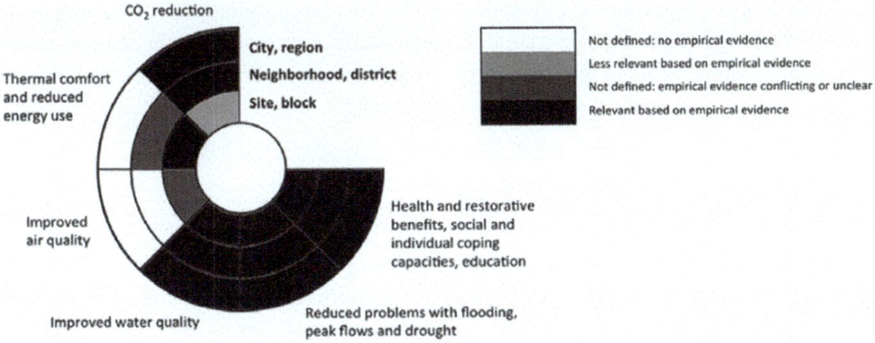

Fig. 9 Relevance of the benefits from green urban infrastructure for climate change adaptation and mitigation on three spatial scales, based on the evidence discussed above. *Source* Demuzere et al. (2014)

Water Cycle Protection

Water flow disturbances (increased runoff, decreased filtration) observed in the water cycle arise mainly due to changes in soil surface conditions in terms of roughness, slope, and permeability towards underground layers (Mahaut, 2009). This makes it crucial to avoid the impermeabilization of soils and plant species, protecting natural zones and soils that are currently highly permeable in order to decrease runoff and boost filtration and the contribution of plant species to rainwater management.

Increasing Biodiversity

Based on the resilience strategies presented by Ahern (2011), biodiversity is an essential element for resilience, through which various solutions can be generated that are associated with natural elements that present diverse responses including increasing rainwater absorption, retention, filtration, and evapotranspiration via plant species.

Strengthening Multifunctionality

The potential of these infrastructures allows for the generation of various functions and solutions to aid flooding risk mitigation, such as by rainwater filtration zones, capture or accumulation zones, etc. In this sense, when there is a broader range of functions and/or solutions, there can be better rainwater management to decrease system vulnerabilities in the face of flooding risks.

Increasing Redundancy

Strengthening the repetition of elements that allow for rainwater filtration and management, such as tanks, filtration zones, or canals, amongst others. In the case that any elements fail or overflow, others can take over their function.

Improving Connectivity

The ability to relate and connect various elements and parts of blue-green infrastructure so they work together as a system. Finally, this strategy lets us aim to generate an adaptive design that can decrease flooding risk, mainly by strengthening adaptive capacity and learning from and transforming the urban system over time.

3 Objectives and Methodology

The general objective of this article is to explore project strategies (possibilities or solutions) for the restoration of preexisting blue connections in the city of Padre Las Casas and the potential of implementing a blue-green infrastructure. Such measures would help mitigate flooding risks and ensure urban ecosystem adaptability as a resilient design measure in the city's peripheral neighborhoods.

3.1 Methodology

- **Bibliographical review** and analysis of authors to help establish the main resilience strategies associated with flooding and waterlogging, including the scope of this concept within the framework of urban planning and design.
- **Review of projects and/or examples** as points of reference to help establish analysis and intervention criteria for resilient design.

- **Analysis of the Padre Las Casas case at various scales** for a general understanding of the natural and urban landscape.
- **Study via design strategies** which allow for adaptive design against flooding hazards.

3.2 Methodological Analysis

The context analysis for the flooding problem in the city of Padre Las Casas was conducted using the main reference of the Andalien ideas contest (Espinosa et al., 2020), whose goal was to generate an intervention proposal around the Andalien River basin that focused on three analysis scales: (a) watershed scale, (b) landscape-city scale and (c) specific site. Analysis for the Padre Las Casas case was divided into three similar analysis scales in order to gain both general and specific understandings of the urban system and the natural landscape of the city.

- **General landscape scale**: Generally associated with the relationship of Padre Las Casas to its natural environment such as existing sloughs and bodies of water, in order to understand this relationship by analyzing different indicators. On this basis, we aim to generate a general planning strategy to implement an urban resilience proposal associated with flooding risk.
- **Intermediate scale**: An approach associated with the direct relationship between watercourses and floodable areas with their closest environment, seeking to focus and go deeper into a representative section of the general strategy established in the previous scale.
- **Specific site scale**: Choosing a defined place based on analysis from the previous scales to design a proposal that allows the aforementioned strategies to be carried out.

4 Multi-scale Analysis: The Case of Padre Las Casas

4.1 General Landscape Scale

By analyzing the general scale, we seek to understand the natural environment of the city and its relationship with this environment so as to understand the components of the existing ecological system. For this, the analysis is divided between two types of indicators, the former associated with green networks and the latter with the blue networks of the city.

Green Network Indicators in the City

Due to accelerated and segregated growth in the city, the old preexisting wetlands were filled in. As shown in Fig. 10, they left traces in the form of a series of vacant lots

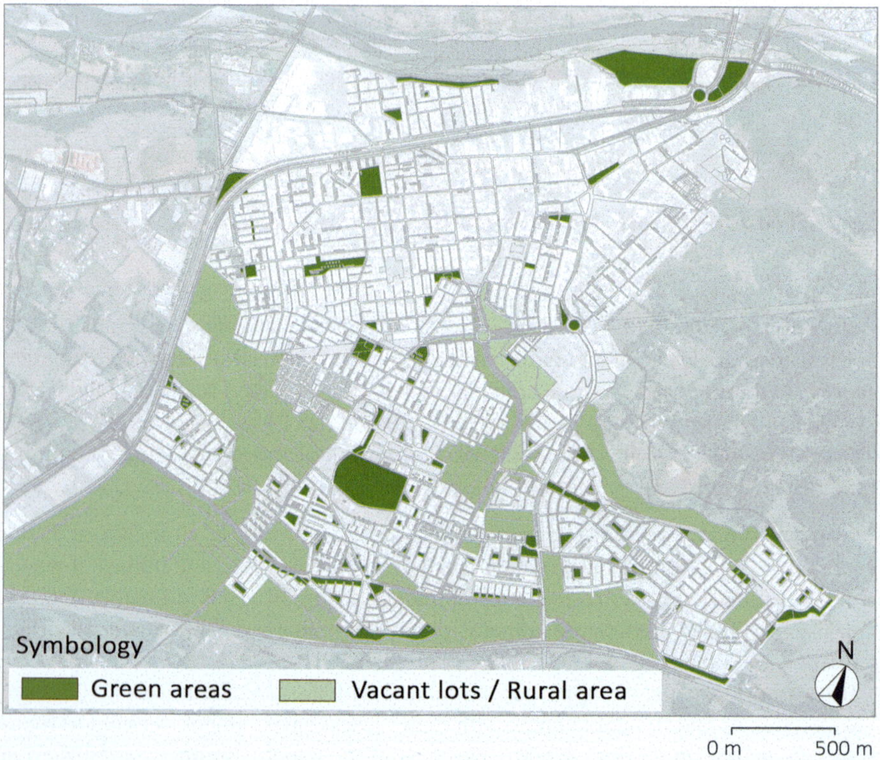

Fig. 10 Vacant lots and green areas in the city. *Source* Author

which maintain the highest percentage of vegetation and biodiversity in the city. The more populated urban areas have mostly lost plant species, as illustrated in Fig. 11.

Blue Network Indicators in the City

There are currently some watercourses with various levels of anthropic intervention in the city, which can be seen in the map in Fig. 12. Over time, these watercourses have become segregated and separated amongst themselves without any direct relationship with the city, and with access to them restricted due to human intervention.

Due to the infill of wetlands south of the city, these zones currently tend to accumulate rainwater in certain seasons, creating significant waterlogging and flooding hazards in various city sectors. To try to address this problem, rainwater collectors have been built in various areas following urban growth, as shown in Fig. 13.

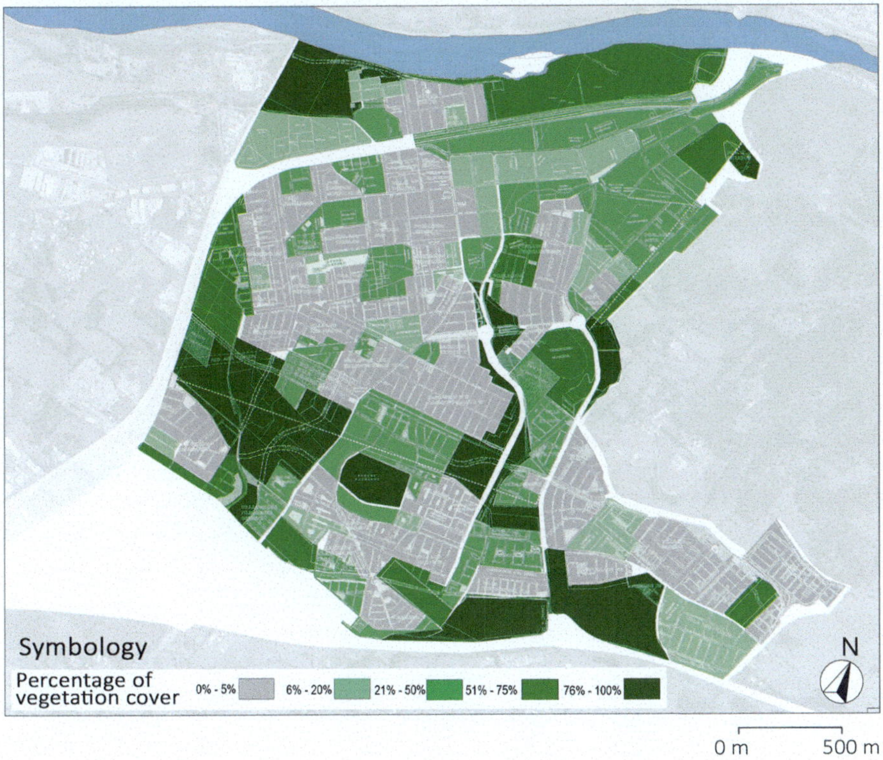

Fig. 11 Percentage of vegetation cover by block. *Source* City corporations atlas, 2017

4.2 Urban Area General Scale

The city can generally be characterized as a bedroom community, given the predominance of residential use and its proximity to the regional capital of Temuco, together with low land-use diversity that has led to low multifunctionality, as illustrated in Fig. 14. The socioeconomic strata shown in Fig. 15 can be characterized by a predominance of lower-middle class groups, with little economic diversity in the city. This encourages growth towards the periphery in search of cheaper lots, and a preference for low-density housing solutions in much of the city.

Despite the urban growth of recent years and continuous real estate expansion, there is still a persistent general housing deficit in the city, with growth into the periphery and natural environments expected to continue as it has in recent years. Together with real estate pressures, this has already driven various housing projects into areas that were previously vacant lots or rural lands.

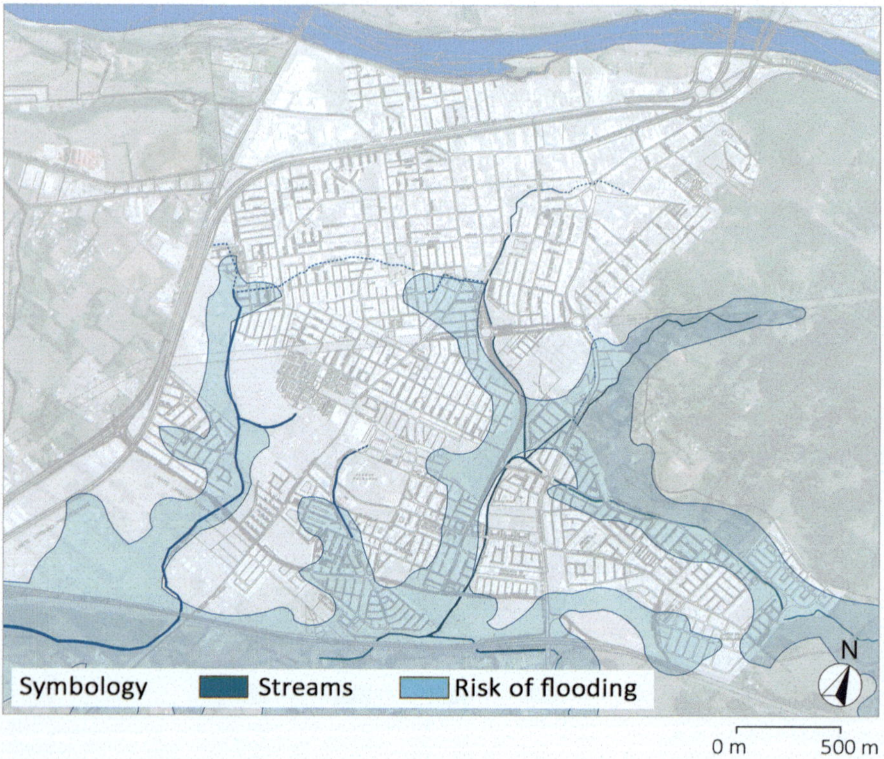

Fig. 12 Streams and flooding hazard zones. *Source* Author

4.3 SWOT Analysis

On the basis of the following indicators, we performed a SWOT analysis (Fig. 16) in order to propose a strategy for overall blue-green restoration.

Strengths

Based on the previous analysis, we can establish some key strengths in the efforts to mitigate flooding risks. The first is the existence of vacant lots, which have the largest capacity for rainwater filtration via soil since they have not undergone any intervention. The higher biodiversity maintained by these sites is another advantage for rainwater management, and makes them less vulnerable to flooding risks. Some of the natural watercourses and streams remain as well as parts of the vegetation and ecosystems associated with this blue fabric of life, which can contribute to decreasing runoff and to managing rainwater.

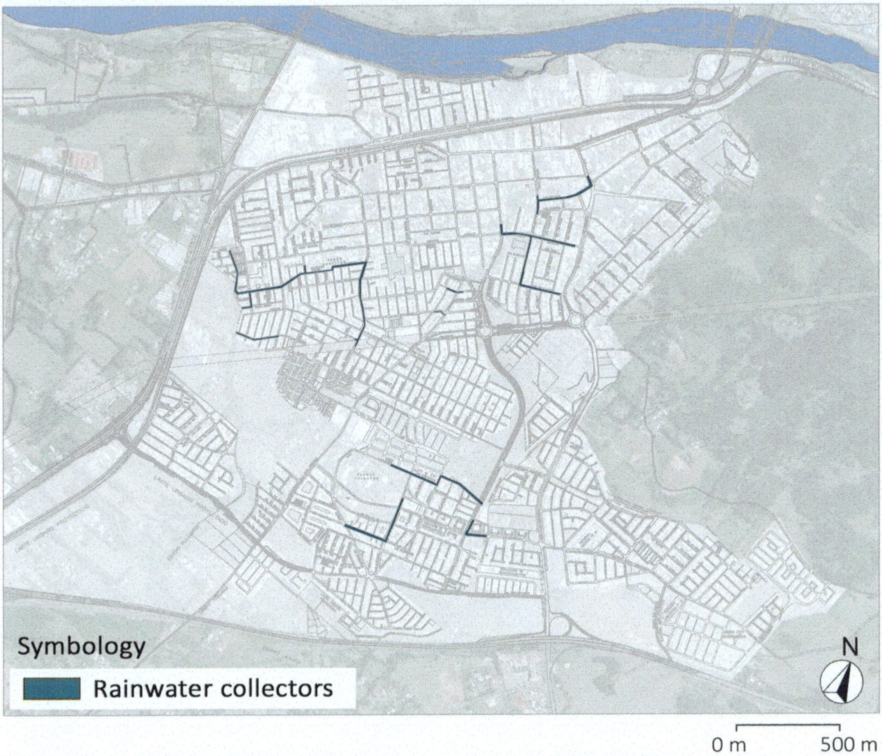

Fig. 13 Rainwater collectors in the city. *Source* MOP, DIRPLAN 2015

Opportunities

Following up on the previous point, we can establish some opportunities associated with existing watercourses and vacant lots to decrease natural flooding risks. Vacant lots can become important sites for catching and draining rainwater from the city, given the biodiversity and permeability they present. They can also become green areas or floodable parks within the urban system, with the goal of protecting these zones and improving general connectivity between different neighborhoods that are currently segregated. The existence of some broad boulevards may drive the design or appearance of green corridors connecting these various vacant lots.

Weaknesses

The weaknesses identified include the inaccessibility of vacant lots, generating fragmentation and segregation between peripheral neighborhoods, decreasing urban permeability, and giving rise to poor citywide connectivity. As mentioned in previous indicators, urban zones have lost much of their vegetation and biodiversity over time, seeing plant cover relegated towards a few green areas or vacant lots. Finally, many

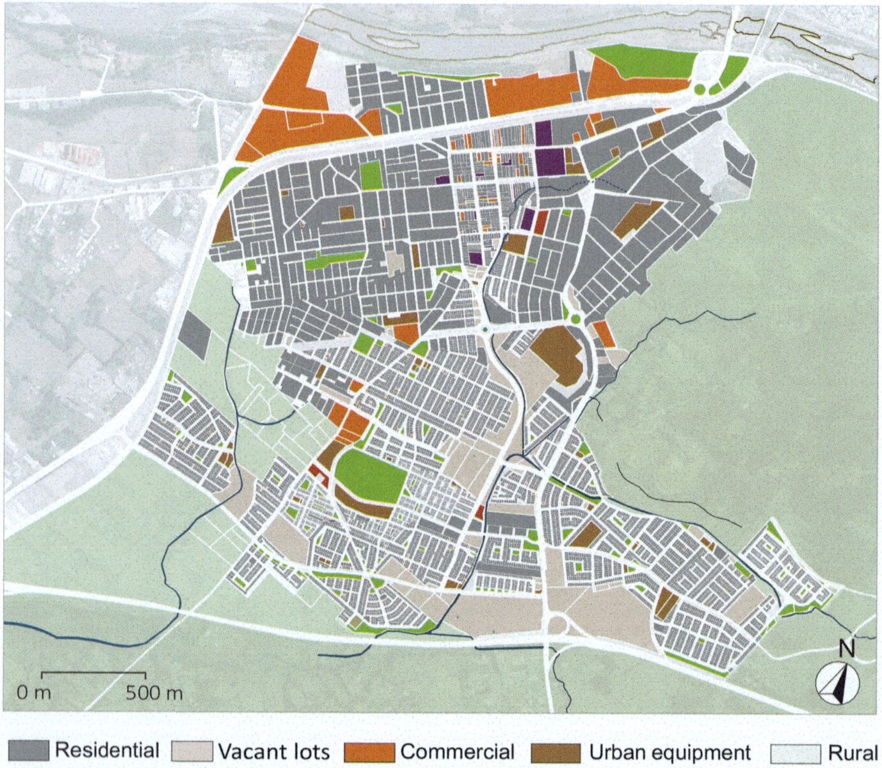

Fig. 14 Land use. *Source* Author, with SII data

of the preexisting streams have become reduced and obstructed due to urban growth, becoming degraded and segregated from the urban environment.

Threats

The main threat present here is the flooding risk due to wetlands infill and increased soil impermeability. This gave rise to the last major flood on the south side of the city in 2015, as seen in Fig. 32. Due to real estate pressure and the present housing deficit, there is also a risk of vacant lots and rural environments being filled in and becoming impermeable due to the need for new construction, which would cause biodiversity loss and increase flooding risk. The present threat to the watercourses is pollution due to garbage dumping or wastewater infiltration, presenting a latent pollution risk.

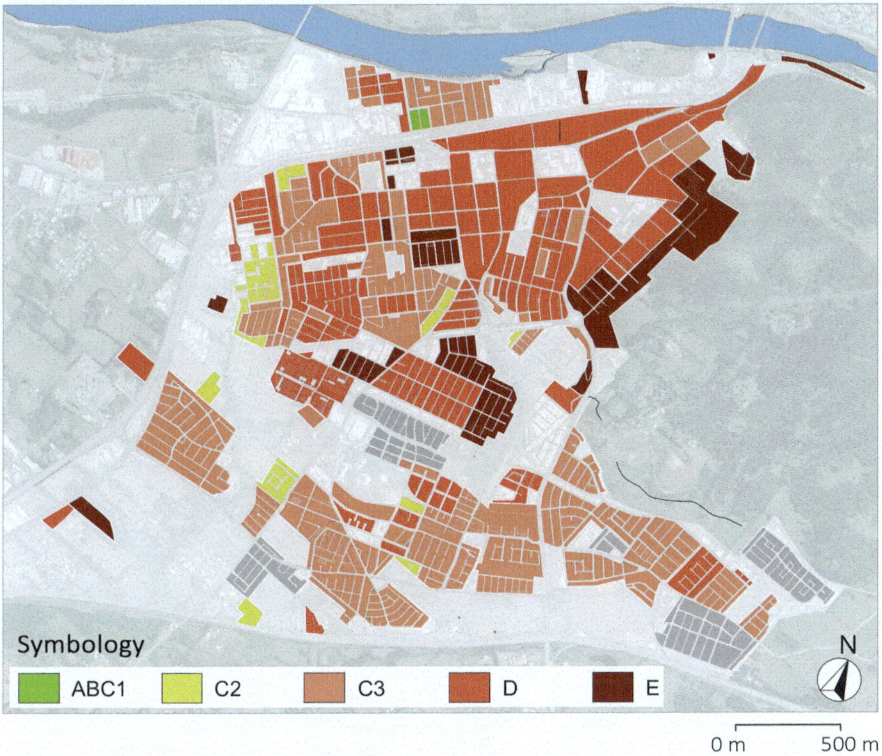

Fig. 15 Socioeconomic strata. *Source* Socioeconomic study and Mapuche community characterization. Universidad Mayor, 2014

4.4 General-Scale Strategy: Lines of Action

Waterway Protection and Recovery (Improvements to Biodiversity and Connectivity)

The first action is to protect existing streams, seeking to decrease the pollution and degradation risks they may face. This will help protect the ecosystems most associated with these watercourses and the *biodiversity* they have. We also seek to restore parts of the streams that have been lost over time, as shown in Fig. 17, interrelating them via new sections that can guarantee *connectivity* in this blue network. In this sense, this reconnection occurs according to the old surface of the preexisting streams, also considering the possibility of intuitive new sections associated with major avenues within the city.

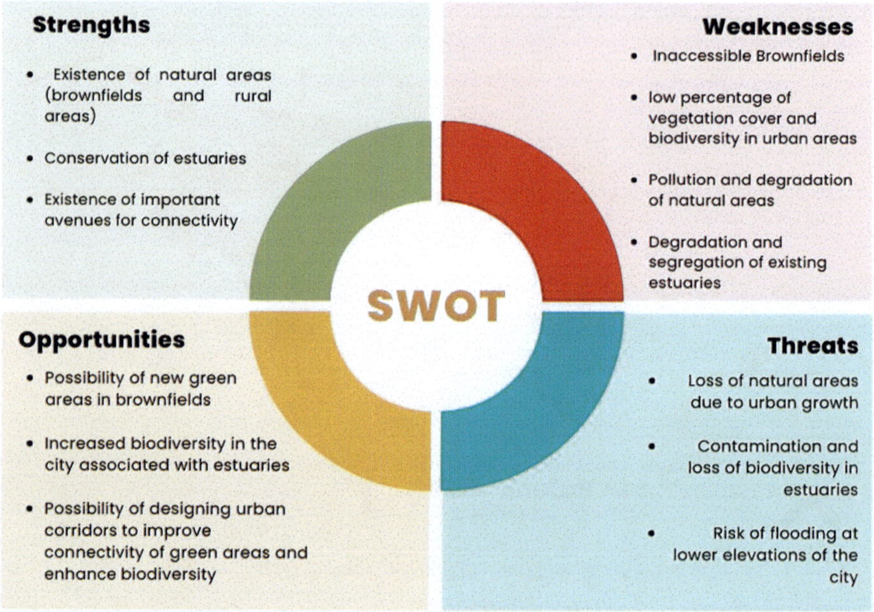

Fig. 16 SWOT analysis. *Source* Author

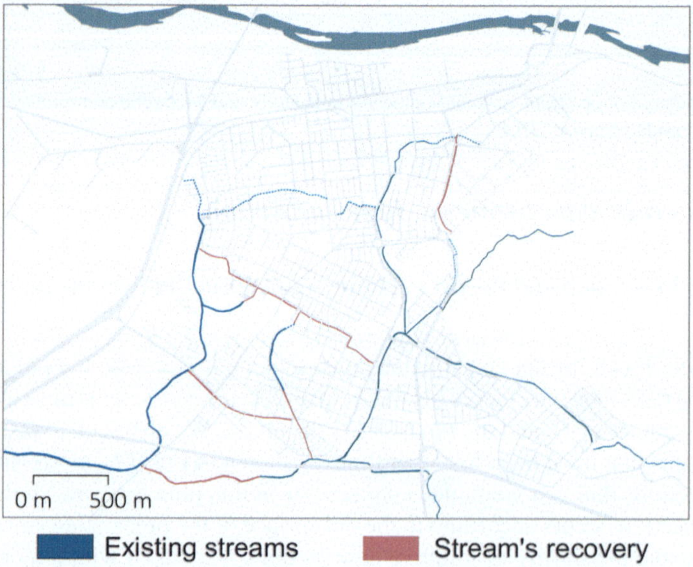

Fig. 17 Blue network recuperation map. *Source* Authors

Design of Floodable Areas in Natural Spaces (Strengthening Multifunctionality and Adaptability)

We can also consider the flooding risk zone as the main area in need of rainwater management, incorporating new floodable areas to allow for rainwater accumulation and filtration in these sectors and adjacent neighborhoods, thus maximizing the advantages of existing soil permeability in these natural spaces, as shown in Fig. 18. In this sense, these floodable areas are built in a multifunctional way based on various programs including floodable parks, plazas, or walkways, which work under normal conditions as well as in the case of flooding or waterlogging, allowing for long-term adaptability.

Protecting Plant Species and Soil Permeability in Natural Spaces (Protecting and Increasing Biodiversity)

The objective of this line of action is to protect vacant lots around watercourses and floodable areas as illustrated in Fig. 19, seeking to protect the soil permeability and biodiversity of these sectors. These can also be considered as green areas limiting urban growth, as well as acting as a second filtration zone between floodable areas and the urban environment. We also aim to recover the connection between various city neighborhoods via the design of these green sites as vacant lots, improving accessibility to watercourses and between various currently segregated sectors south of the city by means of circulations and walkways around the streams and green areas.

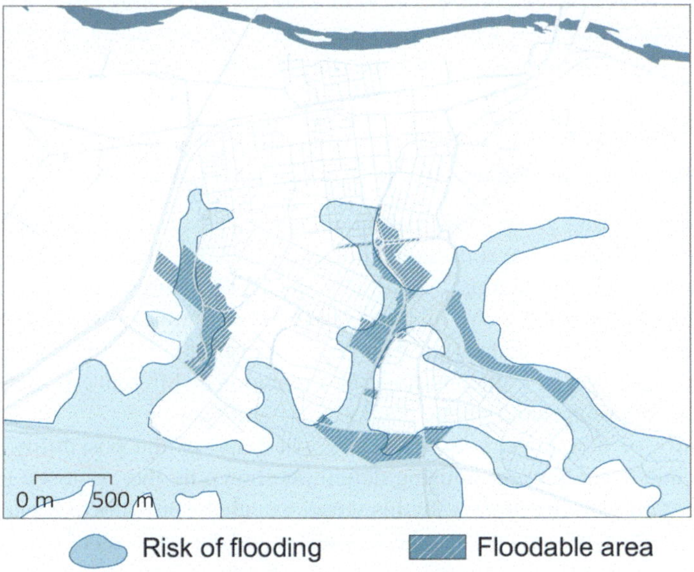

Fig. 18 Cartography of new floodable areas. *Source* Authors

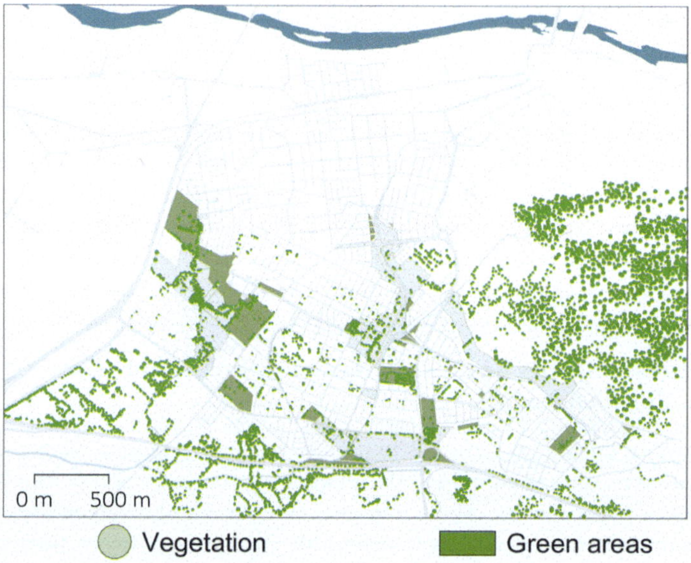

Fig. 19 Map of vegetation and new green areas. *Source* Authors

Urban Corridor Design (Improved Biodiversity and Connectivity)

The urban corridors depicted in Fig. 20 have the main objective of connecting different green areas and natural spaces within the city, and seeking to relate different natural species which exist within them, in order to support urban biodiversity. We also seek to improve connectivity between the various segregated neighborhoods of the city by incorporating walkways around urban corridors that connect the various areas south of the city.

Regeneration of Extant Spaces in the City (Improved Adaptability)

Regenerating existing green areas: As shown in Fig. 21, we propose projecting and improving extant green areas in the city, seeking to take advantage of their better soil permeability, strengthening their biodiversity, and incorporating them into the water management strategy. We also recognize the importance of plant species for rainwater management via absorption and evapotranspiration. In this sense, we seek to increase plant cover by implementing new vegetation in different city areas, mainly associated with avenues and public places.

Possibility of new construction: Finally, we consider the possibility of urban expansion due to the present housing deficit, as shown by the existence of current development projects in progress. In this sense, it could be possible to generate a new relationship between the constructed urban sector and the new web of blue and green infrastructure, preventing the integrity of natural spaces from being compromised and integrating the water landscape into the daily lives of residents.

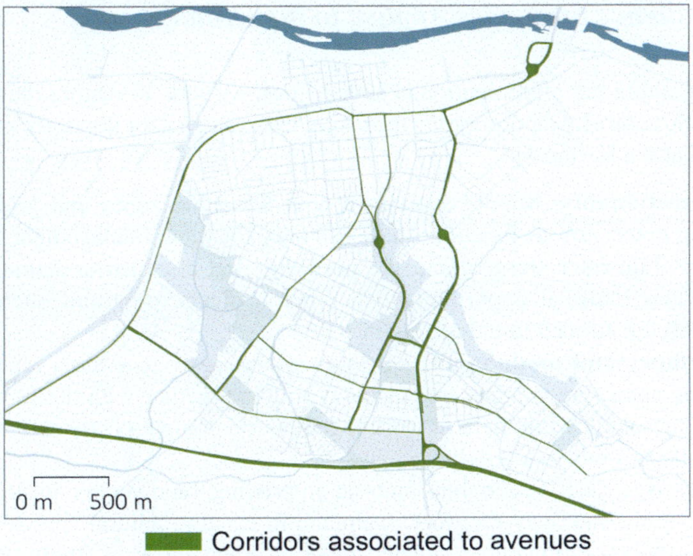

Fig. 20 The urban corridors. *Source* Authors

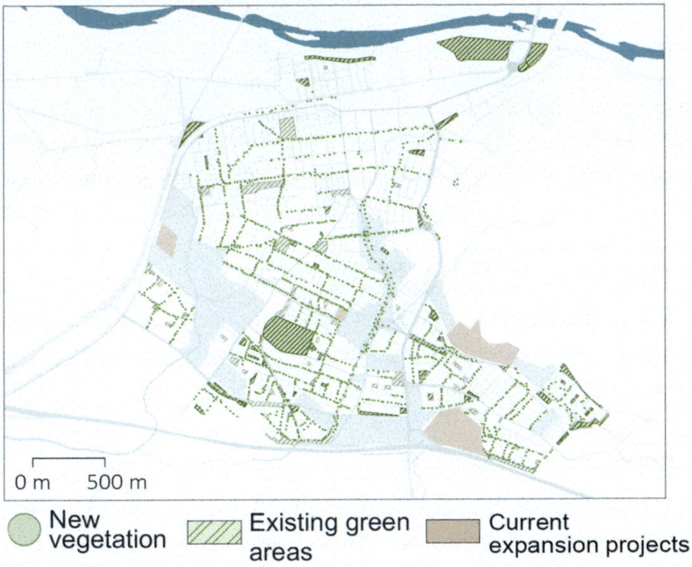

Fig. 21 Adapted elements map. *Source* Authors

4.5 Resilience Strategies Applied to the Urban System

Continuing along the aforementioned action lines, we seek to restore the blue and green infrastructure by incorporating the resilience strategies for flood risk mitigation mentioned at the beginning.

- **Multifunctionality**: Mainly associated with floodable zones and green areas, referring to the various functions these can have for water management, boosting filtration, rainwater storage, and the purifying and conducting actions of this water. There is also an opportunity to incorporate diverse programs, such as those previously mentioned in the action lines.
- **Redundancy and modulation**: This strategy seeks to generate a repetition of floodable areas, green areas, and urban corridors together with the functions of each of these, allowing them to work together as a compact network even if one zone collapses due to a flooding event.
- **Biodiversity**: Each action line aims at improving biodiversity in some way, whether with waterway recovery, maintaining existing natural spaces or green areas, or incorporating new vegetated zones or urban corridors. In this sense, we seek to take advantage of plant species for absorption and evapotranspiration of rainwater.
- **Connectivity**: A key strategy that seeks achievement via waterway restoration and the incorporation of green urban corridors. In this way, every element presented in the action lines is related with the others, creating a network in the city.
- **Adaptability**: Finally, adaptability is sought via the regeneration of extant spaces associated with current green areas as well as neighboring areas that can be adapted for rainwater management, in order to incorporate them as functional elements within this network.

Blue and Green Network Map for Resilience

Based on the six action lines presented, we seek to generate a new way of thinking about urban growth as illustrated in Fig. 22, aiming to recognize preexisting ecosystems associated with the watercourses and vacant lots as a fundamental element for urban resilience against natural flooding risks.

Structural Blue-Green Network Axes

Based on this network developed from the action lines, we can identify four of its structures that are related with different urban zones due to the connectivity arising from the existence of the watercourses and ecological corridors (Fig. 23).

- **Axis A, Pulmahue stream**: Axis A is based around the Pulmahue watercourse and the natural spaces surrounding it. Due to high biodiversity and soil permeability, this estuary has the least anthropic intervention compared to the others, meaning that it has a high amount of vegetation associated with its ecosystem. Nevertheless, this sector has also been affected by pollution and water deterioration due to wastewater leakage in its basin.

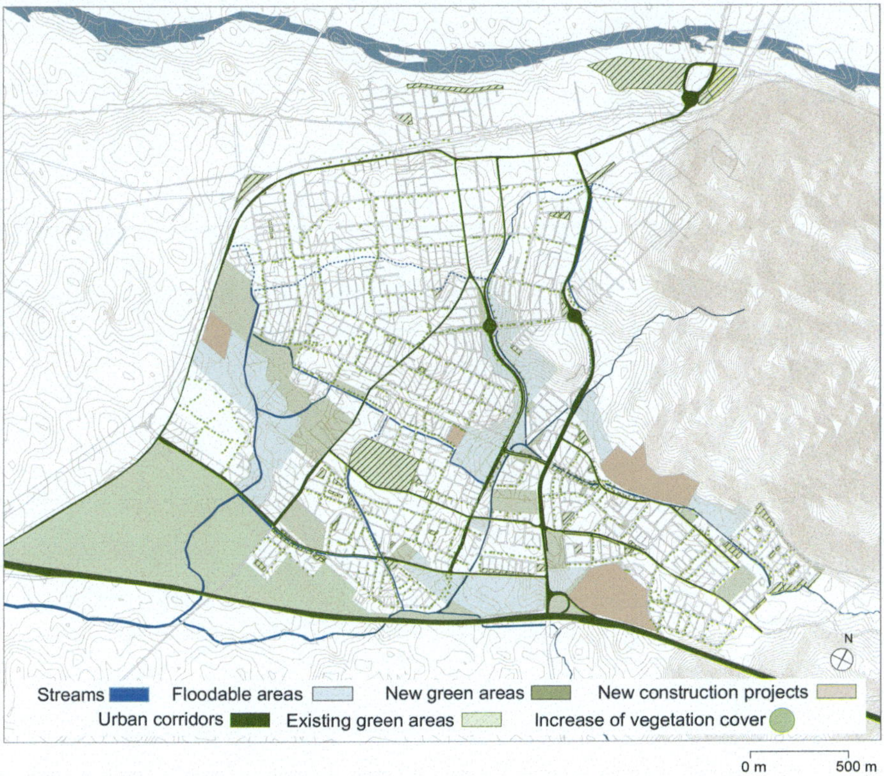

Fig. 22 General map, blue-green network. *Source* Authors

- **Axis B, Llahuallín stream Section 1**: Axis B is mainly associated with a part of Llahuallín stream (T1) that has various relations with its context. It can be connected with floodable areas, residential areas, and vacant lots, and mainly consists of the Llahuallín stream and a main avenue that crosses almost the entire city.
- **Axis C, Huichahue Avenue urban corridor**: Axis C is a major avenue that runs across much of the city and has the potential for implementing one of the urban corridors mentioned in the action lines, connecting different green areas, floodable areas, and residential zones.
- **Axis D, Llahuallín stream Section 2**: Finally, structural Axis D is associated with the second section of the Llahuallín (T2) watercourse. It connects urban areas with the natural environment at the periphery of the city and is associated with different green areas, residential zones, and areas considered at risk of flooding.

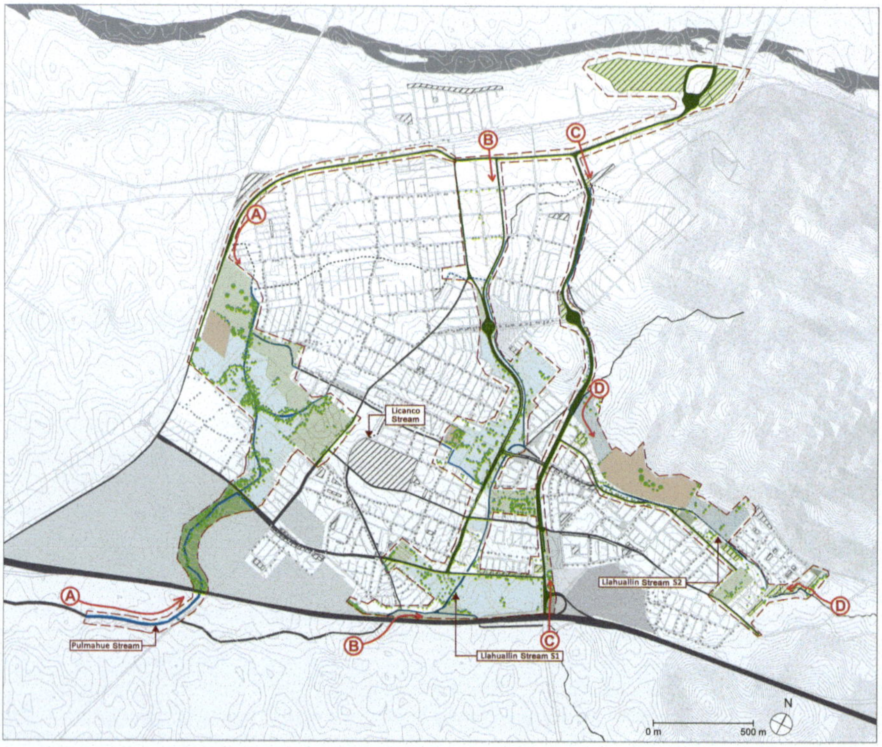

Fig. 23 Llahuallín watercourse structural axis. *Source* Author

Choice of Intermediate Scale

On this basis, we chose structural axis B and its main landmarks as the defined sector to work on hereinafter, so as to provide more detail for the development of the resilience strategies presented on a general scale.

5 Intermediate Scale: Placement Characteristics

5.1 Llahuallín Stream Sector Topography

In the topography of Padre Las Casas, we can identify height variations arising from the existence of old wetlands, as illustrated in the digital elevation model in Fig. 24. These former wetlands are in the lower areas of the city, in a sector containing the axis associated with the Llahuallín stream and La Quebrada Avenue.

On this basis, the topography of the zone lets us intuit the former water flows in this sector as related to wetlands and the stream as shown in Fig. 25, based on local

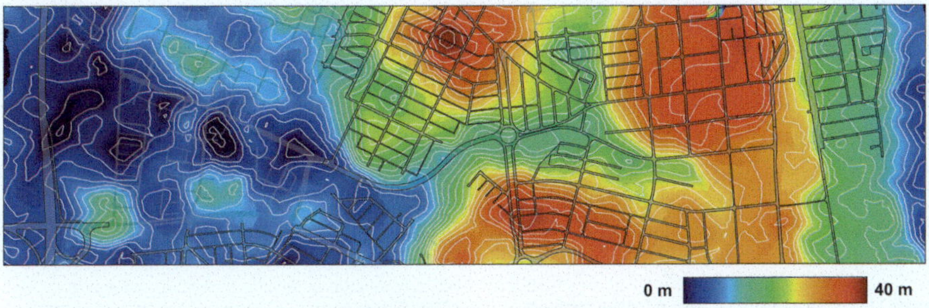

Fig. 24 Digital elevation model (DEM), based on topographical features of the area. *Source* Authors

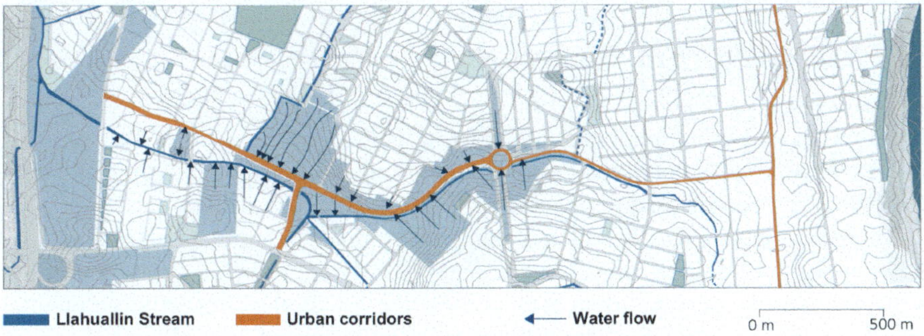

Fig. 25 Water flows, based on local topographical lines. *Source* Authors

topographical lines. By designing this corridor, we can reinterpret these water flows in order to take advantage of their extension and the Llahuallín stream within the city, to capture and manage rainwater and generate multifunctional zones.

5.2 *Natural Spaces Associated with Vacant Lots*

The presence of natural space is of great importance for rainwater management, given the high permeability and biodiversity of these areas. In this sense, the proposed corridor has the potential to connect different natural spaces and generate a continuity of green areas and natural spaces in order to strengthen biodiversity. As illustrated in Fig. 26, the connectivity of natural space crosses much of the city.

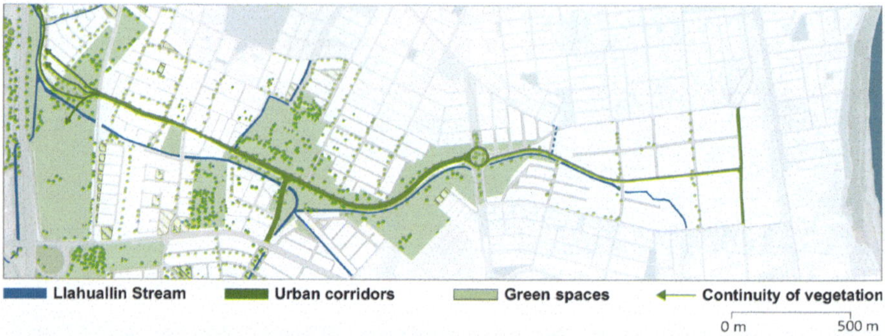

Fig. 26 Connectivity of natural space. *Source* Authors

5.3 Improved Urban Connectivity

Based on current urban permeability, existing natural spaces around the proposed corridor have great potential to become important meeting points within the city, improving connectivity and permeability of the existing urban space.

An agent analysis was conducted using "DepthmapX" simulation software. As shown in Figs. 27 and 28, this software records different possible routes according to the shape of the current urban grid, contrasting it with the connectivity potential of these vacant lots if their accessibility were to be improved, thus allowing them to become important connectivity sites in the city.

Fig. 27 Agent analysis in current urban layout. *Source* Authors

Fig. 28 Agent analysis, potential urban layout. *Source* Authors

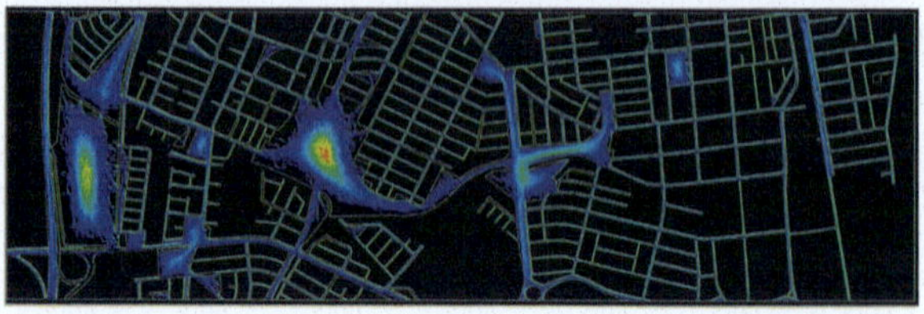

6 Intervention Methodology, Llahuallín Stream Structural Axis

The methodology for the intervention associated with the Llahuallín stream project is grounded in the protection and integration of the ecosystem, seeking to conserve natural environments and existing watercourses associated with vacant lots and extant streams in order to generate a general rainwater management plan associated with five principal aspects:

- **Rainwater capture zones**: Solutions allowing for the gathering and storage of rainwater for future use
- **Rainwater storage**: These help to gather water during the rainy season for reuse in future seasons, whether for household or environmental use
- **Filtration zones**: Places that seek to conserve natural soil permeability by incorporating rainwater gardens in order to avoid runoff and soil surface water accumulation
- **Depuration zones**: Solutions that allow for the filtration of rainwater, removing possible contaminants and sediments dragged by runoff and ultimately permitting the water to return into underground layers or into existing watercourses (Woods et al., 2015).
- **Provision**: Reuse of rainwater in order to maintain vegetation and natural spaces during droughts

Based on this general zoning plan for rainwater management, we seek to incorporate a possible program according to the various characteristics of the sector and its urban environment as shown previously, associated with various land uses and functions.

In this way, as shown in Fig. 29, each design element follows the principle of contributing to rainwater management via the five previously defined aspects. On this basis, we seek to constitute a program and/or possible function within the urban system to contribute to local resilience. For instance, one solution for rainwater accumulation and storage in order to mitigate flooding risk could be to include water pools in green areas in order to avoid runoff and water accumulation in the city. In normal conditions with good weather, these pools can be programmed to be spreading zones that repeat in various sectors and are associated with recreational uses including viewpoints, walking paths, and play areas. This example thus seeks to generate multifunctionality for normal conditions and contribute to redundancy, biodiversity, and adaptive design in case of strong rains.

The graph in Fig. 29 aims to show this methodology for the different solutions presented in the design of the Llahuallín structural axis, part of the blue-green city infrastructure.

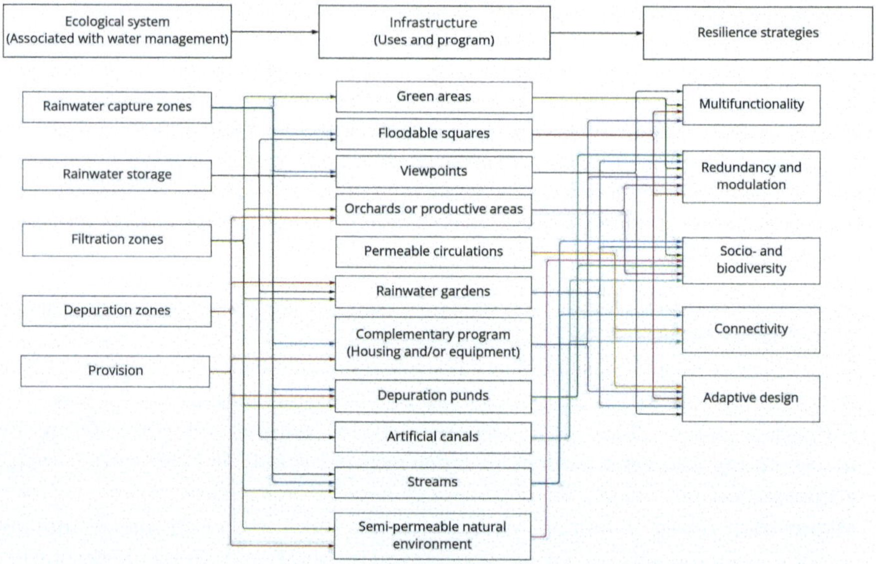

Fig. 29 Methodological schematic for Llahuallín axis intervention. *Source* Authors

7 Intermediate Scale: Llahuallín Stream Structural Axis Project

As illustrated in Fig. 30, the Llahuallín structural axis project seeks to recover the natural spaces that are currently vacant lots in order to use them for rainwater management. The project would take advantage of these spaces' relationship with Llahuallín stream, their biodiversity, high soil permeability, their location in the lowest slopes of the city, and their potential for reconnecting various city neighborhoods, taking advantage of the existence of Avenida La Quebrada to make it an urban corridor.

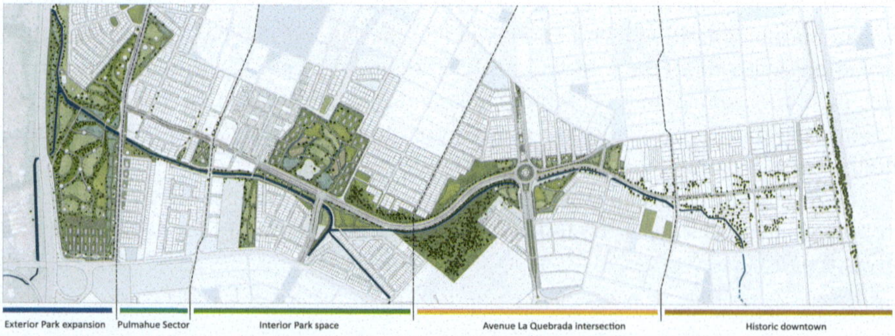

Fig. 30 Llahuallín stream structural axis project. *Source* Author

7.1 Rainwater Management Zoning

Based on the intervention methodology from the Llahuallín structural axis, the first step is planning rainwater management solutions in these sectors, as shown in the water management planning map in Fig. 31. The core goal is to drain waters from residential areas into these natural spaces.

- **Rainwater catchment**: As previously mentioned, water management planning mainly seeks to decrease urban runoff, meaning in this case that it incorporates rainwater capture elements in the natural sectors.
- **Storage**: The objective is to capture and retain rainwater by the different solutions presented and subsequently use it in the different functions associated with the program of each section, as well as for green areas.
- **Infiltration**: We also seek to encourage rainwater infiltration into natural spaces and around Avenida La Quebrada, taking advantage of the high soil permeability in the proposed green areas.
- **Conducting**: The rainwater that cannot be captured and filtered into the soil in the residential neighborhoods around the corridor should be redirected via the avenues, incorporating bioretention strips or other solutions that help evacuate the water from these zones, as well as into Llahuallín stream.
- **Depuration**: For rainwaters that run off into the estuary, the aim is to filter these before they enter the watercourses in order to decrease the possibility of contamination. We thus propose depuration tanks and biofilters around the flow of the Llahuallín stream.
- **Provision**: Finally, we seek to reuse the rainwater captured via the various solutions, as well as the water already in the estuary, for the maintenance of existing green areas and plant species and for domestic purposes associated with the program of each sector.

7.2 Proposed Programmatic Zoning

Historic Downtown

In the first section, the proposal considers the main Street of Avenida La Quebrada, seeking to maintain connectivity between the historic downtown and natural areas. In this sense, there is the possibility to strengthen diverse uses in the place via the intervention of a main avenue, incorporating different types of circulation and resting places.

Avenue La Quebrada Intersection

In this case, the program is associated mainly with recreational uses around Llahuallín stream and this important urban intersection, presenting the possibility of green areas and paths which help protect and maintain local permeability and biodiversity.

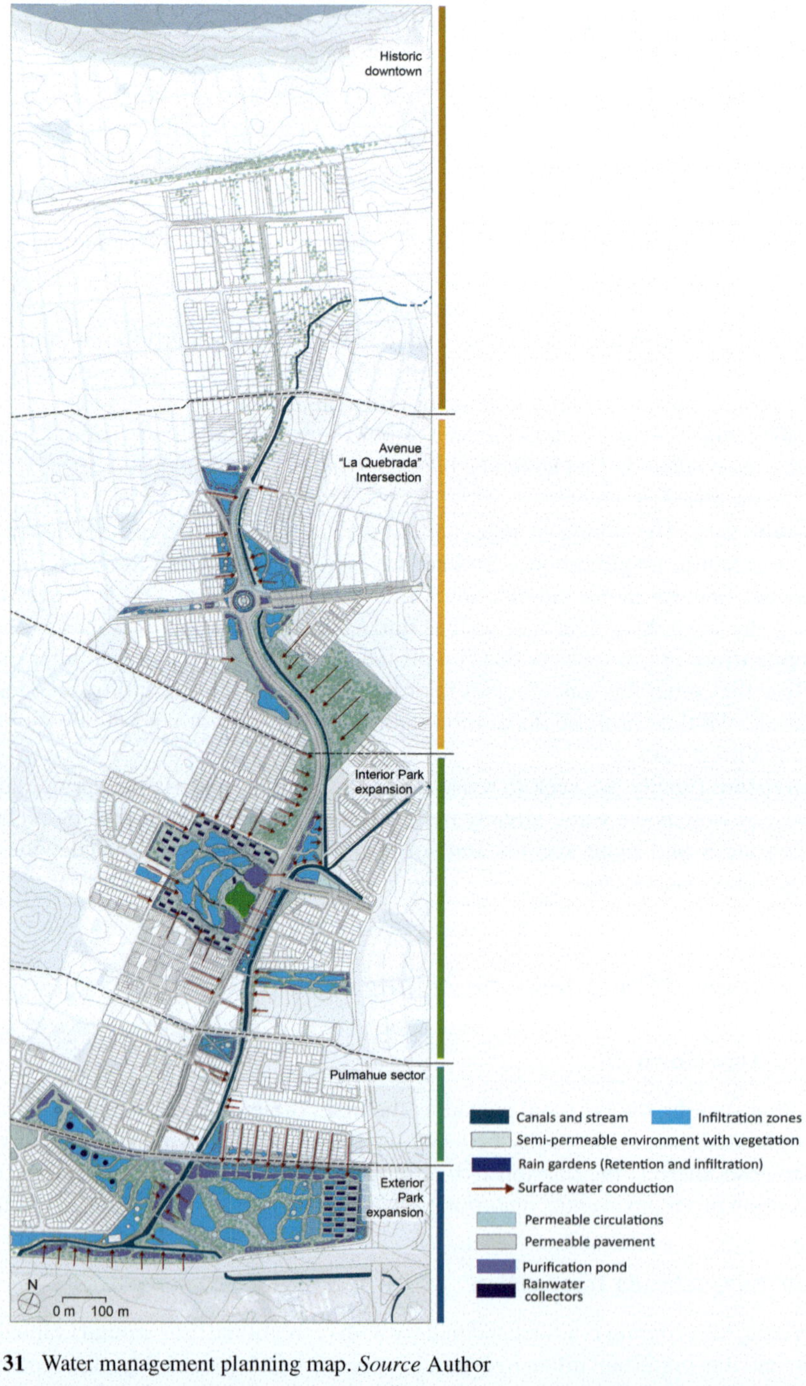

Fig. 31 Water management planning map. *Source* Author

Fig. 32 Proposed programmatic structural axis map for Llahuallín stream. *Source* Author

Interior Park Space

In this section, we seek to activate the preexisting vacant lot by incorporating diverse land use associated with different green areas, plazas, pedestrian malls, and viewpoints. There will also be a complementary housing and equipment program to improve local accessibility.

Pulmahue Sector

In this case, the program is mainly associated with the urban corridor of the avenue, with various circulation types generated (pedestrian, bicycle, and vehicular) in order to improve general corridor connectivity. In addition, extant green areas in this stretch would be regenerated in order to adapt them to rainwater management and for use as recreational spaces.

Exterior Park Expansion

In the last section, we seek to generate an urban park mainly associated with recreational uses and with distribution, featuring a preponderance of green areas and plant species, as well as designing pathways and viewpoints around Llahuallín stream to improve permeability in this urban sector.

We also present the possibility of incorporating a complementary program associated with housing or mainly with equipment in order to activate the sector via diverse land use, generating diverse functioning that could make additional contributions to rainwater management and resilience.

7.3 *Resilience Strategy Planimetry*

We seek to demonstrate the contributions of the presented solutions in rainwater management and programmatic zoning to improving resilience by means of a strategy planimetry. The planimetry shown in Fig. 33 takes into account the five resilience strategies mentioned earlier: multifunctionality, redundancy and modulation, connectivity, biodiversity, and adaptive design.

Multifunctionality

This is related to the various rainwater management solutions presented which fulfill various functions associated with water capture, storage, filtration, depuration, and provision. These give rise to the subsequently proposed program. In this sense, we seek to generate compact zones that can function by generating a direct and close relationship between their components.

Connectivity

We seek to generate connectivity between the different solutions presented via a number of relationship types. To begin, there is a direct relationship by proximity: the water can circulate between different elements, passing from a capture zone towards

Fig. 33 Resilience strategy planimetry. *Source* Author

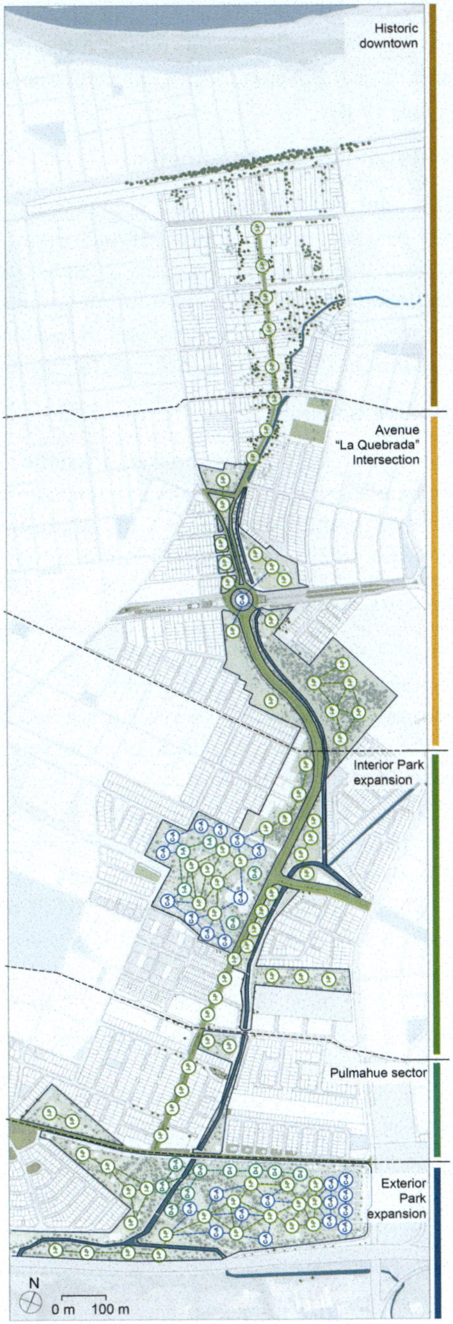

a depuration or filtration zone, or else storing water to supply nature. There is also a larger-scale relationship between the different sections created by the Llahuallín stream or the Avenida La Quebrada corridor, which allows for greater urban-level connectivity.

Redundancy and Modulation

This multifunctionality and connectivity lets us identify modulation between the different solutions, which facilitates their repetition and generates an interconnected system that repeats according to environmental conditions. This redundance also ensures that if one specific solution collapses, there are others to fulfill this function without impacting general system functions.

Modulation between different functions, via a direct relation:

Biodiversity

This manifests via plant conservation through the implementation of gardens, green areas, plazas, and so on. The existence of natural soil and the Llahuallín stream also permits a small- and large-scale relationship between different plant species and natural spaces, contributing to maintaining the green fabric of the city.

Adaptive Design

Finally, adaptive design manifests via the possibility of constituting different water management solutions, as well as the natural environment in a functional program within the city. In this sense, capture and storage devices can be placed in housing, at viewpoints, squares, or other programs. Filtration zones can also be distribution areas, gardens, recreation areas, and more.

This logic repeats with each solution presented, facilitating design adaptation for operations in normal fair-weather conditions, as well as during prolonged rainy periods with flooding or waterlogging risks.

The resilience strategies proposed at this intermediate scale allow us to distinguish between the different sections according to their varying resilience potential, according to the complexity of solutions regarding their multifunctionality and modulation. The most important zones are the exterior park expansion and interior park expansion, which have greater potential and may have a greater positive impact on rainwater management.

On this basis, both sections were taken as examples for design-based research into the possible solutions presented for the place, according to the previously defined strategies.

8 Specific Site Scale

8.1 Specific Section: Exterior Park Expansion

In the first section, shown in Fig. 34, we seek to generate an urban park mainly associated with recreational and distributional uses, which will predominantly have green areas and plant species. In addition, the expansion would involve designing paths and viewpoints around the Llahuallín stream, seeking to improve permeability of the urban area and maintain natural water conditions.

On this basis, we present the possibility of incorporating a complementary program that can be associated with housing or mainly with equipment, in order to activate the sector via diverse land use along the edges of this site. The objective is to generate diverse functions which can contribute to rainwater management and resilience and to balance the need for urban expansion with environmental needs. What follows are three examples of on-site intervention strategies that demonstrate some of the proposed solutions, ranging from an ecological perspective associated with rainwater management to the creation of a possible program.

Capture and Storage Zones

The main water capture and storage zones are infrastructure such as buildings in the case of Z1 (Fig. 35), which can be designed with various programs including high-density housing as well as complementary equipment that increases multi-functionality and vitality within this urban area. In this case, these are three-story buildings with an open ground floor that allows for the possibility of generating a public program and which does not suffer in the case of ground-floor rainwater accumulation. In this sense, water capture can be accomplished via the implementation of green roofs, with the water subsequently stored and used to supply green areas and/or gardens.

The example shown in Z2 is a viewpoint associated with a circulation network, consisting of a tower that allows for the capture and accumulation of water at a high point, which aids in the maintenance of plants during dry periods or droughts.

Filtration and Provision Zones

The filtration zones (Z3 to Z5) are composed of high soil permeability zones which we seek to protect and strengthen by implementing plant species that help with rainwater management via absorption and evapotranspiration, as shown in Fig. 36.

Depuration and Storage Zones (Z6)

Depuration and filtration storage zones (Z6) are composed of bio-retention tanks and/or depuration tanks, as illustrated in Fig. 37, which take advantage of plant species that aid in water filtration in order to return this water into the soil and/or the Llahuallín stream. This temporary water retention also helps control runoff in neighboring spaces, along with the possibility of reusing this water for plants or other functions as needed.

Fig. 34 Exterior park section map. *Source* Authors

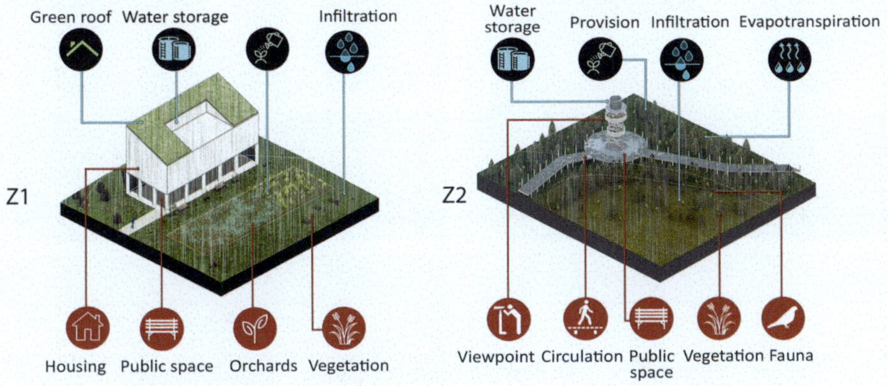

Fig. 35 Exterior park expansion isometrics. *Source* Authors

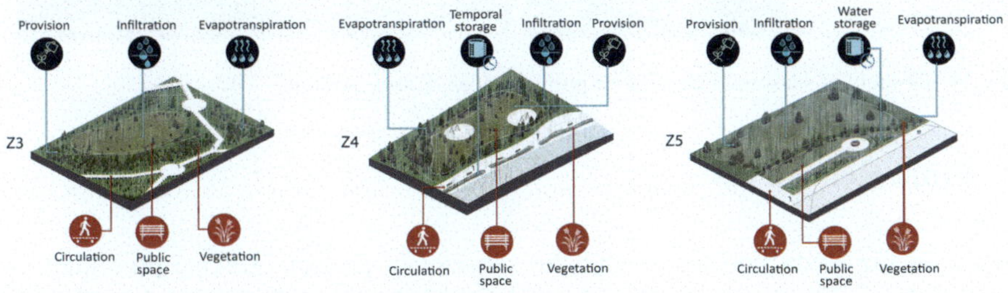

Fig. 36 Exterior park expansion isometrics. *Source* Authors

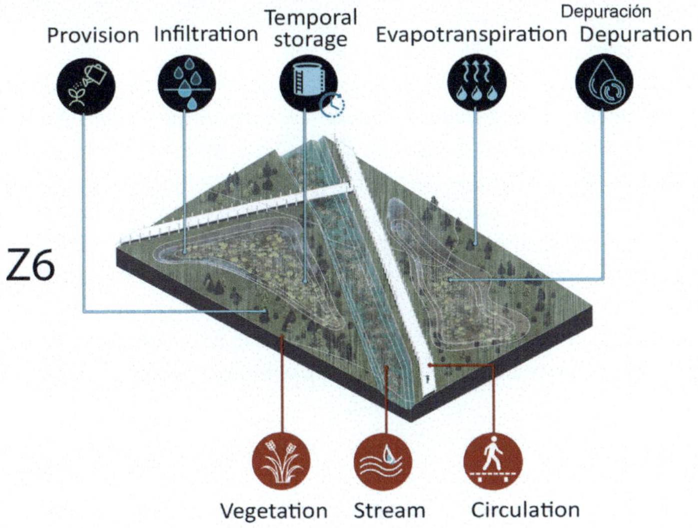

Fig. 37 Exterior park expansion isometrics. *Source* Authors

Fig. 38 Renders of seasonal behavior on public spaces. *Source* Author

9 Final Reflections

Based on the study performed and the subsequent project exercise presented as research by design, it is imperative to understand the dynamics of the natural environment and its importance to the ecosystem, which must be a starting point for urban planning and construction. In this sense, the natural risk that may exist in a given place can be decreased or mitigated significantly via design, intervention and urban planning, depending on the strategies used.

The concept of urban resilience thus supposes a new way of thinking about the city, on the basis of understanding natural systems as something changing and dynamic over time (Espinosa and Agurto, 2021). This leads into the incorporation of the concept of adaptability over time as something fundamental in order to generate a city capable of acting in the face of these environmental transformations.

We can see that the resilience strategies analyzed in this study are not excluding or exclusive for conceiving urban planning and design. On the contrary, they can enrich the creative process by improving final design quality and ultimately human quality of life, whether via natural risk management or as a basis for conceptualizing different programs, functions, and/or urban elements within the city (Fig. 38).

References

Acuti, D., & Bellucci, M. (2019) Resilient cities and regions: Planning, initiatives, and perspectives. In W. Leal Filho, U. Azeiteiro, A. Azul, L. Brandli, P. Özuyar, & T. Wall (Eds.), *Climate action. Encyclopedia of the UN sustainable development goals*. Springer. https://doi.org/10.1007/978-3-319-71063-1_21-1

Ahern, J. (2011). From fail-safe to safe-to-fail: Sustainability and resilience in the new urban world. *Landscape and Urban Planning*, 341–343.

Arenas, F., Lagos, M., & Hidalgo, R. (2010). *Los riesgos naturales en la planificación territorial*. Centro UC Políticas Públicas.

Bruxelles environnement, IBG institut bruxellois pour la gestion de l'environnement. (2012). *Plan de gestion de l'eau, résumé non technique*. Bruxelles.

CCHC. Cámara Regional Temuco. (2019). *Visión ciudad Temuco Padre Las Casas*. Temuco.

Chelleri, L., Waters, J., Olazabal, M., & Minucci, G. (2015). Resilience trade-offs: Addressing multiple scales and temporal aspects of urban resilience. *Enviroment & urbanization*, 181–198.

Christiansen, A. F., & Libby, B. (2017). *Changing cities in a changing climate*. DNV GL Group Technology and Research, Position Paper 1-2017.

Demuzere, M., Orru, K., Heidrich, O., Olazabal, E., Geneletti, D., Orru, H. Faehnle, M. Feliu, E., Mittal, N., & Bhave, A. G. (2014). Mitigating and adapting to climate change: Multi-functional and multi-scale assessment of green urban infrastructure. *Journal of Environmental Management*, 107–115.

Espinosa, P., de Meulder, B., & Ojeda, A. O. (2020). Restauración fluvial como estrategia de diseño urbano. Un diálogo entre investigación y diseño. Concurso río Andalién, Concepción, Chile. AREA, *Agenda de Reflexión en Arquitectura, Diseño y Urbanismo, 26*(1), 1–25.

Espinosa, P., & Agurto, L. (2021). Principios híbridos de diseño urbano (no solo) para adaptarse al cambio climático sino también para combatirlo. El caso del río Andalién en Concepción. In *27FX11 - Aprendizajes e innovación para la resiliencia urbana y riesgo sociocultural Convergencia: ciencia, Estado y ciudadanía Estudios y enfoques para el desarrollo de ciudades resilientes* (págs. 225–243). RIL.

Ghofrani, Z., Sposito, V., & Faggian, R. (2017). A comprehensive review of blue-green infrastructure concepts. *International Journal of Environment and Sustainability, 6*, 15–36. https://doi.org/10.24102/ijes.v6i1.728

Instituto de desarrollo Urbano. (2019). Sistemas urbanos de drenajes sostenibles (SUDS). Foros de sostenibilidad. Bogotá, Colombia.

Laboratorio Cambio Social. (2020). Infraestructura verde-azul y transporte. Balance de transporte justo.

Mahaut, V. (2009). L'eau & la ville, Jardins d'orage. Ecole Polytechnique de Louvain, Departement AUCE - Unité ARCH Architecture & Climat (pp. 55–74).

Ministerio de obras públicas, Dirección de planeamiento (DIRPLAN). (2015). Informe final Temuco y Padre Las Casas. Estudio básico "Análisis brechas de infraestructura urbana MOP en ciudades, etapa I".

Municipio Padre Las Casas. (2019). Informe técnico PLADECO 2020–2025.

Patagua; Legado Chile; Cedeus; Mapa. (2021). Ciudades sensibles al agua. Indaga.

Rojas, P. E. (octubre de 2020). revistaplaneo.cl. Obtenido de. https://revistaplaneo.cl/2020/10/19/del-limite-al-espacio/

Sabatini, F., & Soler, F. (1995). Paradoja de la planificación urbana en Chile. *Revista Eure - Revista De Estudios Urbano Regionales, 21*(62), 61–73.

Territorio Mayor: Centro de estudios ciudad, paisaje y ruralidad. (2015). Estudio socioeconomico y caracterización de comunidades mapuches. *Estudio actualización diagnóstico territorial para modificación del plan regulador*.

Woods, B., Wilson, S., Udale-Clarke, H., Illman, S., Scott, T., Ashley, R., & Kellagher, R. (2015). *The SuDS manual*. CIRIA.

ICELab, a Prototype Module for Environmental Scientists. A Joint Project of the Karlsruhe Institute of Technology (KIT), Germany and the Universidad de Concepción (UdeC), Chile

Renzo Vallebuona and Sophia Schmidt

Abstract The article discusses the challenges of population growth in cities and rural areas and proposes non-invasive architectural strategies that can react to local conditions and enable inhabitation in remote regions. The interdisciplinary team at the Karlsruhe Institute of Technology (KIT) explored the possibilities and limits of building under extreme climatic conditions through case studies in Morocco's Sahara Desert and Tierra del Fuego in Chile. The choice of materials plays a crucial role in meeting complex climatic requirements with simple means, largely dispensing with technical compensation measures. The ICELab project in Patagonia aimed to create a synergy between advanced technologies and local, low-key strategies for optimized adaptation to the specific circumstances of the location. The building serves as both a laboratory and a temporary dwelling for natural scientists, as well as a compact research station for studying changes in the climate. This article highlights the importance of territorial conditions in shaping settlement characteristics and non-invasive building strategies for a particular site, which involve the environment and serve as a catalyst for the development of local construction knowledge.

1 Introduction

Of the approximately eight billion people populating the Earth at the beginning of 2023, almost 60% were living in cities (UN, 2023). All these cities together account for about 1% of the global land mass. Another 3% of the land mass is inhabited by the remaining 40% of the population in rural areas (O'Neill, 2023).

R. Vallebuona (✉) · S. Schmidt
Faculty of Architecture, Karlsruhe Institute of Technology (KIT), Karlsruhe, Germany
e-mail: r.vallebuona@CC-crosscurrents.com; r.vallebuona@vast-architekten.com

S. Schmidt
e-mail: arch.sophiaschmidt@gmail.com

A. Ch. Braun et al. (eds.), *Eco-Industrial Development as an Industrial Strategy*, Green Energy and Technology, https://doi.org/10.1007/978-3-031-73576-9_26

Although these areas are not concentrated in one single place, large portions of the remaining 96% of the Earth's surface have been shown to be unsuitable for settlement. These areas are considered inhospitable or at least unattractive, for example due to extreme temperatures, humidity, or dryness, height above sea level, or distance to water. The planet's constantly growing population continues to be concentrated in urban centers that were established centuries ago in order to allow for interconnectivity and cooperation among the population and thus meet basic needs for shelter, nutrition, social fabric, and education (Bosker & Buringh, 2017).

Although globalization and digitalization increasingly enable participation and networking even without physical contact, the unrestrained rural exodus increasingly causes urban growth beyond the limits of what urban infrastructures and policies are able to bear (Fig. 1). In countries of the Global South in particular, this heterogeneous densification is immanent and leads to precarious conditions and mass migration (Randolph & Storper, 2023; Friesen et al., 2018).

One approach to dealing with this problem is to ameliorate the inhabitability in those regions: not through aggressive urban development, but on a small scale with the help of non-invasive architectural strategies that react to the respective local conditions and learn from them.

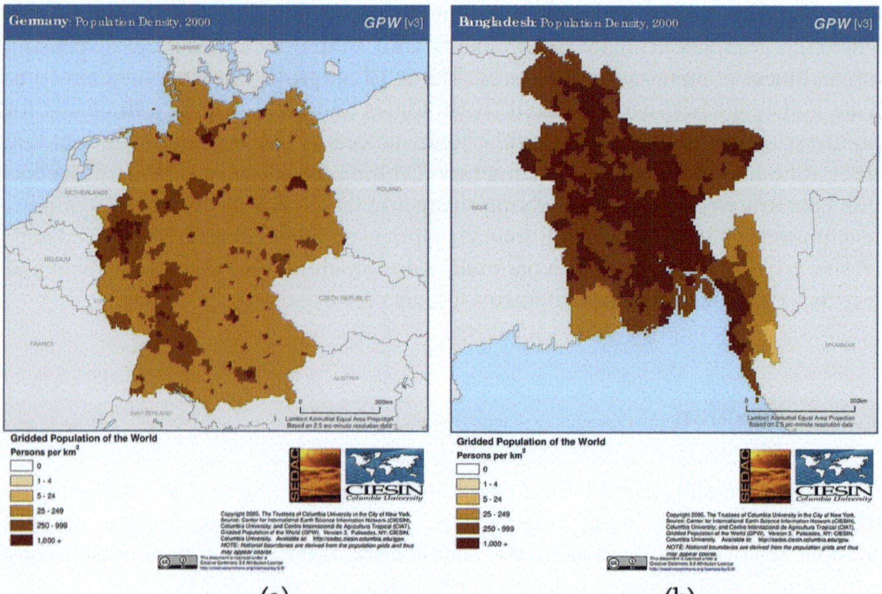

(a) (b)

Fig. 1 Comparison of uncontrolled population density in **a** Germany, and **b** Bangladesh. In NASA SEDAC population estimator, n.d., retrieved February 17, 2022, from https://sedac.ciesin.columbia.edu/mapping/popest/gpw-v4/

In order to find sustainable solutions for remote and newly developed regions, an interdisciplinary group headed by Prof. Vallebuona at the Department of Structural Design Methodology of the Karlsruhe Institute of Technology (KIT) has set itself the task of exploring the possibilities and limits of building under extreme climatic conditions by setting up various case studies. Using a first construction project in M'Hamid in the Moroccan Sahara Desert and a second design project in Tierra del Fuego, southern Chile, non-invasive approaches were investigated for this purpose, where the adverse circumstances of the environment were considered as the determining design parameters.

In general, are territorial conditions such as geographic location, climate, or topography that shape the specific characteristics of a settlement and define the appropriate building strategies for a particular site. The strategies that can be derived from these realities are therefore rarely just self-referential, but are predominantly understood as a collective endeavor and involve the environment.

The resulting constraints serve as a catalyst for the further development of local construction knowledge and existing low-key construction strategies. The clever use of materials leads to the ability to meet complex climatic requirements with very simple means and to largely dispense with technical compensation measures (Fig. 2).

The project in the Sahara Desert was realized by using a local clay deposit near an oasis. The massive structures provide shade and are climate-regulating due to their mass. They store the coolness of the night and release it during the day.

A timber structure was developed for the project in Tierra del Fuego. In contrast to the local availability of the material in the Sahara oasis, building materials in Tierra del Fuego cannot be sourced locally due to nature conservation regulations. In addition to the extreme cold and wind loads, the driving design factor in this region was logistics, which mostly involves deliveries by ship or plane.

The second of these projects was developed within the framework of the cooperation partnerships between KIT and universities of Latin America accompanied by IECO, and could be further developed into full-scale, real project.

(a) (b)

Fig. 2 Use of imported building materials that are not suitable for the climatic conditions **a** Medina of Casablanca, Morocco. Local building technologies **b** Aït-Benhaddou, Morocco

2 Basic Decisions

The basic idea behind the ICELab project is to create a synergy between the use of advanced technologies and high-tech products with local, low-key strategies and optimizing this to the specific circumstances of the location.

While infrastructure and energy self-sufficiency offer advantages in this regard, it was still important for the design of a sustainable future scenario to consider new lifestyles and their impact on constantly changing housing patterns. To this end, it was decided early on to consider a dual program for the design of the module: On the one hand, it is to be used as a laboratory for natural scientists, while also serving as a temporary dwelling for them. This allows for the study of movement in small spaces, temperature variations, functional aspects, etc., in order to obtain results for envisaged future building strategies in Patagonia.

In addition, the building contains a compact research station with which changes in the climate can be studied and evaluated by researchers directly on site. Secondly, however, the building itself represents an experimental object, the monitoring of which provides information about the long-term effects of Patagonia's climatic conditions on the behavior and aging of the materials used in its construction. For this purpose, untreated wood was used as the main construction material and a material newly developed in Germany, 3C composite aluminum panels (Alucobond), was used as wind- and weather-resistant facade cladding.

3 Preparatory Investigations

In parallel, the search for a suitable location within Patagonia for the ICELab took place in cooperation with the Universidad de Concepción (UdeC) and the university's EULA Centre for Environmental Sciences. A suitable site was identified in Caleta Tortel on the Tortel Fjord in Patagonia, southern Chile. Here, the ICELab can be functionally connected directly to the on-site research station for oceanography and glaciology of the Universidad de Concepción.

In its morphological and constructive characteristics, the fishing village of Caleta Tortel reflects the characteristic features of the traditional settlements in the region of Chiloé and Patagonian fjords. The fertile humus soil layer on the predominantly granite topography is mostly only a few centimeters thick. Thus, a vegetation spectrum is defined which represents a unicum of flora and fauna and gives the soil a unique value, culturally speaking.

Construction projects on the shore of the fjord may only be implemented under the condition that the soil is not sealed. Therefore, elevation on stilt bearings is required as a "mediator" between the ground and the building. The removal of soil will be determined on a case-by-case basis, depending on location and exposure to sunlight, so that the vegetation layer is not interrupted by buildings. This strategy is also used for infrastructure such as paths, roads, pipe laying, and other municipal facilities.

Fig. 3 Network of footbridges and wooden stilts as elementary nature protection

Caleta Tortel thus consists of a network of wooden footbridges (Fig. 3) and wooden buildings on stilts, which extends over both the land area and the near-shore waters. This attitude, anchored in the local cultural history, became the constructive concept of the ICELab structure.

Due to the geographical distance, the project determinants were first investigated in Norway, since many areas there have a high climatic similarity to the southern tip of Chile and the physical parameters required for the design could largely be adopted. The data collected there provided the prerequisites for further project work (Figs. 4 and 5).

4 Drafting

To begin, various design concepts were developed in teams, each of which was tested for its performance under the site factors, iterated further, and adapted to the material specifications. Step by step, the findings were merged into a common design for the research unit with a size of approx. 25 m^2 (Fig. 6).

The floor area of the unit is approx. 8.50×3.80 m, while the total height is 4.00 m. Most of the area is used as a laboratory, including a small wet laboratory. The service areas are oriented lengthwise. A shower directly at the entrance allows the scientist to clean his or her equipment and clothing before entering the module.

In the longitudinal direction, the structure is divided into two functional areas. About two-thirds of the building's width is taken up by the drying laboratory, which is also designed to accommodate research staff. Wall shelves running longitudinally along the facade can be rotated 90° into the room and structure it as a room divider if required. The remaining third of the space contains a wet laboratory and all other necessary wet rooms such as a WC, shower, and kitchen. The building services

Fig. 4 Construction of the prototype without the use of large equipment in Caleta Tortel

(a) (b)

Fig. 5 Material research in Norway: weather protection based on high-tech film (**a**); alternative weather protection based on moss (**b**)

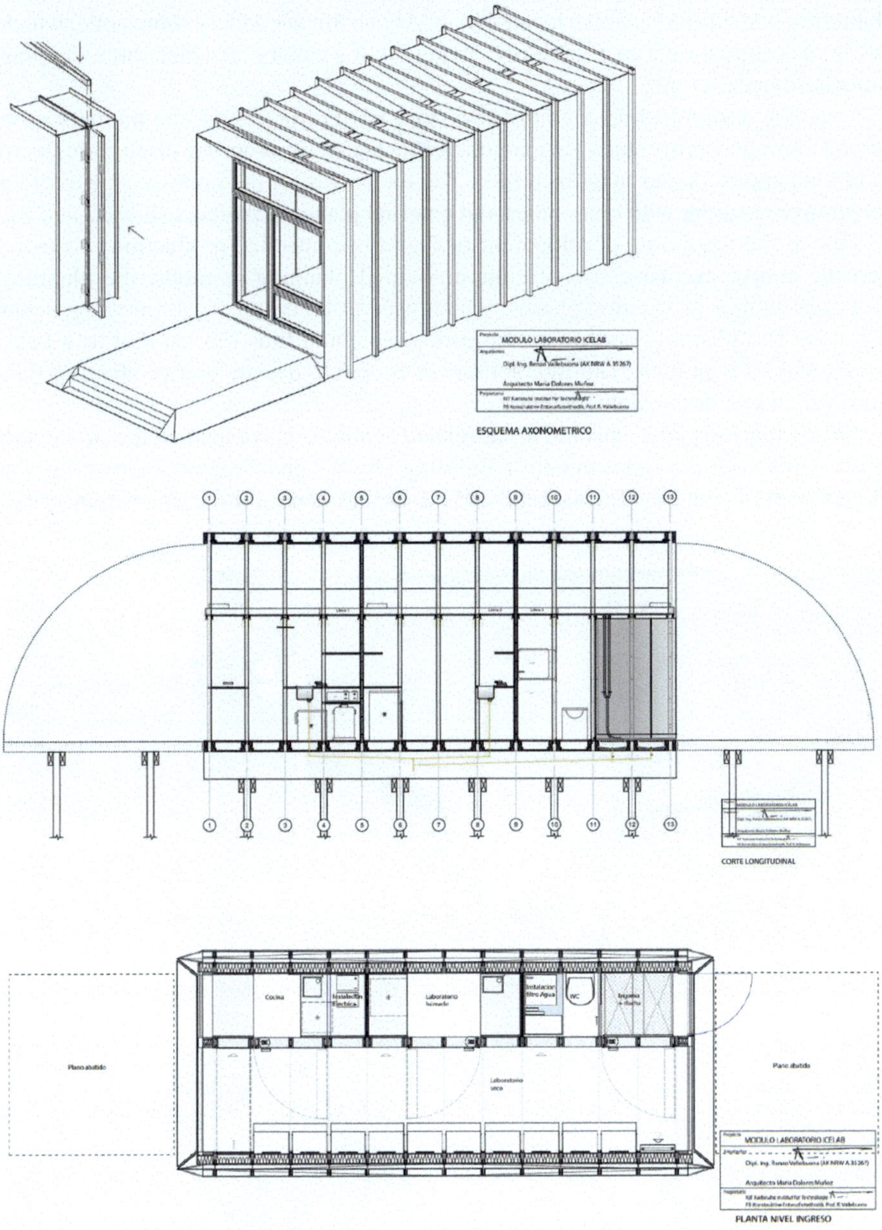

Fig. 6 Floor plans and sections design

(batteries, water filter) are also located here. Above this secondary room zone, which does not occupy the entire available room height, is a gallery on which three sleeping units have been set up.

Once the design had reached the necessary level of precision, the findings were turned into a feasibility study for manufacturing and the appropriate discussions were held with sponsors and manufacturers. The focus was on the premise of enriching integrative teaching with companies and external planning partners (Figs. 7 and 8).

Due to the increasing development of digitally controlled production processes and the almost exclusive use of computer-aided planning methods, the planning and execution or production phases, which used to be far apart, are moving closer together. The planning and execution processes could thus run parallel to a large extent and be kept lean. The interactions of research, design, and production thus grew into a seamless whole.

On the teaching side, specific topic-related seminars accompanied the process in cooperation with the Department of Building Physics and Technical Finishing, the Department of Building Technology, and the Department of Structural Engineering.

Fig. 7 Use of high-tech compound cladding material (left) and laboratory wind tests (right)

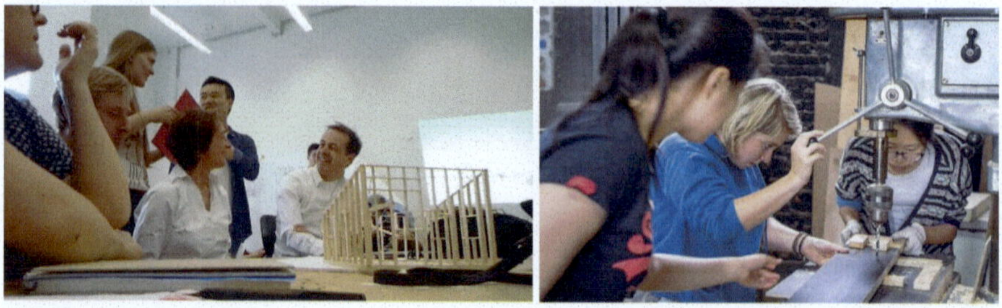

Fig. 8 Integrative teaching

4.1 Prototyping

The prototype "ICELab Pilot Unit" was prefabricated by KIT architecture students in a hangar in Karlsruhe, Germany. The structure was initially erected provisionally, documented and numbered in all parts, then dismantled again and shipped to Chile in a container.

The design of the structure had to take into account the fact that, in addition to truck and cargo ship transport, there were limited means of transport available for the last kilometers to the fjord at Caleta Tortel. Therefore, the constructive elements of the ICELab are dimensioned in such a way that they can all be transported by hand by one or two people, making logistics and assembly independent of infrastructure networks to allow for interventions in rural areas.

Due to the strict requirements in Patagonia, the prototype built in Karlsruhe required a substructure of stilts similar to the local footbridges, which was erected on site a priori by local craftsmen. On this basis, a team of KIT students, alumni, and faculty were able to build the laboratory unit as planned (Figs. 9, 10 and 11).

4.2 Monitoring

The unit itself is the subject of the project, in that it serves to research both the construction method, structural design, and building materials used through long-term monitoring. The laboratory is designed as a wooden skeleton construction in accordance with the local construction technology, so that the results of the monitoring can find local application.

For this purpose, a facade of metal cladding made of aluminum compound elements (Alucobond) was developed, which can be installed or removed without special tools and without special knowledge (Fig. 12).

The roof and facade were designed to collect precipitation water, which after UV filtration is treated for drinking use in the technical room. Waste water is filtered and collected separately in a special tank. The electrical distribution of the wind unit and the batteries is also carried out in the technical room.

The COPAS Sur-Austral scientific group of the Faculty of Natural Sciences and Oceanography (www.sur-austral.cl) of the Universidad de Concepción, led by Prof. Silvio Pantoja, will primarily use the unit as a laboratory in addition to the existing COPAS headquarters in Tortel (Fig. 13). This has resulted in modifications to the originally planned energy supply, rainwater harvesting, and water treatment/natural clarification.

A series of empirical tests are being carried out on site on the ICELab facade as part of upcoming collaborations to investigate a thermal insulator made of high-tech membranes and virtually untreated natural wool (low-key), which is produced in large quantities as industrial waste in the Magellan region of Chile. Other natural materials as well as local renewable resources are also part of this investigation.

Fig. 9 Prototype construction on the site

The results of this study, as well as those of the construction monitoring, will be evaluated and cataloged by KIT in collaboration with the Universidad de Concepción and made available to the local construction planning authority.

Fig. 10 Logistics: transport of the prefabricated components from the storage yard in Germany to the construction site in Caleta Tortel

Fig. 11 The prototype serves as a test bench for various insulation systems made primarily of local industrial waste and resources from the region

Fig. 12 Brackets to hold the Alucobond facade cladding

Fig. 13 ICELab and COPAS

5 Outlook

Donations in kind and know-how from various companies and technical support from experts in all trades have made this project possible. Likewise, the commitment of teachers from both participating universities and the support of IECO were essential for the realization.

Access to the ICELab infrastructure is open to scientists from various national and international institutions as well as KIT. The lab will be part of the activities to improve the visibility of such international collaborations and to disseminate natural and marine sciences in the region. A traveling exhibition is planned to show the results of the monitoring in the Magallanes region and at both participating universities (KIT and UdeC).

References

Bosker, M., & Buringh, E. (2017). City seeds: Geography and the origins of the European city system. *Journal of Urban Economics, 98*, 139–157. https://doi.org/10.1016/j.jue.2015.09.003

Friesen, J., Taubenböck, H., Wurm, M., & Pelz, P. (2018). The similar size of slums. *Habitat International, 73*, 79–88. https://doi.org/10.1016/j.habitatint.2018.02.002

O'Neill, A. (2023). Global share of urban and rural population 1960–2020 | Statista. Published by Statista. https://www.statista.com/statistics/1262483/global-urban-rural-population/, last checked 28 February 2023.

Randolph, G. F., & Storper, M. (2023). Is urbanisation in the Global South fundamentally different? Comparative global urban analysis for the 21st century. *Urban Studies, 60*(1), 3–25. https://doi.org/10.1177/00420980211067926

UN. (2023). Data Portal. Population Division. Hg. v. United Nations. https://population.un.org/dataportal/, last checked 17 February 2023.

Land Use and Industrial Sustainability—Sustainable Land Use—Introduction

Land Use and Industrial Sustainability—Sustainable Land Use—Introduction

Land Use and Sustainability: A Critical Introduction to the Chilean Forestry Context

Andrea Pino Piderit, Ricardo Rozzi, and Andreas Ch. Braun

Abstract This chapter critically examines the interplay between land use and industrial sustainability, focusing on the Chilean forestry sector, providing the context to understand the relevance of the IECO Topic V—"Land Use and Industrial Sustainability". The present introduction chapter highlights the shortcomings of the 2030 Agenda for Sustainable Development, particularly regarding land degradation, soil health, land tenure, and indigenous rights. These aspects are crucial for promoting sustainable land use and social equity, yet they remain inadequately addressed in the Agenda. The chapter illustrates how political and economic factors have driven the expansion of the forestry industry in central Chile, leading to significant environmental impacts, including biotic homogenization due to the establishment of monocultures of fast-growing exotic species like *Pinus* and *Eucalyptus*. The authors advocate for a more nuanced understanding of sustainability that incorporates diverse definitions and dimensions, emphasizing the need for policies that balance competing land uses and respect indigenous land rights. By drawing on ecosystem principles, the chapter proposes strategies for achieving sustainable industrial practices that decouple economic growth from environmental degradation. Ultimately, it calls for a reevaluation of current practices and policies to foster a more equitable and sustainable future for both people and nature in Chile.

A. P. Piderit (✉)
Transdisciplinary Center of Environmental and Sustainable Human Development Studies, Austral University of Chile (CEAM-UACh), Campus Isla Teja, Valdivia, Chile
e-mail: andrea.pino@uach.cl

R. Rozzi
Cape Horn International Center (CHIC), Universidad de Magallanes, University of North Texas, Denton, TX 76205, USA
e-mail: Ricardo.Rozzi@unt.edu

A. Ch. Braun
Kassel Institute for Sustainability, University of Kassel, Mosenthalstr. 8, 34117 Kassel, Germany
e-mail: andreas.braun@uni-kassel.de

A. Ch. Braun et al. (eds.), *Eco-Industrial Development as an Industrial Strategy*,
Green Energy and Technology, https://doi.org/10.1007/978-3-031-73576-9_27

1 Scope of This Contribution

This part of the book, which deals with industrial land use and sustainability, has a slightly different approach than the other parts. This requires explanation, as it may otherwise be unclear how this part fits into the context. While the other Parts are rather "pro industry", this part is explicitly skeptical of the forest industry in Chile.

This book deals with the research results of the German–Chilean Institute for Eco-Industrial Development (IECO) project, which focuses on strategies to make industry more sustainable by imitating ecosystem principles. The book predominantly takes an approach that is positively appreciative of industry and follows a kind of "green growth" logic, namely to keep business and industry oriented towards growth while reducing environmental impacts. This can be achieved, for example, by decoupling growth from emissions and reducing the material footprint. Even if new research shows that the idea of such green growth is not realistic in most countries and has not yet been achieved (Vogel & Hickel, 2023), this research direction is well justifiable. After all, even the most comprehensive Agenda 2030 adheres to such a growth ideology (central in SDG 8—Decent Work and Economic Growth). The chapter by Braun (2024) "Is Eco-Industrial Development Just "More of the Same"? A Hegelian Dialectic View on an Emerging Concept", for example, takes a critical look at the strategy of eco-industrial development.

Especially in the context of Chile, however, this approach must not be maintained without pause. For in Chile there is a very important industrial sector—the forestry industry—which seems to meet precisely such sustainability principles. Many authors regard forest plantations as "planted forests" and thus inherently as an imitation of an ecosystem. Through its own presentation but also in some political discourse, the forest industry is often and willingly presented as a "sustainable eco-industry." The part "Sustainable Material and Energy Transformation of Biomass—Case Studies from Chile and Germany" of this book also contains contributions that address how the forest industry can be further optimized in terms of its sustainability performance through the use of eco-industrial strategies.

However, it should not be forgotten that these considerations often refer to internal aspects. What is often ignored is the connection between industry and land use. The forest industry is extremely space-intensive, with monotypic plantations of exotic, fast-growing tree species, Monterrey pine (*Pinus radiata*) and *Eucalyptus* species especially, covering large portions of the coastal mountains and valleys. Since 1974, Chile's forest industry has been subsidized by Government Decree Law 701, which provides for 75% to 90% of the cost of planting and gives tax exemptions for the products obtained from these plantations (Pastor-Barrué, 2004). The plantations interact and are often in conflict with other forms of land use. There are thus numerous impacts here that are indirectly mediated through land use and are relevant to sustainability (Miranda et al., 2017).

The chapters of this part are dedicated precisely to these problems. They question a supposed "eco-industry" that at first glance at least seems to combine ecological goals with economic sustainability.

2 Introduction: Land Use and the SDGs

When discussing the topic of land use and sustainability, reference should be made to how land use is addressed in the most comprehensive framework on sustainability, the United Nations' 2030 Agenda for Sustainable Development. Adopted in 2015, the 2030 Agenda is a global blueprint for achieving a more sustainable and equitable future. At its core are the 17 Sustainable Development Goals (SDGs) that address a wide range of pressing issues, from poverty eradication to environmental conservation. Among these goals, the role of land use stands out as a critical component, explicitly and implicitly intertwined with several SDG targets. This Section delves into the multifaceted relationship between land use and the 2030 Agenda, highlighting the SDG targets that explicitly and implicitly address land use while also shedding light on critical aspects of land use and sustainability that are not adequately reflected in the Agenda.

Explicit references to land use are found in the following SDGs of the 2030 Agenda:

- SDG 2: Zero Hunger. Target 2.4 explicitly addresses sustainable agriculture, emphasizing resilient agricultural practices and productive land use.
- SDG 6: Clean Water and Sanitation. Target 6.6 aims to protect and restore water-related ecosystems, recognizing the integral role of land use in water quality and availability.
- SDG 11: Sustainable Cities and Communities. Target 11.3 focuses on land use planning, emphasizing inclusive and sustainable urbanization, and reducing the environmental impact of cities.
- SDG 13: Climate Action. Target 13.1 underscores the need for integrated climate change measures into policies, strategies, and planning, which directly influences land use practices.
- SDG 15: Life on Land. Almost all targets within SDG 15 are explicitly tied to land use, as they aim to protect, restore, and sustainably manage terrestrial ecosystems and halt biodiversity loss.

Implicit references to land use in the 2030 Agenda are also ubiquitous:

- SDG 1: No Poverty. While not explicitly mentioning land use, the eradication of poverty often involves access to land for agricultural or housing purposes, making it implicitly linked to land use.
- SDG 7: Affordable and Clean Energy. The expansion of renewable energy sources may require changes in land use, such as the installation of solar panels or wind turbines.
- SDG 9: Industry, Innovation, and Infrastructure. Infrastructure development often involves land conversion, which can impact ecosystems and communities.

However, there are also several critical aspects of land use not adequately addressed in the 2030 Agenda. One significant oversight is the topic of land degradation and soil health, which are critical for sustaining agriculture and ecosystems.

Another neglected topic is land tenure and property rights. The importance of secure land tenure and property rights is not explicitly emphasized in the Agenda, despite their role in promoting sustainable land use and social equity. Moreover, land use trade-offs should have been addressed more explicitly. Balancing competing land uses, such as agriculture, urbanization, and conservation, requires more explicit attention in the Agenda to prevent conflicts and ensure sustainable outcomes. Finally, there is a large regulation gap regarding indigenous land rights and traditional knowledge, topics inadequately addressed in the Agenda. For our Chilean example, the topics of land use tradeoffs and the neglect of indigenous land rights and traditional knowledge are particularly crucial. Therefore, it is imperative that policy makers and stakeholders consider these aspects as they work towards achieving the SDGs by 2030.

3 The Struggle for Definitions

Sustainability is a critical concept in environmental science and land management, emphasizing the need to balance human needs with environmental protection and ecosystem health. This concept has various definitions and interpretations, often influenced by the context in which it is discussed, the scale of land use, and the specific objectives involved.

Given the vast variety of different land uses, it is challenging to provide a meaningful operational definition of land use sustainability. Nonetheless, some key dimensions may be identified. A commonly shared feature of sustainability definitions focuses on the idea of meeting the needs of the present without compromising the ability of future generations to meet their own needs. This definition is derived from the broader concept of sustainable development, as articulated by the Brundtland Commission's 1987 report, "Our Common Future." In the context of land use, this definition implies that land must be utilized in a way that preserves its productivity, biodiversity, and ecological functions over time. In addition, this intertemporal aspect (future generations) is amended by an interspatial aspect of not compromising the needs of people living in other places. Both aspects (intertemporal and interspatial) are very frequently violated in the face of excessive resource use, land grabs, and so on.

3.1 Provision of Ecosystem Services

A more nuanced definition, presented for example by Abson et al. (2014) and Geijzendorffer et al. (2017), emphasizes the importance of ecosystem services in the context of land use sustainability. According to this view, sustainable land use must maintain or enhance the provision of essential services such as water purification, carbon sequestration, soil fertility, and habitat for biodiversity. This approach suggests that sustainable land use requires a comprehensive understanding of the interactions between human activities and ecosystem processes.

3.2 Balances and Tradeoffs Across Scales

Another perspective on land use sustainability, discussed by Agarwal et al. (2002) and DeFries et al. (2004) to name but a few, considers the spatial and temporal scales at which land use decisions are made. This definition stresses the importance of planning and managing land use in a way that is adaptive and resilient to changing conditions. Related studies highlight the need for flexibility in land use policies to accommodate shifting ecological and societal dynamics such as climate change, urbanization, and demographic trends.

3.3 Socio-economic Dimensions of Land Use

In contrast, a socio-economic approach to land use sustainability, as outlined for example by Díaz et al. (2011) and Berkes et al. (2000, 2008), emphasizes the integration of social equity and economic viability into sustainable land use practices. This definition incorporates the concept of sustainable livelihoods, suggesting that land use must support the well-being and economic stability of local communities. It underscores the importance of engaging stakeholders, including local farmers and indigenous groups, in land use planning and decision-making processes to ensure that sustainability goals are met equitably. It is also worth recalling John Rawls' definition of justice (1971), who argued that unequal distribution of goods must be designed in such a way that even the least advantaged must still benefit from this distribution. According to Rawls, there must therefore be no distribution of goods that increases the sum of benefits but completely favors some while only disadvantaging others (separateness of persons argument). A very obvious example where this principle would be violated in terms of land use are so-called sacrifice zones (Scott & Smith, 2017).

3.4 Relevance of Governance Mechanisms

Land use sustainability can also be defined through a legal and governance lens, as proposed by Lambin et al. (2014) and exemplified by Moss (2004). This perspective focuses on the regulatory frameworks and policies that guide land use decisions. In this context, sustainable land use is achieved through robust governance mechanisms that enforce environmental regulations, promote public participation, and ensure transparency in land use planning. This approach underlines the critical role of governments and institutions in shaping sustainable land use practices.

3.5 Resilience

More recent contributions by Cummings (2011) and Schmidt (2022) focus more on the resilience aspect of the spatial systems involved, whereby resilience—just like sustainability—would have to be defined in turn. In our view, the innovative contribution of the resilience concept is that, in contrast to sustainability, it focuses less on the preservation of systems and more on their long-term adaptability to stress and shock. Landscapes—understood as socio-ecological hybrid systems—are resilient if they remain permanently in an integral, at least metastable equilibrium, instead of finding themselves in a degraded equilibrium by crossing tipping points.

3.6 Multispecies Integration

Rupprecht et al. (2020) have aptly reminded sustainability researchers that we are not alone on the planet and that sustainability definitions such as triple bottom line cannot be specifically geared towards humans alone. Landscapes, including utilized landscapes, are inhabited by a variety of organisms that have intrinsic rights to continue to inhabit these landscapes, even if they do *not* provide ecosystem services. In terms of sustainable land use, this means that biodiversity and ecosystem functions must be preserved as an intrinsic value.

One common thread among these definitions is the need for long-term thinking and adaptive management. The dynamic nature of ecosystems and human societies means that land use strategies must be flexible and able to respond to changing conditions. This requires ongoing monitoring, evaluation, and stakeholder engagement to ensure that land use practices remain sustainable over time. In conclusion, land use sustainability encompasses a range of definitions and perspectives, each offering valuable insights into how land can be managed to meet human needs while protecting the environment. Achieving sustainable land use requires a collaborative and adaptive approach, integrating ecological, social, economic, and governance considerations. As land use pressures continue to grow, the importance of a comprehensive and flexible approach to land use sustainability will only increase. In sum, a stipulative definition of land use sustainability could be as follows:

> Land use is sustainable when ecological landscape is used in such a way that ecosystem functions and ecocenoses remain intact in the long term, thus enabling ecosystem services that allow social-ecological systems that are resilient to stress and shocks. These social-ecological systems must be managed in a fair and participatory manner, which means that the capital generated must be distributed in such a way that all stakeholders benefit and that

the use does not transgress at either the spatial or temporal scale in such a way that other social-ecological systems are permanently and adversely affected.

We are aware that although this definition is comprehensive on the basis of the aspects and definitions presented, it is nevertheless very demanding in normative terms. This definition may therefore be rejected for being unrealistic or idealistic. However, we would like to remind you that weaker requirements, such as the triple bottom line, have led to a kind of "sustainability cherry-picking" in which different actors have declared any contribution to one dimension to be sustainable overall.

4 Eco-industry? Impacts of the Forest Industry in Chile

The forest industry in Chile is a very important sector and is often and willingly presented as a "sustainable eco-industry" when it is very far from being so. This industry is extremely space-intensive, with plantations that cover large portions of the coastal mountains and valleys where they interact with and often conflict with other forms of land use. Furthermore, in the forestry industry there is no cooperation between companies as established in the eco-industrial system (Guo & Cui, 2010).

During the pre-European period, the Valdivian Rainforest eco-region (35°–43°30′ S) of Chile covered approx. 11.3 million hectares (ha), but by 2007 this had decreased to 5.7 million ha due to the conversion of native forests to pasturelands, shrublands, and agriculture land (and, since 1974, forest plantations of exotic species), a value that represents 51% of the original coverage (Lara et al., 2012). The principal changes were a drastic decrease in native forest cover and the emergence of large-scale agriculture and forest plantations. While in 1970 there were only 300,000 ha of plantations (Lara & Veblen, 1993), by 2007 this had reached 2.4 million ha, of which 1.8 million were covered with native forest. This was the result of a favorable economic and political context, principally the Decree Law 701 (DL 701) enacted in 1974 (Nahuelhual et al., 2012). Furthermore, this decree benefits the three largest forestry companies: Between the 1970s and early 1990s, 80% of the subsidy payments of DL 701 went to Chile's three largest forestry companies (Nahuelhual et al., 2012).

The establishment of forest plantations has caused significant landscape changes in the areas where they have been established (Bremer & Farley, 2010; Salas, et al., 2016). The forestry incentives can affect the investment decisions of small forest landowners and bring socioeconomic externalities or unintended effects associated with farm management (Bopp et al., 2020).

The temperate forests of Chile are classified as a world hotspot of biodiversity due to their high species diversity and levels of endemism (Armesto et al., 1998; Arroyo et al., 2004; Myers et al., 2000). But in Chile the areas with the greatest species richness of native forests are mainly converted to exotic tree plantations. The Chilean forestry model has proven successful in expanding exotic tree plantations, but so far it has not been compatible with native forest conservation and restoration (Miranda et al., 2017).

In the 1990s, Sala et al. (2000) identified land use change as the main driver of global change causing biodiversity losses. In the 2020s, it was documented again that during the late twentieth century and early twenty-first century land use change has been the dominant direct driver of recent biodiversity loss in terrestrial and freshwater ecosystems worldwide (Díaz et al., 2019; Esse et al., 2021; Jaureguiberry et al., 2022). Land use change has multiple and major impacts on species population and composition as well as on ecosystem structure and function. For example, Barlow et al. (2007), Bernard et al. (2009), Andrade-Núñez and Aide (2010), Bremer and Farley (2010), Pawson et al. (2013), Bergner et al. (2015), Iezzi et al. (2018), and Heilmayr et al. (2020) indicate that the replacement of native forests by non-native plantations decreases diversity and changes the composition of assemblages of terrestrial mammals and ground-dwelling and understory birds in a forest-productive landscape.

The massive replacement of these native forests by exotic monocultures has caused not only severe losses of biodiversity but also erosion and compactness of the soil as well as conditions detrimental to hydrological cycles, causing floods during the winter and droughts during the summer (Aburto et al. 2021; Bathurst et al., 2011a, 2011b, 2022; Birkinshaw et al., 2011; McWethy et al., 2018; Van Holt et al. 2012). In watersheds on the coastal mountain range of central-southern Chile, the replacement of native forest with fast-growing forest plantations reduced water yield and produced more sediments (Alaniz et al., 2021; Alvarez-Garretón et al. 2019; Balocchi et al., 2023; Bonnesoeur et al., 2019; Germer et al., 2009; Jones et al., 2017; Little et al., 2009). This situation, together with others, causes severe water supply problems at the local level and subsequent socioeconomic effects and conflicts (Anderson et al. 2016; Braun, 2021).

In addition, this biophysical degradation has been coupled with a social degradation that has stimulated the migration of the rural population to urban centers due to at least four causes. First, small owners have sold their land to the forest companies, with the subsequent concentration of land ownership entailing not only less access to forest ecosystems, but also the displacement of local communities. Second, forestry plantations require less labor than other agriculture, and labor is required only during intermittent years for planting, thinning, and cutting. Third, most of the labor force is brought with companies from other regions of Chile, and fourth, other multiple uses and values of native forests are eliminated (Andersson et al., 2016; Hofflinger et al., 2021; Rozzi, 2018).

Land cover and land use changes occur in response to economic, social, political, and infrastructural conditions. In the twenty-first century, land-grabbing and other forms of concentration of land ownership have become a major driver of the rapid rates of rural–urban migration (Borras et al., 2011; Borras et al., 2012) . This rural–urban migration has severe consequences for both the native habitats and the well-being of displaced human communities (Grau & Aide, 2007; Hunter, 2005; Majumder & Raham, 2022; Rozzi, 2012; Selod & Shilpi, 2021). For the displaced people, this migration causes a loss of everyday contact with their communities and diverse life habits. In addition, rural to urban migration leads to the loss of ancestral human stewards or custodians of the land. In the cities, displaced people frequently lose their autonomy and lack access to basic services such as food, water, shelter, and sanitary conditions, facing extreme poverty conditions that are rapidly expanding in the marginal neighborhoods of metropolitan areas.

Rural–urban migration and its consequences are not the only problem. The change in land use can impact the price of property and influence the market, which, in turn, favors the hoarding of land by a few owners, impoverishing rural areas and excluding the local population from their habitat and the use and practices they support, monopolizing access and perpetuating the injustice inherited from the colonial era (Berna & Nahuelhual, 2019; Pirard et al., 2017).

Chile has experienced a dramatic acceleration in urban growth, with the percentage of the overall population living in cities growing from 58% in 1950 to 91% in 2020 (Lee, 2011; Rozzi et al., 2015). During the second half of the twentieth century and in the early twenty-first century, Chile's rate of rural–urban migration was significantly higher than the rest of Latin America and the world (Fig. 1). In turn, the expansion of tree plantations in Chile has come at the expense of agricultural activity which has been often the main cause behind migration to urban areas. Younger rural dwellers in particular have been propelled to migrate to cities (Andersson et al., 2016). They migrate first for education and then stay to work, as cities offer more opportunities for salaried jobs with higher pay than that of their parents in rural areas (Andersson et al., 2016). Tree plantations may have an augmented negative effect on local economic opportunities due to the concentration of land ownership by a few companies or families (Alfonso & Miller, 2021; Andersson et al., 2016; Pirard et al., 2017).

Confronted with these drastic social and environmental problems, sustainability policies should be addressed towards the indirect or underlying causes of rural–urban migration, such as the regime that promotes the concentration of land ownership in the hands of a few companies or families, leading to a lack of opportunities for the majority in rural areas. It is necessary to promote balanced land tenure within the framework of territorial planning that contributes to improving the quality of life in both urban and rural areas (Kröger, 2012).

In the same sense, an important point is one of the ethical attributes of sustainability, the transgenerational character that seeks to satisfy the needs of the present generations without compromising the needs of future ones (Thompson, 2007). This transgenerational ethic must be included into decision-making processes on how to use and transform the landscapes and their mosaics of ecosystems. Just actions are those that are fair to present and future generations and maintain the integrity of ecosystems inhabited by them. But, in addition to including a transgenerational perspective within, we must include young people in decision-making processes, and guarantee a prosperous future for them and coming generations.

On the other hand, sustainability requires responsible and equitable use of natural resources. One way to achieve this is to recognize the rights of indigenous peoples to the land, which will allow the perpetuation of traditional knowledge, management practices, and governance systems based on a deep understanding of the nature-community interconnection, and, with this holistic approach, guaranteeing the continuity of life in all its forms. Paradoxically, throughout history, these communities

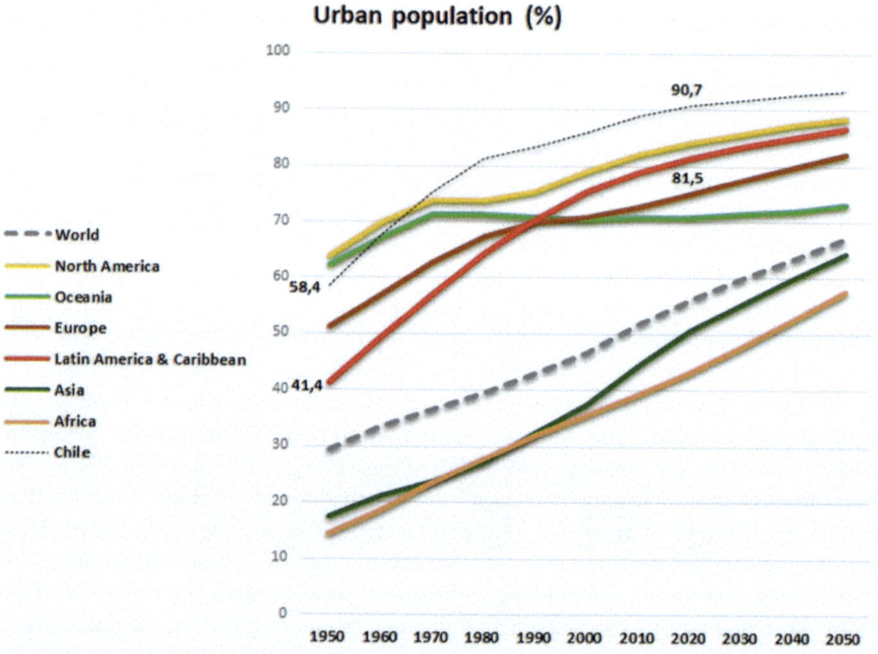

Fig. 1 Relative percentages of urban populations since 1950, and estimated percentages until 2050, in Chile, Latin America and the Caribbean, North America, Europe, Oceania, Asia, Africa, and the world as a whole (modified from Rozzi, 2015)

have faced land dispossession, environmental degradation, and violations of their human and territorial rights.

Today, Mapuche resistance movements aim to stop the transformation of native forests into homogenous plantations of exotic trees (Reinao, 2008). Their demands include recovering access to the land and restoring native forests and their natural water cycles that have been significantly disrupted by monocultures (Torres-Salinas et al., 2016, 2022). The Mapuche movement demonstrates great heterogeneity and includes different forms of political action from armed struggle to the promotion of a non-violent solidarity economy (Schmalz et al., 2022). The non-violent or peaceful solidarity economy movement is based on deep cultural traditions. It is centered on the recovery of ancestral lands, stopping the advance of environmental degradation, and reversing water scarcity associated with forestry monoculture by questioning the one-dimensional capitalist logic of exploiting “natural resources,” and by reintroducing more sustainable ways of managing natural commons valued as communities of co-inhabitants.

This form of the Mapuche environmental justice movement offers an example to the world on how to supersede biocultural homogenization based on monocultures to instead foster biocultural conservation (Rozzi, 2018). On the one hand, this movement stresses the relevance of deconstructing concepts of the coloniality of nature as well as historical, political, spatial, and socio-environmental processes associated with mere production of commodities. On the other hand, the Mapuche worldview offers relational concepts, communities, and territories that orient decolonization struggles to restore lands and ancestral ontologies, which support the human right of coinhabiting the world in different ways.

The conservation of native habitats becomes an ethical imperative because it is a necessary condition for the well-being of human and other-than-human co-inhabitants. Consequently, it has been proposed that the conservation of native habitats and access to them by local communities should be incorporated into sustainability policies as a matter of eco-social justice (Hofflinger et al., 2021; Rozzi, 2013). Biocultural ethics behooves us to consider both the rights of human co-inhabitants and those other-than-human co-inhabitants who share habitats (Rozzi, 2012).

With a systemic perspective it is necessary to assess the social, economic, and environmental consequences of different actions promoted by different public policies. It is important to recognize and incorporate a diversity of agents, involving local communities, experts from various disciplines, environmental organizations, and government agencies. With a participatory approach, different objectives and interests are balanced through deliberative processes that are not free from conflicts. Thus, these social, political, economic, and ethical deliberative processes can promote justice between different biological species, different social classes, and different cultures that coexist in forests and other types of habitats (Rozzi, 2018). In turn, this biocultural conservation approach could foster more responsible and sustainable practices that safeguard both human well-being and ecosystem integrity (Gavin et al., 2015; Sterling et al., 2017).

5 Chapter Outlines

Now we would like to briefly introduce the chapters that follow. This section of the book focuses on sustainable land use by examining three case studies in Chile and Germany. These cases illustrate the issue of land use change as a driving force for both sustainable and unsustainable environmental change. Throughout these chapters, the use of the concept of sustainability in the field of land use change and land use planning has not been uniform due to the multiplicity of definitions, dimensions, and contradictions that accompany it (Moore et al., 2017; Ruggerio, 2021; Virtanen et al., 2020). Consequently, each chapter emphasizes different aspects of sustainability associated with both the nature of the case and the discipline and conceptual framework of the authors.

In the chapter "Relationship between Political-Economic Factors and the Spatial Expansion of the Plantation Industry in the Coastal Region of Central Chile", Braun and Hitzeroth (2024) discuss the relationships between political-economic factors on the one hand and the expansion of the forestry industry on the other. They show how the forestry industry has spread in several phases depending on these factors.

Most importantly, they discuss the impact of the forest industry as a driver of land use change.

Regarding environmental impacts, the article "Regional Biotic Homogenization by Large-Scale Reforestation: Lessons from the Biodiversity Hotspot of Central Chile" by Braun and Dolos (2024) demonstrates how the expansion of the forest industry has led to the biotic homogenization of the coastal mountains. Rather than merely developing biological arguments, however, the authors place the example of Chile into a broader context, highlighting its relevance in current environmental and climate discourse. They report how the deforestation that occurred in this range of central Chile can be largely attributed to the establishment of commercial plantations based on fast-growing exotic species, particularly *Pinus spp.* and *Eucalyptus spp.* (Braun, 2022; Miranda et al., 2017). These forest monocultures are facilitated by the coercion between the country's economic policies, such as D.L 701, the lack of regulations to assess its impact, and the response of the international market, including the growing participation of forestry companies in this world market and the response of these companies to the evolution of global prices.

Finally, Pohl et al. (2024) explain how these correlations might look. In their article "The Contribution of Non-Timber Forest Products (NTFP) Based on the Chilean Hazelnut (*Gevuina avellana*) to the Sustainability of Indigenous Collectors' Livelihoods in Chile," the authors shed light on the realities of another user of space in the coastal mountains and how processes of habitat degradation may be avoided. The livelihood strategies of small farmers and indigenous people who use non-timber forest products are illustrated, but also how these are endangered by the forest industry. This approach is especially appropriate for the subsistence economies of smallholders and forest-dependent rural people whose livelihoods depend on a variety of forest ecosystems' goods and services.

In summary, this section of the book takes a critical look at the supposed sustainability of Chilean forestry industry and proposes eco-industrial strategies that imitate ecosystem principles, with the goal of transforming economic activities across the region, promoting practices that minimize environmental impact, and optimizing economic and social benefits.

References

Abson, D. J., Von Wehrden, H., Baumgärtner, S., Fischer, J., Hanspach, J., Härdtle, W., Heinrichs, H., Klein, A. M., Lang, D. J., Martens, P., & Walmsley, D. (2014). Ecosystem services as a boundary object for sustainability. *Ecological Economics, 103*, 29–37.

Aburto, F., Cartes, E., Mardones, O. & Rubilar, R. 2020. Hillslope soil erosion and mobility in pine plantations and native deciduous forest in the coastal range of south-Central Chile. *Land Degraadation & Development, 2*(1), 453–466.https://doi.org/10.1002/ldr.3700

Alaniz, A. J., Abarzúa, A. M., Martel-Cea, A., Jarpa, L., Hernández, M., Aquino-López, M. A., & Smith-Ramírez, C. (2021). Linking sedimentological and spatial analysis to assess the impact of the forestry industry on soil loss: The case of Lanalhue Basin, Chile. *CATENA, 207*(105660), 105660. https://doi.org/10.1016/j.catena.2021.105660

Afonso, R., & Miller, D. C. (2021). Forest plantations and local economic development: Evidence from Minas Gerais, Brazil. *Forest Policy and Economics, 133*(102618), 102618. https://doi.org/10.1016/j.forpol.2021.102618

Agarwal, C., Green, G. M., Grove, J. M., Evans, T. P., & Schweik, C. M. (2002). A review and assessment of land-use change models: Dynamics of space, time, and human choice. United States Department of Agriculture.

Alvarez-Garreton, C., Lara, A., Boisier, J. P., & Galleguillos, M. (2019). The impacts of native forests and forest plantations on water supply in Chile. *Forests, 10*(6), 473. https://doi.org/10.3390/f10060473

Andersson, K., Lawrence, D., Zavaleta, J., & Guariguata, M. R. (2016). More trees, more poverty? The socioeconomic effects of tree plantations in Chile, 2001–2011. *Environmental Management, 57*, 123–136.

Andrade-Núñez, M. J., & Aide, T. M. (2010). Effects of habitat and landscape characteristics on medium and large mammal species richness and composition in northern Uruguay. *Zoologia (Curitiba, Brazil), 27*(6), 909–917. https://doi.org/10.1590/s1984-46702010000600012

Armesto, J. J., Rozzi, R., Smith-Ramírez, C., & Arroyo, M. T. K. (1998). Effective conservation targets in South American temperate forests. *Science, 282*, 1271–1272.

Arroyo, M. T. K., Marquet, P., Simonetti, J., Marticorena, C., Cavieres, L., & Rozzi R. (2004). Central Chile/Valdivian Patagonian forests. In R. Mittermeier et al. (Eds.), *Hotspots Revisited* (pp. 99–103). CEMEX.

Balocchi, F., Galleguillos, M., Rivera, D., Stehr, A., Arumi, J. L., Pizarro, R., Garcia-Chevesich, P., Iroumé, A., Armesto, J. J., Hervé-Fernández, P., Oyarzún, C., Barría, P., Little, C., Mancilla, G., Yépez, S., Rodriguez, R., White, D. A., Silberstein, R. P., Neary, D. G., & Ramírez de Arellano, P. (2023). Forest hydrology in Chile: Past, present, and future. *Journal of Hydrology, 616*(128681), 128681. https://doi.org/10.1016/j.jhydrol.2022.128681

Barlow, J., Gardner, T. A., Araujo, I. S., Ávila-Pires, T. C., Bonaldo, A. B., Costa, J. E., Esposito, M. C., Ferreira, L. V., Hawes, J., Hernandez, M. I. M., Hoogmoed, M. S., Leite, R. N., Lo-Man-Hung, N. F., Malcolm, J. R., Martins, M. B., Mestre, L. A. M., Miranda-Santos, R., Nunes-Gutjahr, A. L., Overal, W. L., Peres, C. A., et al. (2007). Quantifying the biodiversity value of tropical primary, secondary, and plantation forests. *Proceedings of the National Academy of Sciences of the United States of America, 104*(47), 18555–18560. https://doi.org/10.1073/pnas.0703333104

Bathurst, J. C., Iroumé, A., Cisneros, F., Fallas, J., Iturraspe, R., Novillo, M. G., Urciuolo, A., de Bièvre, B., Borges, V. G., Coello, C., Cisneros, P., Gayoso, J., Miranda, M., & Ramírez, M. (2011a). Forest impact on floods due to extreme rainfall and snowmelt in four Latin American environments 1: Field data analysis. *Journal of Hydrology, 400*(3–4), 281–291. https://doi.org/10.1016/j.jhydrol.2010.11.044

Bathurst, J. C., Birkinshaw, S. J., Cisneros, F., Fallas, J., Iroumé, A., Iturraspe, R., Novillo, M. G., Urciuolo, A., Alvarado, A., Coello, C., Huber, A., Miranda, M., Ramirez, M., & Sarandón, R. (2011b). Forest impact on floods due to extreme rainfall and snowmelt in four Latin American environments 2: Model analysis. *Journal of Hydrology, 400*(3–4), 292–304. https://doi.org/10.1016/j.jhydrol.2010.09.001

Bathurst, J. C., Hagon, H., Hambly Barton, F., Iroumé, A., Kilbride, A., & Kilsby, C. (2022). Partial afforestation has uncertain effect on flood frequency and peak discharge at large catchment scales (100–1000 km^2), south-central Chile. *Hydrological Processes, 36*(5). https://doi.org/10.1002/hyp.14585

Bergner, A., Avcı, M., Eryiğit, H., Jansson, N., Niklasson, M., Westerberg, L., & Milberg, P. (2015). Influences of forest type and habitat structure on bird assemblages of oak (Quercus spp.) and pine (Pinus spp.) stands in southwestern Turkey. *Forest Ecology and Management, 336*, 137–147. https://doi.org/10.1016/j.foreco.2014.10.025

Berkes, F., Folke, C., & Colding, J. (Eds.). (2000). *Linking social and ecological systems: Management practices and social mechanisms for building resilience*. Cambridge University Press.

Berkes, F., Colding, J., & Folke, C. (Eds.). (2008). *Navigating social-ecological systems: Building resilience for complexity and change*. Cambridge University Press.

Berna, F., & Nahuelhual, L. (2019). A trilogy of inequalities: Land ownership, forest cover and ecosystem services distribution. *Land Use Policy, 82*, 247–257. https://doi.org/10.1016/j.landusepol.2018.12.020

Bernard, H., Fjeldså, J., & Mohamed, M. (2009). A case study on the effects of disturbance and conversion of tropical lowland Rain Forest on the non-Volant small mammals in north Borneo: Management implications. *Mammal Study, 34*(2), 85–96. https://doi.org/10.3106/041.034.0204

Birkinshaw, S. J., Bathurst, J. C., Iroumé, A., & Palacios, H. (2011). The effect of forest cover on peak flow and sediment discharge—an integrated field and modelling study in central–southern Chile. *Hydrological Processes, 25*(8), 1284–1297. https://doi.org/10.1002/hyp.7900

Bonnesoeur, V., Locatelli, B., Guariguata, M. R., Ochoa-Tocachi, B. F., Vanacker, V., Mao, Z., Stokes, A., & Mathez-Stiefel, S.-L. (2019). Impacts of forests and forestation on hydrological services in the Andes: A systematic review. *Forest Ecology and Management, 433*, 569–584. https://doi.org/10.1016/j.foreco.2018.11.033

Bopp, C., Engler, A., Jara-Rojas, R., & Arriagada, R. (2020). Are forest plantation subsidies affecting land use change and off-farm income? A farm-level analysis of Chilean small forest landowners. *Land Use Policy, 91*(104308), 104308. https://doi.org/10.1016/j.landusepol.2019.104308

Borras, S. M., Hall, R., Scoones, J., White, B., & Wolford, W. (2011). Towards a better understanding of global land grabbing. *Journal of Peasant Studies, 38*, 209–216.

Borras, S. M., Franco, J. C., Gómez, S., Kay, C., & Spoor, M. (2012). Land grabbing in Latin America and the Caribbean. *Journal of Peasant Studies, 39*, 845–872.

Braun, A. C. (2021). Encroached by pine and eucalyptus? A grounded theory on an environmental conflict between forest industry and smallholder livelihoods in Chile. *Journal of Rural Studies, 82*, 107–120. https://doi.org/10.1016/j.jrurstud.2021.01.029

Braun, A. C. (2022). Deforestation by Afforestation: Land Use Change in the Coastal Range of Chile. *Remote Sensing, 4*(7):1686. https://doi.org/10.3390/rs14071686

Braun, A. C., & Dolos, K. (2024). Regional biotic homogenization by large-scale reforestation: Lessons from the biodiversity hotspot of central Chile. In A.C. Braun, G. Espinosa, D. Tröger, & T. Hirth (Eds.), *Eco-industrial development as an industrial strategy—Contributions from a German-Chilean research partnership*. Springer Nature.

Braun, A. C., & Hitzeroth, M. (2024). Relationship between political-economic factors and the spatial expansion of the plantation industry in the coastal region of Central Chile. In A.C. Braun, G. Espinosa, D. Tröger, & T. Hirth (Eds.), *Eco-industrial development as an industrial strategy—Contributions from a German-Chilean research partnership*. Springer Nature.

Bremer, L. L., & Farley, K. A. (2010). Does plantation forestry restore biodiversity or create green deserts? A synthesis of the effects of land-use transitions on plant species richness. *Biodiversity and Conservation, 19*(14), 3893–3915. https://doi.org/10.1007/s10531-010-9936-4

Contador, T., Rendoll-Cárcamo, J., Gañan, M., Ojeda, J., Kennedy, J., Convey, P., & Rozzi, R. (2023). Underwater with a hand lens: Ecological sciences and environmental ethics to value freshwater biodiversity. In R. Rozzi, A. Tauro, T. Wright, N. Avriel-Avni, & R.H. May Jr. (Eds.), *Field environmental philosophy: Education for biocultural conservation*. Ecology and ethics series (Vol. 5, pp. 53–69). Springer.

Cumming, G. S. (2011). *Spatial resilience in social-ecological systems*. Springer Science & Business Media.

DeFries, R. S., Foley, J. A., & Asner, G. P. (2004). Land-use choices: Balancing human needs and ecosystem function. *Frontiers in Ecology and the Environment, 2*(5), 249–257.

Díaz, S., Settele, J., Brondízio, E. S., Ngo, H. T., Agard, J., Arneth, A., Balvanera, P., Brauman, K. A., Butchart, S. H. M., Chan, K. M. A., Garibaldi, L. A., Ichii, K., Liu, J., Subramanian, S. M., Midgley, G. F., Miloslavich, P., Molnár, Z., Obura, D., Pfaff, A., Polasky , S., & Zayas, C. N. (2019). Pervasive human-driven decline of life on Earth points to the need for transformative change. *Science (New York, N.Y.), 366*(6471). https://doi.org/10.1126/science.aax3100

Díaz, S., Quétier, F., Cáceres, D. M., Trainor, S. F., Pérez-Harguindeguy, N., Bret-Harte, M. S., Finegan, B., Peña-Claros, M., & Poorter, L. (2011). Linking functional diversity and social actor strategies in a framework for interdisciplinary analysis of nature's benefits to society. Proceedings.

Gavin, M. C., McCarter, J., Mead, A., Berkes, F., Stepp, J. R., Peterson, D., & Tang, R. (2015). Defining biocultural approaches to conservation. *Trends in Ecology & Evolution, 30*(3), 140–145. https://doi.org/10.1016/j.tree.2014.12.005

Geijzendorffer, I. R., Cohen-Shacham, E., Cord, A. F., Cramer, W., Guerra, C., & Martín-López, B. (2017). Ecosystem services in global sustainability policies. *Environmental Science & Policy, 74*, 40–48.

Echeverría, C., Coomes, D., Salas, J., Rey-Benayas, J. M., Lara, A., & Newton, A. (2006). Rapid deforestation and fragmentation of Chilean temperate forests. *Biological Conservation, 130*, 481–494. Pastor-Barrué 2004

Esse, C., Ríos, N., Saavedra, P., Fonseca, D., Encina-Montoya, F., Santander-Massa, R., De los Ríos-Escalante, P., Figueroa-Muñoz, G., López-Pérez, A., & Correa-Araneda, F. (2021). Effects of land use change on water availability and water efficiency in the temperate basins of south-central Chile. *Journal of King Saud University. Science, 33*(8), 101650. https://doi.org/10.1016/j.jksus.2021.101650

Germer, S., Neill, C., Vetter, T., Chaves, J., Krusche, A. V., & Elsenbeer, H. (2009). Implications of long-term land-use change for the hydrology and solute budgets of small catchments in Amazonia. *Journal of Hydrology, 364*(3–4), 349–363. https://doi.org/10.1016/j.jhydrol.2008.11.013

Grau, H. R., & Aide, T. M. (2007). Are rural–urban migration and sustainable development compatible in mountain systems? *Mountain Research and Development, 27*(2), 119–123. https://doi.org/10.1659/mrd.0906

Guo, J., & Cui, W. (2010). Research on the stability of Eco-industry chains. *International Journal of Business and Management, 5*(11). https://doi.org/10.5539/ijbm.v5n11p152

Heilmayr, R., Echeverría, C., & Lambin, E. F. (2020). Impacts of Chilean forest subsidies on forest cover, carbon and biodiversity. *Nature Sustainability, 3*(9), 701–709.

Hofflinger, A., Nahuelpan, H., Boso, À., & Millalen, P. (2021). Do large-scale forestry companies generate prosperity in indigenous communities? The socioeconomic impacts of tree plantations in Southern Chile. *Human Ecology*, 1–12.

Hunter, L. M. (2005). Migration and environmental hazards. *Population and Environment, 26*(4), 273–302. https://doi.org/10.1007/s11111-005-3343-x

Iezzi, M. E., Cruz, P., Varela, D., De Angelo, C., & Di Bitetti, M. S. (2018). Tree monocultures in a biodiversity hotspot: Impact of pine plantations on mammal and bird assemblages in the Atlantic Forest. *Forest Ecology and Management, 424*, 216–227. https://doi.org/10.1016/j.foreco.2018.04.049

Jaureguiberry, P., Titeux, N., Wiemers, M., Bowler, D. E., Coscieme, L., Golden, A. S., Guerra, C. A., Jacob, U., Takahashi, Y., Settele, J., Díaz, S., Molnár, Z., & Purvis, A. (2022). The direct drivers of recent global anthropogenic biodiversity loss. *Science Advances, 8*(45). https://doi.org/10.1126/sciadv.abm9982

Jones, J., Almeida, A., Cisneros, F., Iroumé, A., Jobbágy, E., Lara, A., Lima, W. de P., Little, C., Llerena, C., Silveira, L., & Villegas, J. C. (2017). Forests and water in South America. *Hydrological Processes, 31*(5), 972–980. https://doi.org/10.1002/hyp.11035

Pohl, B., Braun, A. C., & Azocar, G. (2024). The contribution of non-timber forest products (NTFP) based on the Chilean Hazelnut (Gevuina avellana) to the sustainability of indigenous collectors' livelihoods in Chile. In A. C. Braun, G. Espinosa, D. Tröger, & T. Hirth (Eds.), *Eco-industrial development as an industrial strategy—Contributions from a German-Chilean research partnership*. Springer Nature.

Kröger, M. (2012). The expansion of industrial tree plantations and dispossession in Brazil. *Development and Change, 43*(4), 947–973. https://doi.org/10.1111/j.1467-7660.2012.01787.x

Lambin, E. F., Meyfroidt, P., Rueda, X., Blackman, A., Börner, J., Cerutti, P. O., Dietsch, T., Jungmann, L., Lamarque, P., Lister, J., Walker, N. F., & Wunder, S. (2014). Effectiveness and synergies of policy instruments for land use governance in tropical regions. *Global Environmental Change, 28*, 129–140.

Lara, A., Solari, M. E., Del Rosario, M., & Peña, M. P. (2012). Reconstrucción de la cobertura de la vegetación y uso del suelo hacia 1550 y sus cambios a 2007 en la ecorregión de los bosques valdivianos lluviosos de Chile (35°–43° 30′ S). *BOSQUE, 33*(1), 13–23. https://doi.org/10.4067/S0717-92002012000100002

Lara, A., & Veblen, T. T. (1993). Forest plantations in Chile: A successful model? In A. Mather (Ed.), *Afforestation* (pp. 118–139). Belhaven Press.

Lee, R. (2011). The outlook for population growth. *Science (New York, N.Y.), 333*(6042), 569–573. https://doi.org/10.1126/science.1208859

Little, C., Lara, A., McPhee, J., & Urrutia, R. (2009). Revealing the impact of forest exotic plantations on water yield in large scale watersheds in South-Central Chile. *Journal of Hydrology, 374*(1–2), 162–170. https://doi.org/10.1016/j.jhydrol.2009.06.011

Majumder, S. C., & Rahman, M. H. (2022). Rural–urban migration and its impact on environment and health: Evidence from Cumilla City Corporation, Bangladesh. *GeoJournal, 88*(3), 3419–3437. https://doi.org/10.1007/s10708-022-10816-z

McWethy, D. B., Pauchard, A., García, R. A., Holz, A., González, M. E., Veblen, T. T., Stahl, J., & Currey, B. (2018). Landscape drivers of recent fire activity (2001–2017) in south-central Chile. *PLoS ONE, 13*(8), e0201195. https://doi.org/10.1371/journal.pone.0201195

Miranda, A., Altamirano, A., Cayuela, L., Lara, A., & González, M. (2017). Native forest loss in the Chilean biodiversity hotspot: Revealing the evidence. *Regional Environmental Change, 17*(1), 285–297. https://doi.org/10.1007/s10113-016-1010-7

Moore, J. E., Mascarenhas, A., Bain, J., & Straus, S. E. (2017). Developing a comprehensive definition of sustainability. *Implementation Science, 12*(1), 1–8.

Moss, T. (2004). The governance of land use in river basins: Prospects for overcoming problems of institutional interplay with the EU water framework directive. *Land Use Policy, 21*(1), 85–94.

Myers, N., Mittermeier, R. A., Mittermeier, C. G., Da Fonseca, G. A., & Kent, J. (2000). Biodiversity hotspots for conservation priorities. *Nature, 403*(6772), 853–858.

Nahuelhual, L., Carmona, A., Lara, A., Echeverría, C., & González, M. E. (2012). Land-cover change to forest plantations: Proximate causes and implications for the landscape in south-central Chile. *Landscape and Urban Planning, 107*(1), 12–20. https://doi.org/10.1016/j.landurbplan.2012.04.006

Pastor-Barrué, M. (2004). *Forêts et développement durable au Chili: Indianité mapuche et mondialisation*. Presses Univ. du Mirail.

Pawson, S. M., Brin, A., Brockerhoff, E. G., Lamb, D., Payn, T. W., Paquette, A., & Parrotta, J. A. (2013). Plantation forests, climate change and biodiversity. *Biodiversity and Conservation, 22*(5), 1203–1227. https://doi.org/10.1007/s10531-013-0458-8

Pirard, R., Petit, H., & Baral, H. (2017). Local impacts of industrial tree plantations: An empirical analysis in Indonesia across plantation types. *Land Use Policy, 60*, 242–253. https://doi.org/10.1016/j.landusepol.2016.10.038

Rawls, J. (1971). *A theory of justice*. Harvard University Press.

Reinao, R. M. (2008). The Mapuche and climate change in the Chilean neoliberal economic system. *Indigenous Affairs, 1*(08), 66–71.

Rozzi, R. (2012). Biocultural ethics: The vital links between the inhabitants, their habits and regional habitats. *Environmental Ethics, 34*, 27–50.

Rozzi, R. (2013). Biocultural ethics: From biocultural homogenization toward biocultural conservation. In R. Rozzi, S. T. A. Pickett, C. Palmer, J. J. Armesto, & J. B. Callicott (Eds.), *Linking ecology and ethics for a changing world: Values, philosophy, and action. Ecology and ethics* (Vol. 1, pp. 9–32). Springer.

Rozzi, R. (2015). Implications of the biocultural ethic for earth stewardship. In R. Rozzi, F. S. Chapin, J. B. Callicott, S. T. A. Pickett, M. E. Power, J. J. Armesto, & R. H. May (Eds.), *Earth stewardship: Linking ecology and ethics in theory and practice. Ecology and ethics* (Vol. 2, pp. 113–136). Springer.

Rozzi, R. (2018). Biocultural homogenization: A wicked problem in the anthropocene. In R. Rozzi, R. H. May, F. S. Chapin, F. Massardo, M. Gavin, I. Klaver, A. Pauchard, M. A. Núñez, & D. Simberloff (Eds.), *From biocultural homogenization to biocultural conservation* (pp. 21–48). Springer.

Ruggerio, C. A. (2021). Sustainability and sustainable development: A review of principles and definitions. *Science of the Total Environment, 786*, 147481.

Rupprecht, C. D., Vervoort, J., Berthelsen, C., Mangnus, A., Osborne, N., Thompson, K., Urushima, A. Y. F., Kóvskaya, M., Spiegelberg, M., Cristiano, S., Springett, J., Marschütz, B., Flies, E. J., McGreevy, S. R., Droz, L., Breed, M. F., Gan, J., Shinkai, R., & Kawai, A. (2020). Multispecies sustainability. *Global Sustainability, 3*, e34.

Sala, O. E., Stuart Chapin, F. I. I. I., Armesto, J. J., Berlow, E., Bloomfield, J., Dirzo, R., Huber-Sanwald, E., Huenneke, L. F., Jackson, R. B., Kinzig, A., Leemans, R., Lodge, D. M., Mooney, H. A., Oesterheld, M., Poff, N. L., Sykes, M. T., Walker, B. H., Walker, M., & Wall, D. H. (2000). Global biodiversity scenarios for the year 2100. *Science, 287*(5459), 1770–1774.

Salas, C., Donoso, P. J., Vargas, R., Arriagada, C. A., Pedraza, R., & Soto, D. (2016). The forest sector in Chile: An overview and current challenges. *Journal of Forestry, 114*, 562–571.

Schmalz, S., Graf, J., Julián-Vejar, D., Sittel, J., & Alister Sanhueza, C. (2022). Challenging the three faces of extractivism: The Mapuche struggle and the forestry industry in Chile. *Globalizations*, 1–19.

Schmidt, C. (2022). *Landscape resilience: Basics, case studies, practical recommendations.* Springer.

Scott, D. N., & Smith, A. A. (2017). "Sacrifice zones" in the green energy economy: Toward an environmental justice framework. *McGill Law Journal, 62*(3), 861–898.

Selod, H., & Shilpi, F. (2021). Rural-urban migration in developing countries: Lessons from the literature. *Regional Science and Urban Economics, 91*(103713), 103713. https://doi.org/10.1016/j.regsciurbeco.2021.103713

Sheather, J., Littler, K., Singh, J. A., & Wright, K. (2023). Ethics, climate change and health—A landscape review. *Wellcome Open Research, 8*, 343. https://doi.org/10.12688/wellcomeopenres.19490.1

Sterling, E. J., Filardi, C., Toomey, A., Sigouin, A., Betley, E., Gazit, N., Newell, J., Albert, S., Alvira, D., Bergamini, N., Blair, M., Boseto, D., Burrows, K., Bynum, N., Caillon, S., Caselle, J. E., Claudet, J., Cullman, G., Dacks, R., Jupiter, S. D., et al. (2017). Biocultural approaches to well-being and sustainability indicators across scales. *Nature Ecology & Evolution, 1*(12), 1798–1806. https://doi.org/10.1038/s41559-017-0349-6

Thompson, I. (2007). The ethics of sustainability. In J. Benson & Roe, M. (2007). *Landscape and sustainability*. Taylor & Francis.

Torres-Salinas, R., García, G. A., Henríquez, N. C., Zambrano-Bigiarini, M., Costa, T., & Bolin, B. (2016). Forestry development, water scarcity, and the Mapuche protest for environmental justice in Chile. *Ambiente & Sociedade, 19*(1), 121–144.

Torres-Salinas, R., Azócar, G., Gallardo, R., & Mendoza, J. (2022). Water extractivism and decolonial struggles in Mapuche territory, Chile. *Water Alternatives, 15*(1), 150–174.

Van Holt, T., Moreno, C. A., Binford, M. W., Portier, K. M., Mulsow, S., & Frazer, T. K. (2012). Influence of landscape change on nearshore fisheries in southernhttps://doi.org/10.1111/j.1365-2486.2012.02674.x Chile. *Global Change Biology, 18*(7), 2147–2160.

Virtanen, P. K., Siragusa, L., & Guttorm, H. (2020). Introduction: Toward more inclusive definitions of sustainability. *Current Opinion in Environmental Sustainability, 43*, 77–82.

Vogel, J., & Hickel, J. (2023). Is green growth happening? An empirical analysis of achieved versus Paris-compliant CO_2–GDP decoupling in high-income countries. *Lancet Planet Health, 7*, e759–e769.

Land Use and Industrial Sustainability—Sustainable Land Use—Case Studies Regarding the Chilean Forest Industry

Relationship Between Political-Economic Factors and the Spatial Expansion of the Plantation Industry in the Coastal Region of Central Chile

Andreas Ch. Braun and Marion Hitzeroth

Abstract The Chilean forest industry is often described as a "green" or even "sustainable" industry, but in many ways it does not live up to this. Many negative impacts relate to the immense land grabs by the forestry industry, which have led to the massive replacement of near-natural forests in the coastal mountains and the displacement of other land uses. In the context of a discourse on alleged "eco-industries," the relationship between industry and spatial development needs to be considered. These considerations expand the discourse around eco-industrial development, which usually focuses on the immediate processes of production within and between companies, but not on their relationship to space. In order to disentangle the differential effects of economic policies and world market signals, relationships between political-economic factors and the spatial effects on land use have to be elaborated. Land-use change drivers are identified and forest losses are attributed to land appropriation by plantations. A prospective analysis is coupled with a retrospective analysis to understand land-use change processes in Chile's Coastal Range. In addition, this analysis aims to disentangle political-economic incentives to spatial developments in Chile. As a result, it is shown that deforestation can largely be attributed to commercial plantation establishment. Moreover, the subsequent results suggest that important incentives in the establishment of these plantations are not only Chilean subsidies but also influences from the development of world forestry markets.

A. Ch. Braun (✉)
Kassel Institute for Sustainability, University of Kassel, Mosenthalstr. 8, 34117 Kassel, Germany
e-mail: Andreas.braun@uni-kassel.de

M. Hitzeroth
Institute of Regional Science (IfR), Karlsruher Institute of Technology (KIT), Kaiserstr. 12, 76131 Karlsruhe, Germany
e-mail: marion.hitzeroth@kit.edu

A. Ch. Braun et al. (eds.), *Eco-Industrial Development as an Industrial Strategy*, Green Energy and Technology, https://doi.org/10.1007/978-3-031-73576-9_28

1 Introduction

In countries under development that are rich in natural resources, the development of extractivist economic models is a frequent phenomenon. In extractivist economies, the exploitation of natural resources with a focus on maximizing economic profit is the major goal (Acosta, 2013; Browder, 1992). Production is frequently concentrated on raw material extraction and commodities (Burchardt & Dietz, 2014; Lopez & Vertiz, 2015; Svampa, 2015), and markets are frequently dominated by oligopolies or even monopolies of a few successful companies (Gudynas, 2018; Pelfini & Mena, 2017). Very frequently, strong environmental impacts and social conflicts result as consequences of extractivist industries (Acuña, 2015; Martinez-Alier & Walter, 2016). Latin America is a region characterized by the development of extractivist economic structures (Petras & Veltmeyer, 2014a; Veltmeyer, 2013). Well-documented examples are the mining industry in Mexico, the fossil fuel industries in Venezuela and Argentina, and agro-industry and mining in Brazil (Azamar & Ponce, 2014; Chiasson-LeBel, 2016; Cleary, 1993; Petras & Veltmeyer, 2014b, 2014c; Renauld, 2016; Tetreault, 2014).

The Chilean plantation industry is a far less-understood example (Clapp, 1998). During the military regime of Augusto Pinochet starting in the 1970s, large-scale plantation industry oligopolies developed (Bull et al., 2006; Torres et al., 2015). As a consequence, commercial plantations of pine and eucalyptus trees are being established at the cost of a pronounced deforestation of near-natural forests (Heilmeyr et al., 2016). These land-use changes have resulted in pronounced environmental impacts and social conflicts that are continuously harshening (Altieri & Rojas, 1999b; Frêne & Nuñez, 2010). Nonetheless, the relationship between economic fostering of the plantation industry and resulting land-use change are far less understood than frequently claimed. Precisely, it is not well documented which political-economic incentives have been decisive in shaping land-use change (Nahuelhual et al., 2012; Niklitschek, 2007; Zamorano-Elgueta et al., 2015). It is well known that the Pinochet regime fostered the plantation industry as an economic sector of strategic importance (Carruthers, 2001; Clapp, 1995; Langenfeld, 2017). The political instrument to do so was DL 701, a decree (rather than a parliamentary law) enacted at the national level (Bitterlich, 2001). With its strongly regulatory nature, DL 701 is an exception to the predominantly (neo) liberal economic turn under the Pinochet government (Clapp, 1995; Langenfeld, 2017; Yates & Bakker, 2014). Firstly, the establishment of commercial plantations on areas "suitable for forestry" was subsidized with 75% of incurring costs (Tit. VI Art. 21 DL 701). Furthermore, omitting to establish forestry on "suitable areas" was progressively sanctioned by charging owners of the areas up to 80% of the estimated fiscal value (Tit. VII Art. 25 DL 701). The notion of "areas suitable for forestry" was legally defined in a manner sufficiently vague that near-natural forests fell under the definition (Tit. I Art. 2 DL 701). In contrast, DL 701 did not provide any incentives to diversify the production spectrum on the one hand, or to regulate environmental or social impacts on the other (cf. DL 701, Bitterlich, 2001). Hence, via DL 701, the government set strong incentives for expanding the plantation industry but did not account for environmental or social consequences (Altieri & Rojas, 1998, 1999a, 1999b; Clapp, 1995; Langenfeld, 2017).

Along with political incentives (subsidies and sanctions in DL 701), there are other influential factors from the (global) economy incentivizing plantation establishment (and consequently, land-use change). Some influential factors may promote an expansion of commercial plantation establishment (e.g. demand on the world market) or hamper it (e.g. the Chilean debt crisis). In total, the coaction of political-economic influential factors determines managerial decisions, which determine an expansion or stagnation of the plantation industry (Bull et al., 2006; Niklitschek, 2007). The resulting expansion or stagnation should become visible in land-use patterns. For this reason, an integrated analysis of land-use change could shed light on how political and economic incentives have coacted (Geist & Lambin, 2002; Geist & McConnell, 2010; Lambin & Geist, 2010; Reid et al., 2010).

The relationship between forest expansion and land-use change has been investigated in two types of studies. The first type consists of several descriptive studies based on remote sensing data (cf. Miranda et al., 2015, 2017; Van Holt et al., 2016 for an overview). Despite the strong agreement between these studies, several desiderata can be identified (Altamirano et al., 2013, related Ramankutty, 2010, Longley, 2002). Firstly, few studies have investigated larger areas (e.g. entire administrative units). The majority has focused on single satellite images, which may lack representativeness. Secondly, there are no studies which analyze short-term intervals. Published studies usually consider intervals of ten to twenty years. However, in such long-term intervals, intermediary usage can be overlooked, which can lead to misinterpretations. For instance, an area covered with native forest in 1990 and commercial plantation in 2010 may not have been deforested in favor of the plantation industry, but in favor of e.g. agriculture, established in 1991 and abandoned in 2000. Hence, in order to attribute land-use change drivers with sufficient certainty, short-term intervals are required. Thirdly, most of the studies name DL 701 as an influential factor, but most of these seem to simply *assume* that DL 701 has been the driver of land-use change, without explicitly elaborating its effects and also taking other influential factors into account.

A second type of studies focuses on forest policies or forest economies without providing a spatially explicit perspective on land use (Bull et al., 2006; Langenfeld, 2017; Manuschevich, 2016; Manuschevich & Beier, 2016; Silva, 2004). Hence, a research need is identified particularly regarding descriptive land-use change analysis (first type of studies) and political-economic influential factors (second type of studies). Such a perspective is provided herein.

The approach of this study is as follows. Regarding land-use, the change processes in two entire administrative regions are analyzed in five-year intervals. Both regions comprise the area predominantly characterized by plantation industry establishment since the 1970s. Using these data, land-use change and replacement processes are clearly established in a prospective and retrospective analysis. Regarding spatial economics, a phase model of plantation industry establishment is developed. Political-economic influential factors are identified from literature and world market prices. Both perspectives are then related to each other to develop a concept explaining the spatial effects of influential factors on land-use change.

The study seeks to firstly provide a methodological approach to how land-use change analysis from remote sensing and spatial economy can be linked. Secondly, it provides a perspective on how political-economic factors have been influential in shaping land use and, consequently, on the environmental and social impacts related to it. The approach is potentially applicable to other situations where resources are extracted above-ground.

The remainder of the study is organized as follows. Section 2 introduces the study region and Sect. 3 explains methods of land-use change analysis and political-economic analysis. Results of the land-use change analysis (Sect. 4.1) serve as the basis for the concluding political-economic analysis (Sect. 4.2) leading to the discussion of the developed phase model (Sect. 5) and concluding remarks (Sect. 6).

2 Study Site

The study site comprises the Coastal Range of the VII. Región del Maule, VIII. Región del Biobío, where deforestation in favor of the plantation industry has been strongest. Both regions belong to the temperate zone of Chile. The climate is a Mediterranean Csb climate (Koeppen–Geiger), with an annual mean temperature of 12 °C and annual precipitation of around 1300 mm (Kottek et al., 2006; Luebert & Pliscoff, 2006). The Coastal Range is a mountain range running from NNE to SSW. To the east, the central depression forms the interior of the country. Towards the Argentinian border, the Andes comprise a high mountain region. The basement of the Coastal Range consists of magmatic and metamorphic rocks of Paleozoic age and younger sediments and volcanic material from the Andes. The two regions cover around 67,358 km^2 and are inhabited by 2.8 million inhabitants. The population density (41 persons per km^2) is among the highest in Chile (INE, 2002). The study site is an area of intensive human modification of the natural environment. During the colonial period, large parts of the area, which is a naturally tree-covered zone, were disturbed by fire clearance in favor of agriculture (Holmgren et al., 2000). Around 1850, the first pine plantations were established presumably to prevent the development of erosive badlands (Camus, 2003). The effectiveness of this measure has recently been questioned (Banfield et al., 2018). Ever since, plantation establishment has intensified (Clapp, 1995, 1998). Due to the original purpose of plantations in 1850 (erosion control), DL 701 is sometimes considered a law for environmental protection in Chile. Due to the well documented and above-mentioned environmental impacts of plantations (Banfield et al., 2018; Braun & Koch, 2016; Braun et al., 2017), this claim must be challenged, however.

3 Materials and Methods

3.1 Land-Use Change Analysis

3.1.1 Datasets and Classification Algorithms

In order to characterize land-use changes in the study site, satellite image data acquired by the Landsat sensor series have been used. The Landsat database has been available since the 1970s, with images available almost globally and with a revisit time of around 14 days. The database is open after registration and data can be used free of charge. Since 1972, several instruments have been orbited, providing data with a reasonably high spatial resolution; due to this continuity, the Landsat series is of paramount value for environmental monitoring (Wulder et al., 2008, 2012). The Landsat data used herein provide an image on land use for the complete study site and in five-year intervals (i.e. 1975, 1980, 1985, …, 2010, 2015).

The information has been retrieved using supervised image classification. In supervised image classification, the human analyst characterizes each land-use system relevant to the study by spatial examples (defining so-called training areas where a certain land-use system's presence on the ground is confirmed). Then, computer algorithms model the characteristics of the land-use systems in the satellite images and, finally, assign each pixel in the images to one land-use class, providing a land-use map of the entire study site (Gorte, 2002).

In order to meet the demanding accuracy requirements for land-use change analysis (Shao & Wu, 2008), an intricate supervised classification algorithm has been developed. It produces sufficiently high accuracy values for the land-use maps (91.5% on average, but only 78.5% for the first Landsat-MSS image in 1975) by (1) introducing textural information regarding the land-use systems (e.g. visual patterns of plantation rows), (2) applying three state-of-the-art image analysis methods, and (3) analyzing the neighborhoods of each pixel (e.g. assigning single pixels within a plantation but misclassified as forests to the plantation class). Its technical details (relevant mainly to remote sensing scientists and image analysts) are provided in Braun et al. (2012, 2014).

It is important to stress that plantation detectability is limited by tree growth. When plantations are established, trees are planted as small saplings into barren soils. During the first years, recently established plantations may be confused with grasslands in Landsat-TM images from 1982 onwards. For earlier Landsat-MSS images (1975 and 1980) they may even be confused with agriculture (which explains the lower accuracy value). Unfortunately, due to the lower spatial and spectral resolution (from 80×80 m instead of 30×30 m, and only four instead of seven or more channels), these uncertainties have to be accepted. Braun (2013) has shown that in the results used herein, plantations become detectable after approx. ten years, when the trees are sufficiently high to be classified as a closed canopy system.

3.1.2 Deforestation Rates

From the land-use maps available for each five-year interval, the percent of land use found within each land-use system is derived. Besides total land use (t_{lu}), change rates such as deforestation rates are important to analyze the effects of political-economic influencing factors. The calculation of deforestation rates has recently been standardized (Puyravaud, 2003). The deforestation rate d_{fr} is calculated as follows:

$$d_{fr} = \frac{1}{t_2 - t_1} \times \ln \frac{A_2}{A_1}$$

where t_2 and t_1 are two points in time and A_2 and A_1 are land-use coverage at t_2 and t_1, respectively.

3.1.3 Prospective and Retrospective Analysis

Besides absolute figures on land-use change (such as total land-use t_{lu} and deforestation rates d_{fr}), drivers of land-use changes between individual classes have to be identified. Particularly, it is important to (1) analyze by which other land-use classes $C_{j \neq i}$ one particular land-use class C_i *has been* replaced, and (2) which other land-use classes $C_{j \neq i}$ one particular land-use class C_i *has* replaced, both between two points in time t_1 and t_2.

The prospective analysis accounts for the first requirement. The prospective analysis looks at the areas covered by a particular land-use class C_i in t_1 that have changed *to other classes* $C_{j \neq i}$ at t_2. It then calculates percentages of $C_{j \neq i}$ at t_2 and thus describes which classes $C_{j \neq i}$ have replaced C_i between the two points in time.

The retrospective analysis accounts for the second requirement. The prospective analysis looks at the areas covered by a particular land use class C_i in t_2 that have changed *from other classes* $C_{j \neq i}$ since t_1. It then calculates percentages of $C_{j \neq i}$ at t_1 and thus describes which classes $C_{j \neq i}$ C_i has replaced between the two points in time.

Two points have to be stressed. Firstly, the results of the prospective and retrospective are not redundant. That a particular class C_1 has been replaced with C_2 by 95% does not necessarily state that C_2 has been established on areas formerly covered with C_1 by 95% also. Instead, it is possible that C_2 has been established on areas formerly covered with C_1 by 10%, with C_3 by 70% and with C_4 by 20%. Combining the perspectives of both analyses, it can be concluded (1) which land-use systems a certain land-use system has vanished for (prospective), and (2) which land-use systems a certain land-use system has replaced (retrospective).

Secondly, both analyses can be applied either for the entire study period (i.e. the forty years between 1975 and 2015) or for each five-year interval. Both types of analyses will be provided in the following section.

3.2 *Analysis of Political-Economic Factors*

In order to better understand the land-use change processes, an analysis of political-economic aspects such as political and economic framework conditions and industrial, trade, and environmental policy changes during 1970–2015 will be attributed to the identified land-use changes.

It has to be noted that the analysis of land-use change comprises two regions, BioBío and Maule, while political-economic aspects are considered on a national level. This is due on the one hand to a lack of regional data, and on the other hand due to Chile's centralized political system, where relevant policies (such as DL 701) account for the entire country. Hence, as e.g. production share of sawnwood—one of the most important products of the Chilean forestry sector—of the two respective regions is about 75% (INFOR, 2017), a chart analysis regarding the impact of framework conditions (economic-political measures based on a relevant literature review) on plantation development and its verbal interpretation is deemed to be appropriate. Regarding investments in forestry, a time lag of up to twenty years has to be considered. This is simply due to the fact that trees take that time from planting to harvest. For that reason, entrepreneurs cannot react to short-term incentives (such as annual fluctuations in market prices) but have to consider long-term economic developments when making decisions to invest in plantation establishment (see e.g. Clapp, 1995). With the aforementioned detectability limits (ten years) in the image data, this introduces some uncertainty to the analysis that has to be taken into account when comparing political-economic incentives and land-use changes, as will be done in the following.

In addition to the analysis of framework conditions in the development of the world forestry market, incentives in terms of approximated global import prices of the two main Chilean export products (sawnwood and chemical wood pulp) and the respective Chilean production and export quantities based on UNECE/FAO-statistical price data from 1964 to 2015 (UNECE, 2018) have been analyzed and integrated into the discussion.

4 Results

4.1 *Results Regarding Land-Use Change*

4.1.1 Land-Use Maps, Deforestation Rates, Total Land-Use Change Between 1975 and 2015

Applying the algorithm outlined in Sect. 3.1.1 (Braun et al., 2012, 2014), a land-use map for each of the nine points in time has been produced. The land-use classes under consideration were: (1) near-natural forests, (2) plantation industry, (3) scrublands,

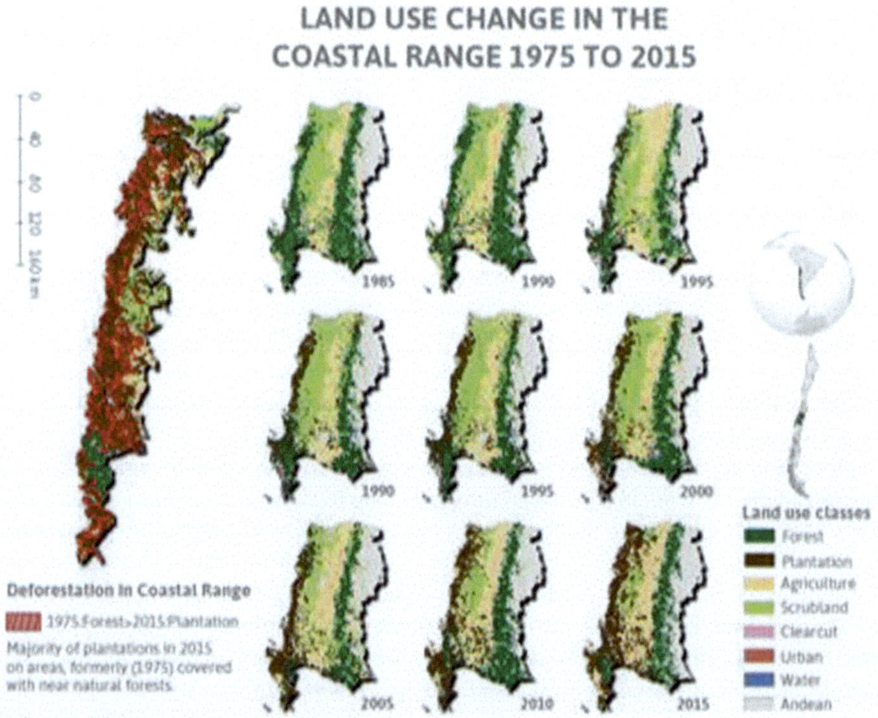

Fig. 1 Land-use change in the Coastal Range from 1975 to 2015

(4) agriculture, (5) urban, (6) open soils/clearcut, (7) water, and (8) barely vegetated Andean areas.

Results are given in Fig. 1. The land-use change between 1975 and 2015 is remarkable. In forty years, the Coastal Range, predominantly covered by near-natural forests in 1975, has altered its land use almost entirely to the plantation industry.

Only a minor portion of areas formerly covered by near-natural forests is now covered by agricultural land use. The disappearance of near-natural forests is predominately present within the Coastal Range; however, a part of the pre-Andean range formerly covered by near-natural forests is today covered by the plantation industry as well. Other land-use change drivers are playing only a subordinate role in the Coastal Range. The plantation industry seems to have been established primarily on former near-native forests and scrublands (and not on open soils/clearcuts, as would be expected from the claim of erosion prevention).

Figure 2a shows that the expansion of the plantation industry becomes visible by 1985, ten years after DL 701. Deforestation becomes visible by 1980. During the first interval (1975–1980), agricultural areas seem to have replaced near-natural forests. Both findings may be related to (1) the properties of the land-use systems and (2) to the properties of the early Landsat-MSS images (which were only four-channel data

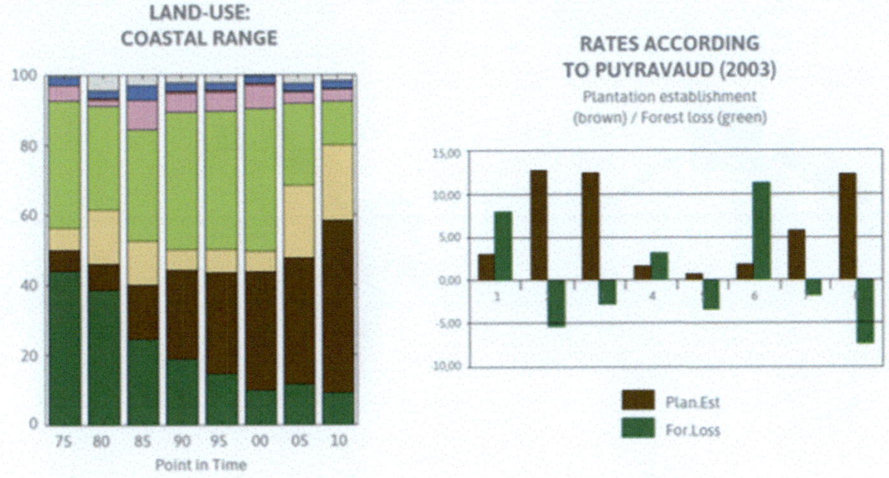

(a) (b)

Fig. 2 **a** Land use: Coastal Range. **b** Rates of plantation establishment according to Puyravaud

with 80 × 80 m resolution). As previously discussed, young plantations in Landsat-MSS images cannot be clearly distinguished from agricultural areas. Hence, it may be possible that parts of the areas identified in 1980 as agricultural are young plantations instead. This would explain why agriculture increases from 1975 to 1980 and then decreases from 1980 to 1985. From 1985 onward, the expansion of the plantation industry and the reduction of near-natural forests become obvious and reach their maximum by 2000. Besides deforestation and plantation industry establishment, a reduction of scrublands also has taken place. Note that scrublands represent more open canopy tree ecosystems similar to native forests (and under comparable land tenure situations). From 2000 onward, an agricultural expansion seems to have taken place, although this is not as spatially extensive as the plantation industry's expansion.

Deforestation rate d_{fr} (and the respective plantation establishment rates) according to the calculus of Sect. 3.1.2 are presented in Fig. 2b. As can be seen, deforestation (assessed by d_{fr}) is not monotonous, but interrupted during several intervals. The same accounts for plantation establishment, which exhibits strong peaks of expansion with stagnation phases in between. Section 4.2 will discuss these phenomena in more detail. It can be stated, however, that assuming that DL 701 has caused monotonous deforestation and expansion of the plantation industry is an oversimplification. In order to attribute land-use changes to spatial drivers, the prospective and retrospective analyses of Sect. 3.1.3 have been applied to the Coastal Range between $t_1 = 1975$ and $t_2 = 2015$. The results are shown in Fig. 3a and b, respectively. Note that the size of the boxes and arrows relates proportionally to absolute areas of land use in the study site.

As the prospective analysis reveals, the most important land-use change process has been deforestation in favor of plantation industry establishment. The majority of

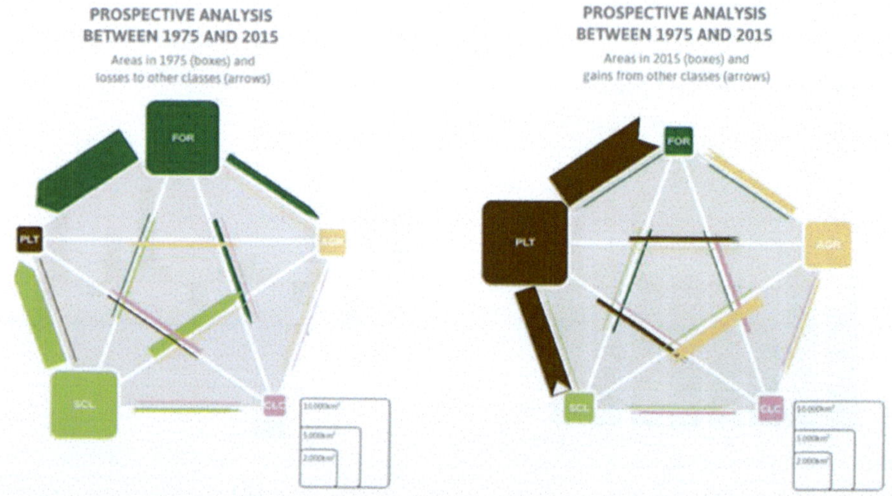

(a) (b)

Fig. 3 **a** Prospective analysis: areas in 1975 and losses to other classes. **b** Prospective analysis: areas in 2015 and gains from other classes

forests has been transformed to industrial plantations; transformation to agricultural areas plays only a subordinate role. The same accounts for a replacement of scrublands, which have also been replaced by plantations and only to a small amount, by agriculture. Note that other land-use change processes play only a marginal role within the Coastal Range. As the retrospective analysis reveals, it is predominately the establishment of industrial plantations that has replaced other land-use systems within the Coastal Range. Plantations have been established to a major part on areas formerly covered by forests and, to a smaller extent, by scrublands. The establishment on former agricultural sites played a minor role between 1975 and 2015 but has become important recently, as Sect. 4.1.2 will show. The agricultural sector has recruited areas formerly covered by other land-use systems as well, but to a much smaller degree than the plantation industry.

From this analysis, it can be assumed that land use in the Coastal Range has changed from a dominance of near-natural forest, scrublands, and small amounts of agriculture in 1975 to a dominance of industrial plantations in 2015.

4.1.2 Land-Use Change and Drivers Between 1975 to 2015

As has been argued in Sect. 3.1.3, in order to clearly attribute total land-use changes to land-use change drivers, a short-term interval analysis has to be performed. In order to attribute deforestation to its drivers clearly, a prospective analysis in five-year intervals has been performed. Figure 4a summarizes the findings, showing into

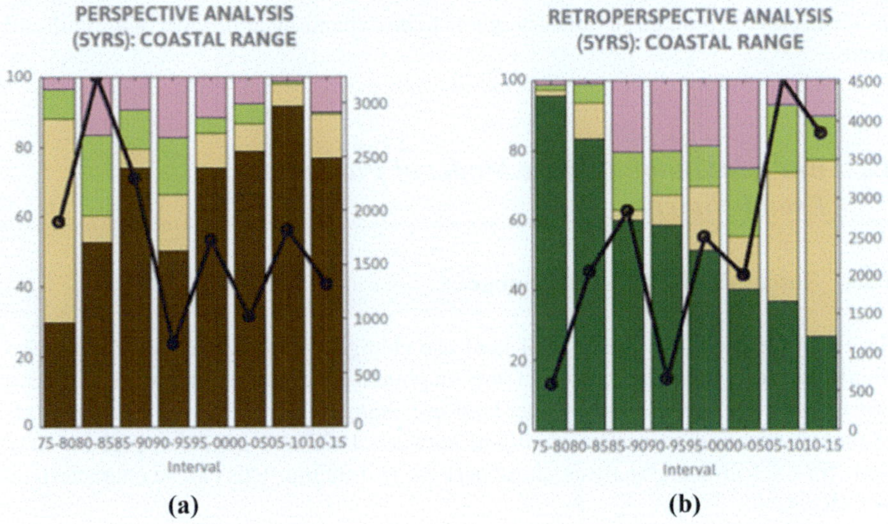

Fig. 4 **a** Perspective analysis; Coastal Range. **b** Retroperspective analysis; Coastal Range

which classes near-natural forests have been replaced in each interval. As can be seen, forests have—with one exception—always been replaced mostly by the plantation industry. The strongest replacement percentage is during the interval of 2005 to 2010, when more than 90% of all forests lost during that interval was due to the establishment of industrial plantations. The only exception is the interval from 1975 to 1980, when replacement by agriculture was greater. However, as discussed in Sect. 4.1.1, confusion between young plantations and agriculture might account for this.

In order to clarify which sites have been preferred for the establishment of plantations, a retrospective analysis in five-year intervals has been performed. Figure 4b shows which land-use systems were found on areas at the beginning of an interval that were being used as plantations by the end of that period. The analysis shows that by 2000, more than 50% of all plantations were established on formerly forested sites. The preference for forested sites for plantation establishment was strongest in the intervals 1975–1980 and 1980–1985, when more than 80% of all new plantations had replaced near-natural forests.

Note that since 1995–2000 another process has begun to occur: plantations are beginning to also draw upon agricultural areas. This may be due to the fact that areas covered with near-natural forests have become scarce since 1995 (cf. Fig. 2a) and in the face of this scarcity, plantation companies are starting strategies to acquire agricultural areas for further expansion.

From this analysis, it can be clearly stated that the plantation industry is principally responsible for the deforestation of Chile's Coastal Range, not because it has had some unintended side effects on forests, but because it has preferred them as the most favored establishment sites while they were available. As of recently this is no longer the case, and the plantation industry seems to be expanding at the cost of agricultural areas.

4.2 *Elaboration of a Phase Model of Plantation Establishment*

4.2.1 Intervals of Land-Use Change

Based on the results of analysis of land-use changes in the regions of BioBío and Maule, the following analysis will focus on plantation development as this has been proven to be responsible for the other land-use changes (see Sect. 4.1.2). Five periods of plantation-building can be identified between 1975 and 2015 (cf. Fig. 5). As a basis for the following analysis, these periods or time intervals can be characterized as follows.

Interval I from 1975 to 1980 can be characterized by a relatively stable situation with plantations' share of land-use still at a low level of around 5%, whereas Ten-Year Interval II (1980–1990) shows a considerable rise of plantations of up to nearly 20%. Between 1990 and 2000 (Interval III) a stabilization on this higher level of plantation land-use share can be observed. Interval IV (2000–2010) is again characterized by a rise in plantation share. In the last Five-Year Interval V (2010–2015), plantation shares nearly doubled from 30% to close to 60%. In Fig. 5 the plantation share rate is also shifted ten years backwards in order to approximate the time of plantation establishment, which is due to the time lag in plantation detectability (cf. Sect. 3.1.1).

4.2.2 Phases of Economic-Political Development

On the basis of the above characterization, the political-economic context on a national level in Chile is summarized and structured with regard to economic conditions and economic and environmental policy conditions. In analogy to the above land-use change, analysis phases of "policy context" are identified (cf. Fig. 5).

Phase A (until 1974) Prior to the military dictatorship, the Chilean forest industry had been in a process of nationalization. At the same time trade, industrial policy was characterized by import substitution and an export ban which had not been problematic as the demand from the world market had still been limited (Clapp, 1995). Nonetheless, after a crisis in the Chilean forestry sector in mid-1960, the government began to promote plantation establishment (Langenfeld, 2017).

Phase B (1974–1982) The political framework conditions changed considerably with the military dictatorship of Pinochet in combination with a 180° turn in industrial and trade policy toward liberalization. Free market economic policy measures entailing privatization of relevant sites and cancellation of export bans were introduced. At

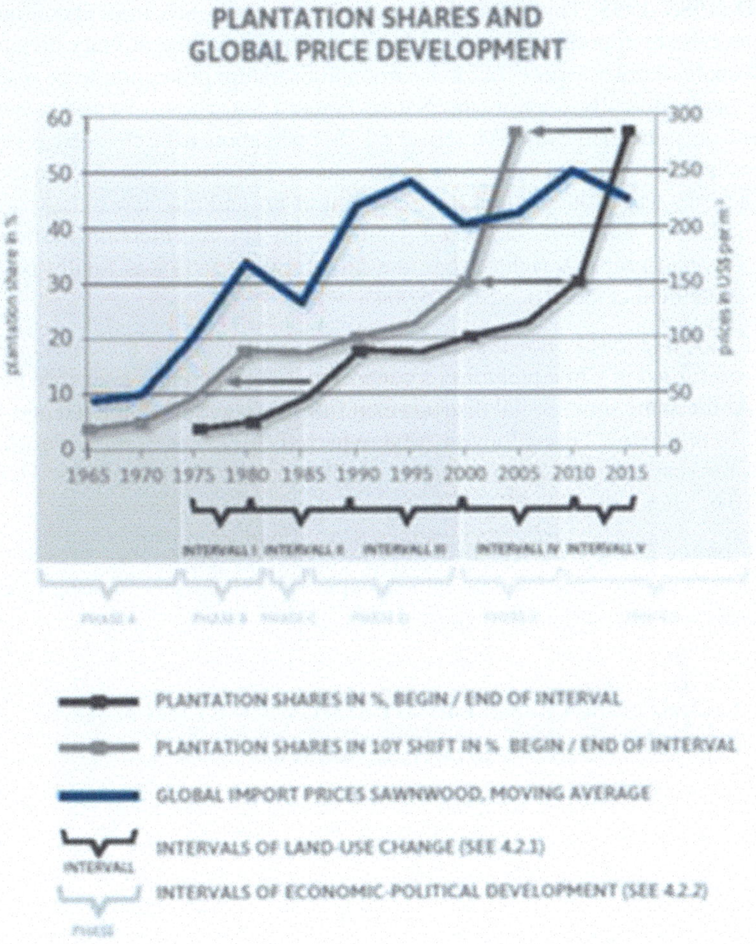

Fig. 5 Phases of economic-political context and import price development to explain land-use changes. *Source* INFOR, 2008, 2017

the same time the DL 701 decree was passed, guaranteeing a high amount of subsidies for plantations even if this was not consistent with the economic ideology of liberalization (Clapp, 1995; Nahuelhual et al., 2012; Niklitschek, 2007; Silva, 2004).

Phase C (1982–1985) Following the Latin American debt crisis, the peso appreciated and Chilean exports became more expensive and continued to fall, initiating a short interventionist intermezzo. After two of the biggest forestry companies fell into significant difficulty and were near collapse, shares of the companies were nationalized again. Tariff rates on imports were raised to 20%. Plantation subsidies were temporarily raised to 90% (Clapp, 1995; Niklitschek, 2007).

Phase D (1985–1998) This phase is characterized by a slightly limited continuation of the liberalization policies started in Phase B. The end of the military dictatorship in 1990 enabled other aspects of environmental or nature protection to be discussed socially and politically, such as the Native Forest Law whose discussion began in 1992 (and only ended in 2007). Moreover DL 701 was about to be changed, restricting subsidies and the rights of large forest companies. But the power of the export-led forest industry was influential on the negotiations around this new bill. Apart from the reduction of subsidies, large companies could enforce their positions on tax exemptions and property rights when the bill (Law 19,561) was finally passed in 1998 (Nahuelhual et al., 2012; Silva, 2004).

Phase E (1998–2008) Following the Asian Crisis of 1998, free trade policies were further reinforced in Chile including a general reduction of tariff rates (Niklitschek, 2007). At the same time, social discourse on forest protection continued and finally resulted in the Native Forest Law in 2008, which includes for example certification systems for sustainable forest management (Heilmayr & Lambin, 2016; Heilmayr et al., 2016; Manuschevich, 2016).

Phase F (since 2008) With the advent of this phase, protection measures of natural forests have become operable but are not expected to have any limiting effects on the extension of plantations (Langenfeld, 2017).

4.2.3 Analysis of the Chilean Forestry Market in the Light of World Market Development

For the purposes of this research, the forestry market has been analyzed with regard to Chilean production and exports (INFOR, 2008, 2017) as well as global import prices (Source: UNECE, 2018, own calculations) for the two largest export products in terms of quantity, namely chemical wood pulp and sawnwood. The development paths of these two products represent different strategies in Chilean forestry market. Sawnwood represents the second-most important Chilean forestry product in terms of production. It can be suggested that, perhaps even more important than DL 701, import prices have been a considerable driver for exports in this time period especially after the export ban has been removed, the consequences of which are quite visible in the export data for this product (cf. Fig. 6a). Until the end of Phase C, the interventionist intermezzo, up to 50% of produced sawnwood has been exported. From 1985 onwards, the export trend goes only slightly upwards, although total Chilean sawnwood production followed the steeply rising global import prices such that a stable export share of only around 30% since the end of 1980s is reached. Perhaps rising global import prices have led to the need for an increase in production to fulfil domestic market needs. Plantation development was relatively stable at that time (Interval III, see Sect. 4.2.1), such that the observed production rise has possibly been within the capacity limit. Therefore, sawnwood seems to have been substituted in the "export portfolio" by other forestry products, with the share of sawnwood of total export value declining from 17% in 1986 to below 10% in 1998 (INFOR, 2017).

This leads into the subsequent discussion of chemical wood pulp, the most important product of Chile's forestry industry.

Contrary to sawnwood, Chilean chemical wood pulp production has always been exported with a share of around 90%, with chemical wood pulp comprising around 45% of the Chilean forestry industry's exports in 2008. Export values only began to rise considerably in 1990 even though the export ban in Chile was removed in 1976 and global import prices more than doubled during that time compared to the beginning of the 1970s (cf. Fig. 6b). This development can possibly be explained by the fact that chemical wood pulp is an industrially processed product requiring further pulp mills that had to be installed before an increase in exports, with three such pulp mills becoming operable in 1990 in the study regions (Clapp, 1995). Since 2001 the export development of this product has followed the trend of global import prices, which typically fluctuate ahead of global economic cycles (Clapp, 1995). Thus, the production and export quantities more than proportionally increased in line with the rising plantation share (see Interval IV) and thus marks the strategy shift in the Chilean forestry industry from domestic to export markets since the beginning of the 1990s.

In sum, this subsection shows that price development on global forestry markets seems to be a relevant stimulus for exports (wood pulp) but also for import substitution (sawnwood), even if the market of the two considered products has turned out to be quite heterogeneous. By way of the above analysis it can be concluded that the end of 1980s marked a strategy shift to explicitly foster industrially processed forestry products such as chemical wood pulp in export markets which have a higher added value compared to unprocessed wood products.

5 Discussion: Integrated Phase Model to Explain Land-Use Changes

The results of the heretofore separated analysis of Sect. 4 have been integrated into a so-called phase model to explain land-use change Intervals I–V in the political-economic context of Phases A-E (Sect. 4.2.2) and the development of the global forestry market (Sect. 4.2.3) considering a time lag of about ten years (see Sect. 3.) This procedure goes beyond previous research, which attributes land-use changes in Chile's forestry sector only to the political-economic context (Langenfeld, 2017; Torres et al., 2015). Figure 5 summarizes the results of Sect. 4 into one phase model diagram. To simplify the comparison, the plantation share rate has been shifted ten years into the past. In addition, as a representative indicator for global forestry market developments, global import prices of sawnwood in a five-year moving average have been added to the diagram, which is explained below. To start with, the relative stability of Interval I with low shares of plantations in both regions below 10% can be explained by nationalization but presumably—and more importantly—by a restrictive trade policy, as demand for Chilean forestry products is to a large extent

(a)

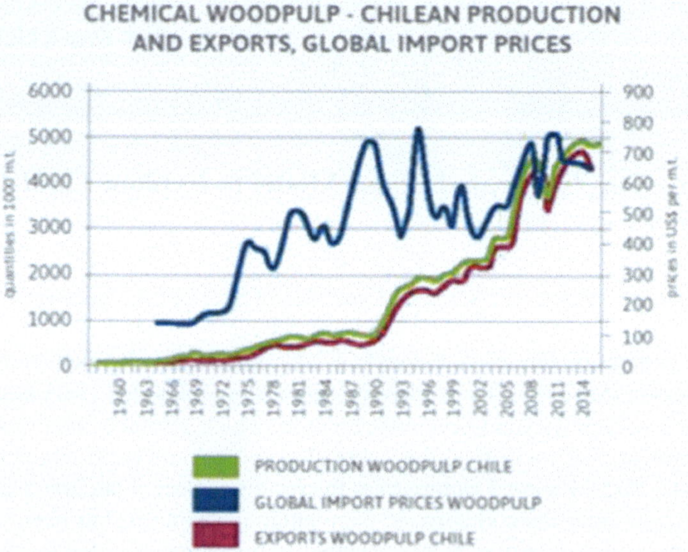

(b)

Fig. 6a Comparison of Chilean production, exports and global import prices of sawnwood. *Source* INFOR (2008, 2017), UNECE (2018). **b** Comparison of Chilean production, exports and global import prices of chemical wood pulp. *Source* INFOR (2008, 2017), UNECE (2018) export-driven (Phase A and Sect. 4.2.1). In the case of chemical wood pulp, which has not yet been listed in INFOR-Statistics in Interval I (INFOR, 2017), there is also a dependency on adequate processing industries.

The considerable rise in plantations during Interval II is to be attributed to two aspects. First, an internal explanation is the start of Chile's military dictatorship and the country's subsequent turn in industrial policy, as well as the subsidies that resulted from the DL 701 decree at the beginning of the 1970s (Phase B). A second, external factor is the continual rise in global import prices of forestry products during that period, serving as a considerable incentive to augment forestry production.

The following ten years of stabilization (1990–2000) that can be observed in Interval III may be explained by the interventionist intermezzo at the beginning of the 1980s (Phase C) when large forestry companies faced significant financial difficulties, thus limiting the expansion of plantations. Exports in the preceding 1980s were also quite stable compared to the previous Interval II, the result being forestry production below capacity limits during this period. Even though Chilean sawnwood production rose in the second half of the 1990s, this development supports the capacity limit argument. A decrease in price development on forestry markets can simultaneously be observed in the mid-1980s, but is possibly less important compared to the internal economic problems in Chile's forestry sector.

Following the interventionist intermezzo, the continuing trend of liberation (albeit a reduced one) of Phase D together with rising (albeit fluctuating) import prices contributed to a rise in plantation shares of up to 30% during Interval IV (2000–2010). Furthermore, since the beginning of the new millennium global import prices have been continuously rising, interrupted only by the Asian crisis in 1998 and the global financial crisis in 2008 with effects of plantation shares that can probably be neglected in the long run. The result is a near-doubling of plantation shares during Interval V to nearly 60% (since 2010).

The effects of a stronger forest protection policy (Phase F) on land-use changes beyond 2015 remain to be seen, but are not expected to be pronounced (Langenfeld, 2017). In this discussion natural limits to land-use shares of industrial plantations have to be considered as well (see Sect. 4.1.2), which is why a stabilization of plantation shares on a high level is expected in the long run.

In conclusion, an integrated analysis of political-economic land-use change drivers shows that internal political and institutional causes are responsible for the first boost in plantation development from 1970 to the end of the 1990s. Since the beginning of the 1990s, the development of world forestry markets seems to be an increasingly important driver of plantation building. During this period, a strategy shift from domestically oriented, unprocessed products such as sawnwood to export-oriented, industrially processed wood pulp can also be observed. For the future, natural capacity limits and potential dampening factors of Chile's New Forest Law have to be awaited.

6 Conclusion

Our analysis aims to disentangle political-economic incentives to actual spatial developments in Chile. Certainly, in order to fully understand these developments, land use data before 1975 would be beneficial, which are simply not available prior to Landsat-MSS's launch in 1972. Nonetheless, it can be concluded that relationships between incentives and the expansion of the forest industry are more complex than frequently assumed, i.e. simply assuming that DL 701 has led to deforestation in a linear and homogeneous fashion. Several changes of trends have been identified, which forced the state and enterprises to change their behavior. Especially since the interventionist intermezzo and following the Chilean debt crisis, forest companies have participated more and more on the world market and reacted to long-term global price developments.

An advantage of an approach like the one presented is that the effects of political-economic incentives are not revealed by assessing entrepreneurial decision making (e.g. by interviews) or via economic indicators but via their spatial concretion. On the one hand, such a spatially explicit perspective is free from any normative notion; on the other hand, it directly reveals effects on environmental assets. By doing so, environmental impacts become immediately apparent and such an approach may complement other approaches such as those based on political economy or ecology, which reveal impacts only immanently while focusing more on power structures. By doing so, the approach of this study can be transferred to other case studies where extractivist land-use changes have caused adverse environmental and social phenomena.

Regarding the regional situation in central Chile, the high rate of subsidization without sufficient regulation on how subsidized areas are used has to be criticized. Forest companies have paved the way for the use of public investment to implement a land use system with pronouncedly negative secondary effects. The loss of natural assets (such as biodiversity and ecosystem services), social conflicts (e.g. with Chilean indigenous populations), impairment of spatial development (by occupying most of the Coastal Range with a single-use system), and natural hazards (such as the 2017 wildfires) are not outweighed by economic benefits, which, at least for Chilean citizens, are low.

Today, the state is hampered in its actions by path dependencies. Since areas and usage rights are sold to private stakeholders, past developments (such as refraining from strong regulations on making plantations sustainable) are difficult to correct. The Chilean government, striving to ensure rule of law and reluctant to threaten the trust of investors, cannot easily apply instruments with strong leverage (such as land reforms). In order to ameliorate the lack of sustainability in plantation forestry, the government has imposed a new "law of native forest recovery and forestry fostering" (Ley 20.283). In order to avoid impacting property rights, the leverage used is harvest. Any harvest is illegal as long as a "management plan" for the area to be harvested is not provided. However, since the requirements for that management plan are markedly vague, it remains to be proven whether path dependencies created by extractive economic policies (e.g. sustainability issues) can be corrected with that instrument.

References

Acosta, A. (2013). Extractivism and neoextractivism: two sides of the same curse. *Beyond Development, 61*, 61–86.

Acuña, R. M. (2015). The politics of extractive governance: Indigenous peoples and socio-environmental conflicts. *The Extractive Industries and Society, 2*(1), 85–92.

Altamirano, A., Aplin, P., Miranda, A., Cayuela, L., Algar, A. C., & Field, R. (2013). High rates of forest loss and turnover obscured by classical landscape measures. *Applied Geography, 40*, 199–211.

Altieri, M., & Rojas, A. (1999a). La tragedia ecológica del milagro neoliberal chileno. *Persona y Sociedad, 1*, 127–141.

Altieri, M. A., & Rojas, A. (1999b). Ecological impacts of Chile's neoliberal policies, with special emphasis on agroecosystems. *Environment, Development and Sustainability, 1*(1), 55–72.

Altieri, M. A., & Rojas, A. (1998). Lessons from a Latin partner: Two decades of free market policies have spurred Chile's economy, but at a steep price. *Alternatives Journal, 24*(4), 24.

Azamar, A., & Ponce, J. I. (2014). Extractivism and development: Mineral resources in Mexico. *Revista Latinoamericana De Economia, 45*(179), 234–242.

Banfield, C. C., Braun, A. C., Barra, R., Castillo, A., & Vogt, J. (2018). Erosion proxies in an exotic tree plantation question the appropriate land use in Central Chile. *CATENA, 161*, 77–84.

Bitterlich, P. F. (2001). Manual de derecho ambiental chileno. Editorial Jurídica de Chile, Santiago de Chile.

Braun, A. C., Troeger, D., Garcia, R., Aguayo, M., Barra, R., & Vogt, J. (2017). Assessing the impact of plantation forestry on plant biodiversity: A comparison of sites in Central Chile and Chilean Patagonia. *Global Ecology and Conservation, 10*, 159–172.

Braun, A. C., & Koch, B. (2016). Estimating impacts of plantation forestry on plant biodiversity in southern Chile—A spatially explicit modelling approach. *Environmental Monitoring and Assessment, 188*(10), 564.

Braun, A. C., Rojas, C., Echeverria, C., Rottensteiner, F., Bahr, H. P., Niemeyer, J., Arias, M. A., Kosov, S., Hinz, S., & Weidner, U. (2014). Design of a spectral–spatial pattern recognition framework for risk assessments using Landsat data—A case study in Chile. *IEEE Journal of Selected Topics in Applied Earth Observations and Remote Sensing, 7*(3), 917–928.

Braun, A. C., Weidner, U., & Hinz, S. (2012). Classification in high-dimensional feature spaces—Assessment using SVM, IVM and RVM with focus on simulated EnMAP data. *IEEE Journal of Selected Topics in Applied Earth Observations and Remote Sensing, 5*(2), 436–443.

Braun, A. C. (2013). *Eine fernerkundungsgestützte geoökologische Prozessanalyse zum Risikozusammenhang zwischen Landnutzung und Biodiversität an einem Beispiel aus Chile.* Ph.D. Thesis, Karlsruhe Institute of Technology (KIT), KIT Scientific Publishing, Karlsruhe.

Browder, J. O. (1992). The limits of extractivism. *BioScience, 42*(3), 174–182.

Bull, G. Q., Bazett, M., Schwab, O., Nilsson, S., White, A., & Maginnis, S. (2006). Industrial forest plantation subsidies: Impacts and implications. *Forest Policy and Economics, 9*(1), 13–31.

Burchardt, H. J., & Dietz, K. (2014). (Neo-) extractivism–a new challenge for development theory from Latin America. *Third World Quarterly, 35*(3), 468–486.

Camus, P. (2003). Federico Albert: Artífice de la gestión de los bosques de Chile. *Revista de Geografía Norte Grande* (30).

Carruthers, D. (2001). Environmental politics in Chile: Legacies of dictatorship and democracy. *Third World Quarterly, 22*(3), 343–358.

Chiasson-LeBel, T. (2016). Neo-extractivism in Venezuela and Ecuador: A weapon of class conflict. *The Extractive Industries and Society, 3*(4), 888–901.
Clapp, R. A. (1998). Waiting for the forest law: Resource-led development and environmental politics in Chile. *Latin American Research Review, 33*(2), 3–36.
Clapp, R. A. (1995). Creating competitive advantage: Forest policy as industrial policy in Chile. *Economic Geography, 71*(3), 273–296.
Cleary, D. (1993). After the frontier: Problems with political economy in the modern Brazilian Amazon. *Journal of Latin American Studies, 25*(2), 331–349.
DL 701. Decreto Ley 701, 15th October, 1974, Ministerio de Agricultura, http://leychile.cl, Last Accessed: 18th of May 2018.
Frêne, C. C., & Nuñez, A., M. (2010). Hacia un nuevo modelo forestal en Chile. *Revista Bosque Nativo, 47*(2), 25–35.
Geist, H. J., & McConnell, W. (2010). Causes and trajectories of land-use/cover change. In E.F. Lambin & H.J. Geist (Eds.), *Land-use and land-cover change—Local processes and global impacts*. Springer.
Geist, H. J., & Lambin, E. F. (2002). Proximate causes and underlying driving forces of tropical deforestation: Tropical forests are disappearing as the result of many pressures, both local and regional, acting in various combinations in different geographical locations. *BioScience, 52*(2), 143–150.
Gorte, B. (2002). Supervised image classification. In A. Stein (Ed.), *Spatial statistics for remote sensing*. Springer.
Gudynas, E. (2018). Extractivisms: Tendencies and consequences. In R. Munck & R.D. Wise (Eds.), *Reframing Latin American development*. Routledge.
Heilmayr, R., Echeverría, C., Fuentes, R., & Lambin, E. F. (2016). A plantation-dominated forest transition in Chile. *Applied Geography, 75*, 71–82.
Heilmayr, R., & Lambin, E. F. (2016). Impacts of nonstate, market-driven governance on Chilean forests. *PNAS, 113*(11), 2910–2915.
Holmgren, M., Avilés, R., Sierralta, L., Segura, A. M., & Fuentes, E. R. (2000). Why have European herbs so successfully invaded the Chilean matorral? Effects of herbivory, soil nutrients, and fire. *Journal of Arid Environments, 44*(2), 197–211.
INE. (2002). Instituto Nacional de Estadisticas, 2002. Censo 2002: Sintesis de Resultados. Santiago de Chile, Chile. Report of Chilean Statistical Authorities.
INFOR. (2017). Anuario Forestal—Chilean Statistical Yearbood of Forestry 2017, Santiago.
INFOR. (2008). Anuario Forestal—Chilean Statistical Yearbood of Forestry 2008, Santiago.
Kottek, M., Grieser, J., Beck, C., Rudolf, B., & Rubel, F. (2006). World map of the Köppen-Geiger climate classification updated. *Meteorologische Zeitschrift, 15*(3), 259–263.
Lambin, E. F., & Geist, H. J. (2010). *Land-use and land-cover change—Local processes and global impacts*. Springer.
Langenfeld, M. (2017). *Waldpolitik in Costa Rica und Chile—Ökosysteme im Spannungsfeld zwischen Markt und Nachhaltigkeit*. Springer VS.
Longley, P. A. (2002). Geographical information systems: Will developments in urban remote sensing and GIS lead to 'better' urban geography? *Progress in Human Geography, 26*(2), 231–239.
Lopez, E., & Vertiz, F. (2015). Extractivism, transnational capital, and subaltern struggles in Latin America. *Latin American Perspectives, 42*(5), 152–168.
Luebert, F., & Pliscoff, P. (2006). *Sinopsis bioclimática y vegetacional de Chile*. Editorial Universitaria.
Martinez-Alier, J., & Walter, M. (2016). Social metabolism and conflicts over extractivism. In F. De Castro, B. Hogenboom, & M. Baud (Eds.), *Environmental governance in Latin America*. Palgrave Macmillan.
Manuschevich, D. (2016). Neoliberalization of forestry discourses in Chile. *Forest Policy and Economics, 69*, 21–30.

Manuschevich, D., & Beier, C. M. (2016). Simulating land use changes under alternative policy scenarios for conservation of native forests in south-central Chile. *Land Use Policy, 51*, 350–362.
Miranda, A., Altamirano, A., Cayuela, L., Lara, A., & González, M. (2017). Native forest loss in the Chilean biodiversity hotspot: Revealing the evidence. *Regional Environmental Change, 17*(1), 285–297.
Miranda, A., Altamirano, A., Cayuela, L., Pincheira, F., & Lara, A. (2015). Different times, same story: Native forest loss and landscape homogenization in three physiographical areas of south-central of Chile. *Applied Geography, 60*, 20–28.
Nahuelhual, L., Carmona, A., Lara, A., Echeverría, C., & González, M. E. (2012). Land-cover change to forest plantations: Proximate causes and implications for the landscape in south-central Chile. *Landscape and Urban Planning, 107*(1), 12–20.
Niklitschek, M. E. (2007). Trade liberalization and land use changes: Explaining the expansion of afforested land in Chile. *Forest Science, 53*(3), 385–394.
Pelfini, A., & Mena, R. (2017). Oligarquization and extractivism: The closed doors on the democratization of environmental policy in Chile. *Perfiles Latinoamericanos, 25*(49).
Petras, J., & Veltmeyer, H. (2014a). *Extractive imperialism in the Americas—Capitalisms new Frontier*. Haymarket Books.
Petras, J., & Veltmeyer, H. (2014b). Imperialist dynamics of US-Venezuela relations. In J. Petras & H. Veltmeyer (Eds.), *Extractive imperialism in the Americas—Capitalisms New Frontier*. Haymarket Books.
Petras, J., & Veltmeyer, H. (2014c). Extractive capitalism and Brazil's great leap backward. In J. Petras & H. Veltmeyer (2014) *Extractive imperialism in the Americas*—Capitalisms new Frontier. Haymarket Books.
Puyravaud, J. P. (2003). Standardizing the calculation of the annual rate of deforestation. *Forest Ecology and Management, 177*(1–3), 593–596.
Ramankutty, N. (2010). Global land-cover change: Recent progress, remaining challenges. In E.F. Lambin & H.J. Geist (Eds.), *Land-use and land-cover change—Local processes and global impacts*. Springer.
Reid, R. S., Tomich, T. P., Xu, J. C., Geist, H. J., & Mather, A. S. (2010). Linking land-change science and policy: Current lessons and future integration. In E.F. Lambin & H.J. Geist (2010) *Land-use and land-cover change—Local processes and global impacts*. Springer.
Renauld, M. (2016). The Esquel effect: Political opportunity structure and adaptation mechanisms in anti-mining mobilisation in Argentine Patagonia. *Canadian Journal of Development Studies/ revue Canadienne D'études Du Développement, 37*(4), 524–540.
Shao, G., & Wu, J. (2008). On the accuracy of landscape pattern analysis using remote sensing data. *Landscape Ecology, 23*(5), 505–511.
Silva, E. (2004). The political economy of forest policy in Mexico and Chile. *Singapore Journal of Tropical Geography, 25*(3), 261–280.
Svampa, M. (2015). Commodities consensus: Neoextractivism and enclosure of the commons in Latin America. *South Atlantic Quarterly, 114*(1), 65–82; Tetreault, D. (2014). Resistance and reform in Mexico's mining sector. In J. Petras & H. Veltmeyer (Eds.), *Extractive imperialism in the Americas—Capitalisms new Frontier*. Haymarket Books.
Tetraulit, D. (2014). Resistance and reform in Mexico's mining sector. In J. Petras & H. Veltmeyer (Eds.), *Extractive imperialism in the Americas*—Capitalisms new Frontier. Haymarket Books.
Torres, R., Azócar, G., Rojas, J., Montecinos, A., & Paredes, P. (2015). Vulnerability and resistance to neoliberal environmental changes: An assessment of agriculture and forestry in the Biobio region of Chile (1974–2014). *Geoforum, 60*, 107–122.
UNECE. (2018). Timber database, historical prices: Export and import unit prices (1964–2015). http://www.unece.org/forests/output/prices.html. 13 July 2018
Van Holt, T., Binford, M. W., Portier, K. M., & Vergara, R. (2016). A stand of trees does not a forest make: Tree plantations and forest transitions. *Land Use Policy, 56*, 147–157.

Veltmeyer, H. (2013). The political economy of natural resource extraction: A new model or extractive imperialism? *Canadian Journal of Development Studies/revue Canadienne D'études Du Développement, 34*(1), 79–95.

Wulder, M. A., Masek, J. G., Cohen, W. B., Loveland, T. R., & Woodcock, C. E. (2012). Opening the archive: How free data has enabled the science and monitoring promise of Landsat. *Remote Sensing of Environment, 122*, 2–10.

Wulder, M. A., White, J. C., Goward, S. N., Masek, J. G., Irons, J. R., Herold, M., Cohen, W. B., Loveland, T. R., & Woodcock, C. E. (2008). Landsat continuity: Issues and opportunities for land cover monitoring. *Remote Sensing of Environment, 112*(3), 955–969.

Yates, J. S., & Bakker, K. (2014). Debating the 'post-neoliberal turn' in Latin America. *Progress in Human Geography, 38*(1), 62–90.

Zamorano-Elgueta, C., Benayas, J. M. R., Cayuela, L., Hantson, S., & Armenteras, D. (2015). Native forest replacement by exotic plantations in southern Chile (1985–2011) and partial compensation by natural regeneration. *Forest Ecology and Management, 345*, 10–20.

Regional Biotic Homogenization by Large-Scale Reforestation: Lessons from the Biodiversity Hotspot of Central Chile

Andreas Ch. Braun and Klara Dolos

Abstract Reforestation for economic purposes but also for climate change mitigation will play an increasing role in the Anthropocene. Economic incentives are suspected to lead to the over-implementation of intensely managed, large-scale forest plantations. Their effects on biodiversity and contribution to biotic homogenization remain controversially discussed. Due to their long existence and strong economic promotion, we argue that the forest plantations of Chile are a unique cautionary tale in this regard. We assess biodiversity conditions in a large (>67,000 km^2) study site on the basis of an exhaustive (195-plot) field assessment. It is shown that an entire landscape has been transformed from ecologically distinct natural forests to plantations with low ecological distinctiveness. We conclude that the Chilean experience may repeat in other countries and outline the importance of considering processes of biotic homogenization in addition to biodiversity losses when planning and implementing reforestation measures.

1 Introduction

Reforestation with planted forests is a phenomenon of utmost importance in the Anthropocene. While the global forest area decreased by 290 million ha between 1990 and 2015, the proportion of planted forests increased by more than 121 million ha according to the Forest Resources Assessment of the Food and Agriculture Organization of the United Nations (FAO), which is why planted forests are said to slow down global deforestation (Payn et al., 2015).

A. Ch. Braun (✉)
Kassel Institute for Sustainability, University of Kassel, Mosenthalstr. 8, 34117 Kassel, Germany
e-mail: Andreas.braun@uni-kassel.de

K. Dolos
Institute of Geography and Geoecology (IfGG), Karlsruher Institute of Technology (KIT), Kaiserstr. 12, 76131 Karlsruhe, Germany
e-mail: klara.dolos@kit.edu

A. Ch. Braun et al. (eds.), *Eco-Industrial Development as an Industrial Strategy*, Green Energy and Technology, https://doi.org/10.1007/978-3-031-73576-9_29

Reforestation is often viewed positively in scientific literature for several reasons (e.g. Paquette & Messier, 2010). On the one hand are economic considerations: planted forests lower the prices of forest products by 24–37% and generate export revenues for operating companies and countries (Buongiorno & Zhu, 2014). On the other hand, ecological expectations for reforestation are relatively high. Studies like Payn et al. (2015), Sedjo and Botkin (1997), and Paquette and Messier (2010) conclude that planted forests will reduce the pressure on semi-natural forests and fulfill a variety of ecosystem services. Reforestation measures are also discussed as a measure of climate change mitigation to generate negative emissions. Doelman et al. (2020) show that negative emissions of up to 410 GtCO_2 could be achieved by 2100. There is thus hope that planted forests can provide both ecological and economic benefits, killing two birds with one stone.

Reforestation activities are likely to increase even more in the future. Global demand for bioenergy is expected to strongly increase prices for many wood products (Buongiorno et al., 2011; Johnston & van Kooten, 2016) and some scenarios expect an almost sixfold increase in demand for fuelwood (Raunikar et al., 2010). The multilateral climate funds provide extensive financial resources ($100 billion USD per year from 2020 to 2025 in the Green Climate Fund) for climate change adaptation and mitigation. Given their low CO_2 storage costs, a substantial amount may flow into reforestation (Honegger et al., 2020). For these reasons, some have predicted an entry into an "age of reforestation" (Yin & Sedjo, 2001).

Of course, different strategies for reforestation exist. One of them is the tree plantation. Despite limited ecological benefits (Lamb et al., 2005; Lewis et al., 2019) tree plantations will continue to play an important role in the future. Lewis et al. (2019) show that almost half of the area already pledged for reforestation by national strategies in the framework of the Paris Agreement is intended to come from such plantations. Doelman et al. (2020) show that more than half of the economically feasible reforestation potential is located in Latin America and Sub-Saharan Africa.

There is a risk that in countries of the Global South, forest policy on the basis of commercial tree plantations is employed as a strategic element of development policy. Environmental concerns are often vastly ignored in this process (Barr & Cossalter, 2004; Clapp, 1995a; Hajdu et al., 2016). It is very important to be aware that even reforestation programs that are explicitly targeted at environmental or climate protection and are internationally subsidized are very easily diverted from their intended purpose by economic interests and political agendas (Barr & Sayer, 2012).

Experience from many countries shows that tree plantations often are not simply established on deforested areas (reforestation). Economic profit expectations often lead to plantations being expanded into near-natural forests and to massive deforestation for the purpose of tree plantation establishment. Examples of this can be found in Indonesia, Malaysia (Austin et al., 2019; Danielsen et al., 2009; Gaveau et al., 2016; Vijay et al., 2016), Cambodia (Davis et al., 2015; Scheidel & Work, 2016), India (Jha et al., 2000), Thailand (Lohmann, 1991, Zambia (Chidumayo, 1989), Ghana (Chiti et al., 2014), Uganda (Hajdu et al., 2016), Kenya (Teucher et al., 2020), Argentina (Zurita et al., 2006), and Chile (Heilmayr et al., 2016; Miranda et al., 2015, 2017).

This poses a challenge for biodiversity hotspots. Biodiversity is defined as the variability among living organisms of all sources, including diversity within species, between species, and of ecosystems (Magurran, 2004). Biodiversity hotspots are also located in regions suitable for reforestation (Brooks et al., 2006; Mittermeier et al., 1998; Myers et al., 2000). As a result, conflicts of objectives arise between economic interests, climate change mitigation, and biodiversity conservation (Caparros & Jacquemont, 2003; Meller et al., 2015; Nunez et al., 2020; Shoyama & Yamagata, 2014). This raises the question of what consequences massive subsidies for reforestation in favor of plantations in the Global South can have on regional biodiversity.

The impacts of plantations on biodiversity remains a controversially discussed issue in the literature (Bremer & Farley, 2010; Brockerhoff et al., 2008; Lindenmayer et al., 2003; Pawson et al., 2013).

Biodiversity conditions in plantations compared to forests may seem irrelevant where plantations are established alongside forests on eroded badlands. However, when plantations replace forests, it is vitally important to understand what this means for biodiversity and how it can be protected.

The effects of plantations on biotic homogenization are even less understood. Biotic homogenization is defined as a "process by which the genetic, taxonomic or functional similarities of regional biotas increase over time" (Olden & Rooney, 2006, p. 113). As Phillips et al. (2017) show, consequences of land-use pressure on biodiversity and biotic homogenization vary markedly across countries, mainly dependent on the type and strength of land-use change. Very few studies have compared processes of biotic homogenization between natural forests and tree plantations. These studies show that the effects vary largely with land-use management, the type of planted forests, and the taxa under investigation. Some studies attest strong impacts on biodiversity and biotic homogenization (Neilan et al., 2019 for birds; Singh et al., 2019 for soil micro- and mesofauna), while some studies document a predominance of biotic homogenization over biodiversity losses (Maas et al., 2009 for birds; Mori et al., 2015 for mites; Manor et al., 2008 for small mammals). Other studies find little effects on biodiversity and biotic homogenization (Lee-Cruz et al., 2013). Mori et al. (2015) correctly claim that biotic homogenization is usually suspected at the global scale and related to processes of biological invasion. Too little attention has been paid to processes of biotic homogenization at regional scales and to mechanisms of biotic homogenization other than biological invasions.

Chile is a country perfectly suited to an investigation into the relationship between forest subsidies, plantation establishment, native forest clearance, and impacts on biodiversity and biotic homogenization. As a consequence of agricultural intensification and non-adapted cultural techniques, central Chile faced heavy erosion towards the end of the 18th century. Forest engineers developed the establishment of *Pinus* plantations on eroded badlands as a strategy to confront erosion (Camus, 2003). Hence, plantation establishment in Chile began with a clear environmental conservation purpose. However, forest policy quickly converted into a strategic development policy. The once environmentally protective measures faded completely into the background with the neoliberal turn under the Pinochet administration, which made plantation expansion a central pillar of economic development (Clapp, 1995a, 1995b, 2001). For more than twenty years (1974–1996) investments in plantation establishment were subsidized by 75% (95% during the debt crisis) by the state (Clapp, 1995a; Ministerio De Agricultura, 1974). Average annual subsidies were as high as $78.6 million USD. Privatization in the forest sector is at 70% and plantations are mainly run by three large companies, which received 80% of all subsidies (Bull et al., 2006; Del Lungo et al., 2006).

The consequences on land use were severe. Several independent studies (summarized in Miranda et al., 2015, 2017) consistently show that due to the lucrative economic incentives, plantations were no longer established on eroded agricultural badlands but by the clearcutting of native forests (cf. Fig. 1). Miranda et al. (2017) estimate that since 1970, as much as 347,816 ha of native forests have been replaced by tree plantations. Today, plantations are found in eight administrative regions of the country. In three of them, plantations have homogenized the landscape, becoming the dominant land-use class (Miranda et al., 2015).

The objectives of this article are to investigate the impact of land-use change in favor of tree plantations on the regional ecology of a large study area in Mediterranean Chile. We are interested in how regional biodiversity has changed and whether a biotic homogenization of the study area has occurred. Thus, our overall objective is

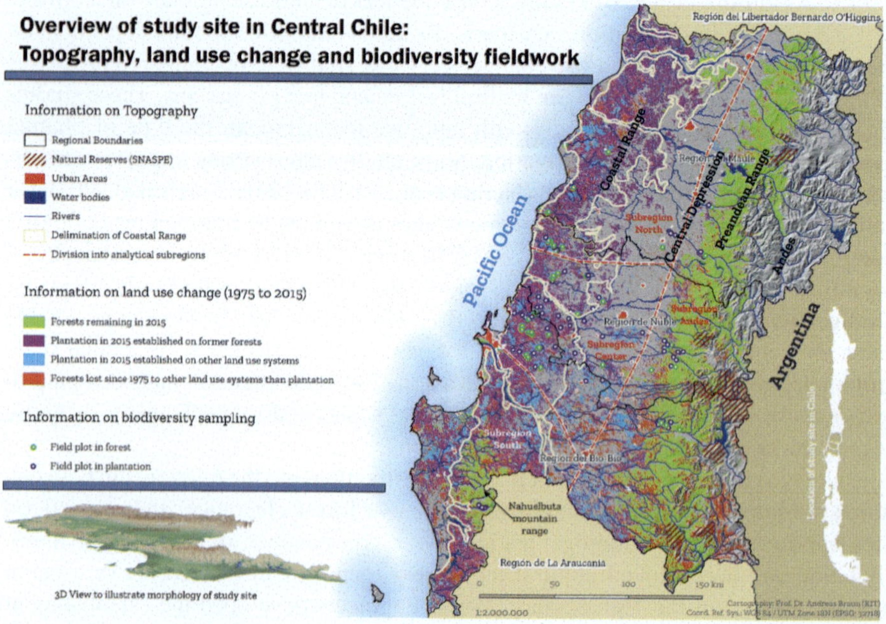

Fig. 1 Overview of the study site. Topographical and elevation data, division into subregions used for homogenization analysis, information on land use change and location of the biodiversity quadrats sampled during fieldwork

to investigate whether massive investments in reforestation measures can lead to a biological impoverishment of the regions if not cautiously implemented.

Our research hypotheses are:

- Biodiversity in terms of species richness, diversity, evenness, and originality of the species present is considerably lower in plantations than in native forests.
- Plantations comprise assemblages that are dominated by a few very abundant species in the tree stratum but also in the understory.
- Community composition in native forest assemblages comprises mainly endemic and native species.
- Community composition in plantation assemblages comprises mainly introduced and invasive species.
- Ecologically distinct forest assemblages in the different subregions of our study site are replaced by ecologically homogenous plantation assemblages.

The remainder of the paper is organized as follows. Section 2 briefly introduces the study site. In Sect. 3 our methods are presented, followed by a presentation of the results on biodiversity and biotic homogenization in Sect. 4. Section 5 discusses the results in the context of the expected global spread of forest plantations. Implications for global reforestation efforts are also derived. Finally, Sect. 6 provides conclusions for both land use research and global forest sector development.

2 Study Site

The VII. Región del Maule, VIII. Región del Biobío, and XVI. Región del Ňuble are part of Mediterranean Chile and extend from 35° to 37° S. The study site covers around 67,358 km^2 and has a population of 2.8 million inhabitants, with a population density of 41 persons per km^2, among the highest in the country (Instituto Nacional de Estadisticas, 2002). Morphologically, the study site is characterized by the Coastal Range, a mountain range running from NNE to SSW. Towards the interior of the study site, the central depression is situated. Eastwards, the Andes form a high mountain region (Fig. 1). The study site has a Mediterranean Csb climate (Koeppen–Geiger), an annual mean temperature of 12 °C, and annual precipitation of 1300 mm (Kottek et al., 2006; Peel et al., 2007). Vegetation is characterized by a latitudinal gradient of sclerophyllous forests in the north, deciduous forests in the center, and coniferous forests in sites of higher elevation, in the south, and in the Andes (Luebert & Pliscoff, 2006; Moreira-Muñoz, 2011). According to Myers et al. (2000), Central Chile hosts 3,429 plant species, of which 1,605 are endemic (0.5% of global endemics).

3 Methods

3.1 Plant Assemblages Studied

We have divided the natural and planted vegetation into main plant assemblages (cf. Fig. 2). All forest vegetation is classified as forest assemblage (FA) and all commercial tree plantations as plantation assemblages (PA). Areas of FA are comprised of typical Chilean primary and secondary forests. They are not used industrially but for firewood extraction, as forest pasture, and for non-timber forest products (NTFP). Exceptions to this are the southern *Araucaria araucana* forests and riparian *Luma apiculata* forests. The plantation assemblages (PA) comprise the commercial tree plantations, which are used industrially as short-rotation plantations harvested by clearcutting (Salas et al., 2016).

To allow for a more precise ecological analysis, the main plant assemblages were divided into eight specific plant assemblages according to the main tree species (three plantation assemblages and five forest assemblages). They are mainly oriented on the vegetation classification of Luebert and Pliscoff (2006). Each of the specific plant assemblages receives an abbreviation. The abbreviation begins with the main plant assemblage to which it is assigned (f for forest assemblages, p for plantation assemblages). This is followed by the first letters of the genus and species name of the dominant tree species.

Deciduous forests dominated by *Nothofagus obliqua* (Mirb.) (fNo) are the typical forest type of the central Coastal Range of the study site, while in the north, sclerophyllous forests dominated by *Cryptocarya alba* (Molina) Looser (fCa) are typical. In the pre-Andean range, deciduous forests dominated by *Nothofagus decidua* (Mirb.) (fNd) are present. Along rivers and streams, riparian forests dominated by *Luma apiculata* (DC.) Burret (fLa) are typical for the study site. In the southern part of our study site, higher than 800 m above sea level, mountainous forests dominated

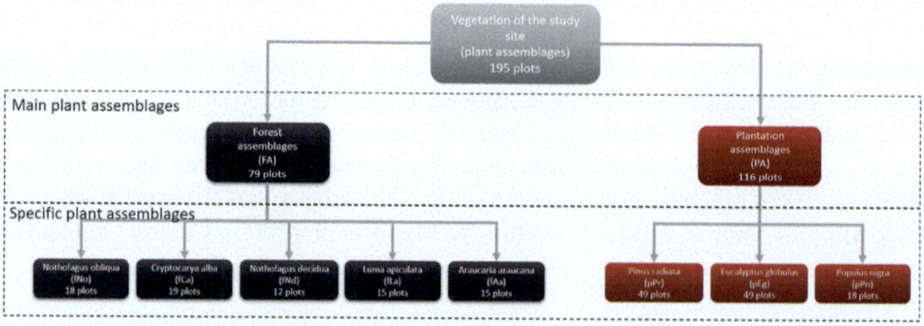

Fig. 2 Structure of the field data. In total, 195 field plots are divided into main plant assemblages, of which 79 consist of forest assemblages and 116 are plantation assemblages. Main plant assemblages are divided into specific plant assemblages (five types of forest assemblages and three types of plantation assemblages)

by *Araucaria araucana* (Molina) K. Koch (fAa) are found. We sampled this type of forest in the Nahuelbuta mountain range (cf. Fig. 1). Within the Coastal Range and the pre-Andean range, plantations of *Pinus radiata* D.Don (pPr) and of *Eucalyptus globulus* Labill. (pEg) are typical. Plantations of *Populus nigra* L. (pPn) occur very rarely, but have been assessed to generalize the findings for pPr and pEg to a third type of plantations.

3.2 Field Sampling and Data Preparation

In subsequent field campaigns (2011, 2012, 2013, 2016), we sampled the plant biodiversity of the eight plant assemblages in terrain according to the sampling methodology of Braun-Blanquet (1964). In total, 195 plots were sampled (cf. Figs. 1 and 2). For the main plant assemblages, we sampled FA = 79, and PA = 116 plot respectively. For the specific plant assemblages, we sampled fNo = 18, fCa = 19, fNd = 12, fLa = 15, fAa = 15, pPr = 49, pEg = 49, pPn = 18 plot. In total, 257 species (trees, shrubs, herbs, epiphytes) were observed. Each plant species present in a plot was recorded and identified. Relative abundance rs is the share of the species in the assemblage in percentage terms. We estimated relative abundance applying the method of Pellissier et al. (2004). The abundance-dominance value ds of each species s was estimated, using Braun-Blanquet's methodology (r = max. 3 individuals, += max. 10 individuals, 1 = 1–5%, 2 = 6–25%, 3 = 26–50%, 4 = 51–75%, 5 = 76–100%). To allow for metrical computation of biodiversity indices, the categorical ds values were transformed into metrical cover values cs, adapting the approach by Pellissier et al. (2004). The following cs values were defined: ds = r → cs = 0.007, ds =+ → cs = 0.015, ds = 1 → cs = 0.030, ds = 2 → cs = 0.150, ds = 3 → cs = 0.375, ds = 4 → cs = 0.625, ds = 5 → cs = 0.875. Relative abundance values rs for each species were computed according to rs = cs/Σcs S, where S is the total number of species found in a plot h.

3.3 Assessment of Biodiversity

3.3.1 Biodiversity Indices for Each Plot

Tuomisto (2010a, 2010b) has provided a systematic nomenclature of biodiversity indices. For biodiversity plots, several key indices can be derived from relative abundance rs data of the species present in the plot. Indices are determined by computing the basic sum Σrs. The basic sum is exponentiated with q (i.e. Σrsq). A different choice of q produces different biodiversity indices. The simplest index is species richness SR = Σrs0 (q = 0), which simply describes the number of species present in the plot. Since rare and common species contribute equally to SR, SR is insensitive to species abundance. Indices sensitive to species abundance use q > 0. For instance, the well-known Simpson index is computed with q = 2, i.e. Σrs2. We follow Tuomisto's

(2010a, 2010b) suggestion to characterize species diversity by the number of effective species. Effective species is the number of equally-common species required to give a particular value of an index. This index for species diversity is produced by inverting the Simpson index, i.e. by computing SD = (Σrs2) − 1. The balance of the frequency distribution of the species of an assemblage can also be well described by the ratio of SR and SD, which we call species evenness SE. SE is calculated as SE = SR/SD. Finally, it is important to record not only the number and abundance of species, but also their origin. Some assemblages are formed almost exclusively by native species and contain many endemic species; other assemblages consist almost exclusively of introduced or even invasive species. We define a multiplier os = +1.0 for endemic, os = +0.5 for native species, os = −0.5 for introduced and os = −1.0 for invasive species. Then, we calculate species originality SO = rs2os. SO indicates to which extent a plot is composed by vegetation originating from the study site. Community composition with respect to origin was assessed by calculating the sum of occurrences of endemic, native, introduced, and invasive species in all plots of a particular plant assemblage and expressing the respective occurrences as percentages of the total number of occurrences within the plots of this plant assemblage.

3.3.2 Rank Abundance Curves of Most Typical Plots

Rank abundance curves (sometimes referred to as Whittaker plots) are used complementary to biodiversity indices, since they outline the contributions of individual species to species assemblages. Shallow curves indicate even communities, while steep curves indicate communities dominated by few species. In order to provide illustrative rank abundance curves for our data, the most typical plot for each specific plant assemblage is identified according to the following method. SR, SD, SE and SO are standardized (z-transformed). The most typical plot for each study system is defined by identifying the plot for which zSR + zSD + zSE + zSO came closest to the arithmetic mean of z-transformed values (i.e. least absolute derivation from μZ = zSR + zSD + zSE + zSO) over all plots belonging to that study system.

3.3.3 Identification of Biodiversity Reductions in Different Systems

In order to assess to which extent one study system A is reduced in terms of species in comparison with another study system B, the following methodology is applied. First a Kruskal–Wallis test is applied to the main plant assemblages as well as specific plant assemblages. Both tests are significant (alpha = 0.05). This allows us to calculate post-hoc tests to find the groups causing this positive test result. However, we apply not only a normal post-hoc test. The SR of all plots of A is multiplied with a factor σ, with σ = [1, …, 0]. We start with $\sigma = 1$ which corresponds to normal post-hoc testing, and then iteratively diminish σ in steps of 0.1. After each multiplication, a one-sided Wilcox test is performed to test whether the SR(A) * σ is significantly greater than SR(B) ($\alpha = 0.05$). The procedure is stopped as soon as the Wilcox test fails to produce a significant result. The last multiplier σ for which the test is significant is considered as the critical value σcr. We interpret a σcr of 0.2 to show that B contains 20% of the species of A (i.e. is reduced in terms of species richness SR by 80%).

3.4 Assessment of Biological Homogenization

In our study site, ecologically different kinds of forests (in the north, south, center of the Coastal Range, and in the Andes), have been replaced by the same kind of plantations (*P. radiata* and *E. globulus*) in the entire study site. Ecologically, this may lead to homogenization of central Chilean ecosystems. In order to assess homogenization, the following analysis is performed. In each of the four subregions, plantation assemblages and forest assemblages are sampled. In the north (N) fCa, pPr and pEg are sampled; in the center (C), we sample fNo, pPr and pEg in the south (S) fAa, pPr and pEg; and in the Andes (A), fNd, pPr and pEg. For each pair of plots [hx hy], x = 1, …, X, y = 1, …, Y, from two different sets (example 1: set X: northern forests vs. set Y: northern *Pinus* plantations, example 2: set X: southern forests vs. set Y: central *Eucalyptus* plantations), pairwise similarity is computed. The pairwise similarity ps index provided by Lennon et al. (2001) is applied as discussed in Tuomisto (2010b). For two plots, $ps = a/(a + \min[b, c])$, where a is the number of species S occurring in both plots and b and c are the numbers of species occurring in one plot but not in the other. Afterwards, ps are averaged over the entire set. Note that pairwise similarity values $\mu ps = 1/(X * Y) * \Sigma ps$ represents the average similarity between the different sets of plots. Forest assemblages of the riparian fLa type are excluded from this analysis, since their distribution depends predominantly on particular morphological and edaphic conditions rather than on geographical latitude/longitude in our study site. Plantation assemblages of the pPn *Populus nigra* type are excluded since they occur rarely and are both widely and randomly distributed in our study site.

4 Results

4.1 Results on Plot Scale Indices

With regard to species richness SR, the main plant assemblages (Fig. 3) show that this is significantly lower for plantation assemblages than for forest assemblages. For forest assemblages a median of SR = 26 species is observed, while plantation assemblages show a median of SR = 10 species. The results on species diversity SD show the same tendency very clearly. The species evenness SE of plantation assemblages is much lower than for forest assemblages. The most significant differences are found in the species originality SO. Here it can be seen that forest assemblages are typically characterized by native species, but plantation assemblages are typically characterized by non-native species; the lowest values of plantation assemblages show that some plots contain hardly any area-typical species.

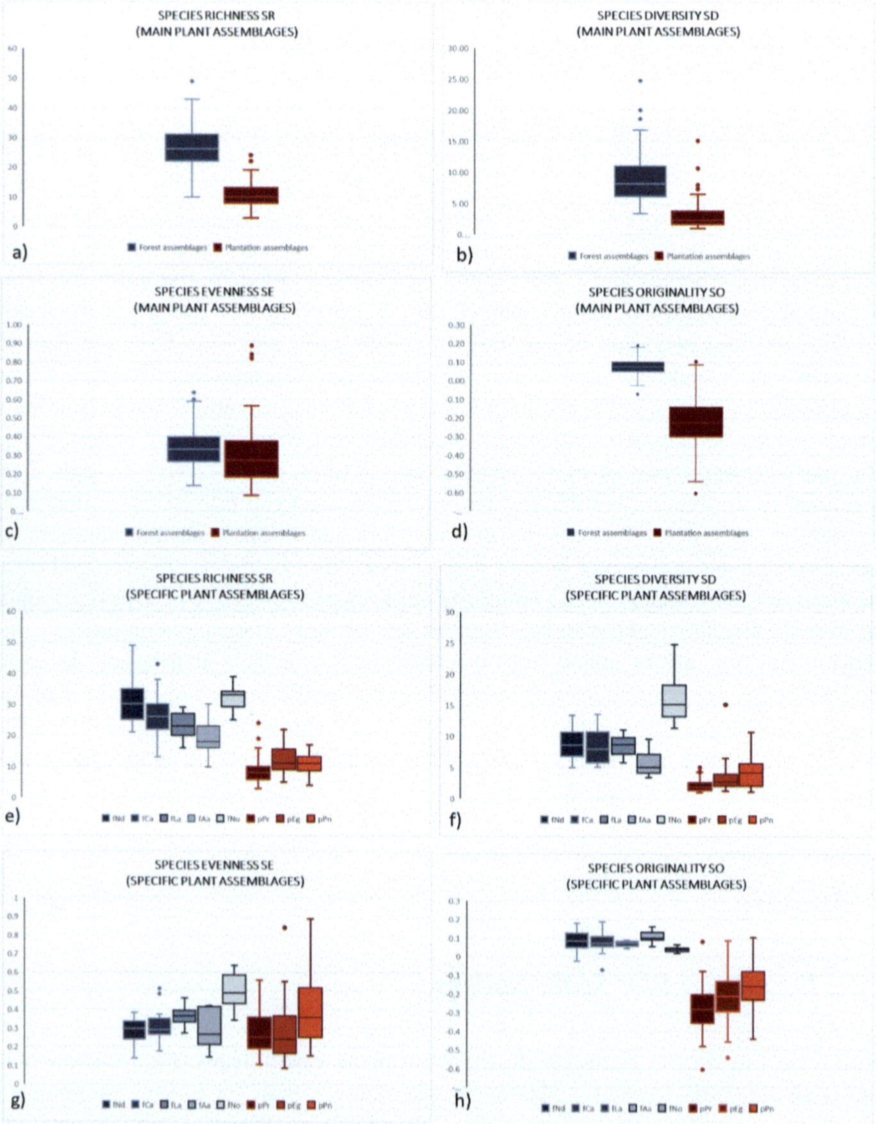

Fig. 3 Boxplot results for plot scale indices species richness SR, species diversity SD, evenness SE and species originality SO for the set of 195 biodiversity quadrats. Top: main plant assemblages (FPA and PPA), Bottom: specific plant assemblages (fNo, fCa, fNd, fLa, fAa, sAc, pPr, pEg, pPn)

The specific plant assemblages show the same tendencies as the main plant assemblages but differentiate them. Within forests, the fNd forests of the pre-Andean range reach the highest species richness SR and species diversity values SD, while the fAa forests of the Nahuelbuta range the lowest. Within the plantation assemblages, the pPr plantations that make up most of the plantations in the terrain (Salas et al., 2016) achieve the lowest species richness, species diversity, and species originality values. The pPr plantation richest in species contains no more species than the fNd forest plot poorest in species that we sampled.

4.2 *Results on Most Typical Plots*

Using the method described in Sect. 3.3.2, we have identified the most typical plot for the main plant assemblages. Figure 4a shows the results for the rank-abundance data. The difference between the most typical forest assemblage (a *N. dombeyi* forest) and the most typical plantation assemblage (a *P. radiata* plantation) is very distinctive. The forest assemblage is characterized by a flat gradient of the rank abundance curve, in which more than fifteen species occur with a high abundance. In contrast, in the plantation assemblage three species dominate, with the rank 1 species having a relative abundance of over 0.6.

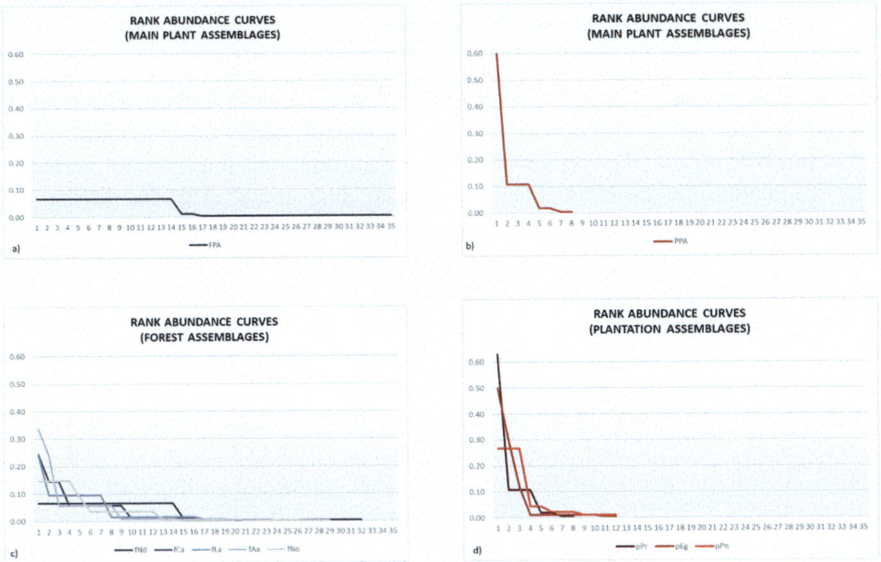

Fig. 4 Rank-abundance curves (Whittaker-Plots) for the most typical quadrats. Species ranks plotted against relative abundance ra values. Top: main plant assemblages. Bottom: specific plant assemblages

Next, we identified the most typical plots for specific plant assemblages (Fig. 4b, c). It becomes clear that the low-ranked species in forest assemblages have relatively low relative abundances r_S. The highest relative abundance of the species with rank 1 reaches the value $r_s = 0.33$ in the plot of the fAa forest. Also, the species with higher ranks reach even higher relative abundances. In contrast, the species with lower ranks in the plantation assemblages achieve significantly higher relative abundances. The relative abundance of the species with rank 1 in the pPr plantations reaches a value of $r_s = 0.6$. The rank abundance curves of the forests are generally flatter, while those of the plantations have steeper slopes.

Figure 5 shows the origins of the species present in the plots. The forest assemblages are dominated by native and endemic species, which together comprise 83% of species occurrences in these plots. In contrast, plantation assemblages are dominated by introduced and invasive species, which comprise 61% of occurrences. The amounts of endemic and native species, respectively, are only half of the values for forest assemblages. A comparison of forest assemblages reveals that results are consistent over all types of forests. The amount of endemic and native species is highest in fAa and fLa forests, which are least affected by human use. All plantation types are characterized by amounts of introduced and invasive species that exceed 56%. The rather uncommon pPn plantation assemblage is most strongly dominated by introduced and invasive species.

4.3 Results on Biodiversity Reduction

We applied the methodology to assess biodiversity reductions described in Sect. 3.3.3. Firstly, we analyzed the differences between main plant formations. Plantation plant formations host 40% of the species of forest plant formations, thus transformation of forests to plantations results in 60% species loss. Next, we assessed the differences between specific plant formations. Figure 6 shows the results regarding the reduction of species richness SR of plantation assemblages in comparison to forest. Reductions span a range from 30% (pPr vs. fAa) to 70% (pPr vs. fNo/fNd). *P. radiata* pPr plantation assemblages show the highest average reduction (62%) in comparison to forest assemblages they have replaced. *E. globulus* pEg planation assemblages are reduced on average by 46% of species present in forest assemblages; the average reduction values for *P. nigra* plantation assemblages are 52%. Calculating the average reduction value of fNo forest assemblages versus any plantation type yields a value of 60% (70% in the case of fNo to pPr, 50% for fNo to pEg, 60% for fNo to pPn). For fCa, the value is 53%, 50% for fLa, 40% for fAa, and finally 63% for fNd. These data show that the most frequent plantation assemblages (pPr) are most reduced in terms of species richness SR. Vice versa, the forest assemblages most strongly replaced by commercial tree plantations (fNo and fNd) are the forest replacements leading to the highest reduction values. The most frequent replacement type fNo/fNd to pPr cause statistically significant reductions of 70% of species richness SR.

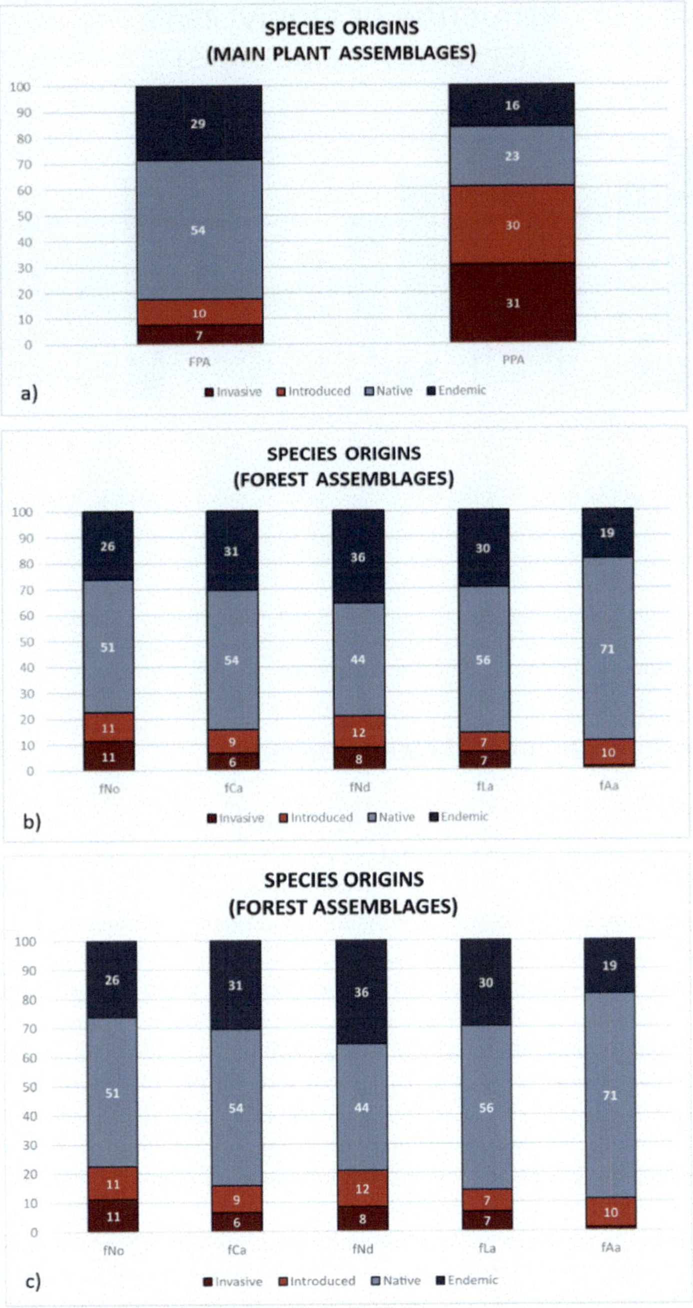

Fig. 5 Species origins: occurences of endemic, native, introduced and invasive species as percentage of total occurences within the plots of a plant assemblage. Left: main plant assemblages, center: forest assemblages, right: plantation assemblages

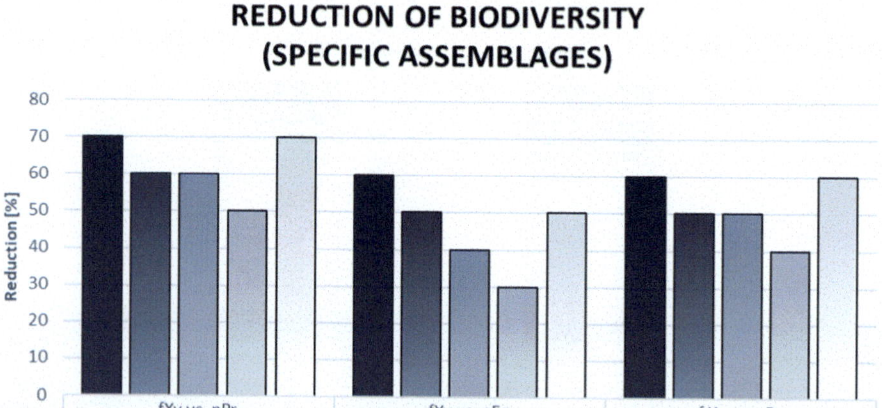

	fXy vs. pPr	fXy vs. pEg	f Xy vs. pPn
■ fNd	70	60	60
■ fCa	60	50	50
■ fLa	60	40	50
□ fAa	50	30	40
□ fNo	70	50	60

Fig. 6 Reductions in biodiversity between specific forest and plantation assemblages. Values indicate the statistically significant ($\alpha = 0.05$) multiplier σ on species richness for the minor group that contains more species, for which a Wilcox test is positive

4.4 Results on Biological Homogenization

Figure 7 visualizes the pairwise similarity within two plantation assemblages (pPr and pEg) and within forest assemblages. Tables 1 and 2 summarize numerical figures and test for significance of the differences. The width of connecting lines is proportional to pairwise similarity values PS between the different plant assemblages. The numbers close to the lines indicate the precise values. The inner (blue) lines provide the results for PS values among each system sampled in four different regions (e.g. a value of 0.39 for pairwise similarity of $pPr_{(N)}$ and $pPr_{(S)}$ denotes a rather high share of common species of northern pine plantations and southern pine plantations, Fig. 7a). The outer (red) lines in Fig. 7a, b indicate the results for plantations of one subregion with forests of the same subregion. In Fig. 7c red lines indicate PS values for forests of a subregion with the less-similar plantation type (either *Pinus* or *Eucalyptus*) of the same subregion. In addition, the value for the more similar plantation assemblage of that subregion is provided in brackets.

Firstly, it is important to analyze the pairwise similarity PS values for the three assemblages within the different subregions (i.e. the forest assemblage typical for the respective subregion fXy, pPr, and pEg). Results for forests (Fig. 7c) show that many comparisons between forests produce low pairwise similarity. The least values are

Table 1 Significance of subregional comparisons. The table shows pairwise similarity values PS values for one study system from different subregions (from north N, center C, south S, or Andes A). Furthermore, it tests for significance of the differences another according to one-way ANOVA (*: $P < 0.05$, **: $P < 0.01$, ***: $P < 0.001$; ****: $P < 0.001$). fXy(R) stands for forests from the same subregion as the plantation, they are tested again (e.g. in the ANOVA of $pPr_{(N)}$ vs. $fXy_{(R)}$, $fXy_{(R)} = fCa_{(N)}$, in the ANOVA of $E_{(C)}$ vs. $fXy_{(R)}$, $fXy_{(R)} = fCa_{(C)}$). The sum Σ indicates the number of significant differences in each row

	$fXy_{(R)}$	$pPr_{(N)}$	$pPr_{(C)}$	$pPr_{(S)}$	$pPr_{(A)}$	Σ
$P_{(N)}$	0.44	–	**0.51***	0.39	**0.48**	2
$P_{(C)}$	0.43	**0.51***	–	**0.46**	**0.47**	3
$P_{(S)}$	0.06	**0.39******	**0.46******	–	**0.41******	3
$P_{(A)}$	0.55	0.48	0.47	0.41	–	0
	$fXy_{(R)}$	$pEg_{(N)}$	$pEg_{(C)}$	$pEg_{(S)}$	$pEg_{(A)}$	Σ
$E_{(N)}$	0.44	–	0.44	0.35	0.41	0
$E_{(C)}$	0.43	**0.44******	–	0.31	0.31	1
$E_{(S)}$	0.08	**0.35******	**0.31******	–	**0.28******	3
$E_{(A)}$	0.55	0.41	0.32	0.28	–	0

Table 2 Significance of subregional cross-comparisons. The table shows the results of tests of pairwise similarity values PS of forests from different subregions (from north N, center C, south S, or Andes A) being significantly different from the pairwise similarity values PS of *Pinus* (column 2) or *Eucalyptus* (column 3) plantations from the same subregions X, Y according to one-way ANOVA (*: $P < 0.05$, **: $P < 0.01$, ***: $P < 0.001$; ****: $P < 0.001$)

Subregional comparison	Forests versus *Pinus* plantations p of $fXy_{(R1)}–fXy_{(R2)}$ versus $pPr_{(R1)}–pPr_{(R2)}$	Forests versus *Eucalyptus* plantations p of $fXy_{(R1)}–fXy_{(R2)}$ versus $pEg_{(R1)}–pEg_{(R2)}$
North to center	0.005**	0.660
North to south	0.000****	0.000****
North to Andes	0.000****	0.000****
Center to south	0.000****	0.000****
Center to Andes	0.004***	0.000****
South to Andes	0.000****	0.000****

observed in comparisons with southern forests from the Nahuelbuta range ($fNd_{(C)}–fAa_{(S)}$: 0.05, $fNo_{(N)}–fAa_{(S)}$: 0.08, $fNo_{(A)}–fAa_{(S)}$: 0.10), but also, the value for $fCa_{(N)}–fAa_{(A)}$: 0.08 is relatively low. On average, pairwise similarity values PS of all forest assemblage comparisons are 0.24. In contrast, Fig. 7a shows that *Pinus* plantation assemblages from all four subregions bear a large similarity to each other, with the lowest pairwise similarity values PS value being 0.39 and the highest 0.51 (the average of all comparisons between *Pinus* plantations 0.45). This result shows that *Pinus* plantation assemblages are determined to a high degree by the same species

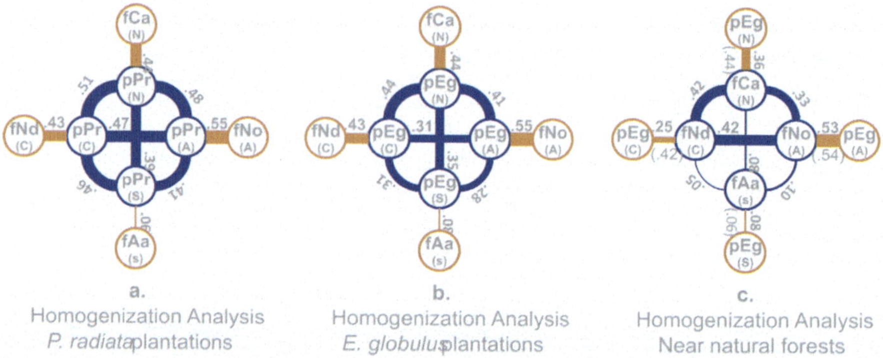

Fig. 7 Results of the homogenization analysis. Average β-similarity values μβSxy for the comparisons of different study sites X, Y. Comparisons of Pinus plantations, Eucalyptus Plantations and forests from different subregions (North N, Center C, South S, Andes A). Forests from different subregions refer to F(N) = fNo, F(C) = fCa, F(S) = fAa, F(A) = fNd

composition over all four subregions. The similarity for *Eucalyptus* plantation assemblages shown in Fig. 7b is less pronounced (average of 0.35) but still relatively high. The lowest pairwise similarity values PS value is for the comparison $E_{(A)}$–$E_{(S)}$ = 0.28.

Secondly, it is interesting to contrast the pairwise similarity values a plantation assemblage yields with the identical plantation assemblages from other subregions with those values, and the plantation assemblage yields with the forest assemblage from its own subregion (i.e. to compare the values of the red vs. blue lines in each of the figures). Figure 7a shows that the similarity of pPr plantation assemblages of most subregions to pPr plantation assemblages of other subregions tends to be higher than pairwise similarity with forests of the same subregion. For instance, $pPr_{(S)}$ are more similar to $pPr_{(N)}$, $pPr_{(C)}$, and $pPr_{(A)}$ than to $fAa_{(S)}$. The same is true for $pPr_{(C)}$ and $pEg_{(S)}$. $pPr_{(N)}$ and $pEg_{(C)}$ are at least more similar to some plantation assemblages of other subregions than to the respective forests of the same subregion. Table 1 summarizes the findings and shows that in most cases, differences are highly significant.

Finally, it is interesting to evaluate whether pairwise similarity between forest assemblages from different subregions is smaller than pairwise similarity values of plantation assemblages from the same subregions. That is, to draw comparison like [$fCa_{(N)}$–$fNd_{(C)}$, $pPr_{(N)}$–$pPr_{(C)}$, $pEg_{(N)}$–$pEg_{(C)}$] or [$fAa_{(S)}$–$fNo_{(A)}$, $pPr_{(S)}$–$pPr_{(A)}$, $pEg_{(S)}$–$pEg_{(A)}$]. For most comparisons, forest assemblages yield the lowest pairwise similarity values PS. The only exception is the comparison between northern and Andean assemblages. There, forests are *Nothofagus*-dominated and many species of the understory appear in both, albeit with different abundances, for which the pairwise similarity index is insensitive. The southern *Araucaria* forests ($fAa_{(S)}$) in particular are floristically different to other forest types. The similarity of different plantations is pronounced, particularly for pPr plantations. The average pairwise similarity values PS values of forests is 0.23. Table 2 shows that the differences observed are statistically significant (with one exception). Hence, it can be observed that plantations of different subregions are significantly more similar than forests in the different subregions.

5 Discussion

5.1 Discussion of Biodiversity Results

Results show reductions of biodiversity in plantations regarding species richness and species diversity. The evenness of plantations varies largely, a finding that has to do with the plantation rotation cycle. Directly after clearcutting and replanting, young plantations have a herbaceous layer, characterized by a set of species with relatively even abundances. In contrast, rank abundance curves in plantation assemblages are steep. Differences in rank abundance data have been related to phenomena of environmental change (Collins et al., 2008; McCarthy et al., 2018). While dominance within the tree stratum of plantations is straightforward, scientific understanding of understory development in plantations is limited (Aubin et al., 2008). In our case, diverse forests with high evenness are replaced by plantations characterized by a few dominant species. Once the canopy closes and plantation litter covers the ground, the plantation tree and the few shrub species present in the understory are the only species with higher abundance. These results are consistent with findings from other areas, such as Sumatra (Rembold et al., 2017) and Patagonia (Paritsis & Aizen, 2008). Results underpin the finding of Aubin et al. (2008) that conifer plantations show a much more limited understory development than deciduous plantations. Results for species originality show that plantations demonstrate a pronounced reduction in native and endemic species and a high abundance of invasive and introduced species, which is consistent with the findings of Sample et al. (2019) for Argentina. The most biodiverse forest assemblages are replaced by the least biodiverse plantation assemblages. Statistically significant biodiversity losses of up to 70% of the species typical for forests of the biodiversity hotspot are the result. Such losses exceed the results for other countries, even in tropical areas (Rembold et al., 2017).

The *Araucaria araucana* forests of the Nahuelbuta range play an interesting role in regional biodiversity. Their contribution to biodiversity at the plot scale is lower than other forests. However, at the landscape scale, they play an important role in the sense that they are ecologically very different from other types of forests. A conservation strategy for *A. araucana* forests is important to prevent losses in landscape scale biodiversity. The recent advances of the forest industry towards higher altitudes of the Nahuelbuta range ought to create concern from a conservational perspective (Braun et al., 2017).

As Tscharntke et al. (2012) and Gámez-Virués et al. (2015) point out, more complex landscapes preserve a landscape-moderated insurance effect, in which biodiversity at the landscape scale insures against ecosystem function decline by species distributed over the landscape. This effect is threatened by the biotic homogenization we observe. This leads to ecologically simplified and monotonous ecosystems which could lose their ability to provide habitat to specialist species (Rédei et al., 2020).

There has been a long debate on the biodiversity impacts of tree plantations (Bremer & Farley, 2010; Brockerhoff et al., 2008; Lindenmayer et al., 2003; Pawson et al., 2013). Most studies focus on comparisons for other taxa rather than plants. Duguid and Ashton (2013) performed a meta-review on ninety-six studies published on biodiversity effects of forests managed for timber, mainly North American temperate forests. Across all studies, they did not find significant effects to understory richness. Kostić et al. (2016) found a moderate reduction of understory plant diversity in *Pseudotsuga menziesii* plantations in comparison to *Fagus moesiaca* forest stands in Serbia. Paritsis and Aizen (2008) found plant diversity in Patagonian conifer plantations to be around 50% lower than in native forests. Instead of profoundly reiterating on the debate about the biodiversity value or impacts of plantations at the plot scale (green deserts or not), it is important to see plantations in their regional context. Paquette and Messier (2010) and Stephens and Wagner (2007) have argued that it is pointless to compare plantation biodiversity to forest biodiversity, favoring instead a focus on the land-use systems that would exist instead of plantations. This is true if plantations are established on sites formerly not covered by trees. In cases of replacement of forests by plantations, understanding to what extent plantations can maintain regional biodiversity is of utmost importance. Such replacement takes place in many (tropical and non-tropical) countries of the Global South. Rembold et al. (2017) compare biodiversity of tropical forests to rubber and oil palm plantations in Sumatra. In their study, the plantations had 59% fewer plant species than the forests. Our results for the Chilean temperate zone are even higher and confirm plot-based findings by Braun et al. (2017) and Heinrichs et al. (2018). Deforestation in favor of plantation establishment tends to replace forests, which causes the highest reductions in biodiversity (*N. obliqua* and *N. dombeyi* forests). Furthermore it tends to establish the *Pinus* plantations which host the lowest biodiversity. Despite the fact that concerns about the biodiversity impacts of plantation forestry have been raised quite early (Frank & Finckh, 1997; Valenzuela et al., 1998), it took a sixteen-year debate before a new law to prevent forest replacement was finally passed in Chile in 2008. Its effectiveness for forest protection and restoration remains in doubt to this day (Manuschevich, 2020).

5.2 *Discussion of Biotic Homogenization*

Neilan et al. (2019) stress the importance of not comparing the effects on biodiversity in human-modified landscapes solely on the basis of assessments at the plot scale. They suggest that landscape-scale effects may cause just as severe consequences for biodiversity. Such effects will only become visible by considering biodiversity assessments among different types of assemblages in dependence of land use. Our results underscore this finding. For bird communities in a fragmented Indonesian landscape dominated by plantations, Maas et al. (2009) did not find biodiversity reductions, but rather biotic homogenization. In contrast, our study documents both phenomena. Plantations in the Chilean temperate zone have effects on biodiversity at the plot scale that are similarly severe as in tropical regions, where forests are usually considered to be more diverse.

Perhaps even more relevant are the processes of biotic homogenization, which are overlooked by analyses based only on plot-scale indicators. Chilean plantations are impoverished in distinct endemic species typical for Chilean forests, and show the consistent occurrence and high abundance of introduced and invasive species (shared by all plantation types in all subregions). The subregional homogenization analysis has shown that ecologically distinct forest types are replaced by ecologically similar plantations. According to our results, plantation forestry homogenizes plant communities and imposes the risk of plant invasions spreading from plantations to the remnants of near-natural native forests (Bustamante & Simonetti, 2005; Simberloff et al., 2010), a process documented in other countries as well (Calviño-Cancela & van Etten, 2018). Our results are consistent with the results of Cifuentes-Croquevielle et al. (2020), who demonstrate biotic homogenization of soil invertebrate diversity in a comparison of native forests and pine plantations in Chile, and also with the results of Šibíková et al. (2019) for *R. pseudoacacia* plantations in Eastern Europe.

Besides their direct impacts on plant biodiversity, processes of biodiversity reduction and biotic homogenization of plant communities are significant due to their cascade effect on other taxa. Manor et al. (2008) analyzes small mammal communities along a land-use intensity gradient ranging from native pine (*Pinus halepensis*) and carob (*Ceratonia siliqua*) stands to *Acacia* plantations. They show that land-use intensity alters plant community composition, which in turn leads to a biotic homogenization of mammal communities (similar to results on moth diversity in Root et al., 2017). Furthermore, besides immediate and mid-term effects during the first rotations, long-term effects are to be expected. For instance, Lee-Cruz et al. (2013) show that after conversion of forests to palm oil plantations in Borneo, no differences in α-diversity of soil bacterial communities were observed and β-diversity was higher in plantation soils. The authors suggest that this finding may reflect preexisting differences of soil chemistry and hydrology before conversion and that biotic homogenization may result in the long term.

5.3 Implications for Reforestation Measures

Favero et al. (2020) show that increased economic incentives e.g. in bioenergy production will create impulses for forest managers to extend and intensify forestry activities—with potential biodiversity losses. Hartley (2002) suggests conserving biodiversity in tree plantations based on a new management scheme. This comprises management options such as polycultures of (partly) native species, effects of plantation size, early thinning, less-destructive harvesting, and longer rotations. Such management schemes are not often employed in the Global South, even in very developed countries such as Chile (Armstrong et al., 1998; Siry et al., 2005). Hence, severe effects on regional biodiversity are to be expected in regions formerly considered biodiversity hotspots (Myers et al., 2000; Smith-Ramírez, 2004).

Ohashi et al. (2019) show negative impacts of climate change mitigation by forest management on biodiversity if the reforestation is conducted with exotic species and without efforts to restore the environment of the original natural forest, as opposed to mixed-species reforestation as suggested by Hulvey et al. (2013), Lewis et al. (2019), and Huang et al. (2018). Heilmayr et al. (2020) have analyzed these circumstances for Chile, showing how economic incentives have led to deforestation and offsetting carbon storage. The Chilean example thus contradicts the position of Williamson (2016) that large-scale tree plantations increase the natural storage of carbon in biomass and forest soil. It confirms the preoccupation of Lewis et al. (2019), which is that forest plantations are far less suited for this purpose than natural forests. Kremen and Merenlender (2018) point out the importance of managing forest landscapes to serve biodiversity, climate change adaptation, as well as people.

The expected benefits from planted forests in Chile are contrasted by severe environmental deterioration. The results of Heilmayr et al. (2020) raise concerns on biodiversity impacts on the basis of surrogate data. This study has demonstrated the actual severity of biodiversity impacts on the basis of an exhaustive field sampling in terrain. Such concerns made Strassburg et al. (2020) rank Chile among the top 20% of priority areas for biodiversity conservation. While impacts on biodiversity are severe, biotic homogenization is the more important effect. Lamb et al. (2005) point out that one of the biggest challenges regarding reforestation will be moving restoration from a site-based activity to a landscape activity. The failure to do so has created monotonous, monofunctional landscapes in central Chile, no longer characterized by ecologically distinct near-natural forests but rather by plantations showing an impoverished and homogenized understory that is relatively independent of the plantation tree and the subregion. Cook-Patton et al. (2020) see large CO_2 sequestration capacities in potential restoration areas in Chile, particularly in deforested interior midlands. This could lead to more plantation forestry there, despite detrimental economic (Andersson et al., 2016; Van Holt et al., 2012) and ecological effects, social conflicts (Braun, 2021; Torres-Salinas et al., 2016), and losses of more sustainable usage of the forest fringe, such as agroecology (Mestmacher & Braun, 2020; and Pohl et al. [in review]). Caparros and Jacquemont (2003) have demonstrated that pricing of carbon storage will not promote forest management, but lead to over-planting of fast-growing introduced species with a potentially negative impact on biodiversity. As shown, biotic homogenization is an additional and underrated threat, to be addressed in reforestation and reforestation measures. In view of the fact that almost 50% of the land currently designated for reforestation is to come from tree plantations (Lewis et al., 2019) and that plantation management is relatively easy to implement and at the same time brings high economic profit margins for corporations (Sedjo, 1999), it must be feared that these severe impacts on regional biodiversity will also occur in other countries.

6 Conclusions

Reforestation measures for economic purposes, climate change mitigation purposes, or both will play an increasingly important role in the future. Economic stimulation and implementation will be mediated from the international, national, and regional levels and involve governments, non-governmental organizations, activist groups, and citizens. Some of these mechanisms will be implemented top-down and make use of incentives that are attractive to large companies. This will be of particular importance in the Global South, where the carbon storage potential from reforestation is situated and where international funds will be directed to. Such mechanisms could induce the widespread establishment of intensely managed plantation monocultures of exotic species, as shown by comprehensive analyses of current policy commitments. This article has contributed to an understanding of which consequences could arise from this scenario.

In terms of ecological consequences, the outcome would be significant reductions in biodiversity. The question of whether plantations are to be considered "green deserts" or not on the basis of plot-scale analyses is less relevant. What is required is a landscape-scale perspective that focuses on biotic homogenization, i.e. conversion into biotically homogenized, monofunctional, industrial tree farms that reduce the insurance function of biodiversity in an ecological sense at the landscape scale and impede implementation of more sustainable forest use, available not only to larger enterprises but also to smallholders such as NTFP collectors and agroforestry/agroecology. Due to the long and continuous presence of industrial plantations in Chile, the country's example provides results on such long-term processes of biotic homogenization. It shows that biotic homogenization results in the long term not only from invasion processes but also from land-use change.

Regarding the potential benefits of planted forests to halt deforestation, a closer look is required. While the assumption that reforestation halts deforestation may be true at the global scale, the regional scale should not be overlooked. In several countries, Chile among them, the establishment of plantations has replaced native forests. Focusing only on the global net values of forest losses and gains can obscure ecologically relevant processes of forest loss at the regional level. More research is still needed to address the issue of biotic homogenization as a consequence of economic incentives in order to inform the practice of forest ecology as well as political stakeholders.

The Chilean example has shown that industrial plantations are a clearly unfavorable option for reforestation activities. Large economic subsidies may lead to the intensification of plantation management and incentivize the replacement of native forests. These processes have heavy impacts on biodiversity and biotic homogenization. Chile is not a crystal ball to the future, of course. It stands as a cautionary tale, however, of what may result if reforestation is not accompanied by effective environmental regulation mechanisms, which must comprise mechanisms for preventing forest clearance in favor of plantation establishment. Such mechanisms will be difficult to implement and control in other contexts in the Global South.

References

Andersson, K., Lawrence, D., Zavaleta, J., & Guariguata, M. R. (2016). More trees, more poverty? The socioeconomic effects of tree plantations in Chile, 2001–2011. *Environmental Management, 57*(1), 123–136.

Armstrong, A. J., Benn, G., Bowland, A. E., Goodman, P. S., Johnson, D. N., Maddock, A. H., & Scott-Shaw, C. R. (1998). Plantation forestry in South Africa and its impact on biodiversity. *Southern African Forestry Journal, 182*(1), 59–65.

Aubin, I., Messier, C., & Bouchard, A. (2008). Can plantations develop understory biological and physical attributes of naturally regenerated forests? *Biological Conservation, 141*(10), 2461–2476.

Austin, K. G., Schwantes, A., Gu, Y., & Kasibhatla, P. S. (2019). What causes deforestation in Indonesia? *Environmental Research Letters, 14*(2), 024007.

Barr, C., & Cossalter, C. (2004). China's development of a plantation-based industry: Government policies, financial incentives, and investment trends. *International Forestry Review, 6*(3–4), 267–281.

Barr, C. M., & Sayer, J. A. (2012). The political economy of reforestation and forest restoration in Asia–Pacific: Critical issues for REDD+. *Biological Conservation, 154*, 9–19.

Braun-Blanquet, J. (1964). *PflanzensoziologiE. Grundzüge der Vegetationskunde*. Springer, Wien.

Braun, A. C. (2021). Encroached by pines and *Eucalyptus*? A grounded theory on plantation forestry impacts on smallholder livelihood in Chile. *Journal of Rural Studies, 82*(2021), 107–120.

Braun, A. C., Troeger, D., Garcia, R., Aguayo, M., Barra, R., & Vogt, J. (2017). Assessing the impact of plantation forestry on plant biodiversity: A comparison of sites in Central Chile and Chilean Patagonia. *Global Ecology and Conservation, 10*, 159–172.

Bremer, L. L., & Farley, K. A. (2010). Does plantation forestry restore biodiversity or create green deserts? A synthesis of the effects of land-use transitions on plant species richness. *Biodiversity and Conservation, 19*(14), 3893–3915.

Brockerhoff, E. G., Jactel, H., Parrotta, J. A., Quine, C. P., & Sayer, J. (2008). Plantation forests and biodiversity: Oxymoron or opportunity? *Biodiversity and Conservation, 17*(5), 925–951.

Brooks, T. M., Mittermeier, R. A., da Fonseca, G. A., Gerlach, J., Hoffmann, M., Lamoreux, J. F., Mittermeier, C. G., Pilgrim, J. D., & Rodrigues, A. S. (2006). Global biodiversity conservation priorities. *Science, 313*(5783), 58–61.

Bull, G. Q., Bazett, M., Schwab, O., Nilsson, S., White, A., & Maginnis, S. (2006). Industrial forest plantation subsidies: Impacts and implications. *Forest Policy and Economics, 9*(1), 13–31.

Buongiorno, J., & Zhu, S. (2014). Assessing the impact of planted forests on the global forest economy. *New Zealand Journal of Forestry Science, 44*(S1), S2. Springer International Publishing.

Buongiorno, J., Raunikar, R., & Zhu, S. (2011). Consequences of increasing bioenergy demand on wood and forests: An application of the global forest products model. *Journal of Forest Economics, 17*(2), 214–229.

Bustamante, R. O., & Simonetti, J. A. (2005). Is *Pinus radiata* invading the native vegetation in central Chile? Demographic responses in a fragmented forest. *Biological Invasions, 7*(2), 243–249.

Calviño-Cancela, M., & van Etten, E. J. (2018). Invasive potential of *Eucalyptus globulus* and *Pinus radiata* into native eucalypt forests in Western Australia. *Forest Ecology and Management, 424*, 246–258.

Camus, P. (2003). Federico Albert: Artífice de la gesti6n de los bosques de ChilE. *Revista De Geografía Norte Grande, 20*(2003), 55–63.

Caparros, A., & Jacquemont, F. (2003). Conflicts between biodiversity and carbon sequestration programs: Economic and legal implications. *Ecological Economics, 46*(1), 143–157.

Chidumayo, E. N. (1989). Land use, deforestation and reforestation in the Zambian Copperbelt. *Land Degradation & Development, 1*(3), 209–216.

Chiti, T., Grieco, E., Perugini, L., Rey, A., & Valentini, R. (2014). Effect of the replacement of tropical forests with tree plantations on soil organic carbon levels in the Jomoro district, Ghana. *Plant and Soil, 375*(1–2), 47–59.

Clapp, R. A. (2001). Tree farming and forest conservation in Chile: Do replacement forests leave any originals behind? *Society & Natural Resources, 14*(4), 341–356.

Clapp, R. A. (1995a). Creating competitive advantage: Forest policy as industrial policy in Chile. *Economic Geography, 71*(3), 273–296.

Clapp, R. A. (1995b). The unnatural history of the Monterey pine. *Geographical Review, 85*(1), 1–19.

Cifuentes-Croquevielle, C., Stanton, D. E., & Armesto, J. J. (2020). Soil invertebrate diversity loss and functional changes in temperate forest soils replaced by exotic pine plantations. *Scientific Reports, 10*(1), 1–11.

Collins, S. L., Suding, K. N., Cleland, E. E., Batty, M., Pennings, S. C., Gross, K. L., Grace, J. B., Gough, L., Fargione, J. E., & Clark, C. M. (2008). Rank clocks and plant community dynamics. *Ecology, 89*(12), 3534–3541.

Cook-Patton, S. C., Leavitt, S. M., Gibbs, D., Harris, N. L., Lister, K., Anderson-Teixeira, K. J., Briggs, R. D., Chazdon, R. L., Crowther, T. W., Ellis, P. W., Griscom, H. P., Herrmann, V., Holl, K. D., Houghton, R. A., Larossa, C., Lomax, G., Lucas, R., Madsen, P., Malhi, Y., … Griscom, B. W. (2020). Mapping carbon accumulation potential from global natural forest regrowth. *Nature, 585*(7826), 545–550.

Danielsen, F., Beukema, H., Burgess, N. D., Parish, F., Brühl, C. A., Donald, P. F., Murdiyarso, D., Phalan, B., Reijnders, L., Struebig, M., & Fitzherbert, E. B. (2009). Biofuel plantations on forested lands: Double jeopardy for biodiversity and climate. *Conservation Biology, 23*(2), 348–358.

Davis, K. F., Yu, K., Rulli, M. C., Pichdara, L., & D'Odorico, P. (2015). Accelerated deforestation driven by large-scale land acquisitions in Cambodia. *Nature Geoscience, 8*(10), 772–775.

Del Lungo, A., Ball, J., & Carle, J. (2006). Global planted forests thematic study—Results and analysis. Food and Agriculture Organisation (FAO), Working Paper FP/38, Rome, Italy

Doelman, J. C., Stehfest, E., van Vuuren, D. P., Tabeau, A., Hof, A. F., Braakhekke, M. C., Gernaat, D. E. H. J., van der Berg, M., van Zeist, W.-J., Daioglou, V., van Meijl, H., & Lucas, P. L. (2020). Reforestation for climate change mitigation: Potentials, risks and trade-offs. *Global Change Biology, 26*(3), 1576–1591.

Duguid, M. C., & Ashton, M. S. (2013). A meta-analysis of the effect of forest management for timber on understory plant species diversity in temperate forests. *Forest Ecology and Management, 303*, 81–90.

Favero, A., Daigneault, A., & Sohngen, B. (2020). Forests: Carbon sequestration, biomass energy, or both? *Science Advances, 6*(13), eaay6792.

Frank, D., & Finckh, M. (1997). Impactos de las plantaciones de pino oregón sobre la vegetación y el suelo en la zona centro-sur de Chile. *Revista Chilena De Historia Natural, 70*, 191–211.

Gámez-Virués, S., Perović, D. J., Gossner, M. M., Börschig, C., Blüthgen, N., De Jong, H., Simons, N. K., Klein, A.-M., Krauss, J., Maier, G., Scherber, C., Steckel, J., Rothenwöhrer, C., Steffan-Dewenter, I., Weiner, C. N., Weissner, W., Werner, M., Tscharntke, T., & Westphal, C. (2015). Landscape simplification filters species traits and drives biotic homogenization. *Nature Communications, 6*(1), 1–8.

Gaveau, D. L., Sheil, D., Salim, M. A., Arjasakusuma, S., Ancrenaz, M., Pacheco, P., & Meijaard, E. (2016). Rapid conversions and avoided deforestation: Examining four decades of industrial plantation expansion in Borneo. *Scientific Reports, 6*(1), 1–13.

Hajdu, F., Penje, O., & Fischer, K. (2016). Questioning the use of 'degradation' in climate mitigation: A case study of a forest carbon CDM project in Uganda. *Land Use Policy, 59*, 412–422.

Hartley, M. J. (2002). Rationale and methods for conserving biodiversity in plantation forests. *Forest Ecology and Management, 155*(1–3), 81–95.

Heilmayr, R., Echeverría, C., & Lambin, E. F. (2020). Impacts of Chilean forest subsidies on forest cover, carbon and biodiversity. *Nature Sustainability, 3*(9), 701–709.

Heilmayr, R., Echeverría, C., Fuentes, R., & Lambin, E. F. (2016). A plantation-dominated forest transition in Chile. *Applied Geography, 75*, 71–82.

Heinrichs, S., Pauchard, A., & Schall, P. (2018). Native plant diversity and composition across a *Pinus radiata* D. Don plantation landscape in south-central Chile—The impact of plantation age, logging roads and alien species. *Forests, 9*(9), 567.

Honegger, M., Michaelowa, A., & Roy, J. (2020). Potential implications of carbon dioxide removal for the sustainable development goals. *Climate Policy*, 1–21.

Huang, Y., Chen, Y., Castro-Izaguirre, N., Baruffol, M., Brezzi, M., Lang, A., Li., Y., Härdtle, W., von Oheimb, G., Yang, X. F., Liu, X. J., Pei, K. Q., Both, D., Yang, B., Eichenberg, D., Assmann, T., Bauhus, J., Behrens, T., Uscot, F., Chen, X. Y., Chesters, D., Ding, B. Y., Durka, W., Erfmeier, A., Fang, J. Y., Fischer, M., Guo, L. D., Guo, D. L., Gutknecht, J. L. M., He, J. S., He, C. L., Hector, A., Hönig, L., Hu, R. Y., Klein, A. M., Kühn, P., Liang, Y., Li, S., Michalski, S., Scherer-Lorenzen, M., Schmidt, K., Scholten, T., Schuldt, A., Shi, X. Z., Tan, M. Z., Tang, Z. Y., Trogisch, S., Wang, Z. W., Welk, E., Wirth, C., Wubet, T., Xiang, W. H., Yu, M. J., Yu, X. D., Zhang, J. Y., Zhang, S. R., Zhang, N. L., Zhou, H. Z., Zhu, C. D., Zhu, L., Bruelheide, H., Ma, K. P., Niklaus, P. A. & Schmid, B. (2018). Impacts of species richness on productivity in a large-scale subtropical forest experiment. *Science, 362*(6410), 80–83.

Hulvey, K. B., Hobbs, R. J., Standish, R. J., Lindenmayer, D. B., Lach, L., & Perring, M. P. (2013). Benefits of tree mixes in carbon plantings. *Nature Climate Change, 3*(10), 869–874.

Instituto Nacional de Estadisticas. (2002). Censo 2002: Sintesis de Resultados. Santiago de Chile, Chile. Report of Chilean Statistical Authorities.

Jha, C. S., Dutt, C. B. S., & Bawa, K. S. (2000). Deforestation and land use changes in Western Ghats, India. *Current Science*, 231–238.

Johnston, C. M., & van Kooten, G. C. (2016). Global trade impacts of increasing Europe's bioenergy demand. *Journal of Forest Economics, 23*, 27–44.

Kostić, O., Jarić, S., Gajić, G., Pavlović, D., Marković, M., Mitrović, M., & Pavlović, P. (2016). The effects of Douglas fir monoculture on stand characteristics in a zone of Montane beech forest. *Archives of Biological Sciences, 68*(4), 753–766.

Kottek, M., Grieser, J., Beck, C., Rudolf, B., & Rubel, F. (2006). World map of the Köppen-Geiger climate classification updated. *Meteorologische Zeitschrift, 15*(3), 259–263.

Kremen, C., & Merenlender, A. M. (2018). Landscapes that work for biodiversity and people. *Science, 362*(6412).

Lamb, D., Erskine, P. D., & Parrotta, J. A. (2005). Restoration of degraded tropical forest landscapes. *Science, 310*(5754), 1628–1632.

Lee-Cruz, L., Edwards, D. P., Tripathi, B. M., & Adams, J. M. (2013). Impact of logging and forest conversion to oil palm plantations on soil bacterial communities in Borneo. *Applied and Environmental Microbiology, 79*(23), 7290–7297.

Lennon, J. J., Koleff, P., Greenwood, J. J. D., & Gaston, K. J. (2001). The geographical structure of British bird distributions: Diversity, spatial turnover and scale. *Journal of Animal Ecology, 70*(6), 966–979.

Lewis, S. L., Wheeler, C. E., Mitchard, E. T., & Koch, A. (2019). Regenerate natural forests to store carbon. *Nature, 568*(7750), 25–28.

Lindenmayer, D. B., Hobbs, R. J., & Salt, D. (2003). Plantation forests and biodiversity conservation. *Australian Forestry, 66*(1), 62–66.

Lohmann, L. (1991). Peasants, plantations, and pulp: The politics of eucalyptus in Thailand. *Bulletin of Concerned Asian Scholars, 23*(4), 3–17.

Luebert, F., & Pliscoff, P. (2006). *Sinopsis bioclimática y vegetacional de Chile*. Editorial Universitaria.

Maas, B., Putra, D. D., Waltert, M., Clough, Y., Tscharntke, T., & Schulze, C. H. (2009). Six years of habitat modification in a tropical rainforest margin of Indonesia do not affect bird diversity but endemic forest species. *Biological Conservation, 142*(11), 2665–2671.

Magurran, A. E. (2004). *Measuring biological diversity*. Blackwell Publishing.

Manor, R., Cohen, O., & Saltz, D. (2008). Community homogenization and the invasiveness of commensal species in Mediterranean afforested landscapes. *Biological Invasions, 10*(4), 507–515.

Manuschevich, D. (2020). Land use as a socio-ecological system: Developing a transdisciplinary approach to studies of land use change in south-central Chile. In *Ecological economic and socio ecological strategies for forest conservation* (pp. 79–97). Springer.

McCarthy, J. K., Mokany, K., Ferrier, S., & Dwyer, J. M. (2018). Predicting community rank-abundance distributions under current and future climates. *Ecography, 41*(9), 1572–1582.

Meller, L., van Vuuren, D. P., & Cabeza, M. (2015). Quantifying biodiversity impacts of climate change and bioenergy: The role of integrated global scenarios. *Regional Environmental Change, 15*(6), 961–971.

Mestmacher, J., & Braun, A. (2020). Women, agroecology and the state: New perspectives on scaling-up agroecology based on a field research in Chile. *Agroecology and Sustainable Food Systems*, 1–26.

Ministerio De Agricultura, (1974). Decreto Ley 701. Fija Regimen Legal de los Terrenos Forestales o Preferenmente Aptos para la Forestacion, y Establece Normas de Fomento sobre la Materia.

Miranda, A., Altamirano, A., Cayuela, L., Pincheira, F., & Lara, A. (2015). Different times, same story: Native forest loss and landscape homogenization in three physiographical areas of south-central of Chile. *Applied Geography, 60*, 20–28.

Miranda, A., Altamirano, A., Cayuela, L., Lara, A., & González, M. (2017). Native forest loss in the Chilean biodiversity hotspot: Revealing the evidence. *Regional Environmental Change, 17*(1), 285–297.

Mittermeier, R. A., Myers, N., Thomsen, J. B., Da Fonseca, G. A., & Olivieri, S. (1998). Biodiversity hotspots and major tropical wilderness areas: Approaches to setting conservation priorities. *Conservation Biology*, 516–520.

Moreira-Muñoz, A. (2011). *Plant geography of Chile*. Springer.

Mori, A. S., Ota, A. T., Fujii, S., Seino, T., Kabeya, D., Okamoto, T., Ito, M., Kaneko, N., & Hasegawa, M. (2015). Biotic homogenization and differentiation of soil faunal communities in the production forest landscape: Taxonomic and functional perspectives. *Oecologia, 177*(2), 533–544.

Myers, N., Mittermeier, R. A., Mittermeier, C. G., Da Fonseca, G. A., & Kent, J. (2000). Biodiversity hotspots for conservation priorities. *Nature, 403*(6772), 853–858.

Neilan, W. L., Barton, P. S., McAlpine, C. A., Wood, J. T., & Lindenmayer, D. B. (2019). Contrasting effects of mosaic structure on alpha and beta diversity of bird assemblages in a human-modified landscape. *Ecography, 42*(1), 173–186.

Nunez, S., Verboom, J., & Alkemade, R. (2020). Assessing land-based mitigation implications for biodiversity. *Environmental Science & Policy, 106*, 68–76.

Ohashi, H., Hasegawa, T., Hirata, A., Fujimori, S., Takahashi, K., Tsuyama, I., Nakao, K., Kominami, Y., Tanaka, N., Hijioka, Y., & Matsui, T. (2019). Biodiversity can benefit from climate stabilization despite adverse side effects of land-based mitigation. *Nature Communications, 10*(1), 1–11.

Olden, J. D., & Rooney, T. P. (2006). On defining and quantifying biotic homogenization. *Global Ecology and Biogeography, 15*(2), 113–120.

Payn, T., Carnus, J. M., Freer-Smith, P., Kimberley, M., Kollert, W., Liu, S., Orazio, C., Rodriguez, L., Silva, L. N., & Wingfield, M. J. (2015). Changes in planted forests and future global implications. *Forest Ecology and Management, 352*, 57–67.

Paquette, A., & Messier, C. (2010). The role of plantations in managing the world's forests in the Anthropocene. *Frontiers in Ecology and the Environment, 8*(1), 27–34.

Paritsis, J., & Aizen, M. A. (2008). Effects of exotic conifer plantations on the biodiversity of understory plants, epigeal beetles and birds in Nothofagus dombeyi forests. *Forest Ecology and Management, 255*(5–6), 1575–1583.

Pawson, S. M., Brin, A., Brockerhoff, E. G., Lamb, D., Payn, T. W., Paquette, A., & Parrotta, J. A. (2013). Plantation forests, climate change and biodiversity. *Biodiversity and Conservation, 22*(5), 1203–1227.

Pohl, B., Azocar, G., & Braun, A. C. (in review). The contribution of non-timber forest products (NTPF) based on the Chilean hazelnut (Gevuina avellana) to the sustainability of indigenous collectors' livelihoods in Chile, Submitted to: Trees, Forests and People.

Peel, M. C., Finlayson, B. L., & McMahon, T. A. (2007). Updated world map of the Köppen-Geiger climate classification. *Hydrology and Earth System Sciences Discussions, 4*(2), 439–473.

Pellissier, V., Gallet, S., & Rozé, F. (2004). Comparison of the vegetation and seed bank on hedge banks of different ages in Brittany, France. *Environmental Management, 34*(1), 52–61.

Phillips, H. R., Newbold, T., & Purvis, A. (2017). Land-use effects on local biodiversity in tropical forests vary between continents. *Biodiversity and Conservation, 26*(9), 2251–2270.

Raunikar, R., Buongiorno, J., Turner, J. A., & Zhu, S. (2010). Global outlook for wood and forests with the bioenergy demand implied by scenarios of the intergovernmental panel on climate change. *Forest Policy and Economics, 12*(1), 48–56.

Rédei, T., Csecserits, A., Lhotsky, B., Barabás, S., Kröel-Dulay, G., Ónodi, G., & Botta-Dukát, Z. (2020). Plantation forests cannot support the richness of forest specialist plants in the forest-steppe zone. *Forest Ecology and Management, 461*, 117964.

Rembold, K., Mangopo, H., Tjitrosoedirdjo, S. S., & Kreft, H. (2017). Plant diversity, forest dependency, and alien plant invasions in tropical agricultural landscapes. *Biological Conservation, 213*, 234–242.

Root, H. T., Verschuyl, J., Stokely, T., Hammond, P., Scherr, M. A., & Betts, M. G. (2017). Plant diversity enhances moth diversity in an intensive forest management experiment. *Ecological Applications, 27*(1), 134–142.

Salas, C., Donoso, P. J., Vargas, R., Arriagada, C. A., Pedraza, R., & Soto, D. P. (2016). The forest sector in Chile: An overview and current challenges. *Journal of Forestry, 114*(5), 562–571.

Sample, M., Aslan, C. E., Policelli, N., Sanford, R. L., Nielsen, E., & Nuñez, M. A. (2019). Increase in nonnative understorey vegetation cover after nonnative conifer removal and passive restoration. *Austral Ecology, 44*(8), 1384–1397.

Scheidel, A., & Work, C. (2016). Large-scale forest plantations for climate change mitigation? New frontiers of deforestation and land grabbing in Cambodia. *Global Governance/Politics, Climate Justice & Agrarian/Social Justice: Linkages and Challenges*, paper no. 11.

Sedjo, R. A. (1999). The potential of high-yield plantation forestry for meeting timber needs. *New Forests, 17*(1999), 339–359.

Sedjo, R. A., & Botkin, D. (1997). Using forest plantations to spare natural forests. *Environment: Science and Policy for Sustainable Development*, *39*(10), 14–30.

Šibíková, M., Jarolímek, I., Hegedüšová, K., Májeková, J., Mikulová, K., Slabejová, D., Škodová, I., Zaliberová, M., & Medvecká, J. (2019). Effect of planting alien *Robinia pseudoacacia* trees on homogenization of Central European forest vegetation. *Science of the Total Environment, 687*, 1164–1175.

Simberloff, D., Nuñez, M. A., Ledgard, N. J., Pauchard, A., Richardson, D. M., Sarasola, M., van Wilgen, B. W., Zalba, S. M., Bustamante, R., Peña, E., & Ziller, S. R. (2010). Spread and impact of introduced conifers in South America: Lessons from other southern hemisphere regions. *Austral Ecology, 35*(5), 489–504.

Singh, D., Slik, J. F., Jeon, Y. S., Tomlinson, K. W., Yang, X., Wang, J., Kerfahi, D., Porazinska, D. L., & Adams, J. M. (2019). Tropical forest conversion to rubber plantation affects soil micro-& mesofaunal community & diversity. *Scientific Reports, 9*(1), 1–13.

Siry, J. P., Cubbage, F. W., & Ahmed, M. R. (2005). Sustainable forest management: Global trends and opportunities. *Forest Policy and Economics, 7*(4), 551–561.

Shoyama, K., & Yamagata, Y. (2014). Predicting land-use change for biodiversity conservation and climate-change mitigation and its effect on ecosystem services in a watershed in Japan. *Ecosystem Services, 8*, 25–34.

Smith-Ramírez, C. (2004). The Chilean coastal range: A vanishing center of biodiversity and endemism in South American temperate rainforests. *Biodiversity & Conservation, 13*, 373–393.
Stephens, S. S., & Wagner, M. R. (2007). Forest plantations and biodiversity: A fresh perspective. *Journal of Forestry, 105*(6), 307–313.
Strassburg, B. B., Iribarrem, A., Beyer, H. L., Cordeiro, C. L., Crouzeilles, R., Jakovac, C. C., Braga Junqueira, A., Lacerda, E., Latawiec, A.E., Balmford, A., Brooks, T. M., Butchart, S. H. M., Chazdon, R. L., Erb, K.-H., Brancalion, P., Buchanan, G., Cooper, D., Diaz, S., Donald, P. F., Kapos, V., Leclere, D., Miles, L., Obersteiner, M., Plutzar, C., Scaramuzza, C. A., Scarano, F. R., & Visconti, P. (2020). Global priority areas for ecosystem restoration. *Nature*, 1–6.
Teucher, M., Schmitt, C. B., Wiese, A., Apfelbeck, B., Maghenda, M., Pellikka, P., Lens, L., & Habel, J. C. (2020). Behind the fog: Forest degradation despite logging bans in an East African cloud forest. *Global Ecology and Conservation, 22*, e01024.
Tscharntke, T., Tylianakis, J. M., Rand, T. A., Didham, R. K., Fahrig, L., Batáry, P., Bengtsson, J., Clough, Y., Crist, T. O., Dormann, C. F., Ewers, R. M., Fründ, J., Holt, R. D., Holzschuh, A., Klein, A. M., Kleijn, D., Kremen, C., Landis, W. L., Lindenmayer, D., … Westphal, C. (2012). Landscape moderation of biodiversity patterns and processes-eight hypotheses. *Biological Reviews, 87*(3), 661–685.
Torres-Salinas, R., García, G. A., Henríquez, N. C., Zambrano-Bigiarini, M., Costa, T., & Bolin, B. (2016). Forestry development, water scarcity, and the Mapuche protest for environmental justice in Chile. *Ambiente & Sociedade, 19*(1), 121–144.
Tuomisto, H. (2010a). A diversity of beta diversities: straightening up a concept gone awry. Part 1. Defining beta diversity as a function of alpha and gamma diversity. *Ecography, 33*(1), 2–22.
Tuomisto, H. (2010b). A diversity of beta diversities: straightening up a concept gone awry. Part 2. Quantifying beta diversity and related phenomena. *Ecography, 33*(1), 23–45.
Valenzuela, E., Moreno, G., Garnica, S., & Ramírez, C. (1998). Micosociología en bosques nativos de Nothofagus y plantaciones de *Pinus radiata* en la X Región de Chile: Diversidad y rol ecológico. *Revista Chilena De Historia Natural, 71*, 133–146.
Van Holt, T., Moreno, C. A., Binford, M. W., Portier, K. M., Mulsow, S., & Frazer, T. K. (2012). Influence of landscape change on nearshore fisheries in southern Chile. *Global Change Biology, 18*(7), 2147–2160.
Vijay, V., Pimm, S. L., Jenkins, C. N., & Smith, S. J. (2016). The impacts of oil palm on recent deforestation and biodiversity loss. *PLoS ONE, 11*(7), e0159668.
Williamson, P. (2016). Emissions reduction: Scrutinize CO_2 removal methods. *Nature, 530*(7589), 153–155.
Yin, R., & Sedjo, R. A. (2001). Is this the age of intensive management? A study of loblolly pine on Georgia's Piedmont. *Journal of Forestry, 99*(12), 10–17.
Zurita, G. A., Rey, N., Varela, D. M., Villagra, M., & Bellocq, M. I. (2006). Conversion of the Atlantic Forest into native and exotic tree plantations: Effects on bird communities from the local and regional perspectives. *Forest Ecology and Management, 235*(1–3), 164–173.

The Contribution of Non-Timber Forest Products (NTFP) Based on the Chilean Hazelnut (*Gevuina Avellana*) to the Sustainability of Indigenous Collectors' Livelihoods in Chile

Bianca Pohl, Andreas Ch. Braun, and Gerardo Azócar

Abstract The Chilean forestry industry oftenly appraises their alleged sustainable practices, but its extensive use of space has sparked socio-environmental concerns. The conversion of central and southern Chilean forests into agricultural areas and monocultures is one of the key issues. Non-timber forest products (NTFPs) offer a potential solution, promoting sustainable development while maintaining biodiversity. This study explores the role of NTFPs, focusing on derivatives of the endemic species *Gevuina avellana*, to foster eco-industrial development in Chile. Using the sustainable livelihoods approach (SLA), the study assesses the perspectives of *G. avellana* collectors and producers in the Biobío Region through a semi-structured survey of experts and indigenous producers. The findings reveal the cultural and economic significance of *G. avellana*, highlighting the unique approaches taken by different producer groups based on social and economic sustainability factors. The research underscores the role of NTFPs in environmental sustainability, particularly from an indigenous perspective, where sustainable practices help prevent overexploitation. These insights point towards innovative strategies for environmental and cultural conservation that can encourage integration and cooperation among diverse cultures. The study also examines the potential disruption caused by "eco-industries" and suggests ways to support sustainable NTFP production.

B. Pohl
Institute of Regional Science (IfR), Karlsruher Institute of Technology (KIT), Kaiserstr. 12, 76131 Karlsruhe, Germany
e-mail: bianca.pohl@student.kit.edu

A. Ch. Braun (✉)
Kassel Institute for Sustainability, University of Kassel, Mosenthalstr. 8, 34117 Kassel, Germany
e-mail: Andreas.braun@uni-kassel.de

G. Azócar
Facultad de Ciencias Ambientales, Universidad de Concepcion, Casilla 160-C, Concepcion, Chile
e-mail: gazocar@udec.cl

A. Ch. Braun et al. (eds.), *Eco-Industrial Development as an Industrial Strategy*, Green Energy and Technology, https://doi.org/10.1007/978-3-031-73576-9_30

1 Introduction

A drastic change in land use can be observed on a global level, especially in the transformation of forests into areas of intense use, often expressed in the form of monocultures (Langenfeld, 2017). This change comes in large part with a variety of environmental damages, which in the long term endanger the fertility and functionality of soils and their surroundings. As a result, more and more social conflicts emerge that incorporate health problems, marginalization, and territorial conflicts, among others, which in turn fosters the manifestation and appearance of poverty and social inequality (Fischer et al., 2002). To counteract this development, in the 1990s an international debate on sustainability began in search of alternatives in land use management. An important measure to stabilize forest production under the goals of sustainability is diversification: firstly, stabilizing the ecosystem and its services (TEEB, 2010) and secondly, offering a variety of income sources to its users which reduces the economic risk of being dependent on one or a few products (Langenfeld, 2017). On this basis, non-timber forest products (NTFPs) are considered to have potential for the diversification of forests (Marshal & Newton, 2003) and to contribute to the diversification of the livelihood strategies of smallholders and forest-dependent rural people with few economic alternatives, for whom NTFPs are often the only source of cash income (Ruiz-Pérez et al., 2004; Belcher & Schreckenberg, 2007). Globally, in many cases the income contribution of NTFPs to rural livelihoods could be described as significant (Ruiz-Pérez et al., 2004; Mahapatra et al., 2005; Lepcha et al., 2023) whether in the form of absolute cash value or as a means to bridge critical years or periods when other income opportunities are scarce (Shackleton, 2007).

NTFP collection activity is often deeply integrated into the traditions and customs of rural people who in many countries belong to indigenous populations. Therefore, NTFP activity also brings with it the potential for cultural conservation (Ruiz-Pérez et al., 2004; Tacón, 2004). These groups of collector-producers possess broad knowledge of the different NTFPs and their characteristics and uses, which often results in the incorporation of traditional rules that regulate the form and quantity of the harvest and thus prevent overexploitation of the NTFPs (Tacón, 2004). As a result, the incorporation of this traditional ecological knowledge (TEK) into planning processes and landscape management approaches like the traditional resource management (TRM) has been discussed (Ticktin & Johns, 2002). Further research will be needed to identify, evaluate, and systemize the TEK (Tacón, 2004) to facilitate its integration into development policies.

Despite this promising image of NTFPs, their implementation into agroforestry systems has been challenging and comes with some social and environmental risks. The rising recognition and income opportunities of NTFPs (Scherr et al., 2004) often lead to increased exploitation which not only threatens the natural resources, it also creates the risk of unequal benefits among the actors along the value chain. The intended beneficiaries, that is to say groups of poor people with low resources and capacities, are often out-competed by actors with access to more forms of capital, such as investment capital, access to resources and know-how, and access to essential business contacts. A decline in the natural resource might then also be equal to a decrease in the livelihood base of poor and forest-dependent people, which in turn makes them even more vulnerable. This becomes especially drastic where negotiated NTFPs are part of significant local practices, own consumption, or medicinal use, risking the loss of cultural traditions and values (Belcher & Schreckenberg, 2007).

The variety of NTFPs on the market in terms of characteristics, uses, origin, values, and position is very high. These differences determine the socio-ecologic impact of the product and strongly limit their potential for comparison and the broader application of findings from individual case studies. In the face of this situation, scientific reviews and meta and multi-case research on NTFP experience and development have been performed to analyze possible patterns and factors that harm or favor the successful implementation of NTFPs to livelihood systems. It has been shown that achievements depend highly on the constitution of favorable local socio-economic and environmental conditions (García-Fernández et al., 2008; Meinhold & Darr, 2019; Mugido & Shackleton, 2018; Stryamets, 2016). Explicitly, a particular need has been identified to implement NTFP supervision and management in the legislation of the respective country (García-Fernández et al., 2008; Silva et al., 2020; Stryamets, 2016), facilitating the inclusion and cooperation of multiple stakeholders (García-Fernández et al., 2008; Guariguata et al., 2010). Consequently, many scientists highlight the need for profound holistic and multidisciplinary inspections of local conditions, enabling the establishment of local conditions more favorable to MFM models and NTFP implementation and the targeting of effective management interventions (Belcher & Schreckenberg, 2007; Guariguata et al., 2010; Meinhold & Darr, 2019; Mugido & Shackleton, 2018; Silva et al., 2020).

Chile has experienced strong socio-environmental conflicts due to the transformation of forests into agricultural areas and, later, into forest monocultures in favor of the wood production and export that has been strongly promoted by the Chilean state in the last forty years (Langenfeld, 2017). Many native peoples are under threat from the exploitation of their forests, pollution (air and soil), and the loss of biodiversity, among others (Torres et al., 2015).

There is a wide variety of Chilean NTFPs and a shift in activity from collection for self-consumption to sale on the national and international markets, with demand continuing to grow. However, information on the origin of the different products, the reality of the producers, and commercial activities along the value chain is scarce and informal, which strongly limits the possibilities for adequate management of the NTFP sector (Calleja, 2007; INFOR, 2019b; Valdebenito et al., 2015). Collectors and producers often belong to marginalized groups with informal organization and a low amount of resources (Calleja, 2007; INFOR, 2014; Tacón, 2004), and without access to the information necessary to participate in trade in an equal manner (Calleja, 2007).

During the last few years, a vertiginous development of the commercial behavior of the NTFP Chilean hazelnut (*G. avellana*), also known as Chile nut, has been noted. Compared to other NTFPs, the hazelnut and its byproducts attain high market values, but due to its relatively recent commercial debut any knowledge about its dynamics is especially scarce (INFOR, 2014; Valdebenito et al., 2015).

Box 1 *Gevuina avellana*, the species of this study. Short biological presentation, NTFP derivation and harvest.

The Chilean hazelnut (*Gevuina avellana* Molina) is an evergreen tree species, endemic to southern Chile and Argentina. The tree is highly tolerant of soil, light, and moisture conditions and occurs between 0 and 1000 m above sea level in the undergrowth of near-natural forests (Hoffmann, 1982). The diversity of the plants and microorganisms in the surrounding ecosystem plays a role in its reproduction (Donoso, 1978; Loewe et al., 2017) and fruit production is dependent on site conditions, especially light. Its fructification cycle is biannual leading to the presence of flowers and ripe fruits at the same time, which makes it vulnerable to rude collection practices (Tacón, 2017). Several NTFPs of the hazelnut are commercialized: the leaves are used for decorative purposes due to their form, glint, and high durability, along with its flowers (Fig. 1). The flowers are the source of a highly reputed honey and the fine-grained wood serves as the basis for quality furniture and instruments. The fruit and its derivatives (oil, flour, coffee, and sweets) are the products with the highest commercial value. Due to its nutrient composition and characteristic flavor, the hazelnut is classified as an important supplement in the diet of many indigenous groups (Tacón, 2004), but also as a delicacy demanded on an international level, specifically in the northern hemisphere, as a so-called superfood, natural or wild, free of suspicion of contamination (FAO, 2018). There is strong interest developing in the extraction of the hazelnut's oil, from the pharmaceutical and cosmetic industry (Valdebenito et al., 2015) in particular, due to its suitability for the production of sunblock (INTEC, 1982) (Fig. 2).

Fig. 1 Flower and leaves of *G. avellana*

Fig. 2 Tree of *G. avellana* with fruits

The fruit (Figs. 3 and 4) is harvested, mostly by hand, by the rural population between March and June. Afterwards it is selected, cleaned, and dried indoors for about a month before the fruit is stewed, peeled, and roasted to make it tasty. The roasted fruit is then crushed and mixed with different percentages of cereal flour to achieve the hazelnut flour that is commercialized on the market (Fig. 4). This hazelnut flour can be mixed with milled coffee to be sold as "café de avellana" or processed into cakes and other sweets. The roasted fruit and its flour is generally sold in small plastic bags of 50 or 100 g at a kilogram price of 10,000 CLP (~$13 USD).

Fig. 3 Peeled and raw nut of *G. avellana*

Fig. 4 Packaged flour of *G. avellana*

Photos: Bianca Pohl (2019).

The growing international demand for NTFPs has different effects at the local level. On the one hand, the demand facilitates the opportunity for economic integration of the collector-producer (Ruiz-Pérez et al., 2004). On the one hand, the development of an uncontrolled market can cause damage to particular groups of actors who, due to their low degree of organization and formalization, do not have the rights and necessary capacities to commercialize their product and receive adequate remuneration for their efforts (Calleja, 2007; Ruiz-Pérez et al., 2004; Tacón, 2004). There is also the risk of overexploitation of the natural resource by some harvesting groups. It should be noted that the need for hazelnut cultivation is expressed in an intensive manner (plantations), whose socio-environmental effects have not yet been estimated.

In this sense, our research seeks to assess who the producers of *G. avellana* NTFPs are in Chile's Biobío Region, an area with greater development of the species. This study examines the local conditions, processes, and trends of the marked dynamic using a holistic investigation model. The aim is to identify the main gaps and needs of *G. avellana* collectors according to the sustainable development of their business in order to achieve a review about the contribution of *G. avellana* to the sustainability of the collectors' livelihoods. The objective is to generate appropriate information for policy and decision makers to promote an achievable, integrated MFM model framing the Chilean NTFP sector and its producers.

2 Applied Method

The research study used the sustainable livelihoods approach (SLA), a holistic research model that has gained importance during the debate on sustainability. SLA allows for a visualization of the complexity of different processes and factors at both the local and macro levels that determine the scope of action of a target group at the local level, including its capacities, commercial activities, and total assets. Subsequently, the degree of resistance to interventions and the capacity of the group to maintain its lifestyle without degrading the available natural resource base can be determined. The further objective of the model is to identify the problems, weaknesses, and needs of the actors as well as their possibilities and strengths, in order to formulate participatory solution strategies (Chambers & Conway, 1992). SLA has proven to be an effective instrument in the planning and implementation of programs and projects of different organizations and actors within international development cooperation.

For the elaboration of the present research design, the SLA structural framework developed by the Department for International Development (DFID) of the Government of the United Kingdom was adapted. DFID (2001) in its SLA framework distinguishes five main components. The characterization of the producers of NTFPs included contents of the analyzed livelihood strategies and livelihood outcomes. The focus of the analysis lies furthermore in the description of the five different livelihood assets which consider the natural, human, social, financial, and physical capital. In addition, components on transforming structures and processes and the vulnerability context are addressed jointly in this paper's analysis due to the significant interrelationship and influence between them.

2.1 Area of Study

The area of study is equivalent to the wild range of the hazelnut tree. The Biobío Region was chosen—specifically the commune of Contulmo located in the Coastal Range, and the commune of Alto Biobío in the Andes—due to the importance of NTFP collection in these areas. The choice of these communities for study was based on data and collector contacts facilitated by the Institute of Agricultural Development (INDAP) in Contulmo and by the University of Concepción (UDEC), Chillán campus. Also important was the information provided by the municipality of Alto Biobío, located in the town of Ralco.

In the Biobío Region, forestry activity has triggered important changes in land use (CONAF, 2019a), with a more relevant advance of plantations in the Coastal Range. Figure 1 highlights the two communes of interest (area of study framed in red) and shows the distribution of native forest as well as forest plantations of small- and medium-scale owners in the year 2016 as possible addressees to establish cooperative relationships according to NTFP incorporation. When including the plantations of large companies such as Forestal Arauco the area reaches 913,173 hectares in total, surpassing the native forests of the region that total 845,552 hectares (INFOR, 2019a). One criterion that guided the selection of both communes was the objective of finding out whether the collectors of NTFPs (*G. avellana*) are affected by the presence of forestry activity (Fig. 5).

The commune of Contulmo is characterized by a decreasing demographic trend and a low population density. In this commune, 41.2% of the people live in rural areas. Of the 6,031 inhabitants (Census 2017), approximately two thousand are Mapuche who live in twelve communities in the Elicura Valley and other neighboring sectors. Contulmo takes advantage of its strategic location for the development of

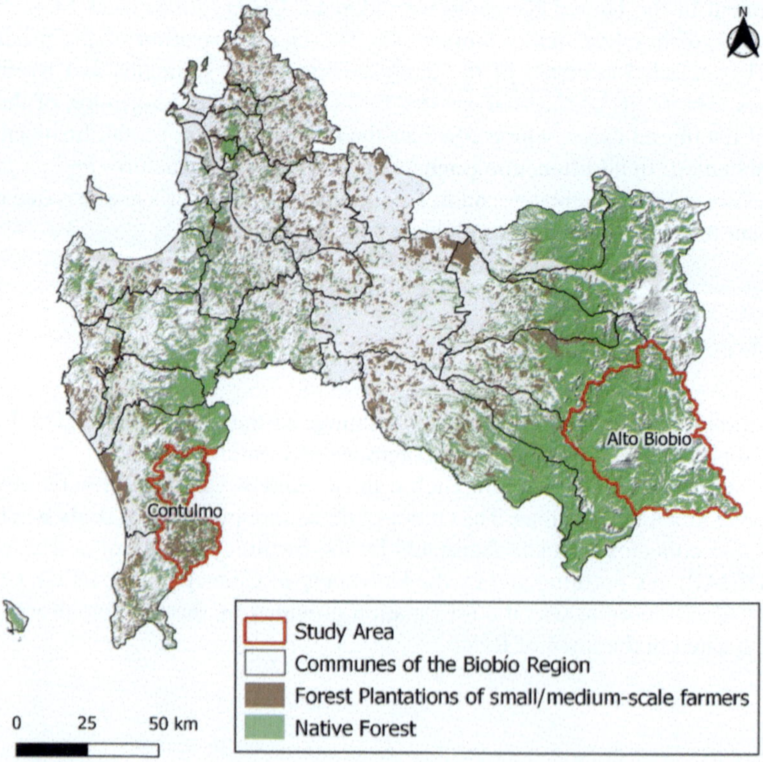

Fig. 5 Distribution of native forest and forest plantations of small- and medium-scale farmers in 2016, Biobío Region. *Source* INFOR (2019a)

tourism, especially for water sports at Lake Lanalhue or as a starting point for visits to Nahuelbuta National Park. The main challenges of community planning are the fight against water pollution (Lake Lanalhue), especially in the mountainous area; the low coordination among local actors; and the lack of cooperation among them with the municipality (SERNATUR, 2014).

The commune of Alto Biobío is the largest commune in the Biobío Region. Located in the southeast, the topography of this mountainous area does not allow for settlements throughout the commune, a situation that determines a very low population density and a percentage of 80% of the population living in rural areas. Of the 5,923 inhabitants (Census 2017), 74.1% are indigenous (Pewenche), which makes Alto Biobío the commune with the greatest ethnic composition at the regional level (Guiñez & Gaete, 2013). A main asphalt road between the town of Ralco and the cities of Santa Barbara and Los Angeles links the commune with the rest of the region. The two secondary, gravel roads to inland locations (Queuco River Valley and Biobío River Valley) can be isolated by landslides and flooding during the winter (ONEMI, 2007).

The most important economic activities of the commune of Alto Biobío are small livestock, tourism, and traditional crops, individual or family, as well as the collection of forest products. Due to the construction of the Ralco Hydroelectric Power Station between 1998 and 2004, which caused strong social conflicts due to environmental and socio-structural changes (SEPADE, 2010), more employment is generated in the areas of electricity, gas, and water, as well as in industry and construction (Social Observatory 2012). Since its creation in 2004, the commune of Alto Biobío has undergone an important transformation with respect to its infrastructure due to public investments and state funds, which has increased the integration of the population into the market and various services. However, significant gaps remain, such as the low level of schooling, a low degree of participation in development decision-making, and poor access to the health system (INDH, 2019).

2.2 *Data Collection and Evaluation Methods*

The interview was chosen as the data collection method: first with the NTFP collector-producers (*G. avellana*) who, according to SLA, are the main actors or subject of the research focus; and second with key informants or experts in the field of collection to include other perspectives related to the subject.

For both groups, the semi-structured qualitative interview was used (cf. Table 3 in the appendix). This method allowed for the collection of a wide range of information in order to evaluate perceptions and areas of action and to understand processes and structural challenges. Formulating the key questions in an open way in turn ensured the structuring of the research, the maintenance of the focus related to the objectives, and the possibility of detecting similarities and patterns among the interviewees (Mayring, 2012). In this research, the key questions were formulated according to the SLA analysis categories.

Sixteen interviews were conducted in total. Twelve of these were with collector-producers, of which 50% came from the three communities of the Alto Biobío commune with an abundance of hazelnuts: El Avellano (3), Pitril (2) and Callaqui (1). The interviewees from the community of El Avellano belong to the eponymous cooperative for the commercialization of NTFPs founded in collaboration with the Foundation for Agrarian Innovation of the University of Concepción (FIAUdeC) in 2017 with the aim of promoting the professional commercialization of the NTFP *G. avellana.* The other six interviewees were collectors-producers from three communities in the commune of Contulmo: Ignacio Huilipan (3), Huenu Mapu (2) and Juana Millahual (1). The main means of contact were the aforementioned institutions. In addition, two interviews were conducted through the snowball system and one through a personal family contact. In most cases, the location of the interview was the domicile of the interviewee. Three interviews were conducted during the Alto Biobío Craft Fair at the Mall del Trébol in Concepción and one interview was held in the office of a manager from the municipality of Ralco.

For the survey of experts, qualified informants were chosen from four different organizations involved in NTFP activities in the Biobío Region: Taller Acción Cultural (TAC), INDAP, the Forestry Institute (INFOR), and the National Forestry Cooperation (CONAF).

The duration of the interviews varied between 23 min and 1 h and 39 min, with an average of 54 min. All interviews were recorded and transcribed. A qualitative content analysis was then conducted in which the information was organized into thematic areas constituting the categories of the SLA structural framework (Appendix 1).

Parallel to the interview, we participated in three meetings of the Regional NTFP Roundtable of the Biobío and Ñuble Regions in Chillán, Arauco, and Concepción (July–September 2019), during which the recent publication of the Cadastre of Collectors of Non-Timber Forest Products (NTFPs) in the Biobío and Ñuble Regions, published by INFOR (2019b), was presented.

To complement the collection of technical data, a consultation was made using the Public Transparency System in Chile to the Servicio Agrícola y Ganadero (SAG), Chile's agricultural inspection service, on the registration of export trades of products of the species *G. avellana* since 2000, including primary products such as foliage and fruits as well as processed products such as extracts or oils.

3 Results

The results are presented in a four-step argumentation. In the first step, we characterize the livelihoods that practice the collection or distribution of NTFPs, especially *G. avellana.* In the second step, we use the SLA framework to investigate the contribution that collection or distribution can make to the sustainability of these livelihoods, breaking down the different levels of capital of SLA complemented by a description of important dynamics and challenges in the sector. Finally, we give a short synthesis of the group's assets and their trends and summarize the strengths and weaknesses of the group of interest.

3.1 Characterization of the Livelihood of the Harvesters

Contulmo's interviewees come from the Mapuche culture, similarly most of the interviewees from the Alto Biobío commune belong to the Pehuenche culture. The vast majority of the interviewees said that they practice collecting with their families or in the community. The average experience for the interviewees in this activity is ten years, with two saying they had only recently started. Two of the interviewees have young hazelnut plantations on their land that are not yet bearing fruit. The El Avellano cooperative reported that they practice management of the hazel trees to increase fruit production, while the other interviewees do not practice any type of management of the trees or the harvesting forest.

Most of the properties are located in rural areas and at long distances from commerce, institutions, and general services, a situation that makes access difficult and transport expensive.

The strategies for obtaining income and maintaining the family group are diverse. In addition to collecting NTFPs, many respondents work in a garden and/or raise animals (poultry, pigs, cattle, sheep, and goats) for self-consumption or exchange of products with neighbors. Some of these products/animals are sold at the local fair; however none of the interviewees gave greater economic relevance to this activity. Rather, some highlighted the lack of economic alternatives. As stated by one interviewee, "[…] There is little work […we are] suffering from poverty, especially in winter." There is a common opinion in Alto Biobío that, in the summer period, tourism is an important source of income, offering accommodation, typical food, and crafts mainly made of NTFPs, wool, cloth, or leather. Most of the interviewees do not have regular income. The only sources of regular income mentioned are a government pension for children, while three interviewees receive a disability pension.

The majority (ten) of the interviewees report collecting different NTFPs. The species mentioned by the Alto Biobío interviewees are: *Araucaria araucana, Ramaria flava, Prunus* spp., *Cyttaria espinosae, Ugni molinae, Rubus* spp., *Cydonia oblonga,* and different medicinal herbs like *Peumus boldus.* In the commune of Contulmo the following species are collected: *Aristotelia chilensis, Ramaria flava,*

Table 1 Typology of NTFP producers

Type of producer		Alto Biobío	Contulmo	Total
Collector		0	0	0
Collector-processor		2	0	2
Collector-intermediary (collects, buys, and processes)		3	5	8
Intermediary	Local	1	1	2
	External	0	0	0
Total		**6**	**6**	**12**

Ugni molinae, Rubus spp, medicinal herbs, and *Cyttaria espinosae.* These goods are marketed without or with added value, for example in the form of jams, sauces, pies, or cosmetics.

The *G. avellana* goods most produced by the interviewees are roasted hazelnut, hazelnut flour (roasted and ground hazelnut with different percentages of wheat flour), and hazelnut coffee (ground coffee mixed with hazelnut flour). These three products have different degrees of added value, which helps to diversify the producer's supply without further complication. Two of the interviewees sell hazelnut seedlings, one sells honey from the *G. avellana* flower, and the other pointed out that he uses the shell as fertilizer.

Table 1 shows a characterization of the interviewees according to their orientation in the market.

The sales strategy is similar in both communes. Most common is the sale at local fairs in nearby cities or direct distribution at the producer's house to the consumer or intermediary. All the interviewees from Alto Biobío indicated that they sell their products at interregional fairs, such as in Concepción and/or Santiago. In addition, sales to traders in cities and through social networks (Facebook) were mentioned.

3.2 *Analysis of Potential and Weaknesses Regarding the Sustainability of Livelihoods*

3.2.1 Natural Capital

The richness of species and ecological productivity of the region require a wide variety of crops that can be used for NTFPs. The focus of this research study, *G. avellana*, is also typical of the region's semi-natural forests. Significant differences in hazelnut abundance between the two harvesting areas are highlighted. Two interviewees in Alto Biobío and all interviewees in Contulmo metioned problems related to low abundances, which they associated with the surrounding forest plantations. Four interviewees reported bad harvesting practices by other collectors, putting the reproduction of the fruit at risk. According to experts, the area of hazelnut tree distribution is more extensive in the Andes while abundance is more focused and fragmented in the Coastal Range, mainly due to the change in land use carried out by forestry companies in the past. However, the area of abundance, in general, is not directly related to hazelnut productivity. The quantity and size of the fruit depends mainly on site conditions such as soil type, the presence of water, and the local climate, but also on the degree of management or supplementation of the *G. avellana* plantation.

One interviewee said she had reached an agreement with the Arauco forestry company for permission to access the company's private land for NTFP collection. However, the collection of hazelnuts in plantations is not possible, in part due to the twelve-year period before the trees bear fruit, which exceeds the duration of forest plantations, and also because of the lack of light in plantations. According to experts, the integration of hazelnut trees, for example, in restoration efforts of secondary forests or as a native edge of plantations for wood production would be possible, "not at the level of forest companies (…), but at the level of small owners" (Interview INFOR, 2019).

3.2.2 Human Capital

In general, NTFP collectors are characterized by a low level of education (INFOR, 2019b). Nevertheless, most interviewees indicated that they provide a profound base of (traditional) knowledge relating to their collection activity, especially to the biological characteristics of the different NTFPs, their harvest, and use. This is also reflected in the high average experience of the collectors. On the other hand, the interviewees indicated a need for knowledge and awareness of commercialization processes due to their recent entry onto the market to sell their NTFPs.

Although the collection of NTFPs in general is a family activity in which different family members and generations participate, a notable predomination of women (66%) in the collection activity of the Biobío Region has been estimated (INFOR, 2019b).

The majority (eight) of the interviewees indicated that they collect only in the community. In Contulmo, four people harvest the fruit on their property while two people report collecting or buying the raw material outside their community, as the hazelnut tree is no longer very abundant in the surrounding area.

Since commercial alternatives are scarce, the importance of trade activities among the neighbors was highlighted by some of the interviewees. When asked which products they or their family produce, some interviewees responded by including the products produced by other members of the community, showing that individuals include the community production in their own production potential.

3.2.3 Social Capital

In this section, we discuss influences on social capital from two perspectives. The first is to work with institutions that can support the livelihoods of collectors and distributors, thus providing support and stabilization. The second is the cultural importance of NTFP collection. This creates opportunities to integrate the traditional, indigenous way of life into market-oriented systems, thus creating potentially beneficial conditions for indigenous people.

The advantage for collectors and distributors is that there are a number of institutions that can support and promote their activities. The most frequently mentioned institutions are INDAP (8), the local municipality (5), the National Corporation for Indigenous Development (CONADI) (4), CONAF (3), the universities (3), and the Pehuén Foundation (3). It is possible to say that most of the interviewees have contact with more than one support institution, which indicates a good perspective of development of the activity for the group interviewed. However, the experiences reported by interviewees were not always positive. Producers stated that they will no longer give their preferred collection sites to other collectors, or that they are less and less willing to engage in group work due to bad experiences. Others mentioned the lack of capacity to take advantage of the support and limits on the institution's authority to establish appropriate promotion programs. Furthermore, there are shortcomings in establishing contact between development institutions and users. One of the limitations of working with collectors is the formal requirements, such as ownership, that institutions establish to accept a collector-producer as a user.

In addition, the contact or relationship is produced by the search for users in rural areas, that is, in the sectors where the programs' beneficiaries reside, by the government institutions. "The collectors in general do not approach [the institution]" and, therefore, the success of the contacts depends greatly on the "willingness of the people who work in the institutions to link these people to the [Regional NTFP] Roundtable, it has been a rather personal effort" (INDAP Interview, 2019). Once a positive relationship has been established, more people become linked either through the "snowball" method or by chance, such as through project meetings on other issues. The coordinator of the Regional NTFP Roundtable and president of TAC mentioned that a broader and more effective link between the collectors and the institutional network would increase the level of formality of the activities and the NTFP market.

According to experts, the most important institution in the organization of the collector-producers, as pointed out by the key informants, is the Regional NTFP Roundtable, since it is the means of communication, exchange, and collaboration between all the institutions linked to the activity and the collectors. For very specific problems that cannot be solved by the roundtable, the importance of the organization of collectors is stressed, and a committee for that particular NTFP must be set up. In addition, it is stressed that the Rural Development National Fund (FNDR) is important for trade promotion, financing of training, and some incentives for NTFP production. CONAF works in the area of harvesting through the Native Forest Research Fund (FIBN), but this seems to be little known by potential users (Monroy-Concha & Pincheira-Ulbrich, 2013).

There is a training structure for the collector-producers. The Regional NTFP Roundtable offers five modules that include facilitating training in organization and cooperation among harvesters, cultural recognition of NTFPs, sustainable harvesting and management of NTFPs, and techniques and means of commercialization. INFOR carries out training in commercialization, technology, finance, and administration. The collectors who have obtained training indicated feeling grateful for it, especially the technical tours. However, the demand for training in the above-mentioned topics generally exceeds the current institutional supply.

According to experts, there is an important relationship between the commercial success of NTFPs and the cultural significance of the activity for the collectors. Interviewees pointed out that they have incorporated the activity of collecting and using different NTFPs into their indigenous traditions for generations. In this sense, culture itself is an important driving force for the participatory construction of their future, and the method to be used to promote the activity and preserve these cultural traditions is the formation of groups and organizations, as experienced by TAC since 1999.

From another point of view, the conservation of the indigenous tradition implies another interesting area of potential, which is the sustainable management with nature. The president of TAC described a special relationship that some indigenous groups often have with nature which differs from the vision of other social groups. This relationship is also expressed in the answers given by the harvester-producers interviewed in response to a question about the significance of forests for them. "Life" was the answer of three interviewees, "wealth," or something "sacred," and they listed its benefits: the forest provides water, diversity, and an environment free from pollution. The importance of the forest is reflected in the following sentence from a Contulmo interviewee: "[...] Today water is more important, let's say, it is a white gold that it has. Without it we could not live." This interviewee distinguished the indigenous vision from the vision of the forestry companies, explaining that, for him, the conservation of the functionality of the land (territory) is more valuable than financial wealth. The forestry companies focus on this last aspect, maximizing the profits of their economic activity without respecting sustainability. It is shown that the high valuation of forests and nature, concepts rooted in the ancestral indigenous culture, still has importance in the population of today's indigenous communities and that this vision implies maintaining careful cultural practices with the use of the soil and the extraction of resources.

Some interviewees pointed out their belief in the connection between commercial success with NTFPs and the will and appreciation of God (chachao). In that sense, their commercial recognition and advances in the NTFP business make them feel cherished by God.

3.2.4 Financial Capital

Most of the interviewees reported selling one kilogram of roasted hazelnut at 10,000 CPL (~$13 USD), regardless of the amount sold. However, some explained that the further away the product is sold (regional or inter-regional fair), the better the price. Estimating the real price received by the producer per kilogram is difficult, as the derivatives are sold in packages of different sizes without awareness of the exact weight of the product.

The same applies to the purchase of raw materials. There are different sack sizes which "are bought full no more" (interviewee of Contulmo, 2019) at prices from 12,000 to 25,000 CPL (~$16 to 33 USD) per 25 kg sack, and from 18,000 to 26,000 CPL (~$24 to 35 USD) for the 50 kg sack. A law has existed since 2005 to protect workers by prescribing a maximum weight of 25 kg per sack (BCN, 2005), but this law still seems poorly implemented. In Alto Biobío its implementation become noticeable due to the presence of the cooperative El Avellano.

The majority of producers (nine out of eleven interviewed) considered the hazelnut business to be profitable. Five of them indicated that, for them and in comparison with other NTFPs and even other items, the hazelnut is more important in terms of income generation because of its high value on the market and the high demand that exceeds supply. They highlighted the importance of this income, explaining that it allows them to pay for food (maintenance) for the winter, their children's education, or to finance the construction of their house.

However, two interviewees stated that the business, in order to be profitable, requires access to property where the hazelnut tree is planted with the aim of becoming independent from the high prices for the raw material. With one exception, all interviewees wanted to expand their activities with *G. avellana*.

The consensus from the experts is that NTFP collection activity has great commercial significance in the Biobío Region. Although the NTFPs of the rosehip, boldo, and other medicinal plants are still more commercially relevant, the hazelnut, especially its oil, has significant development potential, including for sale abroad. As revealed by the SAG consultation, the export volumes of *G. avellana* are still very small. The main destinations are Germany and New Zealand for the seed, and countries such as Japan and the Netherlands for the foliage (SAG, 2019). To meet export requirements, the importance of homogenizing the processing of the product among the harvester-producers is highlighted in order to obtain a consistent and high-quality product, which is tied to the producer's level of experience, their equipment (machinery, drying room, etc.), and their capacity to organize with other producers.

3.2.5 Physical Capital

Most of the interviewees live in wooden houses supplied with water and electricity. Drinking water is stored in a pond supplied by a private company. Access to the internet, an important means of communication and information supply, is limited due to geographical isolation, a situation that affects the possibilities for entry and access to markets. Market access is also limited by the lack of sanitary resolution necessary to formalize the activity. The machinery and tools for harvesting and processing in most households are basic. The El Avellano cooperative represents an exception, with a founded production shed including modern machinery.

3.3 Processes, Structural Changes and Vulnerability

3.3.1 Transition from Self-sufficiency to Market Orientation

The interview partners identify changes in different periods: In the first period, they consider that harvesting has evolved from self-consumption to the sale of NTFPs and, in the second period, they speak of "a revolution in the way [Chileans] eat" (Interview TAC 2019), which is manifested in an increase in the demand for NTFPs in the last fifteen years. In a more recent period, they relate an important fact: the consideration of NTFP collection as a commercial activity, a situation that allows support institutions to establish the most coherent advice for the sustainable promotion of collector-producers.

3.3.2 Valorization and Competence of NTFPs

The biggest changes that the interviewees reported having noticed in the *G. avellana* market in recent years are an increase in the price of the raw material (7), the recognition of the commercial value of the fruit (5), and their perception of increasing competition from the other actors involved (4). This is reflected in a strong increase in the price of the raw material over the last few years, roughly doubling to 26,000 CPL or ~$33 USD for a 25 kg sack. Considering the decrease in the hazelnut supply, it is clear that the scarcity of the resource has also contributed to the increase in its price. In contrast, prices at the fair, as the interviewees pointed out, have not increased. For most of the interviewees, the hazelnut market has an important dynamic due to the general strengthening of the sector, especially the commercialization of hazelnuts. This situation has led to an increase in competition between collectors and producers.

The perception of competition seems to be related to the orientation of the producer. For those producers who have a focus on harvesting (the *collector-processor* and *collector intermediary* with a focus on harvesting), the dynamics that have characterized NTFP harvesting in recent years have been beneficial, because they receive a higher price for their raw hazelnut sack or product. The increased competition on the market due to the presence of other intermediaries thus favors the collector. However, there are other types of producers: those who intensively expand their production, buying sacks of raw hazelnuts from other harvesters (*producer-intermediary* with a focus on buying raw material), and those who only buy raw hazelnuts for processing (*producer-intermediary*). Both types of producers have to compete with higher prices, which are paid by other intermediaries. Thus, they have to invest more for each sack of raw hazelnuts purchased. The presence of other intermediaries then pushes up the price of the hazelnut sack. In addition, the intermediaries interviewed identified other types of intermediaries in the municipalities of Alto Biobío and Contulmo, who they described as "outsiders." These are people from other places, not indigenous, who come to the commune or community to buy raw hazelnuts from the local collectors. Competition with external intermediaries seems to be so significant that two Contulmo interviewees mentioned being forced to change their strategy.

3.3.3 Increasing Challenges to Market Development

The challenges that the development of the *G. avellana* market brings to the collector-producers are also reflected in the answers given about the threats, gaps, and needs they face. The following were mentioned: lack or contamination of water (7), lack of resources (financial, machinery, transport, and land) (6), low abundance of hazelnut tree or its fruit (5), capacity (5), presence of foresters (5), institutional support (through training and/or financial support) (3), unsustainable harvesting by others (3), difficult or dangerous work (3), seasonality of the resource (3), and requirement/achievement of sanitary resolution (3). The interviewees related the lack of water and its contamination to the presence of forest plantations and also connected this to the decrease in hazelnut trees or their fruit. The gaps in the capacity of the harvester-producer are manifested in the lack of knowledge about different business strategies, especially the marketing and sale of their products. In terms of training, access to financing, and obtaining sanitary resolution, the formal requirements and bureaucracy were identified as limiting factors. According to the interviewees, sanitary resolution is a very useful distinction in the promotion of NTFP commerce, since it allows for the expansion of destination markets for the product, such as international export or sale to restaurants.

One interviewee also drew attention to new dependencies faced by the indigenous communities, explaining that previously the family did not need to sell outside the community because they maintained their way of life by exchanging agricultural products among their neighbors. While this exchange continues to be practiced, there is also a need to generate additional income to pay educational expenses for their children, for example. Other interviewees reported significant improvements in their quality of life due to additional income from the sale of NTFPs, making it easier for the family to escape poverty.

The market for *G. avellana* NTFPs, then, seems to be in a phase of growing importance. However, as this development seems to be unexpected, there are no considerations or prognoses about its future development. The recognition of value and unsaturated demand for a high-value product on the market may explain why intermediaries increase their investments in the purchase of NTFPs, why new intermediaries enter the field, or why large producers have started or increased the production of Chilean hazelnut in plantations.

3.4 *Synthesis of the Characterization of Sustainable Livelihoods and Its Dynamic*

All forms of capital show significant weaknesses, indicating an overall low level of livelihood assets. However, there are identified strengths in human and social capital. Furthermore, due to the integration of the institutional network that offers training or financial or marketing support, a growing trend of these two capitals is expected, as well as in physical capital (provision of technology and resources for home improvement and processing rooms). Natural capital shows a decreasing trend due to land use dynamics, difficulties in accessing collection resources, and environmental threats that limit their availability. Due to the lack of data it is not possible to estimate the development trend of financial capital. Strengths and weaknesses of each type of capital are summarized in Table 2.

Table 2 Strengths and weaknesses of the five types of capital

Livelihood assets (capital)	Strengths	Weaknesses
Human	Traditional knowledge of harvesting; family labor	Level of schooling and business training; possibilities for resource management
Natural	NTFP supply from the Biobío Region; a naturally available resource	Environmental threats and abundance (especially in Contulmo); access to land; lack of land for cultivation
Social	Inclusion in the institutional support network The collection activity is compatible with the indigenous worldview and way of life. For this way of life the activity provides a choice for the future	Accessibility of institutional support and selectivity of projects and programs; limits to legal margin and to the provision and structure of the institutional network; non-adopted training; social conflicts at the local level
Financial	*G. avellana* is a potentially good source of income for livelihoods The collection and commercialization of NTFPs has potential regional importance	Availability of resources (financial and property) and land Market access and export opportunities are very limited and subject to obstacles
Physical		Housing standard; technology level

Source Own elaboration complemented with information based on INFOR (2019b)

4 Discussion

4.1 Limitations of Our Study

Our study could focus more on gender relations. Vázquez-García and Ortega-Ortega (2017) show that the use of NTFPs in Mexico can also be subject to gender-related power dynamics, which regulate access rights to NTFPs, for example, but also deprioritize feminine uses. Mestmacher and Braun (2020) show that a feminization of smallholder (agro-ecological) agriculture is taking place in Chile, but is insufficiently supported by the state. Nahuelhual et al. (2008) expect NTFPs (foliage of *G. avellana*, among others) in Chile to contribute to the income of indigenous women, but then show that without further supporting policies the actual contribution will remain rather small.

Uma Shaanker et al. (2004) show that the livelihood benefits of NTFPs always have an eco-logical cost. According to the study, these costs increase with higher dependencies on the re-source, lower level of ecological knowledge about the resource, and the degree to which the market is organized through short-term contracts instead of through permanent participation of local stakeholders. Also, the link between NTFP harvesting and environmental damage is not addressed in this study.

Our study could also more explicitly address the importance of traditional ecological knowledge (TEK), especially with regard to the question of how TEK is related to the sustainability of forest use. As our results indicate, TEK plays an important role and its application in the harvesting of *G. avellana* means little to no damage is caused to the trees. Hunter et al. (2003), for example, were able to show that on China's Hainan Island, indigenous TEK in the use and processing of rattan is a crucial factor in protecting forest edges from further deforestation through rattan NTFPs. He et al. (2009) show how indigenous TEK in China's Yunnan province is successfully integrated into a forest restoration program that implements agroforestry systems based on NTFPs. Jevon and Shackleton (2015) show that TEK is as useful to the local population in assessing plant invasions for their impact on the biodiversity of local species as ecological assessments. Sinha and Bawa (2002) show, however, that indigenous harvesting techniques can also partially damage the plants that produce NTFPs, so that the relationship between TEK and sustainability needs to be examined on a case-by-case basis.

An important aspect not addressed in our study is the possibility of combining NTFP production with measures of ecological restoration and conservation. For example, Robles-Diaz-de-Leon and Kangas (1998) show that riparian buffer zones, which would otherwise be economically unproductive, can be significantly enhanced by indigenous NTFP production. Ashton et al. (2014) show in a special issue that NTFP production by indigenous people is an important factor in the restoration of fragmented tropical forest landscapes.

Our study also deals with only one NTFP; in the future, several NTFPs should be treated synoptically, since the seasonality of NTFPs often makes it difficult to focus a livelihood on a single product (Mechik et al., 2017). Though *G. avellana* is a seasonal product, collectors indicated being able to collect sufficient quantities to earn an income from it throughout the entire year.

4.2 Relationship of the Findings Regarding the Different Forms of Capital to International Experiences

Regarding human capital, it can be stated that the producers of *G. avellana*-based NTFPs are mainly indigenous Mapuche or Pewenche. They live in rural forest fringe communities with limited access. It is mainly women who carry out the NTFP activity (66%). They have a low level of formal education, but a sound knowledge of NTFP characteristics and use through cultural embedding. The families have very low and irregular cash income (only child benefit payments or disability pension, and to a small extent income through local marketing of agricultural products and NTFPs, as well as tourism in Alto Biobío). They are therefore reliant on a diversified production of supply goods and have few alternative sources of income, especially in winter.

The situation of Chilean collectors is comparable to other livelihoods on an international scale incorporating NTFP activities. A predomination of women in NTFP collection activity has been noted because men in male-dominated societies are often involved in other income-generating activities (Heltberg et al., 2000) and because woman often face socio-economic disadvantages due to spending most of their time involved in family care (Hill, 2011), which increases their economic dependency and vulnerability. This female domination in NTFP activity can be reduced for some specific NTFPs in areas where harvesting is prohibited or dangerous due to the presence of wild animals, as was observed in case studies in India (Campbell, 1991; Dash & Behera 2016). However, NTFP activities are one of the few cash-generating opportunities for women in marginalized rural communities (Schreckenberg & Marshall 2006; Abdullah et al., 2013), oftentimes offering greater opportunities for women than their possibilities for employment and thus reducing income inequality (Kassa & Yigezu, 2015).

The level of formal education generally seems to play a role in the degree of involvement in NTFP activity for livelihoods. Studies from Bolivia and Nepal find that with increased formal education the level of engagement with NTFP commercialization declines, the result of better income and employment alternatives beyond the forest sector, for example (Adhikari et al., 2004; Godoy & Contreras, 2001). Dash & Behera (2016), on the other hand, found that households of the Similipal Tiger Reserve in India showed increased interest in participation with NTFPs despite having a higher level of education and assumed this effect is related to NTFPs of elevated value, while those households loose interest in less-valuable NTFPs. Nonetheless, the requirements to participate in NTFP activity depends on experience rather than formal education. The knowledge of practice, forming a fundamental part of the human capital, is accumulated by spending time in the activity which provides insurance or a lowering of risk to livelihoods (Pattanayak & Sills, 2001).

Regarding the natural capital, our study could show that the abundance of *G. avellana* in the study area is relatively high in Alto Biobío but limited in Contulmo. Additionally, the long period of time until fruit formation hampers the economic valorization of the product. Murali et al. (1996) also show that the recruitment and regeneration of the species used by indigenous people in South India is often inadequate, which could lead to supply bottlenecks. This is due not least to the historical overuse of forest ecosystems. Nakazono et al. (2004) are conducting a three-year experiment on the harvesting of *Ischnosiphon polyphyllus*, one of the species used by indigenous people in Central Amazonia for NTFP extraction (production of tools). They investigate the influence of different utilization schemes and show that the long-term possible utilization is variable with time. After the first year of the study, a harvesting limit of 50% would have been proposed, but after three years this figure was found to be too high. As with *G. avellana*, the seasonality of fruit ripening is a problem for NTFP-based livelihoods based on the Brazilian nut in Rodonia (Mechik et al., 2017).

Such results show that the long-term viability of NTFP use is critically dependent on the management scheme and that expansions of use in particular need to be implemented with care. Our results are therefore consistent with the findings of other authors and show that appropriate management is of high importance for the conservation of NTFP resources. Our field experience further shows that the current management schemes are not yet sufficiently based on fundamental ecological knowledge about *G. avellana* to really ensure the conservation of the resource.

In addition, we have been able to show that the forest industry threatens the physical basis of cultivation. Saifullah et al. (2018) show that for indigenous peoples of the Malaysian peninsula, the increasing conversion of forests into forest plantations is becoming a major livelihood risk. Land use pressure is thus a risk factor for the development of sustainable livelihoods based on NTFPs. In Chile, as in other countries, there is a lack of regulation that ensures access and property rights to the necessary resources.

With regard to social capital, we were able to show that cooperation with official, governmental institutions is already taking place, for example through training courses on the use of NTFPs. Elsewhere, cooperation with state institutions is inadequate. For example, there is a lack of pre-financing opportunities, but above all a lack of a legislative framework and management schemes. At the same time, there is a degree of self-organization (NTFP Roundtable), which is very beneficial to the indigenous people. Saifullah et al. (2018) show that indigenous people on the Malaysian peninsula need state support for the management of NTFPs, especially with regard to property rights, in order to not be excluded from access to their resources. Mechik et al. (2017) also show that NTFP use of the Brazilian nut is becoming a successful strategy for indigenous people, particularly through legislative, institutional, and pre-financing support.

De la Torre et al. (2011) show in a comparative study of four Andean countries that often the legislative framework for a legal and sustainable consolidation of indigenous livelihoods through NTFPs is already a problem, for example because access and property rights are not regulated in a goal-oriented manner and forest laws focus primarily on timber extraction. The study by Matias et al. (2018) underlines the importance of the institutional context. In Palawan, Philippines, a community enterprise was founded by indigenous Tagbanua to market NTFP honey. In an integrated value chain analysis, the authors show that it is precisely the downstream actors who benefit, while the upstream actors in this institutional context tend to lose rather than gain. Conversely, in our study, it is the upstream actors who gain from marketing. This makes two things clear. First, the institutional setting in which the use of NTFPs is embedded is of crucial importance. Second, it is difficult to make generalizing statements. Depending on the institutional setting, very different actors can become winners or losers in the value chains.

Furthermore, our studies on social capital have shown that NTFP collection and cultivation have high cultural significance for the indigenous peoples and create a sense of identity. Kim et al. (2012) show that although the use of NTPF for the Sts'ailes Indians in Canada has no economic significance, it fulfills many cultural functions. NTFPs are part of traditional food and medicine and materials, but are also used in spiritual practices. They serve to pass on traditional ecological knowledge and to update cultural identity. Nguyen et al. (2019) show that the consumption of NTFPs in Vietnam correlates with indigeneity and that the traditional indigenous diet contains significantly more calories from NTFPs. Matias et al. (2018), on the other hand, show that the cultural significance of NTFPs is also context-dependent. The Tagbanua honey collectors in their study, who tend to lose rather than gain within the given context, are increasingly abandoning traditional cultural rituals due to economic pressure. It is therefore difficult to make generalized statements about the cultural significance of NTFPs. Again, the extent to which desirable effects (namely the possibility of preserving traditional indigenous identity) are achieved or not is dependent on the institutional framework. In our study the cultural value of NTFPs has been shown to be a crucial motor to the indigenous producers to successfully integrate their activity into the Chilean socio-economic market system.

With regard to financial capital, we were able to show that cultivation on property is considered to be financially profitable by some interviewees. In recent years there has been increasing competitive pressure, from which the collectors but not the distributors have been able to benefit. The latter, consequently, consider changing their strategy to focus on their own production of *G. avellana*. Economic success seems to depend crucially on property rights and on the area under production. Hegde et al. (1996) show that although NTFPs make up 50% of the gross annual income of the indigenous Soliga in southern India, they are not the only source of income. In addition, the income is also high in proportion to the time invested, but NTFPs are not an economically attractive model because the income is too heterogeneous in relation to the various NTFPs and, above all, only minimum income (compared to other activities) is earned (see also Uma Shaanker et al., 2004). Moradi et al. (2017) show that NTFP use in the Zemkan Basin in western Iran is a financially very advantageous business for families in forest-dwelling or forest-fringe villages and creates additional incentives to protect the forests. Saifullah et al. (2018) show that for indigenous people on the Malaysian peninsula, NTFPs account for up to 24% of the average income and are therefore very important. These international results illustrate that the use of NTFPs is potentially promising for Chilean indigenous peoples as well, if it is possible to integrate their use into sustainable livelihoods, as discussed below.

In terms of physical capital, our results show that geographical isolation and lack of access to (market) infrastructure is a major obstacle to the sale of *G. avellana*-based NTFPs. Moradi et al. (2017) also emphasize the importance of management plans for the expansion of use and show that the development of sales and distribution infrastructure is crucial in the transition from local to supra-local markets. Leakey (1998) has shown that lack of access to market infrastructures is a major obstacle for the multi-strata agroforestry of Theobroma cacao into which NTFPs could be integrated. And the focus groups that Horn et al. (2012) surveyed about aguaje (*Mauritia flexuosa*) use in Peru also criticize market access as a major obstacle to development. Our results are thus consistent with international findings and show that it is crucial to improve market access (e.g. through online marketing opportunities, but also transport infrastructures) if NTFP use is to be sustainable in the long term and the actors involved are not to become newly dependent.

4.3 Discussion Regarding the Contribution of NTFPs to the Sustainability of Indigenous Livelihoods

4.3.1 Economic Significance of NTFPs and *G. Avellana* to Producers

The determination of precise economic outcomes from NTFPs in Chile is difficult due to the poor availability of data on indigenous activities, but also on the development of local NTFP markets. Due to these interrelationships, this study approaches the topic using a qualitative approach. From an economic point of view, we observe an at least temporary increase in the sustainability of indigenous livelihoods. Due to the lack of economic perspectives, marketing is considered significant by the majority of producers and experts. Producers report a significant increase in the standard of living (e.g. financing of clothing, supplies for the winter, schooling, and expansion of housing or working space). However, these benefits depend on the type of production (experience, material equipment, quantity and quality of production, networking, etc.). Many producers specialize in the production and marketing of *G. avellana*, since a variety of products can be obtained from *G. avellana* and because *G. avellana* is economically lucrative. Specialization also means developing capacities to produce larger quantities of products and a high degree of processing in order to achieve better quality and prices. Limiting factors here are restrained access to training, which in some cases is due to poor cooperation with authorities, limited ownership, and poor access to the internet, which reduces marketing opportunities when scale is expanded.

It could now be suspected that commercial specialization in a single type of diversification contradicts the sustainability strategies proposed by the SLA approach and development policy. However, we do not think this is true. Producers or collectors as a whole have diversified socio-economic and socio-cultural development strategies, especially Pehuenche and Mapuche. They can maintain their diversified economic activities (livestock breeding, vegetable growing, tourism, and home cooking, etc.) while at the same time complementing them with *G. avellana*-based NTFPs. In addition, the plant can be used to produce a variety of products with or without added value, so the market for sale is also diversified. Contrary to the risks of specialization, factors that inhibit an expansion of the use of *G. avellana* appear to be the greater risk. These are, for example, differences in the appreciation, acceptance, and legitimacy of NTFP use between collectors and landowners. The latter prefer subsidized timber production to NTFP use, while collectors even run the risk of entering unauthorized foreign territories. The specialization in *G. avellana* does not contradict the postulates of the SLA approach at all. Rather, there is a need for recognition, organization, training, and formalization of collecting activities.

If one now looks beyond the economic contributions to individual livelihoods and considers the economic development in the entire NTFP sector, one must above all discuss the increasing pressure of competition and price development. An increase in demand leads to an increase in prices on interregional markets compared to prices on local markets that have remained constant for a long time. Incentives are created to expand the use of NTFPs and new players are emerging along the value chain. This dissolves the monopsony that extern intermediaries originally often had on the demand for *G. avellana* raw materials. The winners of this development are initially the collectors. This firstly helps to redress the social imbalance in the value chain, which originally disadvantaged them. Anyhow, the intermediaries who directly follow the collectors come under pressure. They also belong to the indigenous culture and the marginalized focus group. Some of them therefore no longer see a future in the use of *G. avellana*.

The increasing pressure of competition could lead to a change in social behavior among various indigenous families or groups, manifesting itself in a reduced willingness to cooperate in the procurement and marketing of NTFPs due to negative experiences with other collectors. Since a cooperative economy (exchange of agricultural goods) is both traditionally and currently important in the communities, the question arises as to what consequences this would have for traditional indigenous cultural structures.

4.3.2 Natural Resource Base and Ecological Sustainability

Uma Shaanker et al. (2004) show that the livelihood benefits of NTFPs always have an ecological cost. According to the study, these costs increase with higher dependencies on the resource, lower level of ecological knowledge about the resource, and the more the market is organized through short-term contracts instead of through permanent participation of local stakeholders.

The increased economic value of *G. avellana* intensifies the cultivation and harvesting, resulting in a risk of over-yield or the introduction of bad practices. In addition, actors who lack the TEK on *G. avellana* have been able to enter production. All of this is a risk to the natural resource base, and the stocks ultimately jeopardize economic sustainability.

Another risk to the natural resource base and thus to ecological sustainability is the constant expansion of the forest industry with its commercial plantations. This creates an encroachment on rural livelihoods that are hardly compatible with the presence of the forest industry (Braun, 2021). The use of *G. avellana* is incompatible with forest plantations. Nonetheless, it is possible in near-natural forests and in the form of native edges surrounding forest plantations which could create a win–win situation, enabling plantation owners to protect their land against erosion and for landless and forest-dependent collectors to reobtain access to more valuable forest areas. For example, Robles-Diaz-de-Leon and Kangas (1998) show that riparian buffer zones, which would otherwise be economically unproductive, can be significantly enhanced by indigenous NTFP production. Ashton et al. (2014) show in a special issue that NTFP production by indigenous people is an important factor in the restoration of fragmented tropical forest landscapes. Here, too, environmental governance is required, in which the gatherers receive adequate institutional support as well as access and property rights to be able to pursue their activities on a sustainable basis. However, this is not in sight at present. In these contexts, too, the sustainability of *G. avellana* use for indigenous livelihoods in Chile is endangered.

4.3.3 Institutional Integration and Lack of Support for NTFP Usage

The effects of institutional interventions must be discussed controversially. Although all respondents have some form of contact with state authorities, many have also received financial support. The bureaucratization and formalization of support appears problematic. Support is given to those who are formally recognized as producers. Participation in certain programs determines the inclusion and exclusion of individual producers with regard to their institutional support. In this way, some producers are favored and others disadvantaged. The procedural fairness of access to institutional support is limited. In this sense, formalization creates disparities between formally authorized and non-registered producers. This formalization contradicts the indigenous logic, which often does not find its way around the limitations. It has also been observed that new norms/standards are introduced through institutional intervention. This includes, for example, the introduction of smaller volumes of sacks and benchmarks for prices. In this way, state authorities are very active in intervening in the market. The sustainability of the livelihoods through *G. avellana* use is thus also endangered in institutional terms.

4.3.4 Cultural Relevance and Meaning for Indigenous Livelihoods

In cultural terms, the study shows a change from the autonomous lifestyle of the indigenous groups to integration into the social structures of the Chilean population. These new dependencies can be explained by the decline in indigenous communities' access to natural capital (such as through the expansion of the forestry industry) and therefore explain the increasing importance of the other types of capital they need to cover their basic needs and improve their quality of life. As the opportunities for reclaiming natural capital are limited and subject to strong social and state conflicts (e.g. reclaiming territory, forests, and water), NTFP collection can play an important role in consolidating other types of capital and integrating them into economic structures and activities. This economic recognition that the indigenous collectors receive by selling their products strengthens the self-esteem of a group that, historically, has long been marginalized. This strengthening might be perceived by the group even more intensely due to a perception of their economic success being connected to the approval of God. Harvesting is a traditional activity that ensures important cultural values are kept alive and integrated into the regional and national market and can be a means of improving social and intercultural connectivity.

5 Conclusion

Overall, this results in a heterogeneous assessment of the importance of *G. avellana*-based NTFPs in Chile. NTPFs form a significant part of the socio-economy of indigenous families. There is currently an expansion of use that discriminates against some actors along the value chain while favoring others. The emergence of supra-regional demand in particular offers economic development opportunities arising from *G. avellana* use. However, there are risks with regard to the natural resource base and institutional integration. Whether *G. avellana* will be able to make a sustainable contribution to indigenous livelihoods in the future is greatly dependent, in our view, on environmental governance, which regulates the conservation of the natural resource base through appropriate management techniques, the regulation of land-use conflicts with the forest industry, the organization of collection and use, state support, and access and ownership rights.

As in many other contexts, the question of whether indigenous people can realize the chances of using *G. avellana* depends on certain basic conflicts in the Chilean system. These include the basic conflict over the state's inclusion of indigenous livelihoods, the extent to which Chilean economic policy promotes large corporations on the basis of patronage, the extent to which it deals with small economies, and, last but not least, how fundamental land use conflicts are resolved. Chile's reluctance to amend its constitution to be more environmentally sensitive should be viewed as a missed opportunity to address growing socio-environmental challenges.

Regarding environmental conservation efforts, the findings of this study do not implement concrete proposals to incorporate cultural knowledge into forest management but identify an eligible group to receive governmental and institutional support to develop their environmental sound activity with NTFPs. Since indigenous producers distinguish themselves from "outsider" actors, a focus on the cooperation between both groups would be desirable to reduce mutual rejection and to improve the expansion of the indigenous comprehension and management practices of the natural capital whose destruction tends to be the initial point of the described conflicts.

Another key outcome of the study is the difference in which individuals of the same ethnic group benefit or lose, depending on their strategic position within the value chain. Contrary to many international observations, the actual collectors of *G. avellana* benefit while some intermediaries are disadvantaged. In general, there is a push to reduce benefits received by intermediaries in order to strengthen the primary producer but, in this case, both the intermediaries and primary producers belong to the same vulnerable group. Our findings confirm the necessity of individual, local case studies according to particular NTFPs, ethnicities, and areas of investigation to adapt individual assistance strategies on the basis of international experience.

Appendix 1

See Table 3.

Table 3 Semi-structured interview questions and SLA focus categories

SLA categories		Interview questions (P = Producers; E = Experts)
Livelihood assets	Human capital	P: Who do you collect with? P: How long have you been doing the collection activity? P: Do you practice any kind of forest management or treatment on the hazelnut trees?
	Natural capital	P: Where do you collect? P: Are there problems with the abundance of hazelnut trees and/or their fruit in the area? E: Do you see in danger the forests where you collect? What are the areas of collection of hazelnut in the Region of Biobío and Ñuble? E: What are the main differences between the Andes Mountains and the Coastal Mountains related to the harvesting? E: To your knowledge, the form and/or amount of hazelnut harvesting and production varies between the main regions of abundance Maule, Bío Bío/Ñuble and Araucanía? E: In which areas are there problems or risks with NTFP abundance?
	Social capital	P: Are there any conflicts in the area; which ones? P: Are you in contact with support institutions? P: How is the relation? E: What are the relevant facilities, organizations or institutions in the work with the collectors? E: Do you work exclusively with collectors or also with landowners who produce NTFPs on their property? E: How do you establish contact between institutions and collectors? E: How are composed the trainings for NTFP collectors?
	Financial capital	P: Do you possess land?

(continued)

Table 3 (continued)

SLA categories		Interview questions (P = Producers; E = Experts)
	Physical capital	
Livelihood strategies and returns		P: What products do you get from the hazelnut tree? P: Are there other forest products [NTFPs] that you are harvesting? P: How is the process from purchase/collection to sale? P: Where do you sell? P: What other activities do you and your family receive income from?
	Economic and cultural importance of NTFPs for producers	P: Is the hazelnut business profitable? P: Do you have plans or wishes for the future? P: What does the forest mean to you personally? P: Is the hazelnut part of your or the families diet? E: What role do NTFPs and hazelnut trees play in the Biobio region? E: What are the most important harvest products in the Biobio region and which NTFPs have future potential? E: Do you consider hazelnut products to have potential to exportation?
Changes, structural processes and vulnerability		P: How has changed the NTFP market in recent years? P: Is there competition in the harvest or for your products in the market? E: How has been the hazelnut harvest and market in recent years? E: To your knowledge, how does the industry that processes hazelnuts to make oil and cosmetics affect the work of the harvesters?

(continued)

Table 3 (continued)

SLA categories		Interview questions (P = Producers; E = Experts)
	Gaps and needs of producers	P: What threats are there? P: What are you missing to advance your business? E: What are the main gaps and needs of the collectors or of your institution's work with the collectors? E: What level of formality is there? E: What legal problems limit the activity?

Source Own elaboration

References

Abdulla, A. M. (2013). Non-timber forest products and food security: the case of Yabelo Woreda, Borana Zone, Ethiopia. *Food Science and Quality Management, 22*(12), 9–22.

Ashton, M. S., Goodale, U. M., Bawa, K. S., Ashton, P. S., & Neidel, J. D. (2014). Restoring working forests in human dominated landscapes of tropical South Asia: An introduction. *Forest Ecology and Management, 329*, 335–339.

BCN (2005). Ley 20.001. Regula el peso máximo de carga humana. Biblioteca del Congreso Nacional de Chile (BCN).

Belcher, B., & Schreckenberg, K. (2007). Commercialisation of non-timber forest products: A reality check. *Development Policy Review, 25*(3), 355–377.

Braun, A. C. (2021). Encroached by pine and eucalyptus? A grounded theory on an environmental conflict between forest industry and smallholder livelihoods in Chile. *Journal of Rural Studies, 82*, 107–120.

Calleja, M. (2007). *Non-Timber Forest Products, the Chilean reality and the way ahead.* UMI Research Press & State University of New York. USA.

CONAF (2019a). CONAF actualizó catastro de recursos vegetacionales nativos de región del Biobío. https://www.conaf.cl/conaf-actualizo-catastro-de-recursos-vegetacionales-nativos-de-la-region-del-biobio/ [18.08.2019].

Dash, M., & Behera, B. (2016). Determinants of household collection of non-timber forest products (NTFPs) and alternative livelihood activities in Similipal Tiger Reserve, India. *Forest Policy and Economics, 73*, 215–228.

De la Torre, L., Valencia, R., Altamirano, C., & Ravnborg, H. M. (2011). Legal and administrative regulation of palms and other NTFPs in Colombia, Ecuador, Peru and Bolivia. *The Botanical Review, 77*(4), 327–369.

DFID (The Department for International Development) (2001). *Sustainable livelihoods guidance sheets.* London.

FAO (2018). Non-Wood Forest Products News. http://www.fao.org/forestry/nwfp/84902/en/(Acceso: 2018-07-29)

Fischer, G., van Velthuizen, H., Mahendra Shah, M., & Nachtergaele, S. (2002). Global Agro-ecological Assessment for Agriculture in the 21st Century: Methodology and Results. International Institute for Applied Systems Analysis & FAO. Laxenburg/Rom.

García-Fernández, C., Ruiz-Perez, M., & Wunder, S. (2008). Is multiple-use forest management widely implementable in the tropics?. *Forest Ecology and Management, 256*(7), 1468–1476.

Godoy, R., & Contreras, M. (2001). A comparative study of education and tropical deforestation among lowland Bolivian Amerindians: forest values, environmental externality, and school subsidies.*Economic Development and Cultural Change, 49*(3), 555–574.

Guiñez, D., & Gaete, C. (2013). Diagnostico Alto Biobío. SEREMI de Salud Región de la Región del Biobío.

Guariguata, M. R., García-Fernández, C., Sheil, D., Nasi, R., Herrero-Jauregui, C., Cronkleton, P., & Ingram, V. (2010). Compatibility of timber and non-timber forest product management in natural tropical forests: perspectives, challenges, and opportunities. *Forest Ecology and Management, 259*(3), 237–245.

He, J., Zhou, Z., Weyerhaeuser, H., & Xu, J. (2009). Participatory technology development for incorporating non-timber forest products into forest restoration in Yunnan, Southwest China. *Forest Ecology and Management, 257*(10), 2010–2016.

Hegde, R., Suryaprakash, D., Achoth, D., & Bawa, D. S. (1996). Extraction of non-timber forest products in the forests of Biligiri Rangan Hills, India. 1. Contribution to rural income. *Economic Botany, 50*(3), 243–251.

Hill, C. (2011). Enabling rural women's economic empowerment: Institutions, opportunities and participation. In *Background paper: UN women expert group meeting Accra, Ghana* (pp. 20–23).

Heltberg, R., Arndt, T. C., & Sekhar, N. U. (2000). Fuelwood consumption and forest degradation: a household model for domestic energy substitution in rural India. *Land Economics,* 213–232.

Horn, C. M., Gilmore, M. P., & Endress, B. A. (2012). Ecological and socio-economic factors influencing aguaje (Mauritia flexuosa) resource management in two indigenous communities in the Peruvian Amazon. *Forest Ecology and Management, 267*, 93–103.

Hunter, I., von Hahn, C. G., Zhaohua, Z., & Yanhua, Z. (2003). Stabilising forest margins by growing non-timber forest products: A novel example from Hainan Island, China. *Land Use Policy, 20*(3), 225–230.

INFOR (2019a). Estadísticas Forestales. https://wef.infor.cl/estadisticas_regionales/estadisticasregionales.php

INFOR (2019b). Catastro de Recolectoras y Recolectores de Productos Forestales No Madereros (PFNM) en las Regiones de Biobío y Ñuble. Informe técnico 221.

Jevon, T., & Shackleton, C. M. (2015). Integrating local knowledge and forest surveys to assess *Lantana camara* impacts on indigenous species recruitment in Mazeppa Bay. *South Africa. Human Ecology, 43*(2), 247–254.

Kassa, G., & Yigezu, E. (2015). Women economic empowerment through non timber forest products in Gimbo District, south west Ethiopia. *American Journal of Agriculture and Forestry, 3*(3), 99–104.

Kim, I. A., Trosper, R. L., & Mohs, G. (2012). Cultural uses of non-timber forest products among the Sts'ailes, British Columbia, Canada. *Forest Policy and Economics, 22*, 40–46.

Langenfeld, M. (2017). *Waldpolitik in Costa Rica und Chile.* Springer Fachmedien Wiesbaden.

Leakey, R. R. B. (1998). Agroforestry in the humid lowlands of West Africa: Some reflections on future directions for research. *Agroforestry Systems, 40*(3), 253–262.

Lepcha, L. D., Sarkar, B. C., Shukla, G., & Chakravarty, S. (2023). Non-Timber Forest Products: Diversity, Utilization, and Dependency in Fringe Areas of Jaldapara National Park in Indian Sub-Himalayan Region.

Loewe, V., Delard, C., & Álvarez, A. (2017). Plantaciones de avellano chileno (Gevuina avellana), una alternativa para producir madera de alto valor en Chile. *Ciencia e Investigación Forestal. 23*(1), 49–68.

Matias, D. M. S., Tambo, J. A., Stellmacher, T., Borgemeister, C., & von Wehrden, H. (2018). Commercializing traditional non-timber forest products: An integrated value chain analysis of honey from giant honey bees in Palawan, Philippines. *Forest Policy and Economics, 97*, 223–231.

Mayring, P. (2012). Qualitative Forschungsmethoden in der Psychologie. Differentielle Entwicklungsprozesse. *Psychotherapie & Sozialwissenschaft, 14*(2), 131–135.

Marshal, E. & Newton, A. (2003). Non-timber forest products in the community of el Terrero, Sierra de Manatlan Biosphere Reserve, Mexico: Is their use Sustainable? *Economic Botany, 57*(2), 262–278.

Mahapatra, K., Albers, H. & Robinzon, E. (2005). The impact of NTFP sales on rural households's cash income in India's dry deciduous forest. *Environmental Management35*(3), 258–265.

Mechik, E., von Hauff, M., de Moura, L. H. L., & Held, H. (2017). Analysis of the changes in economic activities of Brazilian forest communities after methodical support and provision of pre-financing capital. *Journal of Tropical Forest Science*, 227–237.

Meinhold, K., & Darr, D. (2019). The processing of non-timber forest products through small and medium enterprises—a review of enabling and constraining factors. *Forests, 10*(11), 1026.

Mestmacher, J., & Braun, A. (2020). Women, agroecology and the state: New perspectives on scaling-up agroecology based on a field research in Chile. *Agroecology and Sustainable Food Systems*, 1–26.

Monsalves-Gavilán, P., Pincheira-Ulbrich, J., & Rojo Mendoza, F. (2013). Climate change and its effects on urban spaces in Chile: A summary of research carried out in the period 2000-2012. *Atmósfera, 26*(4), 547–566.

Mugido, W., & Shackleton, C. M. (2018). Price determination of non-timber forest products in different areas of South Africa. *Ecological Economics, 146*, 597–606.

Moradi, S., Limaei, S. M., Lohmander, P., & Khanmohammadi, M. (2017). Quantitative and financial evaluation of non-timber forest products (case study: Zemkan basin forests, West of Iran). *Journal of Forestry Research, 28*(2), 371–379.

Murali, K. S., Shankar, U., Shaanker, R. U., Ganeshaiah, K. N., & Bawa, K. S. (1996). Extraction of non-timber forest products in the forests of Biligiri Rangan Hills, India. 2. Impact of NTFP extraction on regeneration, population structure, and species composition. *Economic Botany, 50*(3), 252–269.

Nahuelhual, L., Palma, J., Gonzalez, M. E., & Ortiz, K. (2008). Potential for greenery from degraded temperate forests to increase income of indigenous women in Chile. *Agroforestry Systems, 74*(1), 97–109.

Nakazono, E. M., Bruna, E. M., & Mesquita, R. C. (2004). Experimental harvesting of the non-timber forest product Ischnosiphon polyphyllus in central Amazonia. *Forest Ecology and Management, 190*(2–3), 219–225.

Nguyen, T., Lawler, S., & Paul, W. (2019). Socioeconomic and indigeneity determinants of the consumption of non-timber forest products in Vietnam's Bu Gia Map National Park. *International Journal of Sustainable Development & World Ecology, 26*(7), 646–656.

Pattanayak, S., & Sills, E. (2001). Do tropical forests provide natural insurance? The microeconomics of non-timber forest product collection in the Brazilian Amazon. *Land Economics, 77*(4), 595–612.

Pérez, M., Belcher, B., Achdiawan, R., Alexiades M., Aubertin, C., Caballero, J., Campbell, B., Clement, C., Cunningham T., Fantini, A., de Foresta, H., Fernández, C., Gautam, K., Hersch Martínez, P., de Jong, W., Kusters, K., Kutty, G., López, C., Fu, M., Martínez Alfaro, M., Nair, R., Ndoye, O., Rafael Ocampo., Nitin Rai., Martin Ricker, M., Schreckenber, K., Shackleton, S., Shanley, P., Sunderland, T., & Yeo-Chang Youn. (2004). Markets Drive the Specialization Strategies of Forest Peoples. *Ecology and Society, 9*(2), 4.

Robles-Diaz-de-Leon, L. F., & Kangas, P. (1998). Evaluation of potential gross income from non-timber products in a model riparian forest for the Chesapeake Bay watershed. *Agroforestry Systems, 44*(2–3), 215–225.

Saifullah, M. K., Kari, F. B., & Othman, A. (2018). Income dependency on non-timber forest products: An empirical evidence of the indigenous people in Peninsular Malaysia. *Social Indicators Research, 135*(1), 215–231.

SAG (Servicio Agrícola y Ganadero) (2019). Consulta realizada del 19.06.2019 a través del Sistema de Transparencia Pública de Chile.

Schreckenberg, K., Marshall, E., Newton, A., Velde, D. T., Rushton, J., & Edouard, F. (2006). Commercialisation of non-timber forest products: What determines success?

Scherr, S. J., White, A., & Kaimowitz, D. (2004). Making markets work for forest communities. *In Working Forests in the Neotropics: Conservation Through Sustainable Management?* (pp. 130–155). Columbia University Press.

SERNATUR (Servicio Nacional de Turismo) (2014). Licitación N° 1873-19-LE13. Diseño de Implementación Planes Estrategias de la Comuna de Contulmo – Región del Biobío. Gobierno Regional Región del Biobío.

SEPADE (Servicio Evangélico Para el Desarrollo) (2010). Estudio de Capacidades Productivas para la Comuna de Alto Biobío. Union Europea & Evangelischer Entwicklungsdienst.

Shackleton, S., Shanley, P., & Ndoye, O. (2007). Invisible but viable: recognising local markets for non-timber forest products. *International Forestry Review,* 9(3), 697–712.

Sinha, A., & Bawa, K. S. (2002). Harvesting techniques, hemiparasites and fruit production in two non-timber forest tree species in south India. *Forest Ecology and Management, 168*(1–3), 289–300.

Silva, T. C., Araujo, E. C. G., Lins, T. R. D. S., Reis, C. A., Sanquetta, C. R., & Rocha, M. P. D. (2020). Non-Timber Forest Products in Brazil: a bibliometric and a state of the art review. *Sustainability, 12*(17), 7151.

Stryamets, N. (2016). Use and governance of non-wood forest products in market and transition economies. *Acta Universitatis Agriculturae Sueciae,* (2016: 08).

Tacón, A. (2004). Manual de los Productos Forestales No Madereros. *Programa de Fomento para la Conservación de Tierras Privadas de la Décima Región. CIPMA. Valdivia.*

Tacón, A. (2017). Buenas Prácticas de Recolección Sustentable para Productos Forestales No Madereros. Avellano (*Gevuina avellana* Mol.). Cuaderno para recolectoras y recolectores. Fundación para la Innovación Agraria (FIA).

TEEB (The Economics of Ecosystems and Biodiversity) (2010). Mainstreaming the Economics of Nature. A synthesis of the approach, conclusions and recommendations of TEEB.

Ticktin, T., & Johns, T. (2002). Chinanteco management ofAechmea Magdalenae: Implications for the use of TEK and TRM in management plans. *Economic botany, 56*(2), 177–191.

Uma Shaanker, R., Ganeshaiah, K. N., Krishnan, S., Ramya, R., Meera, C., Aravind, N. A., Gauthier, R., et al. (2004). Livelihood gains and ecological costs of non-timber forest product dependence: Assessing the roles of dependence, ecological knowledge and market structure in three contrasting human and ecological settings in South India. *Environmental Conservation, 31*(3), 242–253.

Valdebenito, G., Molina, J., Benedetti, S., Hormazabal, M., & Pavez, C. (2015). Modelos de negocios sustentables de recolección, procesamiento y comercialización de Productos Forestales No Madereros (PFNM) en Chile. Fundación para la Innovación Agraria & Instituto Forestal (INFOR). Santiago de Chile.

Vázquez-García, V., & Ortega-Ortega, T. (2017). Gender, local governance and non timber forest products. The use and management of Satureja macrostema in Oaxaca's central valleys, Mexico. In *Women's studies international forum* (Vol. 65, pp. 47–52). Pergamon.

Summary and Sustainable Reflection

Is Eco-Industrial Development Just "More of the Same"? A Hegelian Dialectic View on an Emerging Concept

Andreas Ch. Braun

Abstract Eco-industrial development (EID) and its sub-disciplines—industrial ecology and circular economy—are often seen as game changers of the economic process in science, business, and politics. These approaches are addressed with the hope that the environmental impact of many economic sectors can be massively reduced through EID while traditional lifestyle and consumption patterns are maintained. This assumption has utopian features. Criticism of EID has therefore rightly been voiced, arguing that EID is just "more of the same." This criticism articulates that although EID discursively postulates that it is a grand redesign of the economic system, it follows the same production and consumption logics as the traditional economy. The article uses a Hegelian dialectic to evaluate the extent to which this criticism is justified, the extent to which it lacks justification, and what a productive synthesis might look like.

1 Introduction

As Korhonen et al. (2018, p. 544) rightly point out, although a significant amount of research has been conducted on the technical possibilities, potentials, boundary conditions, and prerequisites of EID, the values, social structures, cultures, and worldviews that underlie EID have hardly been addressed so far. This is also true of the volume in which this text appears. Across more than 500 pages, various strategies of eco-industrial development (EID), industrial ecology, circular economy, and industrial sustainability as a whole are presented, which have been researched in the

A. Ch. Braun (✉)
Kassel Institute for Sustainability, University of Kassel, Mosenthalstr. 8, 34117 Kassel, Germany
e-mail: Andreas.braun@uni-kassel.de

A. Ch. Braun et al. (eds.), *Eco-Industrial Development as an Industrial Strategy*, Green Energy and Technology, https://doi.org/10.1007/978-3-031-73576-9_31

framework of the IECO—Institute of Eco-Industrial Development project, a major large German-Chilean partnership led by myself (Braun et al., n.d.).[1] Although some contributions in Part F take critical perspectives, they generally endorse the concept of EID—whether implicitly or explicitly—as a useful strategy towards improved sustainability. They do so in the context of a continent (South America) where environmental, social, and land-use conflicts related to extractive industries are both numerous and intense (Petras & Veltmeyer, 2014). Moreover, they do so in the context of a very broad sustainability discourse, in that a wide range of theories and conceptual models about socio-ecological transformations exist. On the one hand, these are based on very different conceptions about the preconditions and character of sustainable development (e.g. efficiency vs. sufficiency, weak vs. strong sustainability, green growth vs. no/post growth) (Braun, 2024). On the other hand, they are based on different ideas about the objectives and the strategies to achieve them. Eco-industrial development, for example, exists in a spectrum of alternatives,[2] which for example includes cleaner production and green chemistry at the corporate level, recycling as an alternative strategy for dealing with industrial residues at the sectoral level, and deindustrialization as an alternative for improving sustainability as a whole at the macroeconomic level.

With respect to either context, the contributions in this volume argue in at least implicitly appreciative terms by presenting and thus recommending precisely EID as a strategy. This is acceptable from the point of view of the goals of this book, because this volume aims to present the results of a joint research partnership that has worked on the topic of EID. However, if one changes the perspective and focuses attention less on the advantages and disadvantages of EID strategies and more on the sustainability discourse as a whole, i.e. if one asks what place EID has in the discourse on socio-ecological transformation as such, then it is unacceptable. Because when seen from this perspective, a volume on a set of strategies must of course take a stand on critical queries towards these very strategies.

The reason for my writing of this chapter is that, over the years the IECO research project has been underway, major critiques of the EID concept and thus of the IECO's goals have sometimes been articulated. A large part of these critiques took place in

[1] Somewhat simplistically, eco-industrial development, industrial ecology, circular economy, and industrial sustainability are subsumed here under the acronym EID. Strictly speaking, this is not correct in many respects, because the strategies in fact mean something different, or behave like upper and lower terms to each other. For the analysis in this article, which aims to develop an external perspective on strategies for the circularization of industrial production and place them in the sustainability discourse as a whole, this vagueness is certainly permissible, especially since relevant literature also fails to precisely delineate the concepts (over 100 definitions are presented in Corvellec et al., 2021) and hence the terms at least very close, e.g. as in Korhonen et al. (2018) and Inigo and Blok (2019).

[2] Thus, "alternative" does not necessarily mean mutual exclusivity. Alternatives can be not only competing, but also complementary. For example, it is perfectly possible to combine green chemistry with EID. "Alternative" means that, from the perspective of industrial sustainability, there are other strategies besides EID whose relative advantages and disadvantages would have to be considered qualitatively and quantitatively if industrial sustainability is to be optimized efficiently in the strict sense.

the context of personal conversations with colleagues, but also with civil society organizations. Some of the criticisms were not even explicitly formulated by other people, but stem from my own reflection on argumentations within the sustainability discourse that I heard during talks or read in papers. Some of these criticisms lacked substance entirely. Others, however, were very justified. I am not concerned with technical criticisms, i.e. with the question of whether this or that implementation mechanism of EID is more useful, more expedient, or more efficient. I am concerned with those substantial and fundamental criticisms which—often formulated from a dependency or world systems theory perspective—can be subsumed under the formula that EID is just "more of the same," which seems to refer loosely to a perpetual model of economic growth and exorbitant consumption that is simply "greenified" by a more circular production.

As the project leader of a collaborative project on EID, I do not see it as my role to defend these concepts unreservedly and skillfully and rhetorically refute every objection. Rather, I see it as my responsibility to deal with these criticisms seriously, discussing them and looking for a synthesis.

What added value do I expect from this venture? First of all, I would like to leave the perspective of the advantages and disadvantages, the possibilities and limitations of EID and turn my attention back to the sustainability discourse as a whole. For this discourse, it is crucial to shed light on the place EID plays in the discourse around sustainable development. This is particularly important in the context of a continent where the impact of industry is massive. Second, I want to clarify the extent to which the criticisms of EID are justified—and the extent to which they are not. Third, I want to attempt to clarify how EID can help in the context of sustainable development.

2 Scientific Approach

2.1 Overview

In order to approach this project as systematically as possible, I will take the following approach. First, I describe my strategy to collect arguments; that is, how critiques against EID were found. Then I describe which of these critiques I selected and by what criteria. Next, I briefly describe the Hegelian dialectic that underlies this contribution as a method—albeit not in its purest, philosophical form. In addition, I will describe some limitations of my approach. In the third section I then try to synthesize hermeneutically what critics mean when they say EID is just "more of the same." In the fourth section, I develop the dialectic in an attempt to take a stand on these criticisms. In the fifth and final section, I summarize the chapter.

2.2 Collection of Arguments

The majority of the arguments listed in Sect. 3 were articulated in the course of personal discussions. Some also stem from my own reflection on other scientific contributions on the topic of sustainability, the core statements of which I related to EID. Thus, the arguments stem from a living—albeit oral—scientific discourse. However, in order to build my argumentation in this paper not only on oral conversations but also align it with written debates and thus make my paper more connectable and broader, I additionally researched literature. First, I conducted a literature search in ISI Web of Science in December 2021 using the following search terms: ["industrial ecology" OR "industrial symbiosis" OR "industrial sustainability" OR "eco-industrial development" OR "eco industrial development" OR "eco-industrial park" OR "eco industrial park" OR "circular economy"] AND ["critique" OR "criticism" OR "contradiction" OR "opposition"]. These searches brought me to the following articles: (Boons, 2021; Corona et al., 2019; Corvellec et al., 2021; Desrochers, 2012; Giampietro & Funtowicz, 2020; Hobson & Lynch, 2016; Hoehn et al., 2021; Inigo & Blok, 2019; Jensen et al., 2011; Keitsch, 2011; Kirchherr & van Santen, 2019; Korhonen et al., 2018; O'Rourke et al., 1996; Sehnem et al., 2019). Furthermore, I added additional literature, for example by specifically searching the publications of some critical authors or by specifically researching certain partial aspects of the discourse (Cohen-Ehrenfeld, 2004, 2007; Rosenthal, 2017). I incorporated the literature thus found into my argument. In doing so, my goal was not to encounter every written argument in the articles I found. Rather, I looked for arguments that underscored and fleshed out in writing what was articulated orally.

2.3 Exclusion and Selection of Arguments

To be able to work reasonably critically, it is not so much a razor-sharp delineation of the central terms associated with EID that is required, but rather a clearly outlined discursive approach. Such an approach can only work if it is limited in terms of the discursively processed critiques, i.e. if it excludes some published aspects. In the following, I ignore intrinsic criticisms, i.e. criticisms that refer to the concept itself (its internal structures and its external application), for example by doubting the fundamental feasibility of the EID approach (e.g. Keitsch, 2011, p. 23). Criticisms that are concerned with the thermodynamic impossibility of the concept are also excluded here (Giampietro & Funtowicz, 2020; Korhonen et al., 2018).

Furthermore, I do not deal with approaches critical of EID for not bring implemented enough in practice (e.g. Kirchherr & van Santen, 2019, p. 1; Corvellec et al., 2021; Haas et al., 2015; Korhonen et al., 2018, p. 547), for the lack of empirical evidence on the usefulness of the concept (Kirchherr & van Santen, 2019, p. 1; Korhonen et al., 2018, p. 549), who diagnose the conceptual inconsistencies in the concepts (Giampietro & Funtowicz, 2020, p. 64), or those who argue that no "one-off," off-the-shelf solutions are possible, but only gradual adaptation instead (Giampietro & Funtowicz, 2020, p. 70). The vagueness of the concept (Corvellec et al., 2021; Korhonen et al., 2018; O'Rourke et al., 1996), technical and other implementation barriers (Corvellec et al., 2021), problems with upscaling (Corvellec et al., 2021) are also less relevant here. Other texts deal with misconceptions, for example the differences between ecology and industrial ecology. There are literally hundreds of studies on this (Corvellec et al., 2021, p. 2), in comparison to this bunch of literature, my study appears rather irrelevant. So what's my point in here?

I am concerned with the fundamental question: the place that EID occupies in the sustainability discourse. Above all, the question of whether EID is actually a step towards more sustainability, i.e. whether EID and sustainability are really compatible, or whether the critics are right who think EID is "just more of the same." This doubt was well expressed by Corvellec et al. (2021, p. 5): "The differences between circular economy and sustainability are often blurred despite the fact that the latter is more holistic. Underpinned by a broader variety of institutional commitments, and as a synonym for a more extensive set of risks and opportunities [...], the questionable conceptual relationship between the two ideas has yet to be thoroughly characterized [...]. [...] The circular economy is presented as the practical solution to the sustainability challenge, but it underestimates the challenge [...]." They conclude that, "It is therefore important to dispel the myth that circular systems are necessarily more environmentally sustainable than linear systems [...]." A single scientist cannot go so far as to "dispel a myth" in a simple anthology, but I would like to point out a way in which this discourse could be conducted and argue why this is important.

2.4 *Hegelian Dialectic*

The Hegelian dialectic (Mueller, 1958) is a framework for understanding the progression of ideas through a process involving contradiction and resolution (cf. Fig. 1). It consists of three stages: a thesis, which is a starting point or assertion; an antithesis, which represents a contradictory or opposing view; and a synthesis, which reconciles the conflict between the thesis and antithesis to form a new, more comprehensive perspective. In the context of eco-industrial development (EID), the thesis might assert that EID is simply an extension of existing industrial practices, merely adding an environmental perspective without fundamentally changing the way industries operate. This viewpoint suggests that EID is "more of the same," adhering to established patterns without offering significant innovation.

The antithesis contradicts this notion, arguing that EID represents a distinct approach that introduces novel concepts and techniques aimed at fostering sustainability and reducing environmental impact. This perspective underscores the transformative potential of EID, highlighting how it redefines industrial processes through recycling, waste reduction, and renewable energy. The synthesis, which reconciles these opposing views, recognizes that while EID builds upon traditional industrial

Fig. 1 Hegel's dialectic. A thesis and an opposing antithesis are brought into dialogue to form a new synthesis. The synthesis encompasses (positive) features of both thesis and antithesis to then become the new thesis (Somnath Mukherjee, on https://www.quora.com/What-is-the-research-term-problem-on-Hegel-s-dialectic, accessed February 10, 2024)

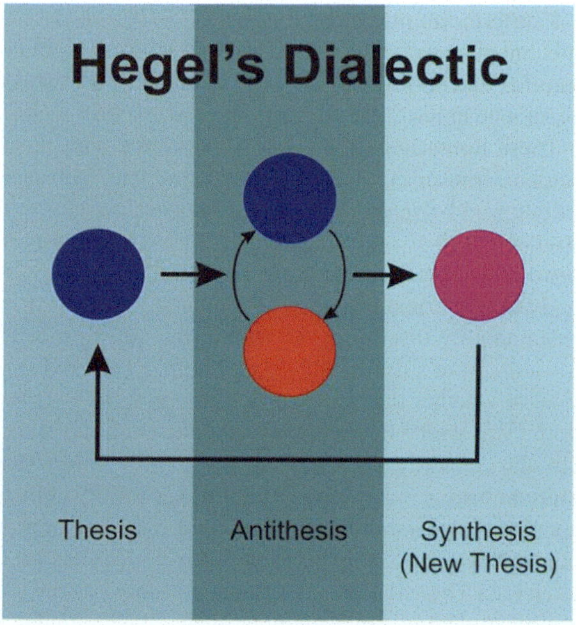

practices, it also introduces unique elements that go beyond simple extensions. The synthesis reveals that EID can lead to significant advancements by integrating environmental sustainability into industrial systems. It acknowledges that EID's innovation comes from its ability to evolve from existing practices while incorporating new approaches that drive genuine environmental benefits. Thus, the Hegelian dialectic demonstrates that the progression of ideas involves a dynamic interplay between conflicting perspectives, ultimately resulting in a more robust understanding that accommodates both the limitations and potentials of each stage. Hence, here the thesis means that EID is just "more of the same;" the antithesis means that EID is not just "more of the same." The synthesis clarifies how EID can help us beyond "more of the same."

2.5 *Limitations*

This contribution is not devoid of a subjective perspective. For one, it is based on my personal hermeneutic interpretation of what the critics of the approach might mean when they say that EID is just "more of the same." Second, the opinion developed here is in no way objectifiable; it is based on argumentative, rhetorical elements that I use to comment on the fundamental critique. Lastly, it is certainly selective, containing some aspects and missing others.

Furthermore, my knowledge of EID is of course limited, for example I am not an economist and therefore only able to assess the economic dimension of EID to a limited extent. I have a well-founded but limited professional perspective on EID, but let's be honest: For which of us would that be different?

These limitations are, in my view, without alternative, because sustainability is a global, transdisciplinary discourse that no scholar can fully grasp. If readers are not deterred by these limitations, they are encouraged to finish reading the rest of the chapter. Afterwards, they are cordially invited to comment on my argumentation, because, as Hegel emphasizes, the synthesis is always to be understood as a new thesis, which can be met dialectically on the basis of an antithesis.

3 Interpretation of "More of the Same"

What do critics mean when they say EID is "more of the same"—what does "the same" mean in the context of sustainable development? As explained at the beginning, I cannot say what was objectively meant, but only what I understood. As I understand the critiques, they encompass several dimensions.

First, they mean that EID is another way of extending the habitual, unsustainable lifestyle into the future by taking actions that gradually reduce, sometimes merely legitimize, the impacts of that lifestyle rather than fundamentally changing the lifestyle itself (cf. Table 1). The unsustainable lifestyle here means that the privileged populations in the Global North in particular consume quantitatively too much, demand qualitatively too high a variety of products, and overall lead lives that exist under the auspices of never-ending need satisfaction and thus ultimately require a growth economy. In this regard, improvements in industrial sustainability are a substitute for the necessary shift toward a lifestyle based on sufficiency and limiting growth.

Moreover, it means a system in which industry always succeeds in securing its own production and sales interests, and in placing those interests above the environmental and social demands of a finite planet (cf. Table 2). What is meant is that industry always succeeds against politics in securing its principle of permanent production—a principle this criticism defines as fundamentally wrong—and in protecting it against massive sanctions and transformations by gradual adjustments of the modes of production.

Lastly, with regard to international relations, "more of the same" means that EID succeeds in continuing to stabilize a system based on exploitation of natural resources in the countries of the Global South (cf. Table 3). Within this system, transnational corporations, or national corporations operating according to neoliberal logics, go on the prowl for raw materials in the Global South, using sustainability strategies such as EID to greenwash their products. In this respect, the discourse around EID is yet another example of an agenda of powerful elites fraught with neocolonial epistemic violence.

Table 1 Dimension one of the criticism expressed. Summarized as: "EID allows us to shirk the necessary lifestyle changes to achieve sustainability." Concretizing aspects with references

Dimension 1	
Abbreviated paraphrase of the criticism expressed	"EID allows us to shirk the necessary lifestyle changes to achieve sustainability"
Concretizing aspects	Literal references
EID follows an anthropocentric concept, instead of overcoming the nature-culture divide. Nature is evaluated according to what is useful about it. EID does not understand the economy as inscribed in nature, but as an external sphere of action that uses nature	Keitsch (2011, p. 25) Giampietro and Funtowicz (2020, p. 64f) Corvellec et al. (2021, p. 4)
EID falls prey to a myth that economic growth and environmental protection are supposedly possible both at once	Korhonen et al. (2018, p. 544)
EID continues to be based on the logic of satisfying needs, which created the problems of the anthropocene in the first place	Keitsch (2011, p. 25)
EID is based on a technocratic and pragmatic conception, but does not treat sustainability fundamentally	Keitsch (2011, p. 30) Giampietro and Funtowicz (2020, p. 69) Corvellec et al. (2021, p. 4)
EID harbors the risk of technological lock-in	Inigo and Blok (2019, p. 285)
EID is a kind of window dressing. A socially manufactured self-assurance that one can continue to consume without limits	Giampietro and Funtowicz (2020, p. 68, 69)
EID follows the old techno-narrative that technology solves the world's sustainability problems	Giampietro and Funtowicz (2020, p. 70)
The image of the "circle" is emotionally alluring, resembles nature, but is already thermodynamically impossible	Corvellec et al. (2021, p. 25)
We can talk ourselves into a good eco-conscience through EID	Valenzuela and Böhm (2017, p. 25–26)
The "green industry" helps us to do this because it satisfies our growing hunger for green products, which helps us to continue consuming	O'Rourke et al. (1996, p. 11)
EID helps us hold on to the harmful belief that we are the guardians of nature	Jensen et al. (2011, p. 683)

Taken together, these three dimensions assert that EID will precisely not lead to greater sustainability but instead continue to allow industry to exploit resources in the Global South, on the one hand fueling economic growth that has been identified since 1972 as fundamentally incompatible with the planet's carrying capacity limits (Meadows et al., 1972), and on the other hand giving us all the opportunity to shirk uncomfortable lifestyle changes. The three dimensions have been fleshed out in various discussions based on diverse aspects. The most important of these aspects

Table 2 Dimension two of the criticism expressed. Summarized as: "EID allows industry to continue to pursue its production interests against (political) sanctions." Concretizing aspects with references

Dimension 2	
Abbreviated paraphrase of the criticism expressed	"EID allows industry to continue to pursue its production interests against (political) sanctions"
Concretizing aspects	Literal references
EID is "economics of technological promises:" The fact that we need industry at all remains unquestioned. Further development (EID) naturalizes the reference point (industry)	Giampietro and Funtowicz (2020, p. 70)
EID also lacks mechanisms to minimize the impact of environmental interventions by industry (e.g. precautionary principle or ecological footprint): improvements are "only" achieved in the technosphere, not in the biosphere	Keitsch (2011, p. 26)
The "technology will save the world" narrative suppresses opposition to technology	Giampietro and Funtowicz (2020, p. 70)
EID also does not question the growth logic of the economy. The approach that the economy can continue to grow by simply decoupling it from additional resource use is an illusion	Giampietro and Funtowicz (2020, p. 65, 68)
EID creates illusions of win–win–win (politics, industry, environment) situations, which are simply myths!	*No literal quotation given*
EID is not radical enough. It does not get to the root of the problem: the way we consume	Hobson and Lynch (2016, p. 3)
This makes EID a reassurance for policymakers	Corvellec et al. (2021, p. 8)
EID is a folk tale to depoliticize the sustainability debate by creating implausible socio-technological visions	Giampietro and Funtowicz (2020, p. 64) Hobson and Lynch (2016, p. 2) Corvellec et al. (2021, p. 7)
EID makes it possible to keep the "sovereignty" over the sustainability discourse with the industry, which outwardly appears to bear responsibility but actually retains control	O'Rourke et al. (1996, p. 7)
Ultimately, EID follows the logic of adapting society so that the economy can continue to grow, rather than adapting the economy to the (socio-ecologically limited) society	Hobson and Lynch (2016, p. 2)
EID reinforces the idea that sustainability can be achieved by industry alone	O'Rourke et al. (1996, p. 12)

Table 3 Dimension three of the criticism expressed. Summarized as: "EID stabilizes a regime of systematic exploitation of the Global South." Concretizing aspects with references

Dimension 3	
Abbreviated paraphrase of the criticism expressed	"EID stabilizes a regime of systematic exploitation of the global South"
Concretizing aspects	Literal references
EID is just another neocolonial intellectual hegemony. It is based on the worldview of the intellectual elites of the Global North, which is superimposed on the actors in the Global South. It does not match the epistemologies of the Global South, which are systematically at disadvantage because they are confronted with the requirement to implement this strategy that is alien to them, and have no chance to establish their own concepts (e.g. buen vivir) as sustainability strategies	Giampietro and Funtowicz (2020, S. 70)
Not only in epistemological terms, but also in terms of power politics, the concept allows for neocolonial hegemonies. It continues to be large transnational corporations that rake in the profits from the resources of the Global South. EID, moreover, further allows them to do so in a purportedly environmentally friendly manner that exacerbates environmental injustices, rather than mitigating them	*No literal quotation given*
What we need is a transformation from an economy of accumulation to an economy of distribution, from an economy of extraction to an economy of restoration, from an economy of efficiency to an economy of sufficiency, from an economy of competition to an economy of cooperation. EID does not do justice to any of these aspects	Banerjee et al. (2021, p. 348)
EID focuses on manufacturing industries that do not apply in the Global South	Kirchherr and van Santen (2019, p. 1)
EID focuses on developed countries and should become a global agenda at the same time!	Kirchherr and van Santen (2019, p. 2)
EID also conceals the fact that in a finite world, economic growth is a zero-sum game. If we grow for a long time in the North, they can no longer do so in the South	Giampietro and Funtowicz (2020, p. 69)
EID also assumes that the world's development problems can be solved through trickle-down and free trade	Giampietro and Funtowicz (2020, p. 69)

(continued)

Table 3 (continued)

Dimension 3	
Abbreviated paraphrase of the criticism expressed	"EID stabilizes a regime of systematic exploitation of the global South"
EID makes sustainability promises that cannot be kept and thus creates new governance conflicts itself	Giampietro and Funtowicz (2020, p. 70)
EID is exported all over the world; indigenous perspectives of the Global South are deliberately ignored	Corvellec et al. (2021, p. 1) Calisto Friant et al. (2020, p. 4)
EID might not fit in the Global South. The technological rebound (efficiency increase leads to increased consumption) could be particularly strong there	Zink and Geyer (2017, p. 20)

are summarized in Table 1 and discussed dialectically below. For reasons of space, I cannot cover all of the aspects mentioned here; instead, I will limit myself to the three most essential ones.

4 An Opinionated Dialectic

Just as the criticisms made are generally based not on data and analysis but on personal interpretations of what sustainability means, my argumentation is also essentially based on discursive arguments. It is therefore a dialectic laden with opinions. I do not intend to implement this approach conclusively here, because I cannot! A conclusive approach would require the cooperation of a wide variety of participating disciplines, of which (industrial) ecology, (political) economics, sustainability science, (economic) sociology, (social) psychological behavioral science, and cultural studies are perhaps only a few of the most important.

My approach is exemplary. I would like to use each of the three dimensions to show how the dialectic of money could be applied in order to criticize EID more constructively and better situate it in the sustainability discourse. The following three sections are dedicated to a thesis, antithesis, and synthesis for each of the three dimensions. In the conclusion that follows, I will attempt to make the findings useful for the sustainability discourse as a whole.

4.1 Dimension 1: On the Relationship Between Industry and Consumers

It seems to me that the first dimension is very clearly about the relationship between the producing industry and the consuming citizen.

Thesis: The fear is that EID is a perpetuation of the anthropocentric, technocentric, growth-believing economic model that we have developed since the Industrial Revolution at the latest. In particular, there are fears that EID will inevitably lead to "greenwashing," which will ultimately only soothe the conscience of consumers, who will then continue to consume unabated instead of changing their lifestyles. Added to this is the fear of a technological lock-in, in which we continue to indulge in the illusion that technological improvements will significantly or even exclusively lead to compliance with planetary boundaries.

Many of these fears can certainly be acknowledged. EID is by no means based on a change in fundamental human-environment relations. There is little to suggest that EID strategies will fundamentally change the assumption that our environment is available to us as a provider of source-sink dynamics. EID is based on ideas of efficiency and consistency and runs the risk of overlooking the importance of ideas of sufficiency. The accusation of techno-centricity is also fundamentally correct; although production should be environmentally optimized, nothing changes in the fundamental production patterns. This certainly implies the risk of technological lock-ins. However, the EID literature gives no reason to assume that it would lead to an overcoming of the belief in growth. The fact that there are indeed other ways of life that seek personal happiness not in material consumption (greener products), but in sustainability, health, solidarity, and sufficiency does not become apparent.

Antithesis: Nevertheless, some objections are also necessary. The discussion must by no means overlook the difference between sustainable production (summarized in SDG 12) and sustainable consumption. The fear that greener production will lead to the continuation of excessive consumption is not an inevitability. It overlooks a certain division of labor between us and production. It is the task of companies to minimize the life cycle impacts of their products as much as possible—even if this means a loss of profit—and even to avoid some forms of production altogether. The task and responsibility of minimizing consumption lies with the consumer. The risk that excessive consumption will continue must not lead logically to the rejection of greener production strategies, but must lead to a minimizing of consumption on the part of consumers through suitable instruments (such as education for sustainable development, consumer protection through eco-labels, etc.). This is not a question of "either/or," but of a logical division of labor in achieving a socio-ecological transformation.

The problem of scaling must also be pointed out again here. For example, so-called LOHAS (lifestyle of health and sustainability) demonstrate that material consumption can very well be minimized for personal happiness. These are very important impulses that really can and should achieve greater social scale. For me, however, the question is to what extent this scaling becomes more likely by rejecting greener production (shirking necessary lifestyle changes). For me, it makes much more sense to expend energy on scaling up such lifestyles (e.g. using Everett Rogers' diffusion of innovation theory).

Another misconception regarding the division of labor lies in the accusation that EID does nothing to change the fundamental belief in growth. While this is factually true, plausible responsibilities must also be distributed here. This accusation can be turned around to the assumption that an abandonment of the belief in growth must come from industry. Although this would be highly desirable, and social entrepreneurs, ecopreneurs, and B-corps are living examples of this, it would still be a risky eutopia to expect significant impetus from industry if there are no external incentives for this. Economic and industrial policy, but also change in demand and changing lifestyles, are impulses that can lead to a post-growth economy. This is therefore a key issue for policymakers, who must be prepared to provide appropriate incentives for both production and consumption.

Finally, technological lock-ins are a risk but not an inevitability. "Keeping your hands off" is hardly a convincing solution to a fear of lock-ins and rebound effects from technological improvements towards greener production. Other instruments are available instead, such as education for sustainable development, socio-psychological interventions, and consumer transparency to raise awareness of the fact that rebound effects exist and must be avoided.

Synthesis: The coining of euphoric images of circularity, which is thermodynamically impossible, certainly implies the risk of the feared eyewash and the resulting avoidance of necessary lifestyle changes. However, this risk in itself is not an argument to refrain from researching and implementing circular strategies. This is not about exclusivity, but about a division of labor: a division of labor in which politics is called upon to enact policies that shape industrial production in such a way that life within planetary boundaries is possible. Industry itself is called upon to make its production as sustainable as possible instead of greenwashing it. Consumers are called upon to change their lifestyles and spread sustainable lifestyles to the maximum. Such a division of labor would go further than ideological trench warfare between green-growth and post-growth economics; on a pragmatic level, it is more about the fact that different subjects are responsible in different ways for both approaches. The consequence of this must be to use civil society and political instruments in such a way that the respective subjects addressed fulfill their respective tasks.

4.2 *Dimension 2: On the Relationship Between Industry, Society, and Policy*

This is clearly about the relationship between industry, society, and, above all, politics. Some of what is criticized in this dimension has already been dealt with in my dialectic in Sect. 4.1, which is why this section is shorter.

Thesis: EID represents a kind of "strategy of lulling" vis-à-vis politics. On the basis of green production strategies, industry puts itself in the driver's seat, so to speak, of socio-ecological transformation. This deceives politicians, who then refrain from more "inconvenient" regulations (CO_2 taxes, strict supply chain laws, etc.). In this way, the industry succeeds in committing itself to sustainability, but stabilizes its adherence to a growth logic overall. To a certain extent, industry and politics are making common cause by seemingly giving in to increasing social pressure for more sustainable development, but hardly changing anything per se.

Although the argument is not without certain Machiavellian traits, it is by no means implausible (e.g. from a Marxist or neo-Marxist perspective).

Antithesis: Even if the described danger exists, I would like to raise several objections. These concern, on the one hand, the *attribution* of the danger to EID per se, and on the other hand, its *inevitability*.

First of all, the attribution: Such "lulling" of politicians *can* and often *does* come from the side of industry, but it is questionable whether it needs to take the detour of EID to do so. After all, there is no shortage of examples where such connections are achieved directly through even more drastic instruments, such as lobbying. Industry enforces political compliance worldwide using completely different mechanisms. Does this eliminate the risk? Of course not, but this does not mean that EID is not a sensible strategy in principle. The possibility of *misuse* of any instrument does not negate the possibility of *sensible* use. The second argument concerns inevitability. The possibility of such misuse may exist. But to the same extent, there is also the possibility that EID is carried out on a reasonable scientific basis within the framework of honest and serious experiments.

Synthesis: Let us assume that EID is part of an honest and serious industrial policy and industrial attempt to minimize the environmental and social impact of the industrial sector. Then many case studies show that this is possible and that production can be organized more sensibly. Once again, it seems to me to be about the division of labor. Preventing the inappropriate use of green production strategies—namely to stabilize unlimited growth economies—is neither in the hands of scientific research nor solely in the hands of the companies that implement these strategies. Just as in the previous section it was the responsibility of industry to produce as sustainably as possible and the responsibility of consumers to minimize their consumption, the same applies here. Industry has a responsibility to implement strategies such as EID with transparent and serious sustainability intentions. Politicians, on the other hand, have a responsibility to rethink and transform the (green) growth-based economic model, creating the necessary economic, industrial, energy, supply chain, and consumption policies (and then some). Furthermore, policymakers have a responsibility to clearly communicate the limitations of planetary boundaries to consumers, instead of misleading them and suggesting a—politically more acceptable—business-as-usual life.

4.3 Dimension 3: On the Relationship Between Industry and Foreign Policy

Dimension 3 is clearly aimed at the importance of industry and foreign policy, or the global political economy. In my view, it is the weakest developed in this respect, as it overlooks numerous actors and focuses too narrowly on the direct relationship between industry and the Global South.

Thesis: Basically, the criticism can be summarized under three aspects. First, in epistemological terms, EID is based on concepts from the Global North which are then "exported" to the Global South but do not fit into this context (e.g. to local industrial systems). Local worldviews would be overlooked. Secondly, that it is based on the same power relations as "traditional industry" and thus reproduces exploitative relationships. And thirdly, that it continues to overlook the fact that within the limits of a finite earth system, growth is always only a zero-sum game, so that growth in the Global North limits growth in the Global South.

I fully agree with the content of this argument. At the same time, in my view it falls short of overcoming what it criticizes.

Antithesis: It is absolutely correct that perspectives from the Global South are also overlooked in sustainability discourses. Other ideas of sustainable development exist in the Global South (buen vivir, sumak kawsay, eco-swaraj, ubuntu, sentirpensar, and pluriversality are just a few examples). Those who claim that these concepts take a back seat to concepts of the Global North (e.g. Brundlandt definition, triple bottom line, etc.) and have too little influence on global agreements such as the Millennium Development Goals (MDGs) or the Sustainable Development Goals (SDGs) are absolutely right. However, anyone who wants to develop a path to change from this justified criticism must also say what could change on the basis of these concepts and, above all, how. Some of the concepts are intended more for the organization of smaller, more local communities (e.g. buen vivir) and it is very important not to stop at the fundamental criticism (Southern concepts are overlooked), but to work on concepts of how these can (better) influence global agreements, for example. There are certainly examples of this, such as the strong implementation of nature's own rights in the constitution and jurisdiction in Ecuador.

A similar argument also applies to the correct assertion that EID is designed more for the industry of the Global North. However, it does not necessarily follow from this that the EID concept is therefore automatically a kind of neo-colonial hegemony. On the one hand, the Global South is possibly even closer to the idea of circularity than the Global North. In any case, several project partners in Chile were able to report from their own experience what life is like on a small farm, where the first principle is to not throw anything away, but to recycle everything. However, EID can also simply be seen as an alternative idea in an arena of alternative ideas on (sustainable) industrial policy, which, if welcomed there in principle, can be adapted in the Global South.

The completely accurate argument of the zero-sum game weighs very heavily. Within the planetary boundaries, consumption in one part of the world limits consumption in another part of the world. However, if the accusation that EID rearticulates oppressive relationships is derived from this without further ado, then much is overlooked. This accusation is made when the division of labor between industry and consumers (Sect. 4.1) and industry and politics (Sect. 4.2) is overlooked. Industry will not be the driving force in overcoming exploitative relationships on its own, not even as an eco-industrial industry. It must be regulated by both consumers and politicians so that these exploitative relationships are not perpetuated. Furthermore, consumers and politicians each have their own tasks to achieve sustainability, which Andy Stirling very aptly equates with "overcoming colonial modernity."

Synthesis: Once again, my aim in the thesis is to acknowledge that many criticisms are justified. In the antithesis, it is important for me to show where these are limited in their scope in order to identify a possible next step in the synthesis. EID in itself will not overcome global political-economic relations and epistemic hegemonies. However, I wonder why anyone would expect it to. To deduce from this that EID strategies in themselves cannot be meaningful and worth researching, however, fundamentally ignores the systemic integration of industry into global power relations, in whose environment there are very many actors who allow, promote, or cannot prevent these exploitative relationships. However, the solution here is not to reject a potentially promising industrial strategy, but to overcome exploitative relationships systemically, in which, for example, international organizations, regimes and agreements, human and international law jurisdictions, political and economic associations in the Global South, civil society actors, and ultimately consumers play a role. In epistemological terms, neocolonial hegemony can be minimized, for example, by actively considering either what EID can look like in the Global South, or why it cannot and should not be a strategy there, and what is adapted there instead.

5 Conclusion

With my contribution, it was important to me to show an approach to what a more consistent, critical way of thinking about EID could look like. First, I would like to turn to the epistemological positioning of EID in the sustainability discourse. On the basis of the dialectic developed here, the question of whether industrial ecology (or EID) can be the science of sustainability (Ehrenfeld, 2004) seems about as problematic as the analogous question of whether oncology can be the science of health.

EID is not sustainability science per se, if only because it covers only one aspect of SDG 12, but not both. The United Nations' 2030 Agenda for Sustainable Development emphasizes how comprehensive and large the necessary socio-ecological dimension is. Many very important aspects are not even touched by EID. Just as oncology addresses a very important health issue, EID addresses a very important sustainability issue: industrial production. It is *one* of many important strategies to which, for example, cleaner production, green chemistry, and energy efficiency could be added. Within *this* framing, however, it is an important and very promising strategy.

More important to me than the question of whether EID is sustainability science per se is the question of *what role EID can play* in the sustainability discourse. Here I have tried to create a certain elucidation via the three Hegel dialectics. I feel reminded of (Hobson & Lynch, 2016, p. 8), who stated that "On paper these seem like interesting and cutting-edge material innovations that are not being critiqued here per se. However, the **social implications of these moves** need to also be thoroughly considered as central to the CE." The most crucial aspect is how to circumvent the problem of a perpetuated production and consumption model that allows industry to use greener strategies to create needs that are satisfied by overconsumption, only to create new needs. If EID is indeed used as a tool to prevent economic regulation on the one hand and changes in consumption patterns on the other, then EID is "just more of the same." If EID is integrated as an efficiency/consistency strategy into a comprehensive socio-ecological strategy, it can support it.

I have tried to show that and why it is less about establishing EID as the "silver bullet" of industry transformation, which is expected to provide "the solution," but rather about clarifying the division of labor. EID on its own does not make more or less sense. The strategy makes sense when it optimizes specific processes within the industrial sector that are synchronized with improvements in lifestyle, consumer behavior, industrial policy, and international relations. The concept must not be normatively overloaded, i.e. it must not be assigned too much transformative potential, nor must it be discarded as a potentially sensible strategy simply because its factual implementation still appears to be full of prerequisites or suboptimal.

So is EID "more of the same"? Indeed, in many respects, there is a risk that this strategy too will only lead to incrementally improved production, which belies the need for more fundamental changes in lifestyle. However, as I have shown, this is neither an inevitability nor an argument for rejecting the strategy itself. To the same extent that it is irresponsible to maintain a production system and lifestyle that is incompatible with the carrying capacity limits of the planet by greenwashing it, it is irresponsible to merely claim the necessity of a major socio-cultural transformation of our way of life and to criticize its nonexistence without presenting any alternative strategies on how to deal with the sustainability deficits of our way of life as long as this socio-cultural transformation of our way of life and production is absent.

It seems to me that a tension is often overlooked in the discourse, one that is important in sustainability science: namely, the tension between material pragmatism and fundamental critique. Fundamental critique attempts to deconstruct social relations of production and reproduction so radically that their intrinsic destructive potential becomes visible and thus also attempts to induce a "great mindshift" (Maja Göpel). Material pragmatism (such as EID), on the other hand, attempts to incrementally improve given material practices, which will not fundamentally transform overnight in such a way that they at least become less destructive. Neither approach has yet been able to convincingly work out how humanity can actually live within planetary boundaries, so the two *must work together*.

References

Banerjee, S. B., Jermier, J. M., Peredo, A. M., Perey, R., & Reichel, A. (2021). Theoretical perspectives on organizations and organizing in a post-growth era. *Organization, 28*, 337–357.

Boons, F. (2021). Freedom versus coercion in industrial ecology: Mind the gap! *Econ Journal Watch, 9*, 100–111.

Braun, A. C. (2024). Suffizienz–Psychische Ressourcen–Transformation. *Organisationsberatung, Supervision, Coaching, 31*(1), 7–28.

Braun, A. C., Espinosa, G., Mantilla, I., Kohler, P., Rümmele, K., & Hirth, T. (n.d.). Internationale scientific cooperation for sustainable development—The German-Chilean Institute of eco-industrial development (IECO). In A. C. Braun & T. Hirth (Eds.), *Eco-industrial development as an industrial sustainability strategy in Latin America: Perspectives from a German-Chilean research cooperation*. Springer.

Calisto Friant, M., Vermeulen, W. J., & Salomone, R. (2020). A typology of circular economy discourses: Navigating the diverse visions of a contested paradigm. *Resources, conservation and recycling, 161*, 104917.

Cohen-Rosenthal, E. (2017). What is eco-industrial development?

Corona, B., Shen, L., Reike, D., Rosales Carreón, J., & Worrell, E. (2019). Towards sustainable development through the circular economy—A review and critical assessment on current circularity metrics. *Resources, Conservation and Recycling, 151*, 104498. https://doi.org/10.1016/j.resconrec.2019.104498

Corvellec, H., Stowell, A. F., & Johansson, N. (2021). Critiques of the circular economy. *Journal of Industrial Ecology*. https://doi.org/10.1111/jiec.13187

Desrochers, P. (2012). Freedom versus coercion in industrial ecology: A reply to boons. *Econ Journal Watch, 9*, 78–99.

Ehrenfeld, J. R. (2004). Can industrial ecology be the "Science of Sustainability"? *Journal of Industrial Ecology, 8*, 1–3. https://doi.org/10.1162/1088198041269364

Ehrenfeld, J. R. (2007). Would industrial ecology exist without sustainability in the background? *Journal of Industrial Ecology, 11*, 73–84. https://doi.org/10.1162/jiec.2007.1177

Giampietro, M., & Funtowicz, S. O. (2020). From elite folk science to the policy legend of the circular economy. *Environmental Science & Policy, 109*, 64–72. https://doi.org/10.1016/j.envsci.2020.04.012

Haas, W., Krausmann, F., Wiedenhofer, D., & Heinz, M. (2015). How circular is the global economy?: An assessment of material flows, Waste production, and recycling in the European Union and the World in 2005. *Journal of Industrial Ecology, 19*, 765–777. https://doi.org/10.1111/jiec.12244

Hobson, K., & Lynch, N. (2016). Diversifying and de-growing the circular economy: Radical social transformation in a resource-scarce world. *Futures, 82*, 15–25. https://doi.org/10.1016/j.futures.2016.05.012

Hoehn, D., Laso, J., Margallo, M., Ruiz-Salmón, I., Amo-Setién, F. J., Abajas-Bustillo, R., Sarabia, C., Quiñones, A., Vázquez-Rowe, I., Bala, A., Batlle-Bayer, L., Fullana-i-Palmer, P., & Aldaco, R. (2021). Introducing a degrowth approach to the circular economy policies of food production, and food loss and waste management: Towards a circular bioeconomy. *Sustainability, 13*, 3379. https://doi.org/10.3390/su13063379

Inigo, E. A., & Blok, V. (2019). Strengthening the socio-ethical foundations of the circular economy: Lessons from responsible research and innovation. *Journal of Cleaner Production, 233*, 280–291. https://doi.org/10.1016/j.jclepro.2019.06.053

Jensen, P. D., Basson, L., & Leach, M. (2011). Reinterpreting industrial ecology. *Journal of Industrial Ecology, 15*, 680–692. https://doi.org/10.1111/j.1530-9290.2011.00377.x

Keitsch, M. (2011). Ethics in industrial ecology. *Problems of Sustainable Development, 6*, 19–31.

Kirchherr, J., & van Santen, R. (2019). Research on the circular economy: A critique of the field. *Resources, Conservation and Recycling, 151*, 104480. https://doi.org/10.1016/j.resconrec.2019.104480

Korhonen, J., Nuur, C., Feldmann, A., & Birkie, S. E. (2018). Circular economy as an essentially contested concept. *Journal of Cleaner Production, 175*, 544–552. https://doi.org/10.1016/j.jclepro.2017.12.111

Meadows, D. H., Meadows, D. L., Randers, J., & Behrens, W. W. (1972). *The limits to growth*. Potomac Associates. Oakland. USA

Mueller, G. E. (1958). The Hegel legend of "Thesis-Antithesis-Synthesis." *Journal of the History of Ideas, 19*, 411. https://doi.org/10.2307/2708045

O'Rourke, D., Conelly, L., & Koshland, C. (1996). Industrial ecology: A critical review. *International Journal of Environment and Pollution, 6*, 89–112.

Petras, J., & Veltmeyer, H. (2014). Extractive imperialism in the Americas: Capitalism's new frontier. Brill.

Sehnem, S., Vazquez-Brust, D., Pereira, S. C. F., & Campos, L. M. S. (2019). Circular economy: Benefits, impacts and overlapping. *SCM, 24*, 784–804. https://doi.org/10.1108/SCM-06-2018-0213

Valenzuela, F., & Böhm, S. (2017). Against wasted politics: A critique of the circular economy. *Ephemera: Theory & Politics in Organization, 17*(1), 23–60.

Zink, T., & Geyer, R. (2017). Circular economy rebound. *Journal of Industrial Ecology, 21*(3), 593–602.

Institute for Eco-Industrial Development (IECO): A Summary and Farewell from the Project Director

Andreas Ch. Braun

Abstract This is not a scientific text. It is the conclusion of this anthology, written from the perspective of the project leader of the Institute of Eco-Industrial Development (IECO). This project was developed, applied for, set up, and managed under my leadership. Many people were involved, both scientifically and administratively. Their work is made evident by the numerous contributions in this volume. In this concluding chapter, I would like to look back on the project and make recommendations that could be important for such projects in the future.

1 Introduction

As explained in the introductory chapter on the IECO and international cooperation, this project, including its initiation phase, ran from 2015 to the end of 2022. It was only possible thanks to the cooperation of many people from Chile and Germany. My role as the leader of this project was a great challenge for me, and the pandemic years of 2020 to 2022 in particular were not exactly a window of opportunity for international cooperation.

On the other hand, the project also helped me take the next step in my career and I am no longer writing this chapter at the Karlsruhe Institute of Technology (KIT), but at the University of Kassel. Above all, however, I was able to personally learn a great deal through my work with the IECO. Over the years, some aspects have crystallized for me that could be important for similar international cooperation projects in the future. The final contribution I can make for the IECO project and the network around it is to document these aspects. These are subjective, of course, but they may still be useful.

My primary concern here is not to summarize the many individual chapters, which are too great in number. Nor would this be helpful, because any summaries would have to remain so superficial that hardly any learnings could be taken from them. Instead, I want to do two things. Firstly, I would like to take a brief look back at what

A. Ch. Braun (✉)
Kassel Institute for Sustainability, University of Kassel, Mosenthalstr. 8, 34117 Kassel, Germany
e-mail: Andreas.braun@uni-kassel.de

A. Ch. Braun et al. (eds.), *Eco-Industrial Development as an Industrial Strategy*, Green Energy and Technology, https://doi.org/10.1007/978-3-031-73576-9_32

has been achieved in terms of the original project objectives and what remains to be done. More importantly, however, I would like to make a few recommendations that could be helpful for scientists, universities, and funding bodies dealing with projects such as the IECO.

2 Review

In this chapter, I would like to take a brief look back at what has been achieved and what is still outstanding, and briefly explain which factors were decisive in each case.

2.1 Achieved Goals

The goals achieved in the research are very adequately documented by the research results published here. In my view, there are also several positive aspects to emphasize.

- **Interdisciplinary thinking**: The IECO has certainly led to the development of levels of discussion between the institutions but also within them, which have enabled a broader reflection on industrial sustainability than the previously existing, more discipline-oriented cooperation formats.
- **Common culture of cooperation**: While I was still somewhat skeptical at the beginning of the project about the prospects of success for German-Chilean collaboration, Chilean cooperation very quickly became a familiar and lived reality. One indicator is certainly KIT's decision in 2018 to make Chile the fourth partner country of its international strategy.
- **Visibility in academic cooperation**: The collaborative aspect of the IECO project also quickly became visible and well-known in the field of German-Chilean cooperation. The German embassy, for example, stated with reference to the IECO that KIT was one of the three major German players in Chile.

2.2 Open Goals

There are many factors behind the unachieved goals, but one of them stands out very clearly: the COVID-19 pandemic between 2020 and 2022. Within a few days in spring 2020, it became apparent that an important and very extensively planned project trip was going to be impossible. At that time, it was not yet foreseeable that it would be virtually impossible to organize international project trips and larger workshops for more than two years. The COVID-19 pandemic hit the IECO hard, from the beginning of the consolidation phase to the end. Throughout this period, it was in fact irrational for participating scientists to focus on international collaboration. For the coordination of the IECO, this meant a "pull path" in which the overarching goal was to maintain motivation and interest in the consortium. However, frequent, long, and exhausting virtual meetings could not compensate for the slowdown in project progress. Even the "IECO Days," which were organized as large virtual conferences, could only somewhat alleviate the severity of the situation. It is against this backdrop that two project objectives could not be fully achieved. Additional factors play a role in this, however, which should be mentioned briefly.

2.2.1 International Graduate School

IECO had originally set itself the goal of establishing a joint German-Chilean graduate school. This did not fully succeed, for which three factors (beyond COVID-19) played a role. Firstly, it became apparent that the framework conditions for post-graduate education in Germany and Chile are too different. For example, doctoral students in Germany are generally paid by the university, but in Chile they pay for their doctorate. There were additional institutional obstacles that were initially unknown, such as the fact that a central graduate school existed at a partner university, alongside which no other, cooperation-specific graduate school could exist. Finally, it does not correspond to the cooperation logic in Chile to work together with other Chilean partners in the area of doctoral training. This was compensated for by a change in strategy towards internationally supervised Cotutelle-Ph.Ds and two very successful IECO Summer Schools.

2.2.2 Coevolutionary Entrepreneurship

Another original project objective was to work together in the area of innovation management and knowledge transfer and possibly even develop strategies for coevolutionary entrepreneurship (using instruments such as close cooperation between the incubators and accelerators of the partner universities). Although transparent and good concepts have been developed in the area of innovation management since 2019, the cooperation remains network-like and not instrumental. In order to deepen the cooperation, more time would have been needed as well as more extensive financial resources, for example to incubate jointly generated knowledge in start-ups and spin-offs. In addition, a parallel project for international innovation cooperation (including with Chile) was created in KIT Innovation and Relationship Management, so that no structures had to or should be duplicated.

3 Recommendations

3.1 Recommendations to Scientists

Of course, scientists are experts in their fields and are completely and rightly convinced of the effectiveness of the sustainability strategies researched there. Nevertheless, two recommendations should be made for international and interdisciplinary cooperation.

3.1.1 Conceptual Opening

The importance of one's own research is not diminished by being open to alternative strategies. It makes sense to be open to technological alternatives, for example, and to learn from them. Interdisciplinary cooperation is based not only on the fundamental recognition of other disciplines, but also on the acceptance that these can sometimes relativize one's own findings. Excessive conceptual competition, which is not based on merit but on academic claims, fails to recognize that we live on a planet that is on the verge of collapse.

3.1.2 Simplification of Coordination

Supposedly, someone once said that coordinating scientists was like herding cats. Even if the collaboration worked well, this person may not have been entirely wrong. For the administrative parts of the team in particular, who cannot claim professorial prestige themselves, it is helpful if emails are answered promptly. It is often simply not apparent what enormous efforts in administrative coordination lie behind such projects. This can be greatly simplified if feedback is given, flexibility in the schedule is organized, and the added value of some administrative requests is acknowledged, which may seem annoying at first glance.

3.2 Recommendations to Universities

For universities, international partnerships are often part of a strategy for internationalization and important instruments for achieving research goals. At the same time, they entail an immense coordination effort for a large number of participants who have to set up, administer and develop these instruments. The necessary travel alone is difficult to manage. Strategic internationalization is difficult to achieve from basic funding and the partnerships are often diffusely formulated at the beginning with an unclear impact. Nevertheless, practically all successful universities strive for good and reliable international partnerships. Three recommendations are also to be made to universities here.

3.2.1 Independence of Persons

International collaborations are often very dependent on the people who establish and lead them. However, the appointment practice in Germany, which makes it almost impossible to be appointed to one's own university, creates a situation in which the most successful people in particular change location more frequently. However, this also creates the risk that a newly established cooperation will collapse, which cannot be in the interests of either the funding body or the universities involved. In this respect, it seems very sensible from the university's point of view to make cooperation as independent of individuals as possible. This is possible, for example, by establishing management structures that are based on more than one person and through the intensive involvement of the international office in the cooperative project.

3.2.2 Recognition in Career Paths

There is still a hierarchy of recognition for academic careers in appointment committees. Scientific publications are rewarded the most, followed by cooperative projects with industry, and teaching and international cooperation at the very bottom. However, it should be noted that these different fields of activity naturally compete with each other. The enormous amount of time invested in establishing international collaborations is of course not spent on research and publication activities. As a result, it only actually makes sense for very well-off scientists to invest in such activities. Younger colleagues in particular risk slowing their career advancement if they take time away from publishing to invest in other activities. This outcome is obviously not desirable. If things are to be built up in a way that benefits the university as a whole, then this must also be rewarded in the CVs of those who work in this way. In this respect, the advice here is to pay attention to the willingness of individuals to establish international cooperation when making appointments.

3.2.3 Durability of Financing

Analogous to the recommendations for funding institutions (Sect. 3.3.3), it is also important that the universities themselves plan very early on how a cooperation that has been established can be made permanent. Even the establishment of coordination over a few years, up to the approval of follow-up projects and so on, can be helpful. Establishing international collaborations is an enormously time-consuming undertaking. The motivation of the consortium can only be strengthened by developing a certain perspective of continuity that rewards this work.

3.3 Recommendations to Funding Institutions

Scientific collaboration is enormously expensive and is not always geared towards being immediately "in the black." A great deal of research is also necessary in the area of lower technology readiness levels (TRLs) until business cases emerge that amortize the social investment, and even increased research cannot and should not lead to tangible effects. Efficiency and profitability criteria must therefore be observed and expenditure is ultimately taxpayers' money, which is why it must remain justifiable to the taxpayer. International cooperation in particular can easily run into a certain justification deficit, because ultimately strategies are often researched—as here—that are clearly geared towards other countries. The direct benefit for Germany can thus be easily called into doubt by many. It is therefore all the more pleasing when German ministries support international cooperation projects as comprehensively as was the case with the IECO (and the other projects in the same call).

Against the backdrop of this initial situation, it therefore seems important to me to make a few recommendations that may be of interest to funding providers.

3.3.1 Financing the Partner Countries

For the reasons mentioned above, it is of course difficult to use taxpayers' money to finance personnel and equipment abroad. Nevertheless, I would like to suggest it for several reasons. If, as in the case of this call for proposals, the call is not mirrored, i.e. there is no call for proposals from the Latin American partner countries, this immediately leads to difficulties in the event of funding. The comprehensively funded German consortium then sets major goals that have to be coordinated in detail and on a regular basis. However, this poses major difficulties for the Latin American partners, as the many coordination meetings and work steps require people who are not funded for this purpose. This creates an asymmetry in which both sides are expected to achieve ambitious goals to the same extent, but only one side is funded. This naturally raises questions of fairness.

Added to this are the operational challenges. This is because Latin American universities are then forced to ensure the coordination of a very complex project by a few volunteer scientists who are not funded for this purpose. This almost inevitably leads to the need to change and rotate these responsibilities. However, this in turn makes it more difficult for the project as a whole to achieve its goals, which have to be regularly discussed with different counterparts. It also almost inevitably implies that the project will be less well perceived by the partner universities. Although this can be compensated for by the prestige of the German university partner, for example, international cooperation should not be based on goodwill. For these reasons, it is in the best interest of the funded country that at least coordinators can be financed and provided with some material resources in the partner country. The effectiveness would therefore be increased by the possibility of financial transfers abroad, which has also been made possible in more recent comparable calls for proposals.

3.3.2 Combination Formats

According to Humboldt's educational ideal, research and teaching are to be understood as a unit. The transfer of knowledge adds a "third mission" for universities. This ideal makes sense and is desirable for a number of reasons. So far, however, it has not been fully reflected in research funding in Germany. The tendency is always to fund either research or teaching or transfer. For international academic cooperation projects in particular, however, it would be desirable to have combined formats. In order to enable long-term cooperation, networking needs to be as multi-layered as possible. Students who are supervised by two international project partners are just as helpful here as another publication. It is also important to recognize that many more people can often be networked much more permanently through teaching than through research. Even in large research projects in many areas,[1] often only a few people can be integrated, while almost any number of people can be integrated into joint degree or doctoral programs with their teaching offerings. This also allows for a more "informal" exchange of perspectives and discussion, which is important for joint teaching. All of this enables long-term networking and should therefore be promoted in calls for proposals together with research.

3.3.3 Durability of Financing

Building something new is always easier than maintaining what already exists. And new things are always more innovative than existing ones. Nevertheless, it would be important to have formats for how international projects such as joint degree programs, joint research institutes, etc. can be financed on a permanent basis. It is quite right that after a comprehensively funded start-up phase, the structures developed will then finance themselves through follow-up projects in research and teaching. What is needed, however, is permanent coordination. Staff are needed to bring the network together on a day-to-day basis and stimulate new activities. If this breaks down, there is a risk that the network will split into smaller sub-networks that already existed before, but that the overall structure will increasingly be forgotten or can simply no longer be operated. Since, as described above, the return on investment often only occurs well after the end of the project term, the effectiveness of the funded measures is at risk here.

3.4 Total Recommendations

Finally, I would like to make a few overall recommendations aimed at all those involved in international cooperation.

[1] The exception is "big science," whereby care must always be taken not to equate "big" with "great." As a rule, it is simply about large research equipment.

3.4.1 Intercultural Competence

One is the importance of intercultural support. In my experience, we engage in international cooperation as if cultural differences don't matter. We travel to partner countries without international communication training, disregarding social, political, and cultural differences and customs. The result is conflicts that arise very easily, in which it is easy to see the cause of the conflict in the respective conflict partner, which is rooted simply in the differences between the systems. For both sides, this entails the risk of rearticulating cultural stereotypes instead of overcoming them. It presupposes the need for culturally sensitive learning, which can be supported by intercultural training such as that offered by the DAAD International Academy.

3.4.2 Linking Back to the Sustainability Discourse as a Whole

Of course, research projects set themselves specific topics that they investigate in depth. From the point of view of policymakers, this carries the risk of exaggerating very prominently communicated projects and strategies. However, as I explained in the previous chapter, it is important to place the projects in an overall context. It should be discussed and made transparent what certain strategies are intended for, what opportunities and limitations they have, and how they compare to other strategies in the sustainability discourse. This requires inter- and transdisciplinary communication, in which it may well be that one's own sacred cow is sometimes slaughtered.

3.4.3 Intrinsic Sustainability

Of course, it is desirable for sustainability projects to try to be sustainable themselves. This applies first of all to the enormously emissions-intensive academic travel activities, which could, if possible, be designed in such a way that travel is minimized overall and reasons for travel to the destination country are bundled. It would also be desirable to at least offset the resulting emissions, which must of course be permitted by the funding body. There are also numerous other aspects.

4 Conclusion and Farewell

Having said this, it only remains for me to emphasize that a great many people at the Federal Ministry of Education and Research (BMBF), the German Aerospace Center (DLR), the administrative and research management of the partner universities, as well as numerous project partners[2] have worked together to make this project possible. I would like to thank them personally once again.

[2] https://ieco-institute.com/network.html.

On the other hand, it is also important for me to remind us all once again what the added value has been beyond individual scientific achievement. We are experiencing an increasingly polarized, unilateralized world. According to the Uppsala Conflict Data Program, 2022 and 2023 will be the most conflict-ridden years since the end of the Cold War. In addition to escalating conflicts, in many countries—including Germany and Latin America—the voices of right-wing extremist politicians and parties that deny pressing sustainability issues such as climate change as well as the importance of openness to the world and international cooperation are on the rise. The question of counter-forces arises. I believe that science, with its open-mindedness, inclusivity, curiosity, and interest in the "new and foreign" can build bridges of cooperation that will outlast such dark times of crisis.

With this hope, all that remains is for me to say goodbye to my project partners at the Universidad de Chile (UCH), Universidad de Concepcion (UdeC), Universidad Austral de Chile (UACh), and Karlsruhe Institute of Technology (KIT), and to wish everyone many more joint successes.

Prof. Dr. Andreas Ch. Braun.

GPSR Compliance

The European Union's (EU) General Product Safety Regulation (GPSR) is a set of rules that requires consumer products to be safe and our obligations to ensure this.

If you have any concerns about our products, you can contact us on ProductSafety@springernature.com

In case Publisher is established outside the EU, the EU authorized representative is:

Springer Nature Customer Service Center GmbH
Europaplatz 3
69115 Heidelberg, Germany

The manufacturer's authorised representative in the EU is Springer Nature Customer Service Centre GmbH, Europaplatz 3, 69115 Heidelberg, Germany. If you have any concerns regarding our products, please contact ProductSafety@springernature.com

Printed and bound by CPI Group (UK) Ltd, Croydon, CR0 4YY
01/10/2026
02225965-0018